ENDORSED BY

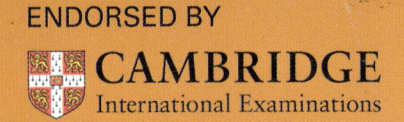

CAMBRIDGE
International Examinations

Essential Mathematics
for Cambridge IGCSE®

Extended

Sue Pemberton

Oxford and Cambridge
leading education together

OXFORD
UNIVERSITY PRESS

OXFORD
UNIVERSITY PRESS

Great Clarendon Street, Oxford OX2 6DP

Oxford University Press is a department of the University of Oxford.
It furthers the University's objective of excellence in research,
scholarship, and education by publishing worldwide in

Oxford   New York

Auckland   Cape Town   Dar es Salaam   Hong Kong   Karachi
Kuala Lumpur   Madrid   Melbourne   Mexico City   Nairobi
New Delhi   Shanghai   Taipei   Toronto

With offices in

Argentina   Austria   Brazil   Chile   Czech Republic   France   Greece
Guatemala   Hungary   Italy   Japan   Poland   Portugal   Singapore
South Korea   Switzerland   Thailand   Turkey   Ukraine   Vietnam

Oxford is a registered trade mark of Oxford University Press
in the UK and in certain other countries

© Oxford University Press 2012

The moral rights of the author have been asserted

Database right Oxford University Press (maker)

First published 2012

All rights reserved. No part of this publication may be reproduced,
stored in a retrieval system, or transmitted, in any form or by any means,
without the prior permission in writing of Oxford University Press, or as
expressly permitted by law, or under terms agreed with the appropriate
reprographics rights organization. Enquiries concerning reproduction
outside the scope of the above should be sent to the Rights Department,
Oxford University Press, at the address above

You must not circulate this book in any other binding or cover
and you must impose this same condition on any acquirer

British Library Cataloguing in Publication Data

Data available

ISBN: 978-0-19-912874-7

10 9 8 7 6

Printed in China by Printplus.

Acknowledgments
® IGCSE is the registered trademark of Cambridge International Examinations.

The publisher would like to thank Cambridge International Examinations for their kind permission to reproduce past paper questions.

Cambridge International Examinations bears no responsibility for the example answers to questions taken from its past question papers which are contained in this publication.

The publishers would like to thank the following for permission to reproduce photographs:

P90: Kuttly/Shutterstock; **P257**: R. Formidable; **P271**: Andreas Meyer/Shutterstock;
P271: Jorgen Mcleman/Shutterstock.

Cover Image: Aguasonic Acoustics / Science Photo Library

Every effort has been made to contact copyright holders of material reproduced in this book.
If notified, the publishers will be pleased to rectify any errors or omissions at the earliest
opportunity.

Contents

	Unit 1		**Unit 2**	
Number	Order of operations	2	Percentages 1	50
	Directed numbers	3	Ratio	52
	Multiples, factors, primes, squares and cubes	5		
	Four rules for fractions	10		
	Significant figures and decimal places	14		
Algebra	Simplifying algebraic expressions	17	Indices 1	58
	Solving linear equations	20	Solving linear inequalities	62
	Constructing formulae	24	Manipulating algebraic fractions	64
	Substitution into formulae	25	The general equation of a straight line	68
	Gradients and straight line graphs	28	Representing linear inequalities on graphs	72
Shape and Space	Angle properties	33	Perimeter and area	76
	Symmetry	36	Pythagoras	78
	Polygons	39	Geometrical constructions	84
			Loci	86
			Area and circumference of a circle	90
Probability and Statistics	Averages and range	41	Displaying data	96
	Frequency tables	44		
	Examination questions	46	Examination questions	100

III

Number

Unit 3		Unit 4	
Standard form	104	Percentages 2	168

Algebra

Simultaneous equations 1	108	Matrix algebra	170
Factorising 1	114	Expanding double brackets	180
Rearranging formulae 1	118	Quadratic graphs	184

Shape and Space

Similar triangles	120	Bearings	190
Reflections, rotations and translations	128	Trigonometry	194
Enlargements	144	Angles of elevation and depression	208
Surface area and volume 1	150		

Probability and Statistics

Probability 1	156	Scatter diagrams	210
Examination questions	164	Examination questions	214

Unit 5

Direct and inverse proportion	218
Increase and decrease in a given ratio	222
Functions	224
Factorising 2	230
Cubic graphs	234
Surface area and volume 2	238
Areas of similar shapes	246
Volumes of similar objects	250
Grouped frequency tables	254
Examination questions	258

Unit 6

Percentages 3	264
Speed, distance and time	268
Sets and Venn diagrams	272
Indices 2	284
Solving quadratic equations by factorisation	288
Reciprocal graphs	292
Circle theorems	294
Stretches and shears	308
Probability 2	316
Examination questions	324

Unit 7

Distance-time graphs	334
Speed-time graphs	338
Rearranging formulae 2	342
Sequences	344
Exponential graphs	354
Matrices and transformation	358
Cumulative frequency	368
Examination questions	376

Number

Algebra

Shape and Space

Probability and Statistics

Unit 8

Rational and irrational
numbers   382

Solving quadratic equations
using the formula   384
Further algebraic fractions   388
Variation   390

Sine and cosine ratios up
to 180°   398
Area of a triangle   400
The sine and cosine rules   404

Histograms   412

Examination questions   416

Unit 9

Upper and lower bounds   422

Simultaneous equations 2   426
Linear programming   428
Solving quadratic equations by
completing the square   432
Using graphs to solve
equations   436

Vectors and vector
geometry   440

Probability 3   452

Examination questions   456

Answers   462
Index   519

About this book

This book has been specially written to help you achieve your highest potential on the Cambridge IGCSE® Mathematics 0580, extended syllabus. It is endorsed by University of Cambridge International Examinations so you know that it is fully up to date and covers the syllabus in depth.

The author is an experienced mathematics teacher and a principal examiner. Packed with carefully chosen examples, helpful hints and plenty of exercises to give you confidence in your abilities, the author's many years of experience ensure that the book is carefully designed to help you succeed.

The contents are organised into nine units and within each unit the sections are grouped into the broad topics: number, algebra, shape and space, and probability and statistics. Each section concludes with a selection of past exam questions. A unit is intended to be covered in around 20 hours; this should leave ample time for revision and exam practice if the syllabus is being taught over two years.

The free CD-ROM which comes with this book contains eighty-five presentations, one for each section of the book, that provide many more fully worked examples.

In all the examination papers it is permissible to use an electronic calculator provided that it is not an algebraic or graphical calculator. Four figure trigonometric tables are also permitted. Opportunities to practise using a calculator arise throughout the book, though in practice many questions will not require the use of a calculator and candidates should be able to use mental and written methods. Indeed certain questions may require evidence of a written method being used.

Note that because this book is written for the extended syllabus a number of core topics – basic arithmetic, measures, time, money, etc. – are not covered explicitly but are woven into the treatment of more advanced topics.

Order of operations

THIS SECTION WILL SHOW YOU HOW TO
- Perform operations in the correct order

When a calculation involves more than one operation it is important to do the operations in the correct order.
1. Work out the **B**rackets first.
2. Work out the **I**ndices next.
3. Work out the **D**ivisions and **M**ultiplications next.
4. Work out the **A**dditions and **S**ubtractions last.

Memory aid

B I D M A S
- **B**rackets
- **I**ndices
- **D**ivision
- **M**ultiplication
- **A**ddition
- **S**ubtraction

EXAMPLE

Calculate $150 - 3^2 \times (8 + 4) + 6 \div 2$

$150 - 3^2 \times (8 + 4) + 6 \div 2$	brackets first
$= 150 - 3^2 \times 12 + 6 \div 2$	indices next, 3^2 means $3 \times 3 = 9$
$= 150 - 9 \times 12 + 6 \div 2$	division and multiplication next
$= 150 - 108 + 3 = 45$	addition and subtraction last

EXERCISE 1.1

Work out:

1. $2 + 3 \times 5$
2. $8 \div 2 + 4$
3. $3 \times 4 - 5 \times 2$
4. $5 \times 6 - 8 \div 2$
5. $7 - 2 \times 3 + 5$
6. $10 - 2 \times 3$
7. $6 + (3 \times 5) - 2$
8. $10 + 3^2 \times 4$
9. $(5 + 4) - 3 \times 2$
10. $(5 \times 4) - (3 \times 2)$
11. $(4 \times 3) \times 2^2$
12. $5 \times (4 - 3) \times 2$
13. $5 \times (16 - 3 \times 2)$
14. $38 \div 2 - 4 \times 3$
15. $56 - (2^3 + 4)$
16. $4 + (3 \times 2)^2$
17. $(4 \times 3 + 2)^2$
18. $(2 \times 3^2) + 4$
19. $4^3 - 5 \times 6$
20. $5 \times (2^3 + 3^2)$

EXERCISE 1.2

Copy these and use brackets (where necessary) to make the statements true.

1. $2 + 3 \times 4 + 5 = 25$
2. $2 \times 3 + 4 \times 5 = 26$
3. $2 + 3 \times 4 + 5 = 29$
4. $2 + 3 \times 4^2 = 80$
5. $2 \times 3 + 4 \times 5 = 46$
6. $5 + 4 \times 3 - 2 = 15$
7. $2 \times 3 + 4 \times 5 = 50$
8. $2 \times 3 + 4 \times 5 = 70$
9. $2 + 3 \times 4 + 5 = 45$
10. $2 + 3 \times 4^2 = 50$

Directed numbers

THIS SECTION WILL SHOW YOU HOW TO
- Perform the four rules on directed numbers

The **positive** and **negative** whole numbers are called **integers**.
They can be shown on a number line.
The number line can be used in practical situations.

NEGATIVE NUMBERS   POSITIVE NUMBERS
−6 −5 −4 −3 −2 −1 0 1 2 3 4 5 6

To find the difference between a temperature of 5 °C and a temperature of −3 °C, you find the gap between these two numbers on the number line.
The difference is 8 °C.

Adding and subtracting directed numbers

The rules for adding and subtracting directed numbers are:

Change	−2 + +5	to	−2 + 5 = 3
Change	−2 + −5	to	−2 − 5 = −7
Change	−2 − +5	to	−2 − 5 = −7
Change	−2 − −5	to	−2 + 5 = 3

EXAMPLE

Work out **a** (−9) − (−3)   **b** (+8) − (+15)   **c** (29) + (−12)

a −9 − −3 = −9 + 3 = −6   change − − to +
b +8 − +15 = +8 − 15 = −7   change − + to −
c 29 + −12 = 29 − 12 = 17   change + − to −

EXERCISE 1.3

Work out:

1 (−3) + (−5)
2 6 − (−4)
3 8 + (−10)

4 2 + (−5)
5 (−4) − (+2)
6 (−15) + (−3)

7 36 + (−8)
8 29 − (+1)
9 (−52) − (−38)

10 (−54) + (−3)
11 (−16) + (−2)
12 (−20) − (−20)

13 (−57) + (+5)
14 41 + (−16)
15 52 − (−3)

16 (−5) − (+10)
17 (−7) − (−14)
18 (−42) + (−5)

19 (−8) + (−2) + (−5)
20 (−6) − (+2) − (−3)
21 7 − (−2) + (−3)

22 (+9) − (−6) + (−6)
23 7 − (+9) + (−3)
24 46 − (−12) + (−5)

Directed numbers   3

Multiplying and dividing directed numbers

The rules for multiplying and dividing directed numbers are:

Multiplication	Division
+ × + = +	+ ÷ + = +
+ × − = −	+ ÷ − = −
− × + = −	− ÷ + = −
− × − = +	− ÷ − = +

If the two signs are the same, the answer will be positive.
If the two signs are different, the answer will be negative.

EXAMPLE

Work out   **a** $(-6) \div (-2)$ **b** $5 \times (-8)$ **c** $(-4)^3$

a $(-6) \div (-2) = 3$ the two signs are the same so the answer is positive

b $5 \times (-8) = -40$ the two signs are different so the answer is negative

c $(-4)^3 = -4 \times -4 \times -4$ first multiply −4 by −4
$ = 16 \times -4$ then multiply by −4 again
$ = -64$

EXERCISE 1.4

Work out:

1 $(-12) \times (-5)$
2 $(-8) \times (+4)$
3 $(+16) \times (-2)$

4 $(-52) \div (-13)$
5 $(-55) \div (+5)$
6 $(-145) \div (-5)$

7 $(+20) \div (-2)$
8 $(-95) \div (-19)$
9 $(-11) \times (-11)$

10 $(-3) \times (-4) \times (-5)$
11 $(-2) \times (+8) \times (-4)$
12 $(+6) \times (-3) \times (-7)$

13 $(-2) \times (-5) \times (+6)$
14 $(-9)^2$
15 $(-15)^2$

16 $(-5)^3$
17 $(-60)^3$
18 $(-4)^3 \times (-1)^3$

19 $(-2)^5 \times (-10)^2$
20 $(-1)^{13}$
21 $(-1)^{15} \times (-1)^{24}$

22 $\dfrac{-6}{-3}$
23 $\dfrac{(-10) \times (+3)}{-15}$
24 $\dfrac{(-12) \times (-5)}{(-2) \times (+10)}$

25 Check your answers to questions **1** to **18** using a calculator.

Find the missing numbers:

26 $\dfrac{\square^2 \times (-12)}{8 \div (-2)} = 48$
27 $\dfrac{(-5) \times \square}{10} = 3$
28 $\dfrac{\square \div 3}{(-2) \times (-6)} = -1$

KEY WORDS
positive
negative
integer

UNIT 1

Multiples, factors, primes, squares and cubes

THIS SECTION WILL SHOW YOU HOW TO
- Identify and use factors, multiples, primes, squares and cubes

Factors

The whole numbers that divide exactly into 15 are called **factors** of 15.
The factors of 15 are 1, 3, 5 and 15.

EXAMPLE

List all the factors of 24.

24 = 1 × 24   write 24 as the product of two factors
24 = 2 × 12   repeat until all pairs have been found
24 = 3 × 8
24 = 4 × 6
Factors of 24 = 1, 2, 3, 4, 6, 8, 12 and 24.

EXAMPLE

Find the **highest common factor (HCF)** of 20 and 36.

Factors of 20 = ①, ②, ④, 5, 10, 20   list the factors of both 20 and 36
Factors of 36 = ①, ②, 3, ④, 6, 9, 12, 18, 36
Common factors of 20 and 36 are 1, 2, and 4.   find the numbers that are in both lists
Highest common factor of 20 and 36 is 4.   select the highest number

Multiples

The **multiples** of 6 are the numbers 6, 12, 18, 24, 30 …

EXAMPLE

Find the **lowest common multiple (LCM)** of 12 and 9.

Multiples of 12 = 12, 24, ㊱, 48, 60, ⑦②, 84 …   list the multiples of 12 and 9
Multiples of 9 = 9, 18, 27, ㊱, 45, 54, 63, ⑦②, 81 …
Common multiples of 12 and 9 are 36, 72 …   find the numbers that are in both lists
Lowest common multiple of 12 and 9 is 36.   select the lowest number

Multiples, factors, primes, squares and cubes   5

EXERCISE 1.5

1. Write down all the factors of:
   - a 10
   - b 15
   - c 9
   - d 17
   - e 60
   - f 80
   - g 100
   - h 64
   - i 125
   - j 90

2. Find the common factors of:
   - a 6 and 8
   - b 10 and 15
   - c 9 and 18
   - d 16 and 20
   - e 20 and 25
   - f 12 and 30
   - g 80 and 100
   - h 42 and 48
   - i 6, 12 and 42

3. Find the highest common factor (HCF) of:
   - a 6 and 8
   - b 10 and 15
   - c 90 and 18
   - d 36 and 45
   - e 23 and 46
   - f 20 and 24
   - g 30 and 45
   - h 42 and 48
   - i 8, 32 and 44

4. List the first six multiples of each of the following numbers.
   - a 10
   - b 6
   - c 9
   - d 18
   - e 25
   - f 40
   - g 100
   - h 12
   - i 27
   - j 121

5. Find the lowest common multiple (LCM) of:
   - a 6 and 8
   - b 5 and 15
   - c 6 and 9
   - d 7 and 8
   - e 4 and 6
   - f 12 and 8
   - g 14 and 21
   - h 11 and 5
   - i 8, 10 and 12

6. A piece of rope can be cut into an exact number of 6 m lengths. The rope could also be cut into an exact number of 8 m lengths. What is the shortest possible length of the rope?

7. A light flashes every 15 minutes. A second light flashes every 18 minutes. Both lights flash together at 2 a.m. What will be the time when they next flash together?

8. A bell rings every 20 seconds. A second bell rings every 25 seconds. A third bell rings every 30 seconds. They all ring together at 8 p.m. How long will it be before all three bells ring together again?

Primes

A **prime** number is a number that has exactly two factors.
5 is a prime number because it has exactly two factors (1 and 5)

EXAMPLE

List the first ten prime numbers

The prime numbers are: 2, 3, 5, 7, 11, 13, 17, 19, 23 and 29

A **prime factor** is a factor that is also a prime number.

EXAMPLE

List the prime factors of 30.

The factors of 30 are: 1, ②, ③, ⑤, 6, 10, 15 and 30   *select the factors that are prime*
The prime factors of 30 are: 2, 3 and 5.

Numbers can be written as the **product of prime factors**.
For example 120 = 2 × 60
= 2 × 2 × 30
= 2 × 2 × 2 × 15
= 2 × 2 × 2 × 3 × 5
= 2^3 × 3 × 5

The next example shows how a factor tree can be used.

EXAMPLE

Write 84 as the product of prime factors.

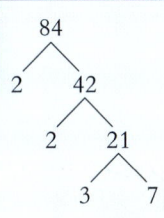

84
2   42
   2   21
      3   7   84 = 2 × 2 × 3 × 7 = 2^2 × 3 × 7

Expressing numbers as the product of prime factors can help you to find highest common factors (HCF) and lowest common multiples (LCM).

EXAMPLE

Find the HCF and the LCM of 270 and 420.

First write 270 and 420 as the product of prime factors.
270 = 2 × 3 × 3 × 3 × 5 and 420 = 2 × 2 × 3 × 5 × 7
Write the prime factors on a diagram.

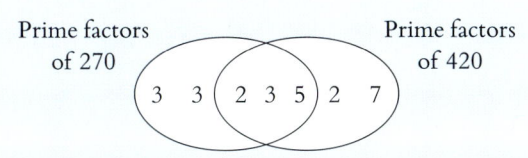

Prime factors   Prime factors
 of 270   of 420
       3   3 (2 3 5) 2   7

The HCF is the product of the numbers in the intersection = 2 × 3 × 5 = 30.
The LCM is the product of all the numbers in the diagram = 3 × 3 × 2 × 3 × 5 × 2 × 7 = 3780

Multiples, factors, primes, squares and cubes

EXERCISE 1.6

1. Which of the following numbers are prime numbers?
   11, 17, 21, 35, 47, 69, 72, 81.

2. Which of the following numbers are prime numbers?
   3, 13, 23, 33, 43, 53, 63, 73.

3. Calculate the value of the following.
   a $2 \times 5 \times 5$
   b $2 \times 3 \times 11$
   c $3 \times 3 \times 3 \times 7$
   d $2 \times 5 \times 7$
   e $2 \times 3 \times 3 \times 13$
   f $2 \times 2 \times 3 \times 11$
   g $3^3 \times 5^2 \times 7$
   h $2^5 \times 3^2 \times 5 \times 11$

4. Write each of the numbers as the product of prime factors.
   a 10
   b 150
   c 81
   d 60
   e 74
   f 100
   g 98
   h 250
   i 1110
   j 275
   k 2004
   l 2210

5. 3, 7, 10, 15, 19, 21, 35
   a Which of these numbers are prime numbers?
   b Which of these numbers are multiples of 3?
   c Which of these numbers are multiples of 5?
   d Which of these numbers are factors of 30?
   e Which of these numbers are factors of 380?

6. List the prime numbers between 80 and 100.

7. Sanjit thinks that 713 is a prime number. Explain why he is wrong.

8. The number 12 can be written as the sum of two prime numbers.
   12 = 5 + 7
   Write each of the following numbers as the sum of two prime numbers.
   a 10
   b 14
   c 25
   d 49
   e 30
   f 20
   g 38
   h 82
   i 36
   j 48

9. Find the HCF and LCM of:
   a 24 and 42
   b 45 and 105
   c 70 and 42
   d 40 and 75

10. Find the HCF and LCM of:
    a 30, 36 and 42
    b 28, 35 and 56
    c 60, 90 and 210

Square and cube numbers

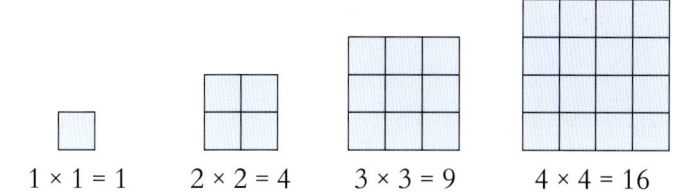

$1 \times 1 = 1$ $2 \times 2 = 4$ $3 \times 3 = 9$ $4 \times 4 = 16$

The numbers 1, 4, 9 and 16 are called **square numbers**.
The number 169 is also a square number because $13 \times 13 = 169$
13×13 can be written as 13^2

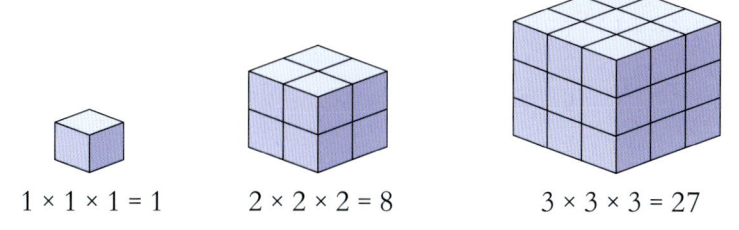

$1 \times 1 \times 1 = 1$ $2 \times 2 \times 2 = 8$ $3 \times 3 \times 3 = 27$

The numbers 1, 8 and 27 are called **cube numbers**.
The number 8000 is a cube number because $20 \times 20 \times 20 = 8000$.
$20 \times 20 \times 20$ can also be written as 20^3.

EXERCISE 1.7

1 Which is biggest 2^3 or 3^2?

2 1, 5, 9, 20, 27, 56, 48, 49, 52, 64, 275, 289, 343, 436, 512.
   a Write down the numbers that are square numbers.
   b Write down the numbers that are cube numbers.

3 The number 64 is a square number because $8 \times 8 = 64$.
It is also a cube number because $4 \times 4 \times 4 = 64$.
Can you find another number (bigger than 1) that is both
a square and a cube number?

KEY WORDS
factor
highest common factor (HCF)
multiple
lowest common multiple (LCM)
prime
prime factor
product of prime factors
square numbers
cube numbers

Multiples, factors, primes, squares and cubes

Four rules for fractions

THIS SECTION WILL SHOW YOU HOW TO
- Use the language of fractions
- Apply the four rules for fractions

The language of fractions

In the fraction $\frac{3}{7}$, the top part is called the **numerator** and the bottom part is called the **denominator**.

The fraction $\frac{8}{3}$ is called an **improper fraction** because the numerator is bigger than the denominator.

The number $2\frac{2}{3}$ is called a **mixed number** because it is made up of a whole number and a fraction.

To simplify a fraction you divide the numerator and denominator by a common factor.

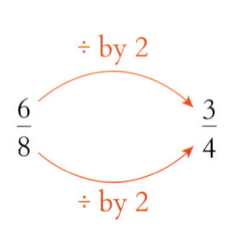

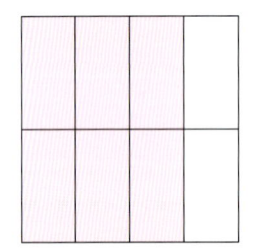

EXERCISE 1.8

Simplify:

1. $\frac{6}{12}$ 2. $\frac{35}{49}$ 3. $\frac{18}{24}$ 4. $\frac{6}{15}$ 5. $\frac{80}{128}$ 6. $\frac{150}{175}$

7. $\frac{36}{63}$ 8. $\frac{60}{75}$ 9. $\frac{42}{48}$ 10. $\frac{36}{132}$ 11. $\frac{51}{60}$ 12. $\frac{84}{144}$

Change to mixed numbers:

13. $\frac{8}{5}$ 14. $\frac{7}{3}$ 15. $\frac{15}{7}$ 16. $\frac{20}{17}$ 17. $\frac{13}{2}$ 18. $\frac{19}{4}$

19. $\frac{17}{12}$ 20. $\frac{47}{15}$ 21. $\frac{15}{8}$ 22. $\frac{22}{21}$ 23. $\frac{17}{5}$ 24. $\frac{30}{9}$

Adding and subtracting fractions

To add or subtract fractions a common denominator is needed as shown below.

EXAMPLE

Calculate a $2\frac{3}{4} + 1\frac{2}{5}$ b $3\frac{1}{5} - 1\frac{2}{3}$

a $2\frac{3}{4} + 1\frac{2}{5}$ write both fractions with a common denominator of 20

$= 2\frac{15}{20} + 1\frac{8}{20}$

$= 3\frac{23}{20}$

$= 4\frac{3}{20}$

b $3\frac{1}{5} - 1\frac{2}{3}$ write both fractions with a common denominator of 15

$= 3\frac{3}{15} - 1\frac{10}{15}$

$= 2\frac{18}{15} - 1\frac{10}{15} = 1\frac{8}{15}$ $3\frac{3}{15} = 2 + 1\frac{3}{15} = 2\frac{18}{15}$

EXERCISE 1.9

Calculate:

1 $1\frac{1}{2} + 2\frac{1}{3}$ 2 $3\frac{1}{4} + 2\frac{1}{5}$ 3 $5\frac{1}{6} + 2\frac{3}{7}$ 4 $3\frac{2}{3} + 3\frac{4}{5}$

5 $1\frac{5}{6} + 2\frac{3}{4}$ 6 $2\frac{5}{6} + 1\frac{2}{5}$ 7 $2\frac{3}{8} + 1\frac{4}{5}$ 8 $5\frac{2}{3} + 9\frac{7}{9}$

9 $3\frac{4}{5} - 1\frac{1}{2}$ 10 $2\frac{5}{8} - 2\frac{1}{3}$ 11 $6\frac{4}{7} - 3\frac{2}{9}$ 12 $4\frac{7}{8} - 2\frac{3}{5}$

13 $5\frac{7}{8} - 2\frac{3}{5}$ 14 $2\frac{2}{5} - 1\frac{5}{6}$ 15 $6\frac{1}{6} - 2\frac{3}{4}$ 16 $7\frac{2}{5} - 3\frac{4}{7}$

17 $-3\frac{4}{5} + 1\frac{1}{2}$ 18 $4\frac{2}{3} - \left(-2\frac{1}{5}\right)$ 19 $2\frac{1}{2} + 3\frac{1}{3} - 1\frac{3}{4}$ 20 $\frac{1}{2} - 2\frac{1}{3} + \frac{5}{6}$

21 $2\frac{1}{5} - 1\frac{4}{5} - 1\frac{2}{3}$ 22 $5 - \left(2\frac{1}{3} + 1\frac{1}{5}\right)$

23 Use a calculator to check your answers to questions **1** to **22**.

Four rules for fractions

24 Omar has some money. He spends $\frac{1}{5}$ of the money on a book, $\frac{3}{7}$ of the money on a CD and he puts the rest of the money in a bank. What fraction of the money does he put in the bank?

25 Fatimah has 3 kg of flour. She uses $1\frac{3}{5}$ kg to make some bread and $\frac{3}{4}$ kg to make a cake. How much flour does she have left?

26 The sum of two fractions is $6\frac{4}{5}$. One of the fractions is $4\frac{9}{10}$. Find the other fraction.

27 The difference of two fractions is $4\frac{2}{7}$. One of the fractions is $2\frac{3}{5}$. Find the other fraction.

28 Find the fraction that is halfway between $-\frac{1}{2}$ and $\frac{1}{3}$.

29 Arrange these fractions in order of size starting with the smallest:

a $\quad \frac{3}{5}, \frac{2}{3}, \frac{13}{30}$ b $\quad \frac{7}{12}, \frac{11}{18}, \frac{5}{9}$

c $\quad \frac{4}{5}, \frac{3}{4}, \frac{7}{10}$ d $\quad \frac{5}{6}, \frac{8}{9}, \frac{11}{12}$

HINT
change the fractions so they have the same common denominators.

Multiplying and dividing fractions

EXAMPLE

Calculate a $2\frac{1}{5} \times 1\frac{3}{4}$ b $3\frac{1}{3} \div 3\frac{1}{2}$

a $2\frac{1}{5} \times 1\frac{3}{4} = \frac{11}{5} \times \frac{7}{4}$ *change to improper fractions*

$\quad = \frac{77}{20}$ *multiply the numerators, multiply the denominators*

$\quad = 3\frac{17}{20}$ *change to a mixed number*

b $3\frac{1}{3} \div 3\frac{1}{2} = \frac{10}{3} \div \frac{7}{2}$ *change to improper fractions*

$\quad = \frac{10}{3} \times \frac{2}{7}$ *dividing by $\frac{7}{2}$ is the same as multiplying by $\frac{2}{7}$*

$\quad = \frac{20}{21}$ *multiply the numerators, multiply the denominators*

EXERCISE 1.10

Calculate:

1. $1\frac{2}{3} \times 2\frac{1}{5}$
2. $2\frac{1}{4} \times 2\frac{1}{4}$
3. $3\frac{2}{5} \times 1\frac{1}{6}$
4. $5\frac{2}{7} \times 3\frac{3}{4}$

5. $4\frac{5}{6} \times 2\frac{1}{3}$
6. $7\frac{3}{10} \times 2\frac{4}{5}$
7. $1\frac{1}{2} \div 2\frac{1}{3}$
8. $3\frac{3}{4} \div 1\frac{1}{2}$

9. $2\frac{5}{6} \div 1\frac{2}{3}$
10. $3\frac{1}{8} \div \frac{4}{5}$
11. $\frac{5}{9} \div 1\frac{1}{2}$
12. $2\frac{4}{5} \div 2\frac{2}{3}$

13. $\left(\frac{3}{5}\right)^2$
14. $\left(2\frac{1}{2}\right)^2$
15. $\left(\frac{3}{20}\right)^3$
16. $\left(1\frac{1}{3}\right)^3$

HINT

$\left(2\frac{1}{2}\right)^2$ means $2\frac{1}{2} \times 2\frac{1}{2}$

17. $\frac{1}{4} + \frac{2}{5} \times \frac{3}{8}$
18. $2\frac{1}{4} - \frac{3}{8} \div \frac{3}{5}$

19. Check your ansewrs to questions **1** to **18** using a calculator.

Find the missing numbers:

20. $\frac{3}{5} \times \square = 1$
21. $\square \times \frac{2}{3} = 1$
22. $\square \div \frac{3}{7} = \frac{5}{8}$
23. $\square \times 2\frac{1}{5} = 5$

24. Tom has six bottles of juice. Each bottle contains $\frac{5}{8}$ of a litre.

    How many litres of juice does he have altogether?

KEY WORDS
numerator
denominator
improper fraction
mixed number

Four rules for fractions

Significant figures and decimal places

THIS SECTION WILL SHOW YOU HOW TO
- Round to a given number of decimal places or significant figures.

It is often useful to give an answer to a number of **decimal places (d.p.)**.
If you are asked to **round** a number to 2 decimal places you must have exactly 2 digits after the decimal point. The following rules are used.

> 1 Count along the digits to the required number of decimal places.
> 2 Look at the next digit.
>   If it is less than 5, leave the digit before it as it is.
>   If it is 5 or more, you must round up the digit before it.

EXAMPLE

a Write 6.34296 correct to 3 decimal places.
b Write 31.3027 correct to 2 decimal places.

a 6.34296 = 6.342⋮96   the next digit is 9 so you must round up
        ↑
     = 6.343 (to 3 d.p.)
b 31.3027 = 31.30⋮27   the next digit is 2 so the digit before it will not change
        ↑
     = 31.30 (to 2 d.p.)   you **must** keep the 0 in your answer

You may also be asked to round to a number of **significant figures (s.f.)**.
To round a number to a given number of significant figures use the following rules.

> 1 Count along the digits to the required number of significant figures.
>   (The most significant figure is the first non-zero figure.)
> 2 Look at the next digit.
>   If it is less than 5, leave the digit before it as it is.
>   If it is 5 or more, you must round up the digit before it.

EXAMPLE

a Write 62934 correct to 3 significant figures.
b Write 0.00047928 correct to 2 significant figures.
c Write 0.08702 correct to 3 significant figures.

a 62934 = 629⋮34   the next digit is 3 so the digit before does not change
        ↑
     = 62900 (to 3 s.f.)
b 0.00047928 = 0.00047⋮928   the next digit is 9 so the digit before it must round up
        ↑
     = 0.00048 (to 2 s.f.)
c 0.08702 = 0.0870⋮2   the next digit is 2 so the digit before does not change
        ↑
     = 0.0870 (to 3.s.f.)   you must keep the 0 after the 7 in your answer

UNIT 1

EXERCISE 1.11

1. Round each of these correct to 1 decimal place (1 d.p.).
   - a 5.347
   - b 8.75
   - c 5.99
   - d 30.021
   - e 0.06
   - f 8.88
   - g 4.2067
   - h 16.255
   - i 643.991
   - j 7.006

2. Round each of these correct to 2 decimal places (2 d.p.).
   - a 6.352
   - b 8.388
   - c 16.1555
   - d 4.2081
   - e 5.6999
   - f 0.0384
   - g 0.0666
   - h 2.994
   - i 3.056
   - j 7.1025

3. Round each of these correct to 3 decimal places (3 d.p.).
   - a 7.2222
   - b 6.1616
   - c 35.6855
   - d 82.0089
   - e 24.7894
   - f 3.00002
   - g 6.0091
   - h 6.66666
   - i 102.1028
   - j 5.0096

4. Round each of these correct to 1 significant figure (1 s.f.).
   - a 2.56
   - b 8.09
   - c 9.7
   - d 48.26
   - e 352
   - f 26700
   - g 51000
   - h 0.0022
   - i 0.087
   - j 0.308

5. Round each of these correct to 2 significant figures (2 s.f.).
   - a 1.7328
   - b 3.094
   - c 5680
   - d 61300
   - e 15.93
   - f 20.07
   - g 103.24
   - h 0.00633
   - i 0.0577
   - j 0.000399

6. Round each of these correct to 3 significant figures (3 s.f.).
   - a 27.348
   - b 6.5137
   - c 2587
   - d 148.8
   - e 16.745
   - f 0.3492
   - g 0.07175
   - h 0.008076
   - i 10.066
   - j 39992

7. In 2010 the population of the world was estimated to be 6 831 000 000. Round this number to 1 s.f.

8. The total area of the Atlantic Ocean is about 106 400 000 square kilometres. Round this number to
   - a 1 s.f.
   - b 2 s.f.
   - c 3 s.f.

9. Mount Everest is the highest mountain in the world. The summit is 8848 metres above sea level. Round this number to
   - a 1 s.f.
   - b 2 s.f.
   - c 3 s.f.

10. An egg has mass of 0.06395 kg. Round this number to
    - a 1 s.f.
    - b 2 s.f.
    - c 3 s.f.
    - d 1 d.p.
    - e 2 d.p.
    - f 3 d.p.

11. The length of the Earth's equator is 40 008.629 km. Round this number to
    - a 1 s.f.
    - b 2 s.f.
    - c 3 s.f.
    - d 4 s.f
    - e 1 d.p.
    - f 2 d.p.

Estimating

It is important that you know how to check if your answer to a calculation is a sensible answer. To do this you need to **estimate** the answer.
To estimate the answer to a calculation use the following steps.

> 1 Round each of the numbers to 1 significant figure.
> 2 Do the calculation using your rounded numbers.

EXAMPLE

Estimate the answers to these calculations.

a $5.94591 \div 2.0738$

b $\dfrac{(19.719)^2 - 2.9513 \times 22.628}{685 \div (5.16 + 1.7694)}$

a $5.94591 \div 2.0738$ *first round each of the numbers to 1 s.f.*
   $\approx 6 \div 2$
   $= 3$ *[the answer using a calculator is 2.86715….]*

b $\dfrac{(19.719)^2 - 2.9513 \times 22.628}{685 \div (5.16 + 1.7694)}$ *first round each of the numbers to 1 s.f.*

   $\approx \dfrac{(20)^2 - 3 \times 20}{700 \div (5+2)}$ *remember to do the operations in the correct order (BIDMAS)*

   $= \dfrac{400 - 60}{700 \div 7} = \dfrac{340}{100} = 3.4$ *[the answer using a calculator is 3.257899….]*

EXERCISE 1.12

1 Estimate the answers to these calculations.
   a 6.4×9.8 b 2.16×7.79 c 289×12.3
   d 49.3×52.1 e 218×372.9 f $17.94 \div 3.15$
   g $22.8 \div 3.89$ h $188.3 \div 42.76$ i $5.76 + 4.73 \times 2.28$

2 Estimate the answers to these calculations.

   a $\dfrac{84.3 + 17.8}{7.16 + 2.947}$ b $\dfrac{8.26 - 1.99}{1.207 + 1.806}$ c $\dfrac{5.16 + (1.94)^2}{7.26 - 3.78}$

3 Use a calculator to find the answers to questions **1** and **2**, give your answers correct to four significant figures. Comment on whether your estimate is either an over- or an under-estimate.

4 Estimate the cost of 2.16 kg of bananas at $3.84 per kg.

5 A theatre sells tickets for a show. They sell 103 tickets at $18.99 per ticket, 321 tickets at $21.50 per ticket and 48 tickets at $27.99 per ticket. Estimate the total amount of money obtained from the sale of the tickets.

> **KEY WORDS**
> round
> decimal places (d.p.)
> significant figures (s.f.)
> estimate

Simplifying algebraic expressions

THIS SECTION WILL SHOW YOU HOW TO
- Collect like terms
- Expand single brackets

You can simplify **expressions** by collecting **like terms**.

EXAMPLE

Simplify these expressions.
a $9x + 5y - 3x - 8y$ **b** $4x^2 - 3x - 7 - 2x^2 - 5x + 1$ **c** $7x^2y - 3xy - 7xy - 2x^2y$

a $9x + 5y - 3x - 8y$ find the like terms
   $= 9x + 5y - 3x - 8y$ put the like terms next to each other
   $= 9x - 3x + 5y - 8y$ $9 - 3 = 6$ and $+5 - 8 = -3$
   $= 6x - 3y$

b $4x^2 - 3x - 7 - 2x^2 - 5x + 1$ find the like terms
   $= 4x^2 - 3x - 7 - 2x^2 - 5x + 1$ put the like terms next to each other
   $= 4x^2 - 2x^2 - 3x - 5x - 7 + 1$ $4 - 2 = 2$ $-3 - 5 = -8$ $-7 + 1 = -6$
   $= 2x^2 - 8x - 6$

c $7x^2y - 3xy - 7xy - 2x^2y$ find the like terms
   $= 7x^2y - 3xy - 7xy - 2x^2y$ put the like terms next to each other
   $= 7x^2y - 2x^2y - 3xy - 7xy$ $7 - 2 = 5$ and $-3 - 7 = -10$
   $= 5x^2y - 10xy$

EXERCISE 1.13

Simplify these expressions.

1. $5x - 8x + 2x$
2. $3y - 4y - 2y$
3. $8xy - 2xy$
4. $6xy - 9yx$
5. $6x - 4y + x - 5y$
6. $3p + 9q + 2p - 4q$
7. $3xy + 5x + 2xy - 8x$
8. $5x^2 - 9x - 6x + 2x^2$
9. $9 - 4x - 2 + 8x$
10. $4xy - 3yx + 2xy - 10yx$
11. $x^2 - 6 - 4x^2 + 15$
12. $7y^2 - 2y^2 + 3y - 5y - 11y^2$
13. $6xy - x + 2xy$
14. $a + 8ab - 12ab$
15. $7ab - 4bc + 3ab - 2bc$
16. $x^2 + x^4 + x^2 - 5 + 3x^2$
17. $6x^2 + 5x^3 - 7x^2 + x$
18. $7fg + 2gh - 5fg - 6gh - 4fg$
19. $6a^2b - 4ab^2 - 12ab^2 - 2a^2b$
20. $8cd^2 - 5c^2d - 4c^2d - 3cd^2$
21. $x + \dfrac{3}{x} + 5x + \dfrac{7}{x}$
22. $\dfrac{8}{x} - \dfrac{2}{y} + \dfrac{3}{x} - \dfrac{5}{y} - \dfrac{2}{x}$
23. $3xy^2 + 7xy - 6x^2y - 5y^2x - 4x^2y + x^2 + 5xy - y^2$
24. $7xy - 5x^2 + 3y^2 - 2x^2 - 9xy + 7x^2 - 15y^2 + 3 + 2xy + 12y^2$

Expressions involving brackets

$2(x + 3)$ means the same as $2 \times (x + 3)$
The diagram below helps to show that $2(x + 3) = 2x + 6$

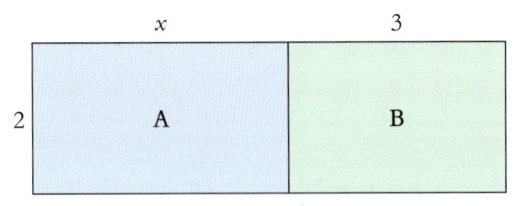

	x	3
2	A	B

The area of rectangle A is $2x$
The area of rectangle B is 6
The area of the whole rectangle is $2x + 6$

When you multiply out the brackets you must multiply each term inside the bracket by the term outside the bracket.

$5(2x + 3y)$ means $5 \times 2x + 5 \times 3y = 10x + 15y$

EXAMPLE

Expand **a** $7(2a - 5b)$ **b** $9x(3x - 4y + 8)$

a $7(2a - 5b)\ \ = 7 \times 2a - 7 \times 5b$
$\qquad\qquad\qquad = 14a - 35b$

b $9x(3x - 4y + 8) = 9x \times 3x - 9x \times 4y + 9x \times 8$
$\qquad\qquad\qquad\qquad = 27x^2 - 36xy + 72x$

You need to be very careful when multiplying by a negative number
To expand $-2(x - 6)$ you multiply both terms in the bracket by -2.
$\qquad -2(x - 6) = (-2 \times x) + (-2 \times -6) = -2x + 12$

EXAMPLE

Expand **a** $-5(3x + 6)$ **b** $-8(7 - 5y)$

a $-5(3x + 6) = (-5 \times 3x) + (-5 \times 6)$
$\qquad\qquad\qquad = -15x + -30$
$\qquad\qquad\qquad = -15x - 30$

b $-8(7 - 5y) = (-8 \times 7) + (-8 \times -5y)$
$\qquad\qquad\qquad = -56 + 40y$

EXAMPLE

Expand and simplify
a $6(2x - 3) + 5(3x - 1)$ **b** $7(2x - 3) - 5(3x - 2)$

a $6(2x - 3) = 12x - 18$ and $+5(3x - 1) = +15x - 5$ *expand each set of brackets*
$\quad 6(2x - 3) + 5(3x - 1) = 12x - 18 + 15x - 5$ *collect like terms*
$\qquad\qquad\qquad\qquad\quad = 27x - 23$

b $7(2x - 3) = 14x - 21$ and $-5(3x - 2) = -15x + 10$ *expand each set of brackets*
$\quad 7(2x - 3) - 5(3x - 2) = 14x - 21 - 15x + 10$ *collect like terms*
$\qquad\qquad\qquad\qquad\quad = -x - 11$

EXERCISE 1.14

1. Freda says that $7(2x + 3) = 14x + 3$.
   Explain why she is wrong.

2. Expand these expressions.
   - a $4(x + 3)$
   - b $6(y + 2)$
   - c $4(x + 5)$
   - d $2(3 - a)$
   - e $7(y - 4)$
   - f $8(x - 9)$
   - g $5(2x + 3)$
   - h $6(3y + 4)$
   - i $7(5a + 6)$
   - j $3(2x + 2y - 1)$
   - k $5(4a + 5b - 3)$
   - l $8(3p - 4q + 7r)$
   - m $\frac{1}{2}(8x - 2)$
   - n $\frac{1}{4}(12x + 8)$
   - o $\frac{1}{3}(15y - 6)$

3. Expand these expressions.
   - a $-3(x + 4)$
   - b $-2(x - 6)$
   - c $-5(6 - 4x)$
   - d $-8(2x + 5)$
   - e $-7(3x - 8)$
   - f $-8(x - 9)$
   - g $-(5x + 4)$
   - h $-(2x - 7)$
   - i $-(3x + 8)$
   - j $-6(3x - 3y + 5)$
   - k $-4(3p + 4q - 5)$
   - l $-3(9x^2 + 3x - 2)$

4. Expand these expressions.
   - a $x(x + 3)$
   - b $y(y - 5)$
   - c $a(7 - a)$
   - d $2x(5 - 3x)$
   - e $5y(y + 8)$
   - f $3x(2x + 4y)$

5. Expand and simplify.
   - a $2(x + 3) + 3(x + 8)$
   - b $5(x - 2) + 6(x + 7)$
   - c $6(y + 1) + 3(y - 2)$
   - d $4(x - 8) + 7(x - 6)$
   - e $6(5x - 2y) + 2(3x + 4y)$
   - f $7(2x^2 - 3x) + 2(5x - 6x^2)$
   - g $x(x + 4) + x(x + 2)$
   - h $x(x - 3) + x(x + 7)$
   - i $2x(3x + 4) + 5x(6 - 3x)$
   - j $2x(3y - 4) + 3y(2x - 5)$
   - k $2x(3x - 5y) + 3y(2x + 7y) + 5x(3y - 2x)$

6. Joe says that $5(2x - 3) - 4(x - 6) = 10x - 15 - 4x - 24 = 6x - 39$.
   Explain why he is wrong.

7. Expand and simplify.
   - a $3(x - 4) - 2(x - 6)$
   - b $5(y + 2) - 3(y + 7)$
   - c $7(a + 2) - 8(3 - 2a)$
   - d $6(7x - 2) - 2(3x + 4)$
   - e $3(x + 4) - (x - 5)$
   - f $7(2 - y) - (y + 3)$
   - g $4(2h - 3g) - 3(3h - 2g)$
   - h $3x(2x + 1) - x(5x + 8)$
   - i $5p(p - 6) - 2p(7 - 3p)$
   - j $2a(3b - 5a + 7) - 5a(7a + 3b - 5) - 3a(2a - 5b - 8)$

8. Expand and simplify.
   - a $10 + 2(x + 3)$
   - b $15 - 3(x - 4)$
   - c $20 - 2(3x - 4y + 6)$
   - d $9 - 5(y + 2)$
   - e $8 - 7(x - 2)$
   - f $17x - 3(x + 5)$
   - g $10y - (2y - x)$
   - h $5x - 2x(x - 3)$

KEY WORDS
expression
like terms
expand

Solving linear equations

THIS SECTION WILL SHOW YOU HOW TO
- Set up simple equations
- Solve linear equations

EXAMPLE

Solve   **a** $6x - 5 = 11$ **b** $9 - 5x = 2$

a $6x - 5 = 11$

$6x - 5 + 5 = 11 + 5$ add 5 to both sides

$6x = 16$

$\dfrac{6x}{6} = \dfrac{16}{6}$ divide both sides by 6

$x = 2\dfrac{2}{3}$ change to a mixed number

b $9 - 5x = 2$

$9 - 5x + 5x = 2 + 5x$ add $5x$ to both sides

$9 = 2 + 5x$

$9 - 2 = 2 + 5x - 2$ subtract 2 from both sides

$7 = 5x$ divide both sides by 5

$\dfrac{7}{5} = \dfrac{5x}{5}$ divide both sides by 5

$x = 1\dfrac{2}{5}$ change to a mixed number

EXAMPLE

Solve   **a** $3(x - 4) = 5(2x + 3)$ **b** $5 - 6y = 2 - 4y$

a $3(x - 4) = 5(2x + 3)$ expand the brackets

$3x - 12 = 10x + 15$

$3x - 12 - 3x = 10x + 15 - 3x$ take $3x$ from both sides

$-12 = 7x + 15$

$-12 - 15 = 7x + 15 - 15$ take 15 from both sides

$-27 = 7x$ divide both sides by 7

$\dfrac{-27}{7} = \dfrac{7x}{7}$

$x = -3\dfrac{6}{7}$ change to a mixed number

b $5 - 6y = 2 - 4y$

$5 - 6y + 6y = 2 - 4y + 6y$ add $6y$ to both sides

$5 = 2 + 2y$

$5 - 2 = 2 + 2y - 2$ take 2 from both sides

$3 = 2y$

$\dfrac{3}{2} = \dfrac{2y}{2}$ divide both sides by 2

$y = 1\dfrac{1}{2}$ change to a mixed number

EXERCISE 1.15

1 Solve these equations.
- a $2x + 3 = 15$
- b $7x - 1 = 62$
- c $5x + 2 = 37$
- d $6x - 2 = 5$
- e $8x + 3 = 15$
- f $4x - 1 = 52$
- g $18 - 2x = 12$
- h $6 - 3x = 8$
- i $10 - 4x = 7$
- j $30 - 8x = 15$
- k $10 - 2x = 15$
- l $6 - 5x = 20$
- m $3x + 4 = 27$
- n $4(y - 5) = 14$
- o $2(x + 3) = 7$
- p $3(2x + 5) = 26$
- q $4(3x - 1) = 80$
- r $5(3x + 8) = 4$
- s $2(13 - x) = 22$
- t $5(10 - 3x) = 8$

2 Solve these equations.
- a $12x + 5 = 8x + 13$
- b $7x + 2 = 3x + 22$
- c $6x - 3 = 5x + 4$
- d $2x - 3 = 6x - 17$
- e $3x + 1 = 2 - 5x$
- f $7x - 9 = 10 + 2x$
- g $3x + 7 = 8 - 2x$
- h $7x - 3 = 9x + 10$
- i $3 - 2x = 8 - x$
- j $10 - 3x = 5 + 2x$
- k $6 - 7x = 3 - 5x$
- l $5 - 2x = 10 - 7x$

3 Solve these equations.
- a $2(x - 3) + 4(2x - 7) = 6$
- b $3x + 11 = 5(x + 1)$
- c $7 - x = 3(5 - x)$
- d $5(2x + 2) = 6(x + 5)$
- e $3(2x + 5) - 2(x - 5) = 51$
- f $3(3x - 2) - 7(x - 2) = 0$
- g $4(1 - 2x) - 2(3 - x) = 0$
- h $6(2x - 5) - 4(x - 2) = 14$
- i $37 - 2(3x + 1) = 30 - 3x$

4 The sum of three consecutive numbers is 171. Find the numbers.

HINT

let the numbers be x, $x + 1$ and $x + 2$.

5 AB is a straight line. Find the value of x.

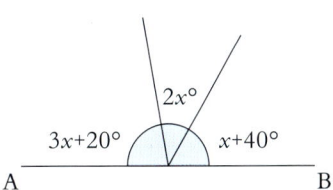

Angles on line AB: $3x+20°$, $2x°$, $x+40°$

HINT

the angles on a straight line add up to 180°.

6 ABCD is a rectangle. Find the value of x.

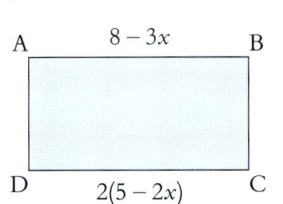

Rectangle ABCD with AB = $8 - 3x$ and DC = $2(5 - 2x)$

HINT

opposite sides of a rectangle are equal.

7 The perimeter of the triangle is 10 cm. Find the value of x.

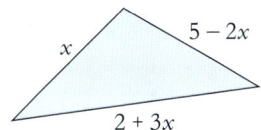

Triangle with sides x, $5 - 2x$, $2 + 3x$

8 Expression A is 18 more than expression B. Find the value of x.

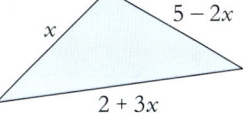

A: $6x - 7$ B: $15 - 2x$

Solving linear equations   21

Solving linear equations involving fractions

EXAMPLE

Solve a $\dfrac{x}{2} + 4 = 12$ b $\dfrac{x-3}{4} = 5$ c $\dfrac{3}{x} - 2 = 8$

a $\dfrac{x}{2} + 4 = 12$ subtract 4 from both sides

$\dfrac{x}{2} = 8$ multiply both sides by 2

$2 \times \dfrac{x}{2} = 2 \times 8$ ➡ **NOTE:** $2 \times \dfrac{x}{2} = x$

$x = 16$

b $\dfrac{x-3}{4} = 5$ multiply both sides by 4

$4 \times \dfrac{(x-3)}{4} = 4 \times 5$ ➡ **NOTE:** $4 \times \dfrac{(x-3)}{4} = x - 3$

$x - 3 = 20$ add 3 to both sides

$x = 23$

c $\dfrac{3}{x} - 2 = 8$ add 2 to both sides

$\dfrac{3}{x} = 10$ multiply both sides by x

$x \times \dfrac{3}{x} = 10 \times x$ ➡ **NOTE:** $x \times \dfrac{3}{x} = 3$

$3 = 10x$ divide both sides by 10

$x = \dfrac{3}{10}$

The next example shows how to solve an equation when there is a single fraction on both sides of the equation.

You multiply both fractions by the lowest common multiple (LCM) of the two denominators.

EXAMPLE

Solve $\dfrac{x-4}{5} = \dfrac{x-2}{6}$

$\dfrac{x-4}{5} = \dfrac{x-2}{6}$ multiply both sides by 30

$30 \times \dfrac{(x-4)}{5} = 30 \times \dfrac{(x-2)}{6}$ $30 \div 5 = 6$ and $30 \div 6 = 5$

$6(x - 4) = 5(x - 2)$ expand the brackets

$6x - 24 = 5x - 10$ add 24 to both sides

$6x = 5x + 14$ subtract $5x$ from both sides

$x = 14$

HINT

you can simplify an equation with a single fraction on each side by 'cross-multiplying'.

$\dfrac{x-4}{5} = \dfrac{x-2}{6}$

$6(x - 4) = 5(x - 2)$

UNIT 1

EXERCISE 1.16

1 Solve these equations.

   a $\dfrac{x}{4} = 5$ b $\dfrac{x}{3} = 20$ c $\dfrac{2x}{5} = 30$ d $\dfrac{3x}{4} = 5$

   e $\dfrac{x}{3} + 2 = 17$ f $\dfrac{x}{4} - 1 = -8$ g $\dfrac{2x}{5} - 3 = 7$ h $\dfrac{3x}{7} + 2 = 6$

   i $\dfrac{x}{2} + 10 = 3$ j $16 - \dfrac{x}{2} = 5$ k $18 - \dfrac{x}{3} = 10$ l $19 - \dfrac{4x}{5} = 3$

2 Solve these equations.

   a $\dfrac{2x+1}{3} = 5$ b $\dfrac{3x+2}{5} = 7$ c $\dfrac{4x-1}{3} = 4$ d $\dfrac{6-2x}{3} = 1$

   e $\dfrac{10-4x}{5} = 4$ f $\dfrac{2(x+4)}{3} = 6$ g $\dfrac{3(2x-1)}{4} = 6$ h $\dfrac{5(2-3x)}{4} = -3$

3 Solve these equations.

   a $\dfrac{24}{x} = 8$ b $24 = \dfrac{12}{x}$ c $4 = \dfrac{7}{x}$ d $5 = \dfrac{3}{2x}$

   e $\dfrac{5}{x} - 2 = 8$ f $\dfrac{2}{x} + 4 = 5$ g $\dfrac{30}{x} - 2 = 4$ h $\dfrac{8}{x} - 7 = -11$

   i $5 - \dfrac{3}{x} = 3$ j $24 - \dfrac{12}{x} = 30$ k $18 - \dfrac{15}{2x} = 15$ l $9 - \dfrac{36}{x} = 18$

   m $\dfrac{8}{x+1} = 2$ n $\dfrac{6}{x-1} = 3$ o $\dfrac{10}{3x+2} = 2$ p $\dfrac{8}{2-5x} = 4$

4 Solve these equations.

   a $\dfrac{3-x}{3} = \dfrac{2+x}{2}$ b $\dfrac{x+1}{4} = \dfrac{8-x}{5}$ c $\dfrac{2x+2}{5} = \dfrac{8-x}{2}$ d $\dfrac{3x-1}{4} = \dfrac{x}{2}$

   e $\dfrac{5x+2}{3} = \dfrac{3x}{4}$ f $\dfrac{x+1}{3} = \dfrac{x-1}{4}$ g $\dfrac{x+5}{4} = \dfrac{3x+1}{5}$ h $\dfrac{6x-3}{5} = \dfrac{4x+5}{2}$

5 Explain why you cannot solve the equation $\dfrac{6x+2}{3} = \dfrac{4x-5}{2}$

6 The perimeter of the equilateral triangle is 15 cm. Find the value of x.

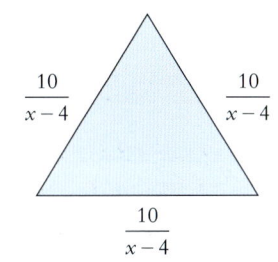

Sides of the triangle: $\dfrac{10}{x-4}$, $\dfrac{10}{x-4}$, $\dfrac{10}{x-4}$

KEY WORDS
equation
solve

Solving linear equations

Constructing formulae

THIS SECTION WILL SHOW YOU HOW TO
- Construct formulae

A **formula** is a rule that shows how **variables** are connected.

EXAMPLE

A chocolate has a mass of 15 grams and a sweet has a mass of 22 grams.
I have x chocolates and $x + 7$ sweets.
Write down a formula for the total mass M, in grams, of the chocolates and sweets.

mass of x chocolates = $15x$
mass of $x + 7$ sweets = $22(x + 7)$
$M = 15x + 22(x + 7)$ *expand brackets*
$M = 15x + 22x + 154$ *collect like terms*
$M = 37x + 154$

➡ **NOTE:** the two variables in this question are x and M.

EXERCISE 1.17

1. A theatre sells 135 tickets at $10 each and n tickets at $13 each
   Write down a formula for the total cost P, in $, of the tickets.

2. I have $100 to spend. I buy n books at $9 each.
   Write down a formula for the amount of money Q, in $, that I have left after buying the books.

3. Light bulbs cost $$b$ each and candles cost $$c$ each.
   Write down a formula for the total cost C, in $, of 4 light bulbs and 7 candles.

4. Rulers cost r cents each, pencils cost p cents each and files cost $$f$ each.
   Write down a formula for the total cost T, in $, of 5 rulers, 6 pencils and 4 files.

5. The cost of hiring a tent is made up of two parts. There is a fixed charge of $30 and then an extra charge of $4 for each day that the tent is rented.
   Write down a formula for the total cost T, in $, when a tent is rented for n days.

6. The instructions for calculating the roasting time for a chicken are given below.

   Allow 45 minutes per kg plus 20 minutes.

   Write down a formula for the total cooking time T, in minutes, to cook a chicken that has a mass of m kg.

KEY WORDS
formula
variable

Substitution into formulae

THIS SECTION WILL SHOW YOU HOW TO
- Replace letters by numbers in an expression
- Replace letters by numbers in a formula

Substitution into an **expression** means replacing the letters in an expression by the given numbers.

EXAMPLE

If $a = 2$, $b = 3$ and $c = -5$ find the value of these expressions.

a $a^2 - bc$ **b** $ac + bc$ **c** $\dfrac{4bc^2 + 5a^2}{2c}$

a $a^2 - bc = 2^2 - (3 \times -5)$ replace the letters with the given numbers
$= 4 - -15 = 19$ $4 - -15 = 4 + 15$

b $ac + bc = (2 \times -5) + (3 \times -5)$ replace the letters with the given numbers
$= -10 + -15 = -25$ $-10 + -15 = -10 - 15$

c $\dfrac{4bc^2 + 5a^2}{2c} = \dfrac{4 \times 3 \times (-5)^2 + 5 \times 2^2}{2 \times -5}$ replace the letters with the given numbers

$\phantom{\dfrac{4bc^2 + 5a^2}{2c}} = \dfrac{300 + 20}{-10}$ $4 \times 3 \times (-5)^2 = 4 \times 3 \times 25$ because $-5 \times -5 = 25$

$\phantom{\dfrac{4bc^2 + 5a^2}{2c}} = \dfrac{320}{-10}$ remember that $+ \div - = -$

$\phantom{\dfrac{4bc^2 + 5a^2}{2c}} = -32$

EXERCISE 1.18

If $a = -4$, $b = 2$ and $c = -3$ find the value of these expressions.

1. $3a + 2b$
2. $5c - 4b$
3. $6a - 3b + 2c$
4. $-5c - 2b$
5. $a^2 + b^2$
6. $c^2 - 3b$
7. $a(b + c)$
8. $7c(2a - b)$
9. $ab + bc + ac$
10. $abc + ab$
11. $3ab^2$
12. $5a^2 bc$
13. $a - (2c)^2$
14. $(3a)^2 - (2b)^2$
15. $b^2 + 5b - 2$
16. $c^2 - 4c + 7$
17. $\dfrac{ac}{b}$
18. $\dfrac{a^2}{b}$
19. $5 - \dfrac{ac}{b}$
20. $\dfrac{b^2 c^2}{a}$
21. $\dfrac{a + b + c}{2}$
22. $\dfrac{2a + 7b}{b}$
23. $\dfrac{a + c}{a + b}$
24. $\dfrac{4ac - 1}{c}$
25. $\dfrac{3b + 2c}{a^2 + b^2}$
26. $\dfrac{a^2 + b^2 + c^2}{a + b + c}$
27. $\dfrac{3a}{b} + \dfrac{4c}{a}$
28. $8 - \dfrac{ac^2}{b}$

Substitution into formulae   25

Substitution into a **formula** means replacing the letters in a formula by the given numbers.

EXAMPLE

The formula $v = u + at$ is used in physics.
Work out the value of v when $u = 20$, $a = 3$ and $t = 5$.

$v = u + at$ replace the letters with the given numbers
$v = 20 + 3 \times 5$ remember to do the operations in the correct order (BIDMAS)
$v = 20 + 15$
$v = 35$

EXAMPLE

The formula $s = ut + \frac{1}{2}at^2$ is also used in physics.
Work out the value of s when $u = 40$, $t = 10$ and $a = -5$.

$s = ut + \frac{1}{2}at^2$ replace the letters with the given numbers

$s = 40 \times 10 + \frac{1}{2} \times -5 \times 10^2$ $40 \times 10 = 400$ and $\frac{1}{2} \times -5 \times 100 = -250$

$s = 400 + \frac{1}{2} \times -5 \times 100$

$s = 400 + -250$ $400 + -250 = 400 - 250$
$s = 150$

You may need to solve an equation after substituting the numbers into the formula.

EXAMPLE

The formula $\frac{1}{f} = \frac{1}{u} + \frac{1}{v}$ is used in physics.
Work out the value of u when $f = 3$ and $v = 4$.

$\frac{1}{f} = \frac{1}{u} + \frac{1}{v}$ replace the letters with the given numbers

$\frac{1}{3} = \frac{1}{u} + \frac{1}{4}$ take $\frac{1}{4}$ from both sides

$\frac{1}{u} = \frac{1}{3} - \frac{1}{4}$ $\frac{1}{3} - \frac{1}{4} = \frac{4}{12} - \frac{3}{12}$

$\frac{1}{u} = \frac{1}{12}$

$u = 12$

EXERCISE 1.19

1. Use the formula $C = \dfrac{5(F-32)}{9}$ to work out the value of C when
   a. $F = 14$ b. $F = 41$ c. $F = -4$ d. $F = 212$

2. Use the formula $s = \dfrac{1}{2}(u + v)t$ to work out the value of s when
   a. $u = 3, v = 7, t = 5$ b. $u = -2, v = 5, t = 6$ c. $u = 8, v = -3, t = 4$

3. Use the formula $f = \dfrac{uv}{u+v}$ to work out the value of f when
   a. $u = 10, v = 15$ b. $u = 3, v = 7$ c. $u = 3.8, v = 1.2$

4. Use the formula $A = \sqrt{\dfrac{hm}{3600}}$ to work out the value of A when
   a. $h = 180, m = 20$ b. $h = 120, m = 43.2$ c. $h = 150, m = 77.76$

5. The cost C, in \$, of hiring a car for n days is given by the formula
   $C = 120 + 45n$
   a. Find C when $n = 7$ b. Find n when $C = 840$

6. The formula for the area A of a trapezium is
   $A = \dfrac{1}{2}(a + b)h$
   a. Find h when $a = 3, b = 7, A = 25$
   b. Find b when $a = 5, h = 4, A = 26$

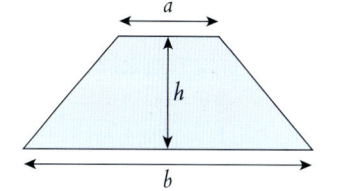

7. The formula for the total surface area A of a cuboid is
   $A = 2lw + 2hw + 2hl$
   where l is the length, w is the width and h is the height.
   a. Find h when $A = 310, w = 5, l = 10$
   b. Find w when $A = 164, l = 3, h = 4$

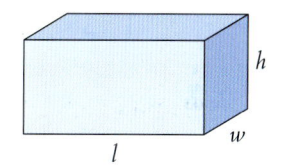

8. The formula for the volume V of a square based pyramid is
   $V = \dfrac{1}{3}x^2 h$
   where x is the length of the sides of the square base and h is the height of the pyramid.
   a. Find h when $V = 300, x = 10$
   b. Find x when $h = 12, V = 256$.

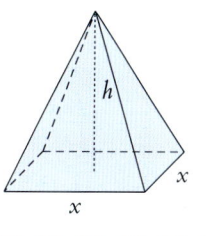

KEY WORDS
substitution
expression

Substitution into formulae

Gradients and straight line graphs

THIS SECTION WILL SHOW YOU HOW TO
- Find the gradient of a straight line
- Draw straight line graphs

The **gradient** of a straight line is a measure of how steep the line is.

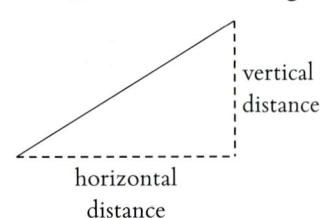

$$\text{gradient} = \frac{\text{vertical distance}}{\text{horizontal distance}}$$

Gradients can be either positive or negative.
This line has positive gradient.   This line has negative gradient.

EXAMPLE

On a graph draw the line joining the points (−2, 6) and (6, 2) and find its gradient.

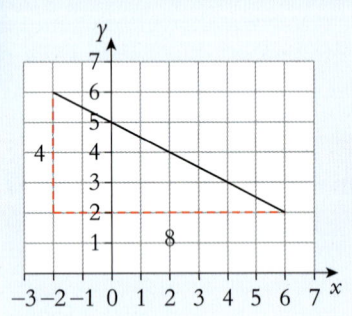

$$\text{gradient} = \frac{\text{vertical distance}}{\text{horizontal distance}}$$

$$\text{gradient} = -\frac{4}{8} = -\frac{1}{2}$$ ➡ **NOTE:** the gradient is negative

The gradient of a line passing through two points can also be calculated using a formula.

The gradient, m, of the line passing through the points (x_1, y_1) and (x_2, y_2) is given by $m = \dfrac{y_2 - y_1}{x_2 - x_1}$

EXAMPLE

Find the gradient of the line joining (−2, −3) and (6, −1).

(−2, −3)   (6, −1)   *first decide which values to use for x_1, y_1, x_2 and y_2.*
 ↑   ↑   ↑   ↑
(x_1, y_1) (x_2, y_2)

$$\text{gradient} = \frac{y_2 - y_1}{x_2 - x_1} = \frac{-1 - -3}{6 - -2}$$

$$= \frac{2}{8} = \frac{1}{4}$$ *simplify the fraction*

EXERCISE 1.20

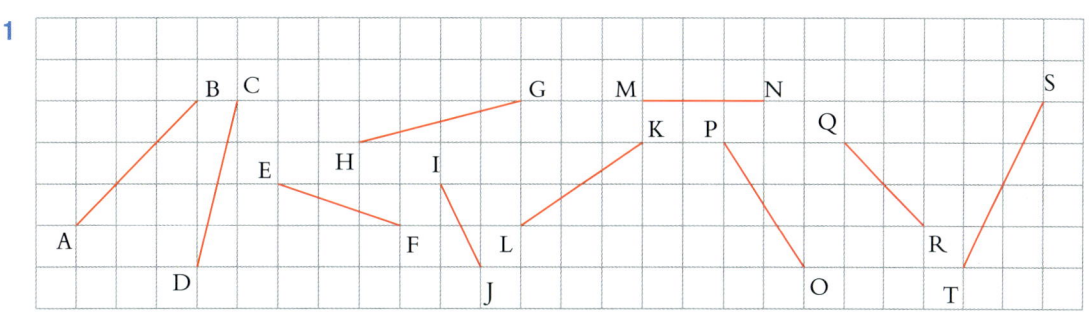

1 Find the gradient of the lines.

a	AB	b	CD	c	EF	d	GH	e	IJ
f	KL	g	MN	h	OP	i	QR	j	ST

2 Use a graph to help you work out the gradient of the line joining these pairs of points.
- a (5, 3), (6, 4)
- b (−2, 4), (6, −5)
- c (−1, 1), (3, 7)
- d (2, −4), (5, −3)
- e (1, 9), (4, −3)
- f (5, 6), (−2, −3)
- g (6, 4), (8, 4)
- h (6, −3), (5, 8)
- i (2, −4), (5, 8)
- j (14, 0), (10, 4)
- k (−7, −8), (−2, −6)
- l (0, −6), (4, −5)

3 Calculate the gradient of the line joining these pairs of points using the formula: gradient = $\dfrac{y_2 - y_1}{x_2 - x_1}$.
- a (2, 3), (5, 6)
- b (−2, 4), (1, 10)
- c (4, 2), (6, 3)
- d (2, −3), (6, 5)
- e (2, 5), (6, 3)
- f (1, 4), (1, −2)
- g (−3, −1), (−6, −4)
- h (6, −3), (5, −3)

4 The line joining (2, 1) to (7, a) has a gradient of 1. Find the value of a.

5 The line joining (4, 2) to (6, b) has a gradient of 3. Find the value of b.

6 The line joining (0, 3) to (6, c) has a gradient of $\dfrac{2}{3}$. Find the value of c.

7 The line joining (3, 6) to (7, d) has a gradient of $\dfrac{1}{2}$. Find the value of d.

8 The line joining (−4, −1) to (−1, e) has a gradient of 2. Find the value of e.

9 The line joining (−2, 1) to (6, 3) is parallel to the line joining (2, −4) to (6, f).
Find the value of f.

10 The line joining (8, 6) to (2g, −g) has a gradient of 2. Find the value of g.

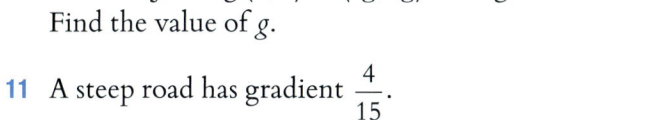

11 A steep road has gradient $\dfrac{4}{15}$.
The road rises by 3 meters. Calculate the horizontal distance x.

The graphs of $x = c$, $y = c$, and $y = x$

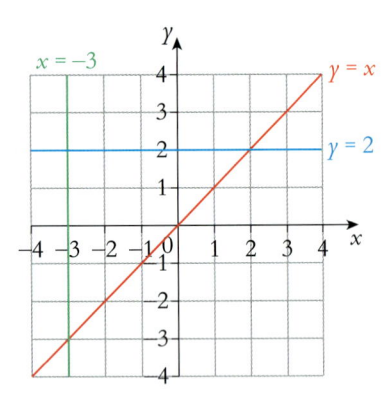

The x and y coordinates are the same at every point on the line $y = x$
eg $(1, 1), (-3, -3), (2, 2) \ldots$

The y coordinate is 2 at every point on the line $y = 2$
eg $(4, 2), (0, 2), (-3, 2) \ldots$

The x coordinate is -3 at every point on the line $x = -3$
eg $(-3, 2), (-3, -1), (-3, 4) \ldots$

➡ **NOTE:** these are important graphs that you should remember

Straight line graphs of the form $y = ax + b$

EXAMPLE

Plot the graph of $y = \frac{1}{2}x + 1$

Choose 3 values of x and work out the corresponding y values.

when $x = 0$, $y = \left(\frac{1}{2} \times 0\right) + 1 = 1$

when $x = 2$, $y = \left(\frac{1}{2} \times 2\right) + 1 = 2$

when $x = 4$, $y = \left(\frac{1}{2} \times 4\right) + 1 = 3$

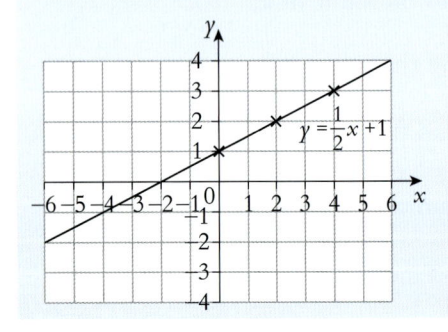

x	0	2	4
y	1	2	3

➡ **NOTE:** to find each y value you halve the x value and then add 1

EXAMPLE

The line $y = ax - 5$ passes through the point (4, 11)
Find the value of a.

$y = ax - 5$ substitute $x = 4$ and $y = 11$ into the equation
$11 = 4a - 5$ add 5 to both sides
$4a = 16$ divide both sides by 4
$a = 4$

EXERCISE 1.21

1. Write down the equations of each of these lines.

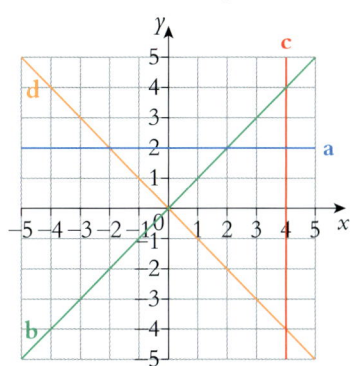

2. For each of the following copy and complete the table of values and then draw a graph of the function. Use coordinate axes with values from −6 to 6.

   a $y = x + 2$

x	0	1	2
y			

   b $y = x - 5$

x	0	1	2
y			

   c $y = 2x - 2$

x	0	1	2
y			

   d $y = 3x - 4$

x	0	1	2
y			

   e $y = \frac{1}{2}x - 1$

x	-2	0	2
y			

   f $y = \frac{1}{3}x - 4$

x	-3	0	3
y			

   g $y = 3 - 2x$

x	0	1	2
y			

   h $y = 5 - 3x$

x	0	1	2
y			

3. Plot these graphs on the same set of axes.

   $y = 3x$ $y = 3x + 2$ $y = 3x - 5$

   What do you notice about the graphs?

4. Plot these graphs on the same set of axes.

   $y = 3x - 2$ $y = 2x + 1$

   Write down the coordinates of the point where the two lines inersect.

5. Plot these graphs on the same set of axes.

   $y = 2x - 3$ $y = \frac{1}{2}x + 3$

   Write down the coordinates of the point where the two lines inersect.

6. The graph $y = ax + 4$ passes through the point (2, 8). Find the value of a.

7. The graph $y = bx - 3$ passes through the point (−2, 6). Find the value of b.

8. The graph $y = 3x + c$ passes through the point (3, −5). Find the value of c.

Gradients and straight line graphs   31

Graphs of the form $ax + by = c$

An easy method to draw or sketch a graph of this form is to find the axis crossing points.
Putting $x = 0$ into the equation will tell you where the graph crosses the y-axis.
Putting $y = 0$ into the equation will tell you where the graph crosses the x-axis.

EXAMPLE

Sketch the graph of $3x + 4y = 12$.

when $x = 0$, $4y = 12$
$y = 3$ the graph crosses the y-axis at $(0, 3)$
when $y = 0$, $3x = 12$
$x = 4$ the graph crosses the x-axis at $(4, 0)$

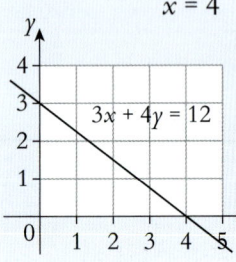

EXERCISE 1.22

Copy and complete the table for each of the following functions and sketch the graph.

1. $2x + y = 4$

x	0	
y		0

2. $3x + 2y = 6$

x	0	
y		0

3. $2x + 5y = 10$

x	0	
y		0

4. $y + 3x = 9$

x	0	
y		0

5. $2x + 3y = -12$

x	0	
y		0

6. $6x + 5y = -30$

x	0	
y		0

7. $x - 2y = 6$

x	0	
y		0

8. $4x - 5y = 20$

x	0	
y		0

9. $2x - 3y = 12$

x	0	
y		0

10. $y - 3x = 15$

x	0	
y		0

11. $5x - 3y = -30$

x	0	
y		0

12. $2x - \frac{1}{3}y = -4$

x	0	
y		0

KEY WORDS

gradient

Angle properties

THIS SECTION WILL SHOW YOU HOW TO
- Use the basic facts about angles

You should already know the following facts.

Isosceles triangle *Equilateral triangle*

Names of angles
acute angles — between 0° and 90°
right angle — 90°
obtuse angle — between 90° and 180°
reflex angle — more than 180°

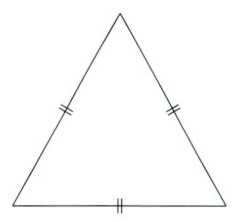

Angle properties

Angles on a straight line

$a + b = 180°$

Angles at a point

$a + b + c = 360°$

Vertically opposite angles

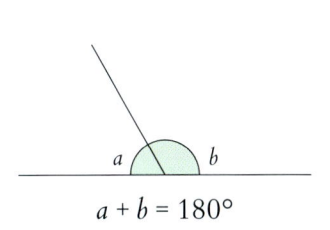

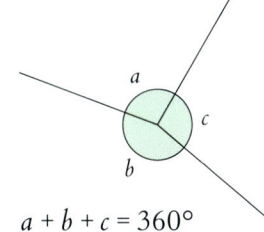

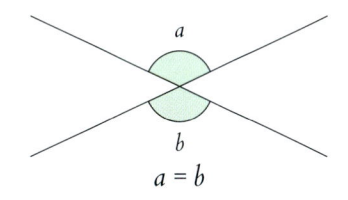

$a = b$

Angles in a triangle

$a + b + c = 180°$

Exterior angles of a triangle

$a + b = c$

Angles in a quadrilateral

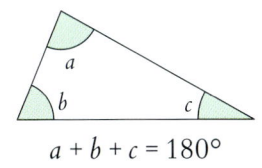

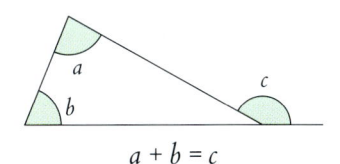

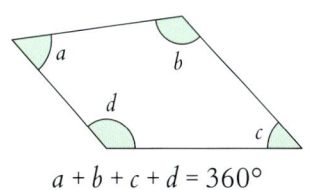

$a + b + c + d = 360°$

Parallel lines

Perpendicular lines

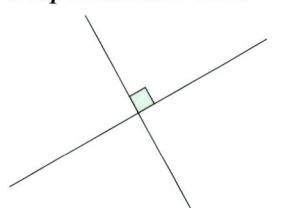

Corresponding angles

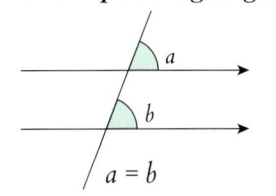

$a = b$

Alternate angles

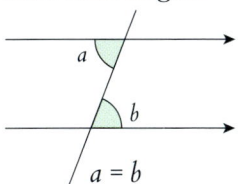

$a = b$

Angle properties 33

EXAMPLE

Find the lettered angles giving reasons for your answers.

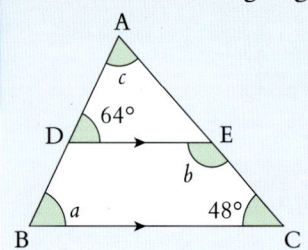

$A\hat{D}E = D\hat{B}C$ $a = 64°$ *corresponding angles*
$A\hat{E}D = E\hat{C}B = 48°$
$C\hat{E}D + A\hat{E}D = 180°$ $b = 132°$ *angles on a straight line*
$D\hat{A}E + A\hat{E}D + E\hat{D}A = 180°$ $c = 68°$ *angles in a triangle*

EXERCISE 1.23

Calculate the size of each lettered angle.

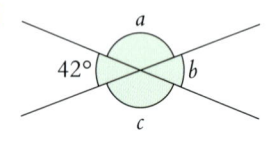

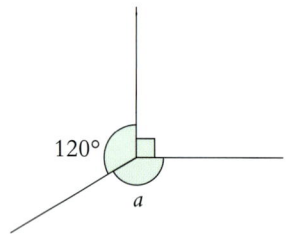

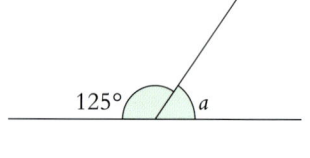

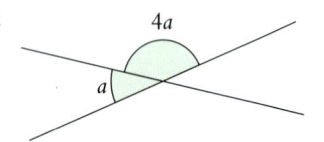

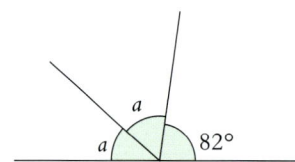

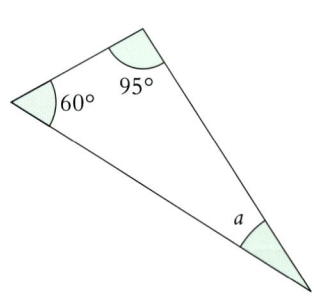

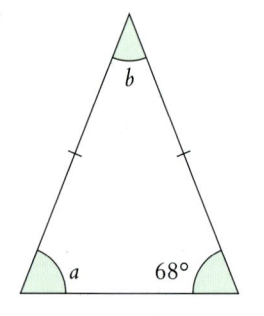

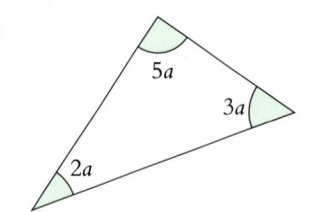

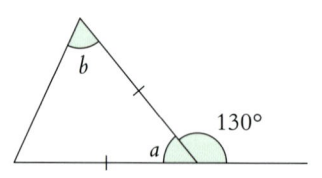

10 — triangle with angle a at top, two equal sides, exterior angle 110°

11 — triangle with angles b and a at top vertex, 70°, with tick marks

12 — right-angled triangle with angle a at top and $a + 30°$ at right

13 — triangle with 20° angle, angle a, and a right angle

14 — quadrilateral with 60°, a, and two right angles, parallel arrows

15 — quadrilateral with 115°, 100°, a, and a right angle

16 — quadrilateral with 120°, $2a$, a, and a right angle

17 — two parallel lines cut by a transversal; angles a, b, 140°

18 — two parallel lines cut by a transversal; angles $3a$ and a

19 — quadrilateral with angles a, b, 80°, 50°; parallel arrows

20 — zig-zag between parallel lines; 40°, a, 25°

21 — zig-zag between parallel lines; 160°, a, 140°

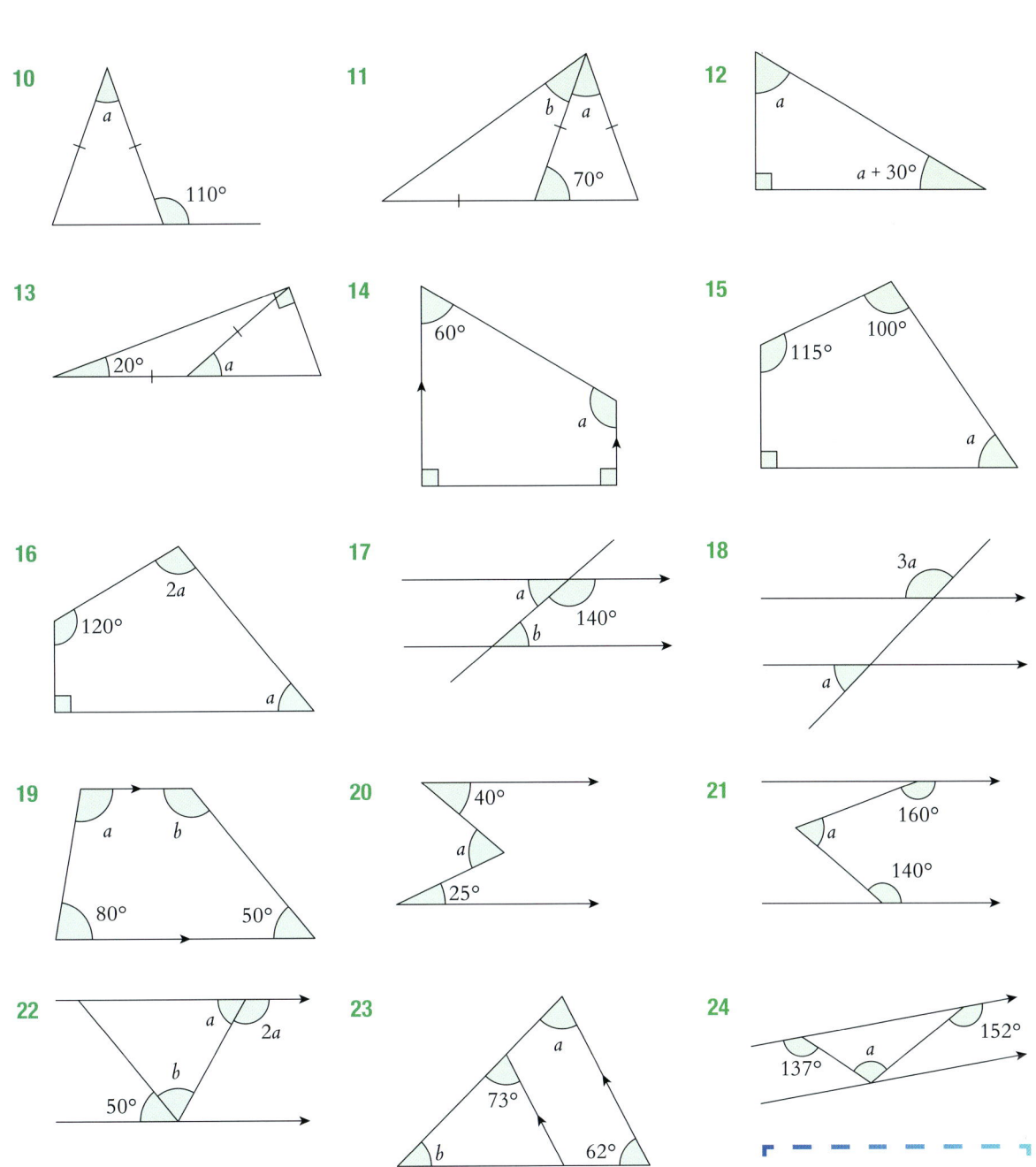

22 — triangle on a line with a, $2a$, b, 50°

23 — triangle with a, 73°, 62°, b and an internal parallel line

24 — triangle between parallel lines with 137°, a, 152°

25 The four angles of a quadrilateral are x, $2x$, $3x$ and $4x + 30°$. Calculate the size of each of the angles.

26

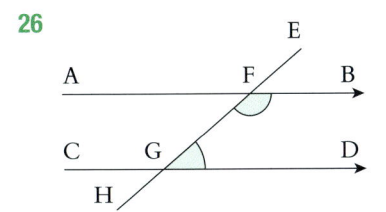

Two parallel lines AB and CD cut by transversal EH passing through F on AB and G on CD.

Prove that the two interior angles $B\hat{F}G$ and $F\hat{G}D$ add up to 180°.

KEY WORDS
acute
obtuse
reflex
right-angle
parallel
perpendicular
corresponding
alternate

Angle properties

Symmetry

THIS SECTION WILL SHOW YOU HOW TO
- Describe line symmetry and rotational symmetry
- Draw planes of symmetry

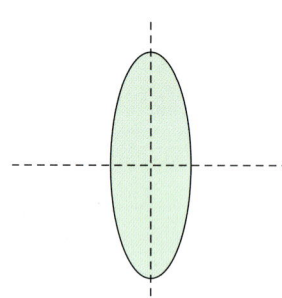

This shape has 2 **lines of symmetry**.

> A line of symmetry divides a shape into two identical parts.

It also has **rotational symmetry** of order 2.

> Rotational symmetry of order 2 means that if you rotate the shape about its centre through one complete turn, it will fit onto itself twice.

EXAMPLE

Describe the symmetry of these two shapes.

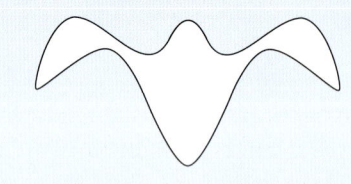

This shape has 0 lines of symmetry and rotational symmetry of order 4.

This shape has 1 (vertical) line of symmetry and rotational symmetry of order 1.

EXERCISE 1.24

1. For each quadrilateral write down the number of lines of symmetry and the order of rotational symmetry.

   a square
   b 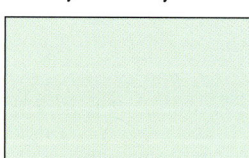 rectangle
   c 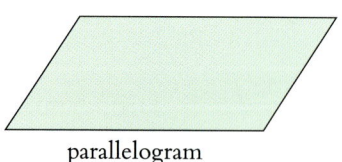 parallelogram
   d 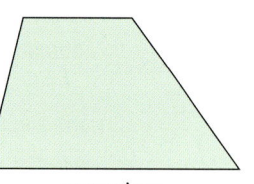 trapezium
   e 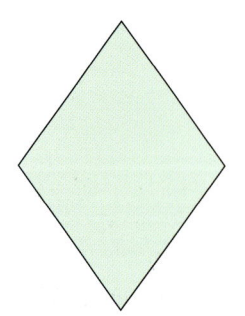 rhombus
   f 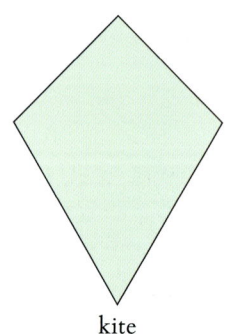 kite

36 UNIT 1

2 For each shape write down the number of lines of symmetry and the order of rotational symmetry.

a 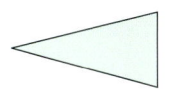   b 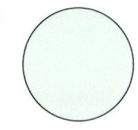   c

d    e    f

g    h 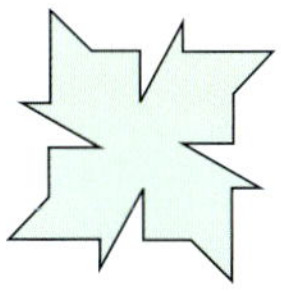   i

j    k    l

3 For each pattern write down the number of lines of symmetry and the order of rotational symmetry.

a 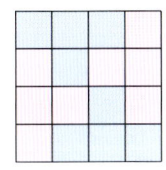   b 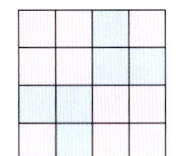   c

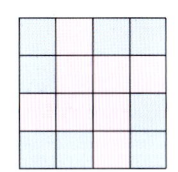

4 Make three copies of this grid. Colour each grid so that it has
   a One line of symmetry and rotational symmetry of order 1
   b Two lines of symmetry and rotational symmetry of order 2
   c No lines of symmetry and rotational symmetry of order 2

5 For each of the following curves write down the number of lines of symmetry and the order of rotational symmetry.

a    b    c    d

6 For an unknown quadrilateral you are told how many lines of symmetry it has, its order of rotational symmetry, whether its diagonals cross at right-angles and whether the diagonals bisect one another. Given this information can you always identify the quadrilateral.

HINT
Start by constructing a table of properties of quadrilaterals.

Symmetry   37

Planes of symmetry

This cuboid has 3 **planes of symmetry**.

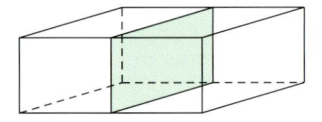

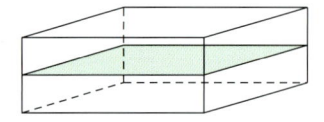

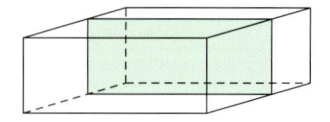

A plane of symmetry cuts a solid into two equal parts.
Each part is a mirror image of the other.

EXAMPLE

How many planes of symmetry does a square based pyramid have?

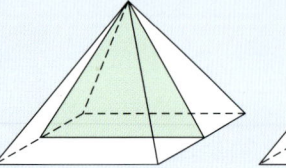

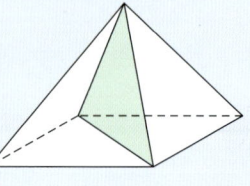

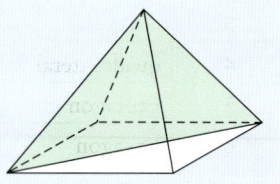

A square based pyramid has 4 planes of symmetry.

EXERCISE 1.25

1 The end face of this prism is an equilateral triangle.
   A plane of symmetry is shown.
   The prism has three more planes of symmetry.
   Draw three diagrams to show these planes of symmetry.

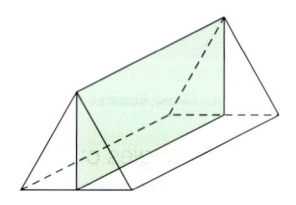

2 How many planes of symmetry are there for each of these solid objects.

   a 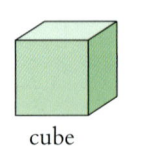 cube
   b  cylinder
   c  sphere
   d  cone
   e 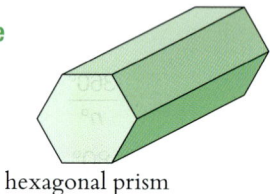 hexagonal prism

3 These solids are made from small cubes.
   How many planes of symmetry does each solid have?

   a 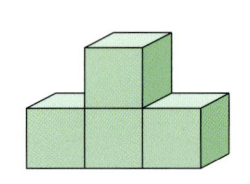   b 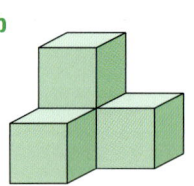   c 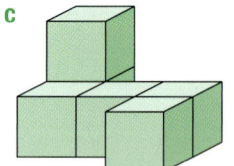   d

4 Draw a prism whose cross section is an isosceles trapezium.
   How many planes of symmetry does the prism have?

KEY WORDS
line of symmetry
order of rotational symmetry
plane of symmetry

Polygons

THIS SECTION WILL SHOW YOU HOW TO
- Find the sum of the interior angles in a polygon
- Find the size of exterior and interior angles in a regular polygon

A **polygon** is a shape enclosed by straight lines.
An n-sided polygon can be divided into $(n-2)$ triangles.

Number of sides	Name of polygon	Sum of interior angles
3	triangle	$(3-2) \times 180° = 180°$
4	quadrilateral	$(4-2) \times 180° = 360°$
5	pentagon	$(5-2) \times 180° = 540°$
6	hexagon	$(6-2) \times 180° = 720°$

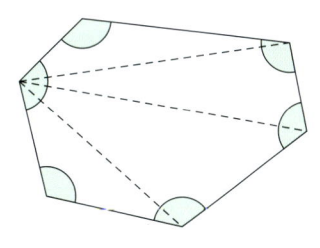

the sum of the **interior angles** in an n-sided polygon $= (n-2) \times 180°$

interior angle + exterior angle = 180°

the exterior angles of any polygon always add up to 360°

A **regular polygon** has all sides equal and all angles the same.

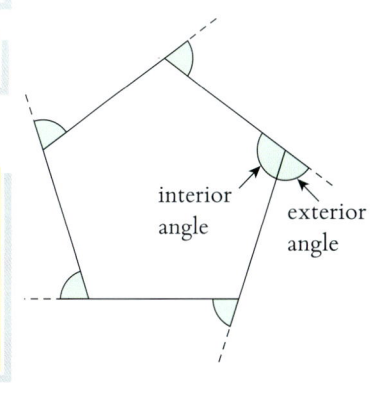

In a regular n-sided polygon
exterior angle $= \dfrac{360°}{n°}$
interior angle $= 180° - \dfrac{360°}{n°}$

EXAMPLE

ABCDE is a regular pentagon.
The lines BA and DE are extended to meet at F.
Calculate the size of angle AFE.

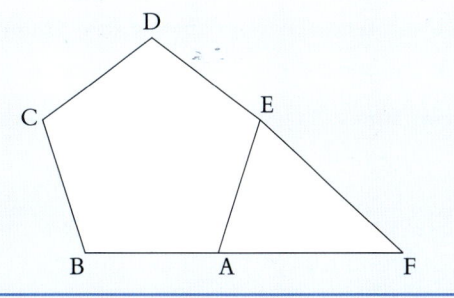

Angle FAE $= \dfrac{360}{5} = 72°$ \qquad angle FAE is an exterior angle of the pentagon
Angle FEA is also 72°
Angle AFE $= 180 - (72 + 72)$ \qquad angles in a triangle add up to 180°
$= 180 - 144 = 36°$

Polygons 39

EXERCISE 1.26

1. Jay says that the sum of the interior angles of a six-sided polygon is 6 × 180° = 1080°. Is he correct? Explain your answer.

2. Four of the angles of a pentagon are 123°, 84°, 113° and 96°. Calculate the fifth angle.

3. A regular polygon has nine sides. Calculate the size of the interior angle.

4. The exterior angle of a regular polygon is 30°. How many sides does the polygon have?

5. The interior angle of a regular polygon is 162°. How many sides does it have?

6. A polygon has 10 sides. Find the sum of the interior angles.

7. Calculate the size of x for each of the diagrams.

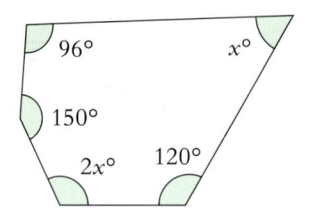

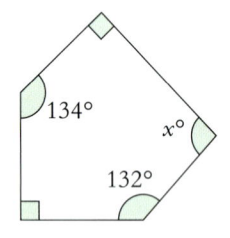

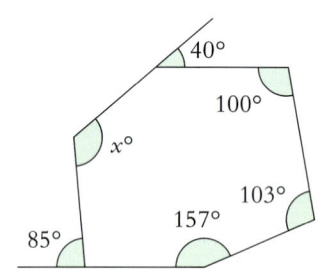

   a. (angles: 120°, 91°, 145°, 165°, 82°, $x°$)

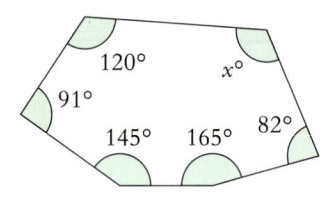

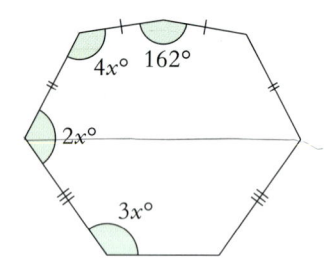

   b. (angles: 96°, $x°$, 150°, $2x°$, 120°)

   c. (angles: $4x°$, 162°, $2x°$, $3x°$)

   d. (angles: 134°, 132°, $x°$, and two right angles)

   e. (angles: 40°, 100°, $x°$, 85°, 157°, 103°)

   f. (angles: 85°, 120°, $x°$, 100°, $2x°$)

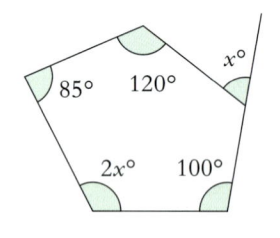

8. The diagram shows a regular octagon and a regular pentagon. Calculate:
   a. angle AED
   b. angle DEF
   c. angle HDE
   d. angle HDG
   e. angle ADE
   f. angle ADG
   g. angle AEF
   h. angle BDA

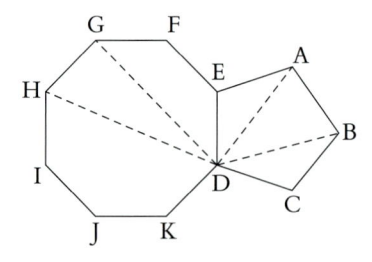

9. The interior angle of a regular polygon is eight times the size of the exterior angle. How many sides does the polygon have?

> **KEY WORDS**
> polygon
> regular polygon
> interior angle
> exterior angle

Averages and range

THIS SECTION WILL SHOW YOU HOW TO
- Find the mean, median, mode and range for a small data set

The three types of **average** that you need to know are

mean	the total of all the values divided by the number of values
median	the middle value when the data are arranged in order
mode	the value (or item) that occurs most often

The **range** is a measure of spread.

range = highest value − smallest value

EXAMPLE

The list shows the times taken by six students to solve a puzzle.

55 sec   48 sec   36 sec   55 sec   29 sec   41 sec

Work out
- **a** the mean
- **b** the median
- **c** the mode
- **d** the range for the times taken.

a   mean = $\dfrac{55+48+36+55+29+41}{6}$

$= \dfrac{264}{6}$

= 44 sec

b   29   36   ㊶   ㊽   55   55   *arrange in order of size first*

There are two middle values.
The median is the number half-way between these values.

median = $\dfrac{41+48}{2}$

= 44.5 sec

c   mode = 55 sec   *the value that occurs most often*

d   range = 55 − 29   *highest value − smallest value*

= 26 sec

Sometimes you might be asked to choose which type of average is most representative for a data set.

EXAMPLE

The list shows the test results of eight students.
   1   17   18   19   19   20   20   20
Find the mean, mode and median.
Comment on each of your averages.

mean = $\frac{1+17+18+19+19+20+20+20}{8} = \frac{134}{8} = 16.75$

The mean is not a representative average because only one student scored less than this.

mode = 20
The mode is not a representative average to use because everyone scored 20 or less.

median = 19
This is the most representative (best) average to use.

EXERCISE 1.27

1. 7, 3, 5, 6, 2, 1, 5

   Raju says that the median of the numbers is 6. Is he correct?
   Explain your answer.

2. Find the mean, median, mode and range for each list of numbers.

   a 8, 10, 11, 10, 6
   b 3, −6, −10, 5, −1, −3
   c 50, 63, 54, 65
   d 3, 1, 6, 5, 2, 3, 5, 4, 3, 1, 2, 3, 1
   e $1\frac{2}{3}, \frac{5}{6}, \frac{1}{2}, 1\frac{1}{3}$

3. The table shows the number of cars sold each month by a car salesman in one year.

Month	Jan	Feb	Mar	Apr	May	Jun	July	Aug	Sept	Oct	Nov	Dec
Number of cars	13	13	12	10	13	12	13	15	14	13	6	10

   Find the mean, median, mode and range for the data.

4. The times (in seconds) for 10 runners in a 100 metre race were

   10.35   10.28   10.05   10.25   10.28   10.65   10.47   10.33   10.28   10.42

   Find the mean, median, mode and range for the data.

5. For the following data sets explain, with reasons, which average you would
   use to describe them.

   a 0, 0, 1, 2, 4, 6, 7, 8, 8, 9
   b 6, 7, 7, 8, 29
   c 1, 2, 3, 4, 5, 11, 12, 13, 14, 15
   d −5, −3, 2, 3, 3

6. a The mean mass of 8 men is 82.4 kg. What is the total mass of the 8 men?
   b The mean length of 7 sticks is 1.02 m. What is the total length of the 7 sticks?

7 The mean height of six women is 1.64 m.
The heights of five of the women are
1.60 m, 1.62 m, 1.64 m, 1.61 m and 1.62 m.
Find the height of the sixth woman.

HINT
first find the total height of the six women.

8 a The mean of four numbers is 59.
The numbers are 52, 57, 58 and p.
Find the value of p.

b The mean of six numbers is 9.
The numbers are 5, 7, 7, x, 12 and 16.
Find the value of x.

c The mean of five numbers is 14.
The numbers are 9, y, 15, 15 and 18.
Find the value of y.

d The mean of seven numbers is 19.
The numbers are 6, 10, 11, 13, 17, 20 and w.
Find the value of w.

9 There are six women and four men in a group of ten people.
The mean age of the six women is 65.
The mean age of the four men is 73.
Find the mean age of the group of ten people.

10 Students from two different schools do an examination paper.
There are 150 students at School A and the mean mark for School A is 66.
There are 250 students at School B and the mean mark for School B is 58.
Calculate the mean mark of the 400 students.

11 A sample of fish is taken from two different lakes.
Seventy fish are taken from Lake A and the mean length of the fish is 35 cm.
Thirty fish are taken from Lake B and the mean length of the fish is 42 cm.
Calculate the mean length of the 100 fish.

KEY WORDS
average
mean
median
mode
range

Frequency tables

THIS SECTION WILL SHOW YOU HOW TO
- Find the mean, median, mode and range for a frequency distribution

Large data sets are usually recorded in a **frequency table**.

EXAMPLE

The list shows the number of absences from a class over a period of 25 days.

0 3 2 2 0 2 2 0 2 2 2 1 0
1 2 1 2 0 1 3 2 1 1 2 2

Arrange the data into a frequency table.
Use your frequency table to work out the
  a mode   b median   c range   d mean.

Number of absences (x)	Frequency (f)
0	5
1	6
2	12
3	2

a Mode = 2 absences   the mode is the number of absences with the highest frequency

b Median = 2 absences   the median is the (25 + 1) ÷ 2 = 13th number. There are 11 numbers up to the end of the 1's, so the 13th number must be a 2.

c Range = 3 absences   highest value − lowest value = 3 − 0

d

Number of absences (x)	Frequency (f)	x × f
0	5	0 × 5 = 0
1	6	1 × 6 = 6
2	12	2 × 12 = 24
3	2	3 × 2 = 6
Total	25	36

A third column is used to find the total number of absences for each row

$$\text{Mean} = \frac{\text{total number of absences}}{\text{total number of days}}$$

$$= \frac{36}{25} = 1.44 \text{ absences}$$

UNIT 1

EXERCISE 1.28

1. a Gina says the mode for the number of fish is 6.
      Is she correct? Explain your answer.
   b Gregor says that the median number of fish is 2.
      Is he correct? Explain your answer.

Number of fish	Frequency
0	5
1	6
2	5
3	3
4	1

2. The table shows the number of people in each car passing a checkpoint.
   a Find the mode for the number of people in a car.
   b Find the median number of people in a car.
   c Copy and complete the table.
      Calculate the mean number of people in a car.

Number of people	Frequency
1	52
2	38
3	8
4	2

Number of people (x)	Frequency (f)	$f \times x$
1	52	1 × 52 =
2	38	
3	8	
4	2	
Total		

3. Twenty children were asked how many pets they own. The table shows the results.

Number of pets	0	1	2	3	4
Frequency	5	9	4	1	1

   Find the mode, median and mean number of pets.

4. The table shows the number of suitcases belonging to each of 100 air passengers.

Number of suitcases	0	1	2	3
Frequency	3	55	38	4

   Find the mode, median and mean number of suitcases.

5. The table shows the scores in a quiz for 40 students.

Quiz score	5	6	7	8	9
Frequency	3	7	17	12	1

   Find the mode, median and mean quiz score.

6. The table shows the test marks for n students. The mean test mark is 4.

Test mark	1	2	3	4	5
Frequency	3	5	6	2	x

   a Find the values of x and n.
   b Find the median test mark.

KEY WORDS
frequency table

Unit 1 Examination questions

1 Write the number 1045.2781 correct to

   (a) 2 decimal places, [1]
   (b) 2 significant figures. [1]

Cambridge IGCSE Mathematics 0580, Paper 21 Q7, June 2008

2 The table gives the average surface temperature (°C) on the following planets.

Planet	Earth	Mercury	Neptune	Pluto	Saturn	Uranus
Average temperature	15	350	−220	−240	−180	−200

   (a) Calculate the range of these temperatures. [1]
   (b) Which planet has a temperature 20°C lower than that of Uranus? [1]

Cambridge IGCSE Mathematics 0580, Paper 2 Q3, June 2006

3 Write down the next two prime numbers after 47. [2]

Cambridge IGCSE Mathematics 0580, Paper 21 Q1, June 2008

4 p is the largest prime number between 50 and 100.
q is the smallest prime number between 50 and 100.

Calculate the value of $p - q$. [2]

Cambridge IGCSE Mathematics 0580, Paper 21 Q3, June 2010

5 Write each number correct to 1 significant figure and estimate the value of the calculation.
You must show your working.

$$2.65 \times 4.1758 + 7.917$$ [2]

Cambridge IGCSE Mathematics 0580, Paper 21 Q1, November 2010

6 Expand the brackets and simplify.

$$\frac{1}{2}(6x - 2) - 3(x - 1)$$ [2]

Cambridge IGCSE Mathematics 0580, Paper 21 Q3, November 2010

7 Two quantities c and d are connected by the formula $c = 2d + 30$.
Find c when $d = -100$. [1]

Cambridge IGCSE Mathematics 0580, Paper 2 Q1, November 2006

8 Angharad has an operation costing $500.
She was in hospital for x days.
The cost of nursing care was $170 for each day she was in hospital.

(a) Write down, in terms of x, an expression for the total cost of her operation and nursing care. [1]
(b) The total cost of her operation and nursing care was $2370.
Work out how many days Angharad was in hospital. [2]

Cambridge IGCSE Mathematics 0580, Paper 2 Q15, June 2006

9

For the diagram above write down

(a) the order of rotational symmetry, [1]
(b) the number of lines of symmetry. [1]

Cambridge IGCSE Mathematics 0580, Paper 21 Q1, November 2009

10

Grade	1	2	3	4	5	6	7
Number of students	1	2	4	7	4	8	2

The table shows the grades by 28 students in a history test.

(i) Write down the mode. [1]
(ii) Find the median. [1]
(iii) Calculate the mean. [3]

Cambridge IGCSE Mathematics 0580, Paper 4 Q2 (a) (i) (ii) (iii), November 2007

Examination questions

11 (a) Shade **one** square in a copy of each diagram so that there is

  (i) one line of symmetry,

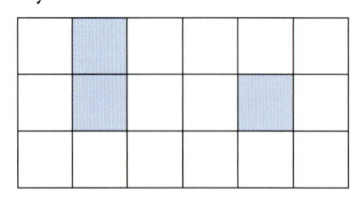

  [1]

  (ii) rotational symmetry of order 2.

  [1]

(b) On a copy of the diagram below, sketch one of the **planes** of symmetry of the cuboid.

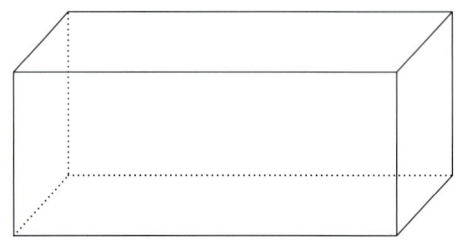

  [1]

(c) Write down the order of rotational symmetry of the equilateral triangular prism about the axis shown.

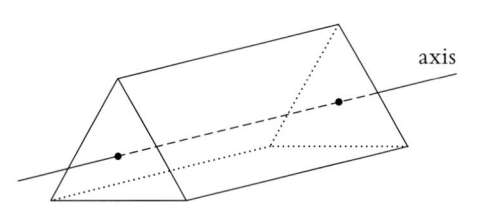

  [1]

Cambridge IGCSE Mathematics 0580, Paper 2 Q21, June 2006

12 A normal die, numbered 1 to 6, is rolled 50 times.

The results are shown in the frequency table.

Score	1	2	3	4	5	6
Frequency	15	10	7	5	6	7

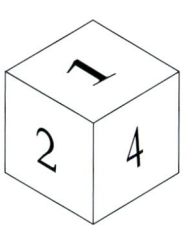

(a) Write down the modal score. [1]
(b) Find the median score. [1]
(c) Calculate the mean score. [2]
(d) The die is then rolled another 10 times.
 The mean score for the 60 rolls is 2.95.
 Calculate the mean score for the extra 10 rolls. [3]

Cambridge IGCSE Mathematics 0580, Paper 4 Q2, June 2009

13 40 students are asked about the number of people in their families.

The table shows the results.

Number of people in family	2	3	4	5	6	7
Frequency	1	1	17	12	6	3

(a) Find
  (i) the mode, [1]
  (ii) the median, [1]
  (iii) the mean. [3]
(b) Another n students are asked about the number of people in their families.
 The mean for the n students is 3.
 Find, in terms of n, an expression for the mean number for all $(40 + n)$ students. [2]

Cambridge IGCSE Mathematics 0580, Paper 41 Q2, June 2010

Percentages 1

THIS SECTION WILL SHOW YOU HOW TO
- Find percentages of a quantity
- Increase and decrease by a given percentage

Percentage means parts of 100 so $37\% = \frac{37}{100}$ $(= 0.37)$

Percentages are often used to compare quantities.

You should know how to convert fractions and decimals to percentages.

> To change a fraction or decimal to a percentage, multiply by 100

$\frac{1}{4} = \frac{1}{4} \times 100\% = 25\%$ $0.03 = 0.03 \times 100\% = 3\%$

> To change a percentage to a fraction or a decimal, divide by 100.

$29\% = 29 \div 100 = \frac{29}{100}$ or 0.29

EXAMPLE

a Find 27% of 540 m   **b** Find 8% of $350

a 27% of 540 m $= \frac{27}{100} \times 540$
$= 145.8$ m

Alternative method:
$0.27 \times 540 = 145.8$ m

b 8% of $350 $= \frac{8}{100} \times 350$
$= \$28$

Alternative method:
$0.08 \times 350 = \$28$

EXAMPLE

An airline increases its ticket prices by 15%.
Calculate the new price of a ticket costing $1300.

Increase in price = 15% of $1300
$= \frac{15}{100} \times 1300$
$= \$195$

New price = $1300 + $195
$= \$1495$

Quicker method:
multiplying factor is 1.15 ($= 1 + 0.15$)
$1.15 \times 1300 = \$1495$

EXAMPLE

A coat costing $180 is reduced by 12% in a sale.
Calculate the price of the coat in the sale.

Decrease in price = 12% of $180
$= \frac{12}{100} \times 180$
$= \$21.60$

New price = $180 − $21.60 = $158.40

Quicker method:
multiplying factor is 0.88 ($= 1 - 0.12$)
$0.88 \times 180 = \$158.40$

50 UNIT 2

EXERCISE 2.1

1. Write these percentages as fractions in their simplest form.
   a 24%   b 80%   c 58%   d 65%   e 30%

2. Write these fractions as percentages.
   a $\frac{3}{5}$ b $\frac{7}{20}$ c $\frac{18}{25}$ d $\frac{37}{50}$ e $\frac{7}{8}$

3. Write these percentages as decimals.
   a 18%   b 70%   c 6%   d 47%   e 2.5%

4. Write these decimals as percentages.
   a 0.28   b 0.8   c 0.08   d 1.2   e 0.755

5. Calculate the following.
   a 15% of $80   b 52% of 600 km   c 27% of 500 kg

6. Which is bigger: 35% of $50 or 30% of $60?

7. a Increase $58 by 12%   b Increase $72 by 24%
   c Increase $136 by 40%   d Increase $230 by 18%
   e Increase $440 by 2.5%   f Increase $90 by 120%

8. a Decrease $64 by 15%   b Decrease $400 by 32%
   c Decrease $250 by 8%   d Decrease $128 by 12.5%
   e Decrease $500 by 3.5%   f Decrease $360 by 0.2%

9. Nadia is paid $700 dollars a week. Her pay is increased by 5%.
   Calculate her new weekly pay.

10. A holiday is advertised as costing $950.
    Raju is given a 7% discount for booking online.
    Calculate how much he pays for the holiday.

11. A new car costs $1750.
    The car depreciates in value by 27% in its first year.
    Find the value of the car after one year.

12. A rare book is bought for $3250. It increases in value by 8%.
    Calculate the new value of the book.

13. Rukhusana and Shahida invested $300 each in company shares.
    Rukhusana's investment increased in value by 5.2% in the first year.
    Shahida's investment decreased in value by 0.8% in the first year.
    At the end of the first year, find how much more Rukhusana's investment is worth than Shahida's.

KEY WORDS
percentage

Ratio

THIS SECTION WILL SHOW YOU HOW TO
- Simplify ratios
- Divide a quantity in a given ratio
- Solve problems using ratios

Simplifying ratios

You can use **ratios** to compare one quantity with another quantity. If there are 4 teachers and 18 students on a school trip then the ratio of teachers to students can be written as

$$\text{teachers : students} = 4 : 18$$

Ratios are simplified in a similar way to fractions.

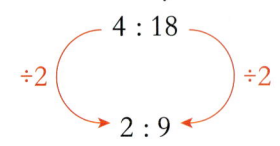

$$4 : 18 \xrightarrow{\div 2} 2 : 9$$

➡ **NOTE:** divide both sides of the ratio by the common factor 2.

This means that for every 2 teachers there are 9 students.

> A ratio in its simplest form (lowest terms) has integer values that cannot be cancelled further.

EXAMPLE

Write these ratios in their simplest form.

a 44 : 36   b 1.7 : 0.4   c $\dfrac{1}{2} : \dfrac{2}{3}$

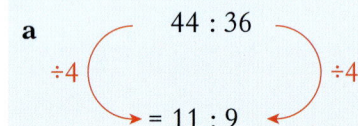

a $44 : 36 \xrightarrow{\div 4} 11 : 9$

b $1.7 : 0.4 \xrightarrow{\times 10} 17 : 4$

c $\dfrac{1}{2} : \dfrac{2}{3} \xrightarrow{\times 6} 3 : 4$

Writing a ratio in the form 1 : n

Map scales are often written as ratios in the form 1 : n

EXAMPLE

Write the ratio 2 cm : 5 km in the form 1 : n.

2 cm : 5 km
= 2 cm : 5000 m   *change 5 km to m*
= 2 cm : 500 000 cm   *change 5000 m to cm*
= 1 : 250 000   *divide both sides of the ratio by 2*

EXERCISE 2.2

1. Write these ratios in their simplest form.
   - a 20 : 30
   - b 18 : 24
   - c 6 : 30
   - d 40 : 25
   - e 29 : 36
   - f 55 : 33
   - g 48 : 42
   - h 63 : 99

2. Write these ratios in their simplest form.
   - a 200 g : 1 kg
   - b 10 cm : 2 m
   - c 5 mm : 2 cm
   - d 1 cm : 50 km
   - e 5 m : 20 cm
   - f 60 g : 3 kg
   - g 80 mm : 2 m
   - h 0.5 kg : 800 g
   - i 6 hours : 1 day

   HINT
   Change to the same units before simplifying.

3. Write these ratios in their simplest form.
   - a 5.6 : 2.4
   - b 0.8 : 0.6
   - c 1.2 : 0.4
   - d 1.5 : 7.5
   - e 0.08 : 0.2
   - f 1.04 : 0.44
   - g 4.9 : 0.7
   - h 3 : 0.02

4. Write these ratios in their lowest terms.
   - a $\frac{1}{2} : \frac{1}{4}$
   - b $\frac{2}{6} : \frac{5}{6}$
   - c $\frac{3}{4} : \frac{2}{3}$
   - d $\frac{1}{5} : \frac{1}{3}$
   - e $\frac{1}{10} : \frac{2}{5}$
   - f $\frac{3}{10} : \frac{1}{3}$
   - g $\frac{3}{8} : \frac{1}{4}$
   - h $2\frac{1}{2} : 1\frac{2}{3}$

5. Write these ratios in their simplest form.
   - a 10 : 40 : 50
   - b 6 : 10 : 8
   - c 5 : 15 : 25
   - d 64 : 24 : 56
   - e 21 : 45 : 77
   - f 0.5 : 2 : 3.5
   - g 1.6 : 2.4 : 0.2
   - h $\frac{1}{2} : \frac{3}{4} : \frac{1}{5}$

6. Write these ratios in the form 1 : n
   - a 2 : 8
   - b 6 : 4
   - c 25 : 4
   - d 8 : 5
   - e 1 cm : 5 km
   - f 5 cm : 20 m
   - g 2 cm : 15 km
   - h 3 mm : 1 km

7. Concrete is made from sand and cement.
   20 kg of concrete contains 15 kg of sand.
   Find the ratio of sand to cement in its simplest form.

8. There are 24 men, 60 women and 44 children on a bus.
   Write the ratio   men : women : children   in its simplest form.

9. Tickets are sold for a school concert.
   72 student tickets and 168 adult tickets are sold.
   Write the ratio   number of student tickets : number of adult tickets
   in its simplest form.

10 1   1   2   3   5   8   13   21   34   …

The rule to find the next term in this sequence is 'add together the two previous terms'.
So the next term = 21 + 34 = 55.
Suri investigates the ratio of adjacent numbers. She writes each ratio in the form $1 : n$.

Ratio of 1st number to 2nd number = 1 : 1
Ratio of 2nd number to 3rd number = 1 : 2
Ratio of 3rd number to 4th number = 2 : 3 = 1 : 1.5
Ratio of 4th number to 5th number = 3 : 5 = 1 : 1.666…

➡ **NOTE:** you can find out more about the golden ratio on the internet.

Write down the next six ratios and describe what happens to the ratio.

Dividing quantities in a given ratio

You can use ratios to share quantities.

Marla wants to share $36 between her two friends Lesley and Barbara in the ratio 3 : 1.
This means that Lesley receives three times as much as Barbara.
The diagram shows how the money is split.

To divide a quantity in a given ratio

- first find the total number of equal parts
- find the value of one part
- multiply to find each share

Lesley ($27)   Barbara ($9)
$9 $9 $9 $9

EXAMPLE

Share $42 between Kurt and Anna in the ratio 5 : 1.

Total number of parts = 5 + 1 = 6
Value of one part = $42 ÷ 6 = $7

Kurt receives 5 parts = 5 × $7 = $35
Anna receives 1 part = 1 × $7 = $7

➡ **CHECK:**
$35 + $7 = $42 ✓

EXAMPLE

Share $2000 between Adam, Jason and Omar in the ratio 1 : 2 : 5.

Total number of parts = 1 + 2 + 5 = 8
Value of one part = $2000 ÷ 8 = $250

Adam receives 1 part = 1 × $250 = $250
Jason receives 2 parts = 2 × $250 = $500
Adam receives 5 parts = 5 × $250 = $1250

➡ **CHECK:**
$250 + $500 + $1250 = $2000 ✓

EXAMPLE

Antonio and Filippo do a sponsored swim to raise money for charity.
The amounts they raise are in the ratio 3 : 7.
Antonio raised $177. How much did Filippo raise?

Antonio raised 3 parts.
Value of one part = $177 ÷ 3 = $59.
Filippo raised 7 parts = 7 × $59 = $413.

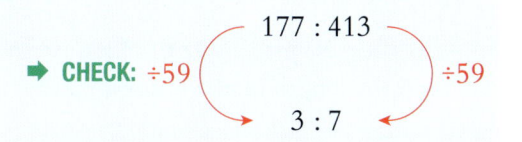

➡ **CHECK:** ÷59   177 : 413   ÷59
                   3 : 7

EXERCISE 2.3

1 Divide each amount in the ratio given.
   a $84 in the ratio 2 : 1
   b 280 m in the ratio 3 : 5
   c $2100 in the ratio 4 : 3
   d 624 km in the ratio 3 : 7
   e 182 kg in the ratio 8 : 5
   f $52.20 in the ratio 4 : 5

2 Divide each amount in the ratio given.
   a $420 in the ratio 1 : 2 : 3
   b 360 m in the ratio 5 : 1 : 2
   c $220 in the ratio 7 : 3 : 1
   d 2850 km in the ratio 6 : 5 : 8
   e 24 000 kg in the ratio 4 : 3 : 5
   f $1078.40 in the ratio 5 : 7 : 8

3 Jenny and Chris share some prize money in the ratio 5 : 3
  Jenny receives $52.25
  How much does Chris receive?

4 The three angles of a triangle are in the ratio 3 : 5 : 7
  Find the largest angle in the triangle.

5 The four angles of a quadrilateral are in the ratio 2 : 5 : 6 : 7
  Find the smallest angle in the quadrilateral.

6 The ratio of boys to girls in a class is 4 : 5
  What fraction of the class are boys?

7 $\frac{4}{7}$ of the children at a swimming club are girls.
  What is the ratio of girls to boys?

8 A bag contains orange sweets and yellow sweets.
  The ratio of orange sweets to yellow sweets is 3 : 2
  There are 51 orange sweets in the bag.
  Find the total number of sweets in the bag.

9 The ratio of adults to children on a train is 7 : 2
  There are 245 adults on the train.
  Find the total number of people on the train.

Ratio

10 At a football match, the ratio of male spectators to female spectators is 11 : 3.
There are 4257 male spectators.
Find the number of female spectators.

11 A length of rope is cut into two pieces in the ratio 4 : 5.
The shortest piece is 92 cm long.
Find the length of the longest piece.

12 Paul is exactly 6 years older than Jane.
The ratio of their ages is 5 : 3.
How old is Paul?

Map scales
Map scales are written as ratios in the form 1: n
A scale of 1 : 2 000 000 means that 1 cm on the map represents 2 000 000 cm on land.
You can convert the units of length as follows

   1 cm : 2 000 000 cm   *divide by 100 to change cm to m*
= 1 cm : 20 000 m   *divide by 1000 to change m to km*
= 1 cm : 20 km

So a map scale of 1 : 2 000 000 means that 1 cm on the map represents 20 km on the land.

EXAMPLE

This map has a scale of 1 : 5 000 000.
Use the map to calculate the actual distance between Buenos Aires and Montevideo.

Distance on map = 4.4 cm.
    1 cm : 5 000 000 cm
= 1 cm : 50 000 m
= 1 cm : 50 km
So 1 cm on the map represents 50 km on land.
Actual distance = 4.4 × 50 = 220 km.

EXAMPLE

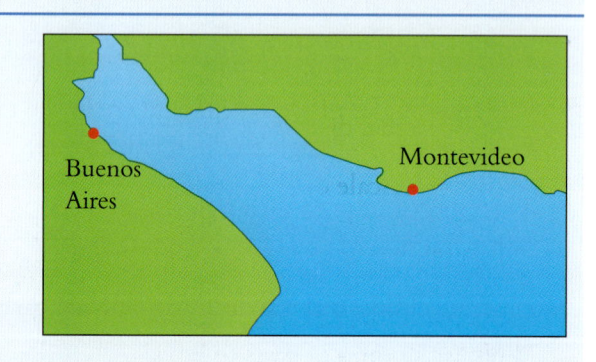

The distance between two villages is 16 km.
How far apart will they be on a map of scale 1 : 200 000?

   1 : 200 000 = 1 cm : 200 000 cm
             = 1 cm : 2000 m
             = 1 cm : 2 km
So each centimetre on the map represents 2 km on land.
Distance between villages on the map = $\frac{16}{2}$ = 8 cm.

EXERCISE 2.4

1. The map has a scale of 1 : 2 000 000.
   a. Use a ruler to measure the distance on the map between
      i. A and B,   ii. B and C,   iii. A and C.
   b. Find the actual distance (in km) between
      i. A and B,   ii. B and C,   iii. A and C.

2. A map has a scale of 1 : 1 000 000
   The distance between two lighthouses on the map is 4 cm.
   What is the actual distance, in kilometres, between the two lighthouses?

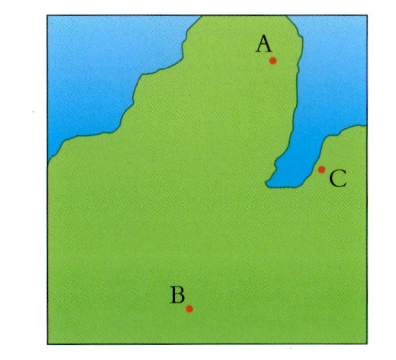

3. The scale of a map is 1 : 500 000
   Find the actual distance, in kilometres, represented by these distances on the map.
   a. 3 cm   b. 6.5 cm   c. 4.2 cm   d. 9 mm

4. The scale of a map is 1 : 200 000
   Find the actual distance, in kilometres, represented by these distances on the map.
   a. 4 cm   b. 2.5 cm   c. 3.8 cm   d. 7 mm

5. A map has a scale of 1 : 50 000
   Find the distance on the map that would represent an actual distance of
   a. 1 km   b. 5 km   c. 2.5 km   d. 4.3 km

6. The scale of a map is 1 : 100 000
   The distance between two houses on the map is 3.2 cm.
   Find the actual distance, in kilometres, between the two houses.

7. A map has a scale of 1 : 50 000. A river is 6 cm long on the map.
   Find the actual length, in kilometres, of the river.

8. A map has a scale of 1 : 500 000. The actual distance between two villages is 30 km.
   Find the distance between the two villages on the map.

9. A map has a scale of 1 : 400 000. Two towns are 6.8 cm apart on the map.
   Find the actual distance between the two towns.

10. A map has a scale of 1 : 10 000 000. The actual distance between two cities is 480 km.
    Find the distance between the two cities on the map.

11. The length of a section of motorway is 39 km. How long will the section of motorway be on a map whose scale is 1 : 500 000?

KEY WORDS
ratio

Ratio

Indices 1

THIS SECTION WILL SHOW YOU HOW TO
- Use simple laws of indices

○ $6 \times 6 \times 6 \times 6 = 6^4$ ← index or power
← base

Multiplying: $2^3 \times 2^4 = (2 \times 2 \times 2) \times (2 \times 2 \times 2 \times 2) = 2^7$

> **RULE 1** $a^m \times a^n = a^{m+n}$

Dividing: $4^5 \div 4^2 = \dfrac{4 \times 4 \times 4 \times \cancel{4} \times \cancel{4}}{\cancel{4} \times \cancel{4}} = 4^3$

> **RULE 2** $a^m \div a^n = a^{m-n}$

Raising to a power: $(3^2)^4 = (3 \times 3) \times (3 \times 3) \times (3 \times 3) \times (3 \times 3) = 3^8$

> **RULE 3** $(a^m)^n = a^{mn}$

EXAMPLE

Simplify   **a** $x^3 \times x^4$ **b** $p^9 \div p^3$ **c** $(q^4)^5$

a $x^3 \times x^4 = x^{3+4}$ add the powers
$= x^7$

b $p^9 \div p^3 = p^{9-3}$ subtract the powers
$= p^6$

c $(q^4)^5 = q^{4 \times 5}$ multiply the powers
$= q^{20}$

EXAMPLE

Simplify   **a** $5x^3 \times 2x^4 y^2 \times 3y^5$ **b** $\dfrac{6x^5 y^8}{3x^2 y^4}$

a $5x^3 \times 2x^4 y^2 \times 3y^5 = 5 \times 2 \times 3 \times x^3 \times x^4 \times y^2 \times y^5$
$= 30 \times x^{3+4} \times y^{2+5}$
$= 30x^7 y^7$

b $\dfrac{6x^5 y^8}{3x^2 y^4} = \dfrac{6}{3} \times (x^5 \div x^2) \times (y^8 \div y^4)$
$= 2 \times x^{5-2} \times y^{8-4}$
$= 2x^3 y^4$

EXERCISE 2.5

1 Sarah says that $3^x \times 3^y = 9^{x+y}$. Is she correct? Explain your answer.

2 Simplify these expressions.
- a $x^2 \times x^7$
- b $b^4 \times b^4$
- c $y^8 \times y^2$
- d $c^3 \times c^9$
- e $a^4 \times a^2 \times a^5$
- f $p^2 \times p^2 \times p^4$
- g $y^3 \times y^5 \times y$
- h $a^2 \times a \times a^4$

HINT $a = a^1$

3
- a $5x^2 \times 3x^4$
- b $2b \times 4b^3$
- c $7y^8 \times 2y^6$
- d $5c^3 \times 5c^3$
- e $2a^2 \times 3a$
- f $7p^4 \times 2p^3 \times 5p$
- g $3y \times 4y^2 \times 5y^3$
- h $3a^2 \times 3a^2 \times 3a^2$

4
- a $2x^2 \times 3y^2 \times x^4$
- b $3x^4 \times 2y \times 5y^2$
- c $5y^2 \times x^2 \times 2y$
- d $x^3 \times 2y^2 \times 3x$
- e $5xy^2 \times 3x^2y$
- f $7xy \times 5x^2$
- g $4x^3y \times xy$
- h $3a^5b^2 \times 2a^3b^2$

5
- a $x^8 \div x^3$
- b $b^7 \div b^6$
- c $a^6 \div a^3$
- d $x^6 \div x$
- e $6x^5 \div x^2$
- f $8y^7 \div 4y^5$
- g $15b^9 \div 3b^2$
- h $21c^8 \div 7c^7$

6
- a $\dfrac{5y^4}{y^2}$
- b $\dfrac{12a^2b}{a}$
- c $\dfrac{2x^2}{8x}$
- d $\dfrac{5y^2}{10y}$
- e $\dfrac{6a^4}{3a^3}$
- f $\dfrac{16a^2b^2c^2}{4abc}$
- g $\dfrac{a^3b^4c^2}{ab^2c}$
- h $\dfrac{(5x^2)^2}{5x^2}$
- i $\dfrac{x^3 \times x^6}{x^4}$
- j $\dfrac{y^8 \times y^2}{y^3}$
- k $\dfrac{3a^4 \times 2a^5}{a^2}$
- l $\dfrac{6x^4 \times 2x^4}{4x^2}$

HINT simplify the numerator first

7
- a $(x^3)^2$
- b $(a^2)^5$
- c $(b^4)^4$
- d $(y^7)^3$
- e $3(a^2)^3$
- f $5(x^5)^4$
- g $(2x^3)^3$
- h $(3y^2)^4$
- i $(2a^2)^3$
- j $(3y^4)^2$
- k $2y^2(3x^2)^2$
- l $5x(2x^2)^3$

8
- a $(3xy^2)^3$
- b $(2a^2b^3)^5$
- c $(5x^4y^3)^2$
- d $(10x^5y^7)^3$

9
- a $\dfrac{3x^2 + 5x^2}{2x^2}$
- b $\dfrac{15x^{10}}{3x^4} + 8x^6$
- c $\dfrac{(2x^2)^3 \times (3x)^2}{2x^5}$
- d $\dfrac{5a^2b^4}{3c} \times \dfrac{abc}{15a^2b^2}$
- e $\dfrac{4a^5b}{cd} \div \dfrac{a}{c^2d^3}$
- f $\dfrac{(4x^8y^3)^3}{2x^5y^2}$
- g $\dfrac{(2x^2y)^6}{(2xy^2)^3}$
- h $\left((2xy^2)^2\right)^2$

Indices 1   59

Negative powers: $4^3 \div 4^5 = \dfrac{\cancel{4} \times \cancel{4} \times \cancel{4}}{4 \times 4 \times \cancel{4} \times \cancel{4} \times \cancel{4}} = \dfrac{1}{4^2}$

and $4^3 \div 4^5 = 4^{3-5} = 4^{-2}$

so $4^{-2} = \dfrac{1}{4^2}$

> **RULE 4** $\quad a^{-m} = \dfrac{1}{a^m}$

Zero powers: $5^4 \div 5^4 = \dfrac{\cancel{5} \times \cancel{5} \times \cancel{5} \times \cancel{5}}{\cancel{5} \times \cancel{5} \times \cancel{5} \times \cancel{5}} = 1$

and $5^4 \div 5^4 = 5^{4-4} = 5^0$

so $5^0 = 1$

> **RULE 5** $\quad a^0 = 1$

Any number raised to the power zero is equal to 1.

EXAMPLE

Work out the value of **a** 5^{-2} **b** 8^0 **c** $(-3)^{-5}$ **d** $\left(\dfrac{3}{5}\right)^{-1}$

a $5^{-2} = \dfrac{1}{5^2} = \dfrac{1}{5 \times 5} = \dfrac{1}{25}$ **b** $8^0 = 1$

c $(-3)^{-5} = \dfrac{1}{(-3)^5} = \dfrac{1}{(-3) \times (-3) \times (-3) \times (-3) \times (-3)} = \dfrac{1}{-243} = -\dfrac{1}{243}$

d $\left(\dfrac{3}{5}\right)^{-1} = \dfrac{1}{\left(\dfrac{3}{5}\right)^1} = 1 \div \dfrac{3}{5} = 1 \times \dfrac{5}{3} = \dfrac{5}{3}$

➡ **NOTE:** it is very useful to know that $\left(\dfrac{3}{5}\right)^{-1}$ is the same as $\left(\dfrac{5}{3}\right)$.

EXAMPLE

Simplify
a $x^9 \times x^{-3} \times x^{-4}$ **b** $5x^{-5} \times 4x^2$ **c** $x^6 \div x^{-8}$ **d** $\left(x^{-2}\right)^3$ **e** $\left(5x^2\right)^{-3}$

a $x^9 \times x^{-3} \times x^{-4} = x^{9+-3+-4}$ add the powers
$= x^{9-3-4}$
$= x^2$

b $5x^{-5} \times 4x^2 = 5 \times 4 \times x^{-5} \times x^2$
$= 20 \times x^{-5+2}$
$= 20x^{-3}$

c $x^6 \div x^{-8} = x^{6--8}$ subtract the powers
$= x^{6+8}$
$= x^{14}$

d $\left(x^{-2}\right)^3 = x^{-2 \times 3}$ multiply the powers
$= x^{-6}$

e $\left(5x^2\right)^{-3} = \dfrac{1}{\left(5x^2\right)^3} = \dfrac{1}{5x^2 \times 5x^2 \times 5x^2}$ or $\left(5x^2\right)^{-3} = (5)^{-3}\left(x^2\right)^{-3}$

$= \dfrac{1}{5 \times 5 \times 5 \times x^2 \times x^2 \times x^2}$ $= 5^{-3} x^{-6}$

$= \dfrac{1}{125x^6}$ or $\dfrac{1}{125}x^{-6}$ $= \dfrac{1}{125}x^{-6}$ or $\dfrac{1}{125x^6}$

EXERCISE 2.6

Work out the value of these.

1.
   a. 3^{-1} b. 5^0 c. 2^{-3} d. 3^{-3} e. 7^0 f. 10^{-3}
   g. 4^{-2} h. 7^{-1} i. 10^{-4} j. 5^{-3} k. 2^{-5} l. 3^{-4}

2.
   a. $(-3)^{-2}$ b. $(-2)^{-3}$ c. $(-6)^{-2}$ d. $(-9)^{-1}$ e. $(-2)^{-4}$ f. $(-10)^{-3}$

3.
   a. $\left(\frac{2}{3}\right)^{-1}$ b. $\left(\frac{2}{5}\right)^{-1}$ c. $\left(\frac{3}{4}\right)^{-1}$ d. $\left(\frac{5}{6}\right)^{-1}$ e. $\left(\frac{2}{7}\right)^{-1}$ f. $\left(\frac{4}{5}\right)^{-1}$

4.
   a. $\left(\frac{2}{3}\right)^{-2}$ b. $\left(\frac{2}{5}\right)^{-2}$ c. $\left(\frac{3}{4}\right)^{-2}$ d. $\left(\frac{5}{6}\right)^{-2}$
   e. $\left(\frac{2}{7}\right)^{-2}$ f. $\left(\frac{4}{5}\right)^{-2}$ g. $\left(\frac{1}{2}\right)^{-3}$ h. $\left(\frac{1}{10}\right)^{-3}$

HINT

$\left(\frac{2}{3}\right)^{-2} = \left(\frac{3}{2}\right)^{2}$

Simplify the following expressions. Give your answers in index form.

5.
   a. $a^6 \times a^{-2}$ b. $b^{-5} \times b^{-4}$ c. $c^{-3} \times c$ d. $d^{-6} \times d^{-7}$
   e. $d^{-3} \times d^{-4} \times d^{10}$ f. $x^2 \times x^5 \times x^{-4}$ g. $y^4 \times y^2 \times y^{-2}$ h. $d^{-3} \times d^{-4} \times d^{10}$
   i. $a^{-3} \times a^3 \div a^{-5}$ j. $b^6 \times b^{-3} \div b^2$ k. $c^4 \times c^{-7} \div c^{-3}$ l. $d^{-3} \times d^{-2} \div d^{-1}$

6.
   a. $\frac{x^2 \times x^4}{x^8}$ b. $\frac{x^{-3} \times x^5}{x^{-3}}$ c. $\frac{4x^5 \times 3x^2}{2x^{-2}}$ d. $\frac{8x^2}{4x^3 \times 2x^4}$

7.
   a. $2x^{-2} \times 4x^3$ b. $5x^{-1} \times 3x^{-2}$ c. $2x^{-2} \times 3x^5$ d. $7x^{-1} \times 4x^{-4}$

8.
   a. $(x^2)^{-1}$ b. $(y^3)^{-2}$ c. $(a^{-4})^5$ d. $(b^{-2})^4$
   e. $(a^{-3})^{-2}$ f. $(y^{-1})^{-2}$ g. $(a^{-2})^{-5}$ h. $(a^{-1})^{-1}$

9.
   a. $(3a^{-1})^2$ b. $(5b^{-1})^3$ c. $(2c^{-2})^4$ d. $(4d^{-2})^2$
   e. $(2x^2)^{-2}$ f. $(3y^2)^{-3}$ g. $(5x^3)^{-2}$ h. $(2x^{-2})^{-2}$

10.
    a. $x^{-1} \div \frac{1}{x^2}$ b. $x^3 \div \frac{2}{x^4}$ c. $y \div \frac{1}{y^{-2}}$ d. $\frac{4}{x^2} + 3x^{-2}$

11. Find the odd one out in these three expressions.

    $5a^3b^{-4} \times a^{-1}b$ $2a^2 + 3b^{-3}$ $\dfrac{5a^{-2}b^{-2}}{a^{-4}b}$

KEY WORDS
index
power
base

Solving linear inequalities

THIS SECTION WILL SHOW YOU HOW TO
- Solve linear inequalities
- Represent solutions to linear inequalities on a number line.

You need to know these **inequality** signs:
> means 'is greater than'   ≥ means 'is greater than or equal to'
< means 'is less than'   ≤ means 'is less than or equal to'

You solve a linear inequality in a similar way to solving linear equations.

$$2x - 3 < 10 \quad \text{add 3 to both sides}$$
$$2x < 13 \quad \text{divide both sides by 2}$$
$$x < 6.5$$

You must be careful if you multiply or divide an inequality by a negative number. You must reverse the inequality sign.
(For example $-3 < 7$, multiplying both sides by -1 gives $3 > -7$)

EXAMPLE

Solve $7 - 2x > 1$ and show your answer on a number line.

Method 1
$7 - 2x > 1$ add $2x$ to both sides
$7 > 1 + 2x$ take 1 from both sides
$6 > 2x$ divide both sides by 2
$3 > x$ so $x < 3$

Method 2
$7 - 2x > 1$ take 7 from both sides
$-2x > -6$ divide both sides by -2
$x < 3$ and reverse the sign

–4 –3 –2 –1 0 1 2 3 4

➡ **NOTE:** a hollow circle is used for < or >.

EXAMPLE

Solve $4(x + 1) \leq 4 - 2(x + 3)$ and show your answer on a number line.

$4(x + 1) \leq 4 - 2(x + 3)$ multiply out the brackets
$4x + 4 \leq 4 - 2x - 6$
$4x + 4 \leq -2 - 2x$ add $2x$ to both sides
$6x + 4 \leq -2$ take 4 from both sides
$6x \leq -6$ divide both sides by 6
$x \leq -1$

–4 –3 –2 –1 0 1 2 3 4

➡ **NOTE:** a solid circle is used for ≤ or ≥.

EXAMPLE

Solve $-1 \leq 2x + 1 < 5$ and show your answer on a number line.

$-1 \leq 2x + 1 < 5$ subtract 1 throughout
$-2 \leq 2x < 4$ divide by 2
$-1 \leq x < 2$

–4 –3 –2 –1 0 1 2 3 4

➡ **NOTE:** if a question asks for the integer values that satisfy the inequality $-1 \leq 2x + 1 < 5$ the final answer will be -1, 0 and 1. (An integer is a whole number.)

UNIT 2

EXERCISE 2.7

Write the inequalities shown on these diagrams.

1. [number line from −3 to 3, closed dot at 1, arrow right]
2. [number line from −3 to 3, open dot at 2, arrow left]
3. [number line from −3 to 3, closed dots at −2 and 2]
4. [number line from −3 to 3, closed dot at −1, open dot at 2]

Solve these inequalities.

5. $2x - 5 \leq 7$
6. $3x + 2 \geq 14$
7. $5x + 4 < 39$
8. $6x - 2 > 10$
9. $8x + 3 \geq 15$
10. $4x - 5 < 19$
11. $2x + 6 \leq 4$
12. $6 > 3x - 15$
13. $10 - 4x \leq 6$
14. $30 - 2x > 15$
15. $4 - 2x \geq 16$
16. $6 - 5x > -4$
17. $27 - 3x \leq 6$
18. $1 - 10x > 3$
19. $7 - 2x < 12$
20. $16 \leq 8 - 2x$
21. $3(2x + 5) \leq 21$
22. $4(y - 5) > 12$
23. $3(2x + 5) \geq 27$
24. $4(3x - 1) < 80$
25. $7 > 2(x + 3)$
26. $-25 \leq 5(3x + 4)$
27. $3(2 - x) \geq -6$
28. $2(13 - x) < 22$
29. $5x - 2 \geq 3x + 4$
30. $8x + 4 < 3x - 31$
31. $2x + 1 \leq 5x - 26$
32. $5 - 2x > 3 - x$
33. $3(x + 1) \leq 4(x - 1)$
34. $2(x + 2) \geq 5(x - 1)$
35. $3(x + 2) - 2(x - 4) \geq 4(x + 2)$
36. $2(3x - 1) - 5(x + 2) > 3(x - 5)$
37. $\dfrac{x-1}{8} \leq 1 - x$
38. $x - 3 \geq \dfrac{x+12}{2}$
39. $\dfrac{x-3}{3} > \dfrac{x-2}{5}$
40. $\dfrac{x+2}{2} \leq \dfrac{x-1}{7}$
41. $\dfrac{x-5}{3} > \dfrac{4-x}{2}$
42. $\dfrac{x-1}{2} > \dfrac{x+2}{5}$
43. $\dfrac{x-3}{5} \geq \dfrac{x+3}{2}$
44. $\dfrac{2x-3}{4} < \dfrac{3x-1}{5}$
45. $5 \leq 3 - \dfrac{2x+1}{5}$

Solve these inequalities and then list the integer solutions.

46. $-6 \leq 2x \leq 10$
47. $-9 \leq 3x < 12$
48. $8 < 4x \leq 36$
49. $-1 < \dfrac{x}{2} \leq 2$
50. $0 < \dfrac{x}{2} - 1 < 1$
51. $-8 \leq 2n - 3 < 8$
52. $3 < 2x + 3 \leq 11$
53. $-2 \leq 2(x - 4) \leq 6$
54. $-3 < 3(x - 4) \leq 15$
55. Write down the smallest integer that satisfies the inequality $3x + 2 \geq 12$
56. Find the largest integer that satisfies $5(3x - 7) < 28$
57. $3x + 2y \leq 6$ and x and y are both positive integers (not including 0).
    List all the possible pairs of values for x and y.

KEY WORDS
inequality

Solving linear inequalities 63

Manipulating algebraic fractions

> **THIS SECTION WILL SHOW YOU HOW TO**
> - Add and subtract algebraic fractions
> - Solve equations involving algebraic fractions

Adding and subtracting algebraic fractions

Reminder

$\dfrac{2}{3} + \dfrac{1}{4}$ Similarly $\dfrac{2x}{3} + \dfrac{x}{4}$ the common denominator is $3 \times 4 = 12$

$= \dfrac{8}{12} + \dfrac{3}{12}$ $= \dfrac{8x}{12} + \dfrac{3x}{12}$ $\dfrac{2}{3} = \dfrac{8}{12}$ and $\dfrac{1}{4} = \dfrac{3}{12}$

$= \dfrac{8+3}{12}$ $= \dfrac{8x+3x}{12}$

$= \dfrac{11}{12}$ $= \dfrac{11x}{12}$

EXAMPLE

Simplify $\dfrac{2x}{3} + \dfrac{x}{4} - \dfrac{3x}{5}$.

$\dfrac{2x}{3} + \dfrac{x}{4} - \dfrac{3x}{5}$ the common denominator is $3 \times 4 \times 5 = 60$

$= \dfrac{40x}{60} + \dfrac{15x}{60} - \dfrac{36x}{60}$ $\dfrac{2}{3} = \dfrac{40}{60}, \dfrac{1}{4} = \dfrac{15}{60}$ and $\dfrac{3}{5} = \dfrac{36}{60}$

$= \dfrac{40x + 15x - 36x}{60}$

$= \dfrac{19x}{60}$

EXAMPLE

Simplify $\dfrac{5(2x-1)}{6} - \dfrac{2(2x-3)}{5}$.

$\dfrac{5(2x-1)}{6} - \dfrac{2(2x-3)}{5}$

$= \dfrac{25(2x-1)}{30} - \dfrac{12(2x-3)}{30}$ the common denominator is $6 \times 5 = 30$

$= \dfrac{25(2x-1) - 12(2x-3)}{30}$ $\dfrac{5}{6} = \dfrac{25}{30}$ and $\dfrac{2}{5} = \dfrac{12}{30}$

$= \dfrac{50x - 25 - 24x + 36}{30}$ remove the brackets and be careful with the signs

$= \dfrac{26x + 11}{30}$ collect like terms in the numerator

EXERCISE 2.8

1 Simplify the following algebraic fractions.

a $\dfrac{3x}{5} + \dfrac{x}{5}$ b $\dfrac{2x}{7} + \dfrac{3x}{7}$ c $\dfrac{x}{4} + \dfrac{x}{4}$

d $\dfrac{x}{8} + \dfrac{3x}{8}$ e $\dfrac{5x}{6} - \dfrac{x}{6}$ f $\dfrac{5x}{8} - \dfrac{x}{8}$

g $\dfrac{7x}{10} - \dfrac{3x}{10}$ h $\dfrac{4x}{5} - \dfrac{2x}{5}$ i $\dfrac{4x}{7} + \dfrac{2x}{7} - \dfrac{x}{7}$

2 Pedro says that $\dfrac{y}{3} + \dfrac{2y}{5} = \dfrac{3y}{8}$.

Explain why he is wrong.

Simplify the following algebraic fractions.

3 a $\dfrac{x}{2} + \dfrac{x}{4}$ b $\dfrac{x}{3} + \dfrac{x}{4}$ c $\dfrac{2x}{5} + \dfrac{x}{3}$

d $\dfrac{3x}{5} + \dfrac{x}{4}$ e $\dfrac{x}{2} - \dfrac{x}{3}$ f $\dfrac{2x}{3} - \dfrac{x}{5}$

g $\dfrac{9x}{10} - \dfrac{2x}{5}$ h $\dfrac{8x}{9} - \dfrac{3x}{4}$ i $\dfrac{x}{2} - \dfrac{3x}{10} + \dfrac{2x}{5}$

4 a $\dfrac{x}{3} + \dfrac{x+2}{4}$ b $\dfrac{x-3}{5} + \dfrac{x+1}{6}$ c $\dfrac{2x-3}{7} + \dfrac{x+4}{5}$

d $\dfrac{6x-3}{7} + \dfrac{x+4}{5}$ e $\dfrac{3(x+1)}{4} + \dfrac{2(x+3)}{3}$ f $\dfrac{8(x-1)}{11} + \dfrac{2(x+4)}{3}$

5 a $\dfrac{6x+1}{3} - \dfrac{x+2}{4}$ b $\dfrac{3x-2}{5} - \dfrac{x-3}{8}$ c $\dfrac{4x+2}{5} - \dfrac{3x+4}{6}$

d $\dfrac{2x+3}{9} - \dfrac{5x-1}{4}$ e $\dfrac{2(3x+4)}{3} - \dfrac{3(2x-1)}{5}$ f $\dfrac{3(x-1)}{4} - \dfrac{2(x+6)}{9}$

6 $\dfrac{2x+1}{4} - \dfrac{3(x-2)}{5} + \dfrac{2(x+1)}{3}$

7 Find the odd one out.

| $\dfrac{2x}{3} - \dfrac{x}{4}$ | $\dfrac{x}{4} + \dfrac{x}{3} - \dfrac{x}{6}$ | $\dfrac{x}{2} - \dfrac{x}{12}$ | $\dfrac{3x}{4} - \dfrac{2x}{3}$ | $\dfrac{x}{4} + \dfrac{x}{6}$ |

Manipulating algebraic fractions

Solving equations

> **EXAMPLE**
>
> Solve $\dfrac{2x}{5} + \dfrac{3x}{2} = 38$
>
> There are two methods for solving this type of equation. Use the method that you prefer.
>
> **Method 1**
>
> $\dfrac{2x}{5} + \dfrac{3x}{2} = 38$ common denominator is 10
>
> $\dfrac{4x}{10} + \dfrac{15x}{10} = 38$
>
> $\dfrac{19x}{10} = 38$ multiply by 10
>
> $19x = 380$ divide by 19
>
> $x = 20$
>
> **Method 2**
>
> $\dfrac{2x}{5} + \dfrac{3x}{2} = 38$ multiply by 10
>
> $^2\cancel{10} \times \dfrac{2x}{\cancel{5}} + {}^5\cancel{10} \times \dfrac{3x}{\cancel{2}} = 10 \times 38$ cancel
>
> $4x + 15x = 380$
>
> $19x = 380$ divide by 19
>
> $x = 20$

> **EXAMPLE**
>
> Solve $\dfrac{2x+8}{3} - \dfrac{x-1}{2} = 4$
>
> **Method 1**
>
> $\dfrac{2x+8}{3} - \dfrac{x-1}{2} = 4$ change fractions to a common denominator of 6
>
> $\dfrac{2(2x+8)}{6} - \dfrac{3(x-1)}{6} = 4$
>
> $\dfrac{2(2x+8) - 3(x-1)}{6} = 4$ remove the brackets and be careful with the signs!
>
> $\dfrac{4x + 16 - 3x + 3}{6} = 4$ collect like terms in the numerator
>
> $\dfrac{x + 19}{6} = 4$ multiply both sides by 6
>
> $x + 19 = 24$ subtract 19 from both sides
>
> $x = 5$
>
> **Method 2**
>
> $\dfrac{2x+8}{3} - \dfrac{x-1}{2} = 4$ multiply both sides by 6
>
> $^2\cancel{6} \times \dfrac{(2x+8)}{\cancel{3}} - {}^3\cancel{6} \times \dfrac{(x-1)}{\cancel{2}} = 6 \times 4$ cancel
>
> $2(2x + 8) - 3(x - 1) = 24$ remove the brackets and be careful with the signs!
>
> $4x + 16 - 3x + 3 = 24$ collect like terms
>
> $x + 19 = 24$ take 19 from both sides
>
> $x = 5$

EXERCISE 2.9

Solve these equations.

1. a $\dfrac{x}{3} - \dfrac{x}{4} = 3$ b $\dfrac{2x}{3} - \dfrac{x}{2} = 3$ c $\dfrac{x}{4} + \dfrac{x}{5} = 18$ d $\dfrac{2x}{3} + \dfrac{3x}{2} = 26$

   e $\dfrac{3x}{5} + \dfrac{x}{4} = 34$ f $\dfrac{2x}{9} - \dfrac{x}{2} = 10$ g $\dfrac{4x}{5} + \dfrac{3x}{2} = 23$ h $\dfrac{6x}{7} - \dfrac{x}{3} = 22$

2. a $\dfrac{x}{5} - \dfrac{x}{6} = \dfrac{1}{2}$ b $\dfrac{3x}{4} - \dfrac{2x}{3} = \dfrac{1}{3}$ c $\dfrac{2x}{5} + \dfrac{x}{4} = 2\dfrac{3}{5}$ d $\dfrac{2x}{9} - \dfrac{x}{8} = \dfrac{7}{12}$

   e $\dfrac{x}{8} - \dfrac{2x}{5} = 1\dfrac{1}{10}$ f $\dfrac{3x}{5} - \dfrac{3x}{8} = \dfrac{3}{20}$ g $\dfrac{x}{4} - \dfrac{x}{7} = \dfrac{9}{14}$ h $\dfrac{x}{8} + \dfrac{3x}{5} = -1\dfrac{9}{20}$

3. a $\dfrac{3x+4}{2} + \dfrac{8-x}{3} = 7$ b $\dfrac{x+1}{4} + \dfrac{2x+1}{5} = 5$ c $\dfrac{4x}{5} + \dfrac{3x-7}{2} = 8$

   d $\dfrac{x+9}{6} + \dfrac{5x-3}{3} = 6$ e $\dfrac{11x+5}{4} + \dfrac{4x-3}{3} = -12$ f $\dfrac{3x+2}{2} + \dfrac{x+6}{3} = 14$

   g $\dfrac{4x-1}{5} + \dfrac{5x-16}{2} = 5$ h $\dfrac{x}{4} + \dfrac{x-5}{3} = 3$ i $\dfrac{3-x}{6} + \dfrac{x+8}{5} = 2$

   j $\dfrac{3x+10}{8} + \dfrac{4x+11}{6} = 1$ k $\dfrac{2x+1}{2} + \dfrac{6x-3}{4} = 6$ l $\dfrac{6x-1}{3} + \dfrac{15x+7}{9} = 1\dfrac{2}{3}$

4. a $\dfrac{x+8}{5} - \dfrac{x+1}{4} = 1$ b $\dfrac{19x+5}{6} - \dfrac{2(x+4)}{5} = 2$ c $\dfrac{5x+6}{3} - \dfrac{5x-1}{7} = 5$

   d $\dfrac{5x-2}{4} - \dfrac{x+6}{3} = 3$ e $\dfrac{4x-1}{3} - \dfrac{20-x}{8} = 3$ f $\dfrac{11x+5}{4} - \dfrac{3x-5}{2} = 5$

   g $\dfrac{2(x-2)}{3} - \dfrac{6x-5}{5} = -3$ h $\dfrac{3x-5}{4} - \dfrac{5(x-4)}{3} = -1$ i $\dfrac{3(3-x)}{2} - \dfrac{2(x+6)}{3} = 7$

   j $\dfrac{3(5-x)}{5} - \dfrac{4(2-x)}{7} = 2$ k $\dfrac{3(4x+1)}{7} - \dfrac{14x-5}{8} = 1$ l $\dfrac{21x+11}{6} - \dfrac{9x-1}{3} = 2\dfrac{1}{3}$

5. $\dfrac{5x-4}{6} - \dfrac{2x+1}{5} = \dfrac{3x-2}{4} - \dfrac{x+1}{3}$

6. The perimeter of this rectangle is 20 cm.
   - a Show that $17x + 31 = 150$
   - b Hence, or otherwise, find the sides of the rectangle.

   (Rectangle with sides $\dfrac{x+2}{3}$ and $\dfrac{4x+7}{5}$)

7. The perimeter of this isosceles triangle is 22 cm. Find the value of x.

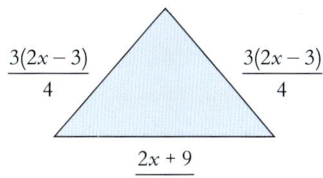

   (Isosceles triangle with equal sides $\dfrac{3(2x-3)}{4}$ and base $\dfrac{2x+9}{2}$)

KEY WORDS

algebraic fraction

The general equation of a straight line

THIS SECTION WILL SHOW YOU HOW TO
- Use the equation $y = mx + c$
- Find the midpoint of a line segment

The diagram shows the graph of $y = 2x + 1$
From the graph you can see that the
- gradient = 2
- **y-intercept** = 1

This important result can be written as

> The graph of $y = mx + c$ is a straight line where m is the gradient and c is the y-intercept.

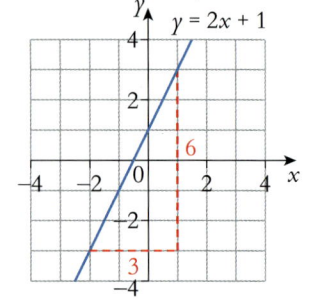

EXAMPLE

Find the gradient and y-intercept for these lines.
a $y = 3x - 5$ b $2y - x = 8$

a $y = 3x - 5$ compare with $y = mx + c$
    $m = 3$ and $c = -5$
    gradient = 3   y-intercept = -5

b $2y - x = 8$ add x to both sides
    $2y = x + 8$ divide both sides by 2
    $y = \frac{1}{2}x + 4$ compare with $y = mx + c$
    $m = \frac{1}{2}$ and $c = 4$
    gradient = $\frac{1}{2}$ y-intercept = 4

Parallel lines

> Parallel lines have the same gradient.

EXAMPLE

A line is parallel to the line $4y = 2x + 8$ and passes through the point (4, 5). Find the equation of the line.

$4y = 2x + 8$ divide both sides by 4
$y = \frac{1}{2}x + 2$ compare with $y = mx + c$

This line has a gradient of $\frac{1}{2}$ so the line parallel to it will also have gradient $\frac{1}{2}$
The equation of the parallel line is $y = \frac{1}{2}x + c$
The line passes through (4, 5) so substitute $x = 4$ and $y = 5$ into $y = \frac{1}{2}x + c$

$5 = \frac{1}{2} \times 4 + c$

$5 = 2 + c$

$c = 3$

The equation of the parallel line is $y = \frac{1}{2}x + 3$

UNIT 2

EXERCISE 2.10

1. Write down the gradient and y-intercept for these lines.
   a. $y = 3x + 2$
   b. $y = 2x - 5$
   c. $y = \frac{1}{2}x + 3$
   d. $y = \frac{2}{3}x - 1$
   e. $y = 3 + 2x$
   f. $y = 4 - 5x$
   g. $y = 7 - \frac{1}{2}x$
   h. $y = \frac{1}{2} + 3x$

2. Find the gradient and y-intercept for these lines.f
   a. $2y = 3x + 4$
   b. $3y = 6x - 2$
   c. $5y = 2x - 3$
   d. $2x + 3y = 5$
   e. $3x + 2x = 6$
   f. $x - 3y = 6$
   g. $4x - 2y = 3$
   h. $3x + 7y = 2$

3. For each of the lines a, b, c and d write down
   i. the gradient
   ii. the y-intercept
   iii. the equation of the line.

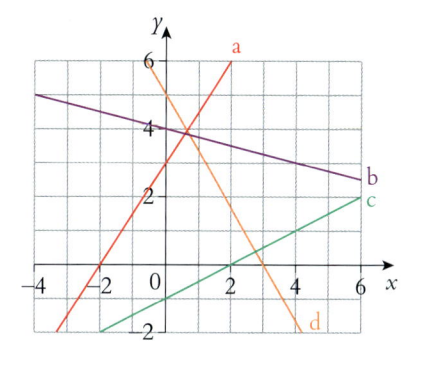

4. Write down the equation of each of these lines.

   Line A
   Gradient = 0
   Passes through (5, −2)

   Line B
   Gradient = 2
   Passes through (0, 0)

   Line C
   Gradient = 3
   Passes through (0, −2)

   Line D
   Parallel to $y = \frac{1}{2}x + 4$
   Passes through (0, 6)

   Line E
   Parallel to $3x + 4y = 12$
   Passes through (2, 5)

   Line F
   Parallel to $2y - x = 6$
   Passes through (−6, 1)

5. Sketch each of these graphs, labelling the point where the graph crosses the y axis.
   a. $y = x + 4$
   b. $y = 3x - 5$
   c. $y = \frac{1}{2}x - 3$
   d. $y = 5 - 2x$
   e. $y + 2x = -4$
   f. $2x + 3y = 12$
   g. $2x - y = 3$
   h. $5x - 2y = 10$

6. A $2y = (x + 8)$ B $y = 3x + 2$ C $x = y + 2$ D $3x + y = 2$ E $x = 2y$

   From the list above write down the lines that
   a. pass through (0, 0)
   b. are parallel
   c. pass through the point (−2, −4)
   d. have negative gradient
   e. are reflections of each other in the y-axis.

The general equation of a straight line

Finding the equation of a line passing through two given points

Some questions give you the y-intercept and another point on the line.

> **EXAMPLE**
>
> Find the equation of the line passing through $(0, -2)$ and $(4, 1)$.
>
> Using a sketch graph:
>
> gradient $= \dfrac{3}{4}$ so $m = \dfrac{3}{4}$ or gradient $= \dfrac{y_2 - y_1}{x_2 - x_1} = \dfrac{1 - -2}{4 - 0} = \dfrac{3}{4}$
>
> y-intercept $= -2$ so $c = -2$
>
> Using $y = mx + c$, $m = \dfrac{3}{4}$ and $c = -2$
>
> The equation of the line is $y = \dfrac{3}{4}x - 2$

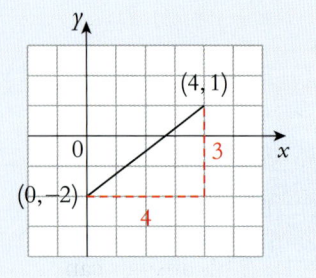

When the two points given do not include the y-intercept it can be more difficult.

> **EXAMPLE**
>
> Find the equation of the line passing through $(-3, 5)$ and $(5, -3)$.
>
> Using a sketch graph: gradient $= -\dfrac{8}{8} = -1$ so $m = -1$
>
> Using $y = mx + c$ and $m = -1$
>
> $y = -x + c$
>
> The line passes through $(-3, 5)$ so substitute $x = -3$ and $y = 5$ into $y = -x + c$
>
> $5 = 3 + c$
>
> $c = 2$
>
> The equation of the line is $y = -x + 2$.

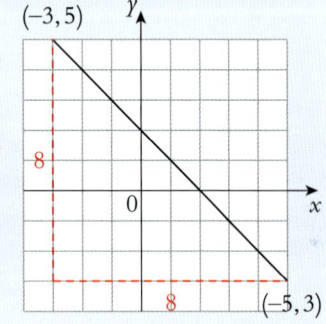

EXERCISE 2.11

Find the equations of the lines passing through these points.

1.
   a. $(0, 6)$ and $(8, 10)$
   b. $(0, -3)$ and $(-2, 1)$
   c. $(0, 1)$ and $(9, 10)$
   d. $(0, -4)$ and $(3, 8)$
   e. $(0, -2)$ and $(-6, 10)$
   f. $(0, 5)$ and $(9, 11)$
   g. $(0, -10)$ and $(5, -13)$
   h. $(0, 8)$ and $(10, 12)$
   i. $(0, 5)$ and $(-4, 2)$

2.
   a. $(2, 3)$ and $(6, 5)$
   b. $(-1, -1)$ and $(1, -7)$
   c. $(-3, 2)$ and $(1, 6)$
   d. $(1, 0)$ and $(-2, 6)$
   e. $(4, -1)$ and $(1, -4)$
   f. $(2, 1)$ and $(5, 3)$
   g. $(-4, 4)$ and $(1, 3)$
   h. $(-4, -6)$ and $(4, -4)$
   i. $(-5, -4)$ and $(-1, -2)$

Finding the midpoint of a line

The **midpoint** of a line segment can be found using a diagram or using the rule
midpoint = (mean of x coordinates, mean of y coordinates)

This rule can be written more formally as

> The midpoint of the line segment joining the points (x_1, y_1) and (x_2, y_2) is given by
> $$\text{midpoint} = \left(\frac{x_1+x_2}{2}, \frac{y_1+y_2}{2}\right)$$

EXAMPLE

Find the midpoint of the line joining the points (1, 2) and (9, 5).

From the graph the midpoint is (5, 3.5).
This answer can also be found using the formula.
First decide which values to use for x_1, y_1, x_2 and y_2.

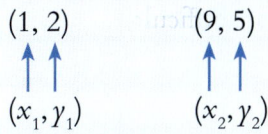

(1, 2)   (9, 5)
↑ ↑   ↑ ↑
(x_1, y_1) (x_2, y_2)

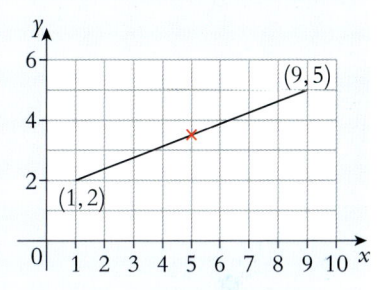

$$\text{midpoint} = \left(\frac{x_1+x_2}{2}, \frac{y_1+y_2}{2}\right) = \left(\frac{1+9}{2}, \frac{2+5}{2}\right) = \left(\frac{10}{2}, \frac{7}{2}\right) = \left(5, 3\frac{1}{2}\right)$$

EXAMPLE

Without drawing the line segment find the midpoint of $(-5, -2)$ and $\left(-2, 2\frac{1}{2}\right)$

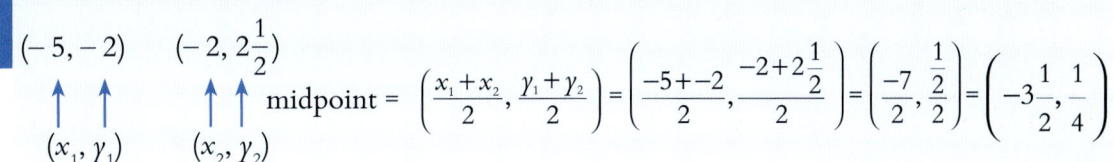

$(-5, -2)$ $\left(-2, 2\frac{1}{2}\right)$
↑ ↑   ↑ ↑
(x_1, y_1) (x_2, y_2)

$$\text{midpoint} = \left(\frac{x_1+x_2}{2}, \frac{y_1+y_2}{2}\right) = \left(\frac{-5+-2}{2}, \frac{-2+2\frac{1}{2}}{2}\right) = \left(\frac{-7}{2}, \frac{\frac{1}{2}}{2}\right) = \left(-3\frac{1}{2}, \frac{1}{4}\right)$$

EXERCISE 2.12

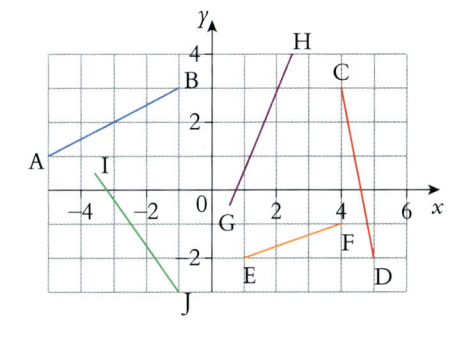

1. Write down the midpoints of the following lines.
   a AB
   b CD
   c EF
   d GH
   e IJ

2. Use the formula to find the midpoint of the lines joining the following points.
   a (2, 3) and (6, 4)
   b (−4, 2) and (0, −2)
   c (1, −4) and (4, 2)
   d (−6, 4) and (−2, 6)
   e (−2, 2) and (−1, 5)
   f (4, −2) and (5, −6)
   g (−6, −6) and (−1, −3)
   h (−5.5, −2.5) and (−2, −2)
   i (1, −1.5) and (4, 3)

KEY WORDS
y-intercept
midpoint

The general equation of a straight line

Representing linear inequalities on graphs

THIS SECTION WILL SHOW YOU HOW TO
- Represent linear inequalities on a graph

You can represent an inequality by a **region** on a graph.
There are some important rules that you must follow.

> If the inequality is ≥ or ≤ the **boundary line** for the region is shown by a solid line.
> If the inequality is > or < the boundary line for the region is shown by a broken (dashed) line.
> You are usually expected to shade the unwanted region.

EXAMPLE

Show the region $x + y \geq 3$ on a graph. Leave the required region unshaded.

First draw the boundary line $x + y = 3$.

x	0	3
y	3	0

The line will be solid because of the ≥ symbol.
The line divides the graph into two regions.
Find the required region using a trial point that is not on the line.
Using a trial point of (3, 2), substitute $x = 3$ and $y = 2$ into $x + y \geq 3$ to see if the inequality is true for the trial point.
$3 + 2 \geq 3$ is true. So (3, 2) is in the required region.

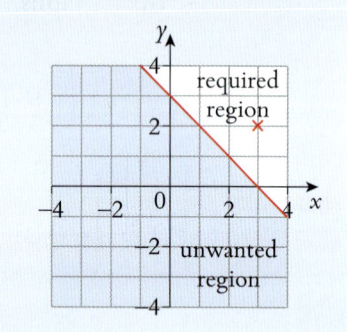

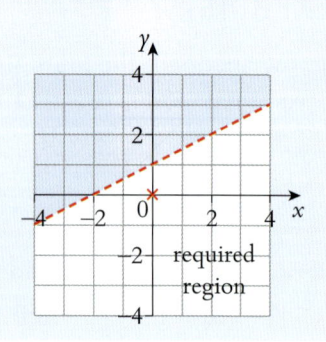

EXAMPLE

Show the region $y < \frac{1}{2}x + 1$ on a graph. Leave the required region unshaded.

x	0	2	4
y	1	2	3

Draw the boundary line $y = \frac{1}{2}x + 1$

The line is broken because of the < symbol.
Using a trial point of (0, 0): $0 < \frac{1}{2} \times 0 + 1$ is true.
So (0, 0) is in the required region.

72 UNIT 2

EXAMPLE

A region, R, contains points whose co-ordinates satisfy the following inequalities:
$-2 < x \leq 1$ $y > -3$ $y \leq x + 1$
On a graph, draw suitable lines and label the region R.

For $-2 < x \leq 1$: draw the lines $x = -2$ (broken) and $x = 1$ (solid)
The required region is between the two lines so shade outside the lines.
For $y > -3$: draw the line $y = -3$ (broken)
The required region is above the line so shade below the line.
For $y \leq x + 1$: draw the line $y = x + 1$ (solid)
The required region is below the line so shade above the line.

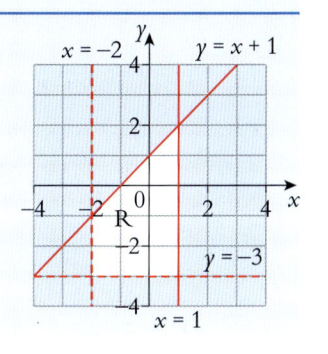

EXERCISE 2.13

1 Write down the inequality that describes the unshaded region for each of these graphs.

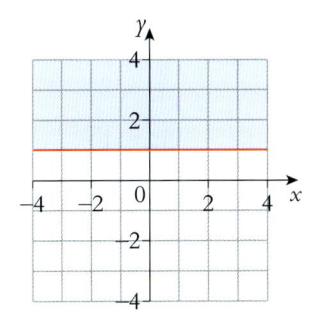

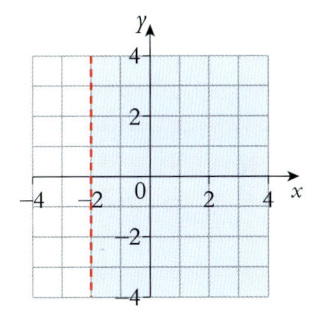

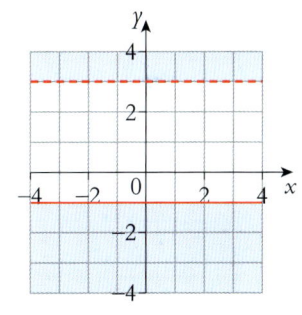

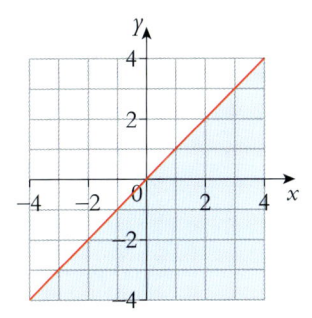

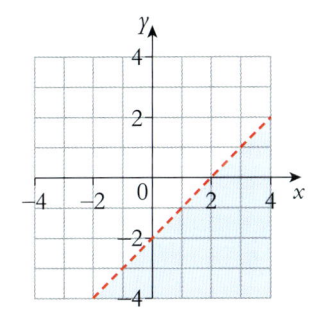

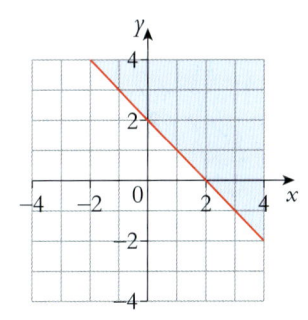

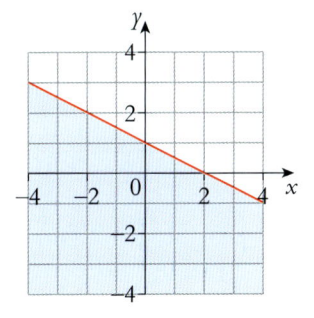

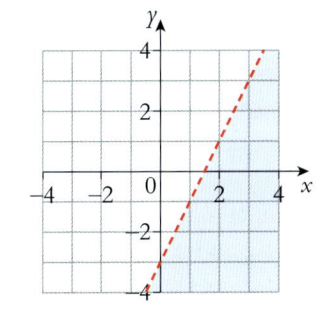

Representing linear inequalities on graphs

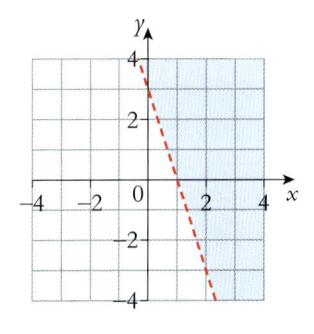

2 Draw a graph to represent each of these inequalities. Leave the required region <u>unshaded</u>.
  a $y < 2$
  b $x \geq 1$
  c $-2 < x \leq 3$
  d $1 < y < 3$
  e $y \leq 2x$
  f $y > \frac{1}{2}x$
  g $y \geq x + 1$
  h $y < x - 2$
  i $x + y \geq 0$
  j $x + y < 4$
  k $x - 2y \geq 4$
  l $x - y < 3$

3 Draw a graph to show the regions defined by each of these inequalities. Leave the required region unshaded.
  a $-2 < y \leq 3$ and $1 < x \leq 2$
  b $-3 < x \leq -1$ and $-2 \leq y \leq 2$

4 The red lines divide the graph into four regions A, B, C and D.

The region D is described by the inequalities:
$y \geq 2$ and $y \geq 2x$
Write down the inequalities that describe
  a Region A
  b Region B
  c Region C.

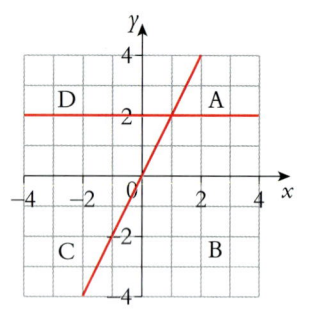

5 Match the regions P, Q and R to the inequality cards A, B and C.

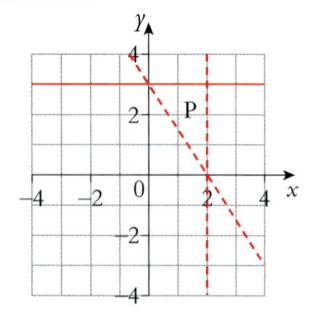

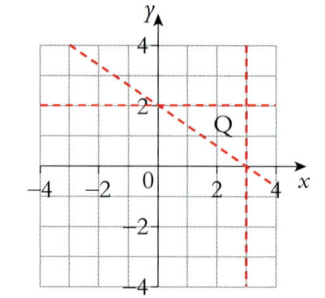

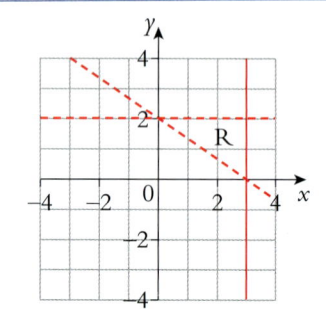

A $y \leq 3, x < 2, 3x + 2y > 6$

B $y < 2, x \leq 3, 2x + 3y > 6$

C $x < 3, y < 2, 2x + 3y < 6$

6 Find the three inequalities which define the unshaded region on each of the graphs.
  a
  b
  c

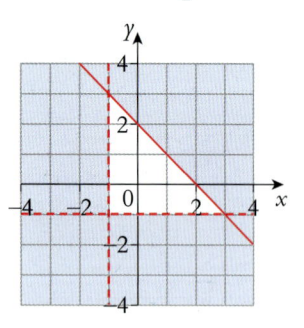

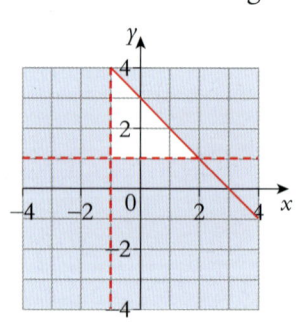

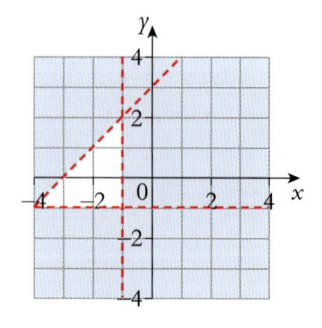

d 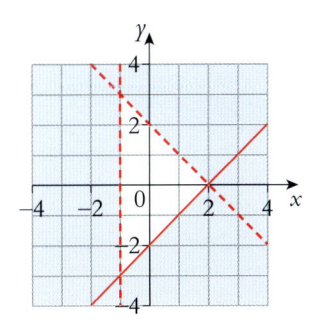
e 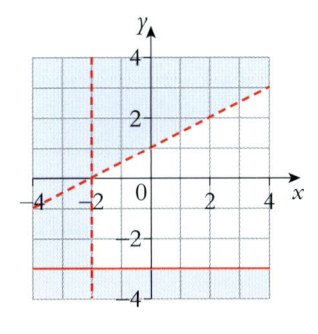
f

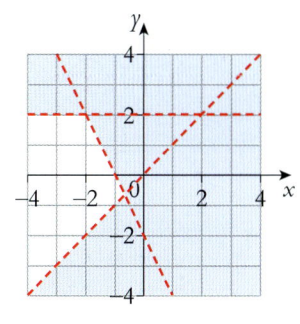

g 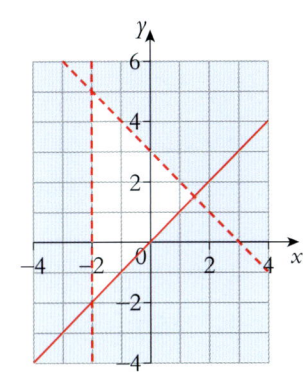
h 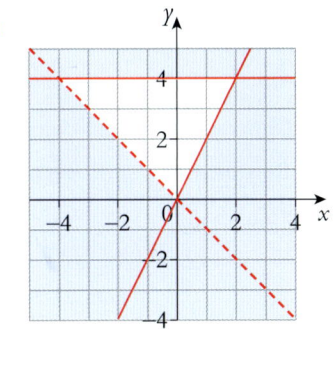
i

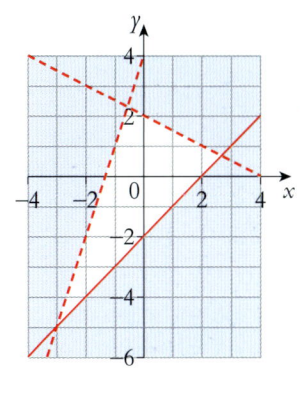

7 A region, R, contains points whose co-ordinates satisfy the inequalities:
$-3 < y \leq 2 \qquad x > -4 \qquad y \geq 2x$
On a graph, draw suitable lines and label the region R.

8 A region, R, contains points whose co-ordinates satisfy the inequalities:
$x + y \leq 2 \qquad y < 2x + 2 \qquad y \geq 0.5x - 1$
On a graph, draw suitable lines and label the region R.

9 A region, R, contains points whose co-ordinates satisfy the inequalities:
$y \leq 4 \qquad x \geq -2 \qquad y \geq x \qquad 2x + y < 4$
On a graph, draw suitable lines and label the region R.

10 Plot the points (0, 3), (4, 3) and (4, 0) on a graph and join the points to make a triangle.
  a Write down the three inequalities that define the region <u>inside</u> the triangle.
  b List all the points <u>inside</u> the triangle that have integer coordinates.

KEY WORDS
region
boundary line

Representing linear inequalities on graphs   75

Perimeter and area

THIS SECTION WILL SHOW YOU HOW TO
- Find perimeters and areas of simple shapes

The distance around an enclosed shape is called the **perimeter**. The amount of space enclosed inside a shape is called the **area**. You should already know these area formulae:

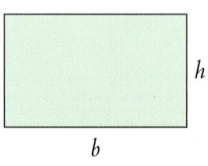

Area of rectangle = base × height

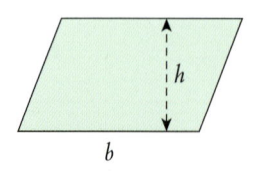

Area of parallelogram = base × height

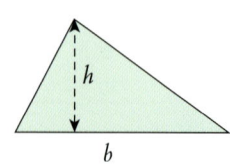

Area of triangle = $\frac{1}{2}$ × base × height

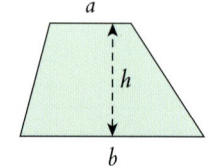

Area of trapezium = $\frac{1}{2}(a + b)h$

EXAMPLE

Find the perimeter and area of this shape.

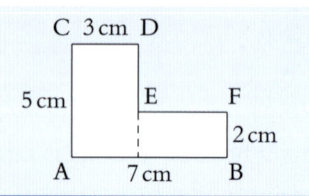

Perimeter = AB + BC + CD + DE + EF + FA
= 7 + 5 + 3 + 3 + 4 + 2 = 24 cm
Area = area of large rectangle + area of small rectangle
= (3 × 5) + (4 × 2) = 23 cm^2

EXAMPLE

Find the area of this trapezium.

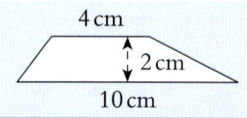

$a = 4$, $b = 10$ and $h = 2$
Area = $\frac{1}{2}(a + b)h$
= $\frac{1}{2}$ × (4 + 10) × 2
= 14 cm^2

To find the area of this rhombus divide it into 2 triangles.

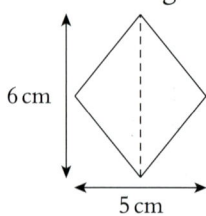

Area of triangle
= $\frac{1}{2}$ × 6 × 2.5 = 7.5
Area of rhombus
= 2 × 7.5 = 15 cm^2

To find the area of this kite divide it into 2 triangles.

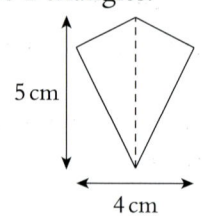

Area of triangle
= $\frac{1}{2}$ × 5 × 2 = 5
Area of kite
= 2 × 5 = 10 cm^2

To convert an area in cm² to an area in mm² you need to remember that:

1 cm² = [10mm × 10mm grid] = 10 × 10 = 100 mm²   1 cm² = 100 mm²
1 m² = 10 000 cm²
1 km² = 1 000 000 m²

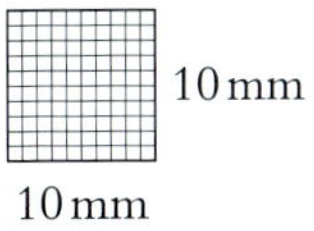

EXERCISE 2.14

1. Find the perimeter and area of these shapes. All lengths are in cm.

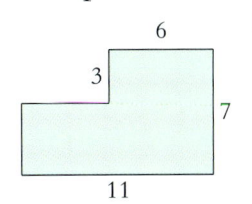

   a, b, 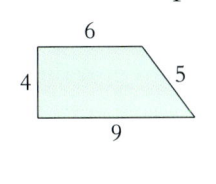 c, 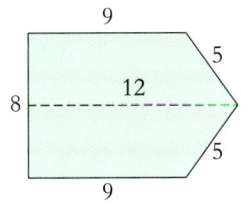 d [shapes with labelled dimensions]

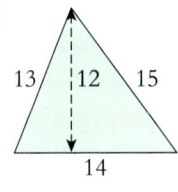

2. Find the area of these parallelograms. All lengths are in cm.

   a, b, c [parallelograms with labelled dimensions]

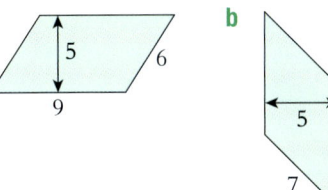

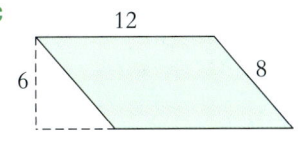

3. Find the area of these trapeziums.

   a, b, 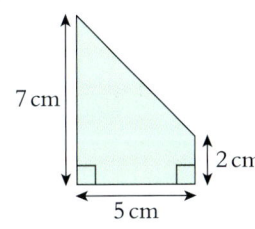 c, 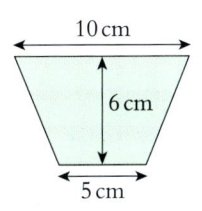 d [trapeziums with labelled dimensions]

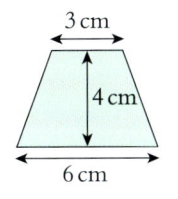

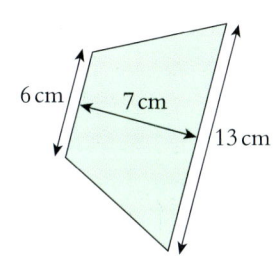

4. Find the area of these shapes. All lengths are in cm.

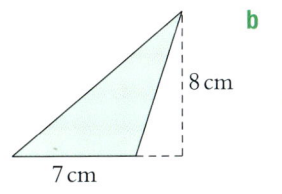

   a, b, 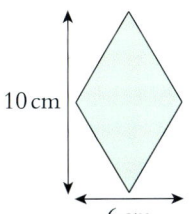 c, 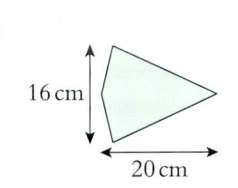 d [shapes with labelled dimensions]

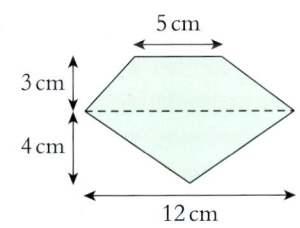

5. A trapezium has an area of 36 cm². The two parallel sides are 3 cm and 6 cm. Find the distance between the two parallel sides.

6. The trapezium has two parallel sides of length x cm and $3x + 2$ cm.
   The distance between the parallel sides is 6 cm.
   The area of the trapezium is 108 cm².
   Find the value of x.

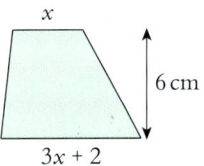

KEY WORDS
area
perimeter

Perimeter and area   77

Pythagoras

THIS SECTION WILL SHOW YOU HOW TO
- Apply Pythagoras' theorem to right-angled triangles
- Find the distance between two co-ordinate points

You can use **Pythagoras' theorem** when you know the length of two sides in a right-angled triangle and you want to find the length of the third side.

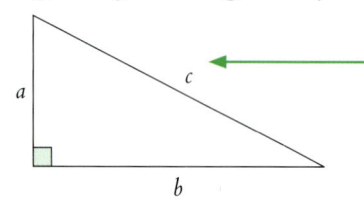

the longest side in a right-angled triangle is called the **hypotenuse**

Pythagoras' theorem for right-angled triangles says that:

The sum of the squares of the lengths of the two shorter sides is equal to the square of the length of the hypotenuse.

$$a^2 + b^2 = c^2$$

EXAMPLE

Find the length of x.

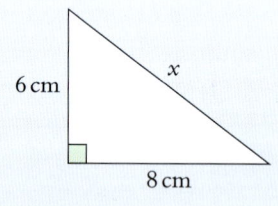

The two shorter sides are 6 cm and 8 cm. The hypotenuse is x cm.
$6^2 + 8^2 = x^2$ square 6 and 8
$36 + 64 = x^2$ $6 \times 6 = 36$ and $8 \times 8 = 64$
$100 = x^2$ square root both sides
$x = \sqrt{100} = 10$ cm

EXAMPLE

Find the length of x.

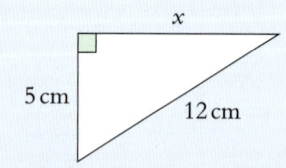

The two shorter sides are x cm and 5 cm. The hypotenuse is 12 cm.
$x^2 + 5^2 = 12^2$ square 5 and 12
$x^2 + 25 = 144$ take 25 from both sides
$x^2 = 119$ square root both sides
$x = \sqrt{119}$
$x = 10.908712…$ round to a suitable degree of accuracy
$x = 10.9$ cm (3 s.f.)

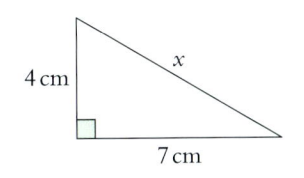

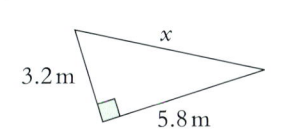

EXERCISE 2.15

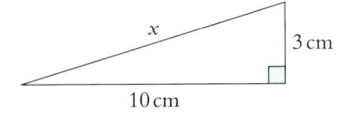

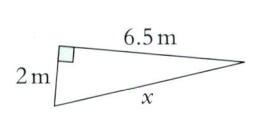

1. Calculate x for each of these triangles.

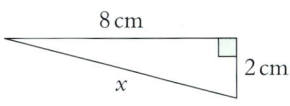

   a. [triangle with legs 4 cm and 7 cm, hypotenuse x]

   b. [triangle with base 10 cm, side 3 cm, hypotenuse x]

   c. [triangle with legs 8 cm and 2 cm, hypotenuse x]

   d. [triangle with legs 3.2 m and 5.8 m, hypotenuse x]

   e. [triangle with legs 2 m and x, hypotenuse 6.5 m]

   f. [triangle with legs x and 5.9 cm, hypotenuse 3.7 cm]

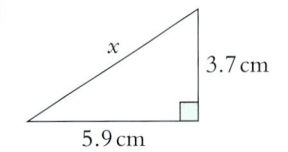

2. Manuel is trying to calculate the value of x for this triangle.
   This is what he writes
   $6^2 + 4^2 = x^2$
   $36 + 16 = x^2$
   $x^2 = 52$
   $x = \sqrt{52}$
   $x = 7.21$ cm
   He has made a mistake. Correct his working.

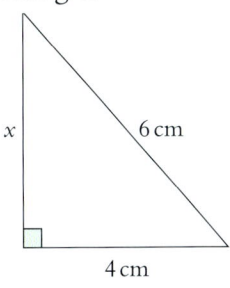

   [triangle with hypotenuse 6 cm, leg 4 cm, and leg x]

3. Calculate x for each of these triangles.

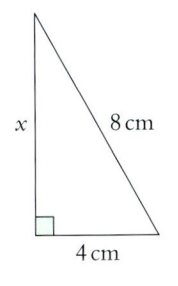

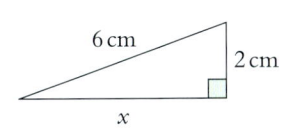

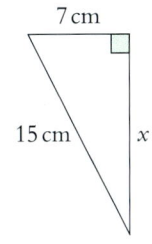

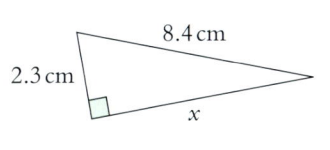

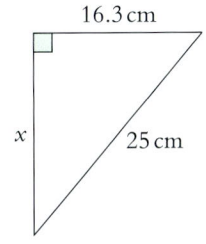

   a. [triangle with legs x and 4 cm, hypotenuse 8 cm]

   b. [triangle with side 6 cm, side 2 cm, base x]

   c. [triangle with legs 7 cm and x, hypotenuse 15 cm]

   d. [triangle with legs 2.3 cm and x, hypotenuse 8.4 cm]

   e. [triangle with legs 16.3 cm and x, hypotenuse 25 cm]

   f. [triangle with legs x and 13.6 cm, hypotenuse 19.8 cm]

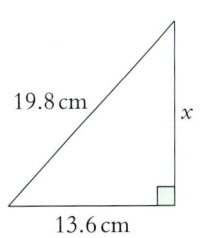

4. A boat sails 15 km due south and then 35 km due east.
   How far is the boat from its starting position?

5. A ladder is 5 m long. It rests against a vertical wall with the bottom of the ladder 1.5 m from the base of the wall. How far up the wall does the ladder reach?

Pythagoras

6 Calculate *x* for these right-angled triangles.

a

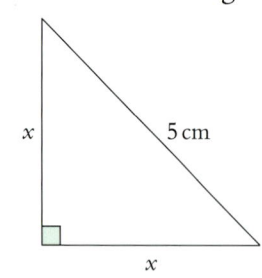

x, 5 cm, *x*

b

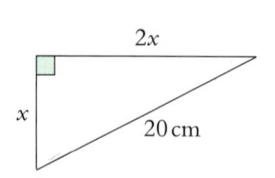

2*x*, *x*, 20 cm

c

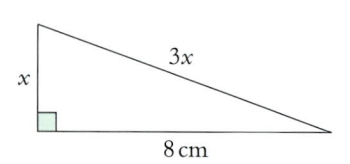

x, 3*x*, 8 cm

7 Calculate AD.

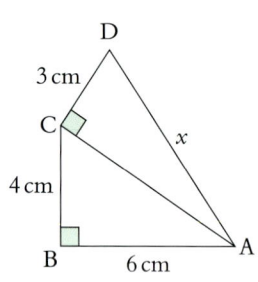

D, 3 cm, C, *x*, 4 cm, B, 6 cm, A

8 Calculate AB.

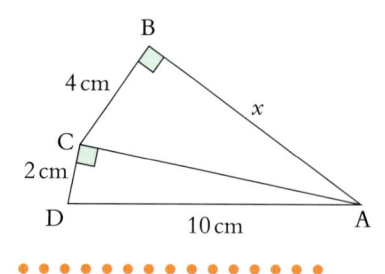

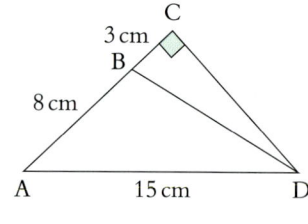

B, 4 cm, *x*, C, 2 cm, D, 10 cm, A

HINT
calculate the length of AC first.

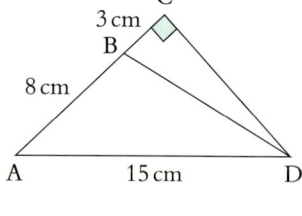

9 Find the length of DB.

C, 3 cm, B, 8 cm, A, 15 cm, D

HINT
calculate the length of CD first.

10 Calculate
  a the length of AD
  b the perimeter of ABCD
  c the area of ABCD
  d the area of triangle ADC.

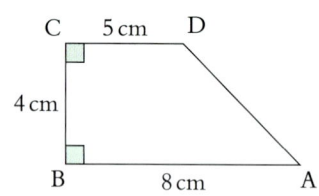

C, 5 cm, D, 4 cm, B, 8 cm, A

11 An equilateral triangle has sides of length 5 cm. Calculate
  a the height of the triangle
  b the area of the triangle.

12 A rhombus has a perimeter of 24 cm. The longest diagonal is 10 cm. Calculate the length of the shortest diagonal.

13 Sungu says this triangle is a right-angled triangle is he correct?

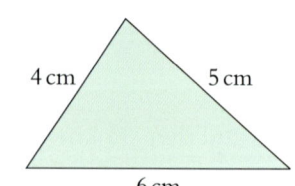

4 cm, 5 cm, 6 cm

HINT
Is $4^2 + 5^2 = 6^2$?

14 A triangle has sides of length 48 cm, 55 cm and 73 cm. Is it a right-angled triangle? Explain your answer.

15 Suri travels 5 miles west, then 4 miles north, then 8 miles east.
How far is she from her starting point?

16 The circle has radius 4 cm.
The vertices of the rectangle lie on the circumference of the circle.
The rectangle has width 6 cm.
Calculate the height of the rectangle.

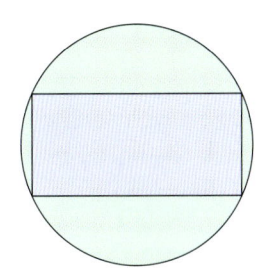

17 A hollow circular cylinder is used for storing thin metal rods.
The cylinder has a height of 15 cm and a base radius 4 cm.
Calculate the length of the longest rod that will fit exactly inside the cylinder.

18 The perpendicular height of a circular cone is twice the length of the diameter of the circular base.
The slant height is 10 cm.
Calculate the diameter of the circular base.

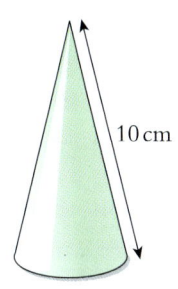

19 The diagram shows a square with sides of length 8 cm.
Inside the square there are four touching circles of radius 2 cm and a smaller circle of radius r cm.
Calculate the radius r.

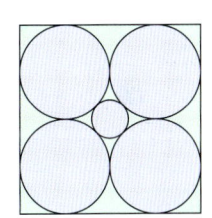

Distance between two points on a graph

You can use Pythagoras' theorem to find the length of a line joining two points on a graph.
To do this you

- Plot the points on a graph
- Make a right-angled triangle
- Mark the horizontal distance and vertical distance on the triangle
- Apply Pythagoras' theorem to the triangle.

EXAMPLE

Find the length of the line joining A (−3, 2) and B (4, 6).

$AB^2 = 7^2 + 4^2$
$AB^2 = 49 + 16$
$AB^2 = 65$
$AB = \sqrt{65}$
$AB = 8.062…$
$AB = 8.06$ (to 3 s.f.)

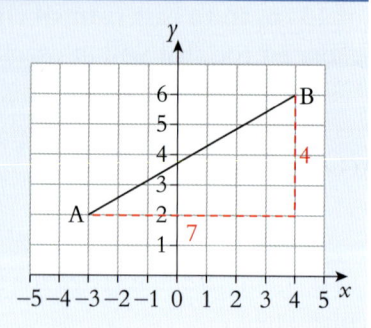

EXERCISE 2.16

1 Work out the length of the following lines.
   a AB
   b CD
   c EF
   d GH

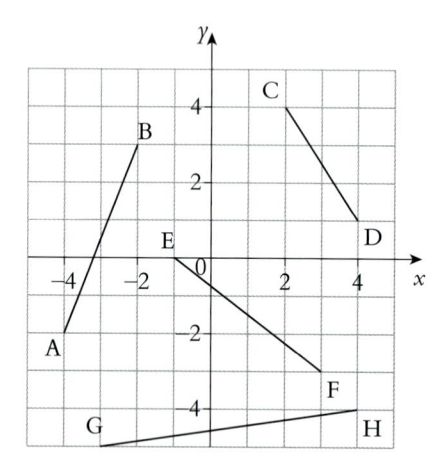

2 Calculate the length of the lines joining the following pairs of points.
   a (2, −4) and (6, 3)
   b (5, 8) and (2, −7)
   c (3, −6) and (−8, 5)
   d (0, −4) and (−3, 4)
   e (−9, 2) and (1, −4)
   f (6, −2) and (−6, 5)
   g (5, −1) and (−4, −4)
   h (−2, −2) and (−5, −1)
   i (−2, −4) and (−6, −3)

Pythagoras in 3-D

To find the length of AG you need to look at triangle AGC.

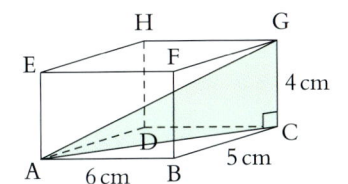

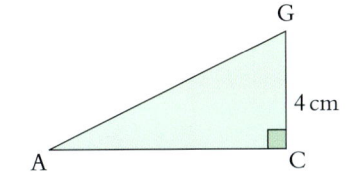

Before you can calculate AG you need to calculate AC using Pythagoras on triangle ABC.

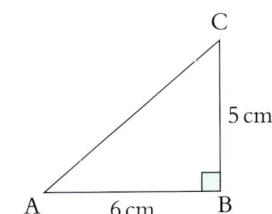

$AC^2 = 6^2 + 5^2$
$AC^2 = 36 + 25$
$AC^2 = 61$
$AC = \sqrt{61}$

Now use Pythagoras on triangle ACG.

$AC^2 + 4^2 = AG^2$
$61 + 16 = AG^2$
$AG^2 = 77$
$AG = 8.77$ cm (to 3 s.f.)

It is useful to remember that AG can be calculated directly using

$AG^2 = 6^2 + 5^2 + 4^2$
$AG^2 = 36 + 25 + 16$
$AG^2 = 77$
$AG = 8.77$ cm

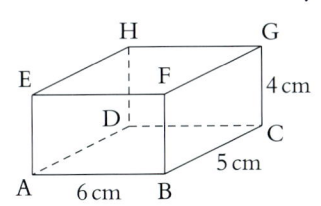

EXERCISE 2.17

1 Find the length of AG for each of these cuboids.

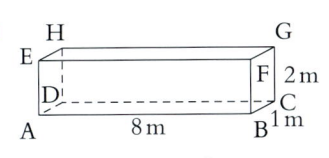

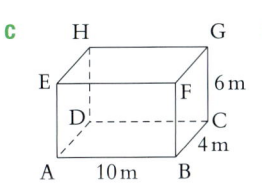

a) 4 m, 2 m, 3 m
b) 8 m, 1 m, 2 m
c) 10 m, 4 m, 6 m
d) 7 m, 7 m, 7 m

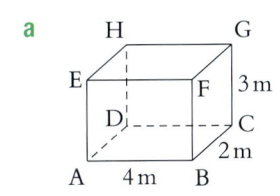

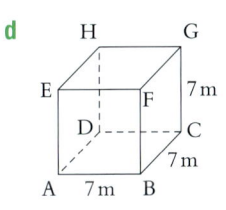

2 Find the length of AC in this triangular prism.

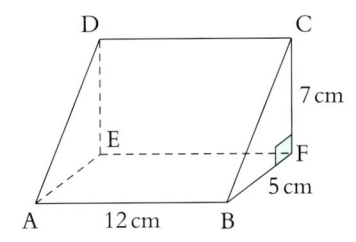

(12 cm, 5 cm, 7 cm)

3 ABCDEFGH is a cuboid.
P is the midpoint of GC.
Calculate the length of AP.

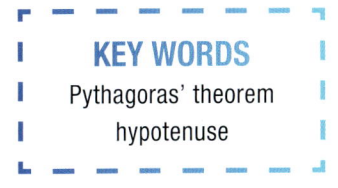

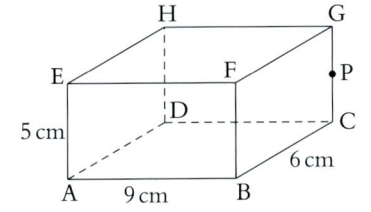

(9 cm, 5 cm, 6 cm)

KEY WORDS
Pythagoras' theorem
hypotenuse

Pythagoras 83

Geometrical constructions

THIS SECTION WILL SHOW YOU HOW TO
- Use a straight edge and compasses to construct a triangle given the lengths of the sides
- Use a straight edge and compasses to construct angle bisectors and perpendicular bisectors

EXAMPLE

Construct triangle ABC in which AB = 6.5 cm, AC = 3.3 cm and BC = 4.5 cm.

- Draw AB = 6.5 cm
- With your compass point on A, draw an arc radius 3.3 cm
- With your compass point on B, draw an arc radius 4.5 cm
- Label the point of intersection of the two arcs as C
- Draw the lines AC and BC

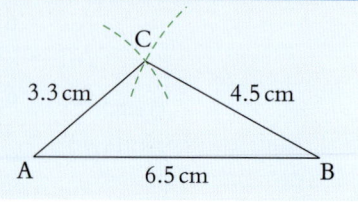

EXAMPLE

Construct an angle of 60° without using a protractor.

- Draw a base line AB
- With your compass point on A, draw an arc to intersect AB at X
- With your compass point on X, draw an arc with the same radius to intersect the first arc at Y
- Draw the line AY   Angle BAY = 60°

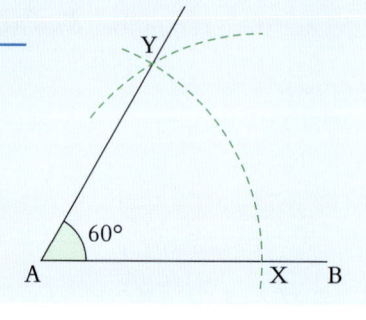

EXAMPLE

Constuct the bisector of angle ABC.

- With your compass point on B, draw an arc to intersect the lines BA and BC at P and Q
- With the same radius, draw arcs from P and Q to intersect each other at R
- Label the point of intersection of the two arcs as R
- Draw the line BR
- Angle ABR = Angle CBR

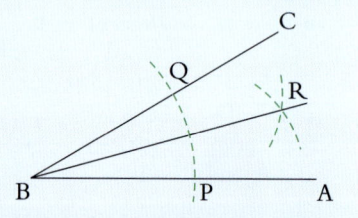

EXAMPLE

Constuct the perpendicular bisector of the line AB.

- With your compass point on A, draw arcs (with the same radius) above and below the line AB
- Using the same radius and your compass point on B, draw arcs above and below the line AB, so that they intersect the first arcs (at P and Q)
- Draw the line PQ
- PQ is the perpendicular bisector of AB

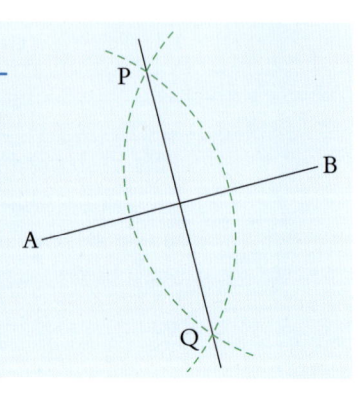

84   UNIT 2

EXERCISE 2.18

Use a straight edge and compasses for each of these constructions.

1. Draw any *acute* angle. Construct the angle bisector.
   Check your results with a protractor.

2. Draw any *obtuse* angle. Construct the angle bisector.
   Check your results with a protractor.

3. Constuct an angle of 60°. Bisect your 60° angle to make two angles of 30°.

4. Draw a line AB that is 10 cm long.
   Construct the perpendicular bisector of your line AB.
   Check your results with a protractor and ruler.

5. Plot the points A(2, 10) and B(8, 1) on a pair of axes.
   Construct the perpendicular bisector of the line joining A to B.
   What are the coordinates of the midpoint of the line AB?

6. Construct an equilateral triangle with sides of length 8 cm.

7. Construct a rhombus whose sides are of length 8 cm and whose shorter diagonal is 5 cm.
   What is the length of the longer diagonal?

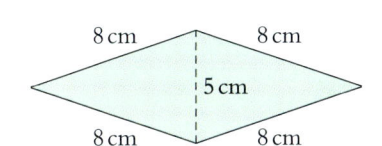

8. Construct a regular hexagon where the length of the line AD is 8 cm.

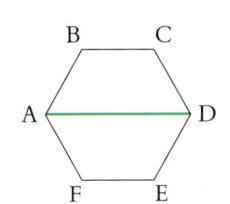

9. Construct a parallelogram with diagonals of length 9 cm and 7 cm that intersect at 60°.
   What is the length of the longest side of the parallelogram?

10. Construct a circle that passes through each of the points A, B and C.
    What are the coordinates of the centre of the circle?

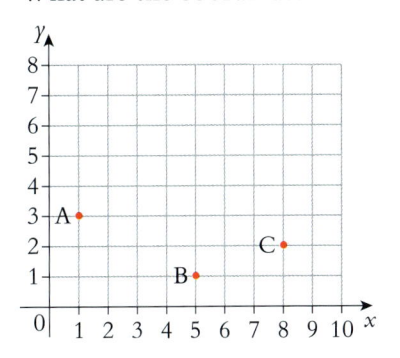

> **HINT**
> construct the perpendicular bisectors of AB and BC.

KEY WORDS
construct
bisector

Geometrical constructions

Loci

THIS SECTION WILL SHOW YOU HOW TO
- Find a set of points that satisfy a given rule

A **locus** is a set of points that satisfy a given rule.
It can be a line, a curve or a region.
Loci is the plural of locus.
These examples show some common loci.

EXAMPLE

Find the locus of points that are 1 cm from the point A.

The locus is a circle of radius 1 cm.

EXAMPLE

Find the locus of points that are 1 cm from the line AB.

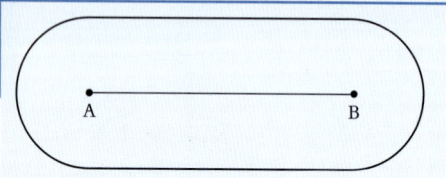

The locus is made from two straight lines and two semicircles.

EXAMPLE

Find the locus of points that are the same distance from the point A as they are from the point B.

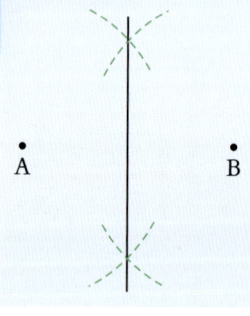

➡ **NOTE:** equidistant can also be used to mean 'the same distance'.

The locus is the perpendicular bisector of the line AB.

EXAMPLE

Find the locus of points that are the same distance from the line AB as they are from the line AC.

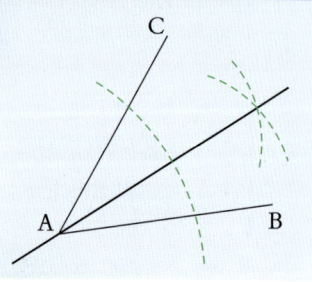

The locus is the bisector of angle BAC.

86 UNIT 2

EXAMPLE

Construct the locus of points inside triangle PQR that are nearer to PQ than they are to PR.

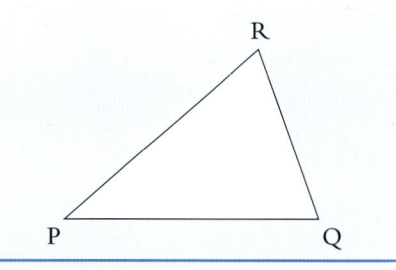

First construct the boundary line where the points are the *same* distance from PQ as they are from PR.
The points that are *nearer* to PQ than they are to PR are below the boundary line.
The required locus is the shaded region on the diagram.

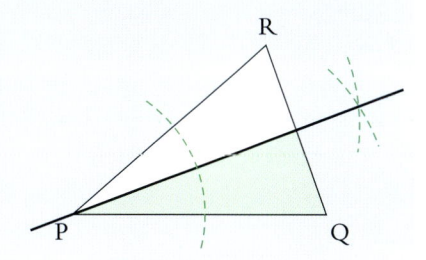

EXERCISE 2.19

For each of these constructions use a ruler and compasses.

1. Construct the locus of points that are 2 cm from a fixed point P.

2. Construct the locus of points that are always less than 1 cm from a fixed point P.

3. Copy the diagram and construct the locus of points that are the same distance from the line PQ as they are from the line PR.

4. Copy the diagram and construct the locus of points that are nearer to the line YZ than the line XY.

5. The points A and B are 8 cm apart.
Draw an accurate diagram and construct the locus of points that are equidistant from point A and point B.

6. The points J and K are 6 cm apart.
Draw an accurate diagram and construct the locus of points that are nearer to J than they are to K.

7. D and E are 7 cm apart. Draw an accurate diagram and construct the locus of points that are nearer to E than to D **and** that are less than 5 cm from D.

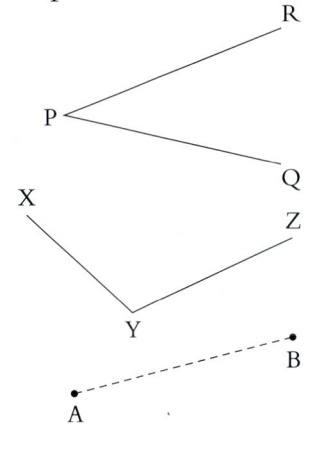

Loci 87

8 Draw a square with sides of length 4 cm.
Construct the locus of all points *outside* the square that are 2 cm from the sides of the square.

9 Draw an equilateral triangle with sides of length 6 cm.
Construct the locus of all points *outside* the triangle that are 2 cm from the sides of the triangle.

10 A camel is tied to a horizontal rail of length 10 m by a rope which is 3 m long.
The rope can slide along the horizontal rail.
Using a scale of 1 cm = 1 m, draw a scale diagram to show all the points that the camel can reach.

11 Draw the square ABCD that has sides of length 6 cm. 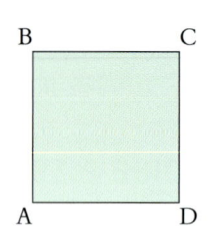
   a Construct the locus of points inside the square that are the same distance from A as from B.
   b Construct the locus of points inside the square that are the same distance from AB as from AD.
   c Shade the region where the points are nearer to B than to A **and** nearer to AB than to AD.

12 Draw the rectangle ABCD that has sides of length 7 cm and 5 cm. 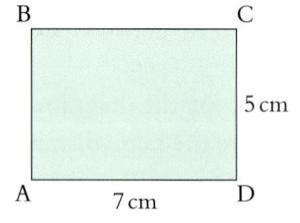
   a Construct the locus of points inside the rectangle that are 4 cm from D.
   b Construct the locus of points inside the rectangle that are equidistant from AB as from BC.
   c Label the point P which lies on both of the loci.

13 Draw the rectangle ABCD that has sides of length 8 cm and 4 cm. 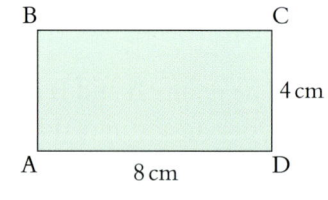
   a Construct the locus of points inside the rectangle that are 3 cm from the midpoint of the line BC.
   b Construct the locus of points inside the rectangle that are the same distance from BC as from AD.
   c Shade the region which is less than 3 cm from the midpoint of the line BC **and** is nearer to AD than to BC.

14 The diagram shows the positions of three villages A, B and C.
A new supermarket is to be built so that it is the same distance from each of the three villages.
Make a scale diagram and use constructions to find the position where the supermarket should be built.
Mark the position with the letter P.

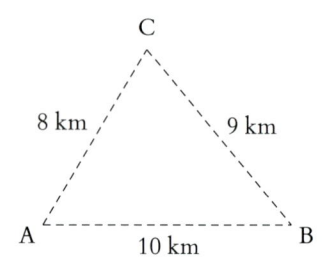

15 The diagram shows the positions of three television transmitters P, Q and R.
The signals from transmitter P have a range of 325 km.
The signals from transmitter Q have a range of 275 km.
The signals from transmitter R have a range of 400 km.
On a scale diagram shade the locus of points where all three signals can be received.

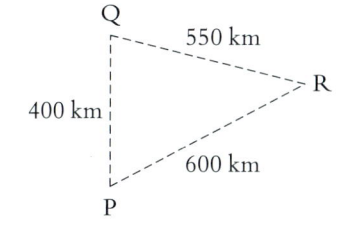

16 The diagram shows a farmer's field ABCD.
The farmer decides to spread fertilizer in the region that is
- nearer to D than to A *and*
- nearer to CD than to AD.

Draw a scale diagram and construct the two loci accurately.
Shade the region where the fertilizer will be spread.

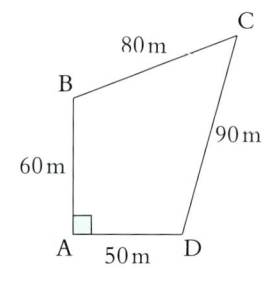

17 Boris wants to pass between the lion and the electric fence.
He walks so that he is always the same distance from the electric fence as he is from the lion.
Use the grid to help you to sketch the path that he must take.

18 A goat is tethered to the corner of a shed by a piece of rope that is 6 m long.
The shed is 8 m long and 4 m wide.
Draw a scale diagram and shade the area that can be reached by the goat.

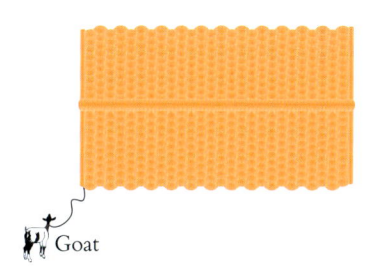

19 The diagram shows a bicycle wheel of radius 36 cm.
The wheel has a reflector (R) attached that is 25 cm from the centre of the wheel (P).
  a Sketch the locus of the point P as the wheel turns though one complete turn.
  b Sketch the locus of the point R as the wheel turns though one complete turn.

KEY WORDS
locus
loci

Loci   **89**

Area and circumference of a circle

THIS SECTION WILL SHOW YOU HOW TO
- Find the circumference of a circle
- Find the area of a circle
- Find the length of an arc
- Find the area of a sector

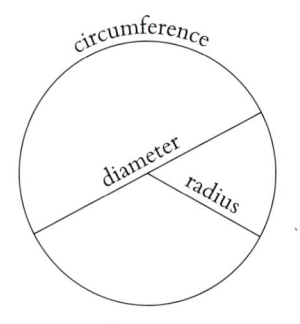

The perimeter of a circle is called the **circumference** (C).
The **diameter** (d) of a circle is twice the **radius** (r).

Circumference = $2\pi r$ (or πd)
Area = πr^2

EXAMPLE

A circular window has a diameter of 1.2 m
Calculate
a the circumference   **b** the area of the window.

a $C = 2\pi r$
$C = 2 \times \pi \times 0.6$
$C = 3.77$ cm (to 3 s.f.)

b $A = \pi r^2$
$A = \pi \times 0.6^2$
$A = 1.13$ cm² (to 3 s.f.)

➡ **NOTE:** use $\pi = 3.142$ or the π key on your calculator.

EXAMPLE

The distance round a CD is 38 cm.
Calculate the diameter of the CD.

$C = 2\pi r$ substitute $C = 38$ in formula
$38 = 2\pi \times r$ divide both sides by 2π
$\dfrac{38}{2\pi} = r$

$r = 6.04788\ldots$ $d = 2r$
$d = 12.1$ cm   give your answer to 3 s.f.

EXAMPLE

A circular clock face has an area of 50 cm².
Calculate the radius of the clock face.

$A = \pi r^2$ substitute $A = 50$ into the formula
$50 = \pi \times r^2$ divide both sides by π
$\dfrac{50}{\pi} = r^2$ square root both sides

$r = \sqrt{\dfrac{50}{\pi}}$

$r = 3.99$ cm   give your answer to 3 s.f.

90 UNIT 2

EXERCISE 2.20

1. Find the circumference and area of these circles. All lengths are in cm.

    a) 6
    b) 15
    c) 4.8
    d) 3.8
    e) 20
    f) 4.7

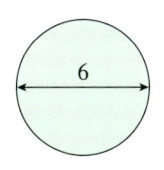

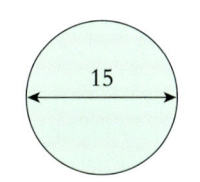

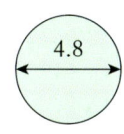

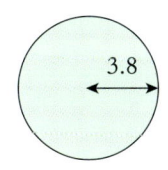

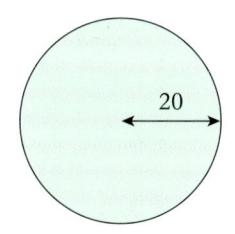

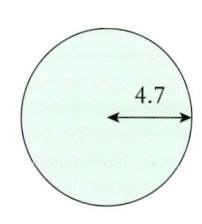

2. The wheels on a car have a diameter of 52 cm.
    a. Calculate the distance travelled by the car when the wheels rotate once.
    b. How many km does the car travel if the wheels rotate 5000 times?

3. A circular table has an area of 1.25 m².
    Calculate   a the radius   b the circumference of the table.

4. The length of the sides of each square is 12 cm.
    The circles fit tightly inside the square.
    Calculate the shaded area for each diagram.

    a   b 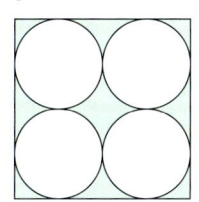   c

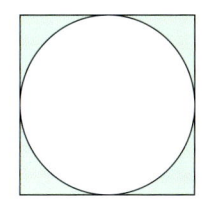

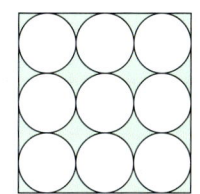

    d   What do you notice about your answers? Can you explain why?

5. The radius of the large circle is 5 cm.
    The radius of the small circle is 3 cm.
    Calculate the shaded area.

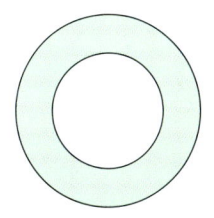

6. The radius of the circle is 10 cm.
    The vertices of the square lie on the circumference of the circle.
    Calculate the shaded area.

Area and circumference of a circle

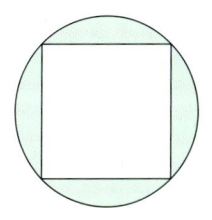

Semicircles and quadrants

A **semicircle** is half a circle. A **quadrant** is quarter of a circle.

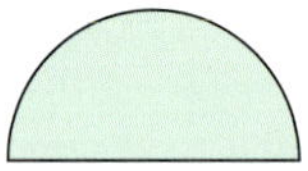

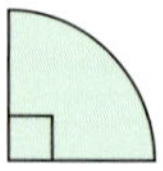

EXAMPLE

Find the perimeter and area of the shape.

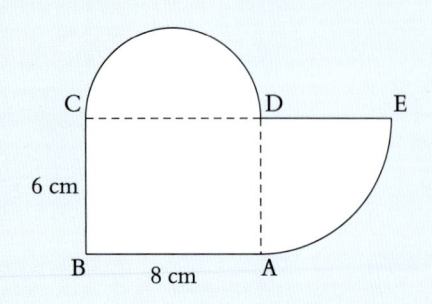

The radius of the semi-circle is 4 cm.
The radius of the quadrant is 6 cm.

Perimeter = AB + BC + arc CD + DE + arc EA

$$= 8 + 6 + \frac{2 \times \pi \times 4}{2} + 6 + \frac{2 \times \pi \times 6}{4}$$

$= 8 + 6 + 4\pi + 6 + 3\pi$
$= 41.99114...$
$= 42.0$ cm (to 3 s.f.)

Area = area of rectangle + area of semicircle + area of quadrant

$$= (8 \times 6) + \frac{\pi \times 4^2}{2} + \frac{\pi \times 6^2}{4}$$

$= 48 + 8\pi + 9\pi$
$= 101.407...$
$= 101$ cm^2 (to 3 s.f.)

EXERCISE 2.21

Calculate the perimeter and area of each of these shaded shapes. All lengths are in cm.

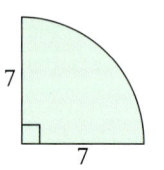

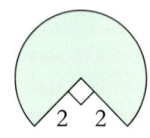

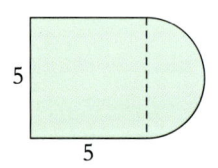

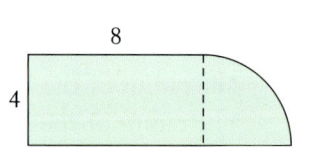

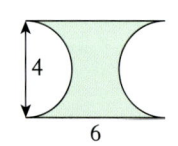

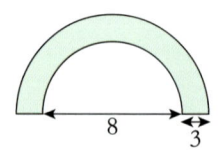

1. (semicircle, diameter 6)

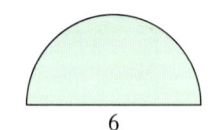

2. (quadrant, 7 by 7)

3. (semicircle with triangular notch, 2 and 2)

4. (rectangle 5 by 5 with semicircle on end)

5. (rectangle 3 by 7 with semicircles on both ends)

6. (rectangle 8 by 4 with quadrant on end)

7. (rectangle 4 by 6 with semicircles cut from sides)

8. (semicircular ring, 8 and 3)

9. (square 10 by 10 with quadrants cut from corners)

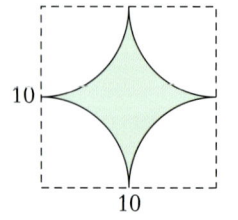

10

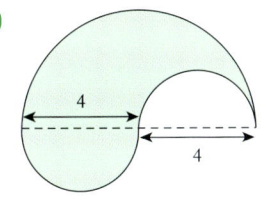

11

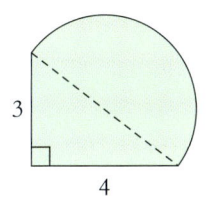

12 The diagram shows four rods of radius 2 cm.
The rods are held tightly together by a piece of wire.
Calculate the length of the wire.

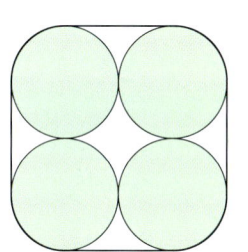

13 a The area of a semicircle is 6 cm². Find the radius and perimeter.
  b The area of a semicircle is 24 cm². Find the radius and perimeter.
  c Compare your answers to parts **a** and **b**. What do you notice?

14 a The area of a quadrant is 5 cm². Find the radius and perimeter.
  b The area of a quadrant is 45 cm². Find the radius and perimeter.
  c Compare your answers to parts **a** and **b**. What do you notice?

15 The perimeter of a quadrant is 24 cm.
Calculate the area of the quadrant.

16 This shape is made from two identical quadrants.
The perimeter of the shape is 40 cm.
Calculate the radius of the quadrants.

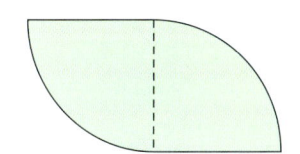

17 A circle has a circumference of x cm and an area of $2x$ cm².
Calculate the radius of the circle.

Arcs and sectors
An **arc** is part of the circumference of a circle.
A **sector** of a circle is a region bounded by an arc and two radii.

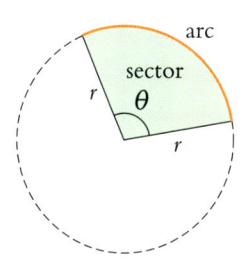

Arc length = $\dfrac{\theta}{360} \times 2\pi r$ Sector area = $\dfrac{\theta}{360} \times \pi r^2$

Area and circumference of a circle   93

EXAMPLE

Find
a   the area of the sector OAB
b   the length of the arc AB
c   the perimeter of the sector AOB.

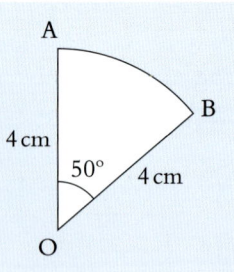

---

a area of sector OAB $= \dfrac{50}{360} \times \pi r^2$

$\phantom{\text{area of sector OAB}} = \dfrac{50}{360} \times \pi \times 4^2$

$\phantom{\text{area of sector OAB}} = 6.98 \text{ cm}^2$ (to 3 s.f.)

b arc length AB $= \dfrac{50}{360} \times 2\pi r$

$\phantom{\text{arc length AB}} = \dfrac{50}{360} \times 2 \times \pi \times 4$

$\phantom{\text{arc length AB}} = 3.49 \text{ cm}$ (to 3 s.f.)

c perimeter of sector OAB = OA + OB + arc AB
$\phantom{\text{perimeter of sector OAB}} = 4 + 4 + 3.49$
$\phantom{\text{perimeter of sector OAB}} = 11.5 \text{ cm}$ (to 3 s.f.)

EXAMPLE

a Arc AB = 7 cm.
Find the value of θ

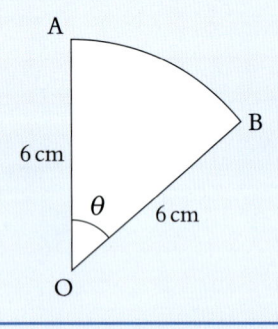

b Area of sector AOB = 50 cm².
Find the value of r.

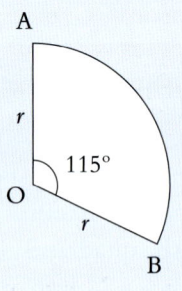

---

a Arc length $= \dfrac{\theta}{360} \times 2\pi r$

$7 = \dfrac{\theta}{360} \times 2 \times \pi \times r$

$7 = \dfrac{\theta}{360} \times 12\pi$

$\theta = \dfrac{7 \times 360}{12\pi}$

$\theta = 66.8°$ (to 3 s.f.)

b Area of sector $= \dfrac{\theta}{360} \times \pi r^2$

$50 = \dfrac{115}{360} \times \pi r^2$

$50 \times 360 = 115\pi \times r^2$

$r = \sqrt{\dfrac{50 \times 360}{115\pi}}$

$r = 7.06 \text{ cm}$ (to 3 s.f.)

EXERCISE 2.22

1. Find the length of the arc AB for each of these.
   - a) A, 7 cm, 30°, 7 cm, B
   - b) 140°, 8 cm
   - c) 4 cm, 88°
   - d) 230°, 8 cm

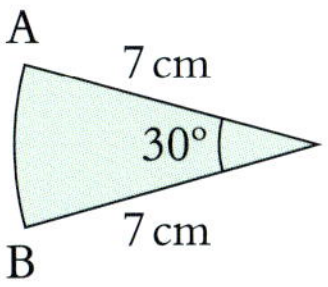

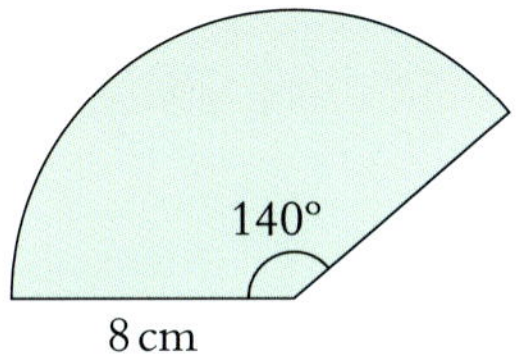

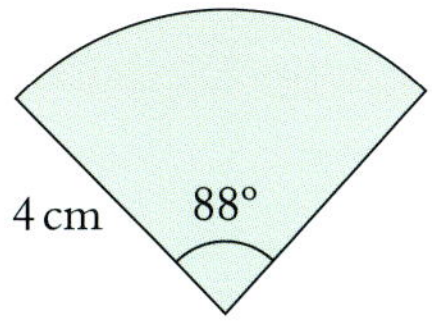

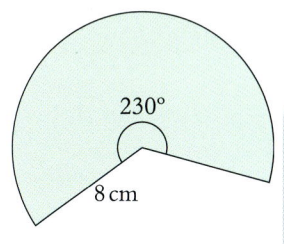

2. Find the area of the sector for each of these.
   - a) 3 cm, 85°
   - b) 5 cm, 102°
   - c) 10 cm, 145°
   - d) 8 cm, 285°

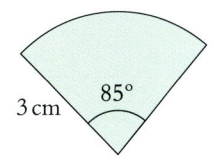

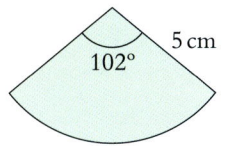

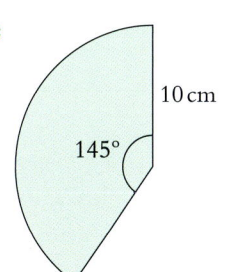

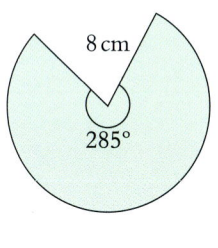

3. Find the perimeter and area of these shapes.
   - a) 2 cm, 3 cm, 40°
   - b) 250°, 6 cm, 2 cm
   - c) 132°, 4 cm, 5 cm
   - d) 100°, 3 cm, 8 cm

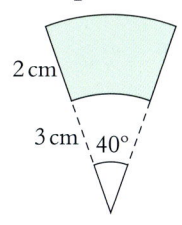

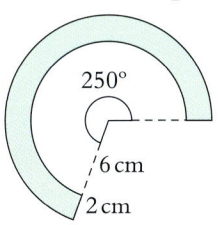

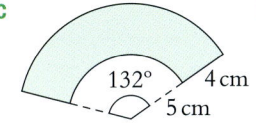

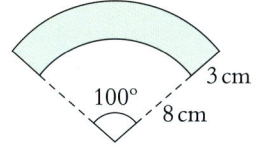

4. The area of the sector is 6 cm². Calculate the value of θ. (3 cm)

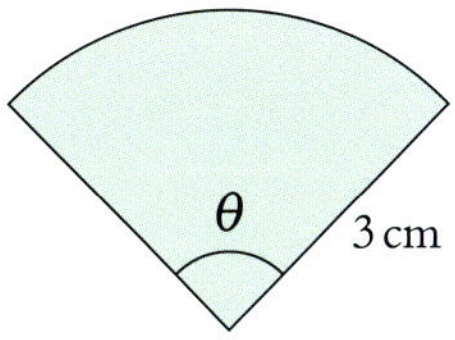

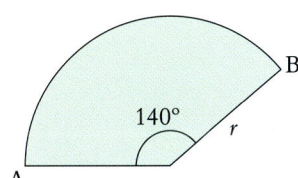

5. The length of arc AB is 10 cm. Calculate the value of r. (140°, B, r, A)

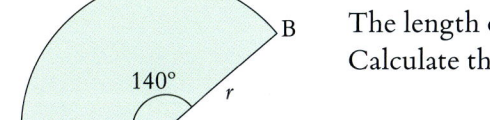

6. This shape is made from a square of length 5 cm and four identical sectors of a circle of radius 5 cm and angle 60°.
   Calculate
   - a) the perimeter
   - b) the area of the shape.

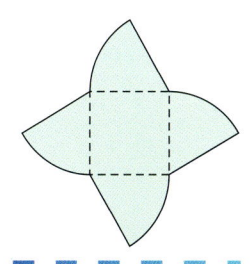

7. The diagram shows three identical coins of radius 1 cm. Calculate the shaded area in the middle of the diagram.

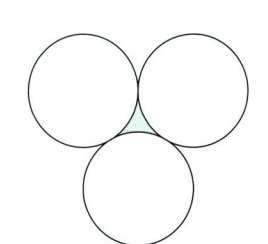

KEY WORDS
circumference
diameter, radius
semicircle, quadrant
arc, sector

Area and circumference of a circle

Displaying data

THIS SECTION WILL SHOW YOU HOW TO
- Display data and interpret data

Types of data

Discrete Data	Continuous Data
• marks in a test • number of people in a boat • number of coins • shoe size	• height of students • mass of sand • time to complete a puzzle • speed of a car

Discrete data can only have exact numbers.

Continuous data can take any value in a particular range.

Bar charts, pictograms and pie charts

A **pictogram** is similar to a **bar chart**, except that the frequencies are represented by a number of identical pictures.

In a **pie chart** the frequency is represented by the angle of the sector of a circle.

EXAMPLE

The table shows the grades obtained by 30 students in an examination. Show this information on
a   a bar chart   b   a pictogram   c   a pie chart.

Grade	A★	A	B	C	D
Number of students	3	12	10	3	2

a

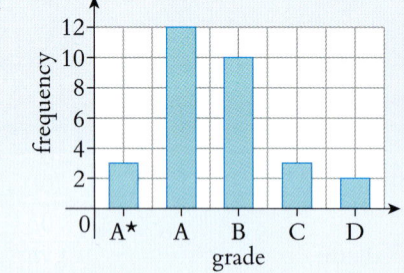

b

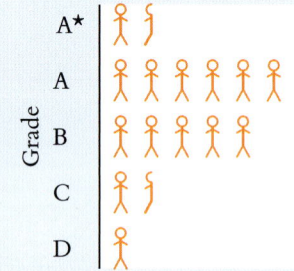

Key: ♁ represents 2 students

c   The pie chart represents 30 students. Each student is represented by 360° ÷ 30 = 12°

Grade	Frequency	Angle
A★	3	3 × 12° = 36°
A	12	12 × 12° = 144°
B	10	10 × 12° = 120°
C	3	3 × 12° = 36°
D	2	2 × 12° = 24°
		360°

Examination grades

96   UNIT 2

Displaying continuous data

Continuous data is often put into a grouped frequency table with equal class widths. A **frequency diagram** can then be drawn to represent the data.

EXAMPLE

2.83	3.22	3.79	2.95	3.88	3.08	4.23	1.74	3.47	2.77
3.56	2.03	3.14	4.11	2.51	2.38	3.98	3.37	3.66	3.33

The table shows the mass (m), in kilograms of 20 newborn babies.
  a Arrange the data into a grouped frequency table using class intervals of $1.5 < m \leq 2.0$, $2.0 < m \leq 2.5$, $2.5 < m \leq 3.0$, $3.0 < m \leq 3.5$, $3.5 < m \leq 4.0$ and $4.0 < m \leq 4.5$.
  b Use the grouped frequency table in part **a** to draw a frequency diagram to represent the data.

a

mass (m) kg	tally	frequency
$1.5 < m \leq 2.0$	|	1
$2.0 < m \leq 2.5$	||	2
$2.5 < m \leq 3.0$	||||	4
$3.0 < m \leq 3.5$	||||		6
$3.5 < m \leq 4.0$	||||	5
$4.0 < m \leq 4.5$	||	2

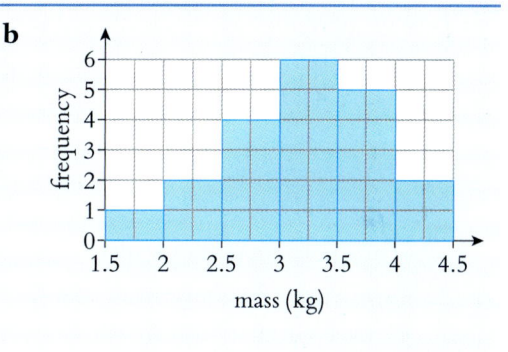

➡ **NOTE:** for continuous data
  • there are no gaps between the bars
  • the horizontal axis is written as a continuous number line

EXERCISE 2.23

1 The table shows the colour of 60 students' mobile phones.

black	silver	blue	other
25	20	10	5

Show this information on
  a a bar chart   b a pictogram   c a pie chart.

2 The table shows the sales of four different flavour packs of crisps.

plain	vinegar	chicken	cheese
16	8	12	4

Show this information on
  a a bar chart   b a pictogram   c a pie chart.

Displaying data   97

3 The table shows how 50 adults travel to work.

walk	cycle	bus	car	train
15	5	10	18	2

Show this information on
a a bar chart   b a pictogram   c a pie chart.

4 The pie chart shows the ingredients used to make a cake.
The cake mixture has a total mass of 900 g.
Measure each of the angles on the pie chart and then calculate the mass of each of the ingredients used to make the cake.

5 Two dice are thrown 40 times and the scores on each die are added together to give a total score. These are the results.

5	7	4	8	10	6	7	3	9	6
6	9	8	7	5	2	11	10	11	9
7	3	8	5	12	6	7	4	8	6
10	6	4	11	7	7	8	5	9	4

a Construct a frequency table to show the total scores.
b Draw a bar chart to show the total scores.

6 The table shows information about the ages of people in a swimming pool.
a Draw a frequency diagram to show the information in the table.
b Find the percentage of people that are over 30 years old.

Age (x years)	Number of people
$0 < x \leq 10$	4
$10 < x \leq 20$	20
$20 < x \leq 30$	15
$30 < x \leq 40$	6
$40 < x \leq 50$	4
$50 < x \leq 60$	1

7 The table shows information about the times taken by 32 teachers to travel to work.
a Draw a frequency diagram to show the information in the table.
b Find the percentage of teachers that take more than 5 minutes.

Time (t minutes)	Frequency
$0 < t \leq 5$	8
$5 < t \leq 10$	12
$10 < t \leq 15$	6
$15 < t \leq 20$	4
$20 < t \leq 25$	2

Two-way tables
A **two-way table** shows two or more sets of data at the same time.

EXERCISE 2.24

1 The table shows information about the number of boys and girls in Class 5A who did and did not do their homework.

	Boys	Girls
Homework	13	15
No homework	4	1

  a How many boys are in Class 5A?
  b How many students did not do their homework?
  c How many students are there altogether in Class 5A?

2 In a survey 100 students are asked if they are left-handed or right handed.
   The table shows the results.
   Copy and complete the table.

	Boys	Girls	Total
Left-handed	7		18
Right-handed		41	
Total			100

3 Students in a small school must attend one lunchtime club.
   They have a choice of Athletics Club, Drama Club or Music Club.
   The table shows information about the number of students attending each club.

	Boys	Girls	Total
Athletics	56	42	
Drama	58	37	
Music	16		
Total			250

  a Copy and complete the table.
  b What percentage of the students attend the Music Club?

4 The table shows information about the times taken by a group of people to solve a puzzle.

		TIME (t minutes)		
		$0 < t \leq 10$	$10 < t \leq 20$	$20 < t \leq 30$
AGE (x years)	$10 < x \leq 20$	5	9	23
	$20 < x \leq 30$	15	10	7
	$30 < x \leq 40$	13	12	15
	$40 < x \leq 50$	10	15	16

  a How many people are there altogether?
  b How many people took more than 10 minutes?
  c What percentage of the people took 10 minutes or less?

KEY WORDS
discrete data
continuous data
bar chart
pictogram
pie chart
frequency diagram
two-way table

Displaying data

Unit 2 Examination questions

1. Simplify $\dfrac{x}{3} + \dfrac{5x}{9} - \dfrac{5x}{18}$. [2]

   Cambridge IGCSE Mathematics 0580, Paper 21 Q2, June 2008

2. Solve the equation
   $$\dfrac{x-2}{4} = \dfrac{2x+5}{3}.$$ [3]

   Cambridge IGCSE Mathematics 0580, Paper 2 Q13, June 2006

3. (a) $\dfrac{2}{3} + \dfrac{5}{6} = \dfrac{x}{2}.$

   Find the value of x. [1]

   (b) $\dfrac{5}{3} \div \dfrac{3}{y} = \dfrac{40}{9}.$

   Find the value of y. [1]

   Cambridge IGCSE Mathematics 0580, Paper 2 Q2, November 2006

4. Solve the inequality $6(2-3x) - 4(1-2x) \leq 0$. [3]

   Cambridge IGCSE Mathematics 0580, Paper 21 Q13, November 2009

5. Solve the inequality $\dfrac{2-5x}{7} < \dfrac{2}{5}$. [3]

   Cambridge IGCSE Mathematics 0580, Paper 21 Q12, November 2008

6. Solve the inequality $\dfrac{2x-5}{8} > \dfrac{x+4}{3}$. [3]

   Cambridge IGCSE Mathematics 0580 Paper 21 Q13, June 2008

7. Solve the inequality $2x + 5 < \dfrac{x-1}{4}$. [3]

   Cambridge IGCSE Mathematics 0580 Paper 21 Q13, November 2010

8. Find the length of the line joining the points $A(-4, 8)$ and $B(-1, 4)$. [2]

   Cambridge IGCSE Mathematics 0580, Paper 21 Q8, November 2009

9 A straight line passes through two points with co-ordinates (6, 8) and (0, 5).
Work out the equation of the line. [3]

Cambridge IGCSE Mathematics 0580, Paper 21 Q9, June 2008

10 Find the co-ordinates of the mid-point of the line joining
the points $A(2, -5)$ and $B(6, 9)$. [2]

Cambridge IGCSE Mathematics 0580, Paper 21 Q7, June 2009

11

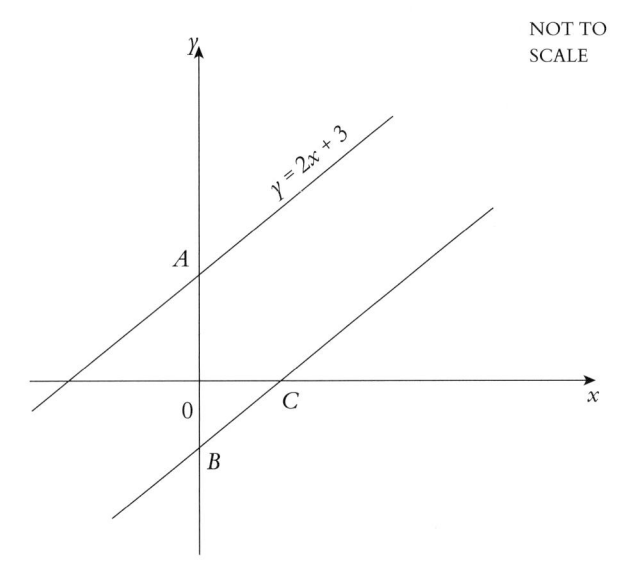

NOT TO SCALE

The distance AB is 7 units.

(a) Write down the equation of the line through B which is parallel to $y = 2x + 3$. [2]
(b) Find the co-ordinates of the point C were this line crosses the x-axis. [1]

Cambridge IGCSE Mathematics 0580, Paper 21 Q9, November 2008

12

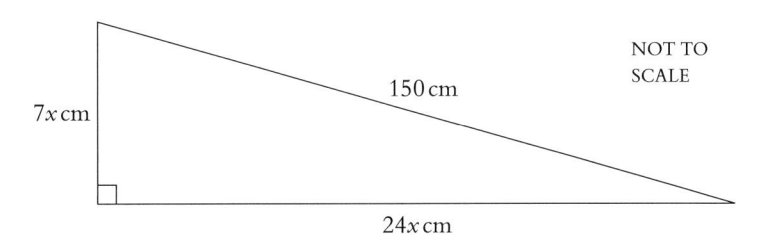

NOT TO SCALE

The right-angled triangle in the diagram has sides of length $7x$ cm, $24x$ cm and 150 cm.

(a) Show that $x^2 = 36$. [2]
(b) Calculate the perimeter of the triangle. [1]

Cambridge IGCSE Mathematics 0580, Paper 2 Q10, November 2006

13

0.8 m

1.4 m

NOT TO SCALE

The top of a desk is made from a rectangle and a quarter circle.
The rectangle measures 0.8m by 1.4m.

Calculate the surface area of the top of the desk. [3]

Cambridge IGCSE Mathematics 0580, Paper 21 Q7, June 2010

14

1.5 m

3.5 m

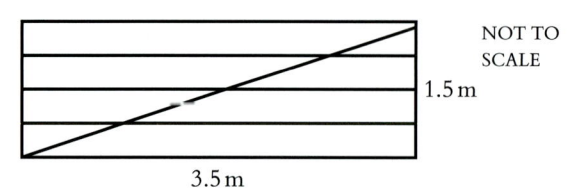

NOT TO SCALE

The diagram represents a rectangular gate measuring 1.5 m by 3.5 m.
It is made from eight lengths of wood.

Calculate the total length of wood needed to make the gate. [3]

Cambridge IGCSE Mathematics 0580, Paper 21 Q10, June 2010

15

8 cm

6 cm

6 cm

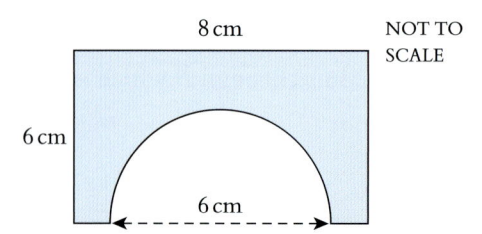

NOT TO SCALE

The semicircle of diameter 6 cm is cut from a rectangle with sides 6 cm and 8 cm.

Calculate the perimeter of the shaded shape, correct to 1 decimal place. [3]

Cambridge IGCSE Mathematics 0580, Paper 21 Q15, November 2010

16

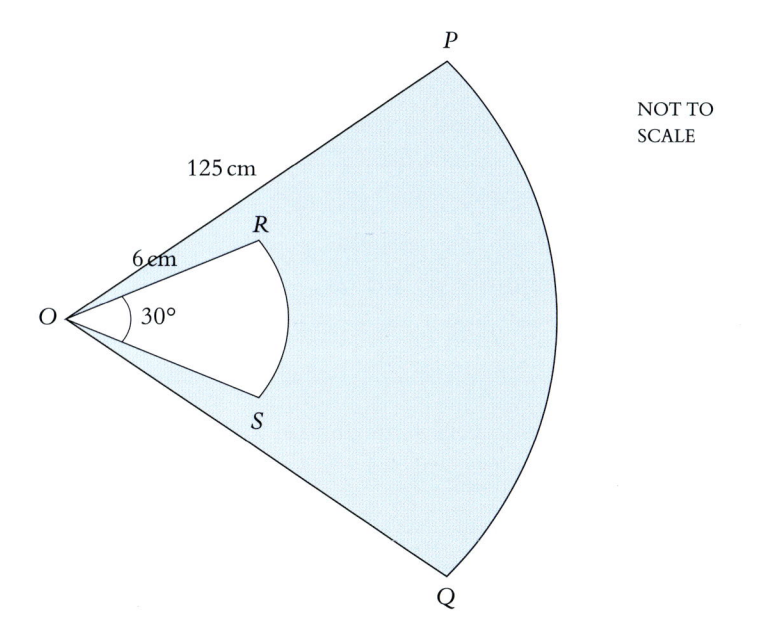

OPQ is a sector of a circle, radius 12 cm, centre *O*. Angle *POQ* = 50°.

ORS is a sector of a circle, radius 6 cm, also centre *O*. Angle *ROS* = 30°.

(a) Calculate the shaded area. [3]

(b) Calculate the perimeter of the shaded area, *PORSOQP*. [3]

Cambridge IGCSE Mathematics 0580, Paper 21 Q19, November 2008

Standard form

THIS SECTION WILL SHOW YOU HOW TO
- Write numbers in standard form
- Solve problems and do calculations in standard form

A number in standard form must be written as

$A \times 10^n$ where $1 \leq A < 10$ and n is an integer

It is a very useful way of writing very large and very small numbers.

Large numbers

EXAMPLE

Write these numbers in standard form
a 8 000 000   b 30 500

a $8\,000\,000 = 8 \times 10^6$ b $30\,500 = 3.05 \times 10^4$

> **NOTE:**
> $10^1 = 10$
> $10^2 = 100$
> $10^3 = 1000$
> $10^4 = 10\,000$
> $10^5 = 100\,000$
> $10^6 = 1\,000\,000$

EXAMPLE

Write these as ordinary numbers
a 1.36×10^5 b 7×10^8

a $1.36 \times 10^5 = 136\,000$ b $7 \times 10^8 = 700\,000\,000$

EXAMPLE

$a = 4.5 \times 10^8$ and $b = 3 \times 10^5$
Calculate the following giving your answers in standard form.
a $a + b$ b $a \times b$ c $a \div b$

a $a + b = (4.5 \times 10^8) + (3 \times 10^5)$ *change so they are both $\times 10^5$*
$= (4500 \times 10^5) + (3 \times 10^5)$
$= 4503 \times 10^5$ *this is not in standard form so replace 4503 by 4.503×10^3*
$= 4.503 \times 10^3 \times 10^5$ *$10^3 \times 10^5 = 10^{3+5} = 10^8$*
$= 4.503 \times 10^8$

b $a \times b = (4.5 \times 10^8) \times (3 \times 10^5)$ *multiply 4.5 by 3 and 10^8 by 10^5*
$= 13.5 \times 10^{8+5}$
$= 13.5 \times 10^{13}$ *this is not in standard form so replace 13.5 by 1.35×10^1*
$= 1.35 \times 10^1 \times 10^{13}$
$= 1.35 \times 10^{14}$

c $a \div b = (4.5 \times 10^8) \div (3 \times 10^5)$ *divide 4.5 by 3 and 10^8 by 10^5*
$= 1.5 \times 10^{8-5}$
$= 1.5 \times 10^3$

EXERCISE 3.1

1. Which of these numbers are written in standard form?
   - a 4×10^2
   - b 15×10^6
   - c 9.5×10^4
   - d 10×10^3
   - e 23.4×10^7
   - f 1×10^8
   - g 7.07×10^5
   - h 50.03×10^8

2. Write these numbers in standard form.
   - a 54 000
   - b 200
   - c 700 000
   - d 5330
   - e 82 000 000
   - f 3020
   - g 666 000
   - h 36.36

3. Write these as ordinary numbers.
   - a 4.8×10^5
   - b 3.7×10^2
   - c 5.64×10^8
   - d 7.06×10^7
   - e 2×10^4
   - f 1.1×10^6
   - g 3.001×10^2
   - h 9.5×10^9

4. Calculate the following, giving your answer in standard form.
   - a $(4 \times 10^3) + (2 \times 10^3)$
   - b $(4 \times 10^5) + (5 \times 10^4)$
   - c $(8 \times 10^4) - (5 \times 10^4)$
   - d $(4.3 \times 10^6) - (2 \times 10^5)$
   - e $(4 \times 10^3) \times (8 \times 10^5)$
   - f $(2.1 \times 10^5) \times (3 \times 10^2)$
   - g $(8 \times 10^7) \div (2 \times 10^4)$
   - h $(7.5 \times 10^9) \div (3 \times 10^2)$
   - i $(4 \times 10^6)^2$
   - j $\dfrac{(4 \times 10^8) + (5 \times 10^7)}{9 \times 10^3}$
   - k $\dfrac{(2.1 \times 10^3) \times (4 \times 10^5)}{2 \times 10^6}$
   - l $\dfrac{(4.9 \times 10^8) - (3 \times 10^6)}{(7 \times 10^5) + (3 \times 10^5)}$

5. Check your answers to question 4 using a calculator.

6. $p = 6 \times 10^9$ and $q = 2 \times 10^8$
   Calculate the following giving your answers in standard form.
   - a $p + q$
   - b $p - q$
   - c $2p - 5q$
   - d $p \times q$
   - e $p \div q$
   - f p^2
   - g $\dfrac{p+q}{q}$
   - h $\dfrac{p^2}{2q}$

7. Simplify the ratios.
   - a $6 \times 10^5 : 2 \times 10^4$
   - b $2.5 \times 10^6 : 1 \times 10^8$
   - c $5.6 \times 10^7 : 11.2 \times 10^4$

8. Solve the equations.
   - a $5x + 3 \times 10^4 = 5.1 \times 10^5$
   - b $2(x - 6 \times 10^3) = 3 \times 10^4$
   - c $3(x - 2 \times 10^4) = 2(x + 7 \times 10^5)$

9. Simplify, giving your answer in standard form.
   - a $\sqrt{4 \times 10^8}$
   - b $\sqrt{9 \times 10^4}$
   - c $\sqrt{3.6 \times 10^7}$
   - d $\sqrt{4.9 \times 10^9}$

10. Write these in order of size, smallest first.
    2.2×10^{14} 2.04×10^{14} 5.8×10^{13} 2.13×10^4

11. The distance between the Earth and the Sun is 149 million kilometres.
    Write this number in standard form.

12. The area of land on the Earth is 1.5×10^8
    The area of ocean on the Earth is 3.6×10^8
    Write the ratio of land to ocean in its simplest form.

13. The population of China is approximately 1.34×10^9
    The population of Argentina is approximately 4.05×10^7.
    Calculate the difference in population, giving your answer in standard form.

Standard form

Small numbers

EXAMPLE

Write these numbers in standard form
a 0.0007
b 0.0000061

a $0.0007 = 7 \times 10^{-4}$
b $0.0000061 = 6.1 \times 10^{-6}$

➡ **NOTE:**
$10^{-1} = 0.1$
$10^{-2} = 0.01$
$10^{-3} = 0.001$
$10^{-4} = 0.0001$
$10^{-5} = 0.00001$
$10^{-6} = 0.000001$

EXAMPLE

Write these as ordinary numbers
a 4.23×10^{-2}
b 9×10^{-4}

a $4.23 \times 10^{-2} = 0.0423$
b $9 \times 10^{-4} = 0.0009$

EXAMPLE

$a = 3 \times 10^5$ $b = 4 \times 10^{-3}$ $c = 2 \times 10^{-4}$
Calculate the following giving your answers in standard form.
a $a \times b$ b $b \times c$ c $a \div c$ d $c \div b$ e $b - c$

a $a \times b = (3 \times 10^5) \times (4 \times 10^{-3})$ multiply 3 by 4 and 10^5 by 10^{-3}
$= 12 \times 10^2$ this is not in standard form so replace 12 by 1.2×10^1
$= 1.2 \times 10^1 \times 10^2$ $10^1 \times 10^2 = 10^{1+2} = 10^3$
$= 1.2 \times 10^3$

b $b \times c = (4 \times 10^{-3}) \times (2 \times 10^{-4})$ multiply 4 by 2 and 10^{-3} by 10^{-4}, $10^{-3} \times 10^{-4} = 10^{-3+-4} = 10^{-7}$
$= 8 \times 10^{-7}$

c $a \div c = (3 \times 10^5) \div (2 \times 10^{-4})$ divide 3 by 2 and 10^5 by 10^{-4}, $10^5 \div 10^{-4} = 10^{5--4} = 10^9$
$= 1.5 \times 10^9$

d $c \div b = (2 \times 10^{-4}) \div (4 \times 10^{-3})$ divide 2 by 4 and 10^{-4} by 10^{-3}, $10^{-4} \div 10^{-3} = 10^{-4--3} = 10^{-1}$
$= 0.5 \times 10^{-1}$ this is not in standard form so replace 0.5 by 5×10^{-1}
$= 5 \times 10^{-1} \times 10^{-1}$ $10^{-1} \times 10^{-1} = 10^{-1+-1} = 10^{-2}$
$= 5 \times 10^{-2}$

e **Method 1**
$b - c = (4 \times 10^{-3}) - (2 \times 10^{-4})$
$= (4 \times 10^{-3}) - (0.2 \times 10^{-3})$
$= 3.8 \times 10^{-3}$

Method 2
$b - c = (4 \times 10^{-3}) - (2 \times 10^{-4})$
$= 0.004 - 0.0002$
$= 0.0038$
$= 3.8 \times 10^{-3}$

EXERCISE 3.2

1 Write these numbers in standard form.
   a 0.015
   b 0.00067
   c 0.0009
   d 0.000606
   e 3 200 000
   f 0.00052
   g 0.44
   h 53 080 000
   i 0.000 000 057
   j 0.000613
   k 700 200
   l 0.00089

2 Write these as ordinary numbers.
   a 8.8×10^{-5} b 2.3×10^{-7} c 6×10^{-3}
   d 5.11×10^{-2} e 9.9×10^{4} f 1.04×10^{-6}
   g 6.8×10^{-4} h 6×10^{8} i 3.8×10^{-1}
   j 2.2×10^{1} k 4.08×10^{5} l 9.5×10^{-6}

3 Calculate the following, giving your answer in standard form.
   a $(4 \times 10^{-2}) + (2 \times 10^{-2})$ b $(7 \times 10^{-5}) + (2 \times 10^{-6})$
   c $(8 \times 10^{-8}) - (3 \times 10^{-8})$ d $(6.2 \times 10^{-4}) - (2 \times 10^{-5})$
   e $(5 \times 10^{-5}) \times (6 \times 10^{-5})$ f $(3.2 \times 10^{-3}) \times (2 \times 10^{-4})$
   g $(9 \times 10^{7}) \times (3 \times 10^{-5})$ h $(4.8 \times 10^{-6}) \div (2 \times 10^{-4})$
   i $(3 \times 10^{-5})^2$ j $\dfrac{(5 \times 10^{4}) + (2 \times 10^{4})}{3.5 \times 10^{-2}}$
   k $\dfrac{(1.8 \times 10^{-2}) \times (3 \times 10^{8})}{2 \times 10^{3}}$ l $\dfrac{(9 \times 10^{-4}) + (7 \times 10^{-4})}{(4 \times 10^{-5}) \times (2 \times 10^{9})}$

4 Check your answers to question 3 using a calculator.

5 Simplify, giving your answer in standard form.

   a $\sqrt{9 \times 10^{-6}}$ b $\sqrt{4 \times 10^{-10}}$
   c $\sqrt{6.4 \times 10^{-5}}$ d $\sqrt{8.1 \times 10^{-13}}$

6 $p = 6 \times 10^{-8}$ and $q = 3 \times 10^{-9}$
   Calculate the following, giving your answers in standard form.
   a $p + q$ b $p - q$ c $3p + 2q$ d $p \times q$
   e $p \div q$ f p^2 g q^2 h $\dfrac{5p^2}{4q}$

7 The mass, in grams, of a single hydrogen atom is 1.67×10^{-24} g.
   Find the total mass of 500 hydrogen atoms.
   Give your answer in standard form.

8 The mass, in grams, of a single oxygen atom is 2.67×10^{-23} g.
   Calculate the number of oxygen atoms in 5 grams of oxygen.
   Give your answer in standard form.

9 The time taken by light to travel one metre is approximately 3×10^{-9} seconds.
   How long does it take light to travel 5 kilometres?
   Give your answer in standard form.

10 The mass, in kilograms, of a grain of sand is 3.5×10^{-10} kg.
    Find the total mass of 75 000 grains of sand.
    Give your answer in standard form.

KEY WORDS
standard form

Standard form

Simultaneous equations 1

THIS SECTION WILL SHOW YOU HOW TO
- Solve simultaneous equations using a graph
- Solve simultaneous equations using the elimination method

Simultaneous equations are a set of equations that are all satisfied by the same values of the variables. Simultaneous equations can be solved using a graph or by using algebra.

EXAMPLE

Solve these simultaneous equations using a graph.
$x + 2y = 4$
$x - y = 1$

First, make a table of values for each equation.

$x + 2y = 4$

x	0	4
y	2	0

$x - y = 1$

x	0	1
y	-1	0

Then draw both lines on one set of axes.
The lines intersect at the point (2, 1)
The solution to the simultaneous equations is $x = 2$, $y = 1$
➡ **CHECK:** substitute the values for x and y into the original equations:
$2 + (2 \times 1) = 4$ ✓
$2 - 1 = 1$ ✓

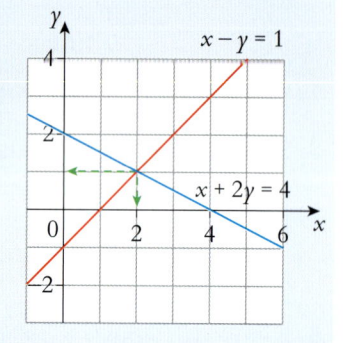

EXERCISE 3.3

1. Copy and complete these tables for the given equations.

   $y = x + 2$

x	0	2	4
y			

   $y = 4 - x$

x	0	2	4
y			

   Draw both graphs on one set of axes.
   Use your graph to solve the simultaneous equations $y = x + 2$ and $y = 4 - x$

2. Copy and complete these tables for the given equations.

   $y = \dfrac{1}{2}x + 1$

x	0	2	4
y			

   $5x + 4y = 20$

x	0	
y		0

   Draw both graphs on one set of axes.
   Use your graph to solve the simultaneous equations $y = \dfrac{1}{2}x + 1$ and $5x + 4y = 20$

Solving simultaneous equations using a graph takes time and can be inaccurate if the coordinates of the point of intersection are not integers. It is quicker and easier to use an algebraic method.

Elimination method

The elimination method involves adding or subtracting the equations (or multiples of them) to **eliminate** one of the variables.

EXAMPLE

Solve these simultaneous equations. $3x + 2y = 19$
$x + 2y = 13$

$3x + 2y = 19$ (1)
$x + 2y = 13$ (2)

Subtracting equation (2) from equation (1) will eliminate the y terms
$2x = 6$
$x = 3$

To find y substitute $x = 3$ in equation (1) and then solve.
$9 + 2y = 19$
$2y = 10$
$y = 5$

The solution is $x = 3$, $y = 5$.

➡ **CHECK:**
$(3 \times 3) + (2 \times 5) = 19$ ✓
$3 + (2 \times 5) = 13$ ✓

EXAMPLE

Solve these simultaneous equations. $4x + 3y = 1$
$2x - 3y = 14$

$4x + 3y = 1$ (1)
$2x - 3y = 14$ (2)

Adding equations (1) and (2) will eliminate the y terms
$6x = 15$
$x = 2\frac{1}{2}$

Substitute $x = 2\frac{1}{2}$ in equation (1) and then solve to find y:
$10 + 3y = 1$
$3y = -9$
$y = -3$

The solution is $x = 2\frac{1}{2}$, $y = -3$.

➡ **CHECK:**
$\left(4 \times 2\frac{1}{2}\right) + (3 \times -3) = 10 - 9 = 1$ ✓

$\left(2 \times 2\frac{1}{2}\right) - (3 \times -3) = 5 - -9 = 14$ ✓

Simultaneous equations 1

EXERCISE 3.4

Solve these simultaneous equations.

1.
   a. $2x + y = 10$
   $x + y = 9$

   b. $3x + 2y = 19$
   $x + 2y = 9$

   c. $2x + y = 19$
   $2x + 3y = 15$

   d. $4x + 3y = 28$
   $4x + y = 20$

   e. $5x + 2y = 8$
   $3x + 2y = 4$

   f. $7x + y = -17$
   $5x + y = -11$

   g. $4x + 5y = -34$
   $4x + 2y = -28$

   h. $3x + 8y = -1$
   $5x + 8y = 9$

2.
   a. $5x - 2y = 2$
   $3x + 2y = 14$

   b. $3x + 5y = 52$
   $4x - 5y = 11$

   c. $4x - 3y = -22$
   $5x + 3y = 40$

   d. $2x - 9y = -13$
   $2x + 9y = 41$

   e. $4x + 9y = 2$
   $3x - 9y = 33$

   f. $5x - 4y = -31$
   $2x + 4y = 10$

   g. $7x + 2y = 39$
   $3x - 2y = 31$

   h. $6x + y = -21$
   $10x - y = -19$

3.
   a. $3x - 7y = 4$
   $x - 7y = -8$

   b. $-2x + 5y = 9$
   $-2x + 3y = 3$

   c. $8x - y = 42$
   $6x - y = 32$

   d. $4x - 3y = -3$
   $8x - 3y = 3$

   e. $3x + 2y = 17$
   $-3x + 4y = 22$

   f. $2x - 3y = 0$
   $5x - 3y = 6$

   g. $8x - 4y = 1$
   $3x - 4y = 6$

   h. $7x - 2y = -35$
   $3x - 2y = -19$

4. The sum of two numbers is 15 and the difference is 4. Let the two numbers be x and y. Write down a pair of simultaneous equations involving x and y. Solve the equations.

5. Find two numbers with
   a. a mean of 27 and a difference of 24
   b. a mean of 4 and a difference of 22

6. Find the values of x and y.

   (Rectangle: top side $x + 2y$, right side 7, bottom 15, left side $x - 2y$)

7. Find the values of x and y.

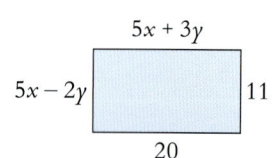

   (Rectangle: top side $5x + 3y$, right side 11, bottom 20, left side $5x - 2y$)

8. Seven pens and two pencils cost $6.59. Seven pens and 5 pencils cost $7.55. Find the cost of a pencil and the cost of a pen.

9. The line $y = mx + c$ passes through the points $(-2, 1)$ and $(4, 4)$.
   Write down a pair of simultaneous equations involving m and c.
   Solve the simultaneous equations.

10. The cost of a taxi journey is made of two parts.
    There is a fixed charge of $\$x$ and then an extra charge of $\$y$ per km.
    The total cost for a 5 km journey is $14.60. The total cost for a 12 km journey is $23.84.
    Find the total cost of a 24 km journey.

To eliminate one of the variables you might first need to multiply one or both of the equations by a suitable number before adding or subtracting.

EXAMPLE

Solve these simultaneous equations. $8x + 3y = 7$
$3x + 5y = -9$

$8x + 3y = 7$ (1)
$3x + 5y = -9$ (2)

Multiply equation (1) by 5 and equation (2) by 3.
$40x + 15y = 35$
$9x + 15y = -27$

Subtracting these two equations will eliminate the y terms.
$31x = 62$
$x = 2$

Substitute $x = 2$ in equation (1) and then solve to find y:
$16 + 3y = 7$
$3y = -9$
$y = -3$.

The solution is $x = 2$, $y = -3$.

➡ **CHECK:**
$(8 \times 2) + (3 \times -3) = 16 + -9 = 7$ ✓
$(3 \times 2) + (5 \times -3) = 6 + -15 = -9$ ✓

Some simultaneous equations need to be re-arranged before solving them.

EXAMPLE

Solve these simultaneous equations. $5x + 2y - 25 = 0$
$2x - 1 = y$

First re-arrange the equations into the form $ax + by = c$
$5x + 2y = 25$ (1)
$2x - y = 1$ (2)

Multiply equation (2) by 2.
$5x + 2y = 25$
$4x - 2y = 2$

Adding the equations will eliminate the y terms
$9x = 27$
$x = 3$

Substitute $x = 3$ in equation (1) and then solve to find y:
$15 + 2y = 25$
$2y = 10$
$y = 5$

The solution is $x = 3$, $y = 5$.

➡ **CHECK:**
$(5 \times 3) + (2 \times 5) = 25$ ✓
$(2 \times 3) - 5 = 1$ ✓

Simultaneous equations 1

EXERCISE 3.5

Solve these simultaneous equations.

1. a. $2x + 3y = 21$
      $5x - y = 10$
   b. $8x + 2y = 34$
      $3x + y = 13$
   c. $4x + 3y = 36$
      $2x - y = 8$
   d. $3x + 7y = 58$
      $x + 2y = 17$

   e. $5x + 2y = 21$
      $x + y = 3$
   f. $6x + 5y = 45$
      $2x - 3y = 29$
   g. $4x + 2y = 22$
      $6x - y = 2$
   h. $7x + 8y = 1$
      $x + 4y = -7$

2. a. $3x + 8y = 68$
      $5x - 6y = -22$
   b. $4x + 3y = 18$
      $3x - 2y = 5$
   c. $5x - 2y = 19$
      $2x - 3y = 1$
   d. $4x + 5y = 18$
      $6x + 2y = 16$

   e. $3x - 4y = 0$
      $2x + 5y = 46$
   f. $9x + 7y = 127$
      $3x + 2y = 41$
   g. $2x + 3y = 33$
      $6x - 5y = -27$
   h. $8x + 5y = 43$
      $5x + 3y = 26$

3. a. $10x - 2y = -13$
      $4x - 3y = -14$
   b. $2x + 7y = -29$
      $3x - 4y = 29$
   c. $5x + 6y = 32$
      $3x + 9y = 57$
   d. $8x - 5y = -14$
      $3x - 4y = -1$

   e. $6x - 7y = -6$
      $8x + 2y = 26$
   f. $2x - 8y = 0$
      $3x + 12y = 36$
   g. $8x + 7y = -7$
      $6x + 5y = -4$
   h. $2x - 3y = -30$
      $3x - 2y = -35$

4. a. $x = 10 - 3y$
      $2x - 3y = 2$
   b. $y = x - 1$
      $3x - y = 5$
   c. $x + 2y - 8 = 0$
      $2x - 3y - 14 = 0$
   d. $3x + 2y = 11$
      $y - 2x = 3$

   e. $y - 3x = -9$
      $x + 2y = -4$
   f. $y = x + 3$
      $3y + x = 5$
   g. $3y - 5x = 13$
      $3x + y + 5 = 0$
   h. $3y = 2x + 4$
      $2y = 7 - 3x$

5. a. $\frac{x}{4} + \frac{y}{3} = 8$
      $\frac{x}{2} - \frac{y}{5} = 3$
   b. $\frac{x}{2} + \frac{y}{4} = 6$
      $\frac{x}{5} - \frac{y}{2} = 12$
   c. $\frac{x}{6} - \frac{y}{2} = -8$
      $\frac{x}{3} - \frac{y}{5} = -8$
   d. $\frac{x}{8} + \frac{y}{3} = 1$
      $\frac{x}{12} + \frac{y}{2} = -1$

6. Four times one number added to three times another number gives 33. The sum of the two numbers is 10. Find the two numbers.

7. Five times one number take away two times another number gives 15. The sum of the two numbers is 6.5. Find the two numbers.

8. A bag contains 36 coins. The coins are a mixture of one-dollar coins and 50-cent coins. The total value of the coins in the bag is $26.50. How many one-dollar coins are there?

9. Cinema tickets cost $8.50 per child and $14 per adult. One evening the cinema sells 201 tickets for a total of $2550. How many adult tickets were sold?

10 Eight bottles of fruit juice and three bottles of water cost $17.89
Five bottles of fruit juice and four bottles of water cost $13.37
Find the cost of a bottle of fruit juice.

11 Four lemons and three oranges cost $5.45
Seven lemons and four oranges cost $9.30
Find the cost of an orange.

12 The straight line $ax + by = 15$ passes through the points $(-5, 6)$ and $(15, -6)$.
Find the values of a and b.

13 The straight line $ax + by = 9$ passes through the points $(12, 5)$ and $(-6, -7)$.
Find the values of a and b.

14 For the rectangle ABCD, find
  a the values of x and y
  b the perimeter of the rectangle
  c the area of the rectangle.

CD = $4x + 2y$, CB = $2x + 2y$, DA = $x - y$, BA = $x - 3y + 4$

15 For the square ABCD, find
  a the values of x and y
  b the perimeter of the square
  c the area of the square.

CD = $2y - 2x$, CB = $8x - y - 3$, DA = $y + 1$, BA = $4x - 1$

16 For the equilateral triangle ABC, find
  a the values of x and y
  b the perimeter of the triangle
  c the area of the triangle.

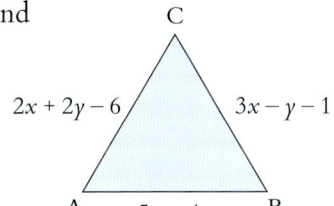

CA = $2x + 2y - 6$, CB = $3x - y - 1$, AB = $5x - 4y$

KEY WORDS
simultaneous equations
eliminate

Factorising 1

THIS SECTION WILL SHOW YOU HOW TO
- Factorise simple expressions
- Use factorisation to simplify fractions

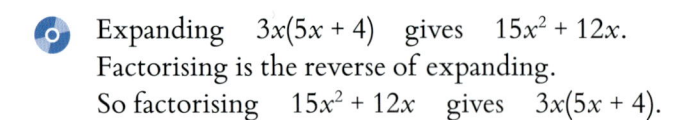

Expanding $3x(5x + 4)$ gives $15x^2 + 12x$.
Factorising is the reverse of expanding.
So factorising $15x^2 + 12x$ gives $3x(5x + 4)$.

EXAMPLE

Factorise $6x + 8$

$6x + 8$
$= 2(3x + 4)$

2 is the highest common factor (HCF) of the two terms
$6x = 2 \times 3x$ and $8 = 2 \times 4$
It is sensible to check your answer by expanding the brackets.
$2(3x + 4) = 2 \times 3x + 2 \times 4 = 6x + 8$ ✓

EXAMPLE

Factorise $5x^2 + 10xy$

$5x^2 + 10xy$
$= 5x(x + 2y)$

$5x$ is the HCF of the two terms
$5x^2 = 5x \times x$ and $10xy = 5x \times 2y$

EXAMPLE

Factorise $xy^2 + x^3y^4$

$xy^2 + x^3y^4$
$= xy^2(1 + x^2y^2)$

xy^2 is the HCF of the two terms
$xy^2 = xy^2 \times 1$ and $x^3y^4 = xy^2 \times x^2y^2$

EXAMPLE

Factorise $x(x + 2) + 5(x + 2)$

$x(x + 2) + 5(x + 2)$
$= (x + 2)(x + 5)$

$x + 2$ is the HCF of the two terms
➡ NOTE: $(x + 5)(x + 2)$ is also correct

EXAMPLE

Factorise $xy - 2x + 3y - 6$

$xy - 2x + 3y - 6$
$= x(y - 2) + 3(y - 2)$
$= (y - 2)(x + 3)$

factorise the first two terms and the last two terms
$y - 2$ is the HCF of these two terms
➡ NOTE: $(x + 3)(y - 2)$ is also correct

EXERCISE 3.6

1. Abdullah has got his factorising homework wrong. Copy out each question and correct his mistakes.

> a) $6x + 12y$ b) $5xy - 10xy^3$ c) $4x + 6y$
>
> $= 2(3x + 6y)$ $= 5x(y - y^2)$ $= 4(x + 1\frac{1}{2}y)$

Factorise these expressions completely.

2.
   a. $4x + 8$
   b. $3x - 6$
   c. $8x + 2$
   d. $10x - 15$
   e. $7y - 14$
   f. $9y + 12$
   g. $20 - 8x$
   h. $36 - 12y$
   i. $42 - 7x$
   j. $18y - 4$
   k. $24x + 16$
   l. $22y - 55$

3.
   a. $x^2 + 5x$
   b. $x^2 - 8x$
   c. $y^2 - 4y$
   d. $a^2 + 9a$
   e. $4x^2 - 8x$
   f. $3y^2 - 9y$
   g. $10x^2 - 4x$
   h. $8y - 2y^2$
   i. $15y^2 + 5y$
   j. $14x - 7x^2$
   k. $6x - 4x^2$
   l. $12y^2 - 9y$

4.
   a. $8\pi + 4$
   b. $15 - 9\pi$
   c. $\pi r^2 + \pi rl$
   d. $2\pi r^2 + 2\pi rh$

5.
   a. $6x^2y - 3xy$
   b. $10x^4y^2 - 5x^4y$
   c. $8xy^3 - 32x^3y$
   d. $15x^3y^2 - 12x^2y^4$
   e. $12x^2y - 15xy$
   f. $28pq^2 - 35p^2q$
   g. $6a^2b + 9ab$
   h. $21x^5y - 28x^2y$
   i. $56a^4b^7 - 35ab^5$

6.
   a. $3a + 3b - 3c$
   b. $5x + 5y + 5x^2$
   c. $2ax - ay - 3ay^2$
   d. $x^3 + 4x^2 - 2x$
   e. $y^4 + 6y^3 - 6y^2$
   f. $p^2q^2r + pqr + pq^2r$
   g. $20a^2b^2c^2 - 8a^2b^2 + 16abc^2$
   h. $15x^3y^2 - 27x^2y + 21x^2y^2$
   i. $8p^7q^5r^3 + 6p^4q^3r^2 - 4p^3q^3r^3$

7.
   a. $x(y + 2) + 3(y + 2)$
   b. $a(b - 4) + 2(b - 4)$
   c. $p(q + 3) - 5(q + 3)$
   d. $x(y - 5) - 3(y - 5)$
   e. $x(x + y)^2 + y(x + y)^2$
   f. $(a + b)^3 - (a + b)^2$
   g. $(x - y)^2 - (x - y)^3$
   h. $5(x + y) + (x + y)^2$
   i. $(x - y)^3 - 4(x - y)^2$

8.
   a. $xy + y + 3x + 3$
   b. $5y - 10 + xy - 2x$
   c. $4a + 4b + ac + bc$
   d. $pq + 2pr + 3q + 6r$
   e. $3x + 12 - xy - 4y$
   f. $xy - 3x - 4y + 12$

You can use factorisation to simplify algebraic fractions.

EXAMPLE

Simplify $\dfrac{x^2+4x}{x}$

$\dfrac{x^2+4x}{x}$ factorise the numerator

$= \dfrac{\overset{1}{\cancel{x}}(x+4)}{\underset{1}{\cancel{x}}}$ cancel

$= x + 4$

EXAMPLE

Simplify $\dfrac{6x-12}{18}$

$\dfrac{6x-12}{18}$ factorise the numerator

$= \dfrac{\overset{1}{\cancel{6}}(x-2)}{\underset{3}{\cancel{18}}}$ cancel

$= \dfrac{x-2}{3}$

EXAMPLE

Simplify $\dfrac{x^2-xy}{7x^2-7xy}$

$\dfrac{x^2-xy}{7x^2-7xy}$ factorise the numerator and denominator

$= \dfrac{\overset{1}{\cancel{x}}(\cancel{x-y})^{1}}{7\underset{1}{\cancel{x}}(\cancel{x-y})_{1}}$ cancel

$= \dfrac{1}{7}$

EXAMPLE

Simplify $\dfrac{3x+3y}{x^2-xy} \times \dfrac{5x-5y}{2x+2y}$

$= \dfrac{3(x+y)}{x(x-y)} \times \dfrac{5(x-y)}{2(x+y)}$ factorise both numerators and denominators

$= \dfrac{3(\cancel{x+y})^{1}}{x(\cancel{x-y})_{1}} \times \dfrac{5(\cancel{x-y})^{1}}{2(\cancel{x+y})_{1}}$ cancel

$= \dfrac{15}{2x}$

EXERCISE 3.7

1 Sheeba has got her simplifying algebraic fractions homework wrong.
   Copy out each question and correct her mistakes.

   a) $\dfrac{4x-8}{4} = 4x - 2$

   b) $\dfrac{x^2 + 2x}{4x} = \dfrac{x^2 + 2}{4}$

   c) $\dfrac{6x + 8}{10} = \dfrac{2(3x + 8)}{10} = \dfrac{(3x + 8)}{5}$

Simplify these algebraic fractions.

2 a $\dfrac{3x+6}{3}$ b $\dfrac{8y+4}{2}$ c $\dfrac{5x+10}{5}$ d $\dfrac{12y+8}{4}$

   e $\dfrac{6x-12}{8}$ f $\dfrac{9x+6}{12}$ g $\dfrac{15-20y}{10}$ h $\dfrac{24-16a}{20}$

3 a $\dfrac{x^2+9x}{x}$ b $\dfrac{y^2-8y}{y}$ c $\dfrac{x^2-3x}{x^2}$ d $\dfrac{x-x^3}{x}$

   e $\dfrac{8x+4y}{4z}$ f $\dfrac{5x+5y}{5z}$ g $\dfrac{x^2+xy}{xy}$ h $\dfrac{x^2-2x}{xy}$

4 a $\dfrac{3x+3y}{x+y}$ b $\dfrac{5a-5b}{a-b}$ c $\dfrac{x-y}{6x-6y}$ d $\dfrac{2x+2y}{10x+10y}$

   e $\dfrac{2-2y}{5-5y}$ f $\dfrac{3x+3xy}{7+7y}$ g $\dfrac{xy+x}{yz+z}$ h $\dfrac{10x-10y}{15x-15y}$

5 a $\dfrac{x^2+x}{x+1}$ b $\dfrac{5x}{x^2-5x}$ c $\dfrac{x^2-3x}{5x-15}$ d $\dfrac{4y-6y^2}{10-15y}$

   e $\dfrac{x^2-x}{(x-1)^2}$ f $\dfrac{3y-6}{(y-2)^3}$ g $\dfrac{x^2+3x}{x+3}$ h $\dfrac{3x^2y-6xy}{6x-12}$

6 a $\dfrac{4p+4q}{2p-2q} \times \dfrac{5p-5q}{3p+3q}$ b $\dfrac{7a+14b}{a-b} \times \dfrac{2a-2b}{3a+6b}$ c $\dfrac{x^2-3x}{y} \times \dfrac{x}{xy-3y}$

   d $\dfrac{x^2+x}{y^2+y} \times \dfrac{y}{x}$ e $\dfrac{7x-7y}{xy} \times \dfrac{x^2+xy}{x-y}$ f $\dfrac{f^2-6f}{2g} \times \dfrac{4g}{fg-6g}$

7 a $\dfrac{2x-4y}{6xy} \div \dfrac{4x-8y}{5x^2y^2}$ b $\dfrac{3x^2-6x}{6y-2y^2} \div \dfrac{5x^2-10x}{9y-3y^2}$

> **KEY WORDS**
> factorise

Rearranging formulae 1

> **THIS SECTION WILL SHOW YOU HOW TO**
> - Change the subject of a formula

In the formula $P = 2x + 2y$, P is called the **subject** of the formula. The formula can be rearranged to make x the subject as follows:

$$P = 2x + 2y \quad \text{take } 2y \text{ from both sides}$$
$$P - 2y = 2x \quad \text{divide both sides by 2}$$
$$\frac{P - 2y}{2} = x$$

EXAMPLE

Rearrange the formula $y = \frac{m-n}{2}$ to make m the subject.

$$y = \frac{m-n}{2} \quad \text{multiply both sides by 2}$$
$$2y = m - n \quad \text{add } n \text{ to both sides}$$
$$2y + n = m$$

EXAMPLE

Rearrange the formula $x = \frac{y^2}{3} + z$ to make y the subject.

$$x = \frac{y^2}{3} + z \quad \text{take } z \text{ from both sides}$$
$$x - z = \frac{y^2}{3} \quad \text{multiply both sides by 3}$$
$$3(x - z) = y^2 \quad \text{square root both sides}$$
$$y = \pm\sqrt{3(x-z)}$$

➡ **NOTE:** remember for example that both $5^2 = 25$ and $(-5)^2 = 25$

EXAMPLE

Rearrange the formula $T = 2\pi\sqrt{\frac{l}{g}}$ to make l the subject.

$$T = 2\pi\sqrt{\frac{l}{g}} \quad \text{divide both sides by } 2\pi$$
$$\frac{T}{2\pi} = \sqrt{\frac{l}{g}} \quad \text{square both sides}$$
$$\frac{T^2}{4\pi^2} = \frac{l}{g} \quad \text{multiply both sides by } g$$
$$\frac{T^2 g}{4\pi^2} = l$$

EXERCISE 3.8

1. Hassan has got his rearranging formulae homework wrong. Copy out each question and correct his mistakes.

   a) $p = 3x - q$
   $p - q = 3x$
   $\dfrac{p-q}{3} = 3$

   b) $2x + a = 8b$
   $x + a = 4b$
   $x = 4b - a$

   c) $y = \sqrt{x} + b$
   $y^2 = x + b^2$
   $y^2 - b^2 = x$

Make x the subject of these formulae.

2. a $x + 3 = y$ b $x - 6 = 5y$ c $x + y = 4$ d $x - a = b$
   e $y = x - 8$ f $x + 2y - 5 = 0$ g $x - 4y + 8 = 2y$ h $y - x = 10$

3. a $6x = y$ b $9x = 3y$ c $3x = y - 4$ d $ax = 7 - 5y$
   e $\dfrac{x}{2} = p$ f $\dfrac{x}{5} = 3 + d$ g $\dfrac{x}{2} = y - 8$ h $\dfrac{x}{a} = 10 - 4y$
   i $\dfrac{2x}{3} = y$ j $\dfrac{4x}{5} = 3y$ k $\dfrac{3x}{7} = a + b$ l $\dfrac{ax}{b} = c + d$

4. a $2x + c = d$ b $5x - y = 2z$ c $4x + 3y = 6$ d $8x - 3b = c + d$
   e $ax + b = d$ f $bx - c = cd$ g $px - q = a + b$ h $y - xy = 8$

5. a $x(y + z) = 2$ b $p = (2q + 3)x$ c $a = 3(2x + y)$ d $p = q(x - r)$

6. a $\dfrac{x+a}{b} = c$ b $\dfrac{x-2f}{3g} = h$ c $\dfrac{ax+by}{c} = d$ d $\dfrac{p-qx}{r} = m$
   e $\dfrac{p}{x} = q$ f $f = \dfrac{g}{x}$ g $\dfrac{p+q}{x} = r$ h $5 = \dfrac{p-q}{x}$

7. a $ax^2 = b$ b $px^2 = q$ c $abx^2 = c$ d $ax^2 = bc$
   e $x^2 + a = b$ f $x^2 - d = e$ g $ax^2 + b = c$ h $px^2 - 5q = 4r$
   i $\dfrac{x^2}{4} - a = b$ j $\dfrac{x^2}{a} + bc = d$ k $f - \dfrac{x^2}{g} = 3h$ l $3fg - \dfrac{x^2}{2} = g$

8. a $\sqrt{x+a} = b$ b $\sqrt{x-p} = qr$ c $\sqrt{2x-y} = 4$ d $2\sqrt{x-y} = z$

9. Sofia and Karl are asked to make h the subject of the formula $A = 2\pi r(r + h)$

   Sofia writes:
   $A = 2\pi r(r + h)$
   $A = 2\pi r^2 + 2\pi rh$
   $A - 2\pi r^2 = 2\pi rh$
   $\dfrac{A - 2\pi r^2}{2\pi r} = h$

   Karl writes:
   $A = 2\pi r(r + h)$
   $\dfrac{A}{2\pi r} = r + h$
   $\dfrac{A}{2\pi r} - r = h$

   KEY WORDS
   subject

   Who is correct? Explain your answer.

Similar triangles

THIS SECTION WILL SHOW YOU HOW TO
- Recognise congruent and similar shapes
- Use the properties of similar shapes to find missing lengths and angles

Congruency

Two shapes are **congruent** if one of the shapes fits exactly on top of the other shape.

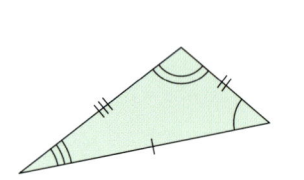

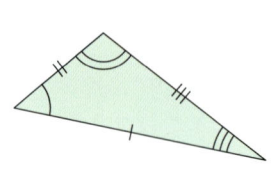

These three triangles are all congruent.

In congruent shapes

- corresponding angles are equal
- corresponding lengths are equal.

To prove that two triangles are congruent you must show that they satisfy one of the following four sets of conditions.

SSS three sides are equal.

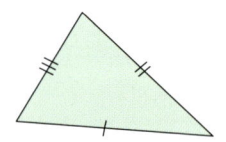

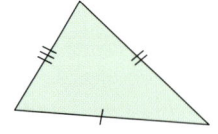

SAS two sides and the included angle are the same.

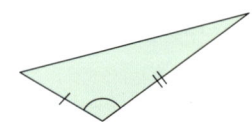

ASA two angles and the included side are the same.

RHS right-angled triangles with hypotenuse and one other side the same.

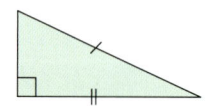

EXAMPLE

Explain why the following pairs of triangles are congruent.

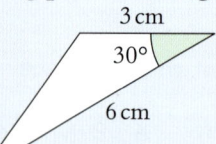

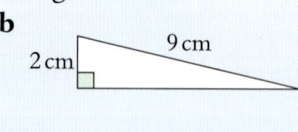

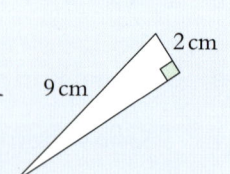

a 6 cm, 3 cm, 30°; 3 cm, 30°, 6 cm

b 2 cm, 9 cm; 9 cm, 2 cm

a   Two sides and the included angle are the same. (SAS)
b   The triangle is right-angled (R), the hypotenuses (H) are the same and one other side (S) is the same. (RHS)

EXERCISE 3.9

1. Using axes from 0 to 10, draw each pair of triangles.
   Which pairs of triangles are congruent?

a	Triangle A: (0, 5) (0, 9) (1, 6)	Triangle B: (3, 10) (7, 10) (5, 9)
b	Triangle C: (0, 2) (1, 0) (3, 2)	Triangle D: (1, 7) (1, 10) (3, 8)
c	Triangle E: (4, 4) (6, 6) (3, 7)	Triangle F: (7, 7) (10, 8) (8, 10)
d	Triangle G: (2, 3) (6, 3) (7, 4)	Triangle H: (6, 5) (9, 5) (10, 6)
e	Triangle I: (0, 4) (1, 3) (2, 7)	Triangle J: (9, 0) (10, 1) (8, 3)
f	Triangle K: (3, 9) (6, 8) (7, 6)	Triangle L: (5, 0) (7, 1) (8, 4)

2. Explain why each of the following pairs of triangles are congruent.

   a

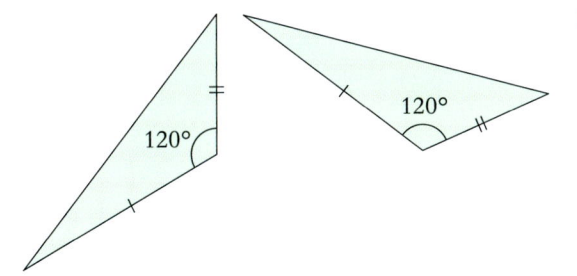

   b

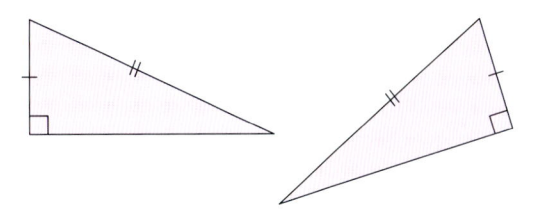

   c

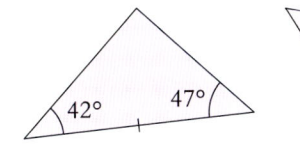

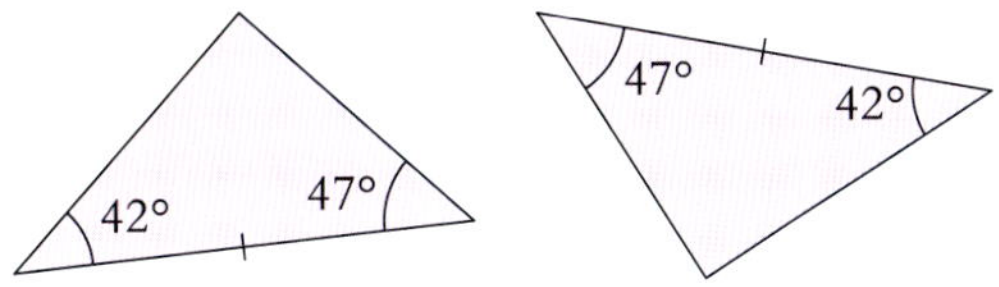

   d

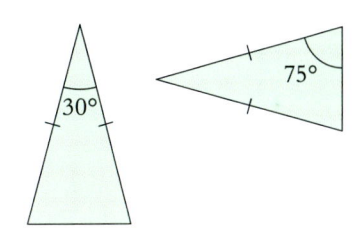

   e

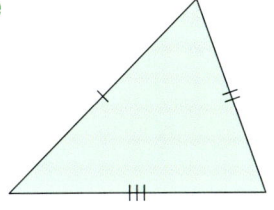

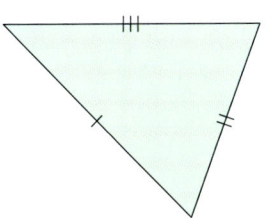

   f

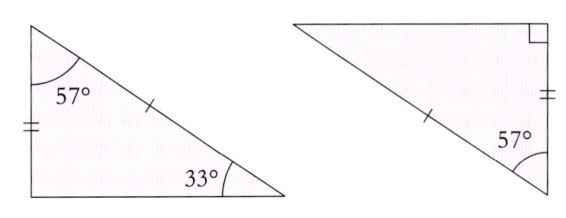

Similar triangles 121

3 ABCD is a parallelogram.
Explain why triangles ABC and ACD are congruent.

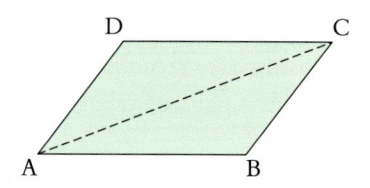

4 PQRS is a rectangle.
Explain why triangles PQS and QSR are congruent.

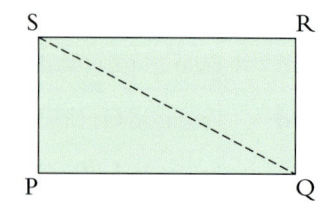

5 Explain why triangles ABC and PQR are congruent.

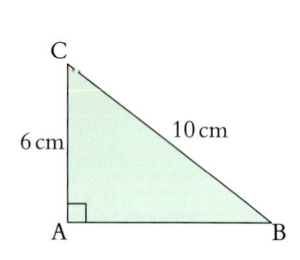

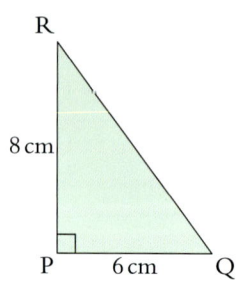

Similar shapes

These two quadrilaterals are **similar**.

$$\frac{PQ}{AB} = \frac{QR}{BC} = \frac{RS}{CD} = \frac{SP}{DA} = 2$$

In similar shapes
- Corresponding angles are equal
- Corresponding sides are in the same ratio.

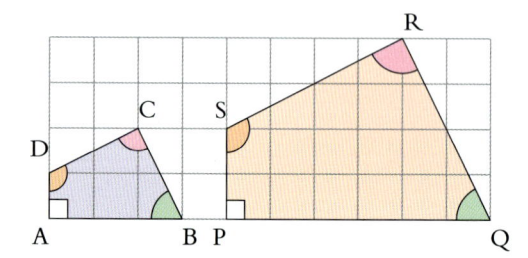

Similar triangles

To show that two triangles are similar it is sufficient to show that just *one* of the above conditions is satisfied.

EXAMPLE

Are these triangles similar?

For the triangles to be similar, corresponding angles must be equal.
Triangle A: third angle = 180 − (58 + 78)
             = 44   *Angles are: 58° 78° 44°.*
Triangle B: third angle = 180 − (78 + 44)
             = 58   *Angles are: 58° 78° 44°.*
Triangles A and B are similar.

EXAMPLE

Are these triangles similar?

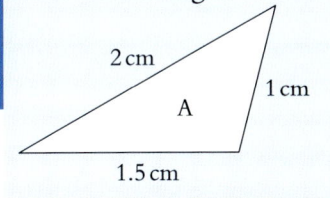

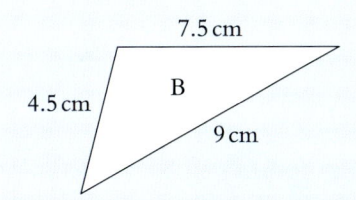

For the triangles to be similar, corresponding sides must be in the same ratio.

Ratio of longest sides = $\frac{9}{2}$ = 4.5

Ratio of shortest sides = $\frac{4.5}{1}$ = 4.5

Ratio of remaining sides = $\frac{7.5}{1.5}$ = 5

Triangles A and B are not similar.

EXAMPLE

Find x and y.

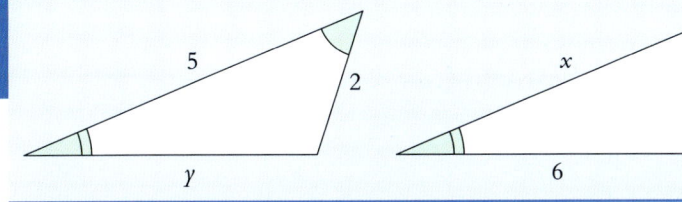

Using ratios of corresponding sides $\frac{3}{2} = \frac{x}{5} = \frac{6}{y}$

separating gives $\frac{3}{2} = \frac{x}{5}$ and $\frac{3}{2} = \frac{6}{y}$

'cross multiply' 15 = 2x 3y = 12
              x = 7.5 y = 4

Alternative method:
using scale factors
      3 = 2 × 1.5
So x = 5 × 1.5 = 7.5
and y = 6 ÷ 1.5 = 4

Similar triangles   123

EXERCISE 3.10

1 Which of these triangles are similar in shape?

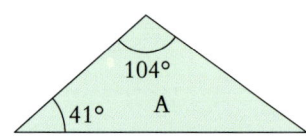

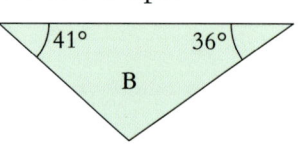

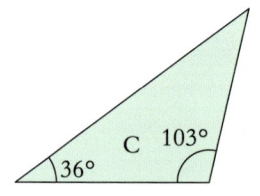

2 Which of these triangles are similar in shape?

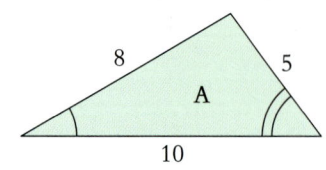

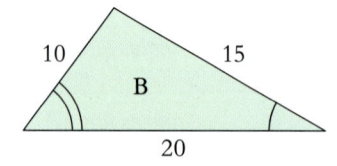

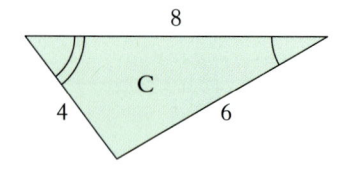

3 The following pairs of shapes are similar. Find the unknown lengths.

a

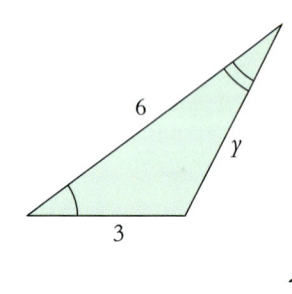

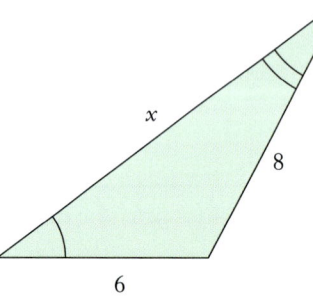

b

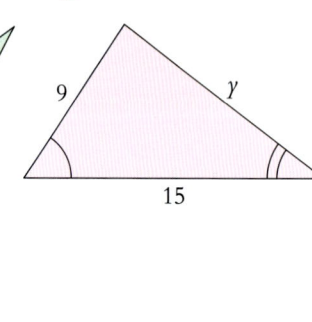

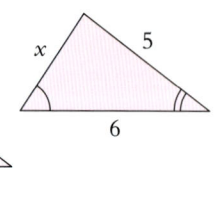

c

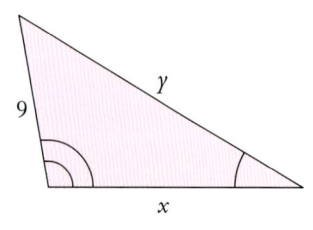

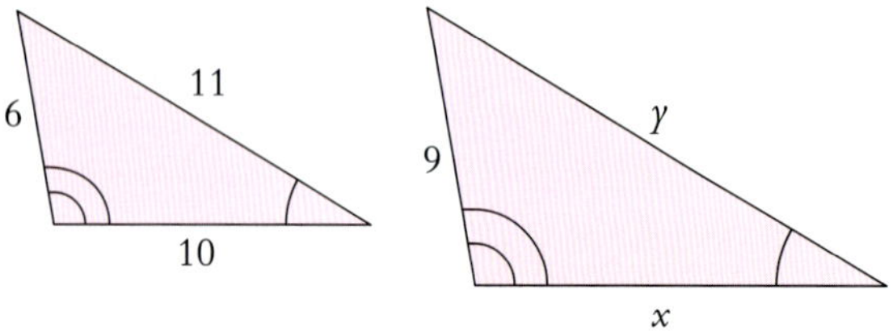

d

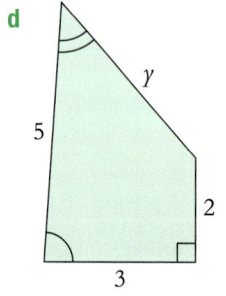

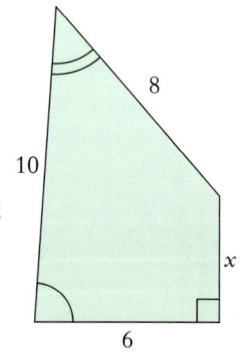

e

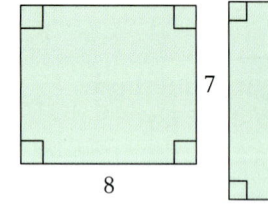

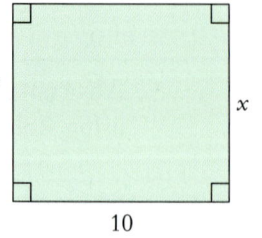

f

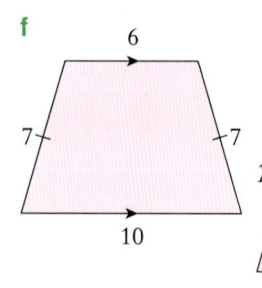

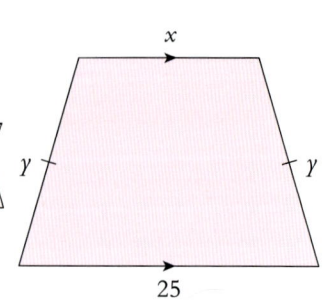

EXAMPLE

Triangles ABC and PQR are similar. Find the values of x and y.

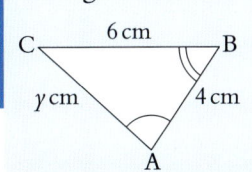

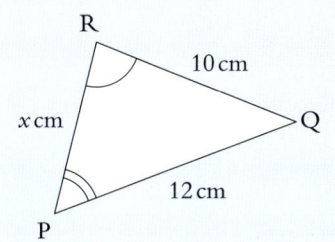

Rotating the second triangle makes it easier to see which are the corresponding sides.

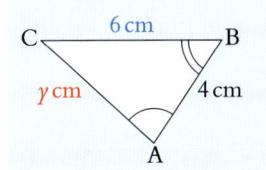

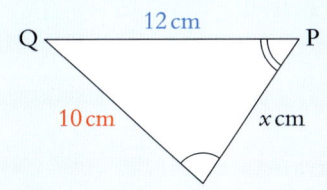

Using ratios of corresponding sides:
$$\frac{12}{6} = \frac{x}{4} = \frac{10}{y}$$

so $\frac{12}{6} = \frac{x}{4}$ and $\frac{12}{6} = \frac{10}{y}$

$2 = \frac{x}{4}$ $\qquad$ $2 = \frac{10}{y}$

$x = 8$ $\qquad$ $y = 5$

Alternative method:
using scale factors
$12 = 6 \times 2$
So $x = 4 \times 2 = 8$
and $y = 10 \div 2 = 5$

EXAMPLE

AB is parallel to DE.
AB = 14 cm, AD = 6 cm, CD = 18 cm and CE = 12 cm.
BE = x cm and DE = y cm.
Find the values of x and y.

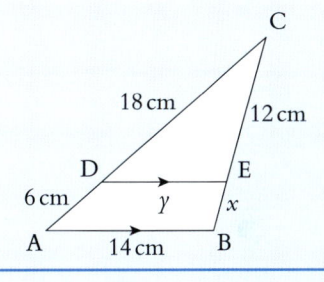

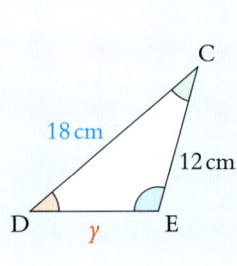

Triangles ABC and DEC are similar because:
∠DCE = ∠ACB (common angle)
∠CDE = ∠CAB (corresponding angles)
∠CED = ∠CBA (corresponding angles)
(Draw the two triangles separately to help you to identify the corresponding sides.)
Using ratios of corresponding sides:
$$\frac{24}{18} = \frac{x+12}{12} = \frac{14}{y}$$ 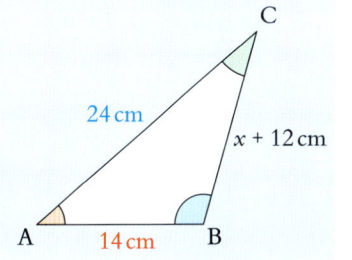

so $\frac{24}{18} = \frac{x+12}{12}$ and $\frac{24}{18} = \frac{14}{y}$

$288 = 18x + 216$ $\qquad$ $24y = 252$ $\quad$ 'cross multiplying'

$18x = 72$ $\qquad$ $y = 10.5$

$x = 4$

Similar triangles

EXERCISE 3.11

1 Find the values of x and y for each of the following diagrams.

   a

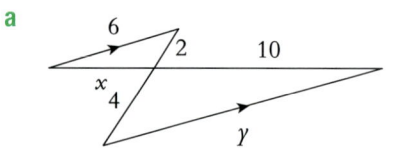

   b

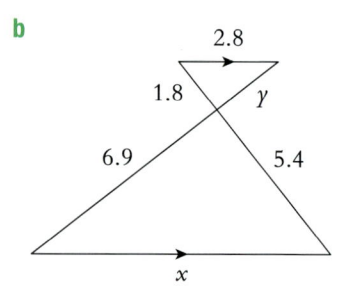

2 The width of a river can be estimated without having to cross over the river. Use the diagram below to calculate the width (w) of this river.

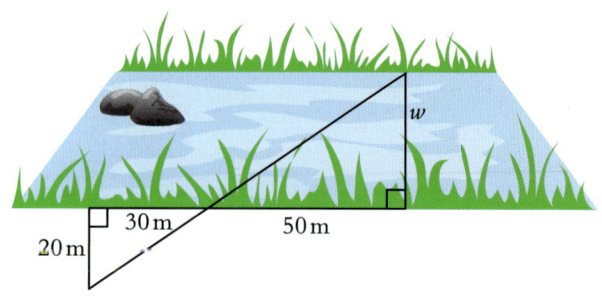

3 Find the value of x for each of the following diagrams.

   a

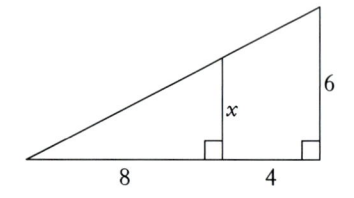

   b

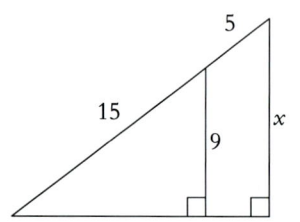

   c

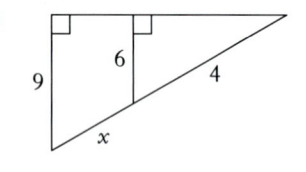

   d

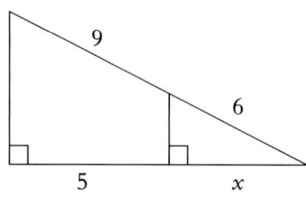

4 At midday the length of the shadow from a building is 50 m and the length of the shadow from a man is 1.65 m.
   The height of the man is 1.8 m.
   Calculate the height of the building.

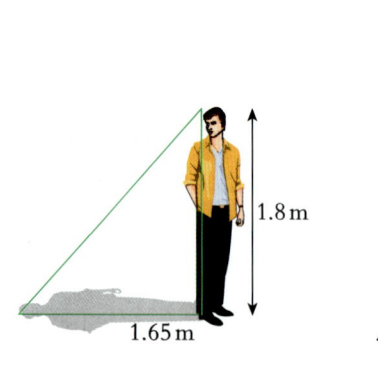

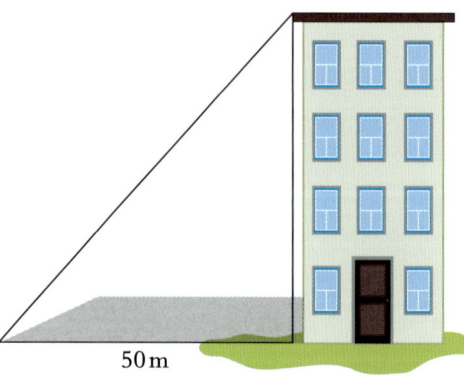

5 Find the values of *x* and *y* for each of the following diagrams.

a

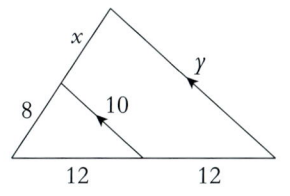

b

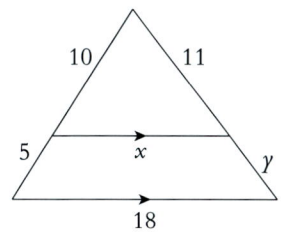

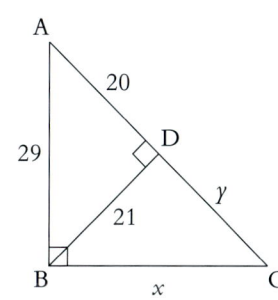

6 DE is parallel to AC.
AD = $\frac{1}{4}$ AB.
   a   Find the coordinates of D and E.
   b   Find the coordinates of the midpoint of the line DE.

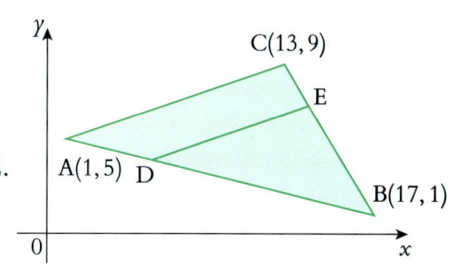

7 A is the point (3, −7) and B is the point (13, −2).
P and Q are points that lie on the line segment joining A to B.
   a   Find the coordinates of the midpoint of the line AB.
   b   AP : PB = 3 : 2. Find the coordinates of the point P.
   c   AQ : AB = 1 : 5. Find the coordinates of the point Q.

8 Find the values of *x* and *y* for each of the following diagrams.

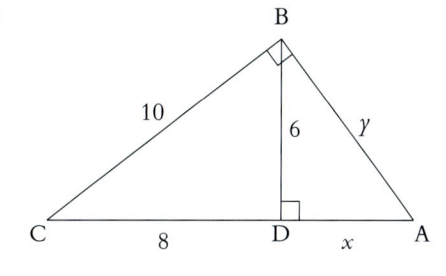

HINT
∠BCD = ∠ABD
in each of these triangles.

a

b

9 Find the values of *x* and *y* for each of the following diagrams.
   a   ∠CAD = ∠DCB 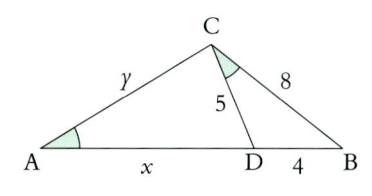
   b   ∠DAB = ∠CBD

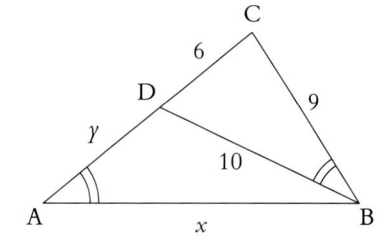

KEY WORDS
congruent
similar

Similar triangles   127

Reflections, rotations and translations

THIS SECTION WILL SHOW YOU HOW TO
- Reflect, rotate and translate a shape
- Use a combination of these transformations

A **transformation** changes the position or size of a shape.
The original shape is called the **object** and the final shape is called the **image**.
Reflections, **rotations** and **translations** are transformations that change only the position of a shape.
The object and image are congruent for reflections, rotations and translations.

Reflections
A reflection is a transformation that involves a mirror line.

In the diagram, triangle A′B′C′ is the image of triangle ABC after a reflection in the line $x = 4$.

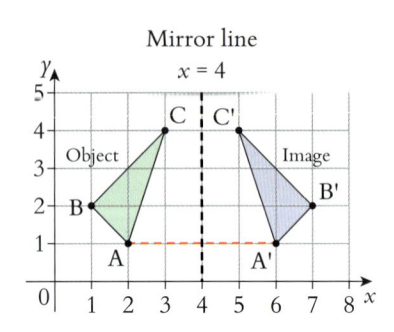

The line joining the point A to its image A′ is at right angles to the mirror line **and** A is the same distance in front of the mirror line as A′ is behind the mirror line.

EXAMPLE

On a graph plot the points P(3, 1), Q(7, 3) and R(7, 7).
Reflect triangle PQR in the line $y = x$. Label the image P′Q′R′.

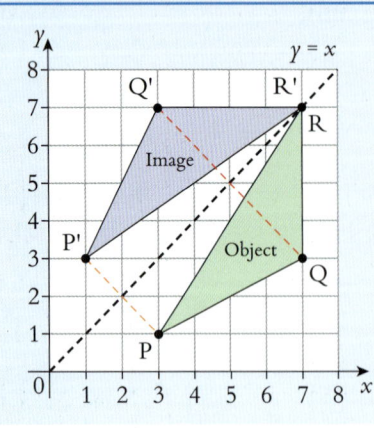

The point R lies on the line $y = x$ so it does not move when reflected in the line. It is called an **invariant** point.

The lines PP′ and QQ′ are at right angles to the mirror line $y = x$.

128 UNIT 3

EXERCISE 3.12

1 Copy the diagrams and draw the image of each shape when reflected in the mirror line.

a b 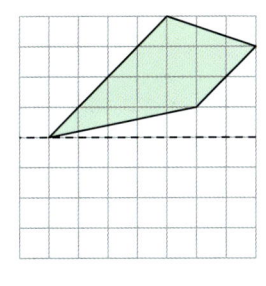 c

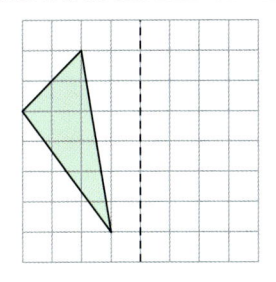

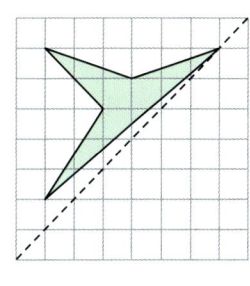

d 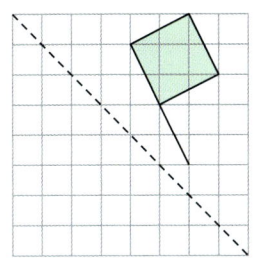 e 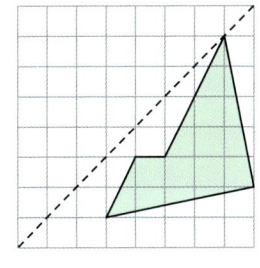 f

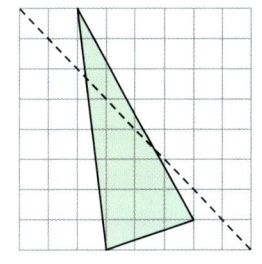

2 Write the equation of the mirror line for each of these reflections.
- a A onto B
- b B onto C
- c B onto D
- d E onto F
- e E onto G
- f E onto I
- g F onto H
- h G onto H

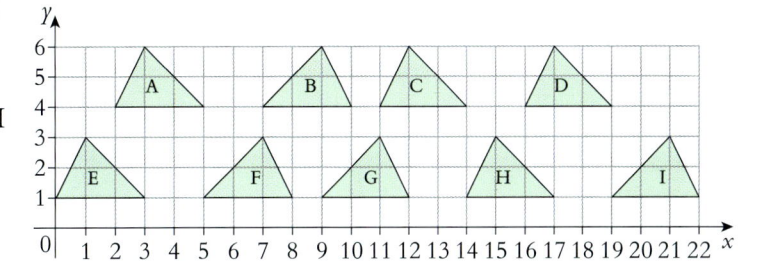

3 Name the mirror line for each of these reflections.
- a A onto B
- b A onto C
- c D onto F
- d E onto F
- e G onto H
- f H onto I

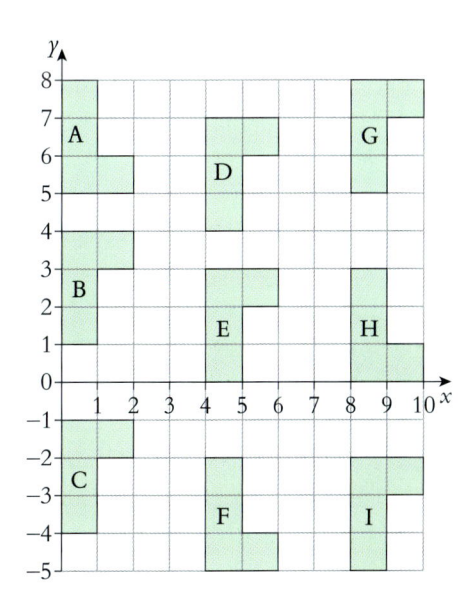

Reflections, rotations and translations

4 Name the mirror line for each of these reflections.
   a A onto B
   b A onto D
   c B onto C

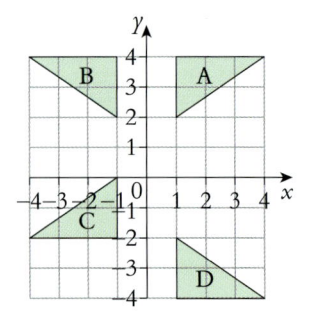

5 Name the mirror line for each of these reflections.
   a A onto B
   b A onto C
   c C onto E
   d B onto E
   e D onto E

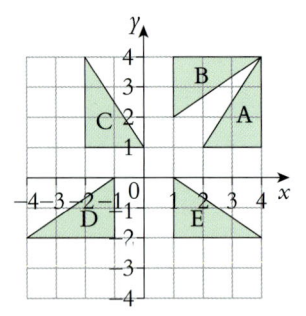

6 Name the mirror line for each of these reflections.
   a A onto D   b F onto G
   c C onto E   d B onto H
   e C onto D   f C onto H
   g D onto G   h A onto E
   i H onto G   j F onto B

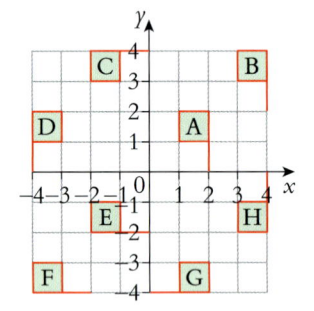

7 Copy the diagram onto graph paper.
   a Reflect shape A in the line $x = 0$ and label the image B.
   b Reflect shape A in the line $y = 1$ and label the image C.
   c Reflect shape A in the line $y = x$ and label the image D.

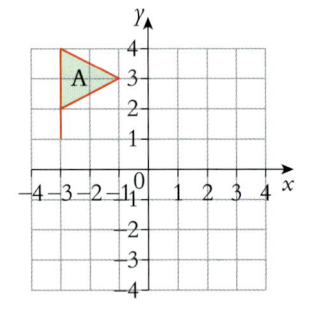

8 Copy the diagram onto graph paper.
   a Reflect shape A in the line $y = 2$ and label the image B.
   b Reflect shape A in the line $y = -1$ and label the image C.
   c Reflect shape A in the line $x = 0$ and label the image D.
   d Reflect shape A in the line $y = -x$ and label the image E.

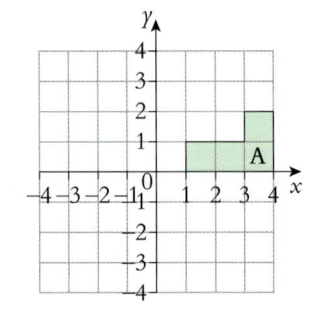

130 UNIT 3

9 Copy the diagram onto graph paper. 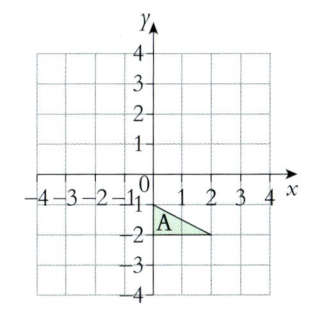
   a Reflect shape A in the line $y = 1$ and label the image B.
   b Reflect shape A in the line $x = -1$ and label the image C.
   c Reflect shape A in the line $x = 2$ and label the image D.
   d Reflect shape A in the line $y = x$ and label the image E.
   e Reflect shape A in the line $y = x + 2$ and label the image F.
   f Reflect shape A in the line $x + y = 1$ and label the image G.

10 Copy the diagram onto graph paper. 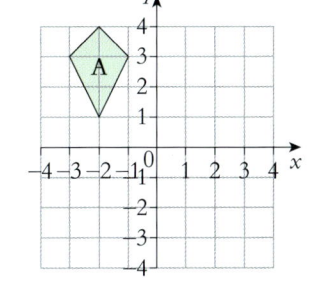
   a Reflect shape A in the x-axis and label the image B.
   b Reflect shape A in the line $x = 0.5$ and label the image C.
   c Reflect shape A in the line $y = x$ and label the image D.
   d Reflect shape A in the line $y = x + 3$ and label the image E.

11 Use axes from −4 to 4.
   Draw the pentagon whose vertices are $(-1, -1)$, $(2, 3)$, $(1, 4)$, $(-2, 3)$ and $(-3, 1)$.
   Reflect the pentagon in the line $y = x$.
   The object and image are congruent.
   Mark the sides and angles to
   show which are equal to each other.

12 Copy the diagram onto graph paper. 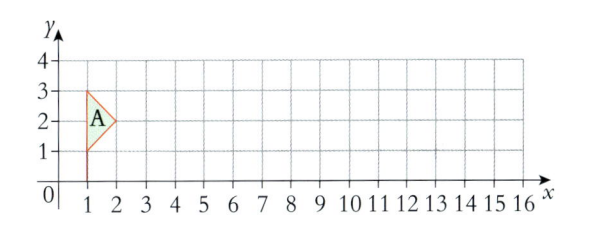
   a Reflect shape A in the line $x = 4$ and label the image B.
   b Reflect shape B in the line $x = 8$ and label the image C.
   c Reflect shape C in the line $x = 10.5$ and label the image D.
   d Shape A can be transformed onto shape D by a **single** reflection.
      Name the mirror line for this single reflection.

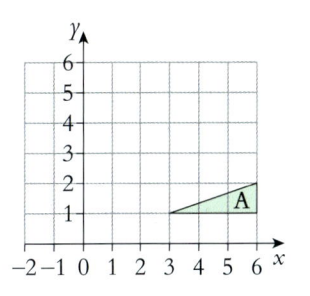

13 Copy the diagram onto graph paper.
   a Reflect shape A in the line $y = x$ and label the image B.
   b Reflect shape B in the line $x + y = 4$ and label the image C.
   c Reflect shape C in the line $x = 2$ and label the image D.
   d Shape A can be transformed onto shape D by a **single** reflection. Name the mirror line for this single reflection.

Reflections, rotations and translations

Rotations

A **rotation** is a transformation that turns a shape about a fixed point.

The fixed point is called the **centre of rotation**. The shape can be rotated in a clockwise (↻) or anticlockwise (↺) direction.

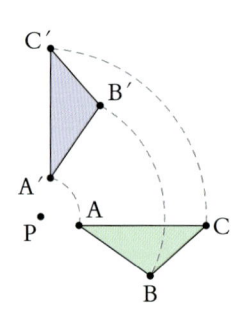

To describe a rotation fully, you need to give
- the angle of rotation
- the direction of rotation
- the centre of rotation.

In the diagram above, triangle A′B′C′ is the image of triangle ABC after a rotation of 90° anticlockwise about the point P.

➡ **NOTE:** you can use tracing paper to help you to do rotations.

➡ **NOTE:** you can use compasses to check the paths traced out by the points.

EXAMPLE

Rotate shape A 180° about the point (8, 2).

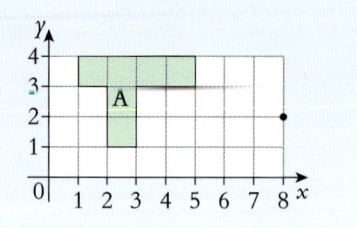

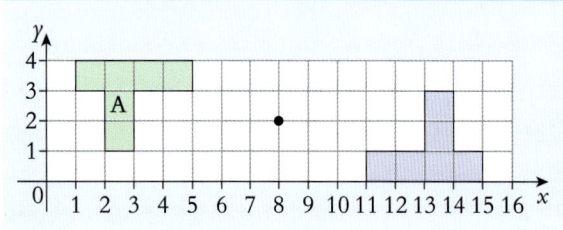

A rotation of 180° clockwise is the same as a rotation of 180° anticlockwise so the direction does not need to be given.

EXERCISE 3.13

1 Copy these shapes on graph paper.
   Draw the image of each shape for the given rotations.
   Use the marked point as the centre of rotation.

   a 180°
   b 90° anticlockwise
   c 90° clockwise
   c 90° anticlockwise
   d 180°
   e 90° clockwise

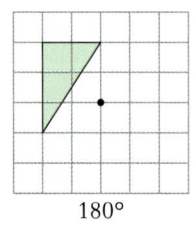

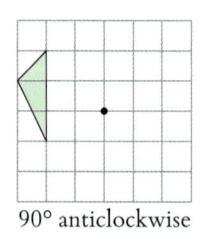

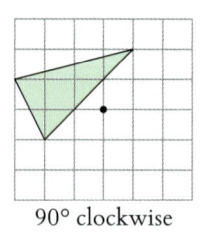

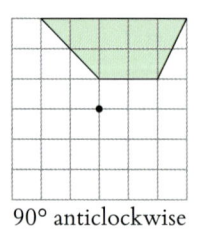

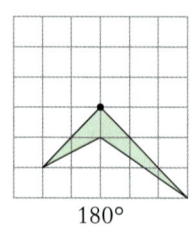

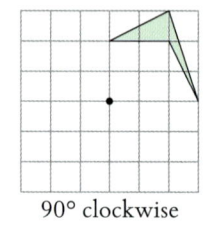

132   UNIT 3

2 Use axes from −4 to 4.
   a Draw the triangle with vertices (−4, 1), (−4, 4) and (−3, 1).
      Label the triangle A.
   b Rotate A through 90° clockwise about the origin.
      Label the image B.
   c Rotate A through 180° about the origin.
      Label the image C.
   d Rotate A through 90° anticlockwise about the origin.
      Label the image D.

3 Use axes from −4 to 4.
   a Draw the triangle with vertices (3, 1), (3, 4) and (2, 3).
      Label the triangle A.
   b Rotate A through 180° about the point (2, 0).
      Label the image B.
   c Rotate A through 180° about the point (0, 1).
      Label the image C.
   d Rotate A through 90° clockwise about the point (1, 3).
      Label the image D.

4 Use axes from −4 to 4.
   a Draw the kite with vertices (1, 0), (2, 2), (1, 3) and (0, 2).
      Label the kite A.
   b Rotate A through 90° anticlockwise about the point (−1, 1).
      Label the image B.
   c Rotate A through 180° about the point (−1, 1).
      Label the image C.
   d Rotate A through 90° clockwise about the point (−1, 1).
      Label the image D.

5 Use axes from −6 to 6.
   a Draw the triangle with vertices (3, −4), (6, −1) and (5, 1).
      Label the triangle A.
   b Rotate A through 180° about the point (2, −1).
      Label the image B.
   c Rotate B through 90° clockwise about the point (0, 3).
      Label the image C.
   d Rotate C through 90° anticlockwise about the point (−2, 1).
      Label the image D.

Reflections, rotations and translations

Describing a rotation

EXAMPLE

Describe fully the single transformation that takes triangle ABC onto triangle A′B′C′.

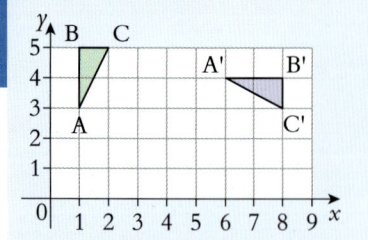

The line AB is vertical and the line A′B′ is horizontal so the angle of rotation is 90°.
The shape has been rotated clockwise.
There are two methods for finding the centre of rotation.

Method 1 (By trial and error)

Use tracing paper to try to find the centre of rotation.
Through trial and error you should find that the centre of rotation is at the point (4, 1).

Method 2 (By construction)

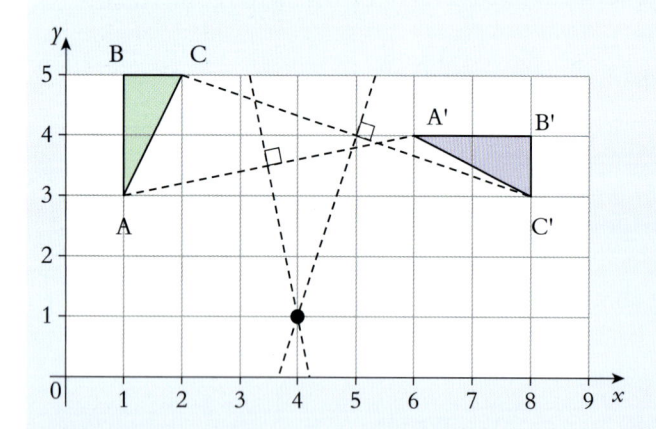

➡ **NOTE:** only two perpendicular bisectors need to be drawn to locate the centre of rotation.

Join A to A′ and draw the perpendicular bisector of the line AA′
Join C to C′ and draw the perpendicular bisector of the line CC′
The point where the two perpendicular bisectors meet is the centre of rotation.
The transformation that takes triangle ABC onto triangle A′B′C′ is a rotation 90° clockwise about the point (4, 1).

EXERCISE 3.14

1 Describe fully the single transformation that takes shape A onto shape B for each of the following diagrams.

a

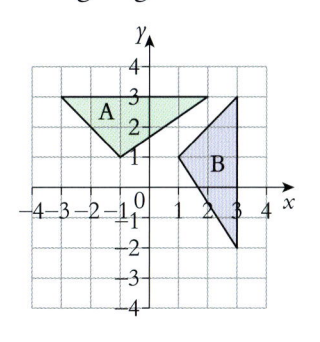

b

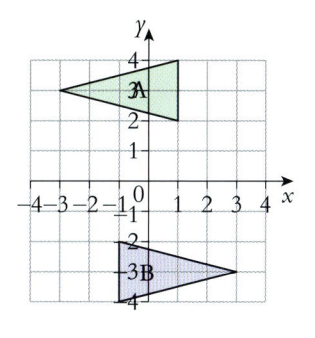

c

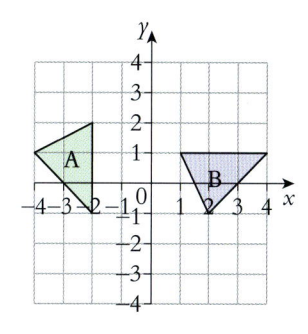

d

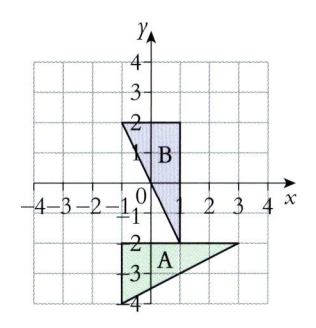

e

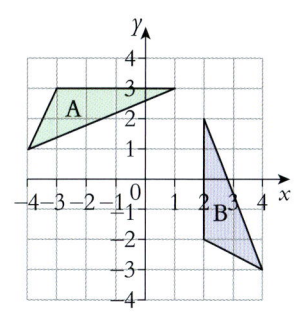

f

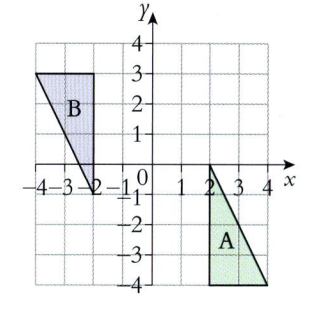

2

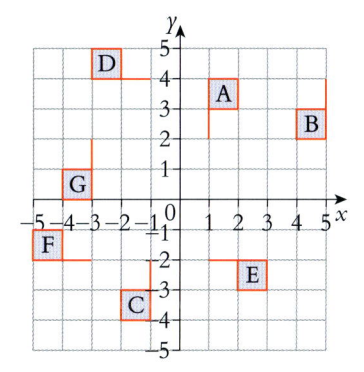

Describe fully the single transformation that moves shape:
a A onto B
b A onto C
c A onto D
d A onto E
e A onto F
f A onto G
g C onto D
h C onto E
i C onto F
j D onto E
k D onto G
l E onto G.

Translations

A **translation** is a transformation that 'slides' a shape from one place to another. In a translation every point on the original shape moves the same distance and in the same direction.

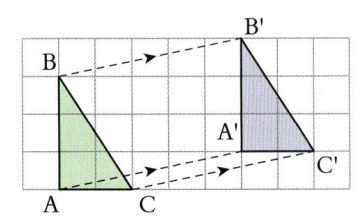

This is a translation of 5 squares to the right and 1 square up.
The translation can be described using a **column vector**:

This is a translation of $\begin{pmatrix} 5 \\ 1 \end{pmatrix}$ ← The top number is the movement in the x direction.
← The bottom number is the movement in the y direction.

For the top number: a movement to the right is positive
a movement to the left is negative.

For the bottom number: a movement up is positive
a movement down is negative.

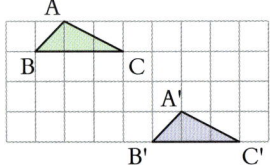

This is a translation of $\begin{pmatrix} 4 \\ -3 \end{pmatrix}$

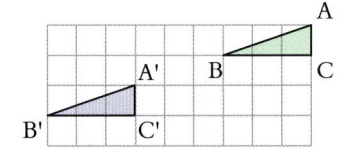

This is a translation of $\begin{pmatrix} -6 \\ -2 \end{pmatrix}$

EXAMPLE

Translate this shape by the vector $\begin{pmatrix} -6 \\ 1 \end{pmatrix}$.

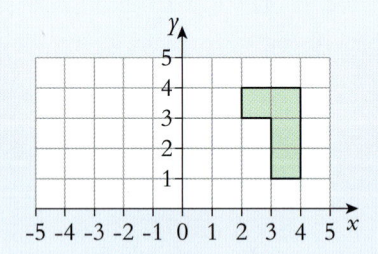

$\begin{pmatrix} -6 \\ 1 \end{pmatrix}$ means:

- move 6 squares to the left and
- move 1 square up.

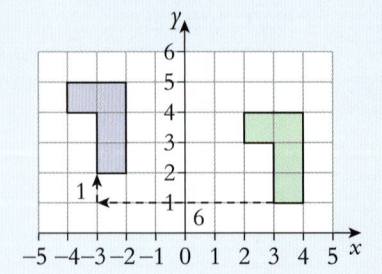

136 UNIT 3

EXERCISE 3.15

1 Write down the vector for the translation that maps shape A onto shape B.

a, b, 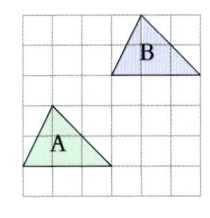 c, 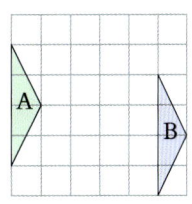 d, 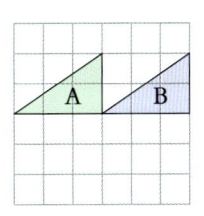 e, 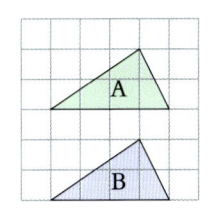 f

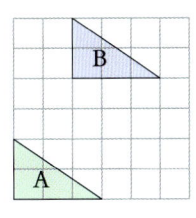

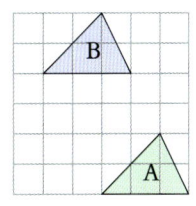

2 Copy these shapes onto squared paper. Translate each shape by the given vector.

a $\begin{pmatrix} 5 \\ 2 \end{pmatrix}$ b $\begin{pmatrix} -4 \\ 3 \end{pmatrix}$ c $\begin{pmatrix} 0 \\ -3 \end{pmatrix}$

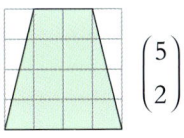

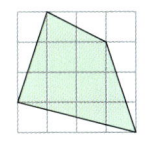

d $\begin{pmatrix} 3 \\ -4 \end{pmatrix}$ e $\begin{pmatrix} -3 \\ -4 \end{pmatrix}$ f $\begin{pmatrix} -4 \\ 0 \end{pmatrix}$

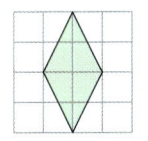

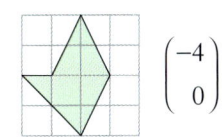

3 Copy and complete the following table.

	Point	Transformation	Image
a	(3, 6)	Translation of $\begin{pmatrix} 5 \\ 2 \end{pmatrix}$	
b	(1, 0)	Translation of $\begin{pmatrix} 4 \\ -2 \end{pmatrix}$	
c	(2, −5)	Translation of $\begin{pmatrix} -6 \\ 4 \end{pmatrix}$	
d	(−2, 8)	Translation of $\begin{pmatrix} -7 \\ -2 \end{pmatrix}$	
e	(−10, −4)	Translation of $\begin{pmatrix} 6 \\ -8 \end{pmatrix}$	
f	(0, −3)	Translation of $\begin{pmatrix} 7 \\ -5 \end{pmatrix}$	

Reflections, rotations and translations

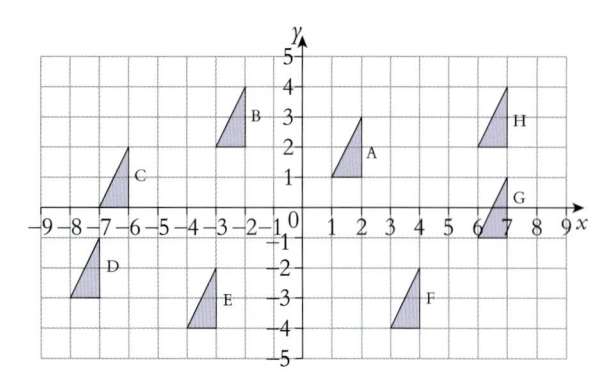

4   Describe fully the single transformation that moves triangle:
   a   A onto B
   b   A onto C
   c   A onto D
   d   A onto E
   e   A onto F
   f   A onto G
   g   A onto H
   h   H onto G
   i   F onto E
   j   B onto H
   k   C onto F
   l   G onto D.

5 a   Translate shape A by the vector $\begin{pmatrix} 3 \\ 3 \end{pmatrix}$
       Label the image B.
   b   Translate shape B by the vector $\begin{pmatrix} 4 \\ -1 \end{pmatrix}$
       Label the image C.
   c   Describe the single transformation that moves shape A onto shape C.

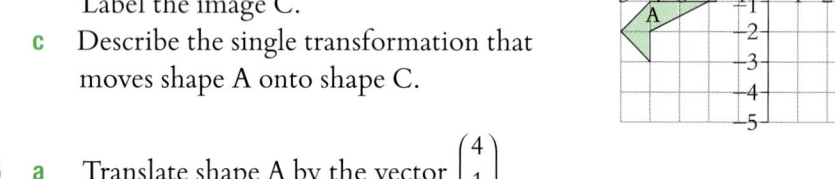

6 a   Translate shape A by the vector $\begin{pmatrix} 4 \\ 1 \end{pmatrix}$
       Label the image B.
   b   Translate shape B by the vector $\begin{pmatrix} -8 \\ -3 \end{pmatrix}$
       Label the image C.
   c   Describe the single transformation that moves shape A onto shape C.

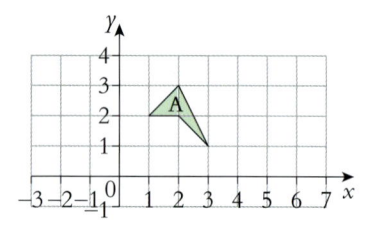

7   Triangle ABC is translated onto triangle A′B′C′.
    Copy and complete the table.

A = (1, 1)	B = (3, 2)	C = (…, …)
A′ = (4, −4)	B′ = (…, …)	C′ = (4, −1)

8   Quadrilateral ABCD is translated onto quadrilateral A′B′C′D′.
    Copy and complete the table.

A = (…, …)	B = (2, 2)	C = (…, …)	D = (1, 2)
A′ = (2, −3)	B′ = (0, −2)	C′ = (−3, −1)	D′ = (…, …)

138   UNIT 3

EXERCISE 3.16

1

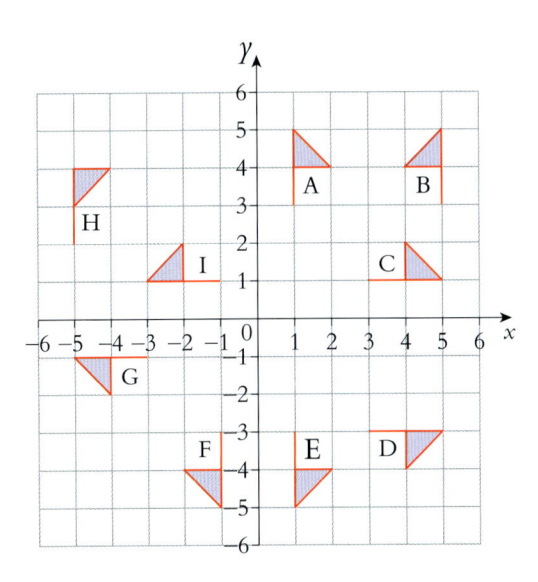

Describe fully the single transformation that moves shape:
- **a** A onto B
- **b** A onto C
- **c** A onto E
- **d** A onto F
- **e** A onto G
- **f** A onto H
- **g** A onto I
- **h** G onto F
- **i** E onto G
- **j** C onto D
- **k** E onto F
- **l** C onto I.

2

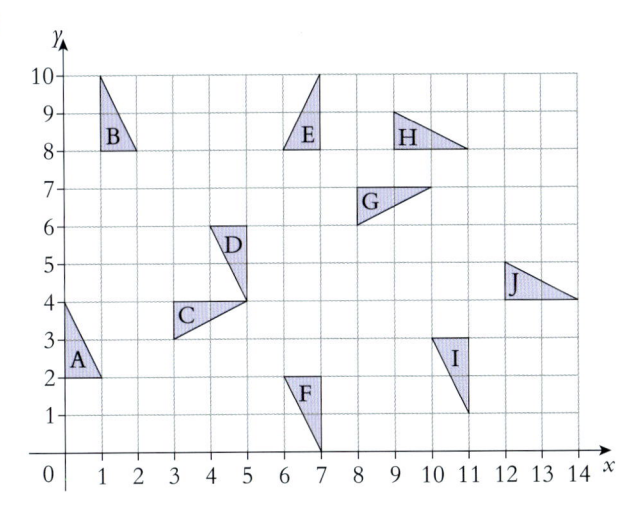

Describe fully the single transformation that moves triangle:
- **a** A onto B
- **b** A onto F
- **c** B onto E
- **d** B onto D
- **e** C onto D
- **f** C onto G
- **g** D onto I
- **h** E onto F
- **i** I onto J
- **j** J onto H
- **k** E onto H
- **l** F onto C.

Reflections, rotations and translations

Combined transformations

Transformations can be applied one after another. A combination of transformations can sometimes be described using a single transformation.

EXAMPLE

a Draw triangle A whose vertices are (2, 1), (4, 1) and (2, 2)
b Reflect triangle A in the line $y = x$. Label the image B.
c Rotate triangle B 90° anticlockwise about the point (0, 0). Label the image C.
d Describe fully the single transformation that maps triangle A onto triangle C.

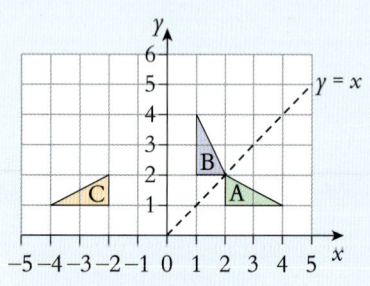

The single transformation that maps triangle A onto triangle C is a reflection in the y-axis.

EXAMPLE

a Draw triangle A whose vertices are (−6, 1), (−2, 1) and (−5, 3)
b Reflect triangle A in the line $x = -1$. Label the image B.
c Reflect triangle B in the line $y = -1$. Label the image C.
d Describe fully the single transformation that maps triangle A onto triangle C.

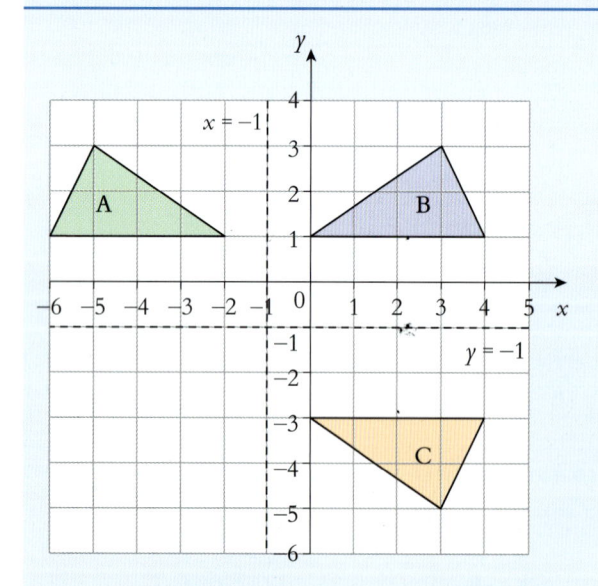

The single transformation that maps triangle A onto triangle C is a rotation of 180° about the point (−1, −1).

EXAMPLE

a Draw triangle P whose vertices are (1, 0), (0, 1) and (2, 3).
b Reflect triangle P in the line $x = 3$. Label the image Q.
c Reflect triangle Q in the line $x = 9$. Label the image R.
d Describe fully the single transformation that maps triangle P onto triangle R.

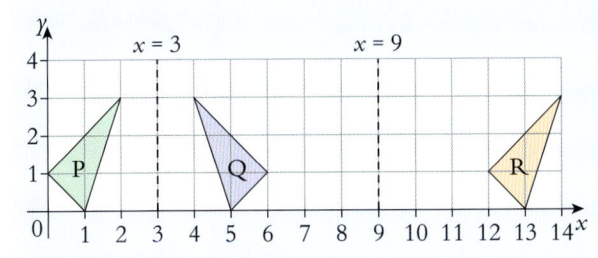

The single transformation that maps triangle P onto triangle R is a translation of $\begin{pmatrix} 12 \\ 0 \end{pmatrix}$.

EXERCISE 3.17

1 Draw axes from −6 to 6.
   a Draw triangle P whose vertices are (6, 1), (1, 1) and (2, 5).
   b Reflect triangle P in the x-axis. Label the image Q.
   c Reflect triangle Q in the y-axis. Label the image R.
   d Describe fully the single transformation that maps triangle P onto triangle R.

2 Draw axes from −6 to 6.
   a Draw triangle P whose vertices are (1, 1), (1, 6) and (3, 6).
   b Rotate triangle P 180° about the point (0, 0). Label the image Q.
   c Translate triangle Q by the vector $\begin{pmatrix} 6 \\ 3 \end{pmatrix}$. Label the image R.
   d Describe fully the single transformation that maps triangle P onto triangle R.

3 Draw axes from −6 to 6.
   a Draw triangle P whose vertices are (4, 1), (2, 2) and (4, 5).
   b Reflect triangle P in the line $y = 0$. Label the image Q.
   c Rotate triangle Q 180° about the point (−1, 0). Label the image R.
   d Describe fully the single transformation that maps triangle P onto triangle R.

Reflections, rotations and translations

4   Draw axes from −6 to 6.
    a   Draw triangle P whose vertices are (1, 3), (2, 6) and (4, 6)
    b   Reflect triangle P in the line $y = 2$.
        Label the image Q.
    c   Reflect triangle Q in the line $y = -1$.
        Label the image R.
    d   Describe fully the single transformation that maps triangle P onto triangle R.

5   Draw axes from −6 to 6.
    a   Draw triangle P whose vertices are (1, 3), (2, 6) and (4, 6)
    b   Reflect triangle P in the line $y = 0$.
        Label the image Q.
    c   Rotate triangle Q 90° clockwise about the point (0, 0).
        Label the image R.
    d   Describe fully the single transformation that maps triangle P onto triangle R.

6   Draw axes from −6 to 6.
    a   Draw triangle P whose vertices are (2, 3), (5, 3) and (5, 5)
    b   Rotate triangle P 180° about the point (4, 1).
        Label the image Q.
    c   Rotate triangle Q 180° about the point (1, −3).
        Label the image R.
    d   Describe fully the single transformation that maps triangle P onto triangle R.

7   Which single transformation is the same as:
    a   A translation of $\begin{pmatrix} 4 \\ 3 \end{pmatrix}$ followed by a translation of $\begin{pmatrix} 2 \\ 7 \end{pmatrix}$
    b   A translation of $\begin{pmatrix} -5 \\ 6 \end{pmatrix}$ followed by a translation of $\begin{pmatrix} 8 \\ 4 \end{pmatrix}$

8   Which single transformation is the same as:
    a   A reflection in $y = 2$ followed by a reflection in $x = 3$
    b   A reflection in $y = 2$ followed by a reflection in $y = 5$

9   Which single transformation is the same as:
    a   A rotation of 180° about the point (1, 2) followed by a rotation of 180° about the point (3, 4).
    b   A rotation of 180° about the point (5, 0) followed by a rotation of 180° about the point (2, 5).

If you translate an object by the vector $\begin{pmatrix} 5 \\ 2 \end{pmatrix}$ and then translate the image by the vector $\begin{pmatrix} -5 \\ -2 \end{pmatrix}$, the image will be transformed back to the original object.

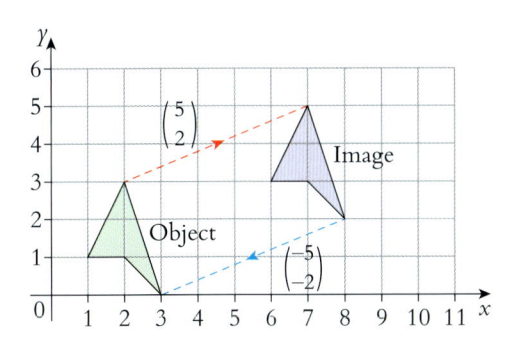

A translation of $\begin{pmatrix} 5 \\ 2 \end{pmatrix}$ and a translation of $\begin{pmatrix} -5 \\ -2 \end{pmatrix}$ are called **inverse transformations**.

EXERCISE 3.18

Write down the inverse of each of these transformations.

1 A translation of $\begin{pmatrix} 2 \\ -4 \end{pmatrix}$

2 A translation of $\begin{pmatrix} 0 \\ 6 \end{pmatrix}$

3 A translation of $\begin{pmatrix} -5 \\ 7 \end{pmatrix}$

4 A translation of $\begin{pmatrix} -8 \\ 0 \end{pmatrix}$

5 A reflection in the x-axis.

6 A reflection in the y-axis.

7 A reflection in the line $y = 3$

8 A reflection in the line $x = -5$

9 A reflection in the line $x + y = 4$

10 A rotation of 90° clockwise about the point (0, 0)

11 A rotation of 90° anticlockwise about the point (0, 0)

12 A rotation of 180° about the point (0, 0)

13 A rotation of 90° clockwise about the point (5, −2)

14 A rotation of 90° anticlockwise about the point (−4, 7)

KEY WORDS
transformation
object
image
reflection
invariant
rotation
translation
centre of rotation
column vector
inverse transformation

Reflections, rotations and translations

Enlargements

THIS SECTION WILL SHOW YOU HOW TO
- Find the centre and scale factor of an enlargement
- Enlarge a shape

An **enlargement** is a transformation where
- Corresponding angles are the same
- Lengths of corresponding sides are in the same ratio.

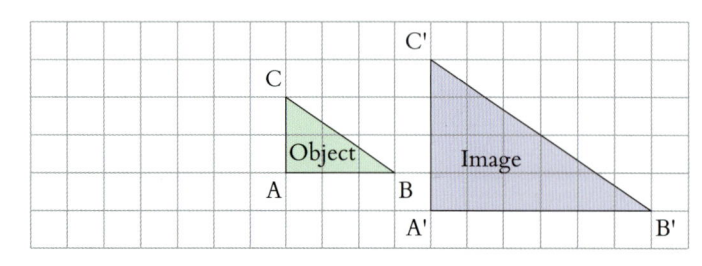

➡ **NOTE:** the object and image are similar in shape.

Triangle ABC has been enlarged with **scale factor** 2 to give triangle A'B'C'.
(The scale factor is 2 because the lengths of the sides of the triangle have doubled.)

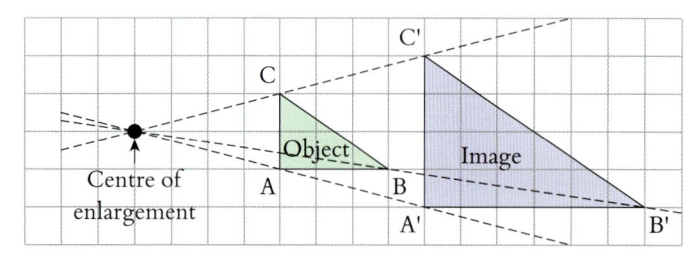

The position of the centre of enlargement is found by drawing lines through A and A', B and B' and C and C'. The point where these lines intersect is called the **centre of enlargement**.

EXAMPLE

Describe fully the transformation that maps triangle PQR onto triangle P'Q'R'.

The lengths of all the sides of the triangle have trebled so the scale factor of the enlargement is 3.
To find the centre of enlargement draw the lines through P and P', Q and Q' and R and R'. These lines intersect at the point (1, 6).
The transformation is an enlargement scale factor 3, centre (1, 6).

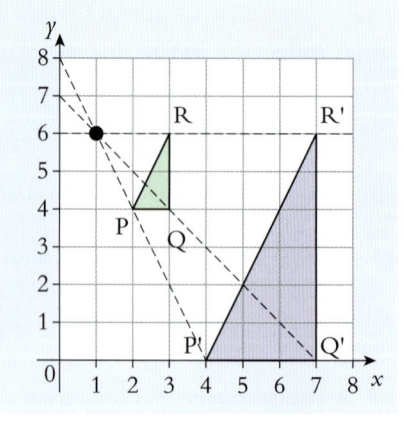

To describe an enlargement fully you must state the scale factor and the centre of enlargement.

EXERCISE 3.19

1 In each diagram shape A has been enlarged to make shape B.
   State the scale factor and the centre of enlargement for each diagram.

   a   b 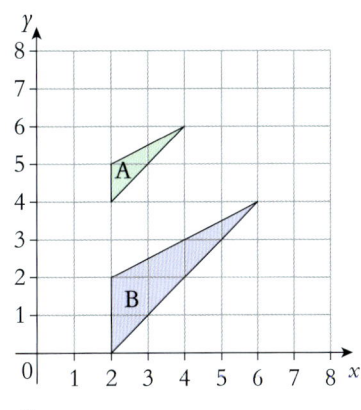
   c 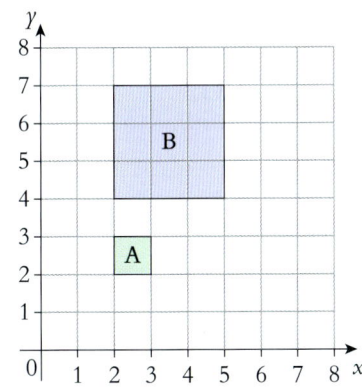   d 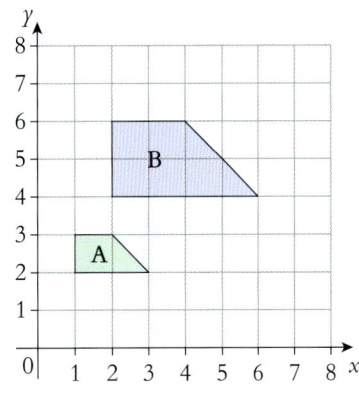
   e 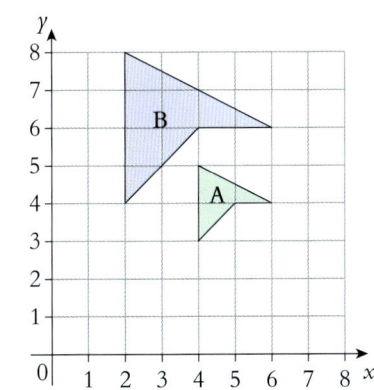   f

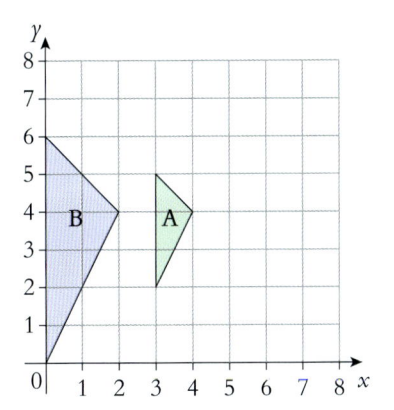

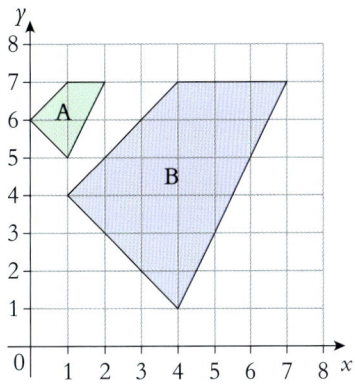

2 Describe fully the single transformation that maps shape:
   a A onto C
   b A onto D
   c D onto F
   d B onto E
   e B onto C
   f B onto D

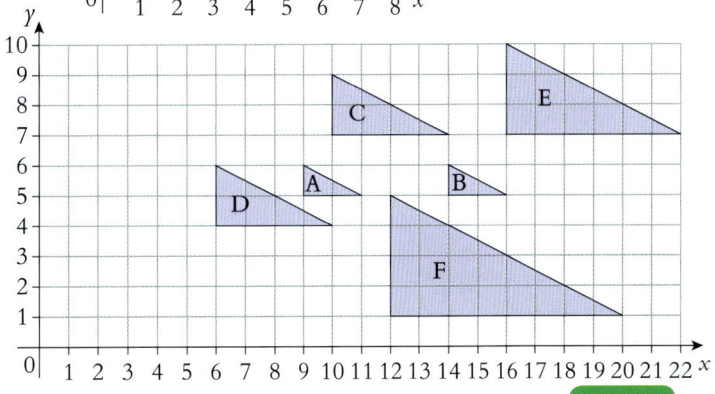

Enlargements

EXAMPLE

Enlarge triangle ABC using a scale factor of 2 and centre of enlargement P(1, 2).

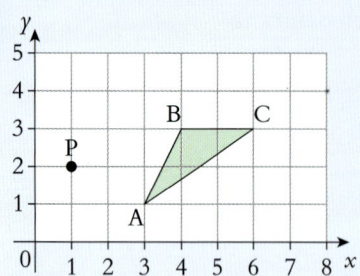

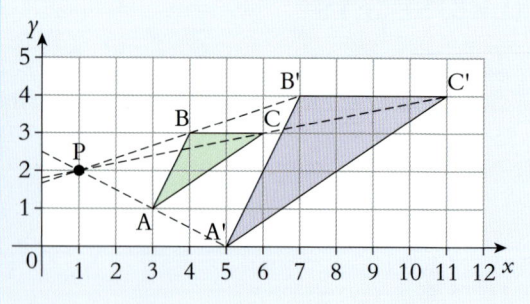

PA′ = 2 × PA
PB′ = 2 × PB
PC′ = 2 × PC

EXAMPLE

Enlarge quadrilateral ABCD using a scale factor of $\frac{1}{2}$ and centre of enlargement P(11, 3).

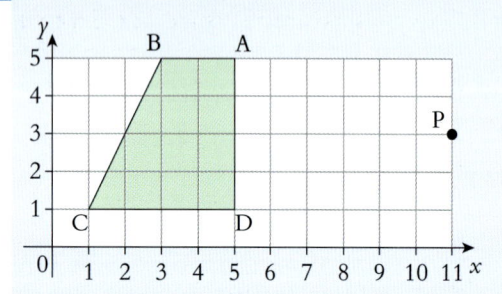

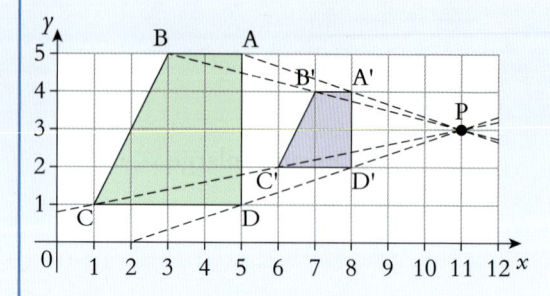

PA′ = $\frac{1}{2}$ × PA
PB′ = $\frac{1}{2}$ × PB
PC′ = $\frac{1}{2}$ × PC
PD′ = $\frac{1}{2}$ × PD

An enlargement scale factor $\frac{1}{2}$ makes the shape smaller.

EXAMPLE

Enlarge triangle ABC using a scale factor of −2 and centre of enlargement P(4, 2).

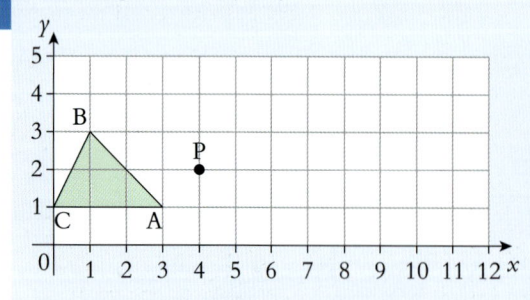

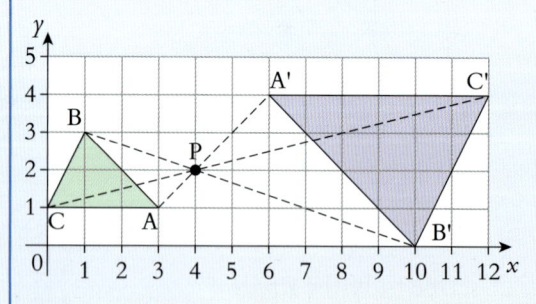

PA′ = 2 × PA
PB′ = 2 × PB
PC′ = 2 × PC

For a negative enlargement the object and image are on opposite sides of the centre of enlargement.

146 UNIT 3

EXERCISE 3.20

1. Copy each diagram and enlarge with scale factor 2. Use point P as the centre of enlargement.

   a   b 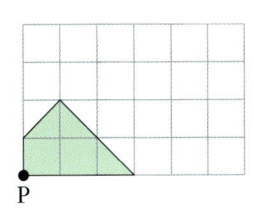   c

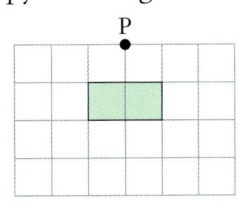

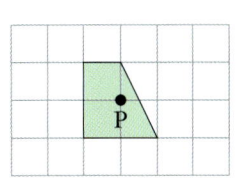

2. Copy each diagram and enlarge with scale factor $\frac{1}{2}$. Use point P as the centre of enlargement.

   a   b 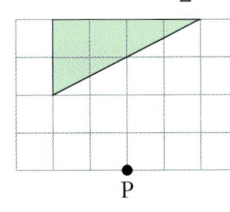   c

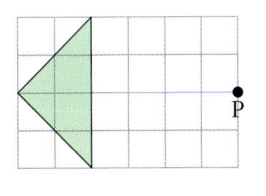

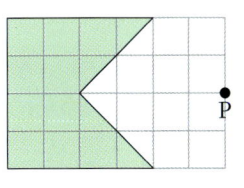

3. Copy each diagram and enlarge with scale factor −2. Use point P as the centre of enlargement.

   a   b 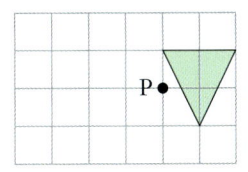   c

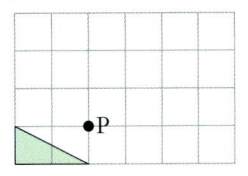

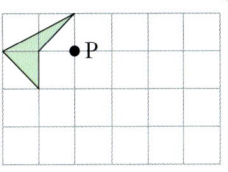

4. Copy the following diagrams. Enlarge each shape using the given scale factor and the point P as the centre of enlargement.

   a   scale factor 3

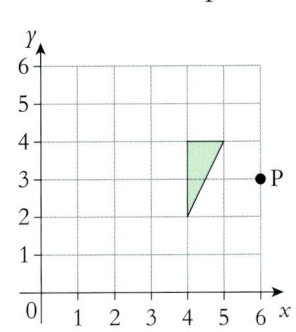

   b   scale factor 2

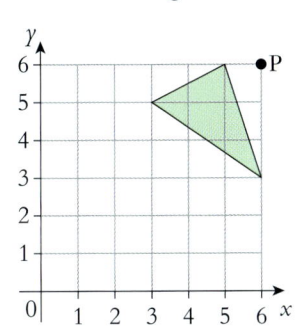

   c   scale factor $\frac{1}{2}$

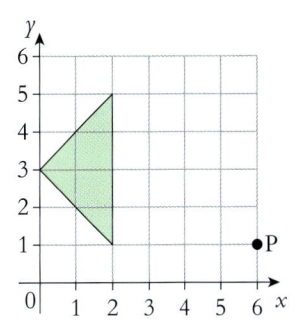

   d   scale factor −2

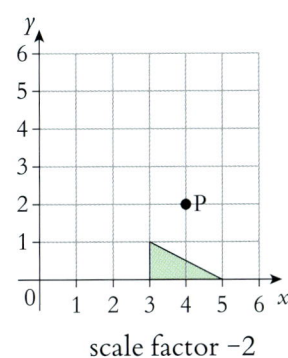

   e   scale factor −1

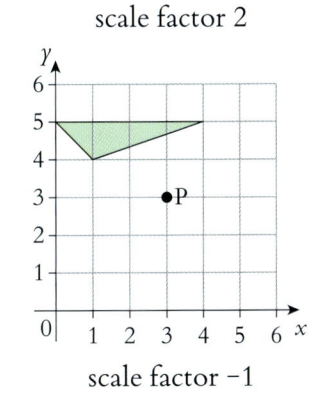

   f   scale factor $-\frac{1}{3}$

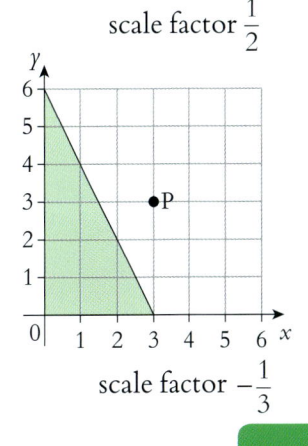

Enlargements

5 Draw axes from −7 to 7.
   Draw triangle T whose vertices are (2, 0), (0, 2) and (−2, 2).
   Draw the image of triangle T after each of these enlargements.
   Label your images A, B, C, and D.

Scale factor	Centre of enlargement	Image
2	(−2, 0)	A
$\frac{1}{2}$	(0, −4)	B
−1	(−2, 1)	C
−2	(0, −1)	D

6 Draw axes from −6 to 6.
   Draw triangle T whose vertices are (4, 2), (6, 2) and (4, 6).
   Draw the image of triangle T after each of these enlargements.
   Label your images P, Q, R and S.

Scale factor	Centre of enlargement	Image
$\frac{1}{2}$	(0, 0)	P
$\frac{1}{2}$	(−6, 4)	Q
−1	(4, 0)	R
−2	(2, 3)	S

7 Draw axes from −6 to 6.
   Draw triangle T whose vertices are (−6, −2), (−4, −4) and (−2, −4).
   Enlarge triangle T with scale factor $2\frac{1}{2}$, centre (−6, −6).

8 Draw axes from −6 to 6.
   Draw triangle T whose vertices are (1, 0), (−3, 4) and (0, −2).
   Enlarge triangle T with scale factor $1\frac{1}{3}$, centre (6, 1).

9 Draw axes from −7 to 7.
   Draw triangle T whose vertices are (5, −2), (5, 2) and (−3, 2).
   Enlarge triangle T with scale factor $-1\frac{1}{2}$, centre (1, 0).

EXAMPLE

Enlarge triangle P with scale factor 3, centre (0, 0). Label the image Q.
Enlarge triangle Q with scale factor $\frac{1}{3}$, centre (−9, 6). Label the image R.
Describe fully the single transformation that maps triangle P onto triangle R.

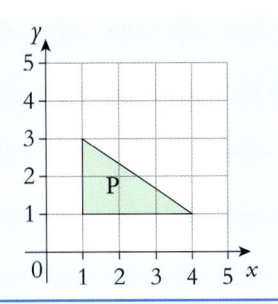

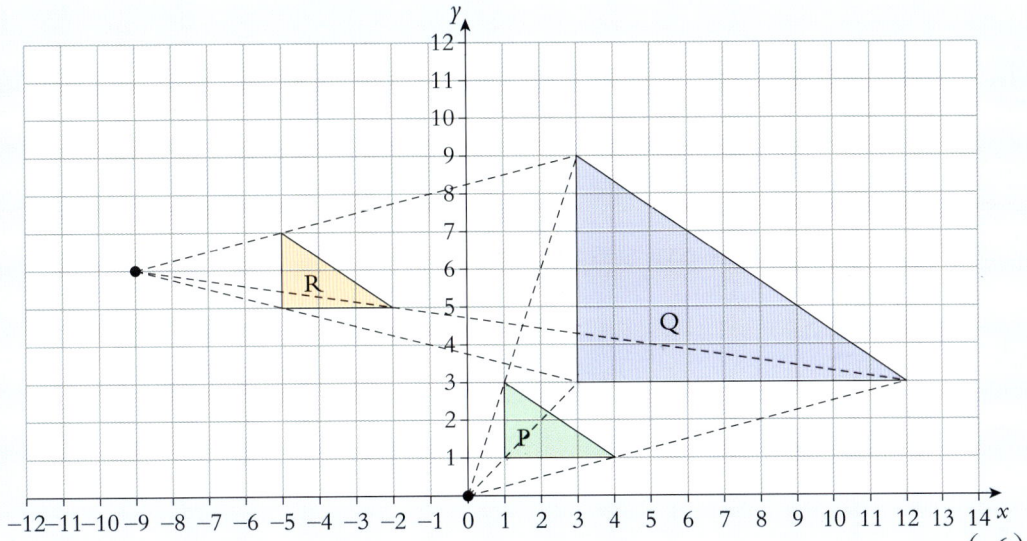

The single transformation that maps triangle P onto triangle R is a translation of $\begin{pmatrix} -6 \\ 4 \end{pmatrix}$

EXERCISE 3.21

1. Draw axes from −6 to 6
   a. Draw triangle P with vertices (1, 0), (0, 1) and (−2, −1)
   b. Enlarge triangle P with scale factor 2, centre (2, −2). Label the image Q.
   c. Enlarge triangle Q with scale factor $\frac{1}{2}$, centre (−6, −6). Label the image R.
   d. Describe fully the single transformation that maps triangle P onto triangle R.

2. Draw axes from −6 to 6
   a. Draw triangle P with vertices (−3, 0), (−6, 0) and (−5, −1)
   b. Enlarge triangle P with scale factor −2, centre (−3, −2). Label the image Q.
   c. Enlarge triangle Q with scale factor −2, centre (0, −2). Label the image R.
   d. Describe fully the single transformation that maps triangle P onto triangle R.

3. Draw axes from −6 to 6
   a. Draw triangle P with vertices (−4, 1), (−4, 3) and (−3, 1).
   b. Enlarge triangle P with scale factor 4, centre (−6, 2). Label the image Q.
   c. Enlarge triangle Q with scale factor $\frac{1}{2}$, centre (−6, −2). Label the image R.
   d. Describe fully the single transformation that maps triangle P onto triangle R.

KEY WORDS
enlargement
scale factor
centre of enlargement

Enlargements

Surface area and volume 1

THIS SECTION WILL SHOW YOU HOW TO
- Find the volume and surface area of a prism

Volume of a cuboid

Volume of a cuboid = length × width × height

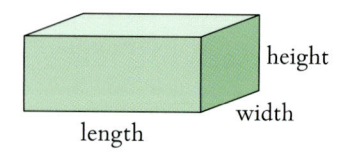

Volume of a prism
A **prism** is a three-dimensional shape with a uniform **cross-section**.

Volume of a prism = area of cross-section × length

➡ **NOTE:** a cuboid is prism with a rectangular cross-section.

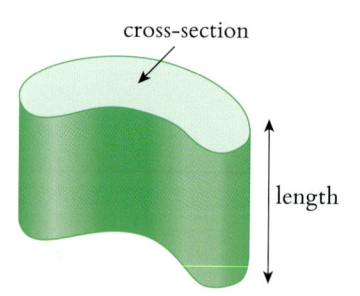

A **cylinder** is a prism with a circular cross-section.
Volume of cylinder = area of cross-section × length
$= \pi r^2 \times h$

Volume of cylinder = $\pi r^2 h$

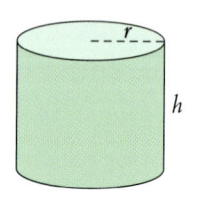

Converting units of volume

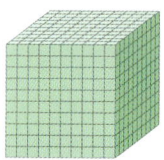

$1 \text{ cm}^3 = 10 \times 10 \times 10 \text{ mm}^3$

Similarly,
$1 \text{ m}^3 = 100 \times 100 \times 100 \text{ cm}^3$

$1 \text{ cm}^3 = 1000 \text{ mm}^3$

$1 \text{ m}^3 = 1\,000\,000 \text{ cm}^3$

Capacity
The **capacity** is the amount that a container can hold.
Units for capacity are litres (ℓ) and millilitres (mℓ).

1 litre = 1000 millilitres

1 cm^3 contains 1 mℓ

'The bottle has a capacity of 200 mℓ' means that the bottle has a volume of 200 cm³

UNIT 3

EXAMPLE

Find the volume of the prism.

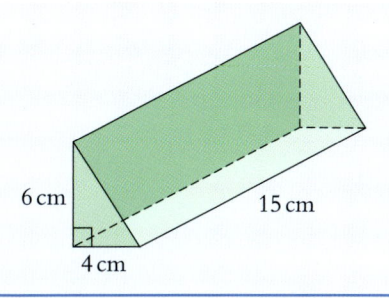

Area of cross-section = $\frac{1}{2} \times 4 \times 6 = 12$ cm²
Volume of prism = area of cross-section × length
  = 12 × 15
  = 180 cm³

EXAMPLE

a Find the volume of the cylinder
b Find the capacity of the cylinder in litres.

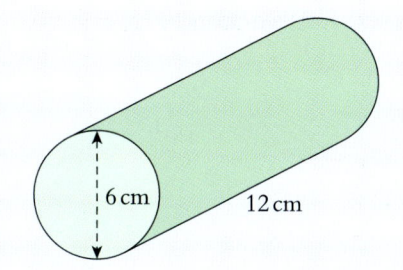

a Volume of cylinder = $\pi r^2 h$
  = $\pi \times 3^2 \times 12$
  = 339 cm³ (3 s.f.)
b Capacity = 339 mℓ
  = 0.339 litres

Remember: 1 cm³ = 1 mℓ and 1000 ml = 1 litre

EXERCISE 3.22

1 Find the volume of these cuboids.

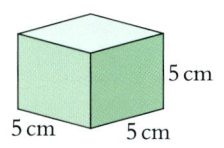

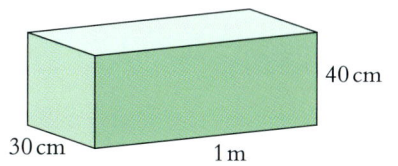

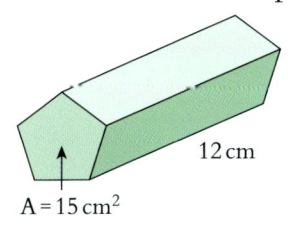

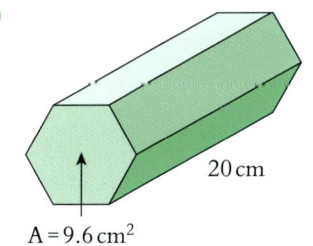

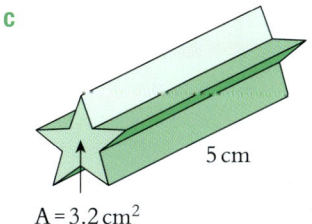

   a 5 cm × 5 cm × 5 cm
   b 8 cm × 6 cm × 2 cm
   c 30 cm × 1 m × 40 cm

2 Find the volume of these prisms.

   a A = 15 cm², 12 cm
   b A = 9.6 cm², 20 cm
   c A = 3.2 cm², 5 cm

Surface area and volume 1   151

3 Find the volume of these prisms. (All lengths are in cm.)

a, b, 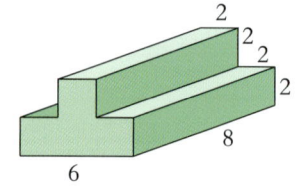 c, 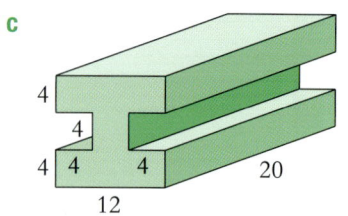 d, 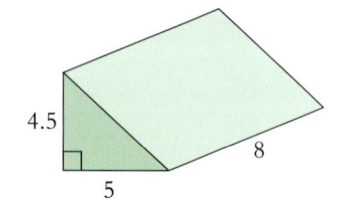 e, 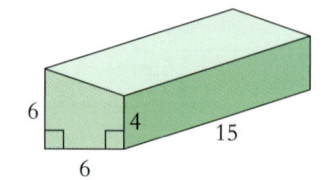 f

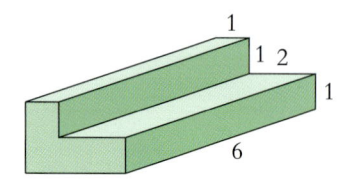

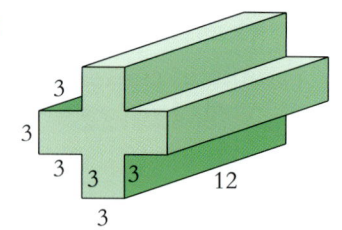

Dimensions shown:
- a: 1, 1, 2, 1, 6
- b: 2, 2, 2, 2, 6, 8
- c: 4, 4, 4, 4, 12, 20
- d: 4.5, 5, 8
- e: 6, 4, 6, 15
- f: 3, 3, 3, 3, 3, 12

4 Find the volumes of these cylinders.

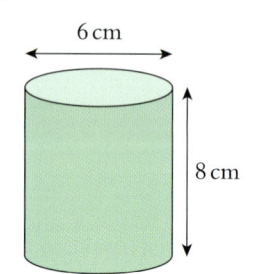

a: 2 m diameter, 1 m height
b: 6 cm diameter, 8 cm height
c: 5 cm diameter, 12 cm length

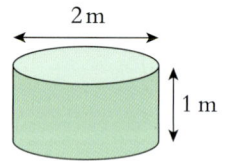

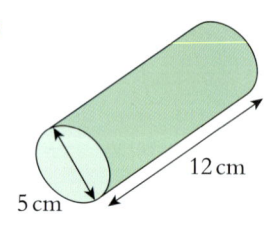

5 Find the volume of these hollow cylinders.

a: 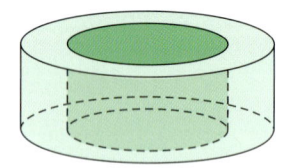
outer diameter = 10 cm
inner diameter = 6 cm
length = 3 cm

b: 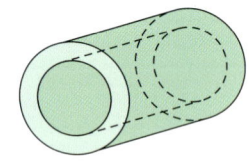
outer diameter = 8 cm
inner diameter = 5 cm
length = 16 cm

c: 
outer diameter = 50 cm
inner diameter = 30 cm
length = 20 cm

6 A carton contains 1 litre of mango juice.
The length of the carton is 10 cm and the width is 5 cm.
Calculate the height of the carton.

5 cm, 10 cm

7 This container is a prism.
The cross-section of the prism is a trapezium.
  a Find the volume of the container in m³.
  b Find the volume of the container in cm³.
  c How many litres of water does the container hold when it is full?

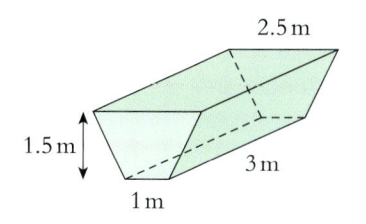

8 This piece of cheese is a prism.
The cross-section is a sector of a circle of radius 14 cm and angle 25°
The height of the piece of cheese is 6 cm.
Calculate the volume.

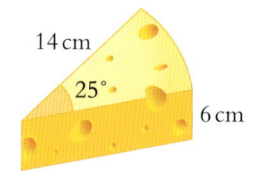

9 A cylindrical can contains 1.5 litres of coconut milk.
The height of the can is 15 cm.
Calculate the radius of the can.

10 This water trough is a prism.
The cross-section is a semi-circle of diameter 50 cm.
When full the water trough holds 300 litres of water.
Calculate the length of the water trough.

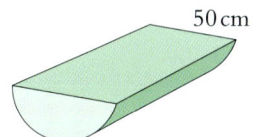

11 The diagram shows a cube with a circular hole.
The cube has sides of length 5 cm.
The circular hole has radius 1 cm.
Calculate the volume of the solid.

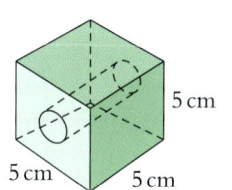

12 The cross-section of this solid prism is a regular hexagon with sides of length 8 cm.
The length of the prism is 25 cm.

  a Find the area of the regular hexagon.
  b Find the volume of the prism.
The prism is made of wood.
1 cm³ of wood has a mass of 0.65 g.
  c Find the mass of the hexagonal prism

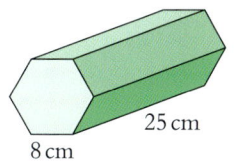

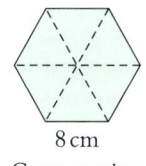

HINT
The hexagon is made from six equilateral triangles. Find the area of one of the triangles and multiply by six.

Surface area and volume 1

Surface area of a prism

> **Surface area** of prism = sum of area of all the faces

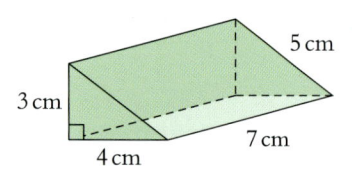

Area of rectangle A = 3 × 7 = 21 cm²
Area of rectangle B = 4 × 7 = 28 cm²
Area of rectangle C = 5 × 7 = 35 cm²
Area of triangle D = $\frac{1}{2}$ × 3 × 4 = 6 cm²
Area of triangle E = $\frac{1}{2}$ × 3 × 4 = 6 cm²
Total surface area = 96 cm²

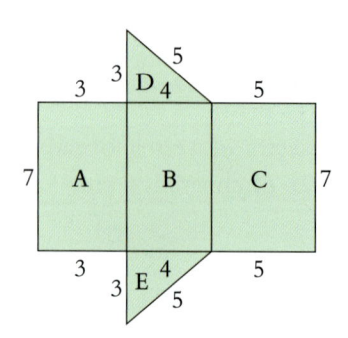

Surface area of a cylinder
The surface of a solid cylinder consists of two circles and a rectangle.

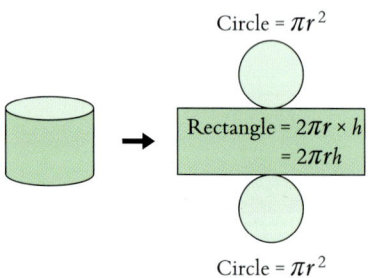

Circle = πr^2
Rectangle = $2\pi r \times h$ = $2\pi rh$
Circle = πr^2

The two circles are flat surfaces.
The rectangle is made from the **curved surface** of the cylinder.

> Curved surface area of cylinder = $2\pi rh$

> Total surface area of cylinder = $2\pi rh + 2\pi r^2$

EXAMPLE

A solid cylinder has radius 3 cm and length 10 cm.
Calculate the total surface area of the cylinder

Total surface area of cylinder = $2\pi rh + 2\pi r^2$
$= 2 \times \pi \times 3 \times 10 + 2 \times \pi \times 3^2$
$= 60\pi + 18\pi$
$= 78\pi$
$= 245$ cm² to 3 s.f.

EXERCISE 3.23

1 Calculate the surface area of these prisms.

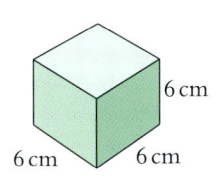

   a 6 cm, 6 cm, 6 cm   **b** 8 cm, 5 cm, 4 cm   **c** 7 cm, 20 cm, 12 cm

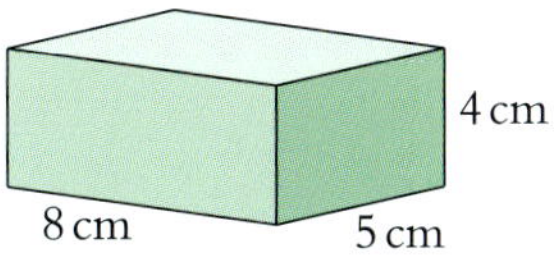

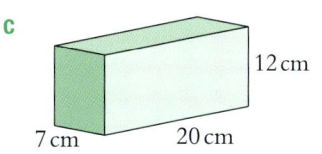

2 Find the surface area of these prisms. (All lengths are in cm.)

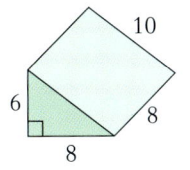

   a 10, 6, 8, 8   **b** 1, 1, 2, 2, 8   **c** 2, 2, 2, 2, 6, 7

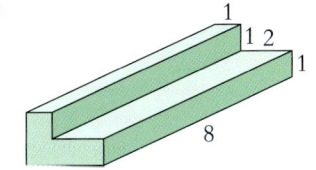

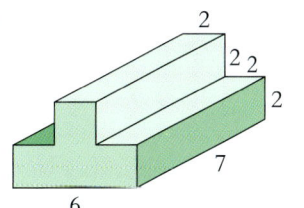

3 Find the surface area of these prisms. (All lengths are in cm.)

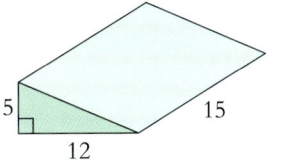

   a 5, 12, 15   **b** 6, 4, 10, 3

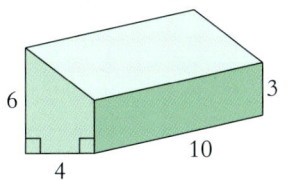

> **HINT**
> You will need to use Pythagoras to find the length of the sloping edges.

4 Find the **curved** surface areas of these cylinders.

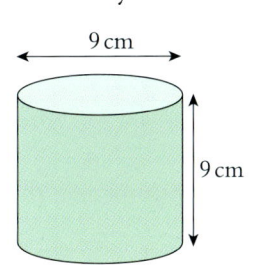

   a 8 cm diameter, 4 cm height   **b** 9 cm diameter, 9 cm height   **c** 4 cm diameter, 14 cm length

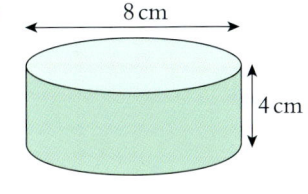

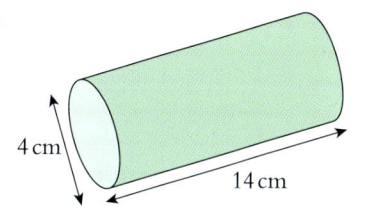

5 Find the **total** surface area of these solid cylinders.

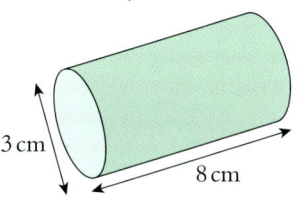

   a 4 cm diameter, 4 cm height   **b** 3 cm diameter, 8 cm length   **c** 4 m diameter, 2 m height

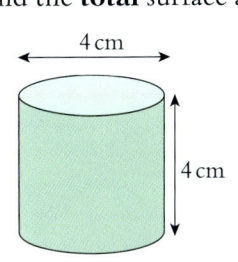

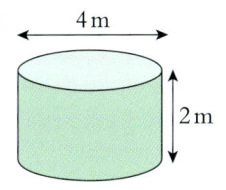

6 This piece of cheese is a prism.
The cross-section is a sector of a circle of radius 12 cm and angle 32°.
The height of the piece of cheese is 5 cm.
Calculate the total surface area.

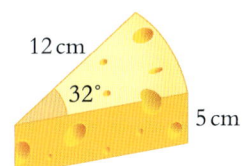

> **KEY WORDS**
> prism, cross-section
> cylinder, capacity
> volume, surface area
> curved surface

Probability 1

THIS SECTION WILL SHOW YOU HOW TO
- Understand and use the probability scale from 0 to 1
- Calculate the probability of a single event
- Use relative frequency
- Use possibility diagrams to calculate the probability of simple combined events

Probability is a measure of how likely an **event** is to happen.
An event can be made up of one or more possible **outcomes**.
For example, when a dice is rolled then the possible outcomes are 1, 2, 3, 4, 5 and 6.
If the event A is 'the number on the dice is greater than 4' then there are two desired outcomes for event A which are 5 and 6.

Probability is measured on a scale from 0 to 1

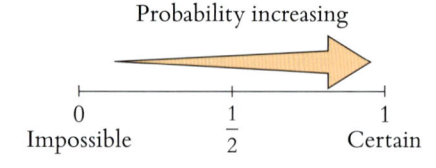

Probability increasing

0 — Impossible
$\frac{1}{2}$
1 — Certain

➡ **NOTE:** you can write probability as a fraction, decimal or a percentage.

If all possible outcomes are equally likely, you can calculate the theoretical probability, P(A), of the event happening.

$$P(A) = \frac{\text{number of desirable outcomes}}{\text{total number of possible outcomes}}$$

So, when a **fair** dice is rolled the probability the number on the dice is greater than $4 = \frac{2}{6} = \frac{1}{3}$

A fair (or **unbiased**) dice means the possible outcomes are equally likely.
A **biased** dice means the possible outcomes are not equally likely.

EXAMPLE

The spinner has eight coloured sections. When the spinner is spun it is equally likely that it stops on any one of the sections.
Find the probability that it stops on
a a green section,
b a yellow section,
c a blue section,
d a red section,
e a section that is not green,
f a blue or green section.

a P(green) = $\frac{1}{8}$
b P(yellow) = $\frac{3}{8}$
c P(blue) = $\frac{4}{8} = \frac{1}{2}$

d P(red) = 0
e P(not green) = $\frac{7}{8}$
f P(blue or green) = $\frac{4+1}{8} = \frac{5}{8}$

UNIT 3

In the last example   P(green) = $\frac{1}{8}$ and   P(not green) = $\frac{7}{8}$

This is an example of the rule $P(\bar{A}) = 1 - P(A)$ where $\bar{A}$ means 'not A'.

EXAMPLE

The probability that Omar is late for school is 0.17.
Find is the probability that he is not late to school.

P(not late) = 1 − P(late) = 1 − 0.17 = 0.83

EXERCISE 3.24

1. An unbiased eight-sided die is thrown. Find the probability that the die
   a. lands on a 5
   b. lands on an even number
   c. lands on a square number
   d. lands on a number greater than 4
   e. lands on a 6 or a 7
   e. lands on a number smaller than 9

2. One of the five cards is chosen at random.
   Find the probability of choosing a card with
   a number that is   [3] [6] [6] [8] [9]
   a. a six
   b. an odd number
   c. not prime
   d. square
   e. a multiple of 3
   f. a factor of 72

3. A box contains 5 orange balls, 4 yellow balls and 1 green ball.
   A ball is picked out of the box at random. Find the probability that it is
   a. green
   b. orange
   c. yellow
   d. not green
   e. orange or green
   f. green or yellow
   g. blue
   h. not blue

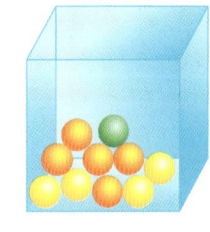

4. [A] [C] [B] [C] [B] [A] [B] [C] [A]

   One of the nine cards is chosen at random. Find the probability that
   a. the letter on the card is an A
   b. the letter on the card is not an A
   c. a green card is chosen
   d. the card is orange with a letter B

5. The table shows the number of boys and girls in Years 8
   and 9 at a school.
   a. A student is chosen at random from Year 8.
      Find the probability that the student is a girl.
   b. A student is chosen at random from Year 9.
      Find the probability that the student is a boy.

	Year 8	Year 9
Boys	52	48
Girls	68	52

Probability 1

EXAMPLE

One of these six cards is chosen at random. The first card is **not** replaced before a second card is chosen at random.

| 7 | 8 | 8 | 7 | 7 | 9 |

a Find the probability that the first card chosen is a
   i 7   **ii** 8   **iii** 9

b If the first card chosen is a 8, find the probability that the second card chosen is a
   i 7   **ii** 8   **iii** 9

---

a i $P(7) = \frac{3}{6} = \frac{1}{2}$ because there are **three** cards with a number 7 and **six** cards to chose from

   ii $P(8) = \frac{2}{6} = \frac{1}{3}$ because there are **two** cards with a number 8 and **six** cards to chose from

   iii $P(9) = \frac{1}{6}$ because there is **one** card with a number 9 and **six** cards to chose from

b If the first card chosen was a 8 then the cards remaining are | 7 | 8 | 7 | 7 | 9 |

   i $P(7) = \frac{3}{5}$ because there are **three** cards with a number 7 and **five** cards to chose from

   ii $P(8) = \frac{1}{5}$ because there is now only **one** card with a number 8 and **five** cards to chose from

   iii $P(9) = \frac{1}{5}$ because there is **one** card with a number 9 and **five** cards to chose from

Sometimes the probabilities of all the outcomes are put into a table.
The probabilities for part **a** in the example above could be displayed as:

Number	7	8	9
Probability	$\frac{1}{2}$	$\frac{1}{3}$	$\frac{1}{6}$

➡ **NOTE:** the bottom row of the table will always add up to 1.

EXAMPLE

Score	1	2	3	4	5
Probability	0.1	x	0.2	0.3	0.15

The table shows the probabilities of getting scores on a biased 5-sided spinner.
a Find the value of x
b Find the probability of a score greater than 2
c Find the probability of not scoring a 3

---

a The probabilities of all the possible outcomes add up to give 1.
   So $0.1 + x + 0.2 + 0.3 + 0.15 = 1$
        $x + 0.75 = 1 \Rightarrow x = 0.25$

b $P(\text{score} > 2) = P(3) + P(4) + P(5)$
                        $= 0.2 + 0.3 + 0.15 = 0.65$

c $P(\text{not } 3) = 1 - P(3)$
                   $= 1 - 0.2 = 0.8$

158   UNIT 3

EXERCISE 3.25

1. The circular board is divided into eight numbered sections. When the arrow is spun it is equally likely to stop in any of the eight sections.

    a. Copy and complete the table which shows the probability of the arrow stopping at each number.
    b. The arrow is spun once. Find
        i. the probability of a number smaller than 3
        ii. the probability of an odd number
        iii. the probability of a prime number
        iv. the probability the number is a 4

Number	1	2	3
Probability		$\frac{1}{8}$	

2. There are five red sweets and three yellow sweets in a bag. Rosa takes a sweet at random from the bag and eats it. Rosa then takes a second sweet at random from the bag.
    a. Find the probability that the first sweet that Rosa took from the bag was
        i. red ii. yellow
    b. If the first sweet that Rosa took from the bag was red, find the probability that the second sweet she takes from the bag is
        i. red ii. yellow
    c. If the first sweet that Rosa took from the bag was yellow, find the probability that the second sweet she takes from the bag is
        i. red ii. yellow.

3. The table shows the probability of the spinner landing on each of the four colours.

Colour	Red	Yellow	Green	Blue
Probability	$\frac{1}{6}$	$\frac{1}{4}$	$\frac{1}{3}$	x

    a. Find the value of x.
    b. What is the most likely colour?
    c. Find the probability of not getting a red.

4.
Number	1	2	3	4	5	6
Probability	0.1	0.15	0.05	0.1	$2x$	x

    The table shows the probabilities of getting scores on a biased 6-sided dice.
    a. Find the value of x.
    b. What is the most likely score?
    c. Find the probability of a score greater than 2.

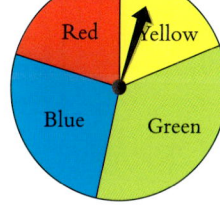

Possibility diagrams

When a dice is rolled and a one-dollar coin is tossed, the possible outcomes can be recorded in a possibility diagram. Let H mean heads and T mean tails

		\multicolumn{6}{c}{Dice}
		1	2	3	4	5	6
Coin	H	H1	H2	H3	H4	H5	H6
	T	T1	T2	T3	T4	T5	T6

The possibility diagram shows that there are 12 equally likely outcomes.

The probability of each of the outcomes = $\frac{1}{12}$.

There are three places on the diagram where you can get a head and an even number; they are H2, H4 and H6.

So, the probability of getting a head and an even number = $\frac{3}{12} = \frac{1}{4}$

EXAMPLE

Two unbiased six-sided dice are rolled. The scores are added together.
a Draw a possibility diagram to show all the outcomes.
b What is the most likely total score?
c Use the diagram in part **a** to find the probability of
   i a total score of 9   ii a total score less than 4

a
Red dice

		1	2	3	4	5	6
Blue dice	1	2	3	4	5	6	7
	2	3	4	5	6	7	8
	3	4	5	6	7	8	9
	4	5	6	7	8	9	10
	5	6	7	8	9	10	11
	6	7	8	9	10	11	12

➡ **NOTE:** there are 36 possible outcomes.
The probability of each outcome = $\frac{1}{36}$

b The most likely total score is 7   *there are more outcomes with a total of 7 than any other number*

c i   P(total score of 9) = $\frac{4}{36} = \frac{1}{9}$ *there are 4 scores of 9 on the diagram*

   ii P(total score less than 4) = $\frac{3}{36} = \frac{1}{12}$ *there are 3 scores less than 4 on the diagram*

EXERCISE 3.26

1

	1	2	3	4	5
1					
2				6	
3					
4					
5			8		

An unbiased five-sided spinner is spun twice and the scores are added.
- **a** Copy and complete the possibility diagram.
- **b** Use the diagram to find the probability that the total score is
  - **i** 9
  - **ii** 3
  - **iii** 1
  - **iv** an odd number
  - **v** a square number,
  - **vi** a prime number.

2

	1	2	3	4
1	2	3		
2				
3				
4			7	

Two unbiased four-sided dice are thrown and their scores are added together.
- **a** Copy and complete the possibility diagram.
- **b** Use the diagram to find the probability the total score is
  - **i** 7
  - **ii** 5
  - **iii** even
  - **iv** less than 5.

3

	2	4	6	8
1	2			
2		8		
3				24

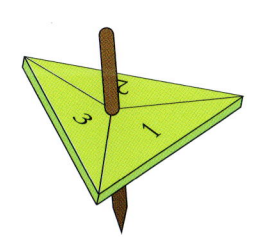

An unbiased three-sided spinner numbered 1, 2, 3 and an unbiased four-sided spinner numbered 2, 4, 6, 8 are spun.
The scores are multiplied together.
- **a** Copy and complete the possibility diagram.
- **b** Use the diagram to find the probability the product of the scores is
  - **i** 8
  - **ii** odd
  - **iii** even
  - **iv** smaller than 7.

Probability 1

Experimental probability

You can use data from a survey or experiment to estimate **experimental probability**. The more trials that are done, then the more reliable the estimate will be.

Relative frequency is an estimate of experimental probability.

$$\text{Relative frequency} = \frac{\text{number of times an event occurs}}{\text{total number of trials}}$$

EXAMPLE

Sami thinks that a particular spinner is biased. To test his theory he spins it 100 times. The results are shown in the table.

Score	1	2	3	4	5
Frequency	13	15	16	42	14

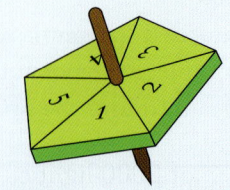

a Calculate the relative frequency for each number on the spinner.
b Compare the relative frequencies with the theoretical probabilities to decide if the spinner is biased.
c How could the experiment be improved?

a

Score	1	2	3	4	5
Frequency	13	15	16	42	14
Relative frequency	$\frac{13}{100}$ = 0.13	$\frac{15}{100}$ = 0.15	$\frac{16}{100}$ = 0.16	$\frac{42}{100}$ = 0.42	$\frac{14}{100}$ = 0.14

b The theoretical probability for obtaining each number is $\frac{1}{5}$ = 0.2.

The figures suggest that the spinner is biased towards landing on the number 4.

c The experiment could be improved by doing more trials.

You may need to work out the **expected number** of times that an event might happen.

$$\text{Expected number} = \text{total number} \times \text{probability}$$

EXAMPLE

A biased coin is tossed 150 times. It lands on heads 51 times.
a Estimate the probability that the coin lands on a head on the next throw.
b Estimate the number of heads you would expect when you toss the coin 650 times.

a P(Head) = $\frac{51}{150} = \frac{17}{50}$

b Expected number of heads = total number × probability

$$= 650 \times \frac{17}{50} = 221$$

UNIT 3

EXERCISE 3.27

1. The probability that a biased coin will land on heads is 0.72
   The coin is tossed 400 times.
   Estimate the number of times the coin will land on a head.

2. The probability that a biased dice will land on a five is 0.29
   The dice is rolled 500 times.
   Estimate the number of times the dice will land on a five.

3. A fair die is rolled 900 times.
   Estimate the number of times the dice will land on a five.

4. The probability that a light bulb is defective is 0.015
   Find the number of defective light bulbs that you would expect to find in a batch of 2000 light bulbs.

5. The probability that a drawing pin lands point down when it is dropped is 0.82.
   The drawing pin is dropped 600 times.
   Estimate the number of times it will land point down.

6. The table shows the results when the three-sided spinner is spun 200 times.

Score	1	2	3
Frequency	51	84	65

   a Copy and complete the table.

Score	1	2	3
Relative frequency			

   b Estimate the number of times the spinner will land on the number 3 if the spinner is spun 720 times.

7. The table shows the results when the circular spinner is spun 200 times.

Colour	Red	Yellow	Green	Blue
Frequency	36	72	54	38

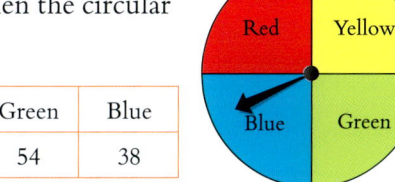

   a Copy and complete the table.

Colour	Red	Yellow	Green	Blue
Relative frequency				

   b Estimate the number of times the spinner will land on each of the four colours if the spinner is spun 900 times.

KEY WORDS
probability
event
outcome
fair
unbiased
biased
possibility diagram
experimental probability
relative frequency
expected number

Probability 1

Unit 3 Examination questions

1 The planet Neptune is 4 496 000 000 kilometers from the Sun.
Write this distance in standard form. [1]

Cambridge IGCSE Mathematics 0580, Paper 2 Q1, June 2006

2 Work out $\dfrac{240^2}{5 \times 10^6}$.
Give your answer in standard form. [2]

Cambridge IGCSE Mathematics 0580, Paper 21 Q6, November 2010

3 **(a)** Write 23 860 000 in standard form. [1]
(b) Calculate $8.29 \div (3.54 \times 10^2)$, giving your answer in standard form. [2]

Examination style question

4 Calculate the value of $5(6 \times 10^3 + 400)$, giving your answer in standard form. [2]

Cambridge IGCSE Mathematics 0580, Paper 21 Q5, June 2010

5 **(a)** There are 10^9 nanoseconds in 1 second.
Find the number of nanoseconds in 5 minutes, giving your answer in standard form. [2]
(b) Solve the equation $5(x + 3 \times 10^6) = 4 \times 10^7$. [2]

Cambridge IGCSE Mathematics 0580, Paper 21 Q14, June 2009

6 A light on a computer comes on for 26 700 microseconds.

One microsecond is 10^{-6} seconds.

Work out the length of time, in seconds, that the light is on

(a) in standard form, [1]
(b) as a decimal. [1]

Cambridge IGCSE Mathematics 0580, Paper 21 Q4, November 2008

7 Solve the simultaneous equations
$$6x + 18y = 57,$$
$$2x - 3y = -8.$$
[3]

Cambridge IGCSE Mathematics 0580, Paper 21 Q9, November 2009

8 Solve the simultaneous equations
$$0.4x + 2y = 10,$$
$$0.3x + 5y = 18.$$ [3]

Cambridge IGCSE Mathematics 0580, Paper 2 Q12, June 2006

9 Solve these simultaneous equations.
$$x + 2y - 18 = 0$$
$$3x - 4y - 4 = 0$$ [3]

Cambridge IGCSE Mathematics 0580, Paper 21 Q10, November 2008

10 Solve the simultaneous equations.
$$\frac{2x + y}{2} = 7$$
$$\frac{2x - y}{2} = 17$$ [3]

Cambridge IGCSE Mathematics 0580, Paper 21 Q13, June 2010

11 Factorise completely.
$$6xy - 2yz$$ [2]

Examination style question

12 Make x the subject of the formula.
$$y = \frac{x}{2} + 7$$ [2]

Examination style question

13 Make d the subject of the formula $c = \frac{5d + 4w}{2w}$. [3]

Cambridge IGCSE Mathematics 0580, Paper 21 Q11, June 2010

14 Rearrange the formula to make y the subject.
$$x + \frac{\sqrt{y}}{9} = 1$$ [3]

Cambridge IGCSE Mathematics 0580, Paper 21 Q9, June 2009

15

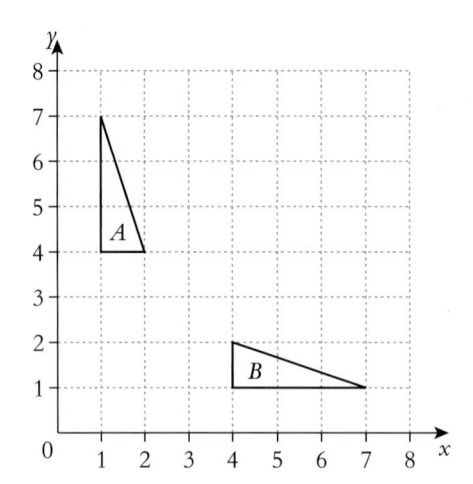

(a) Describe fully the **single** transformation which maps triangle A onto triangle B. [2]
(b) On a copy of the grid, draw the image of triangle A after rotation by 90° clockwise about the point (4, 4). [2]

Cambridge IGCSE Mathematics 0580, Paper 21 Q17, June 2009

16

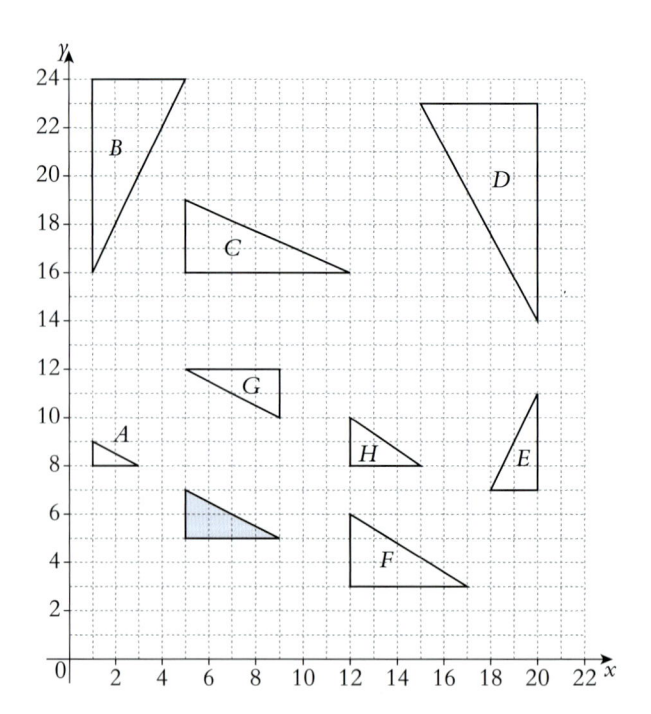

Write down the letters of all the triangles which are

(a) congruent to the shaded triangle. [2]
(b) similar, but not congruent, to the shaded triangle. [2]

Cambridge IGCSE Mathematics 0580, Paper 21 Q18, June 2010

17 (a) Draw and label x and y axes from −6 to 6, using a scale of 1 cm to 1 unit. [1]
    (b) Draw triangle ABC with $A(2, 1)$, $B(3, 3)$ and $C(5, 1)$. [1]
    (c) Draw the reflection of triangle ABC in the line $y = x$. Label this $A_1B_1C_1$. [2]
    (d) Rotate **triangle $A_1B_1C_1$** about $(0, 0)$ through 90° anti-clockwise. Label this $A_2B_2C_2$. [2]
    (e) Describe fully the single transformation which maps triangle ABC onto triangle $A_2B_2C_2$. [2]

Cambridge IGCSE Mathematics 0580, Paper 4 Q2 (a) (b) (c) (d) (e), June 2007

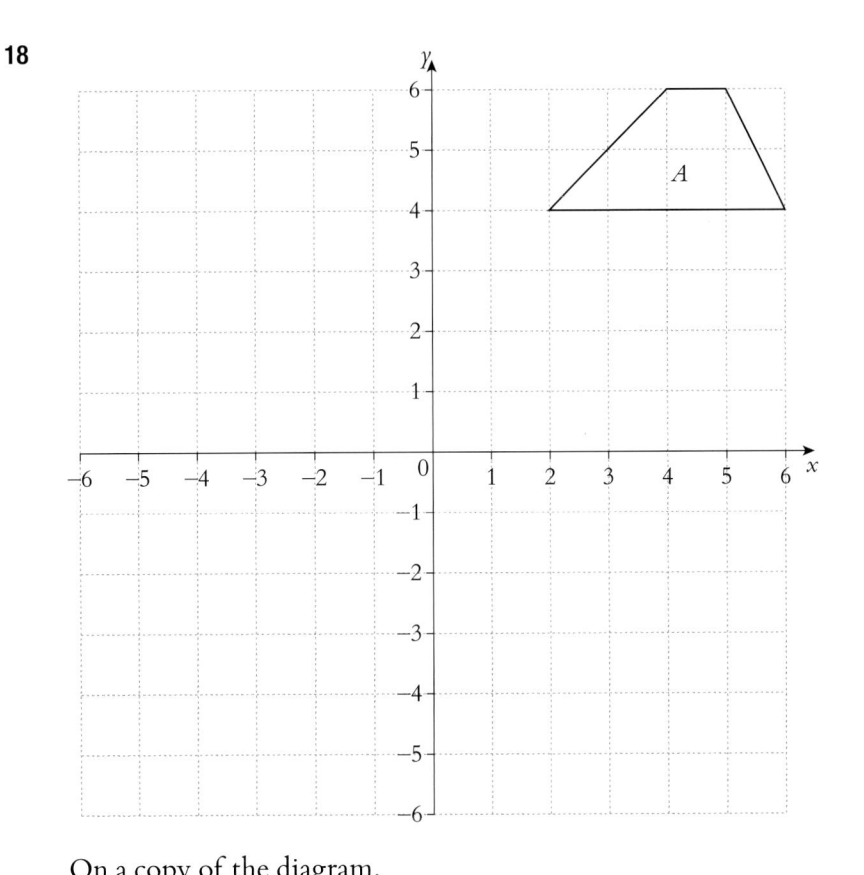

18

On a copy of the diagram.
   (a) (i)   Draw the reflection of shape A in the x-axis. Label the image B. [2]
       (ii) Draw the rotation of **shape B** 180° about $(0, 0)$. Label the image C. [2]
       (iii) Describe fully the **single** transformation that maps shape C onto shape A. [2]
   (b) (i)   Draw the enlargement of shape A, centre $(0, 0)$, scale factor $\frac{1}{2}$. [2]

Examination style question

19 A cylinder has a height of 14 cm and a volume of 850 cm³.
    Calculate the radius of the base of the cylinder. [3]

Examination style question

Percentages 2

THIS SECTION WILL SHOW YOU HOW TO
- Express one quantity as a percentage of another quantity
- Calculate percentage increase and decrease
- Calculate percentage profit and loss

Expressing one quantity as a percentage of a second quantity

To write one quantity as a percentage of a second quantity
- Write the first quantity as a fraction of the second quantity
- Multiply by 100 to change the fraction into a percentage

EXAMPLE

Write 52 as a percentage of 65.

$$\text{Percentage} = \frac{\text{first quantity}}{\text{second quantity}} \times 100\% = \frac{52}{65} \times 100\% = 80\%$$

Percentage increase and percentage decrease

$$\%\text{ increase} = \frac{\text{increase}}{\text{original value}} \times 100\% \quad \%\text{ decrease} = \frac{\text{decrease}}{\text{original value}} \times 100\%$$

EXAMPLE

The population of a village increases from 2400 to 2520. Calculate the percentage increase.

Increase = 2520 − 2400 = 120

$$\%\text{ increase} = \frac{\text{increase}}{\text{original value}} \times 100\%$$

$$= \frac{120}{2400} \times 100\% = 5\%$$

Percentage profit and loss

When a question involves money you are often asked to work out the **percentage profit** or the **percentage loss**.

$$\%\text{ profit} = \frac{\text{profit}}{\text{original cost}} \times 100\% \quad \%\text{ loss} = \frac{\text{loss}}{\text{original cost}} \times 100\%$$

EXAMPLE

A car bought for $5500 is sold for $5280. Calculate the percentage loss.

Loss = 5500 − 5280 = 220

$$\%\text{ loss} = \frac{\text{loss}}{\text{original cost}} \times 100\%$$

$$= \frac{220}{5500} \times 100\% = 4\%$$

EXERCISE 4.1

1. Jarred scores 69 out of 120 in a test. Write this as a percentage.

2. Express $44.40 as a percentage of $120.

3. Eugenio buys a washing machine for $845 and then pays an installation charge of $30. Express the installation charge as a percentage of the washing machine price.

4. In a sale the price of a bag is reduced from $50 to $45.
   Find the percentage decrease.

5. A factory produces 55 000 torches one year and 88 000 torches the next year.
   Find the percentage increase.

6. A set of kitchen scales records a mass of 2.1 kg when the true mass is 2 kg.
   Calculate the percentage error.

7. The number 3.1 is used as an approximation for π.
   Calculate the percentage error.

8. A computer is bought for $1358 and is then sold for $1400.
   Calculate the percentage profit.

9. A clock valued at $600 is sold for $570.
   Calculate the percentage loss.

10. Cecilia's income increases from $70 000 a year to $85 000 a year.
    Calculate the percentage increase.

11. The number of students in a school decreases from 640 to 620.
    Calculate the percentage decrease.

12. In 2005 the population of Malaysia was 2.56×10^7.
    In 2009 the population of Malaysia was 2.75×10^7.
    Calculate the percentage increase in the population.

13. Fernando is a tax payer. He earns $60 000 a year.
    - The first $12 000 of his earnings is tax free.
    - He pays 10% tax on the next $8 000 of his income.
    - He pays 22% tax on the rest of his earnings.

    Express the total amount of tax that he pays as a percentage of the $60 000.

KEY WORDS

percentage profit

percentage loss

Matrix algebra

THIS SECTION WILL SHOW YOU HOW TO
- Describe the size of a matrix
- Find the sum and product of two matrices
- Calculate the determinant of a matrix
- Find the inverse of a matrix

These tables show the sales of TVs and radios over three days in shop A and shop B.

SHOP A	TV	Radio
DAY 1	5	3
DAY 2	7	8
DAY 3	4	5

SHOP B	TV	Radio
DAY 1	9	4
DAY 2	8	5
DAY 3	6	3

These can be combined to give the total sales of TVs and radios for shops A and B.
This can be written in **matrix** form as follows.

TOTAL	TV	Radio
DAY 1	14	7
DAY 2	15	13
DAY 3	10	8

$$\begin{pmatrix} 5 & 3 \\ 7 & 8 \\ 4 & 5 \end{pmatrix} + \begin{pmatrix} 9 & 4 \\ 8 & 5 \\ 6 & 3 \end{pmatrix} = \begin{pmatrix} 14 & 7 \\ 15 & 13 \\ 10 & 8 \end{pmatrix}$$

Two **matrices** can be added together if they are the same size.
The size of a matrix is called the **order** of a matrix.

The matrix $\begin{pmatrix} 5 & 3 \\ 7 & 8 \\ 4 & 5 \end{pmatrix}$ has order 3×2. (3 **rows** and 2 **columns**)

➡ **NOTE:** the number of rows is always written first.

EXAMPLE

Write down the order of this matrix $\begin{pmatrix} 2 & 5 & -1 & 6 & -3 \\ 3 & 0 & 1 & 5 & -1 \end{pmatrix}$

$\begin{pmatrix} 2 & 5 & -1 & 6 & -3 \\ 3 & 0 & 1 & 5 & -1 \end{pmatrix}$

The matrix has 2 rows and 5 columns.
The order is 2×5

EXAMPLE

$A = \begin{pmatrix} 1 & 6 & 2 \\ 0 & -4 & 3 \end{pmatrix}$ $B = \begin{pmatrix} -7 & 10 & -3 \\ 2 & 5 & 7 \end{pmatrix}$.

Find **a** $A + B$ **b** $A - B$

a $A + B = \begin{pmatrix} 1+-7 & 6+10 & 2+-3 \\ 0+2 & -4+5 & 3+7 \end{pmatrix} = \begin{pmatrix} -6 & 16 & -1 \\ 2 & 1 & 10 \end{pmatrix}$

b $A - B = \begin{pmatrix} 1--7 & 6-10 & 2--3 \\ 0-2 & -4-5 & 3-7 \end{pmatrix} = \begin{pmatrix} 8 & -4 & 5 \\ -2 & -9 & -4 \end{pmatrix}$

EXERCISE 4.2

1 Write down the order of these matrices.

a $(3\ 2\ 5\ 1)$ b (2) c $(4\ -1\ 0)$ d $(5\ 5)$

e $\begin{pmatrix} 2 & 5 \\ 1 & -3 \end{pmatrix}$ f $\begin{pmatrix} 6 & -2 & -5 \\ 0 & -1 & 3 \end{pmatrix}$ g $\begin{pmatrix} 5 \\ -4 \end{pmatrix}$ h $\begin{pmatrix} 1 & 7 & -3 & 6 \\ 2 & -5 & 4 & 7 \end{pmatrix}$

i $\begin{pmatrix} 4 \\ 3 \\ -1 \end{pmatrix}$ j $\begin{pmatrix} 2 & 3 & 4 \\ 5 & 6 & 7 \\ 8 & 9 & 10 \end{pmatrix}$ k $\begin{pmatrix} -1 & 2 \\ 4 & 0 \\ 6 & 9 \end{pmatrix}$ l $\begin{pmatrix} 5 & 3 & 5 \\ 2 & 0 & -1 \\ -5 & 7 & 2 \\ 0 & 6 & 0 \end{pmatrix}$

2 $A = \begin{pmatrix} 5 & 4 \\ 3 & 8 \end{pmatrix}$ $B = \begin{pmatrix} 2 & 0 \\ 1 & 5 \end{pmatrix}$ Find a $A + B$ b $A - B$

3 $A = (5\ 3\ 7)$ $B = (3\ 8\ -2)$ Find a $A + B$ b $A - B$

4 $A = \begin{pmatrix} 5 & 1 & 7 \\ -2 & 0 & 6 \end{pmatrix}$ $B = \begin{pmatrix} 7 & 3 & -1 \\ 0 & -2 & 5 \end{pmatrix}$ Find a $A + B$ b $A - B$

5 $A = \begin{pmatrix} 6 & 2 \\ -8 & 0 \\ 2 & -5 \\ 8 & -4 \end{pmatrix}$ $B = \begin{pmatrix} 10 & 5 \\ -8 & 0 \\ -2 & 5 \\ -6 & -3 \end{pmatrix}$ Find a $A + B$ b $A - B$

6 $A = \begin{pmatrix} 5 & 1 \\ -3 & 2 \end{pmatrix}$ $B = \begin{pmatrix} 2 & 9 \\ -7 & 0 \end{pmatrix}$ $C = \begin{pmatrix} 6 & -2 \\ 2 & -4 \end{pmatrix}$ Find

a $A + B$ b $A + C$ c $A - C$ d $B - C$
e $A + B + C$ f $A + B - C$ g $A - (B - C)$ h $A - (B + C)$

7 $A = \begin{pmatrix} 3 & 1 \\ -4 & 2 \\ 7 & 8 \end{pmatrix}$ $B = \begin{pmatrix} 0 & -8 \\ 4 & 0 \\ 7 & 5 \end{pmatrix}$ $C = \begin{pmatrix} -1 & 2 \\ 6 & 4 \\ 3 & -9 \end{pmatrix}$ Find

a $A + B$ b $A + C$ c $B + C$ d $B - C$
e $A - B$ f $A + B - C$ g $A - (B - C)$ h $B - (A + C)$

8 $\begin{pmatrix} x & 6 \\ -2 & y \end{pmatrix} + \begin{pmatrix} -3 & 4 \\ 0 & 7 \end{pmatrix} = \begin{pmatrix} 4 & 10 \\ -2 & -1 \end{pmatrix}$ Find the values of x and y.

9 $\begin{pmatrix} 2 & x \\ -5 & 4 \\ 0 & -3 \end{pmatrix} + \begin{pmatrix} 7 & 1 \\ 8 & y \\ 4 & 0 \end{pmatrix} = \begin{pmatrix} 9 & -6 \\ 3 & -1 \\ 4 & -3 \end{pmatrix}$ Find the values of x and y.

Matrix algebra   171

Multiplying a matrix by a number

If $A = \begin{pmatrix} 1 & 2 \\ 4 & -3 \end{pmatrix}$ then $2A = A + A = \begin{pmatrix} 1 & 2 \\ 4 & -3 \end{pmatrix} + \begin{pmatrix} 1 & 2 \\ 4 & -3 \end{pmatrix} = \begin{pmatrix} 2 & 4 \\ 8 & -6 \end{pmatrix}$

So 2A is the same as multiplying each term in the matrix A by the number 2.

EXAMPLE

$A = \begin{pmatrix} 5 & -1 & 0 \\ -3 & 4 & 2 \end{pmatrix}$ Find 4A.

$4A = \begin{pmatrix} 4 \times 5 & 4 \times -1 & 4 \times 0 \\ 4 \times -3 & 4 \times 4 & 4 \times 2 \end{pmatrix} = \begin{pmatrix} 20 & -4 & 0 \\ -12 & 16 & 8 \end{pmatrix}$

Multiplying a row matrix by a column matrix

If a TV costs $500 and a radio costs $40, you can work out the amount of money shop A takes for the TVs and radios on day 1 as follows

$5 \times 500 + 3 \times 40 = 2500 + 120 = 2620$

You can write this in matrix form as $\begin{pmatrix} 5 & 3 \end{pmatrix} \begin{pmatrix} 500 \\ 40 \end{pmatrix} = (2620)$

Similarly for day 2 $\begin{pmatrix} 7 & 8 \end{pmatrix} \begin{pmatrix} 500 \\ 40 \end{pmatrix} = (3820)$

Similarly for day 3 $\begin{pmatrix} 4 & 5 \end{pmatrix} \begin{pmatrix} 500 \\ 40 \end{pmatrix} = (2200)$

SHOP A	TV	Radio
DAY 1	5	3
DAY 2	7	8
DAY 3	4	5

EXAMPLE

Work out **a** $\begin{pmatrix} 2 & 5 & 3 \end{pmatrix} \begin{pmatrix} 4 \\ 6 \\ 1 \end{pmatrix}$ **b** $\begin{pmatrix} -3 & 0 & 1 & 2 \end{pmatrix} \begin{pmatrix} 4 \\ -2 \\ 1 \\ 5 \end{pmatrix}$

a $\begin{pmatrix} 2 & 5 & 3 \end{pmatrix} \begin{pmatrix} 4 \\ 6 \\ 1 \end{pmatrix} = (2 \times 4 + 5 \times 6 + 3 \times 1) = (41)$

b $\begin{pmatrix} -3 & 0 & 1 & 2 \end{pmatrix} \begin{pmatrix} 4 \\ -2 \\ 1 \\ 5 \end{pmatrix} = (-3 \times 4 + 0 \times -2 + 1 \times 1 + 2 \times 5) = (-1)$

172   UNIT 4

EXERCISE 4.3

1. $A = (3\ -4)$ Find $4A$

2. $A = \begin{pmatrix} 4 & -1 \\ 0 & 2 \end{pmatrix}$ Find $\frac{1}{2}A$

3. $A = \begin{pmatrix} 6 & -3 & 2 \\ 0 & 1 & -1 \end{pmatrix}$ Find $8A$

4. $A = \begin{pmatrix} -2 & 1 & 6 & -5 \\ 1 & 0 & 2 & 1 \end{pmatrix}$ Find $-3A$

5. $A = \begin{pmatrix} 2 & -3 \\ 1 & 1 \end{pmatrix}$ $B = \begin{pmatrix} 5 & 0 \\ -2 & 1 \end{pmatrix}$ $C = \begin{pmatrix} -4 & -6 \\ 0 & 2 \end{pmatrix}$ Find

   a $2A$ b $5C$ c $3B$ d $\frac{1}{2}C$

   e $-3A$ f $2A + 3B$ g $5C - 3B$ h $2A + \frac{1}{2}C$

6. Calculate the following.

   a $(1\ \ 5)\begin{pmatrix} 2 \\ 3 \end{pmatrix}$ b $(-1\ \ 2)\begin{pmatrix} 0 \\ 7 \end{pmatrix}$ c $(6\ \ 6)\begin{pmatrix} -6 \\ -6 \end{pmatrix}$

   d $(2\ \ 5\ \ 3)\begin{pmatrix} 1 \\ 2 \\ 1 \end{pmatrix}$ e $(4\ \ -1\ \ 0)\begin{pmatrix} 2 \\ 0 \\ 1 \end{pmatrix}$ f $(-2\ \ 2\ \ -2)\begin{pmatrix} -1 \\ 1 \\ -1 \end{pmatrix}$

   g $(6\ \ 2\ \ 1\ \ 3)\begin{pmatrix} 5 \\ 1 \\ 3 \\ 3 \end{pmatrix}$ h $(5\ \ -2\ \ 0\ \ 4)\begin{pmatrix} -2 \\ 0 \\ 3 \\ 1 \end{pmatrix}$ i $(0\ \ 2\ \ 6\ \ 2)\begin{pmatrix} 8 \\ 2 \\ 1 \\ 4 \end{pmatrix}$

7. $(x\ \ 3)\begin{pmatrix} 2 \\ -1 \end{pmatrix} = (15)$ Find the value of x.

8. $(1\ \ x\ \ 3)\begin{pmatrix} 2 \\ 4 \\ 2 \end{pmatrix} = (36)$ Find the value of x.

9. $(2\ \ -3\ \ -1\ \ -4)\begin{pmatrix} 1 \\ -6 \\ x \\ 3 \end{pmatrix} = (10)$ Find the value of x.

HINT

Form an equation in x and then solve.

Matrix algebra

Multiplying matrices

In the last section you saw that when a TV costs $500 and a radio costs $40

DAY 1: $(5 \quad 3)\begin{pmatrix} 500 \\ 40 \end{pmatrix} = (2620)$ DAY 2: $(7 \quad 8)\begin{pmatrix} 500 \\ 40 \end{pmatrix} = (3820)$

SHOP A	TV	Radio
DAY 1	5	3
DAY 2	7	8
DAY 3	4	5

DAY 3: $(4 \quad 5)\begin{pmatrix} 500 \\ 40 \end{pmatrix} = (2200)$

These can be combined into just one matrix equation.

$$\begin{pmatrix} 5 & 3 \\ 7 & 8 \\ 4 & 5 \end{pmatrix} \begin{pmatrix} 500 \\ 40 \end{pmatrix} = \begin{pmatrix} 2620 \\ 3820 \\ 2200 \end{pmatrix}$$

order 3×2 2×1 3×1

same

It is only possible to multiply two matrices when the number of columns in the first matrix is the same as the number of rows in the second matrix.

In general   First matrix   Second matrix
Order $a \times b$ $c \times d$

You can multiply the matrices if $b = c$.
The product of the two matrices will be of order $a \times d$.

EXAMPLE

Work out $\begin{pmatrix} 2 & 1 \\ 3 & 4 \end{pmatrix} \begin{pmatrix} 5 \\ 8 \end{pmatrix}$

First consider the order of the matrices: 2×2 2×1

The numbers in the middle are the same so it is possible to multiply the matrices. The answer will be a 2×1 matrix.

$\begin{pmatrix} 2 & 1 \\ 3 & 4 \end{pmatrix}\begin{pmatrix} 5 \\ 8 \end{pmatrix} = \begin{pmatrix} \star \\ \star \end{pmatrix}$ You must now multiply each row in the first matrix with the column in the second matrix to find the missing numbers.

$2 \times 5 + 1 \times 8 = 18$
$3 \times 5 + 4 \times 8 = 47$

$\begin{pmatrix} 2 & 1 \\ 3 & 4 \end{pmatrix}\begin{pmatrix} 5 \\ 8 \end{pmatrix} = \begin{pmatrix} 18 \\ 47 \end{pmatrix}$

EXAMPLE

Work out $\begin{pmatrix} 2 & 1 & 4 \\ 1 & 0 & 3 \\ 5 & 3 & 2 \end{pmatrix} \begin{pmatrix} 3 & 4 \\ 1 & 2 \\ 2 & 1 \end{pmatrix}$

First consider the order of the matrices: 3 × 3   3 × 2   The answer will be a 3 × 2 matrix.

$\begin{pmatrix} 2 & 1 & 4 \\ 1 & 0 & 3 \\ 5 & 3 & 2 \end{pmatrix} \begin{pmatrix} 3 & 4 \\ 1 & 2 \\ 2 & 1 \end{pmatrix} = \begin{pmatrix} \star & \star \\ \star & \star \\ \star & \star \end{pmatrix}$

You must now multiply each row in the first matrix with each column in the second matrix to find the missing numbers.

$2 \times 3 + 1 \times 1 + 4 \times 2 = 15$ $2 \times 4 + 1 \times 2 + 4 \times 1 = 14$
$1 \times 3 + 0 \times 1 + 3 \times 2 = 9$ $1 \times 4 + 0 \times 2 + 3 \times 1 = 7$
$5 \times 3 + 3 \times 1 + 2 \times 2 = 22$ $5 \times 4 + 3 \times 2 + 2 \times 1 = 28$

$\begin{pmatrix} 2 & 1 & 4 \\ 1 & 0 & 3 \\ 5 & 3 & 2 \end{pmatrix} \begin{pmatrix} 3 & 4 \\ 1 & 2 \\ 2 & 1 \end{pmatrix} = \begin{pmatrix} 15 & 14 \\ 9 & 7 \\ 22 & 28 \end{pmatrix}$

EXERCISE 4.4

1   Multiply these matrices.

a $\begin{pmatrix} 1 & 4 \\ 5 & 2 \end{pmatrix} \begin{pmatrix} 3 & 1 \\ 0 & 4 \end{pmatrix}$

b $\begin{pmatrix} 2 & 0 \\ 1 & 1 \end{pmatrix} \begin{pmatrix} 5 & 1 & 2 \\ 0 & 4 & 1 \end{pmatrix}$

c $\begin{pmatrix} 6 & 2 \\ 1 & 0 \end{pmatrix} \begin{pmatrix} 5 \\ 3 \end{pmatrix}$

d $\begin{pmatrix} 6 & 1 \\ 0 & 8 \end{pmatrix} \begin{pmatrix} 5 & 3 \\ 2 & 1 \end{pmatrix}$

e $\begin{pmatrix} 5 & 1 & 2 \\ 4 & 0 & 1 \end{pmatrix} \begin{pmatrix} 2 & 1 \\ 0 & -1 \\ 3 & 0 \end{pmatrix}$

f $\begin{pmatrix} 1 & 6 \end{pmatrix} \begin{pmatrix} 7 \\ 2 \end{pmatrix}$

g $\begin{pmatrix} 5 & 1 & 2 & 3 \\ 0 & 1 & -2 & 4 \end{pmatrix} \begin{pmatrix} 5 & -1 \\ 0 & 2 \\ 1 & 0 \\ 3 & 1 \end{pmatrix}$

h $\begin{pmatrix} 5 & 1 \\ 1 & 2 \\ -3 & 0 \\ 6 & 4 \end{pmatrix} \begin{pmatrix} 3 & 0 & 1 & 2 \\ 4 & 3 & 2 & 1 \end{pmatrix}$

2   $A = \begin{pmatrix} 2 & 8 \\ -5 & 3 \\ 2 & -1 \end{pmatrix}$ and $B = \begin{pmatrix} 6 & 1 & 7 & 3 \\ 2 & 0 & 5 & 4 \end{pmatrix}$

a   The order of the matrix AB is $x \times y$. Write down the values of x and y.
b   Explain why it is impossible to calculate the matrix BA.

3   Work out $\begin{pmatrix} 4 & 2 & -1 \\ 3 & 0 & 5 \\ 2 & -6 & 4 \end{pmatrix} \begin{pmatrix} 3 \\ -2 \\ 1 \end{pmatrix}$

Matrix algebra

2 × 2 matrices

$$\begin{pmatrix} 0 & 0 \\ 0 & 0 \end{pmatrix} \begin{pmatrix} 2 & 3 \\ -4 & 1 \end{pmatrix} = \begin{pmatrix} 0 & 0 \\ 0 & 0 \end{pmatrix} \text{ and } \begin{pmatrix} 2 & 3 \\ -4 & 1 \end{pmatrix} \begin{pmatrix} 0 & 0 \\ 0 & 0 \end{pmatrix} = \begin{pmatrix} 0 & 0 \\ 0 & 0 \end{pmatrix}$$

The matrix $\begin{pmatrix} 0 & 0 \\ 0 & 0 \end{pmatrix}$ is called the **zero** matrix.

$$\begin{pmatrix} 1 & 0 \\ 0 & 1 \end{pmatrix} \begin{pmatrix} 2 & 3 \\ -4 & 1 \end{pmatrix} = \begin{pmatrix} 2 & 3 \\ -4 & 1 \end{pmatrix} \text{ and } \begin{pmatrix} 2 & 3 \\ -4 & 1 \end{pmatrix} \begin{pmatrix} 1 & 0 \\ 0 & 1 \end{pmatrix} = \begin{pmatrix} 2 & 3 \\ -4 & 1 \end{pmatrix}$$

The matrix $\begin{pmatrix} 1 & 0 \\ 0 & 1 \end{pmatrix}$ is called the **identity** matrix for multiplication and is denoted by the letter I.

When you multiply a matrix by the identity matrix (I) the numbers in the matrix stay the same.

IA = A and AI = A

If $A = \begin{pmatrix} a & b \\ c & d \end{pmatrix}$, the **determinant** of matrix A (denoted by $|A|$) is defined as follows.

$|A|$ = ad – bc

The steps for finding the determinant of the matrix $A = \begin{pmatrix} 6 & 7 \\ 2 & 3 \end{pmatrix}$ are:

STEP 1: $\begin{pmatrix} 6 & 7 \\ 2 & 3 \end{pmatrix}$ Multiply the numbers on the leading diagonal. 6 × 3

STEP 2:   Multiply the numbers on the other diagonal. 2 × 7

STEP 3:   Subtract $|A|$ = 6 × 3 − 2 × 7
$|A|$ = 18 − 14
$|A|$ = 4

EXAMPLE

$A = \begin{pmatrix} -2 & 4 \\ -2 & 3 \end{pmatrix}$ Find $|A|$.

$\begin{pmatrix} -2 & 4 \\ -2 & 3 \end{pmatrix}$ $|A|$ = −2 × 3 − 4 × −2
$|A|$ = −6 − −8
$|A|$ = 2

EXERCISE 4.5

1 Work out the determinant of each of these matrices.

a $\begin{pmatrix} 1 & 1 \\ 3 & 4 \end{pmatrix}$
b $\begin{pmatrix} 5 & 3 \\ 3 & 2 \end{pmatrix}$
c $\begin{pmatrix} 9 & -2 \\ -4 & 1 \end{pmatrix}$
d $\begin{pmatrix} 3 & -2 \\ -2 & -1 \end{pmatrix}$

e $\begin{pmatrix} 4 & 11 \\ 3 & 8 \end{pmatrix}$
f $\begin{pmatrix} 0.5 & 0.5 \\ 6 & 10 \end{pmatrix}$
g $\begin{pmatrix} 2 & 5 \\ -2 & -3 \end{pmatrix}$
h $\begin{pmatrix} 2 & 4 \\ 3 & 6 \end{pmatrix}$

2 The determinant of the matrix $\begin{pmatrix} 2 & -1 \\ 3 & x \end{pmatrix}$ is 10. Find the value of x.

3 The determinant of the matrix $\begin{pmatrix} 2 & -4 \\ x & 3 \end{pmatrix}$ is 2. Find the value of x.

> **HINT**
>
> AB means $\begin{pmatrix} 3 & 0 \\ 2 & -1 \end{pmatrix}\begin{pmatrix} -1 & 4 \\ 0 & 2 \end{pmatrix}$
>
> BA means $\begin{pmatrix} -1 & 4 \\ 0 & 2 \end{pmatrix}\begin{pmatrix} 3 & 0 \\ 2 & -1 \end{pmatrix}$

4 $A = \begin{pmatrix} 3 & 0 \\ 2 & -1 \end{pmatrix}$ $B = \begin{pmatrix} -1 & 4 \\ 0 & 2 \end{pmatrix}$

Find   a AB   b BA

5 $A = \begin{pmatrix} 5 & 1 \\ 1 & -2 \end{pmatrix}$ $B = \begin{pmatrix} -3 & 1 \\ 2 & 3 \end{pmatrix}$

Find   a AB   b BA   c A^2 d B^2

> **HINT**
>
> A^2 means $\begin{pmatrix} 5 & 1 \\ 1 & -2 \end{pmatrix}\begin{pmatrix} 5 & 1 \\ 1 & -2 \end{pmatrix}$

6 $A = \begin{pmatrix} 2 & -1 \\ -3 & 2 \end{pmatrix}$ and $B = \begin{pmatrix} 2 & 3 \\ 1 & 4 \end{pmatrix}$ Find

a 3A   b 5B   c A − B   d 3A + 5B
e AB   f BA   g A^2 h B^2
i |A| j |B|

7 $A = \begin{pmatrix} 1 & -1 \\ 1 & 2 \end{pmatrix}$ and $B = \begin{pmatrix} 5 & -3 \\ 1 & -1 \end{pmatrix}$ Find

a 2A   b 4B   c 2A − 4B   d AB
e BA   f A^2 g B^2 h $(A - B)^2$
i |A| j |B|

8 $A = \begin{pmatrix} 5 & 3 \\ 2 & 1 \end{pmatrix}$. Find the 2 × 2 matrix B such that $A + B = \begin{pmatrix} 0 & 0 \\ 0 & 0 \end{pmatrix}$.

Matrix algebra

Inverse matrices

If AB = I, where I is the identity matrix then B is called the **inverse** of matrix A.
The inverse of matrix A is denoted by A^{-1}.
The inverse matrix is calculated using the following formula.

$$\text{If } A = \begin{pmatrix} a & b \\ c & d \end{pmatrix} \text{ then } A^{-1} = \frac{1}{|A|}\begin{pmatrix} d & -b \\ -c & a \end{pmatrix}$$

The steps for finding the inverse of the matrix $A = \begin{pmatrix} 4 & 6 \\ 1 & 2 \end{pmatrix}$ are:

STEP 1: Calculate the determinant. $|A| = 4 \times 2 - 1 \times 6 = 8 - 6 = 2$

STEP 2: Swap the numbers on the leading diagonal to give $\begin{pmatrix} 2 & 6 \\ 1 & 4 \end{pmatrix}$

STEP 3: Change the signs of the numbers on the other diagonal to give $\begin{pmatrix} 2 & -6 \\ -1 & 4 \end{pmatrix}$

STEP 4: Divide by the determinant. $A^{-1} = \begin{pmatrix} 1 & -3 \\ -\frac{1}{2} & 2 \end{pmatrix}$ or $A^{-1} = \frac{1}{2}\begin{pmatrix} 2 & -6 \\ -1 & 4 \end{pmatrix}$

EXAMPLE

Find the inverse of $A = \begin{pmatrix} 1 & 7 \\ -1 & -4 \end{pmatrix}$

$A = \begin{pmatrix} 1 & 7 \\ -1 & -4 \end{pmatrix}$

$|A| = 1 \times -4 - 7 \times -1 = -4 - -7 = 3$

$A^{-1} = \frac{1}{3}\begin{pmatrix} -4 & -7 \\ 1 & 1 \end{pmatrix}$

$A^{-1} = \frac{1}{3}\begin{pmatrix} -4 & -7 \\ 1 & 1 \end{pmatrix}$ or $A^{-1} = \begin{pmatrix} -1\frac{1}{3} & -2\frac{1}{3} \\ \frac{1}{3} & \frac{1}{3} \end{pmatrix}$

- swap the numbers on the leading diagonal
- change the signs of the numbers on the other diagonal
- divide by the determinant

➡ **CHECK:** $A^{-1}A = \frac{1}{3}\begin{pmatrix} -4 & -7 \\ 1 & 1 \end{pmatrix}\begin{pmatrix} 1 & 7 \\ -1 & -4 \end{pmatrix} = \frac{1}{3}\begin{pmatrix} 3 & 0 \\ 0 & 3 \end{pmatrix} = \begin{pmatrix} 1 & 0 \\ 0 & 1 \end{pmatrix}$ ✓

A matrix has no inverse when the determinant is zero because you cannot divide by zero.
A matrix with no inverse is called a **singular** matrix.
A matrix with an inverse is called a **non-singular** matrix.

EXERCISE 4.6

1 Find the inverse matrix for each of these matrices.

a $\begin{pmatrix} 2 & 5 \\ 1 & 3 \end{pmatrix}$
b $\begin{pmatrix} 3 & 4 \\ 2 & 3 \end{pmatrix}$
c $\begin{pmatrix} 5 & 2 \\ 4 & 2 \end{pmatrix}$
d $\begin{pmatrix} -1 & 1 \\ 1 & -2 \end{pmatrix}$

e $\begin{pmatrix} 3 & 2 \\ -2 & -1 \end{pmatrix}$
f $\begin{pmatrix} 6 & -1 \\ -4 & 1 \end{pmatrix}$
g $\begin{pmatrix} 2 & 4 \\ -3 & -7 \end{pmatrix}$
h $\begin{pmatrix} 1 & 1 \\ -3 & -4 \end{pmatrix}$

i $\begin{pmatrix} -9 & -1 \\ -7 & -1 \end{pmatrix}$
j $\begin{pmatrix} 3 & -5 \\ -2 & 4 \end{pmatrix}$
k $\begin{pmatrix} 4 & -1 \\ -1 & 0.5 \end{pmatrix}$
l $\begin{pmatrix} 1 & -2 \\ 1 & -1 \end{pmatrix}$

2 Explain why $\begin{pmatrix} 4 & -8 \\ -2 & 4 \end{pmatrix}$ does not have an inverse.

3 The matrix $\begin{pmatrix} x & 5 \\ -2 & 2 \end{pmatrix}$ has no inverse. Find the value of x.

4 The matrix $\begin{pmatrix} 2 & x \\ 4 & 3 \end{pmatrix}$ has no inverse. Find the value of x.

5 $X = \begin{pmatrix} 7 & 3 \\ 2 & 1 \end{pmatrix}$ and $XY = I$. Find the matrix Y.

6 $Y = \begin{pmatrix} -2 & 1 \\ 8 & 3 \end{pmatrix}$ and $XY = I$. Find the matrix X.

7 $A = \begin{pmatrix} 5 & 3 \\ 2 & 1 \end{pmatrix}$ Find the 2 × 2 matrix C such that $AC = \begin{pmatrix} 1 & 0 \\ 0 & 1 \end{pmatrix}$.

8 $A = \begin{pmatrix} 3 & 4 \\ 1 & 1 \end{pmatrix}$ Find

a 6A
b A^2
c $A^2 + 2A$
d $(3A)^2$
e A^3
f $|A|$
g A^{-1}
h $A^{-1}A$

9 $XY = \begin{pmatrix} 7 & -1 \\ 11 & -5 \end{pmatrix}$ and $X = \begin{pmatrix} 2 & -1 \\ 4 & -1 \end{pmatrix}$ Find the matrix Y.

10 $XY = \begin{pmatrix} 12 & 1 \\ 14 & -3 \end{pmatrix}$ and $Y = \begin{pmatrix} 2 & 1 \\ 5 & 0 \end{pmatrix}$ Find the matrix X.

KEY WORDS
matrix
matrices
order
rows
columns
zero matrix
identity matrix
inverse matrix
singular matrix
non-singular matrix

Matrix algebra   179

Expanding double brackets

THIS SECTION WILL SHOW YOU HOW TO
- Expand and simplify expressions with two brackets

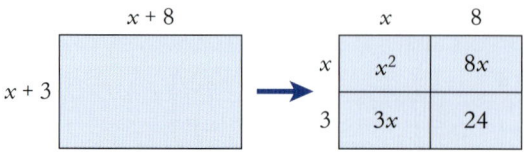

The area of a rectangle that has sides of length $x + 3$ and length $x + 8$ can be found by splitting the rectangle into four parts.

Area $= (x + 3)(x + 8)$ 　　Area $= x^2 + 8x + 3x + 24$

So 　　$(x + 3)(x + 8) = x^2 + 8x + 3x + 24$
　　　　　　　　　　　$= x^2 + 11x + 24$

EXAMPLE

Use a diagram to expand $(x + 2)(x + 7)$

$(x + 2)(x + 7) = x^2 + 7x + 2x + 14$
　　　　　　　　$= x^2 + 9x + 14$

EXERCISE 4.7

1. Copy and complete the following.

   $(x + 3)(x + 10) = \ldots\ldots + 10x + \ldots\ldots + \ldots\ldots$
   　　　　　　　$=$

Use diagrams to help you expand these brackets.

2. $(x + 3)(x + 4)$
3. $(x + 7)(x + 3)$
4. $(x + 4)(x + 2)$
5. $(x + 1)(x + 5)$
6. $(x + 8)(x + 7)$
7. $(x + 2)(x + 6)$
8. $(x + 6)(x + 7)$
9. $(x + 4)(x + 4)$
10. $(x + 15)(x + 4)$
11. $(x + 6)(x + 12)$
12. $(2x + 3)(x + 5)$
13. $(3x + 2)(x + 5)$

First – Outside – Inside – Last

You can expand double brackets without drawing a diagram.

> To expand double brackets you multiply each term in the first bracket by each term in the second bracket.

The mnemonic **FOIL** will help you remember what to do when expanding double brackets.

F	**f**irst
O	**o**utside
I	**i**nside
L	**l**ast

$(x + 2)(x + 7) = x^2 + 7x + 2x + 14 = x^2 + 9x + 14$

EXAMPLE

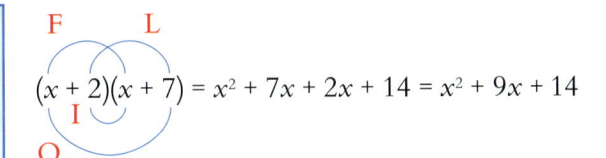

Expand and simplify **a** $(x + 8)(x + 7)$ **b** $(x + 4)(x - 5)$ **c** $(x - 3)(x - 10)$

a $(x + 8)(x + 7)$ F $x \times x = x^2$ O $x \times 7 = 7x$ I $8 \times x = 8x$ L $8 \times 7 = 56$
  $= x^2 + 7x + 8x + 56$
  $= x^2 + 15x + 56$

b $(x + 4)(x - 5)$ F $x \times x = x^2$ O $x \times -5 = -5x$ I $4 \times x = 4x$ L $4 \times -5 = -20$
  $= x^2 - 5x + 4x - 20$
  $= x^2 - x - 20$

c $(x - 3)(x - 10)$ F $x \times x = x^2$ O $x \times -10 = -10x$ I $-3 \times x = -3x$ L $-3 \times -10 = 30$
  $= x^2 - 10x - 3x + 30$
  $= x^2 - 13x + 30$

➡ **NOTE:** remember to look out for double negatives: $-3 \times -10 = +30$

You need to be careful when squaring a bracket.
$(x + 6)^2$ means $(x + 6)(x + 6)$
  $= x^2 + 6x + 6x + 36$
  $= x^2 + 12x + 36$

EXAMPLE

Expand and simplify **a** $(x - 7)^2$ **b** $(3x + 4)(2x - 5)$

a $(x - 7)^2$
  $= (x - 7)(x - 7)$ F $x \times x = x^2$ O $x \times -7 = -7x$ I $-7 \times x = -7x$ L $-7 \times -7 = 49$
  $= x^2 - 7x - 7x + 49$
  $= x^2 - 14x + 49$

b $(3x + 4)(2x - 5)$ F $3x \times 2x = 6x^2$ O $3x \times -5 = -15x$ I $4 \times 2x = 8x$ L $4 \times -5 = -20$
  $= 6x^2 - 15x + 8x - 20$
  $= 6x^2 - 7x - 20$

Expanding double brackets   181

EXERCISE 4.8

1. Rose has made mistakes in her expanding brackets homework.
   Copy out each question and correct her mistakes.

   a $(x + 3)(x + 4) = x^2 + 4x + 3x + 7 = x^2 + 7x + 7$
   b $(x + 4)(x - 6) = x^2 - 6x + 4x - 2 = x^2 - 10x - 2$
   c $(x - 5)(x - 4) = x^2 - 4x - 5x - 20 = x^2 - 9x - 20$
   d $(x + 4)^2 = x^2 + 16$

Expand and simplify:

2. a $(x + 1)(x + 2)$ b $(x + 5)(x + 7)$ c $(x + 3)(x + 8)$ d $(x + 4)(x + 2)$
   e $(x + 4)(x + 5)$ f $(x + 3)(x + 1)$ g $(x + 9)(x + 3)$ h $(x + 5)(x + 6)$
   i $(x + 4)(x + 3)$ j $(x + 2)(x + 3)$ k $(x + 8)(x + 2)$ l $(1 + x)(4 + x)$

3. a $(x - 1)(x + 2)$ b $(x - 2)(x + 6)$ c $(x + 6)(x - 3)$ d $(x + 4)(x - 1)$
   e $(x - 9)(x + 7)$ f $(x - 2)(x + 8)$ g $(x - 8)(x + 4)$ h $(x - 7)(x + 2)$
   i $(x + 2)(x - 5)$ j $(x - 3)(x + 8)$ k $(x - 3)(x + 1)$ l $(9 + x)(4 - x)$

4. a $(x + 3)(x - 3)$ b $(x + 8)(x - 8)$ c $(x - 6)(x + 6)$ d $(x + 10)(x - 10)$
   e $(x - 4)(x + 4)$ f $(x - 9)(x + 9)$ g $(x - 1)(x + 1)$ h $(12 - x)(12 + x)$

5. a $(x - 5)(x - 2)$ b $(x - 4)(x - 1)$ c $(x - 3)(x - 8)$ d $(x - 6)(x - 2)$
   e $(x - 7)(x - 7)$ f $(x - 6)(x - 3)$ g $(x - 5)(x - 4)$ h $(x - 4)(x - 7)$
   i $(x - 9)(x - 4)$ j $(x - 1)(x - 1)$ k $(x - 5)(x - 1)$ l $(2 - x)(2 - x)$

6. a $(x + 3)^2$ b $(y + 5)^2$ c $(x - 4)^2$ d $(x + 7)^2$
   e $(x + 8)^2$ f $(a - 10)^2$ g $(6 - x)^2$ h $(2 + x)^2$

7. a $(2y + 1)(y - 3)$ b $(4x - 3)(x + 2)$ c $(5x + 2)(x + 4)$ d $(7y - 2)(y + 5)$
   e $(4y + 3)(2y - 7)$ f $(2x - 11)(5x - 3)$ g $(5x - 6)(3x - 4)$ h $(9y - 2)(3y - 10)$
   i $(8y - 5)(3y - 4)$ j $(3a - 7)(2a + 3)$ k $(1 - 3a)(3 - 4a)$ l $3(5x - 3)(2x + 1)$

8. a $(2x + 1)(2x - 1)$ b $(3x + 2)(3x - 2)$ c $(5x + 7)(5x - 7)$ d $(5x - 2)(5x + 2)$

9. a $(2x - 1)^2$ b $(5x + 2)^2$ c $(3x + 2y)^2$ d $(8 - 3y)^2$

10. a $(x + 1)(x - 9) + (x + 4)(x + 7)$
    b $(2x + 3)(x - 1) + (3x - 5)(2x + 7)$
    c $(2x + 1)(2x - 5) - (3x - 1)(x + 2)$
    d $(4x - 3)(3x + 2) + (2x - 9)(x - 1)$
    e $(3x + 1)^2 + (2x - 5)^2$
    f $(2x - 3)^2 + (2x + 1)^2$
    g $(6x + 5)^2 - (2x - 3)^2$
    h $2(3x + 1)^2 - 3(2x + 5)^2$

EXERCISE 4.9

1. Solve these equations.
   a. $(x + 3)(x + 4) = (x - 2)(x + 5)$
   b. $(3x + 2)(x + 4) = 3x(x + 6)$
   c. $(x + 2)^2 = (x - 5)(x + 2)$

2. Find x for each of these right angled triangles.

   a. [Triangle with legs 12 and x, hypotenuse $x + 2$]

   b. [Triangle with legs $x - 2$ and 20, hypotenuse $x + 6$]

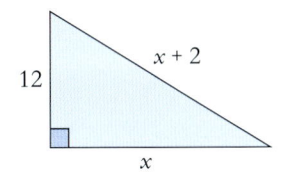

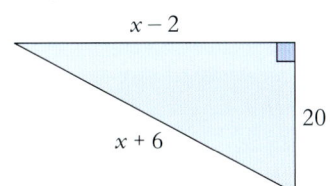

3. | A $x - 2$ | B $x + 4$ | C $x + 2$ | D $x - 3$ | E $x - 4$ |

   Which two cards when multiplied together give:
   a. $x^2 + 2x - 8$ b. $x^2 - 7x + 12$ c. $x^2 - 16$ d. $x^2 + 6x + 8$
   e. $x^2 + x - 12$ f. $x^2 - 4$ g. $x^2 - 5x + 6$ h. $x^2 - 6x + 8$

4. These rectangles have the same area. Find the value of x.

   [Rectangle with sides $x + 5$ and $x - 2$; rectangle with sides $x + 2$ and x]

5. Write down and simplify an expression for
   a. the volume of the cuboid
   b. the surface area of the cuboid.

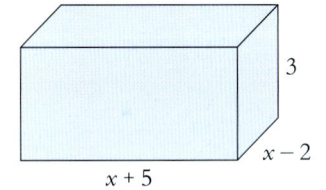

   [Cuboid with dimensions $x + 5$, $x - 2$, and 3]

6. Show that $(n - 2)^2 + (n - 1)^2 + n^2 + (n + 1)^2 + (n + 2)^2$ simplifies to $5(n^2 + 2)$.

7. a. Write down an expression for the area of the white square in terms of c.
   b. Write down an expression for the area of the large square in terms of a and b.
   c. Write down an expression for the area of all four of the shaded right-angled triangles in terms of a and b.
   d. Use your answers from parts a, b and c to prove that $a^2 + b^2 = c^2$.

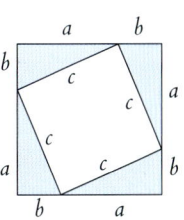

KEY WORDS
FOIL (first, outside, inside, last)

Quadratic graphs

THIS SECTION WILL SHOW YOU HOW TO
- Recognise quadratic graphs
- Draw quadratic graphs
- Use quadratic graphs in practical situations

$y = ax^2 + bx + c$ is called a **quadratic function**.
The simplest quadratic function is $y = x^2$.

The graph of $y = x^2$ is a smooth ∪-shaped curve.
The graph is symmetrical about the y-axis and it passes through the point (0, 0)

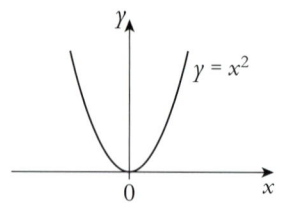

EXAMPLE

a Complete the table of values for $y = x^2 - 5x + 4$
b Draw the graph of $y = x^2 - 5x + 4$
c Name the line of symmetry of the curve.
d Use your graph to find the values of x when $y = 2$

x	0	1	2	3	4	5
y		0		-2		4

a

x	0	1	2	3	4	5	2.5
y	4	0	-2	-2	0	4	-2.25

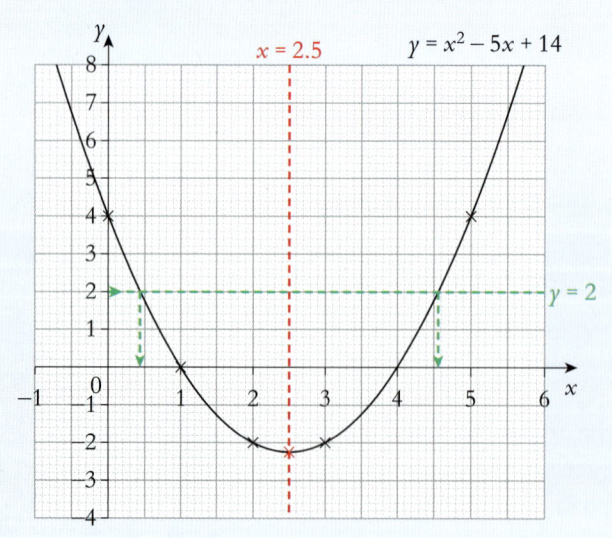

The extra column in the table is needed to find where the curve turns

When $x = 0$, $y = (0)^2 - (5 \times 0) + 4 = 0 - 0 + 4 = 4$
When $x = 2$, $y = (2)^2 - (5 \times 2) + 4 = 4 - 10 + 4 = -2$
When $x = 4$, $y = (4)^2 - (5 \times 4) + 4 = 16 - 20 + 4 = 0$
When $x = 2.5$, $y = (2.5)^2 - (5 \times 2.5) + 4$
$= 6.25 - 12.5 + 4 = -2.25$

b

The points should be marked with a cross.

The points must be joined with a smooth curve.
They must not be joined with straight lines.

The axes and the curve should be labelled.

c The line of symmetry is $x = 2.5$
d When $y = 2$, $x \approx 0.4$ or $x \approx 4.6$

The symbol '$\approx$' means 'is approximately equal to'.

184 UNIT 4

EXERCISE 4.10

1 Copy and complete the table for each of these quadratic functions.
Draw the graph of each function.
State the equation of the line of symmetry for each graph.

a $y = x^2 + 1$

x	-3	-2	-1	0	1	2	3
y						5	

b $y = x^2 - 2$

x	-3	-2	-1	0	1	2	3
y	7						

c $y = -x^2$

x	-3	-2	-1	0	1	2	3
y						-4	

d $y = 2x^2 - 5$

x	-3	-2	-1	0	1	2	3
y			-3				

e $y = \frac{1}{2}x^2 - 2$

x	-3	-2	-1	0	1	2	3
y		0					

f $y = x^2 + 2x$

x	-4	-3	-2	-1	0	1	2
y		3					

g $y = x^2 - 3x$

x	-1	0	1	2	3	4
y				-2		

h $y = 3x - x^2$

x	-1	0	1	2	3	4
y				2		

2 a Copy and complete the table of values for $y = x^2 - 4x + 3$

x	-1	0	1	2	3	4	5
y	8						

b Draw the graph of $y = x^2 - 4x + 3$
c Use your graph to find the values of x when $y = 7$

3 a Copy and complete the table of values for $y = x^2 - x - 2$

x	-2	-1	0	1	2	3
y				-2		

b Draw the graph of $y = x^2 - x - 2$
c Use your graph to find the values of x when $y = 2$

4 a Copy and complete the table of values for $y = 2 + x - x^2$

x	-2	-1	0	1	2	3
y	-4					

b Draw the graph of $y = 2 + x - x^2$
c Use your graph to find the values of x when $y = 1$

5 a Copy and complete the table of values for $y = 2x^2 - 5x + 1$

x	-1	0	1	2	3	4
y				-1		

b Draw the graph of $y = 2x^2 - 5x + 1$
c Use your graph to find the values of x when $y = 0$.

HINT

$2x^2$ means $2 \times x^2$
You must square the number and then multiply by 2.

Quadratic graphs

In the last exercise you should have found that the graphs of quadratic functions are always ∪-shaped or ∩-shaped.

$y = ax^2 + bx + c$
- if a is positive then the graph is ∪-shaped
- if a is negative then the graph is ∩-shaped
- c is the intercept on the y-axis

Gradient of a curve

The gradient of a curve changes as you move along it.
To estimate the gradient of a curve at a point you draw a **tangent** to the curve at the point.
(A tangent is a straight line, which just touches the curve at the point.)
You then work out the gradient of the tangent.

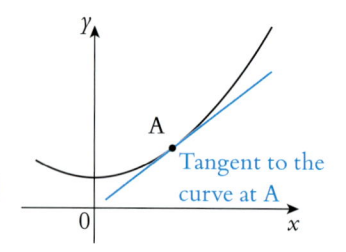

EXAMPLE

By drawing tangents to the graph, estimate the gradient of the curve when
a $x = 5$ b $x = 1$

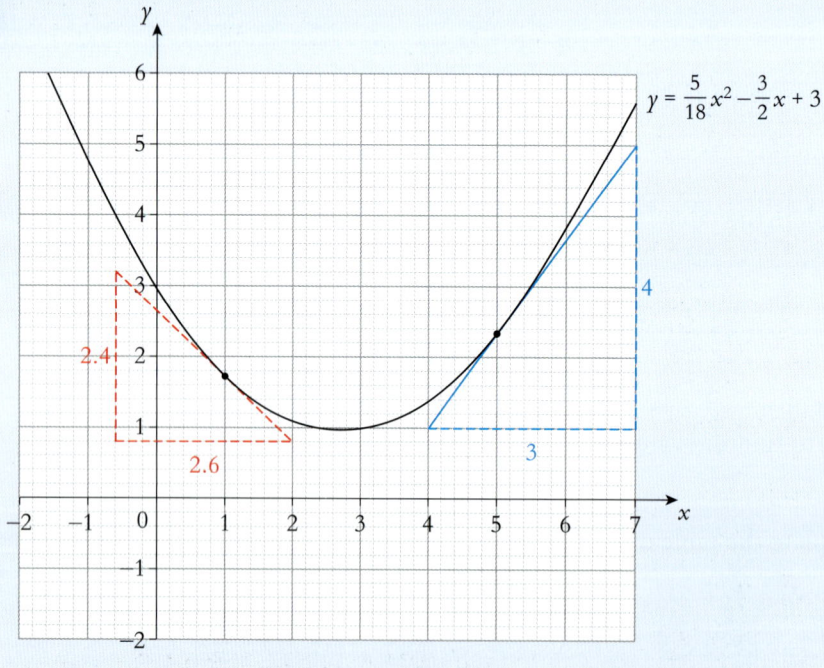

$y = \frac{5}{18}x^2 - \frac{3}{2}x + 3$

a   When $x = 5$, the gradient of the graph $\approx \frac{4}{3} = 1.33$ (3 s.f.)

b   When $x = 1$, the gradient of the graph $\approx -\frac{2.4}{2.6} = -0.923$ (3 s.f.)

EXERCISE 4.11

1 Match the quadratic functions to the correct graph.

a   b 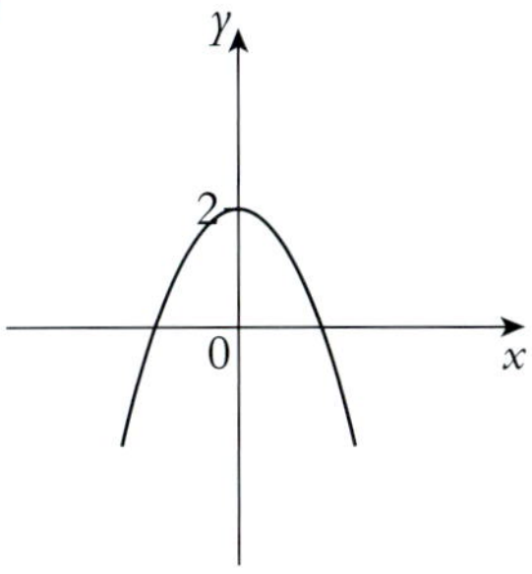   c

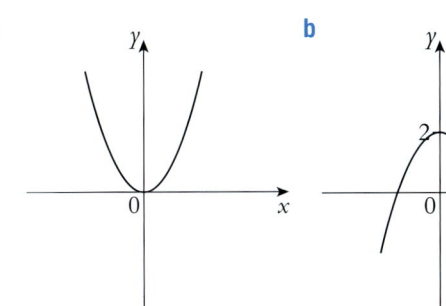

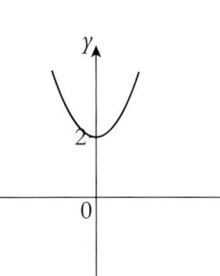

d 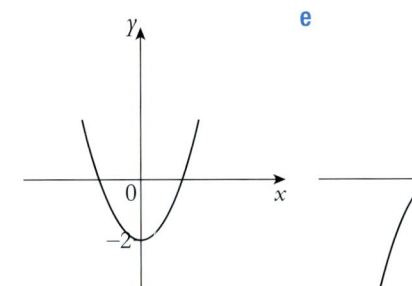   e 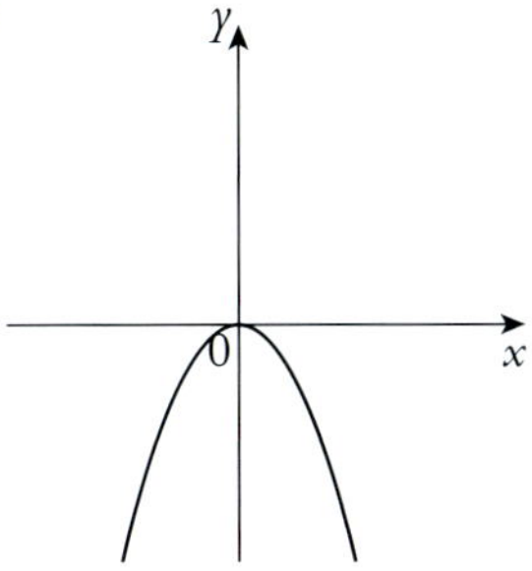   f

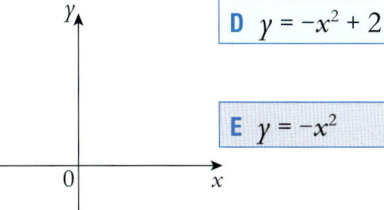

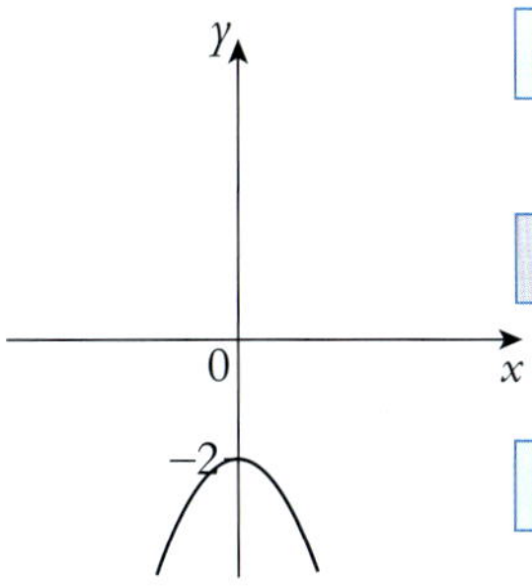

A $y = -x^2 - 2$

B $y = x^2 + 2$

C $y = x^2$

D $y = -x^2 + 2$

E $y = -x^2$

F $y = x^2 - 2$

2 $y = 2x - \frac{1}{8}x^2$

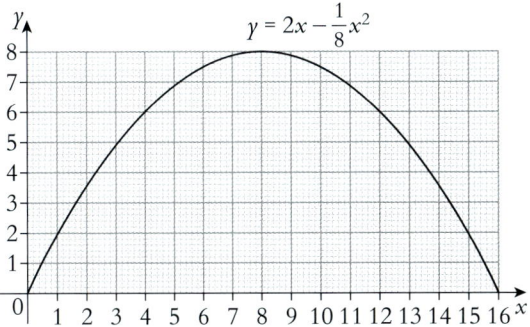

Copy the graph. Draw tangents to estimate the gradient of the graph when
a $x = 2$ b $x = 4$ c $x = 6$ d $x = 8$ e $x = 10$ f $x = 12$ g $x = 14$
Comment on your results.

3 $y = \frac{1}{2}x^2 - 4x - 2$

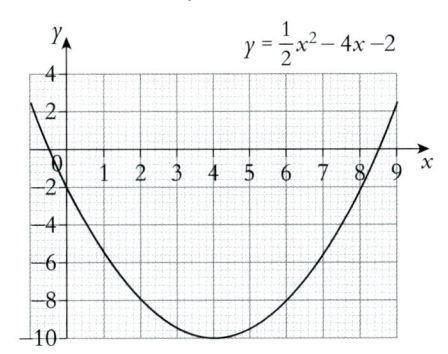

➡ **NOTE:** the scales are not the same on the x and y-axis. You must consider this when calculating the gradient.

Copy the graph. Draw tangents to estimate the gradient of the graph when
a $x = 0$ b $x = 2$ c $x = 4$ d $x = 6$ e $x = 8$
Comment on your results.

Quadratic graphs

Practical applications of quadratic graphs

EXAMPLE

Anu wants to make a rectangular enclosure for his donkey. He uses 100 m of fencing.
a If the length of the rectangle is x m, explain why the width is $(50 - x)$ m.
b Show that the area, A m², of the rectangular enclosure is given by the formula $A = 50x - x^2$.
c Copy and complete the table below.

x	0	10	20	30	40	50
A					400	

d Draw the graph of $A = 50x - x^2$ for $0 \leq x \leq 50$.
e What is the largest possible area for the rectangular enclosure?

a Perimeter of rectangle = 100
   $2x + 2y = 100$ *subtract 2x from both sides*
   $2y = 100 - 2x$ *divide both sides by 2*
   $y = 50 - x$
   width = $50 - x$

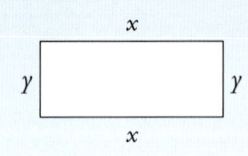

b Area of rectangle = length × width
                    = $x(50 - x)$
                    = $50x - x^2$

c
x	0	10	20	30	40	50	25
A	0	400	600	600	400	0	625

➡ **NOTE:** the extra column in the table is needed to find where the curve turns.

When $x = 25$, $A = 50 \times 25 - 25^2 = 1250 - 625 = 625$

d

e The largest possible area for the rectangular enclosure is 625 m².

EXERCISE 4.12

1. The water tank has a rectangular base with sides of length $(x + 1)$ m and x m. The height of the tank is 1 m.
   a. Show that the volume, V m³, of the tank is given by the formula $V = x^2 + x$
   b. Copy and complete the table below.

x	0	0.2	0.4	0.6	0.8	1
V		0.24				

   c. Draw the graph of $V = x^2 + x$ for $0 \leq x \leq 1$
   d. What value of x gives a volume of 1.3 m³?

2. A stone is projected vertically upwards with a speed of 25 m/s.
   Its height, h m, above the ground after t s is given by the formula $h = 25t - 5t^2$.
   a. Copy and complete the table below.

t	0	1	2	3	4	5
h				30		

   b. Draw the graph of $h = 25t - 5t^2$ for $0 \leq t \leq 5$
   c. What is the maximum height reached by the stone?
   d. What values of t give a height of 26 m?

3. A ball is kicked on horizontal ground.
   The equation for the path of the ball is $y = x - 0.02x^2$ where y m is the height of the ball above the ground and x m is the horizontal distance travelled by the ball.
   a. Copy and complete the table below.

x	0	10	20	30	40	50
y				12		

   b. Draw the graph of $y = x - 0.02x^2$ for $0 \leq x \leq 50$.
   c. What is the maximum height reached by the ball?
   d. Between what distances is the ball at least 10 m above the ground?

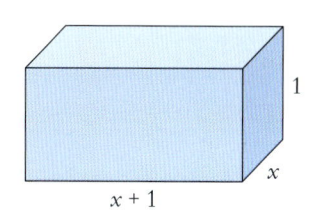

> **KEY WORDS**
> quadratic function
> tangent

Quadratic graphs

Bearings

THIS SECTION WILL SHOW YOU HOW TO
- Use bearings to describe directions

Bearings are used to describe directions.
In this diagram the bearing of B from A is 100°

> Bearings are always measured **clockwise** from the **north** line.

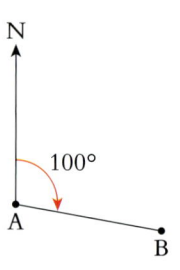

EXAMPLE

Draw diagrams to show these bearings.
a The bearing of P from Q is 068°
b The bearing of Y from X is 250°
c The bearing of C from D is 300°

➡ **NOTE:** bearings are always written as three-figure numbers.
e.g. 068° not 68°

EXAMPLE

The bearing of B from A is 060°
Draw a diagram to show the positions of A and B.
Use the diagram to work out the bearing of A from B.

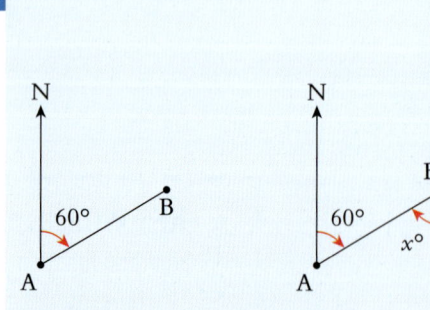

To find the bearing of A from B you must draw a north line at B and find the angle measured clockwise from the north line at B.

$x = 60°$ alternate angles

Bearing of A from B = 180° + 60° = 240°

EXAMPLE

B is 5 km due east of A. C is 5 km from A on a bearing of 060°.
Find the bearing of C from B.

First draw a diagram to show the positions of A, B and C.
∠CAB = 90° − 60° = 30°

Next draw a north line at B and indicate the required angle.

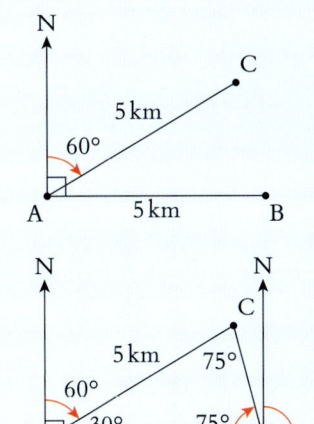

Triangle ABC is isosceles.

∠ACB = ∠ABC = $\begin{pmatrix} 6 & 2 \\ -8 & 0 \end{pmatrix}$ = 75°

Bearing of C from B = 270° + 75°
= 345°

EXERCISE 4.13

1. Draw diagrams to show these bearings:
   a. The bearing of B from A is 056°
   b. The bearing of F from G is 270°
   c. The bearing of N from M is 135°
   d. The bearing of P from Q is 340°

2. Write down the bearing for each of these compass directions.
   a. south east
   b. due south
   c. due west
   d. south west
   e. north east
   f. north west

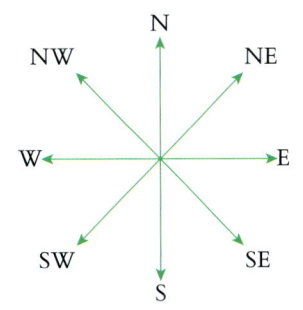

3. Write down the bearing of
   a. Dubai from Doha
   b. Doha from Kuwait City
   c. Shiraz from Doha
   d. Kuwait City from Dubai.

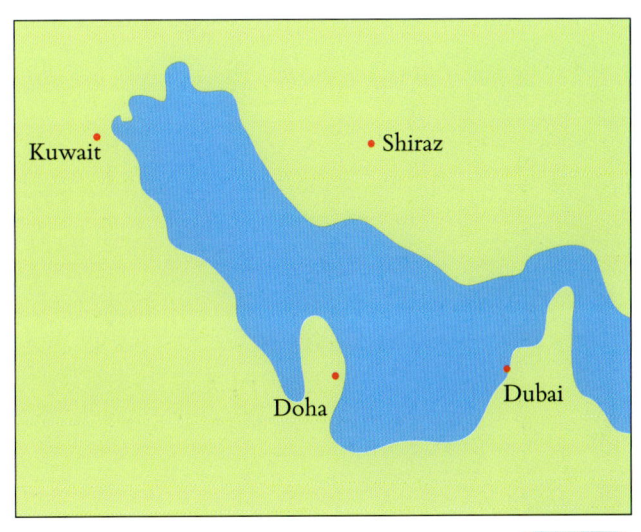

Bearings

4 Bermuda is 1650 km from Orlando on a bearing of 077°
   Jamaica is 1990 km from Bermuda on a bearing of 220°
   Draw a scale diagram to show the positions of Bermuda, Orlando and Jamaica.

5 A boat sets sail from port A and sails 16 km due east to port B.
   The boat then sails 12 km on a bearing of 140° to port C.
   a Using a scale of 1 cm to represent 2 km draw a scale diagram to show the journey.
   b Use your scale diagram to find the bearing of port C from port A.
   c Use your scale diagram to find the distance between port A and port C.

6 Work out the bearing of A from B for these diagrams.

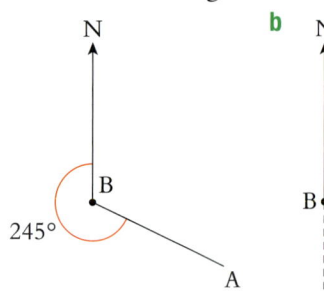

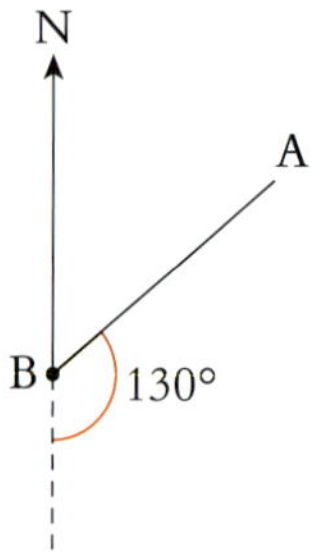

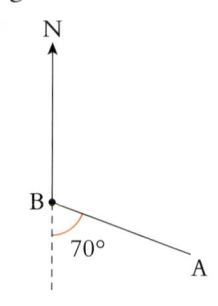

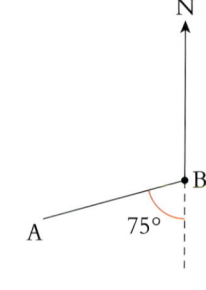

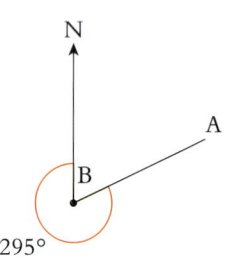

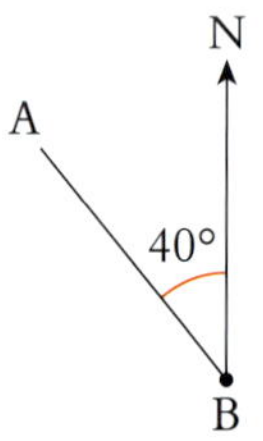

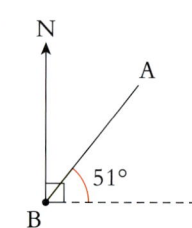

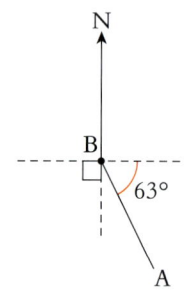

7 B is 8 km due east of A.
   C is 8 km from A on a bearing of 130°.
   Find the bearing of C from B.

8 Work out the bearing of P from Q for these diagrams:

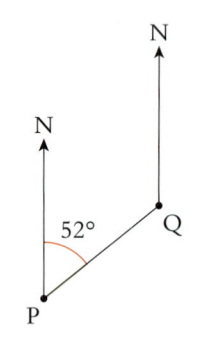

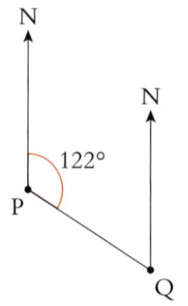

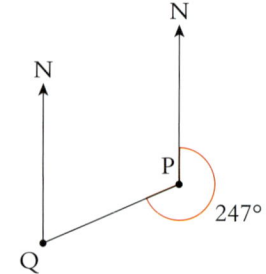

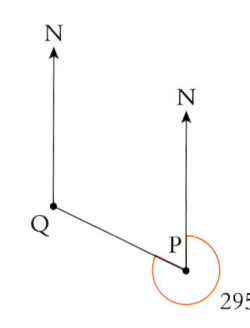

9 The diagram shows the positions of three towns A, B and C.
   Find the bearing of
   a B from C
   b C from B
   c A from B
   d B from A
   e A from C
   f C from A.

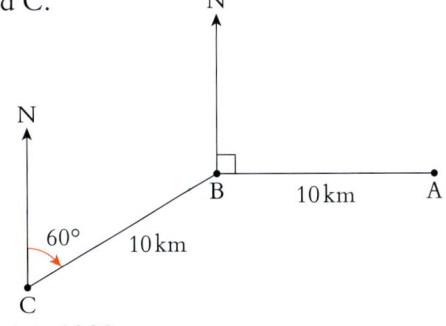

10 The bearing of B from A is 070°, the bearing of C from A is 120°
   and the bearing of C from B is 125°.
   a Find the bearing of A from C.
   b Find the bearing of B from C.

11 Two runners, Deepika and Khadeeja, set off from the point A.
   Deepika runs 8 km on a bearing of 095° to the point B.
   Khadeeja runs 8 km on a bearing of 155° to the point C.
   a Find the distance between B and C.
   b Find the bearing of B from C.
   c Find the bearing of C from B.

12 B is 5 km from A on a bearing of 038°.
   C is 12 km from B on a bearing of 128°.
   Find the distance between A and C.

13 A ship sails on a bearing of 110° from a port P.
   A lighthouse L is 8 km south east of port P.
   a Draw a scale diagram to show the positions of P and L and the path of the ship.
   b Mark clearly on your scale diagram the locus of points where the ship is less than 4 km
      from the lighthouse.

14 The bearing of Q from P is 120°
   The bearing of R from Q is 240°
   The bearing of P from R is 020°
   Find the values of x, y and z.

15 The bearing of B from A is 140°
   The bearing of C from B is 275°
   The bearing of A from C is 043°
   Find the values of x, y and z.

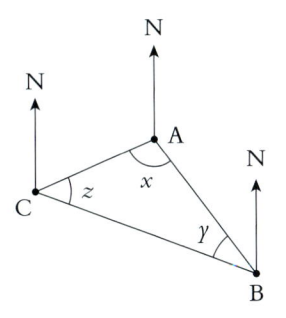

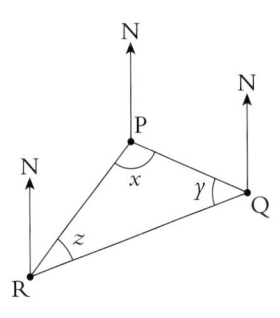

KEY WORDS
bearing

Bearings   193

Trigonometry

THIS SECTION WILL SHOW YOU HOW TO
- Use the tangent, sine and cosine ratios to find sides and angles

Trigonometry is used to find lengths and angles in a right-angled triangle without using scale drawings.
The sides of a right-angled triangle are given special names.
The longest side is called the **hypotenuse**.
The side opposite the angle $x°$ is called the **opposite** side.
The side next to the angle $x°$ is called the **adjacent** side.

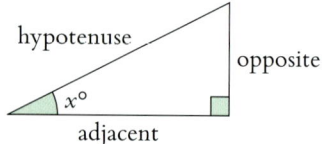

The different relationships or ratio of the sides to each other in a triangle make up trigonometric formulae. The three trigonometrical ratios that you have to know are tangent, cosine and sine.

The tangent ratio

$$\tan x = \frac{\text{opposite}}{\text{adjacent}}$$

Calculating sides

The following examples show you how to use the tangent ratio to find the length of a side.

EXAMPLE

Find the length of x.

(adj) 4 cm, 63°, x (opp)

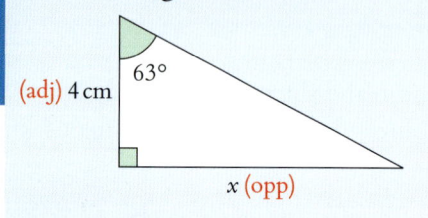

$\tan 63° = \dfrac{\text{opp}}{\text{adj}}$ replace *opp* by x and *adj* by 4

$\tan 63° = \dfrac{x}{4}$ multiply both sides by 4

$x = 4 \tan 63°$

$= 7.85$ cm (to 3 s.f.)

EXAMPLE

Find the length of x.

(adj) x, 48°, 6 cm (opp)

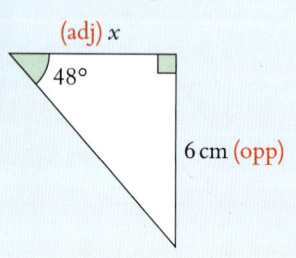

$\tan 48° = \dfrac{\text{opp}}{\text{adj}}$ replace *opp* by 6 and *adj* by x

$\tan 48° = \dfrac{6}{x}$ multiply both sides by x

$x \times \tan 48° = 6$ divide both sides by $\tan 48°$

$x = \dfrac{6}{\tan 48°}$

$= 5.40$ cm (to 3 s.f.)

EXERCISE 4.14

In this exercise all lengths are in cm.

1. Find the length x for each triangle.

a, b, 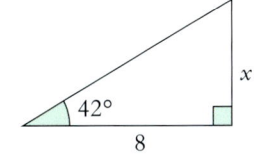 c, 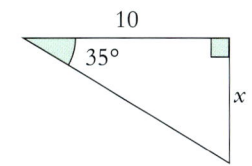 d, 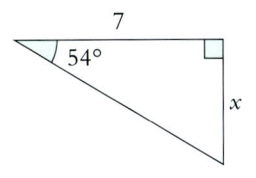 e, 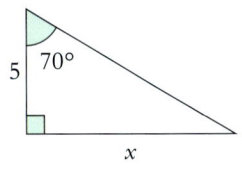 f, 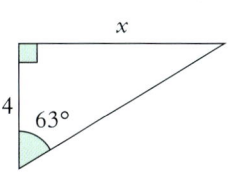 g, 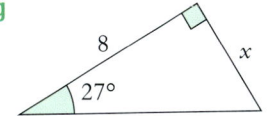 h, 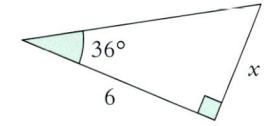 i, 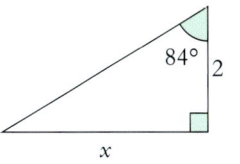 j, 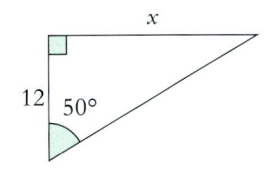 k, 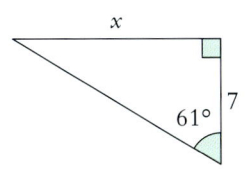 l, 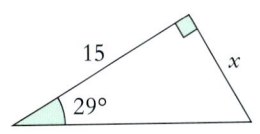 m, 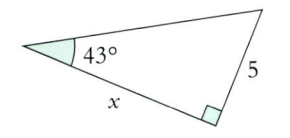 n, 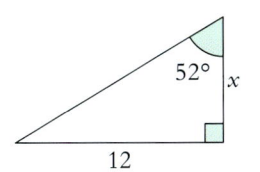 o

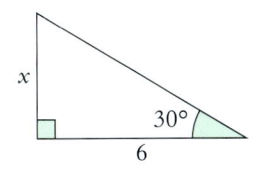

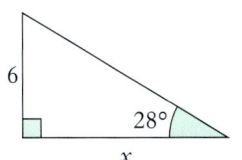

2. Find the lengths of x and y.

a, b, 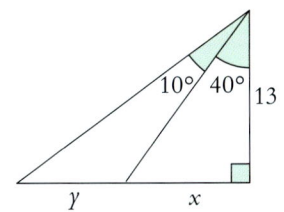 c, 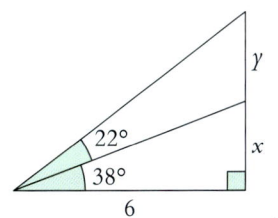 d, 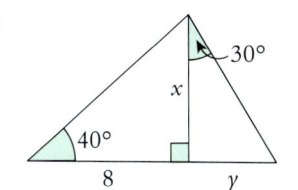 e, 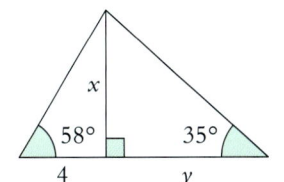 f

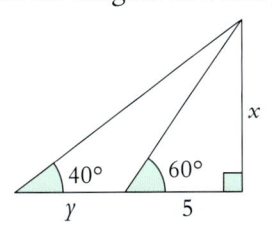

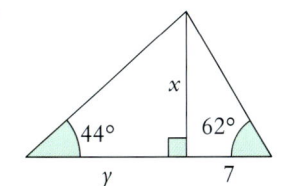

Calculating angles using the tangent ratio

If you know both the opposite and adjacent sides in a right-angled triangle, you can calculate the size of an angle inside the triangle.

To do this you must use the 'inverse tan' button on your calculator. (On most calculators this is labelled tan^{-1})

EXAMPLE

Find the size of angle x.

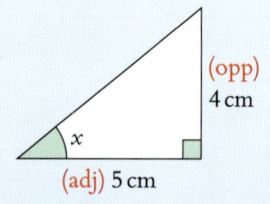

(opp) 4 cm
(adj) 5 cm

$\tan x = \dfrac{\text{opp}}{\text{adj}}$ *replace opp by 4 and adj by 5*

$\tan x = \dfrac{4}{5}$ *to find x use the tan^{-1} button on your calculator*

$x = \tan^{-1}\left(\dfrac{4}{5}\right)$

$x = 38.7°$ (to 1 d.p.) ➡ **NOTE:** you should give angle answers correct to 1 d.p.

EXAMPLE

ABCD is a rectangle. Calculate the size of angle x.

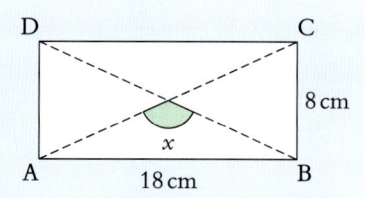

8 cm, 18 cm

The tangent ratio can only be used for right-angled triangles, so split the bottom triangle into two identical right-angled triangles.

$\tan y = \dfrac{\text{opp}}{\text{adj}}$ *replace opp by 9 and adj by 4*

$\tan y = \dfrac{9}{4}$ *to find y use the tan^{-1} button on your calculator*

$y = \tan^{-1}\left(\dfrac{9}{4}\right)$

$y = 66.0375.....$
$x = 2 \times y$
   $= 132.1°$ (to 1 d.p.)

EXERCISE 4.15

In this exercise all lengths are in cm.

1 Find the size of angle x for each triangle.

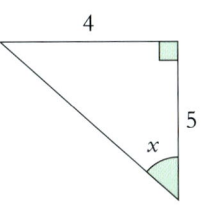

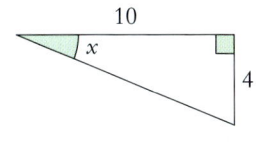

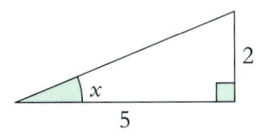

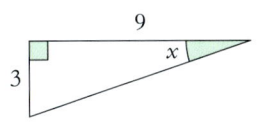

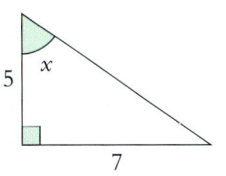

a) triangle with sides 3, 8 and angle x

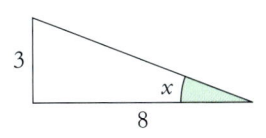

b) triangle with sides 4, 5 and angle x

c) triangle with sides 10, 4 and angle x

d) triangle with sides 2, 5 and angle x

e) triangle with sides 9, 3 and angle x

f) triangle with sides 5, 7 and angle x

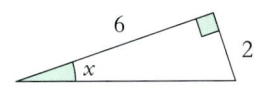

g) triangle with sides 6, 2 and angle x

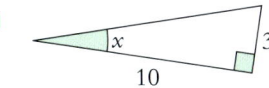

h) triangle with sides 10, 3 and angle x

i) triangle with sides 6, 4 and angle x

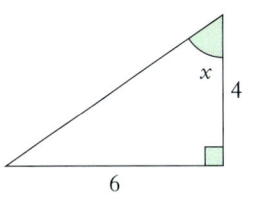

2 A hospital H is 7 km East and 5 km North of a school S. Calculate the bearing of
   a H from S,   b S from H.

3 ABCD is a rectangle. Find the size of angles x and y.

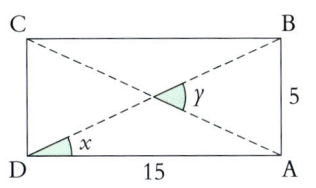

Rectangle with DA = 15, AB = 5, diagonals meeting with angles x at D and y at intersection.

4 Find angle BAD for each of these diagrams.

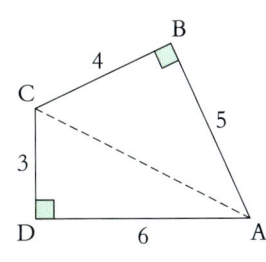

a) Quadrilateral with BC = 4, BA = 5, CD = 3, DA = 6, right angle at D and B

b) Triangle with BC = 8, CD = 7, DA = 15, right angles at C and D

c) Parallelogram with BA = 8, CD = 6, angle BCD = 52°, right angle at BDC

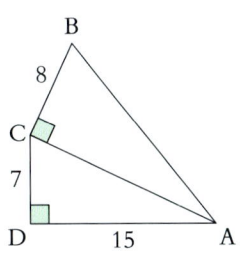

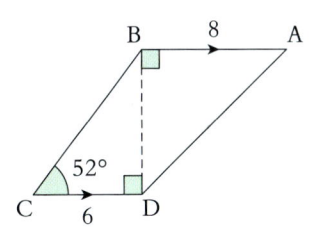

5 Find the size of angles x and for each of these diagrams.

a) Triangle with 4, 3, 6 and angles x, y

b) Triangle with 4, 4, 5 and angles x, y

c) Triangle with 16, 7, 4 and angles x, y

Trigonometry   197

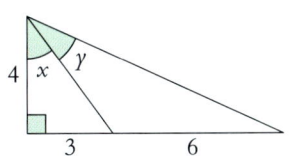

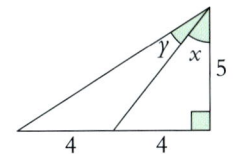

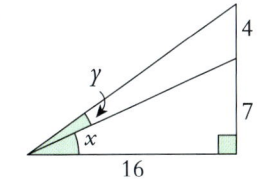

The sine and cosine ratios:

$$\sin x = \frac{\text{opposite}}{\text{hypotenuse}} \qquad \cos x = \frac{\text{adjacent}}{\text{hypotenuse}}$$

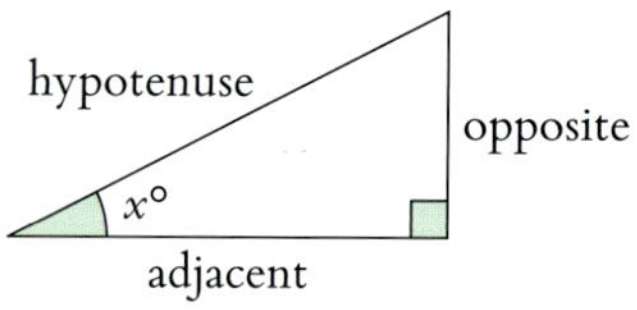

The following examples show you how to use the sine and cosine ratios to find the length of a side.

EXAMPLE

Find the length of x.

(hyp) 6 cm, 25°, x (opp)

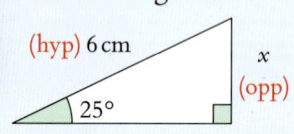

$\sin 25° = \dfrac{\text{opp}}{\text{hyp}}$ replace *opp* by x and *hyp* by 6

$\sin 25° = \dfrac{x}{6}$ multiply both sides by 6

$x = 6 \times \sin 25°$

$x = 2.54 \text{ cm (to 3 s.f.)}$

HINT

It helps to label the sides opp, adj and hyp when solving problems.

EXAMPLE

Find the length of x.

x (adj), 34°, 15 cm (hyp)

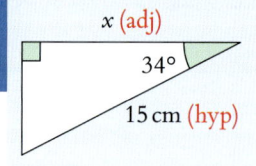

$\cos 34° = \dfrac{\text{adj}}{\text{hyp}}$ replace *adj* by x and *hyp* by 15

$\cos 34° = \dfrac{x}{15}$ multiply both sides by 15

$x = 15 \times \cos 34°$

$x = 12.4 \text{ cm (to 3 s.f.)}$

EXAMPLE

Find the length of x.

(adj) 3 cm, 65°, x (hyp)

$\cos 65° = \dfrac{\text{adj}}{\text{hyp}}$ replace *adj* by 3 and *hyp* by x

$\cos 65° = \dfrac{3}{x}$ multiply both sides by x

$x \times \cos 65° = 3$ divide both sides by $\cos 65°$

$x = \dfrac{3}{\cos 65°}$

$x = 7.10 \text{ cm (to 3 s.f.)}$

EXERCISE 4.16

In this exercise all lengths are in cm.

1 Find the length of x for each triangle.

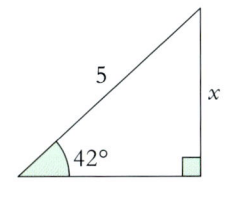

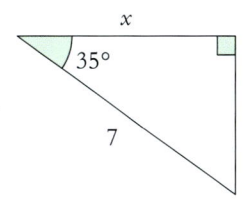

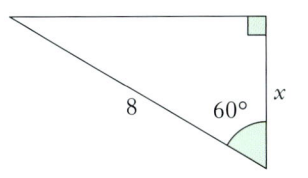

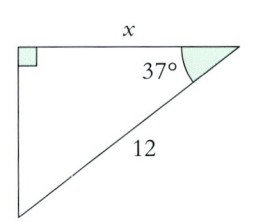

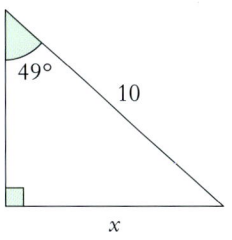

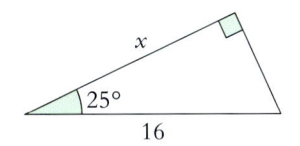

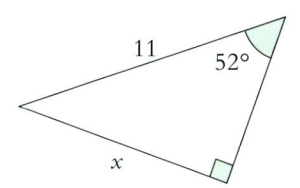

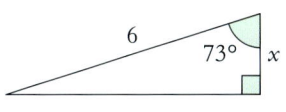

a) triangle with hypotenuse 16, angle 24°, opposite side x

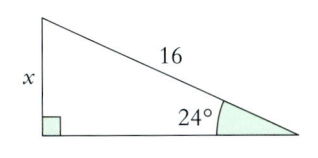

b) triangle with hypotenuse 5, angle 42°, opposite side x

c) triangle with angle 35°, adjacent side x, opposite side 7

d) triangle with side 8, angle 60°, side x

e) triangle with angle 37°, adjacent side x, hypotenuse 12

f) triangle with angle 49°, hypotenuse 10, opposite side x

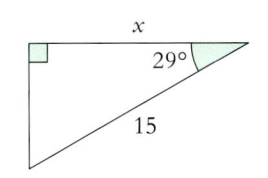

g) triangle with angle 25°, side 16, side x

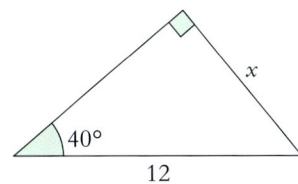

h) triangle with side 11, angle 52°, side x

i) triangle with side 6, angle 73°, side x

j) triangle with angle 29°, side 15, side x

k) triangle with angle 40°, side 12, side x

l) triangle with angle 38°, side 20, side x

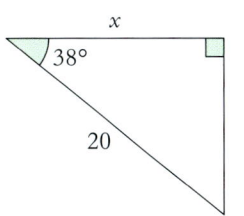

2 Find the length of x for each of these triangles.

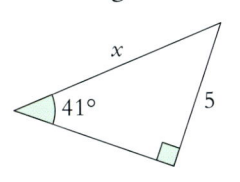

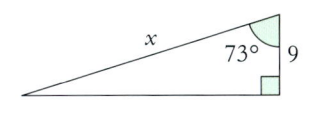

a) triangle with angle 41°, side 5, side x

b) triangle with angle 73°, side 9, side x

c) triangle with angle 56°, side 7, side x

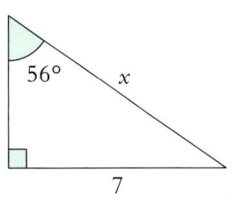

3 Calculate the height and area of each of these triangles.

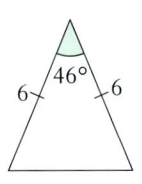

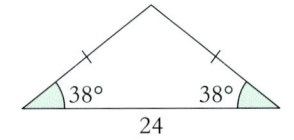

a) isosceles triangle with two sides of 6 and apex angle 46°

b) isosceles triangle with base 24 and base angles 38°

c) equilateral triangle with side 8

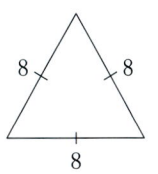

4 Calculate the value of x and the area of this trapezium.

trapezium with parallel sides 8 and (bottom), slant side 5 at 44°, side x

Trigonometry 199

Calculating angles using the sine and cosine ratios

EXAMPLE

Find the size of angle x

(hyp) 8 cm, 4 cm (opp)

$\sin x = \dfrac{\text{opp}}{\text{hyp}}$ replace *opp* by 4 and *hyp* by 8

$\sin x = \dfrac{4}{8}$ to find x use the $\sin^{-1}$ button on your calculator

$x = \sin^{-1}\left(\dfrac{4}{8}\right)$

$x = 30°$

EXERCISE 4.17

In this exercise all lengths are in cm.
Find the size of angle x for each triangle.

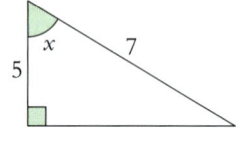

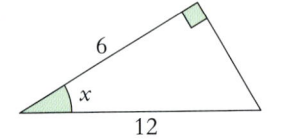

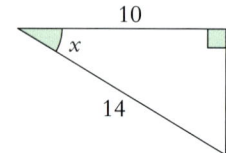

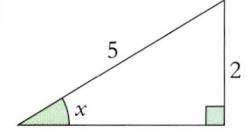

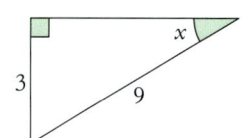

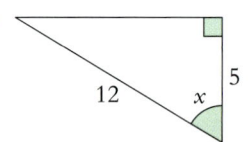

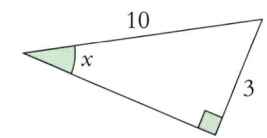

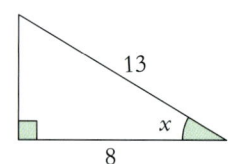

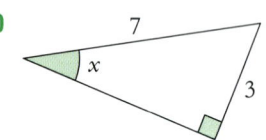

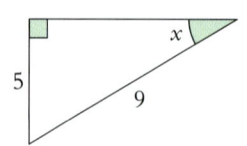

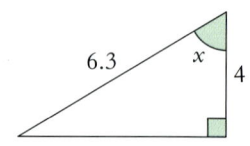

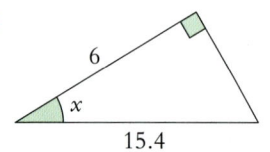

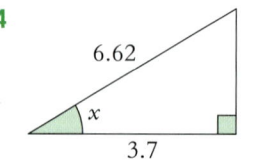

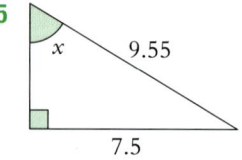

1 x, 7, 5

2 6, x, 12

3 10, x, 14

4 5, 2, x

5 3, 9, x

6 6, x, 4

7 12, x, 5

8 10, x, 3

9 13, x, 8

10 7, x, 3

11 5, 9, x

12 6.3, x, 4

13 6, x, 15.4

14 6.62, x, 3.7

15 x, 9.55, 7.5

Summary of rules for right-angled triangles

The three trigonometric ratios are:

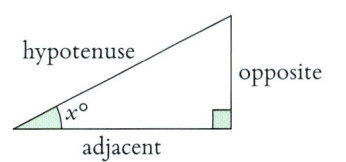

$$\text{Sin } x = \frac{\text{Opposite}}{\text{Hypotenuse}} \quad \text{Cos } x = \frac{\text{Adjacent}}{\text{Hypotenuse}} \quad \text{Tan } x = \frac{\text{Opposite}}{\text{Adjacent}}$$

A useful memory aid for remembering these three ratios is

SOHCAHTOA

Pythagoras' theorem also applies to right-angled triangles.

Pythagoras' $a^2 + b^2 = c^2$

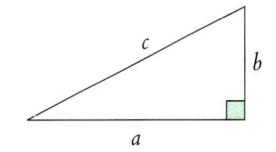

Sometimes problems involve both Pythagoras' theorem and trigonometry.

EXAMPLE

a Calculate the length of AC.
b Calculate the length of DC.
c Calculate the size of angle x.

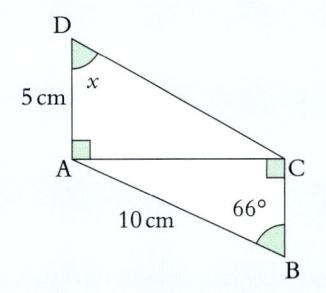

---

a Using triangle ABC:
$\sin 66° = \frac{AC}{10}$
AC = $10 \times \sin 66°$
AC = 9.135...
AC = 9.14 cm (to 3 s.f.)

b Using Pythagoras on triangle ACD:
$DC^2 = AC^2 + AD^2$
$DC^2 = 9.135^2 + 5^2$
$DC^2 = 108.4...$
DC = 10.4 cm (to 3 s.f.)

c Using triangle ACD:
$\tan x = \frac{AC}{5} = \frac{9.135...}{5}$
$x = \tan^{-1}\left(\frac{9.135...}{5}\right)$
$x = 61.3°$ (to 1 d.p.)

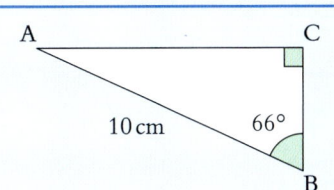

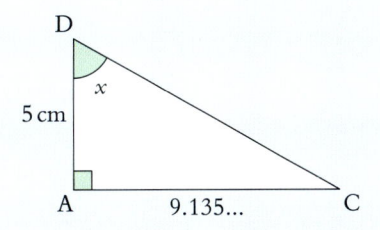

➡ **NOTE:** when using a previous answer in another calculation you should always use the unrounded value.

Trigonometry

EXERCISE 4.18

In this exercise all lengths are in cm.

1 Find the value of x for each triangle.

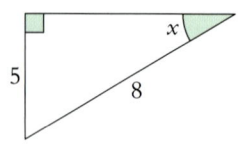

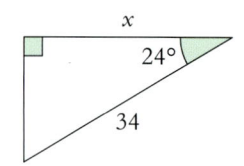

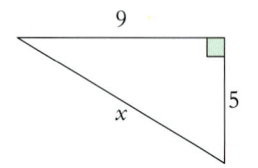

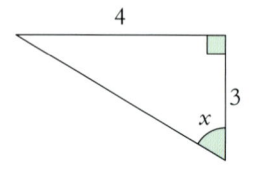

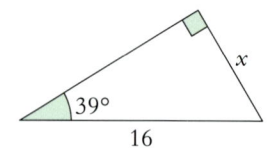

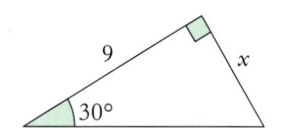

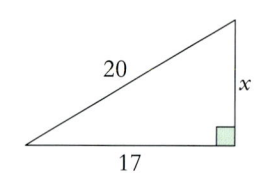

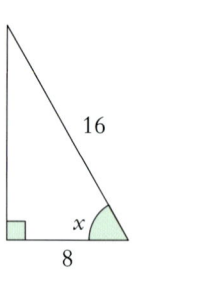

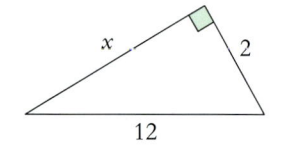

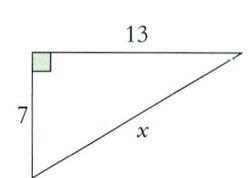

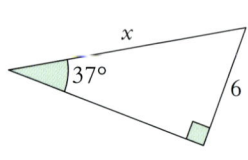

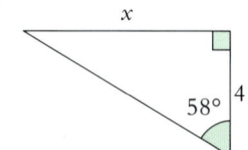

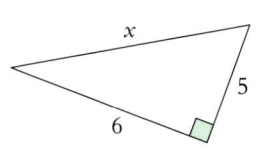

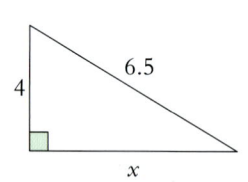

a. (right triangle: vertical leg 6, angle 41°, horizontal leg x)

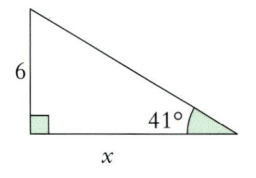

b. (right triangle: leg 5, hypotenuse 8, angle x)

c. (right triangle: hypotenuse 34, angle 24°, leg x)

d. (right triangle: leg 9, leg 5, hypotenuse x)

e. (right triangle: leg 4, leg 3, angle x)

f. (right triangle: angle 39°, adjacent 16, opposite x)

g. (right triangle: leg 9, angle 30°, leg x)

h. (right triangle: hypotenuse 20, leg 17, leg x)

i. (right triangle: hypotenuse 16, leg 8, angle x)

j. (right triangle: leg 2, hypotenuse 12, leg x)

k. (right triangle: leg 7, hypotenuse 13, leg x)

l. (right triangle: angle 37°, leg 6, hypotenuse x)

m. (right triangle: leg 4, angle 58°, leg x)

n. (right triangle: leg 5, leg 6, hypotenuse x)

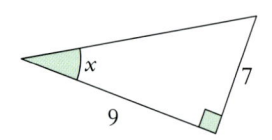

o. (right triangle: leg 4, hypotenuse 6.5, leg x)

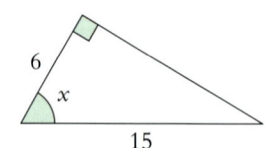

p. (right triangle: leg 9, hypotenuse 7, angle x)

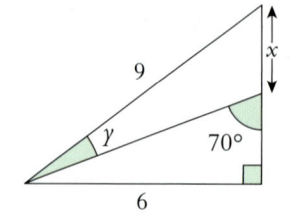

q. (right triangle: leg 6, hypotenuse 15, angle x)

r. (right triangle: hypotenuse 8, leg 3, angle x)

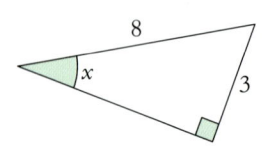

2 Find the values of x and y in each of these triangles.

a. (triangle: hypotenuse 15, angle 65°, base 4, leg x, angle y)

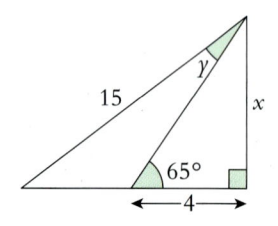

b. (triangle with sides 9 and 6, angle 70°, x, y)

c. (triangle: angle 40°, base segments 5 and 3, x, y)

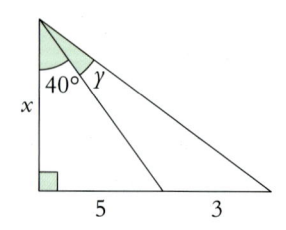

3 Find the values of *x*, *y* and *z* in each of these triangles.

   a   **b** 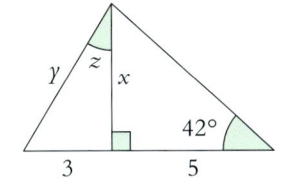   **c**

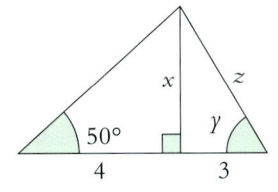

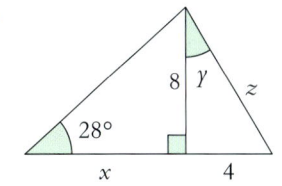

4 Calculate the size of angle *x* in this trapezium.

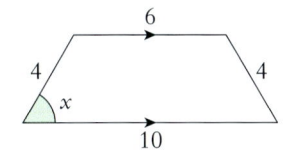

5 Find the values of *x* and *y* in these diagrams.

   a   **b**

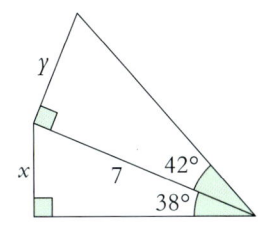

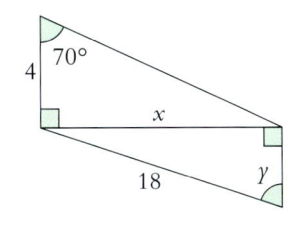

6 Find the values of *x*, *y* and *z* in this diagram.

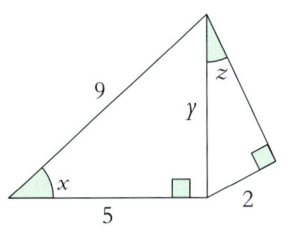

7 Find the area of each of these regular polygons.

   a   **b** 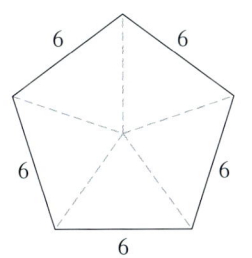   **c**

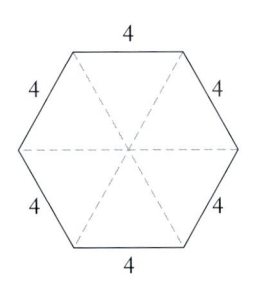

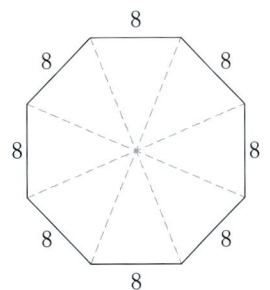

8 The cross-section of this solid prism is a regular pentagon.
The length of each side of the pentagon is 8 cm.
The length of the prism is 20 cm.
Calculate:
  a the surface area of the prism,
  b the volume of the prism.

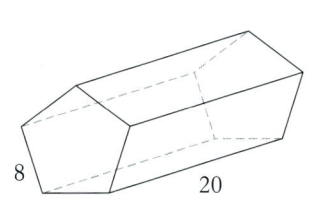

Trigonometry

Practical applications of trigonometry and Pythagoras

EXERCISE 4.19

1. The diagram shows a bridge over a river.
   A boat B is 100 m vertically below the bridge.
   Angle QPB = 56° and angle PQB = 65°.
   Calculate the length of the bridge PQ.

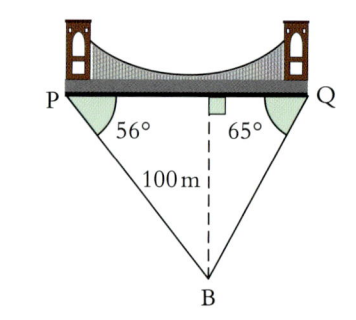

2. ABCDE is the cross-section of a tent.
   AB = DE = 1.7 m   BC = CD = 1.3 m
   Angle EAB = angle AED = 75°
   Angle CBD = angle CDB = 22°
   Calculate
   a   the length of BD
   b   the length of AE
   c   the height (h) of the tent.

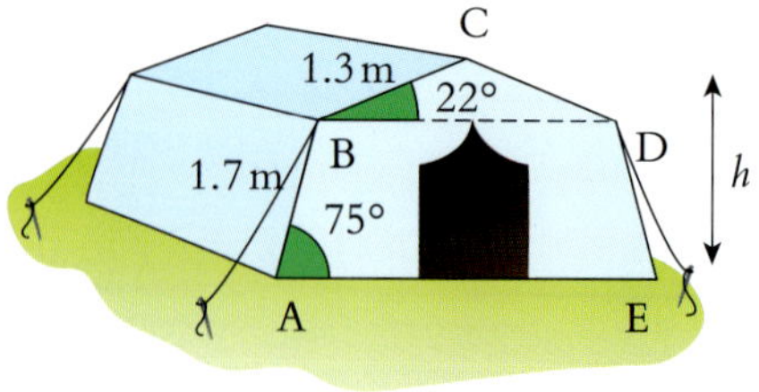

3. The diagram shows a 25 m high radio mast PQ.
   Two wires PA and PB secure the mast.
   PA = 30 m and PB = 35 m
   Calculate
   a   angle PAB
   b   angle PBA
   c   the distance between A and B.

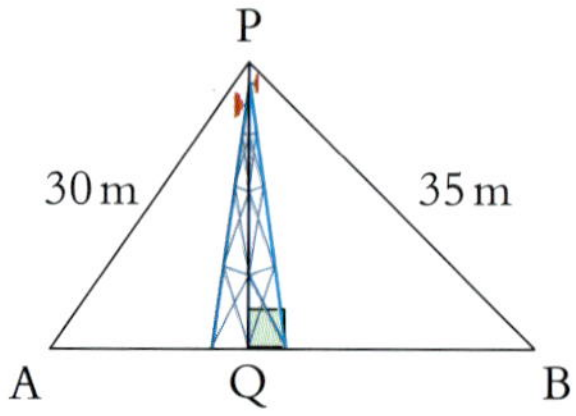

4. A cable car travels at 50° to the horizontal for 70 m and then 35° to the horizontal for 100 m. Calculate the vertical height (h) gained when travelling from A to C.

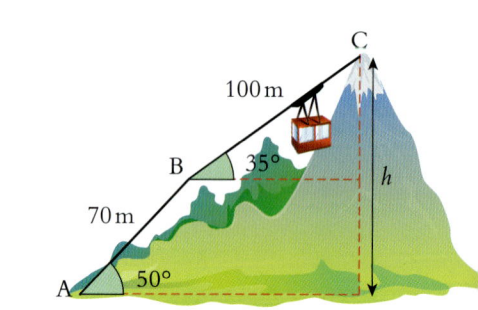

5. ABCD is a field.
   Calculate
   a   the length of AC
   b   the length of CD
   c   angle BAD
   d   angle BCD
   e   the area of the field.

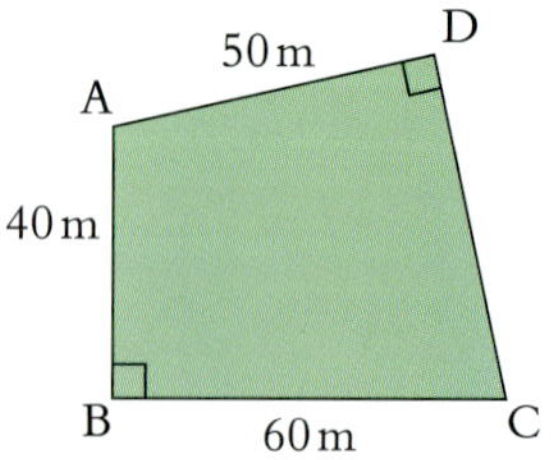

6. The vertices of a regular pentagon lie on the circumference of a circle radius 10 cm.
   Calculate
   a   the lengths of the sides of the pentagon
   b   the difference in area between the circle and the pentagon.

7 The Tower of Pisa leans at an angle of 4° to the vertical.
   The top of the tower stands 56.7 m above the ground.
   a Calculate the height of the tower.
   b The tower used to lean at an angle of 5.5° to the vertical.
      How much lower did the top of the tower used to stand
      above the ground?

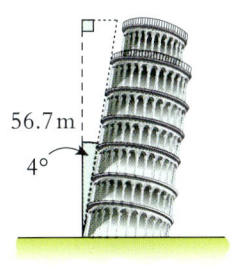

8 A boat sails 50 km on a bearing of 080°.
   Calculate:
   a How far north it travels,
   b How far east it travels.

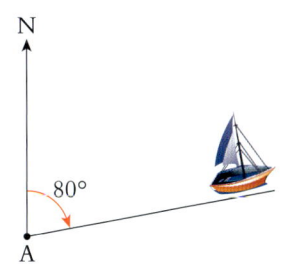

9 A hot air balloon flies 3000 km on a bearing of 245°.
   Calculate:
   a How far south it travels
   b How far west it travels.

10 An aircraft flies 500 km from an airport A on a bearing of 070°
   to arrive at airport B.
   The plane then flies 800 km from airport B on a bearing of 035°
   to arrive at airport C.
   a Calculate how far north C is from A
   b Calculate how far east C is from A
   c Calculate the direct distance between A and C
   d Calculate the bearing of C from A
   e What is the bearing of A from C?

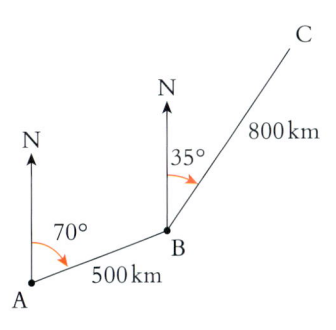

Trigonometry 205

Three-dimensional trigonometry

To find the angle between the line AB and the plane PQRS:
- drop a perpendicular from the point B to meet the plane at X
- draw the right-angled triangle BAX
- the angle between AB and the plane is angle BAX

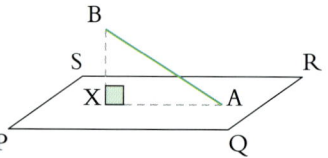

When doing a 3D trigonometry question it is important to draw clearly the triangle (usually right-angled) that you are going to use in your calculations. On your triangle show all the known lengths and angles.

EXAMPLE

ABCDEFGH is a cuboid.
AB = 7 cm, BC = 4 cm and CG = 6 cm.
Calculate
a the length of BD
b the angle between the line BH and the base ABCD.

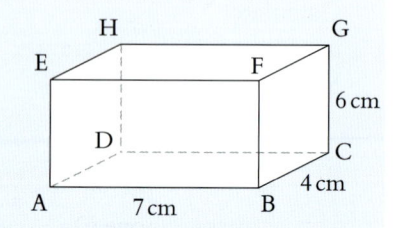

a Draw the rectangular base of the cuboid and on it show the line BD.
Use Pythagoras on the right-angled triangle DAB.
$$BD^2 = 7^2 + 4^2$$
$$BD^2 = 65$$
$$BD = \sqrt{65}$$
$$BD = 8.062...$$
$$BD = 8.06 \text{ cm (3 s.f.)}$$

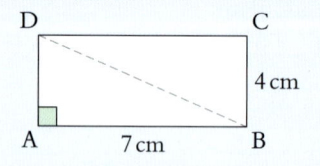

b Drop a perpendicular from H to meet the base ABCD.
The perpendicular meets the base at the point D.
The angle between the line BH and the base ABCD is angle HBD.
Draw triangle HBD. (It has a right-angle at D.) 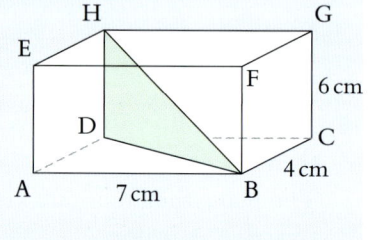
$$\tan x = \frac{\text{opp}}{\text{adj}}$$
$$\tan x = \frac{6}{\sqrt{65}}$$
$$x = \tan^{-1}\left(\frac{6}{\sqrt{65}}\right)$$
$$x = 36.7° \text{ (to 1 d.p.)}$$

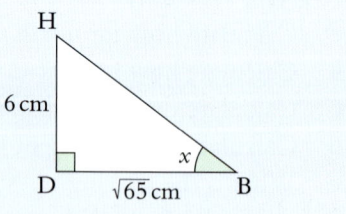

EXERCISE 4.20

1 ABCDEFGH is a cube.
   Calculate:
   a the length of AC,
   b the length of AG,
   c the angle between AG and the base ABCD.

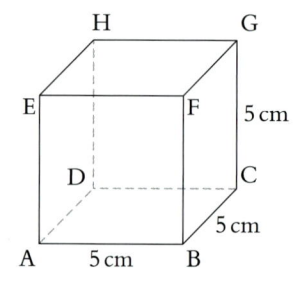

2 ABCDEFGH is a cuboid.
Calculate:
- **a** the length of AC,
- **b** the length of AG,
- **c** the angle between AG and the base ABCD.

3 ABCDEF is a triangular prism.
Calculate:
- **a** the length of BD,
- **b** the length of DE,
- **c** the angle between DE and the base ABCD.

4 ABCDV is a square-based pyramid.
AB = 4 cm and VC = 5 cm.
O is the middle of the square base.
V is vertically above O.
X is the midpoint of AD.
Y is the midpoint of BC.
Calculate:
- **a** the lengths of AC, VO and VY,
- **b** the angle between VC and the base ABCD,
- **c** angle XVY.

5 ABCDV is a rectangular-based pyramid.
AB = 12 cm, BC = 9 cm and VO = 6 cm.
O is the middle of the rectangular base.
V is vertically above O.
X is the midpoint of AD.
Y is the midpoint of BC.
Calculate:
- **a** the lengths of OB, VB and VY,
- **b** the angle between VB and the base ABCD,
- **c** angle XVY.

6 ABCDEF is a triangular prism.
Triangle ABE is equilateral.
AB = 5 cm and BC = 15 cm.
Calculate the angle between AF and the base ABCD.

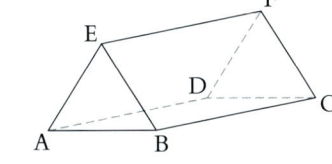

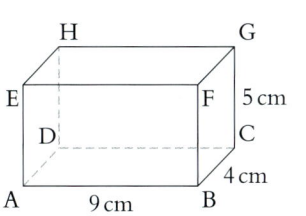

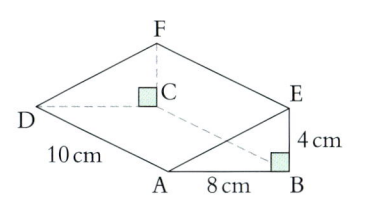

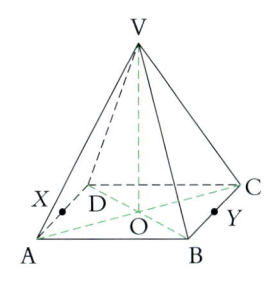

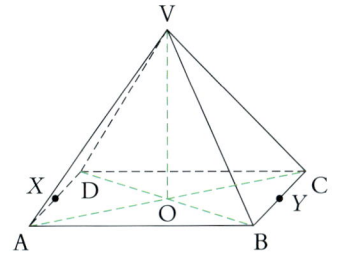

KEY WORDS
tangent
sine
cosine
adjacent
opposite
hypotenuse

Trigonometry

Angles of elevation and depression

THIS SECTION WILL SHOW YOU HOW TO
- Use angles of elevation and depression

The **angle of elevation** is the angle measured upwards from the horizontal.

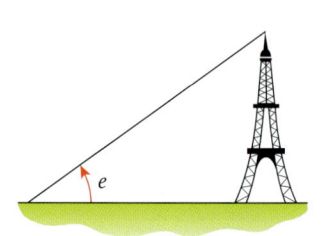

The **angle of depression** is the angle measured downwards from the horizontal.

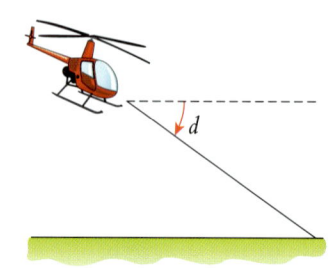

e = angle of elevation

d = angle of depression

EXAMPLE

A radio mast stands on horizontal ground. The angle of elevation of the top of the mast from a point 200 m away from the base of the mast is 17°. Calculate the height, *h*, of the radio mast.

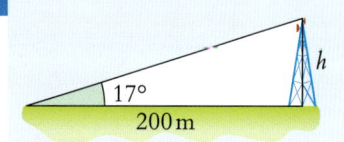

$$\tan 17° = \frac{h}{200}$$
$$h = 200 \times \tan 17°$$
$$h = 61.1 \text{ m}$$

EXERCISE 4.21

1. An electricity pylon stands on horizontal ground. The angle of elevation of the top of the pylon from a point 150 m away from the base of the pylon is 44°. Calculate the height of the pylon.

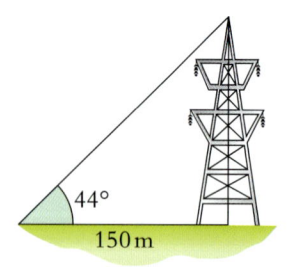

2. A tree stands on horizontal ground. The angle of elevation of the top of the tree from a point 50 m away from the base of the tree is 11.3°. Calculate the height of the tree.

3. The picture shows the CN tower in Canada. The angle of elevation of the top of the tower from a point 100 m away from the base of the tower is 79.75°. Calculate the height of the CN tower.

4 The picture shows the Burj Khalifa tower in Dubai.
The angle of elevation of the top of the tower from a point 100 m away from the base of the tower is 83.11°.
Calculate the height of the Burj Khalifa tower.

5 Use your answers to questions 3 and 4 to find the difference in heights between the CN tower and the Burj Khalifa tower.

6 An aeroplane is flying at a height of 1000 m.
The angle of depression of airport A from the plane is 15°.
Calculate the horizontal distance between the aeroplane and the airport.

7 The lighthouse is 36 m high. Two boats A and B are due South of the lighthouse.
From the top of the lighthouse the angles of depression of boats A and B are 22° and 15°.
Calculate the distance AB.

8 A hot air balloon is 70 m above the sea.
Two buoys A and B are due East of the balloon.
The angle of depression from the balloon to buoy A is 60°.
The angle of depression from the balloon to buoy B is 40°.
Calculate the distance AB.

9 The diagram shows a tower.
The point A is 30 m from the base of the tower.
On top of the tower is a flagpole PQ.
The angle of elevation of P from the point A is 53°.
The angle of elevation of Q from the point A is 50°.
Calculate the length of PQ.

10 PQRS is a rectangular horizontal car park.
PT is a vertical lamp-post.
PS = 50 m and PQ = 40 m.
The angle of elevation of T from S is 8°.
Calculate:
  a PT and TR,
  b the angle of elevation of T from Q,
  c the angle of depression of R from T.

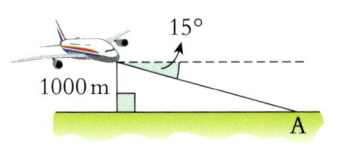

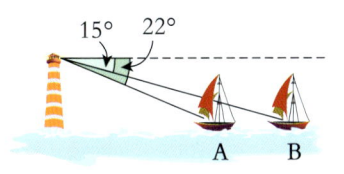

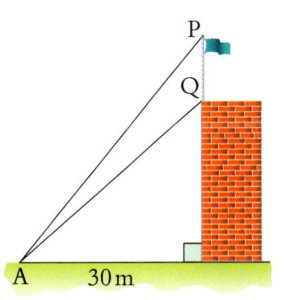

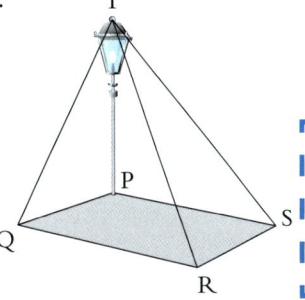

KEY WORDS
angle of elevation
angle of depression

Angles of elevation and depression   209

Scatter diagrams

THIS SECTION WILL SHOW YOU HOW TO
- Draw scatter diagrams and lines of best fit
- Understand positive, negative and zero correlation

You can use a **scatter diagram** to represent paired **data**.

EXAMPLE

The table shows the heights and masses of ten adult males.

Height (cm)	160	166	170	175	180	185	188	190	195	200
Mass (kg)	66	76	70	90	82	100	94	90	110	104

Draw a scatter diagram to show this information.

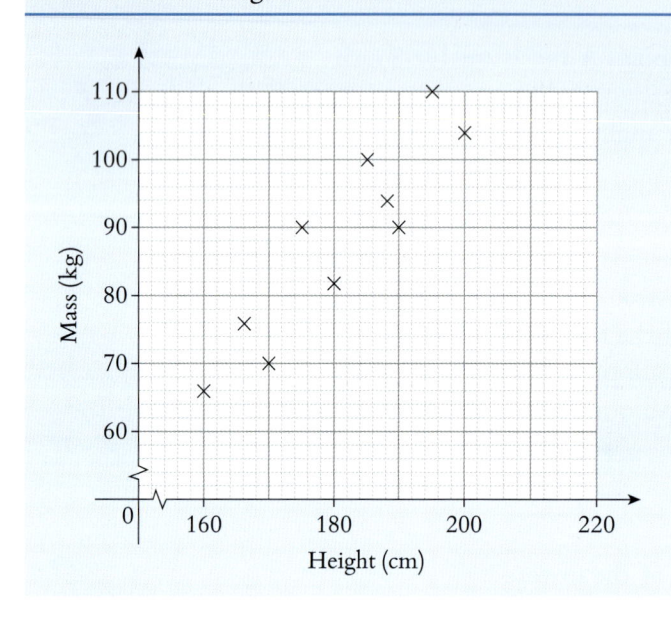

➡ **NOTE:** the information is plotted as coordinate points: (160, 66), (166, 76), (170, 70)...

➡ **NOTE:** a jagged line shows that the scale does not start at zero.

Correlation

A scatter diagram shows you if there is **correlation** (a connection) between the two variables.

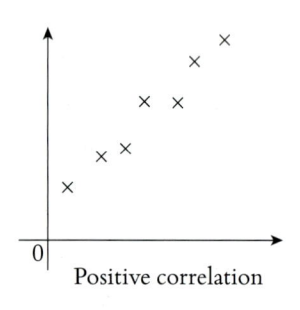

Positive correlation 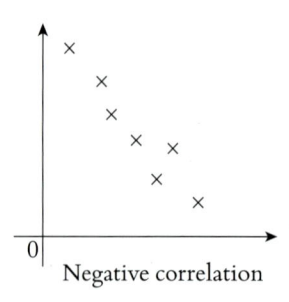   Negative correlation 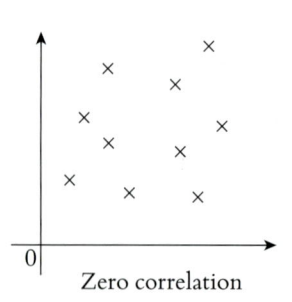   Zero correlation

In the example above, the scatter diagram shows that there is positive correlation between the two variables.

210 UNIT 4

EXERCISE 4.22

1 Name the type of correlation shown by each of these scatter diagrams.

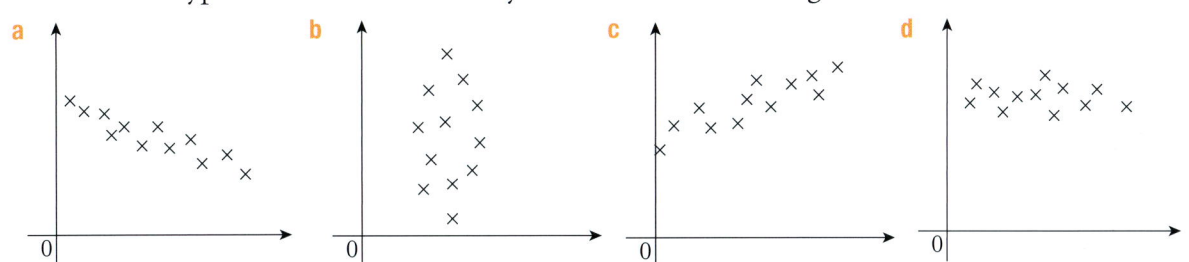

a, b, c, d

2 The table shows the test results for eight students in mathematics and art.

Mathematics	5	8	8	10	13	15	17	18
Art	10	6	18	13	13	19	15	11

a Show this information on a scatter diagram.
b Name the type of correlation shown.

3 The table shows information collected by a café owner for eight days.

Midday temperature (°C)	11	30	18	24	25	12	17	22
Number of hot drinks sold	24	4	11	8	11	20	18	15

a Show this information on a scatter diagram.
b Name the type of correlation shown.

4 The table shows the test results of ten students in chemistry and physics.

Chemistry	6	8	9	9	11	12	13	15	16	19
Physics	6	7	9	11	15	15	17	14	16	18

a Show this information on a scatter diagram.
b Name the type of correlation shown.

5 The table shows the age and value of eight cars.
a Show this information on a scatter diagram.
b Name the type of correlation shown.

Age (years)	Value ($)
0	18 000
1	15 000
2	12 000
4	8 000
4	10 000
5	7 000
5	8 000
6	5 000

6 Name the type of correlation that you would expect for each of these:
a The number of hours of sunshine in a day and the number of cold drinks sold.
b The midday temperature and the number of coats sold.
c The distance travelled by a car and the amount of money spent on petrol.
d The height of an adult and the IQ of the adult.
e The age of a child and the mass of the child.

Scatter diagrams

Lines of best fit

When a scatter diagram shows positive or negative (linear) correlation you can draw a **line of best fit**.

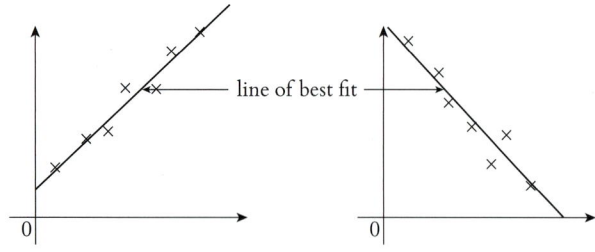

A line of best fit is usually drawn 'by eye' and should
- follow the general trend of the points
- have a similar number of points above and below the line.

The closer the points are to the line of best fit, the stronger the correlation.
The line of best fit can be used to predict data values within the range of data collected.
If you are asked to calculate the mean of the two data sets, then the mean point can also be plotted on the scatter graph and the line of best fit drawn through it.

EXAMPLE

The table shows the daily amount of rainfall (mm) and the number of sun hats sold in a shop over a period of ten days.

Rainfall (mm)	0	1	3	5	2	6	8	4	2	6
Number of sunhats sold	26	23	18	11	21	9	2	15	17	5

a Calculate the mean rainfall for the ten days.
b Calculate the mean number of sunhats sold for the ten days.
c Show the information from the table on a scatter diagram and draw the line of best fit.
d Name the type of correlation shown by the diagram.
e Use your line of best fit to predict the number of sunhats sold on a day with 7 mm of rainfall.

a Mean rainfall = $\dfrac{0+1+3+5+2+6+8+4+2+15}{10} = \dfrac{37}{10} = 3.7$

b Mean number of hats sold = $\dfrac{26+23+18+11+21+9+2+15+17+5}{10} = \dfrac{147}{10} = 14.7$

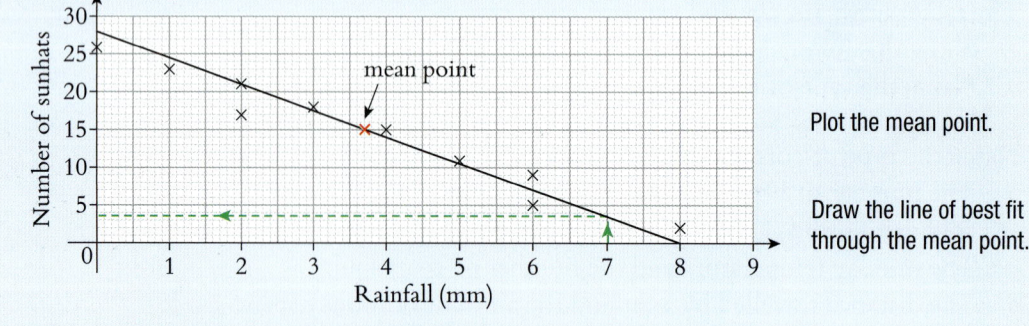

c

Plot the mean point.

Draw the line of best fit through the mean point.

d The scatter graph shows negative correlation
e From the graph the predicted number of sunhats is approximately 4.

EXERCISE 4.23

1. The scatter diagram shows information about the marks earned by a group of students in two examinations.
   a. Use the line of best fit to predict the economics mark for a student who scored 60% in the mathematics examination.
   b. Use the line of best fit to predict the mathematics mark for a student who scored 40% in the economics examination.

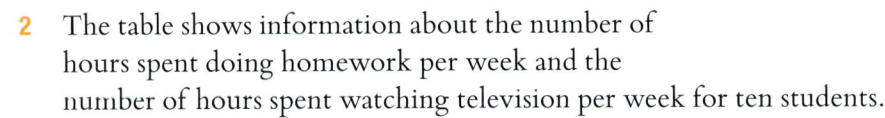

2. The table shows information about the number of hours spent doing homework per week and the number of hours spent watching television per week for ten students.

Homework (hours)	3	13	10	6	1	5	8	14	12	11
Television (hours)	9	3	7	8	12	9	8	2	5	5

   a. Calculate the mean number of hours spent doing homework per week.
   b. Calculate the mean number of hours spent watching television per week.
   c. Show the information from the table on a scatter diagram and draw the line of best fit.
   d. Name the type of correlation shown by the diagram.
   e. A student watched six hours of television in one week. Use your line of best fit to predict the number of hours this student spent doing homework that week.

3. The table shows information about the hand span and the height of ten adult females.

Hand span (cm)	20.5	18	22.4	22.7	23.5	19	19.7	21.5	22.2	21
Height (m)	170	160	186	183	190	167	171	177	181	177

   a. Calculate the mean hand span.
   b. Calculate the mean height.
   c. Show the information from the table on a scatter diagram and draw the line of best fit.
   d. Name the type of correlation shown by the diagram.
   e. An adult female has a hand span of 20.7 cm. Use your line of best fit to predict the height of this female.

KEY WORDS
scatter diagram
correlation
line of best fit

Scatter diagrams

Unit 4 Examination questions

1. Lin scored 18 marks in a test and Jon scored 12 marks.
   Calculate Lin's mark as a percentage of Jon's mark. [2]

   Cambridge IGCSE Mathematics 0580, Paper 21 Q3, June 2008

2. In 1970 the population of China was 8.2×10^8.
   In 2007 the population of China was 1.322×10^9.
   Calculate the population in 2007 as a percentage of the population in 1970. [2]

   Cambridge IGCSE Mathematics 0580, Paper 21 Q5, November 2009

3. Work out $\begin{pmatrix} 2 & -1 & 4 \\ 6 & 5 & 0 \\ -2 & 1 & 3 \end{pmatrix} \begin{pmatrix} 0 \\ 1 \\ 2 \end{pmatrix}$. [3]

   Examination style question

4. $\begin{pmatrix} 1 & -2 \\ 0 & 1 \\ 5 & 5 \end{pmatrix} \begin{pmatrix} 3 & 4 & 8 & 7 \\ 1 & 1 & 3 & 3 \end{pmatrix}$

   The answer to this matrix multiplication is of order $a \times b$.

   Find the values of a and b. [2]

   Cambridge IGCSE Mathematics 0580, Paper 21 Q2, November 2008

5. $\mathbf{A} = \begin{pmatrix} -2 & 3 \\ -4 & 5 \end{pmatrix}$

   Find $\mathbf{A}^{-1}$, the inverse of the matrix $\mathbf{A}$. [2]

   Cambridge IGCSE Mathematics 0580, Paper 21 Q5, June 2009

6. $\mathbf{A} = \begin{pmatrix} 0 & 1 \\ -8 & -4 \end{pmatrix}$ $\mathbf{B} = \begin{pmatrix} 7 & 1 \\ 0 & -5 \end{pmatrix}$

   Calculate the value of $5\,|\mathbf{A}| + |\mathbf{B}|$, where $|\mathbf{A}|$ and $|\mathbf{B}|$ are the determinants of $\mathbf{A}$ and $\mathbf{B}$. [2]

   Cambridge IGCSE Mathematics 0580, Paper 21 Q6, November 2009

7 (a) Find $(3 \quad 4)\begin{pmatrix} 5 \\ 2 \end{pmatrix}$. [2]

(b) $\begin{pmatrix} 7 \\ 3 \end{pmatrix}(x \quad y) = \begin{pmatrix} 28 & 42 \\ 12 & 12 \end{pmatrix}$. Find the values of x and y. [2]

(c) Explain why $\begin{pmatrix} 15 & 20 \\ 6 & 8 \end{pmatrix}$ does not have an inverse. [1]

Cambridge IGCSE Mathematics 0580, Paper 2 Q19, November 2006

8 (a) **A** is a (2×4) matrix, **B** is a (3×2) matrix and **C** is a (1×3) matrix.
Which two of the following matrix products is it possible to work out?
A²   **B**²   **C**²   **AB**   **AC**   **BA**   **BC**   **CA**   **CB** [2]

(b) Find the inverse of $\begin{pmatrix} \frac{1}{2} & \frac{3}{4} \\ \frac{1}{8} & \frac{1}{4} \end{pmatrix}$.

Simplify your answer as far as possible. [3]

(c) Explain why the matrix $\begin{pmatrix} 4 & 2 \\ 6 & 3 \end{pmatrix}$ does not have an inverse. [1]

Cambridge IGCSE Mathematics 0580, Paper 21 Q21, June 2010

9 Simplify $16 - 4(3x - 2)^2$. [3]

Cambridge IGCSE Mathematics 0580, Paper 21 Q12, November 2009

10

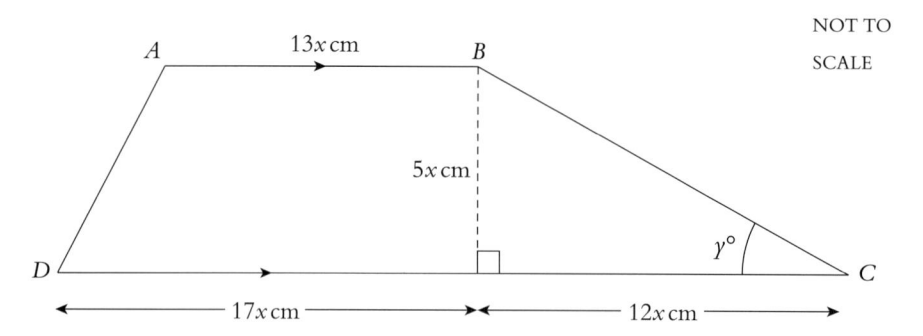

ABCD is a trapezium.

(a) Find the area of the trapezium in terms of x and simplify your answer. [2]

(b) Angle $BCD = y°$. Calculate the value of y. [2]

Cambridge IGCSE Mathematics 0580, Paper 2 Q16, November 2006

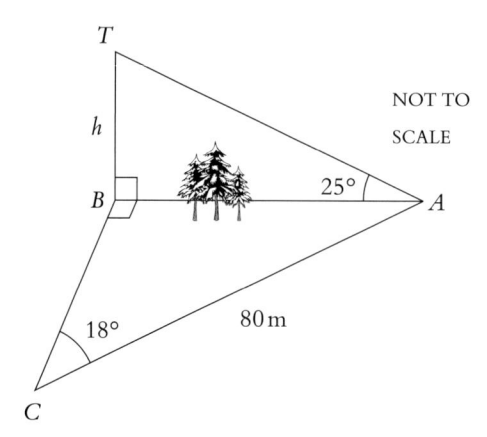

11

Mahmoud is working out the height, h meters, of a tower BT which stands of level ground.
He measures the angle TAB as 25°.
He cannot measure the distance AB and so he walks 80 m from A to C,
where angle $ACB = 18°$ and angle $ABC = 90°$.

Calculate

(a) the distance AB, [2]

(b) the height of the tower, BT. [2]

Cambridge IGCSE Mathematics 0580, Paper 21 Q15, June 2009

12

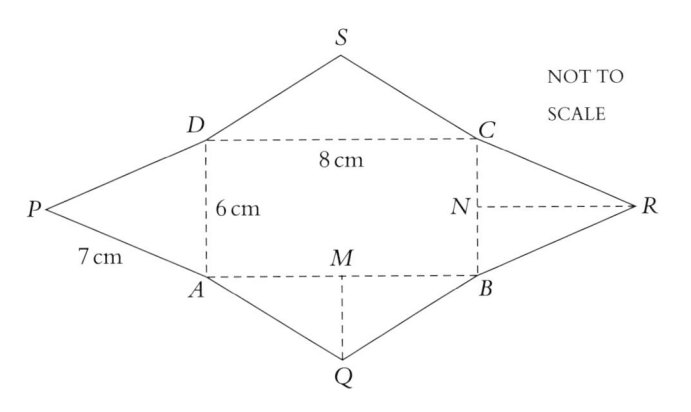

The diagram above shows the net of a pyramid.
The base *ABCD* is a rectangle 8 cm by 6 cm.
All the sloping edges of the pyramid are of length 7 cm.
M is the mid-point of *AB* and *N* is the mid-point of *BC*.

(a) Calculate the length of
   (i) *QM*, [2]
   (ii) *RN*. [1]
(b) Calculate the surface area of the pyramid. [2]
(c)

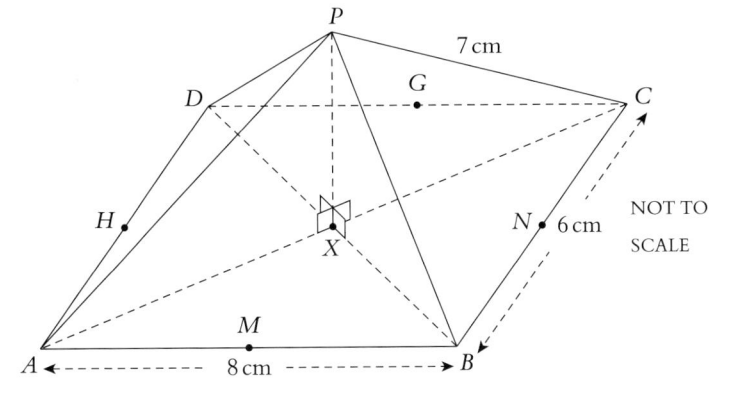

The net is made into a pyramid, with *P*, *Q*, *R* and *S* meeting at *P*.
The mid-point of *CD* is *G* and the mid-point of *DA* is *H*.
The diagonals of the rectangle *ABCD* meet at *X*.
   (i) Show that the height, *PX*, of the pyramid is 4.90 cm, correct to 2 decimal places. [2]
   (ii) Calculate angle *PNX*. [2]
   (iii) Calculate angle *HPN*. [2]
   (iv) Calculate the angle between the edge *PA* and the base *ABCD*. [3]
   (v) Write down the vertices of a triangle which is a plane of symmetry of the pyramid. [1]

Cambridge IGCSE Mathematics 0580, Paper 4 Q5, November 2007

Direct and inverse proportion

THIS SECTION WILL SHOW YOU HOW TO
- Solve problems involving direct and inverse proportion
- Calculate money conversions

Direct proportion

Two quantities are in **direct proportion** if the graph of one quantity against the other quantity is a straight line through the origin. For example, if one quantity doubles then the other quantity doubles.

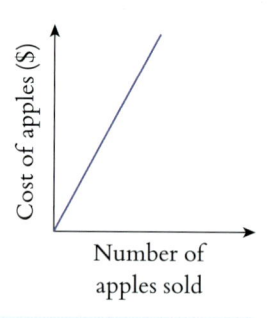

EXAMPLE

The cost of three apples is $2.37
What is the cost of five apples?

3 apples cost $2.37
1 apple costs $2.37 ÷ 3 = $0.79
5 apples cost $0.79 × 5 = $3.95

Inverse proportion

Two quantities are in **inverse proportion** if one quantity increases at the same rate as the other quantity decreases. For example, if one quantity doubles then the other quantity halves.

EXAMPLE

Seven men build a wall in 3 hours.
How long will it take two men to build the same wall?

7 men take 3 hours
1 man will take 7 × 3 = 21 hours
2 men will take 21 ÷ 2 = 10.5 hours

EXAMPLE

A bag of birdseed will feed 30 birds for 12 days.
a   For how many days will the bag of birdseed feed 40 birds?
b   How many birds will the bag of birdseed feed for 15 days?

a   30 birds for 12 days
    1 bird for 30 × 12 = 360 days
    40 birds for 360 ÷ 40 = 9 days

b   30 birds for 12 days
    1 bird for 30 × 12 = 360 days
    360 ÷ 15 = 24 birds

EXERCISE 5.1

1. Twelve metres of material costs $31.56
   Calculate the cost of five metres of the same material.

2. Three kilograms of fish costs $11.25
   Calculate the cost of eight kilograms of fish.

3. Jarred earns $101.50 for working 7 hours in a café.
   Calculate how much he will be paid for working 10 hours.

4. A car uses 20 litres of petrol to travel 120 kilometres.
   a How far will the car travel if it uses 25 litres of petrol?
   b How many litres of petrol are needed for a journey of 315 kilometres?

5. The recipe for pancakes will make 8 pancakes.
   Anna wants to make 60 pancakes.
   Calculate the amount of each ingredient needed.

   RECIPE:
   200g flour
   480ml milk
   2g salt
   2 eggs

6. A machine fills 600 bottles in 4 minutes.
   How long does the machine take to fill 40 bottles?
   Give your answer in seconds.

7. Three people take 4 hours to paint a fence.
   How long will it take five people?
   Give your answer in hours and minutes.

8. Six people take 5 days to pick a crop of oranges.
   How long would it take 4 people to pick the same crop?

9. A bag of sweets is shared between 8 children so that they each receive 9 sweets.
   How many sweets would each child receive if there were 12 children?

10. It takes 48 hours to fill a swimming pool with water using 3 identical pumps.
    a How long would it take to fill the pool if there were eight of these pumps?
    b How many pumps would be needed to fill the pool in 12 hours?

11. A hotel has 120 rooms.
    It takes 6 cleaners 4 hours to clean all the rooms.
    How many cleaners would be needed to clean all the rooms in 3 hours?

12. One sack of rice will feed 100 people for 10 days.
    a For how many days will the sack of rice feed 250 people?
    b How many people will the sack of rice feed for 40 days?

Direct and inverse proportion

Converting money from one currency to another currency

You can use a conversion graph to convert one currency to another currency.

The graph below can be used to convert between US dollars ($) and South African Rand (ZAR).

The blue dashed lines show how to convert 500 ZAR into US dollars ($).

500 ZAR = $73.50

The red dashed lines show how to convert $40 into South African Rand (ZAR).

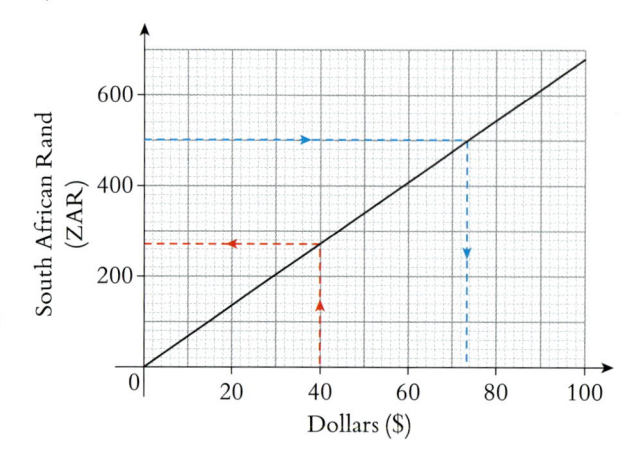

$40 = 272 ZAR

If you know the exchange rate you can use proportion to convert money from one currency to another.

EXAMPLE

Rafael changed $500 into Japanese Yen when the exchange rate was $1 = 83.7 Yen.
How many Japanese Yen did he receive?

$1 = 83.7 Yen
$500 = 500 × 83.7 Yen
$500 = 41850 Yen

EXAMPLE

Jani changed $350 into pounds (£) when the exchange rate was £1 = $1.315
How many pounds did he receive?
Give your answer correct to 2 decimal places.

$1.315 = £1

$1 = £$\frac{1}{1.315}$

$350 = 350 × £$\frac{1}{1.315}$ = £266.159 ... = £266.16

EXERCISE 5.2

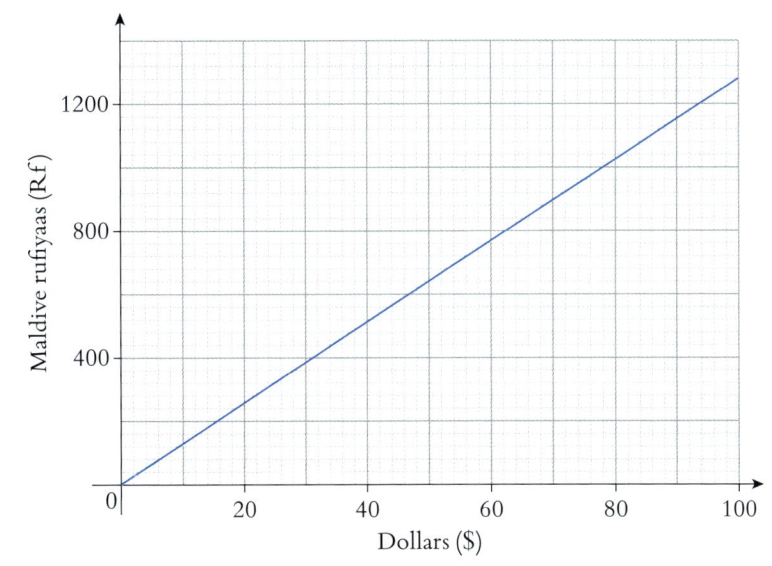

1. The graph can be used to convert between US dollars ($) and Maldives Rufiyaa (Rf).
   a   Use the graph to convert the amounts in dollars into Maldives Rufiyaa (Rf).
       i $25   ii $54   iii $81   iv $200
   b   Use the graph to convert the amounts in Maldives Rufiyaa (Rf) into dollars.
       i   600 Rf   ii   1100 Rf   iii   240 Rf   iv   2000 Rf

2. If the exchange rate between US dollars and Argentine Peso (ARS) is $1 = 3.98 ARS,
   a   Change the following amounts of money into Argentine Peso (ARS).
       i $50   ii $300   iii $72   iv $2000
   b   Change the following amounts of money into US dollars ($).
       i   40 ARS   ii   82 ARS   iii   500 ARS   iv   36.4 ARS

3. If the exchange rate between Malaysian Ringgit (MYR) and US dollars ($) is 1 MYR = $0.319,
   a   change the following amounts of money into Malaysian Ringgit (MYR).
       i $80   ii $20   iii $74   iv $500
   b   change the following amounts of money into US dollars ($).
       i   60 MYR   ii   18 MYR   iii   300 MYR   iv   470 MYR

4. Zara changed 800 US dollars ($) into Thai Baht (THB) when the exchange rate was 1 THB = $0.032
   One month later she changed all the Thai Baht back into US dollars when the exchange rate was 1 THB = $0.035
   Calculate how much profit she made.
   Give your answer in dollars.

KEY WORDS

direct proportion

inverse proportion

Increase and decrease in a given ratio

THIS SECTION WILL SHOW YOU HOW TO
- Increase or decrease a quantity in a given ratio

If you want to increase $60 in the ratio 5 : 4 there are two possible methods.

Method 1
4 parts = $60
1 part = $60 ÷ 4 = $15
5 parts = 5 × $15 = $75
So the answer is $75.

Method 2
$60 × $\frac{5}{4}$ = $75

EXAMPLE

Decrease 540 kg in the ratio 17 : 20

Method 1
20 parts = 540 kg
1 part = 540 kg ÷ 20 = 27 kg
17 parts = 17 × 27 kg = 459 kg
So the answer is 459 kg.

Method 2
540 kg × $\frac{17}{20}$ = 459 kg

EXAMPLE

10 cm
15 cm

A photograph measures 15 cm by 10 cm.
The photograph is enlarged in the ratio 3 : 2
a Calculate the measurements of the enlarged photograph.
b Calculate   **i**   the area of the original photograph
             ii   the area of the enlarged photograph.
c Write down and simplify the following ratio
    area of enlarged photograph : area of original photograph

a   15 cm × $\frac{3}{2}$ = 22.5 cm   and   10 cm × $\frac{3}{2}$ = 15 cm

Enlarged photograph is 22.5 cm by 15 cm.

b   **i**   Area of original photograph = 15 × 10 = 150 cm^2
     ii   Area of enlarged photograph = 22.5 × 15 = 337.5 cm^2

c   area of enlarged photograph : area of original photograph
    = 337.5 : 150   *divide both sides by 37.5*
    = 9 : 4

➡ **NOTE:** it is important to notice that 9 : 4 = 3^2 : 2^2

EXERCISE 5.3

1. Increase $60 in the following ratios
   a 5 : 3   b 7 : 4   c 13 : 12   d 11 : 8

2. Decrease 45 kg in the following ratios
   a 4 : 9   b 11 : 15   c 2 : 3   d 4 : 5

3. One year a farmer grows 5000 kg of potatoes.
   The next year he uses a fertiliser to increase the total mass of potatoes that he grows.
   The mass of potatoes that he grows increases in the ratio 5 : 4
   Calculate the new mass of potatoes that he grows.

4. Robert runs in an 800-metres race and he takes 2 minutes 15 seconds.
   He trains hard for his next competition to try to improve the time that he takes.
   He decreases the time he takes in the ratio 13 : 15
   Calculate the new time that he takes to complete the race.

5. A car manufacturer produces 480 cars a week.
   The demand for new cars increases so the car manufacturer increases the number of cars that are produced each week in the ratio 7 : 6
   Calculate the new number of cars produced each week.

6. Daisy has a mass of 120 kg.
   She is told that she must diet.
   She diets and decrease her mass in the ratio 7 : 8
   Calculate her new mass.

7. A shop has a sale and reduces the cost of a computer in the ratio 23 : 25
   The original cost of the computer was $1800.
   Calculate the sale price of the computer.

8. A rectangle measures 6 cm by 8 cm.
   The lengths of the sides of the rectangle are increased in the ratio 5 : 4
   a What are the lengths of the new rectangle?
   b What is the area of   i   the new rectangle   ii   the original rectangle?
   c Write down and simplify the following ratio
      the area of the new rectangle : the area of the original rectangle.

9. A cuboid measures 6 cm by 15 cm by 21 cm.
   The lengths of the sides of the cuboid are increased in the ratio 4 : 3
   a What are the measurements of the new cuboid?
   b What is the volume of   i   the new cuboid   ii   the original cuboid?
   c Write down and simplify the following ratio
      the volume of the new cuboid : the volume of the original cuboid.

Functions

THIS SECTION WILL SHOW YOU HOW TO
- Use function notation
- Form composite and inverse functions

Function notation

A **mapping** transforms one set of numbers into a different set of numbers.
A **function** is a special mapping that maps each number of a set to only one number in another set.

$$\begin{matrix} 1 \\ 2 \\ 3 \\ 4 \end{matrix} \xrightarrow{f} \begin{matrix} 4 \\ 5 \\ 6 \\ 7 \end{matrix}$$

The function f for the diagram above is 'add 3'.
This can be written using function notion in two ways:

$f : x \rightarrow x + 3$ which is read as f is such that x is mapped to $x + 3$
or $f(x) = x + 3$ which is read as f of $x = x + 3$

$f(1) = 4$ means that the function f maps the number 1 to the number 4.
$f(2) = 5$ means that the function f maps the number 2 to the number 5.

EXAMPLE

If $f(x) = x^2 + 5x$ Find   **a** $f(6)$ **b** $f(-4)$

a f is the function that squares the number and then adds 5 lots of the number, so
$f(6) = 6^2 + 5 \times 6$
$= 36 + 30$
$= 66$

b $f(-4) = (-4)^2 + (5 \times -4)$
$= 16 + -20$
$= -4$

Sometimes you have to work in reverse.

EXAMPLE

If $f(x) = 3x + 4$.   Find the value of x for which $f(x) = 25$.

Replace $f(x)$ in the function $f(x) = 3x + 4$ by the number 25.
$3x + 4 = 25$
$3x = 21$
$x = 7$

➡ **CHECK:**
$3 \times 7 + 4 = 25$ ✓

EXERCISE 5.4

1. $f(x) = 2x + 3$. Find the value of
   a. $f(4)$ b. $f(-6)$ c. $f\left(\dfrac{1}{2}\right)$ d. $f(0)$

2. $f(x) = 5 - 3x$. Find the value of
   a. $f(8)$ b. $f(-4)$ c. $f\left(\dfrac{1}{2}\right)$ d. $f\left(-\dfrac{1}{2}\right)$

3. $f(x) = x^2 + 3x$. Find the value of
   a. $f(2)$ b. $f(-3)$ c. $f(0)$ d. $f\left(\dfrac{1}{2}\right)$

4. $f(x) = 2x^2 + x - 5$. Find the value of
   a. $f(3)$ b. $f(-1)$ c. $f\left(\dfrac{1}{2}\right)$ d. $f(0)$

5. $f(x) = x^3 + 2$. Find the value of
   a. $f(4)$ b. $f(-2)$ c. $f\left(\dfrac{1}{2}\right)$ d. $f(0)$

6. $f(x) = \dfrac{3x^2 + 1}{4}$. Find the value of
   a. $f(4)$ b. $f(-6)$ c. $f\left(\dfrac{1}{2}\right)$ d. $f(0)$

7. $g(x) = 5x - 2$. Find the value of x when $g(x) = 13$

8. $f(x) = x^2 + 2$. Find the values of x when $f(x) = 38$

9. $f(x) = \dfrac{1}{2x + 4}$. Find the value of x when $f(x) = 2$

10. $f(x) = \dfrac{5}{2x + 3}$. Find the value of x when $f(x) = -2$

11. $f(x) = 3x - 2$ and $g(x) = 4x + 1$.
    Find the value of x such that $f(x) = g(x)$.

    HINT
    Put the two functions equal to each other and solve to find x

12. $f(x) = x^2 + 7x$ and $g(x) = x^2 - x + 4$.
    Find the value of x such that $f(x) = g(x)$.

13. $f(x) = x^2 + 3x - 8$ and $g(x) = 3x + 1$. Find the values of x such that $f(x) = g(x)$.

14. $f(x) = \dfrac{1}{x + 2}$ and $g(x) = \dfrac{3}{2x - 1}$. Find the value of x such that $f(x) = g(x)$.

Functions

Composite functions

When one function is followed by another function, the resulting function is called a **composite function**.

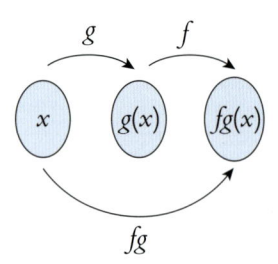

$fg(x)$ means the function g acts on x first, then f acts on the result.

EXAMPLE

$f(x) = x^2 \quad g(x) = x + 5$.
Find **a** $fg(3)$ **b** $fg(-7)$ **c** $fg(x)$.

a $fg(3)$ 　　　　　　　　　g acts on 3 first and $g(3) = 3 + 5 = 8$
　$= f(8)$ 　　　　　　　　　　f is the function 'square'
　$= 8^2$
　$= 64$

b $fg(-7)$ 　　　　　　　　g acts on -7 first and $g(-7) = -7 + 5 = -2$
　$= f(-2)$ 　　　　　　　　　f is the function 'square'
　$= (-2)^2 = 4$

c $fg(x)$ 　　　　　　　　　g acts on x first and $g(x) = x + 5$
　$= f(x + 5)$ 　　　　　　　f is the function 'square'
　$= (x + 5)^2$

EXAMPLE

$f(x) = 2x + 1 \quad g(x) = x^2 + 3$
Find **a** $fg(x)$ **b** $gf(x)$ **c** $ff(x)$

a $fg(x)$ 　　　　　　　　　g acts on x first and $g(x) = x^2 + 3$
　$= f(x^2 + 3)$ 　　　　　　f is the function 'double and add 1'
　$= 2(x^2 + 3) + 1$
　$= 2x^2 + 7$

b $gf(x)$ 　　　　　　　　　f acts on x first and $f(x) = 2x + 1$
　$= g(2x + 1)$ 　　　　　　g is the function 'square and add 3'
　$= (2x + 1)^2 + 3$ 　　　$(2x + 1)^2 = (2x + 1)(2x + 1) = 4x^2 + 2x + 2x + 1$
　$= 4x^2 + 4x + 1 + 3$
　$= 4x^2 + 4x + 4$

c $ff(x)$ 　　　　　　　　　f acts on x first and $f(x) = 2x + 1$
　$= f(2x + 1)$ 　　　　　　f is the function 'double and add 1'
　$= 2(2x + 1) + 1$
　$= 4x + 3$

EXERCISE 5.5

1. $f(x) = 2x$ and $g(x) = x - 8$ Find   **a** $fg(5)$ **b** $gf(5)$

2. $f(x) = 3x + 1$ and $g(x) = 2 - 3x$ Find   **a** $fg(4)$ **b** $gf(4)$

3. $f(x) = x - 3$ and $g(x) = x^2$ Find   **a** $fg(-2)$ **b** $gf(-2)$

4. $f(x) = 5 + \dfrac{6}{x}$ and $g(x) = x - 4$ Find   **a** $fg(1)$ **b** $gf(1)$

5. $f(x) = \dfrac{4}{x+1}$ and $g(x) = \dfrac{1}{x-3}$ Find   **a** $fg(3)$ **b** $gf(3)$

6. $f(x) = (x + 1)^2$, $g(x) = 3 - x$ and $h(x) = \dfrac{1}{x-4}$
   Find   **a** $f(-4)$ **b** $gf(2)$ **c** $fh(6)$ **d** $fgh(5)$ **e** $ff(-2)$

7. $p : x \to x - 5$, $q : x \to \dfrac{x}{2}$ and $r : x \to x^2 + 3$
   Find   **a** $q(-12)$ **b** $qr(5)$ **c** $pq(-4)$ **d** $pqr(6)$ **e** $qq(3)$

8. If $f(x) = 2x + 3$ and $g(x) = 2x - 3$, match these cards

   | **a** $fg(x)$ | **b** $gf(x)$ | **c** $gg(x)$ | **d** $ff(x)$ |

   | **A** $4x + 3$ | **B** $4x - 9$ | **C** $4x + 9$ | **D** $4x - 3$ |

9. $f(x) = x - 3$ and $g(x) = x + 8$
   Find   **a** $fg(x)$ **b** $gf(x)$ **c** $ff(x)$

10. $p(x) = 2x + 1$ and $q(x) = x - 5$
    Find   **a** $qp(x)$ **b** $pq(x)$ **c** $pp(x)$

11. $f(x) = x^2 + 3$ and $g(x) = 5x$
    Find   **a** $gf(x)$ **b** $fg(x)$ **c** $gg(x)$

12. $p(x) = \dfrac{3}{x-2}$ and $q(x) = x + 1$
    Find   **a** $pq(x)$ **b** $qp(x)$

13. $f(x) = x + 6$ and $g(x) = 3x$
    Find x if $fg(x) = 27$

14. $f(x) = 2x - 3$ and $g(x) = \dfrac{x}{5}$
    Find x if $gf(x) = 6$

15. $f(x) = x^2 + 2$ and $g(x) = \dfrac{1}{x-2}$
    Find x if $gf(x) = 4$

Functions

Inverse functions

The inverse of a function $f(x)$ is the function that undoes what $f(x)$ has done.
The inverse of the function $f(x)$ is written as $f^{-1}(x)$.

If $f(x) = x + 4$ then $f^{-1}(x) = x - 4$.
If $f(x) = x - 5$ then $f^{-1}(x) = x + 5$.
If $f(x) = 2x$ then $f^{-1}(x) = \dfrac{x}{2}$.

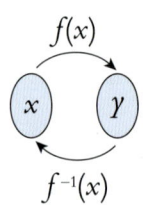

For many straightforward functions the inverse can be found using a flow chart.
The flow chart for finding the inverse of the function $f(x) = \dfrac{3x+1}{4}$ is

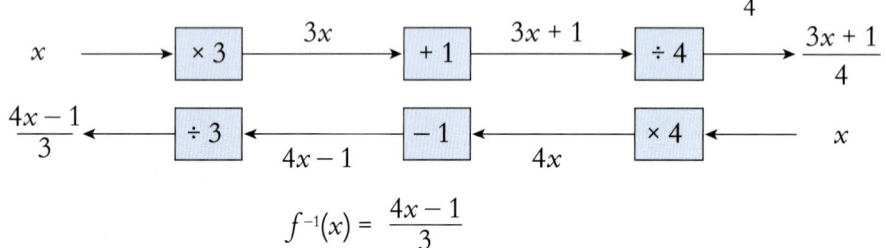

$x \longrightarrow \boxed{\times 3} \xrightarrow{3x} \boxed{+1} \xrightarrow{3x+1} \boxed{\div 4} \longrightarrow \dfrac{3x+1}{4}$

$\dfrac{4x-1}{3} \longleftarrow \boxed{\div 3} \xleftarrow{4x-1} \boxed{-1} \xleftarrow{4x} \boxed{\times 4} \longleftarrow x$

$$f^{-1}(x) = \dfrac{4x-1}{3}$$

You can also find the inverse of the function $f(x) = \dfrac{3x+1}{4}$ using an algebraic method:

Step 1: Write the function as $y = \quad\longrightarrow\quad y = \dfrac{3x+1}{4}$

Step 2: Interchange the x and y variables. $\longrightarrow\quad x = \dfrac{3y+1}{4}$

Step 3: Rearrange to make y the subject. $\longrightarrow\quad x = \dfrac{3y+1}{4}$

$$4x = 3y + 1$$
$$4x - 1 = 3y$$

➡ **CHECK:**

$f(5) = \dfrac{3 \times 5 + 1}{4} = \dfrac{15+1}{4} = 4$

$f^{-1}(4) = \dfrac{4 \times 4 - 1}{3} = \dfrac{16-1}{3} = 5$

$$y = \dfrac{4x-1}{3}$$
$$f^{-1}(x) = \dfrac{4x-1}{3}$$

EXAMPLE

Find the inverse of the function $f(x) = \dfrac{5}{x-2}$, $x \neq 2$.

Step 1: $y = \dfrac{5}{x-2}$ $\longrightarrow$ Step 2: $x = \dfrac{5}{y-2}$ $\longrightarrow$ Step 3: $x(y-2) = 5$
$xy - 2x = 5$
$xy = 2x + 5$

➡ **CHECK:**

$f(7) = \dfrac{5}{7-2} = \dfrac{5}{5} = 1$

$f^{-1}(1) = \dfrac{2 \times 1 + 5}{1} = \dfrac{7}{1} = 7$

$$y = \dfrac{2x+5}{x}$$
$$f^{-1}(x) = \dfrac{2x+5}{x}$$

EXERCISE 5.6

1 Write down the inverse of these functions.
- **a** $f(x) = x + 2$
- **b** $f(x) = x - 7$
- **c** $f(x) = 22 + x$
- **d** $f(x) = x - 0.5$
- **e** $f(x) = 3x$
- **f** $f(x) = 8x$
- **g** $f(x) = \dfrac{x}{4}$
- **h** $f(x) = \dfrac{x}{6}$

2 Find the inverse of these functions.
- **a** $f(x) = 2x - 5$
- **b** $f(x) = 3x + 8$
- **c** $f(x) = 5x - 2$
- **d** $f(x) = 3 + 4x$
- **e** $f(x) = 3(x + 2)$
- **f** $f(x) = 4(x - 7)$
- **g** $f(x) = 2(x - 10)$
- **h** $f(x) = 2(5 + x)$
- **i** $f(x) = 2(5x + 4)$
- **j** $f(x) = 3(5 - 2x)$
- **k** $f(x) = 4(7 - 3x)$
- **l** $f(x) = 3(2x - 9)$
- **m** $f(x) = \dfrac{x - 2}{4}$
- **n** $f(x) = \dfrac{x + 6}{5}$
- **o** $f(x) = \dfrac{2x - 3}{4}$
- **p** $f(x) = \dfrac{5 - 3x}{6}$

3 Find the inverse of these functions.
- **a** $f(x) = \dfrac{5}{x} - 3$
- **b** $f(x) = \dfrac{3}{x} + 2$
- **c** $f(x) = \dfrac{6}{x} - 7$
- **d** $f(x) = 4 - \dfrac{5}{x}$
- **e** $f(x) = \dfrac{2}{3 + x}$
- **f** $f(x) = \dfrac{4}{2x + 1}$
- **g** $f(x) = \dfrac{3}{2x - 5}$
- **h** $f(x) = \dfrac{6}{3 - 2x}$
- **i** $f(x) = \dfrac{2x + 1}{x - 3}$
- **j** $f(x) = \dfrac{x - 7}{2x + 5}$
- **k** $f(x) = \dfrac{x + 6}{x - 8}$
- **l** $f(x) = \dfrac{1 - 2x}{3 - 4x}$

4 Find the inverse of these functions.
- **a** $f(x) = (x + 3)^3$
- **b** $f(x) = (x - 5)^3$
- **c** $f(x) = (2x + 7)^3$
- **d** $f(x) = (5 - 3x)^3$

5 If $f(x) = 2x + 3$, find
- **a** $f^{-1}(0)$
- **b** $f^{-1}(7)$
- **c** $f^{-1}(2)$

6 If $f(x) = 6 - 5x$, find
- **a** $f^{-1}(4)$
- **b** $f^{-1}(-2)$
- **c** $f^{-1}(6)$

7 If $f(x) = 2x - 1$, solve the equation $f(x) = f^{-1}(x)$

8 If $f(x) = 3x + 2$, solve the equation $f(x) = f^{-1}(x)$

9 An important result to remember is: $f^{-1}f(x) = ff^{-1}(x) = x$. This is because f^{-1} undoes what f has done. So if you have the function $f(x) = 3x - 1$ and you are asked to write down the value of $f^{-1}f(5)$ the answer is 5. Similarly $ff^{-1}(2) = 2$

- **a** If $f(x) = 2x - 9$, write down the value of $f^{-1}f(3)$
- **b** If $f(x) = 2(5x - 11)$, write down the value of $f^{-1}f(2.5)$
- **c** If $f(x) = \dfrac{2x - 9}{3x + 4}$, write down the value of $ff^{-1}(-29)$

KEY WORDS

mapping
function
composite function
inverse function

Factorising 2

> **THIS SECTION WILL SHOW YOU HOW TO**
> - Factorise a quadratic expression

A **quadratic** expression is an expression of the form $ax^2 + bx + c$ where a, b and c are numbers and $a \neq 0$.
Examples of quadratic expressions are $\quad 3x^2 - 5x + 4 \quad 2x^2 - 6x \quad x^2 - 9$

Factorising quadratic expressions of the form $ax^2 + bx$
Reminder: $\quad x^2 + 3x = x(x + 3)$
$\qquad\qquad 5x - x^2 = x(5 - x)$
$\qquad\qquad 6x^2 - 4x = 2x(3x - 2)$

Factorising quadratic expressions of the form $x^2 + bx + c$
Factorising is the reverse of expanding.
To factorise you need to find two numbers that multiply to give the **constant** and also add to give the **coefficient** of x.

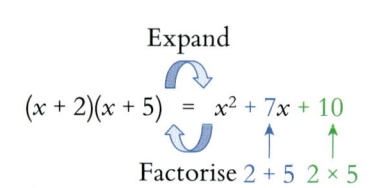

Expand
$(x + 2)(x + 5) = x^2 + 7x + 10$
Factorise $\quad 2 + 5 \quad 2 \times 5$

EXAMPLE

Factorise   **a** $x^2 + 10x + 16$ **b** $x^2 - 2x - 15$ **c** $x^2 - 5x + 6$

a $x^2 + 10x + 16$
List the pairs of numbers that **multiply** to give 16 and then check to see which pair **add** to give 10.
1 and 16   ⬭2 and 8⬭   4 and 4
so
$x^2 + 10x + 16 = (x + 2)(x + 8)$
➡ **CHECK:**
$(x + 2)(x + 8) = x^2 + 8x + 2x + 16 = x^2 + 10x + 16$ ✓

b $x^2 + 2x - 15$
List the pairs of numbers that **multiply** to give −15 and then check to see which pair **add** to give −2.
1 and −15   ⬭3 and −5⬭   −1 and 15   −3 and 5
so
$x^2 - 2x - 15 = (x + 3)(x - 5)$
➡ **CHECK:**
$(x + 3)(x - 5) = x^2 - 5x + 3x - 15 = x^2 - 2x - 15$ ✓

c $x^2 - 5x + 6$
List the pairs of numbers that **multiply** to give 6 and then check to see which pair **add** to give −5.
1 and 6   2 and 3   −1 and −6   ⬭−2 and −3⬭
so
$x^2 - 5x + 6 = (x - 2)(x - 3)$
➡ **CHECK:**
$(x - 2)(x - 3) = x^2 - 3x - 2x + 6 = x^2 - 5x + 6$ ✓

EXERCISE 5.7

1 Factorise these expressions.
- **a** $x^2 + 5x$
- **b** $x^2 - 7x$
- **c** $3x^2 - 15x$
- **d** $12x - 18x^2$

2 Factorise these expressions.
- **a** $x^2 + 4x + 3$
- **b** $x^2 + 7x + 10$
- **c** $x^2 + 2x + 1$
- **d** $x^2 + 9x + 18$
- **e** $x^2 + 6x + 8$
- **f** $x^2 + 6x + 5$
- **g** $x^2 + 5x + 6$
- **h** $x^2 + 20x + 36$
- **i** $x^2 + 10x + 25$
- **j** $x^2 + 13x + 36$
- **k** $x^2 + 9x + 20$
- **l** $x^2 + 9x + 14$
- **m** $x^2 + 10x + 21$
- **n** $x^2 + 11x + 28$
- **o** $x^2 + 22x + 21$
- **p** $x^2 + 8x + 15$

3 Factorise these expressions.
- **a** $x^2 + x - 6$
- **b** $x^2 - 5x - 6$
- **c** $x^2 + 4x - 21$
- **d** $x^2 - 6x - 16$
- **e** $x^2 + 6x - 16$
- **f** $x^2 - 15x - 16$
- **g** $x^2 + 6x - 7$
- **h** $x^2 - 3x - 40$
- **i** $x^2 + 17x - 38$
- **j** $x^2 - 8x - 33$
- **k** $x^2 + 7x - 30$
- **l** $x^2 - 2x - 15$
- **m** $x^2 + x - 20$
- **n** $x^2 - 19x - 20$
- **o** $x^2 + 20x - 21$
- **p** $x^2 - 6x - 7$

4 Factorise these expressions.
- **a** $x^2 - 7x + 10$
- **b** $x^2 - 16x + 15$
- **c** $x^2 - 10x + 21$
- **d** $x^2 - 8x + 12$
- **e** $x^2 - 11x + 10$
- **f** $x^2 - 7x + 12$
- **g** $x^2 - 8x + 7$
- **h** $x^2 - 6x + 9$
- **i** $x^2 - 11x + 30$
- **j** $x^2 - 11x + 28$
- **k** $x^2 - 22x + 21$
- **l** $x^2 - 8x + 15$
- **m** $x^2 - 9x + 14$
- **n** $x^2 - 10x + 24$
- **o** $x^2 - 14x + 13$
- **p** $x^2 - 15x + 26$

5

> You may need to take out a common factor before using double brackets.
>
> $2x^2 - 10x - 72 \longrightarrow 2(x^2 - 5x - 36) \longrightarrow 2(x - 9)(x + 4)$
>
> Common factor   Double brackets

Factorise these expressions.
- **a** $2x^2 + 12x + 16$
- **b** $3x^2 - 3x - 16$
- **c** $2x^2 - 8x - 10$
- **d** $3x^2 - 36x + 33$
- **e** $4x^2 + 16x + 12$
- **f** $2x^2 + 20x + 18$
- **g** $2x^2 - 12x + 18$
- **h** $3x^2 - 15x - 108$
- **i** $5x^2 - 20x - 25$
- **j** $4x^2 + 32x + 60$
- **k** $7x^2 + 21x + 14$
- **l** $2x^2 + 4x - 70$

6 To simplify algebraic fractions you must first factorise and then cancel any common factors.

$$\frac{x^2 + 9x + 14}{x + 7} = \frac{\cancel{(x+7)}(x+2)}{\cancel{x+7}} = x + 2$$

Simplify these algebraic fractions.
- **a** $\dfrac{x^2 + 4x + 3}{x + 1}$
- **b** $\dfrac{x^2 + 8x + 15}{x + 5}$
- **c** $\dfrac{x^2 + 11x + 18}{x + 2}$
- **d** $\dfrac{x^2 + 13x + 22}{x + 2}$
- **e** $\dfrac{x^2 + 2x - 3}{x + 3}$
- **f** $\dfrac{x^2 - 5x - 36}{x - 9}$
- **g** $\dfrac{x - 1}{x^2 + 23x - 24}$
- **h** $\dfrac{x^2 + 5x + 6}{x^2 - 2x - 8}$
- **i** $\dfrac{x^2 - 3x - 40}{x^2 + 2x - 15}$
- **j** $\dfrac{x^2 + 3x - 28}{x^2 - 9x + 20}$
- **k** $\dfrac{x^2 - 5x}{x^2 + x - 30}$
- **j** $\dfrac{x^2 + x - 2}{x^2 + 2x}$

Factorising 2

The difference of two squares

$$(x+5)(x-5) \xrightleftharpoons[\text{Factorise}]{\text{Expand}} x^2 - 25$$

> In general $x^2 - a^2 = (x + a)(x - a)$
> This result is called the **difference of two squares**.

EXAMPLE

Factorise these expressions   **a** $x^2 - 81$ **b** $25x^2 - 36$

a $x^2 - 81 = (x + 9)(x - 9)$
  ➡ **CHECK:**
  $(x + 9)(x - 9) = x^2 - 9x + 9x - 81 = x^2 - 81$ ✓

b $25x^2 - 36 = (5x + 6)(5x - 6)$
  ➡ **CHECK:**
  $(5x + 6)(5x - 6) = 25x^2 - 30x + 30x - 36$
  $ = 25x^2 - 36$ ✓

Factorising expressions of the form $ax^2 + bx + c$

Factorising quadratic expressions with more than one x^2 is difficult.
To factorise $2x^2 + 7x + 3$:
- first look at the x^2 term, $2x^2$ must come from $2x \times x$, so use $(2x + ?)(x + ?)$
- next look at the constant term, 3 must come from 3×1 or 1×3
- expand both options, $(2x + 3)(x + 1)$ and $(2x + 1)(x + 3)$, to find which is correct.
$(2x + 3)(x + 1) = 2x^2 + 2x + 3x + 3 = x^2 + 5x + 3$ ✗
$(2x + 1)(x + 3) = 2x^2 + 6x + x + 3 = x^2 + 7x + 3$ ✓
So $2x^2 + 7x + 3 = (2x + 1)(x + 3)$

EXAMPLE

Factorise $3x^2 - 10x - 80$

$3x^2$ comes from $3x \times x$, so use $(3x + ?)(x + ?)$
-8 comes from 1×-8 or -1×8 or 2×-4 or -2×4
$(3x + 1)(x - 8) = 3x^2 - 24x + x - 8 = 3x^2 - 23x - 8$ ✗
$(3x - 1)(x + 8) = 3x^2 + 24x - x - 8 = 3x^2 + 23x - 8$ ✗
$(3x + 2)(x - 4) = 3x^2 - 12x + 2x - 8 = 3x^2 - 10x - 8$ ✓
$(3x - 2)(x + 4) = 3x^2 + 12x - 2x - 8 = 3x^2 + 10x - 8$ ✗
So $3x^2 - 10x - 8 = (3x + 2)(x - 4)$

EXERCISE 5.8

1 Factorise these expressions.
- **a** $x^2 - 81$
- **b** $y^2 - 1$
- **c** $x^2 - 64$
- **d** $a^2 - 36$
- **e** $y^2 - 100$
- **f** $16 - a^2$
- **g** $x^2 - y^2$
- **h** $144 - x^2$
- **i** $4x^2 - 9$
- **j** $16y^2 - 1$
- **k** $25x^2 - 36$
- **l** $9 - 16x^2$
- **m** $100x^2 - 81y^2$
- **n** $25a^2 - 49b^2$
- **o** $81x^2 - 49y^2$
- **p** $121x^2 - 100$

2 Factorise these expressions.
- **a** $2x^2 - 50$
- **b** $2y^2 - 8$
- **c** $3x^2 - 3$
- **d** $5y^2 - 80$
- **e** $3y^2 - 192$
- **f** $4x^2 - 4y^2$
- **g** $2x^2 - 242$
- **h** $6n^2 - 600$
- **i** $75y^2 - 108$

> **HINT**
> Take out a common factor first.

3 Factorise $x^4 - y^4$

> **HINT**
> You will need to factorise twice.

4 Factorise these expressions.
- **a** $2x^2 + 11x + 5$
- **b** $2x^2 + 5x + 3$
- **c** $2x^2 + 15x + 7$
- **d** $2x^2 + 13x + 11$
- **e** $3x^2 + 5x + 2$
- **f** $3x^2 + 16x + 5$
- **g** $3x^2 + 10x + 3$
- **h** $3x^2 + 10x + 7$
- **i** $2x^2 + 5x - 3$
- **j** $3x^2 + 2x - 5$
- **k** $3x^2 - 22x + 7$
- **l** $5x^2 - 2x - 3$
- **m** $7x^2 - 9x + 2$
- **n** $5x^2 + 3x - 2$
- **o** $11x^2 - 21x - 2$
- **p** $2x^2 - 15x + 13$

5 Factorise these expressions.
- **a** $2x^2 + 11x + 12$
- **b** $2x^2 + 9x + 10$
- **c** $3x^2 + 14x + 8$
- **d** $7x^2 + 38x + 15$
- **e** $6x^2 + 7x + 2$
- **f** $6x^2 + 11x + 5$
- **g** $8x^2 + 14x + 3$
- **h** $10x^2 + 71x + 7$
- **i** $3x^2 + 7x - 6$
- **j** $5x^2 + 22x + 8$
- **k** $2x^2 - 11x + 12$
- **l** $7x^2 + 26x - 8$
- **m** $5x^2 + 21x - 20$
- **n** $3x^2 + x - 10$
- **o** $2x^2 - 9x + 10$
- **p** $3x^2 + 14x + 16$

6 Factorise these expressions.
- **a** $4x^2 + 2x - 42$
- **b** $6x^2 + 21x + 9$
- **c** $6x^2 + 16x + 8$
- **d** $12x^2 + 28x + 8$
- **e** $16x^2y - 20xy - 6y$
- **f** $6x^3 - 10x^2 - 4x$

> **HINT**
> Take out a common factor first.

7 Simplify these algebraic fractions.
- **a** $\dfrac{2x^2 + 7x + 3}{x^2 + 3x}$
- **b** $\dfrac{3x^2 + 5x - 2}{x^2 - 4}$
- **c** $\dfrac{2x^2 + 3x - 20}{x^2 + 6x + 8}$
- **d** $\dfrac{4x^2 - 9}{2x^2 - x - 3}$
- **e** $\dfrac{x^2 - 5x}{2x^2 - 7x - 15}$
- **f** $\dfrac{2x^2 - 50}{3x^2 + 7x - 40}$

> **HINT**
> Factorise the numerator and the denominator before cancelling factors.

8 Factorise $(3x + 2)^4 - 4(3x + 2)^2$

> **KEY WORDS**
> quadratic
> coefficient, constant
> difference of two squares

Factorising 2 233

Cubic graphs

THIS SECTION WILL SHOW YOU HOW TO
- Recognise cubic graphs
- Draw cubic graphs
- Use cubic graphs in practical situations

$y = ax^3 + bx^2 + cx + d$ is called a **cubic function** where a, b, c and d are constants and $a \neq 0$.
The simplest cubic function is $y = x^3$.
The curve has rotational symmetry of order 2 about the origin.

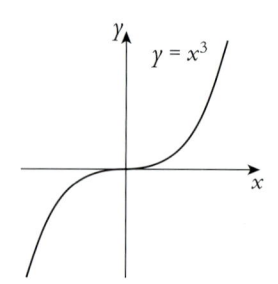

EXAMPLE

Draw the graph of $y = x^3 - 5x + 3$ for $-3 \leq x \leq 3$.
Use your graph to find the values of x when $y = 5$.

First work out a table of values for the graph.
When $x = -3$, $y = (-3)^3 - (5 \times -3) + 3$
$y = -27 - -15 + 3$
$y = -27 + 15 + 3$
$y = -9$ etc

x	-3	-2	-1	0	1	2	3
y	-9	5	7	3	-1	1	15

Use your table of values to draw the graph of the function.

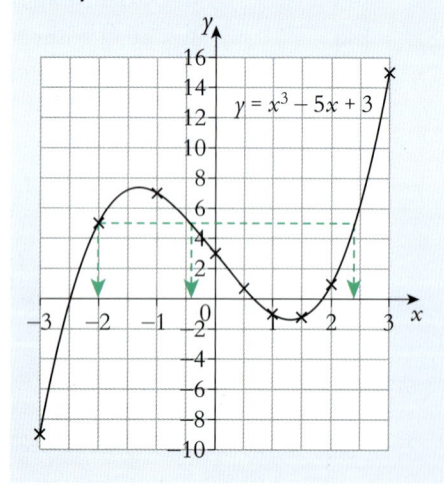

To find the values of x when $y = 5$ you need to draw the line $y = 5$ and then find the x coordinates of the three points where the line intersects the curve.

The graphs intersect in three places.
$x = -2$ $x = -0.4$ $x = 2.4$

The general shape of a cubic graph depends on the value of a

a is positive a is negative

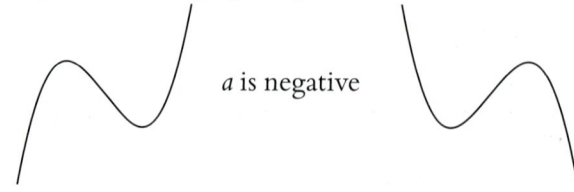

EXERCISE 5.9

1 Copy and complete the table for each of these cubic functions.
Draw the graph of each function.

a $y = x^3 + 1$

x	−3	−2	−1	0	1	2	3
y						9	

b $y = 2 − x^3$

x	−3	−2	−1	0	1	2	3
y			3				

c $y = −x^3$

x	−3	−2	−1	0	1	2	3
y						−8	

d $y = x^3 + 2x$

x	−3	−2	−1	0	1	2	3
y					3		

e $y = 3x^2 − x^3$

x	−2	−1	0	1	2	3	4
y					4		

f $y = x^3 − 2x^2 − 4x$

x	−2	−1	0	1	2	3	4
y		1					

g $y = x^3 − 4x^2 + 2$

x	−1	0	1	2	3	4
y				−6		

h $y = \frac{1}{2}x^3 − 3x^2 − 2$

x	−6	−5	−4	−3	−2	−1	0	1	2
y				10					

2 a Copy and complete the table of values for
$y = x^3 + 2$

x	−3	−2	−1	0	1	2	3
y			1				

b Draw the graph of $y = x^3 + 2$
c Use your graph to find the value of x when $y = 7$
d By drawing a tangent to the graph estimate the gradient of the curve when $x = 1$.

3 a Copy and complete the table of values for
$y = \frac{1}{2}x^3 − 2x^2 + 6$

x	−2	−1	0	1	2	3	4
y					2		

b Draw the graph of $y = \frac{1}{2}x^3 − 2x^2 + 6$
c Use your graph to find the values of x when $y = 4$
d By drawing a tangent to the graph estimate the gradient of the curve when $x = 2$.

4 The function $f(x) = \frac{1}{2}x^3$, draw the graph of $y = f(x)$ for $−3 \leq x \leq 3$
On the same axes draw the graph of $y = 2x$
Write down the coordinates of the points of intersection of the curve and the line.

5 Draw the graph of $y = x^3 − 4x^2 + 3$ for $−1 \leq x \leq 4$
On the same axes draw the graph of $x + y = 1$
Write down the coordinates of the points of intersection of the curve and the line.

6 The function $f(x) = x^3 − 3x^2$, draw the graph of $y = f(x)$ for $−2 \leq x \leq 4$
Write down the values of x where the graph has negative gradient.

Cubic graphs

Practical applications of cubic graphs

EXAMPLE

A square piece of card has sides of length 20 cm.
A square of length x cm is removed from each of the four corners of the card.
The remaining piece of card is then folded to make an open box as shown below.

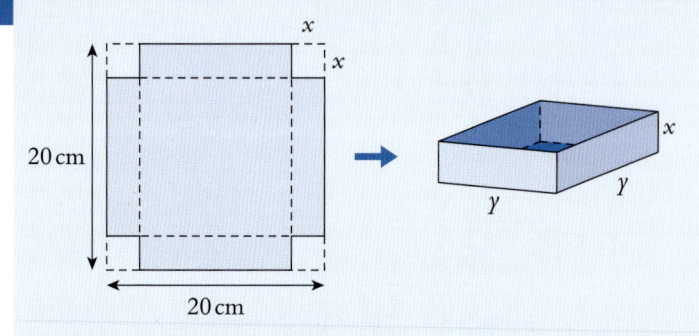

a Show that the volume, V cm³, of the box is given by the formula
$V = 400x - 80x^2 + 4x^3$
b Draw the graph of $V = 400x - 80x^2 + 4x^3$ for $0 \leq x \leq 10$.
c Use your graph to estimate the maximum possible volume of the box and the value of x that gives the maximum volume.

---

a $y = 20 - 2x$

$$\begin{aligned}\text{Volume} &= y \times y \times x \\ &= (20 - 2x)(20 - 2x)x \quad \text{expand brackets} \\ &= (400 - 40x - 40x + 4x^2)x \\ &= 400x - 80x^2 + 4x^3\end{aligned}$$

b

x	0	1	2	3	4	5	6	7	8	9	10
V	0	324	512	588	576	500	384	252	128	36	0

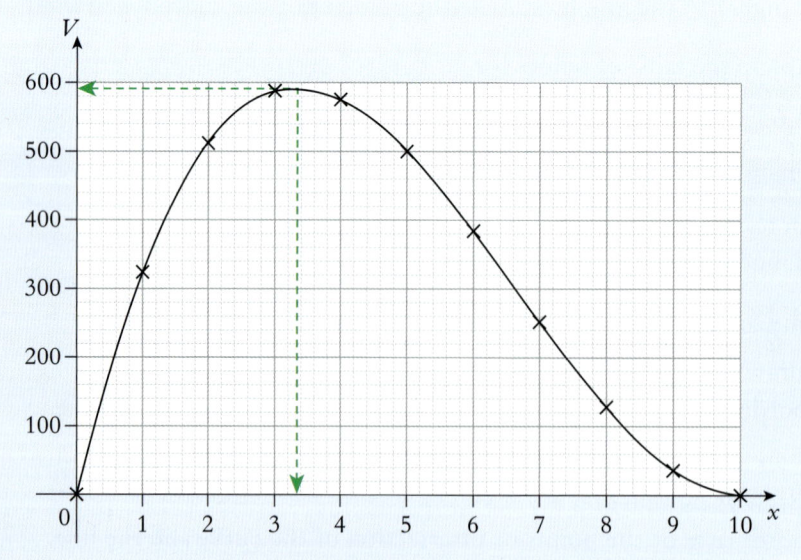

c Maximum volume ≈ 590 when $x ≈ 3.3$

EXERCISE 5.10

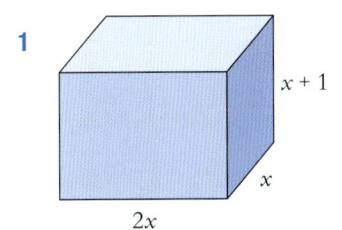

1

The rectangular tank has sides of length $2x$, x and $(x+1)$ m.

a Show that the volume, V cm^3, of the tank is given by the formula
$$V = 2x^3 + 2x^2$$

b Copy and complete the table of values.

x	0	0.2	0.4	0.6	0.8	1
V	0					

c Draw the graph of $V = 2x^3 + 2x^2$ for $0 \leq x \leq 1$

d Use your graph to estimate the value of x that gives a volume of 2 m^3.

2 A rectangular piece of card has sides of length 20 cm and 16 cm.
A square of length x cm is removed from each of the four corners of the card.
The remaining piece of card is then folded to make an open box as shown below.

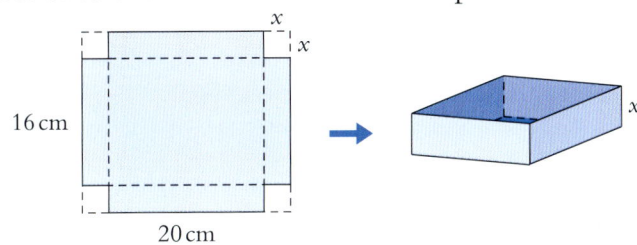

a Show that the volume, V cm^3, of the box is given by the formula
$$V = 320x - 72x^2 + 4x^3$$

b Copy and complete the table of values.

x	0	1	2	3	4	5	6	7	8
V	0								

c Draw the graph of $V = 320x - 72x^2 + 4x^3$ for $0 \leq x \leq 8$

d Use your graph to estimate the maximum possible volume of the box and the value of x that gives the maximum volume.

KEY WORDS
cubic function

Surface area and volume 2

THIS SECTION WILL SHOW YOU HOW TO
- Find the surface area and volume of pyramids, cones and spheres

Pyramids

Volume of a pyramid = $\frac{1}{3}$ × base area × perpendicular height
Surface area = sum of areas of all the faces of the pyramid

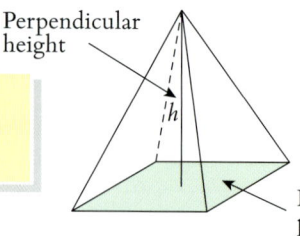

EXAMPLE

ABCDE is a square-based pyramid.
AB = BC = 6 cm.
The perpendicular height of the pyramid is 4 cm.
P is the mid-point of BC and O is the middle of the base.
a Calculate the volume of the pyramid.
b Calculate the length of EP.
c Calculate the area of △ BCE.
d Calculate the surface area of the pyramid.

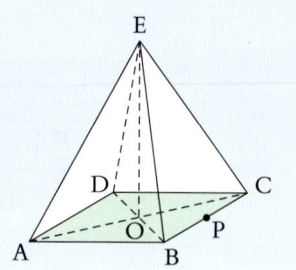

a Volume of pyramid = $\frac{1}{3}$ × base area × perpendicular height

= $\frac{1}{3}$ × (6 × 6) × 4

= 48 cm³

b Using Pythagoras on △ EOP: $EP^2 = 4^2 + 3^2$
$EP^2 = 25$
$EP = 5$ cm

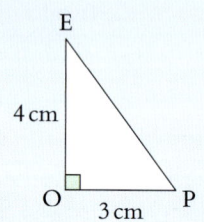

c Area of △ BCE = $\frac{1}{2}$ × base × height

= $\frac{1}{2}$ × 6 × 5 = 15 cm²

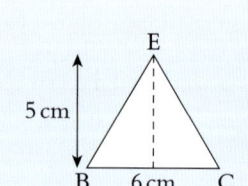

d Surface area = sum of all the faces of the pyramid
= area of base + area of four triangular faces
= 36 + (4 × 15)
= 96 cm²

238 UNIT 5

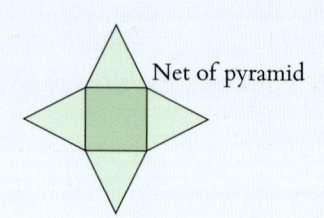

EXERCISE 5.11

1. Find the volume of each of these pyramids.
   The base area (A cm²) and the perpendicular height (h cm) are given for each pyramid.

   a) $A = 15, h = 8$
   b) $A = 11, h = 9$
   c) $A = 28, h = 6$
   d) $A = 12, h = 5$

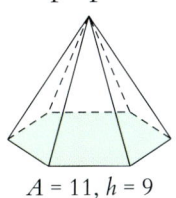

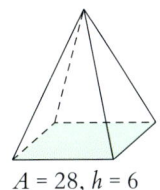

2. A pyramid has a square base of side 8 cm.
   The perpendicular height is 6 cm.
   Calculate the volume of the pyramid.

3. A pyramid has a rectangular base of side 6 cm and 10 cm.
   The perpendicular height is 7 cm.
   Calculate the volume of the pyramid.

4. A pyramid has an equilateral triangular base of side 5 cm.
   The perpendicular height is 4 cm.
   Calculate the volume of the pyramid.

5. A pyramid has a square base of side x cm.
   The perpendicular height is x cm.
   The volume of the pyramid is 72 cm³. Find the value of x.

6. The Great Pyramid of Khutu in Egypt is a square-based pyramid.
   The square base has sides of length 241 m.
   The perpendicular height of the pyramid is 153 m.
   Calculate the volume of the pyramid.
   Give your answer in standard form correct to 3 s.f.

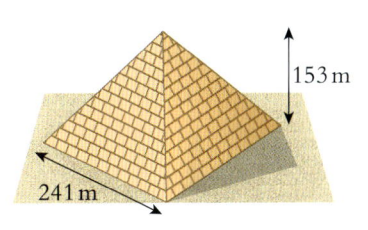

7. The Louvre Pyramid in France is a large glass and metal square-based pyramid.
   The square base has sides of length 35.42 m.
   The perpendicular height of the pyramid is 21.64 m.
   Calculate the volume of the pyramid.
   Give your answer in standard form correct to 3 s.f.

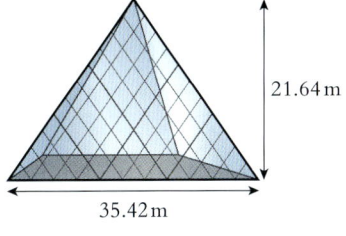

8. A crystal is in the form of a **regular octahedron**.
   The edges of the octahedron are all 4 mm long.
   Calculate the volume of the crystal.

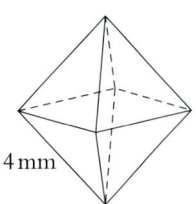

9 A square-based pyramid has a base of side 8 cm and a height of 12 cm.
A square-based pyramid of height 6 cm is removed by cutting parallel to the base, leaving the solid shown. Calculate the volume of the **truncated pyramid**.

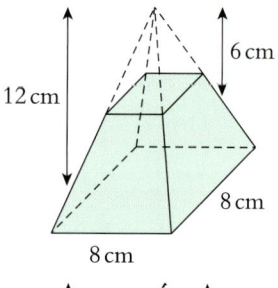

10 A rectangular-based pyramid has a base of sides 12 cm and 9 cm and a height of 15 cm.
A rectangular-based pyramid of height 5 cm is removed by cutting parallel to the base, leaving the solid shown. Calculate the volume of the truncated pyramid.

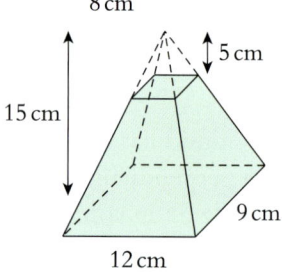

11 Find the surface area of each of these pyramids.

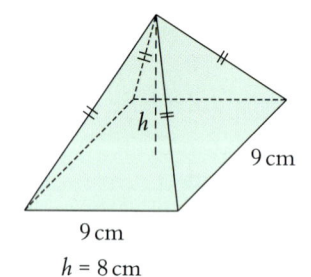

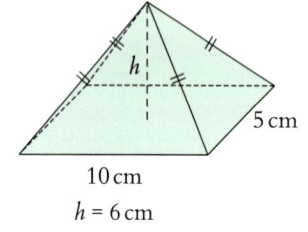

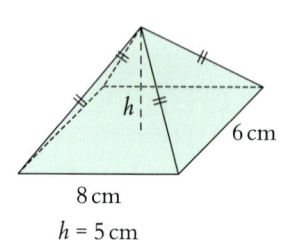

a 5 cm, 5 cm, $h = 6$ cm

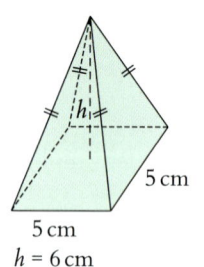

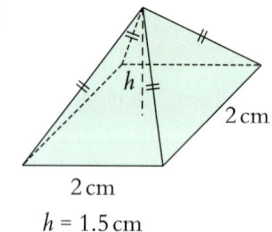

b 9 cm, 9 cm, $h = 8$ cm

c 2 cm, 2 cm, $h = 1.5$ cm

d 10 cm, 5 cm, $h = 6$ cm

e 8 cm, 6 cm, $h = 5$ cm

f 15 cm, 9 cm, $h = 10$ cm

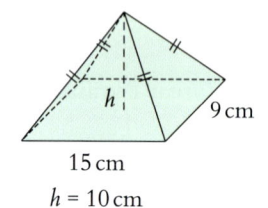

12 Find the surface area of these **regular tetrahedrons**.

a 10 cm

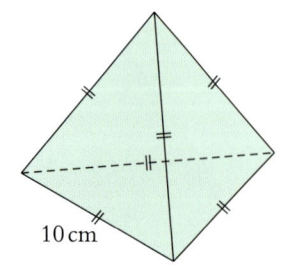

b 4.8 cm

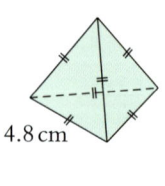

HINT

Tetrahedron is the name for a triangular-based pyramid (tetra- for 4 faces).
Regular means that all the sides are the same length so each face is an equilateral triangle.

13 Find the volume of each regular tetrahedron in question 12.

Cones

A **cone** is a pyramid with a circular base.

Volume of a pyramid = $\frac{1}{3}$ × base area × perpendicular height

So, Volume of a cone = $\frac{1}{3} \times \pi r^2 \times h$

The surface of a cone is made from a flat circular base and a curved surface.
The curved surface is made from a sector of a circle.

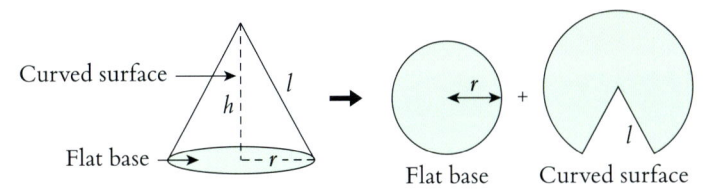

> Volume of a cone = $\frac{1}{3}\pi r^2 h$
> Curved surface area of a cone = $\pi r l$, where l is the slant height
> Total surface area of cone = $\pi r^2 + \pi r l$

When you make a cut parallel to the base of a cone and remove the top part, the part that is left is called a **frustum**.

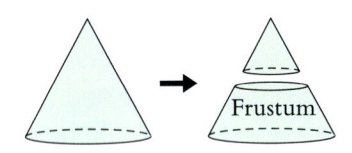

> Volume of frustum = volume of whole cone − volume of smaller cone

EXAMPLE

The cone has base radius 3 cm and perpendicular height 4 cm. 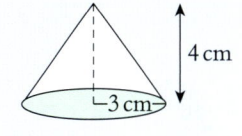
a Calculate the volume of the cone.
b Calculate the curved surface area of the cone.
c Calculate the total surface area of the cone

a Volume = $\frac{1}{3} \times \pi r^2 h = \frac{1}{3} \times \pi \times 3^2 \times 4 = 37.7$ cm³

b Curved surface area = $\pi r l$
   (so you must first calculate the slant height, l)
   Using Pythagoras: $l^2 = 3^2 + 4^2$
   $l^2 = 25$
   $l = 5$ 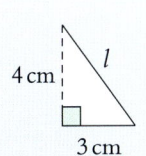
   Curved surface area = $\pi r l = \pi \times 3 \times 5 = 47.1$ cm²

c Total surface area = $\pi r^2 + \pi r l = \pi \times 3^2 + 47.1 = 75.4$ cm²

Surface area and volume 2

EXERCISE 5.12

1. Calculate the volume of these cones.

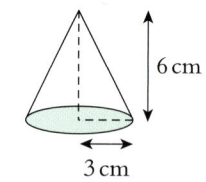

   a 7 cm height, 4 cm radius

   b 6 cm height, 3 cm radius

   c 9 cm height, 5 cm radius

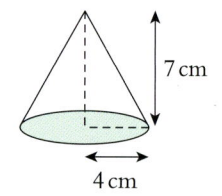

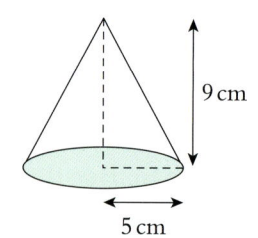

2. A solid cone has a base radius of 4 cm.
   The volume of the cone is 100 cm³.
   Calculate the height of the cone.

3. A solid cone has a height of 12 cm.
   The volume of the cone is 250 cm³.
   Calculate the radius of the base of the cone.

4. The diagram shows a toy rocket made from a cylinder and a cone.
   Calculate the volume of the toy rocket.

   (Cone: 8 cm high; Cylinder: 10 cm high, 4 cm diameter)

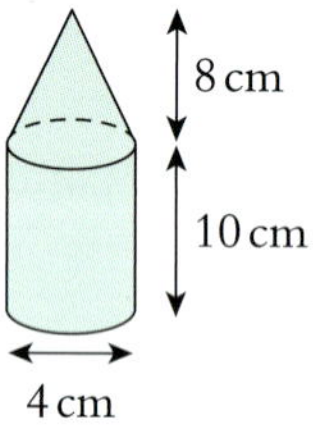

5. A drinking glass is a truncated cone.
   Calculate the volume of the drinking glass.

   (Top diameter 10 cm, 15 cm height of glass, 15 cm to apex)

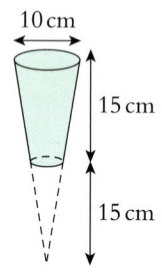

6. Calculate the volume of each of these frustums.

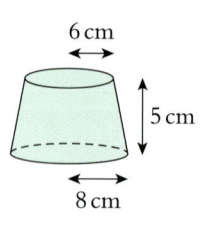

   a top 3 cm, height 10 cm, base 6 cm

   b top 6 cm, height 5 cm, base 8 cm

   c top 3 cm, height 6 cm, base 12 cm

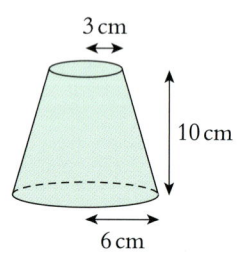

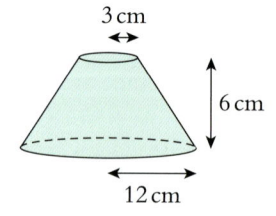

7. The diagram shows a frustum of a cone.
   Find an expression in terms of r and h for the volume of the frustum.

   (Top radius r, height h, base diameter $2r$)

242 UNIT 5

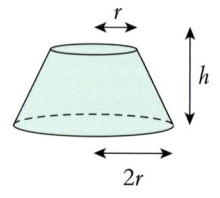

8 Calculate the **curved** surface area of each of these cones.

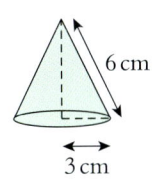

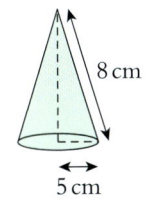

a) 6 cm slant, 3 cm radius
b) 8 cm slant, 5 cm radius
c) 9 cm slant, 4 cm radius

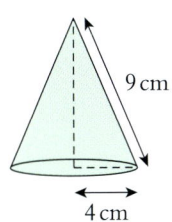

9 Calculate the **total** surface area of each of these cones.

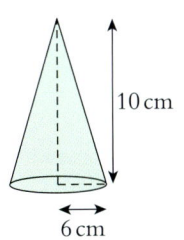

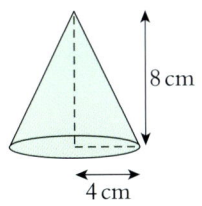

a) 10 cm height, 6 cm radius
b) 8 cm height, 4 cm radius
c) 15 cm height, 9 cm radius

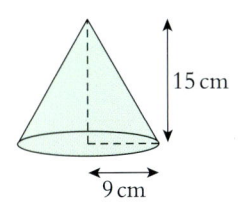

10 Each of these sectors forms the curved surface of a cone.
Calculate the curved surface area of each cone.

a) 160°, 2 cm
b) 250°, 5 cm
c) 143°, 8 cm

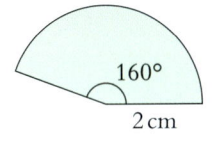

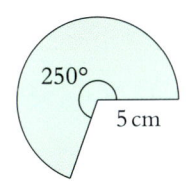

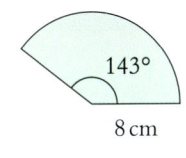

11 The straight edges of each of these sectors are joined together to make a cone.
Calculate: i   the curved surface area of each cone,
ii the radius of the base of each cone,
iii the height of each cone.

a) 138°, 6 cm
b) 165°, 14 cm
c) 242°, 25 cm

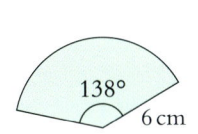

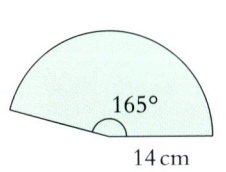

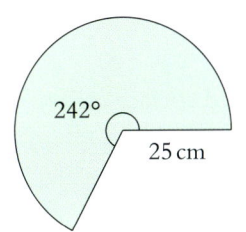

12 Calculate the **total** surface area of each of these solid frustums.

a) top 5 cm, bottom 10 cm, height 4 cm
b) top 4 cm, bottom 12 cm, height 12 cm
c) top 6 cm, bottom 9 cm, height 10 cm

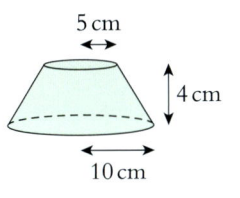

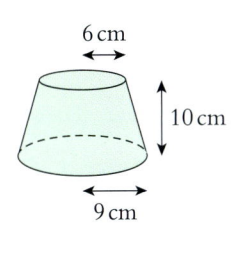

Surface area and volume 2   243

Spheres

Volume of a **sphere** = $\frac{4}{3}\pi r^3$
Surface area of a sphere = $4\pi r^2$

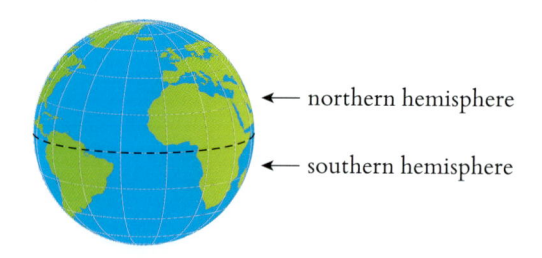

← northern hemisphere
← southern hemisphere

A **hemisphere** is half a sphere.
The earth is divided into two hemispheres.

EXAMPLE

A football has a radius of 11.1 cm.
Calculate:
a the volume
b the surface area of the football.

a Volume of football = $\frac{4}{3}\pi r^3 = \frac{4}{3} \times \pi \times 11.1^3 = 5730 \text{ cm}^3$ (to 3 s.f.)

b Surface area of football = $4\pi r^2 = 4 \times \pi \times 11.1^2 = 1550 \text{ cm}^2$ (to 3 s.f.)

EXAMPLE

A solid hemisphere has radius 6 cm.
Calculate:
a the volume
b the total surface area of the hemisphere.

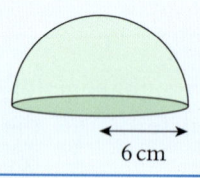

6 cm

a Volume = $\frac{1}{2} \times \frac{4}{3}\pi r^3 = \frac{1}{2} \times \frac{4}{3} \times \pi \times 6^3 = 452 \text{ cm}^3$

b Area of base = $\pi r^2 = \pi \times 6^2 = 36\pi$
Curved surface area = $\frac{1}{2} \times 4\pi r^2 = \frac{1}{2} \times 4\pi \times 36 = 72\pi$
Total surface area = area of base + curved surface area
= $36\pi + 72\pi = 108\pi = 339 \text{ cm}^2$

EXAMPLE

The diagram shows a toy that is made from a cone
on top of a hemisphere.
Calculate the volume of the toy.

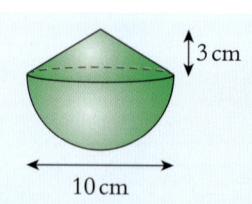

3 cm

10 cm

Volume of hemisphere = $\frac{1}{2} \times \frac{4}{3}\pi r^3 = \frac{1}{2} \times \frac{4}{3}\pi \times 5^3 = 261.8 \text{ cm}^3$

Volume of cone = $\frac{1}{3}\pi r^2 h = \frac{1}{3} \times \pi \times 5^2 \times 3 = 78.54 \text{ cm}^3$

Volume of toy = 261.8 + 78.54 = 340 cm³ (to 3 s.f.)

EXERCISE 5.13

1. Find the volume of each of these spheres.

    a) 3 cm   b) 8 cm   c) 5.7 cm

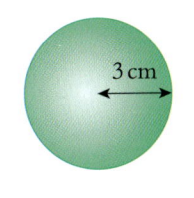

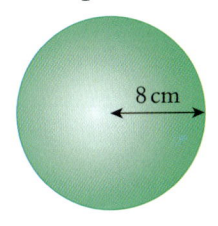

2. Find the surface area of each sphere in question **1**.

3. Find the volume of each of these solid hemispheres.

    a) 4 cm   b) 16 cm   c) 10 cm

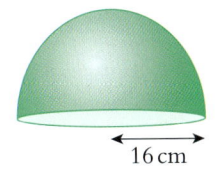

4. Find the total surface area of each of the solid hemispheres in question **3**.

5. The volume of a sphere is 432 cm^3. Calculate the radius of the sphere.

6. The surface area of a sphere is 500 cm^2. Calculate the radius of the sphere.

7. A cylindrical tube contains two tennis balls.
   The radius of each tennis ball is 3.75 cm.
   The tennis balls touch the sides, top and bottom of the tube.
   Calculate the volume of empty space inside the tube.

8. A cricket ball is dropped into a cylindrical tank of water.
   The ball sinks to the bottom of the tank.
   The diameter of the cricket ball is 7.2 cm
   The diameter of the cylindrical tank is 15 cm.
   Calculate the increase in height of the water in the tank.

9. A solid metal cube of side 8 cm is melted and then recast to make a solid sphere.
   Calculate the radius of the sphere.

10. Calculate the volume of these solid shapes.

    a) Hemisphere; Cylinder height 8 cm, radius 6 cm
    b) Cone height 10 cm, radius 14 cm; Hemisphere

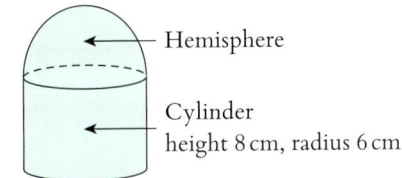

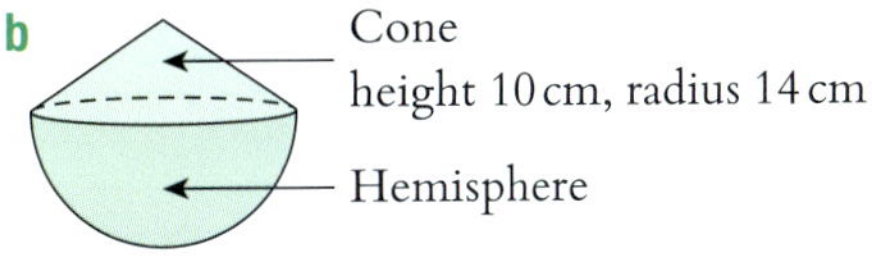

11. Calculate the total surface area of the solid shapes in question **10**.

> **KEY WORDS**
> pyramid
> truncated pyramid
> regular octahedron
> regular tetrahedron
> cone, frustum, sphere
> hemisphere

Areas of similar shapes

THIS SECTION WILL SHOW YOU HOW TO:
- Find lengths and areas of similar shapes using scale factors

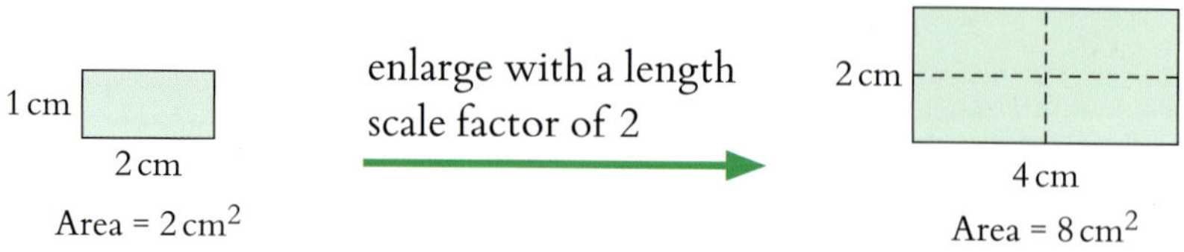

1 cm | 2 cm — enlarge with a length scale factor of 2 → 2 cm | 4 cm
Area = 2 cm²   Area = 8 cm²

The diagram shows that if the **length scale factor** is 2, then the **area scale factor** is 4.

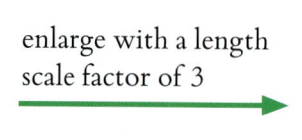

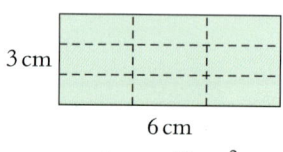

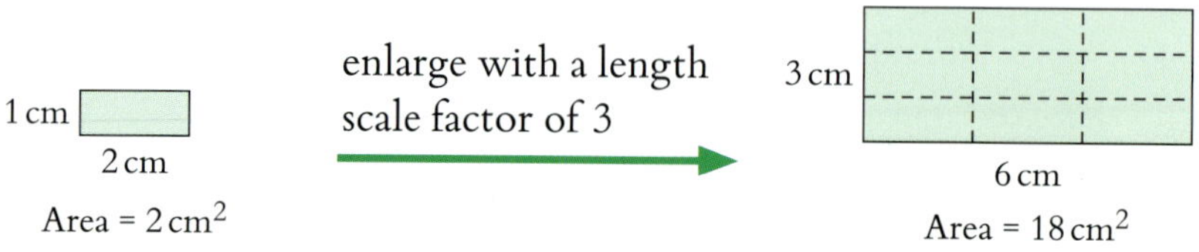

1 cm | 2 cm — enlarge with a length scale factor of 3 → 3 cm | 6 cm
Area = 2 cm²   Area = 18 cm²

The diagram shows that if the length scale factor is 3, then the area scale factor is 9.

> General rule: If the length scale factor is k, then the area scale factor is k^2.

EXAMPLE

The two quadrilaterals are similar.
The area of the smaller quadrilateral is 5 cm².
Find the area of the larger quadrilateral.

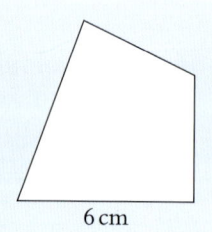

2 cm   6 cm

The length scale factor $k = \dfrac{6}{2} = 3$.

The area scale factor $k^2 = 3^2$.

So the area of the larger quadrilateral = $3^2 \times 5 = 45$ cm².

EXAMPLE

The two triangles are similar.
The area of the larger triangle is 10.8 cm².
Find the area of the smaller triangle.

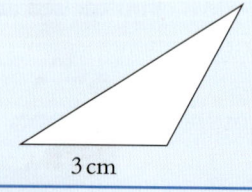

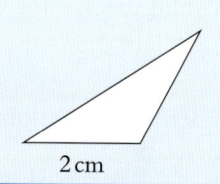

3 cm   2 cm

The length scale factor $k = \dfrac{2}{3}$

The area scale factor $k^2 = \left(\dfrac{2}{3}\right)^2$

So the area of the smaller triangle = $\left(\dfrac{2}{3}\right)^2 \times 10.8 = 4.8$ cm²

EXAMPLE

The two shapes are similar.
The area of the larger shape is 32 cm².
The area of the smaller shape is 8 cm².
Find the value of x.

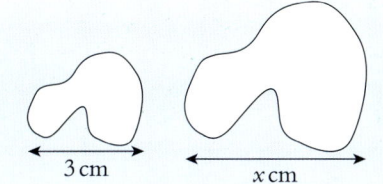

The area scale factor $k^2 = \dfrac{32}{8} = 4$.

The length scale factor $k = \sqrt{4} = 2$.
So $x = 2 \times 3 = 6$ cm.

EXERCISE 5.14

In each question the pair of shapes are similar.
The area of each shape is written inside the shape.
Find the value of x for each question.

1 x cm², 4 cm ; 5 cm², 2 cm

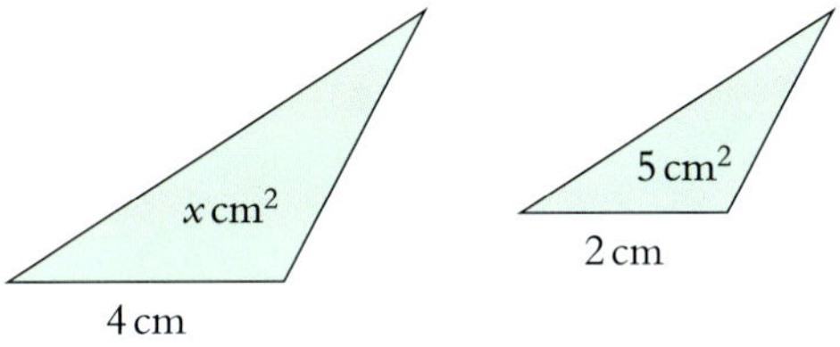

2 15 cm, x cm² ; 5 cm, 10 cm²

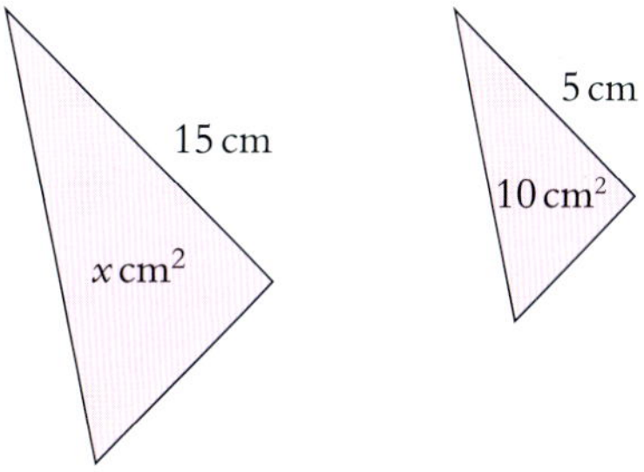

3 x cm², 6 cm ; 12 cm², 4 cm

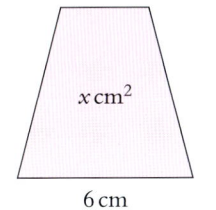

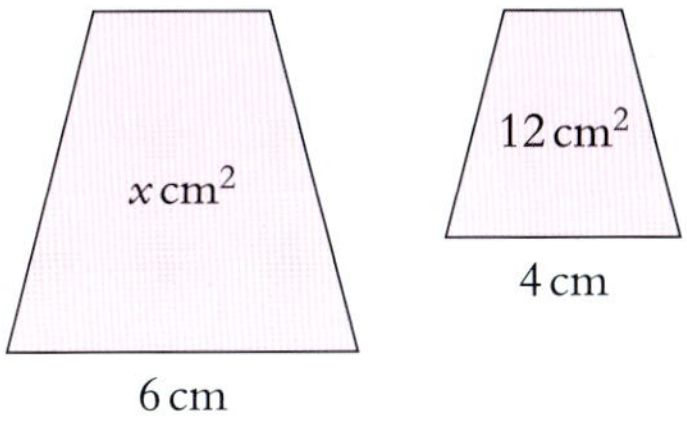

4 120 cm², 12 cm ; x cm², 6 cm

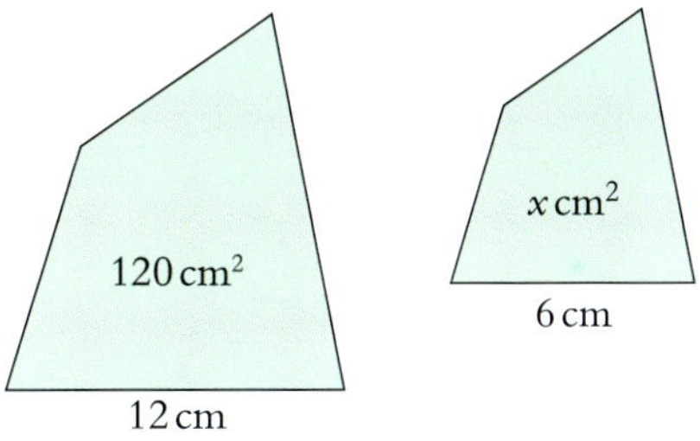

5 x cm², 3 cm ; 99 cm², 9 cm

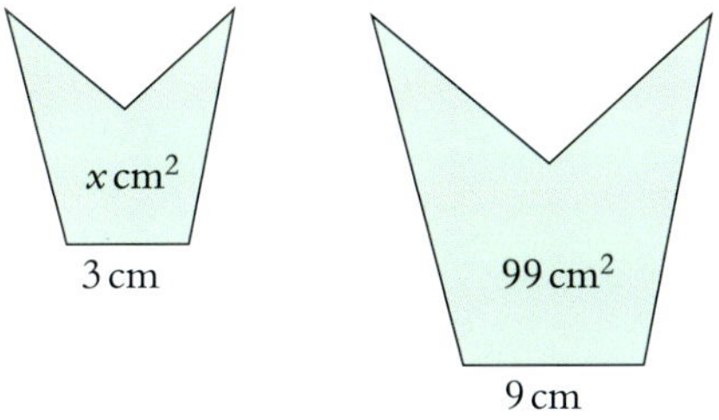

6 x cm², 20 cm ; 12 cm², 8 cm

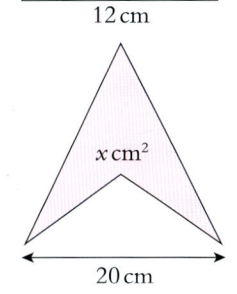

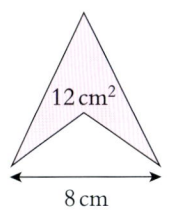

7 108 cm², 9 cm ; x cm², 4.5 cm

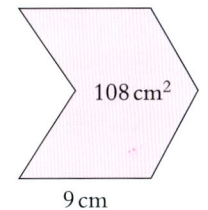

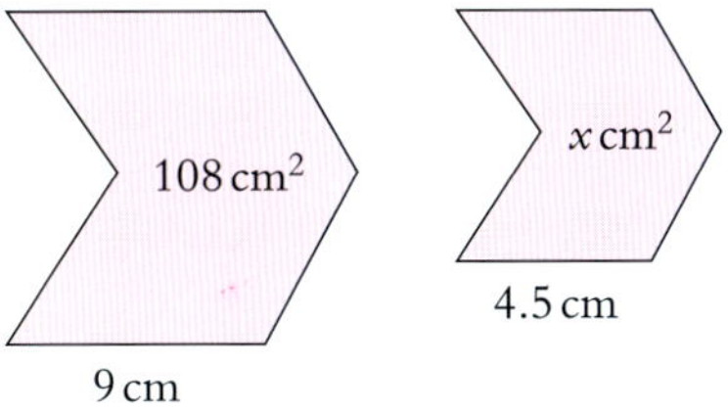

8 4 cm, x cm² ; 8 cm, 80 cm²

Areas of similar shapes   247

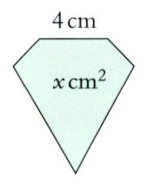

9 x cm, 28 cm², 3 cm, 7 cm²

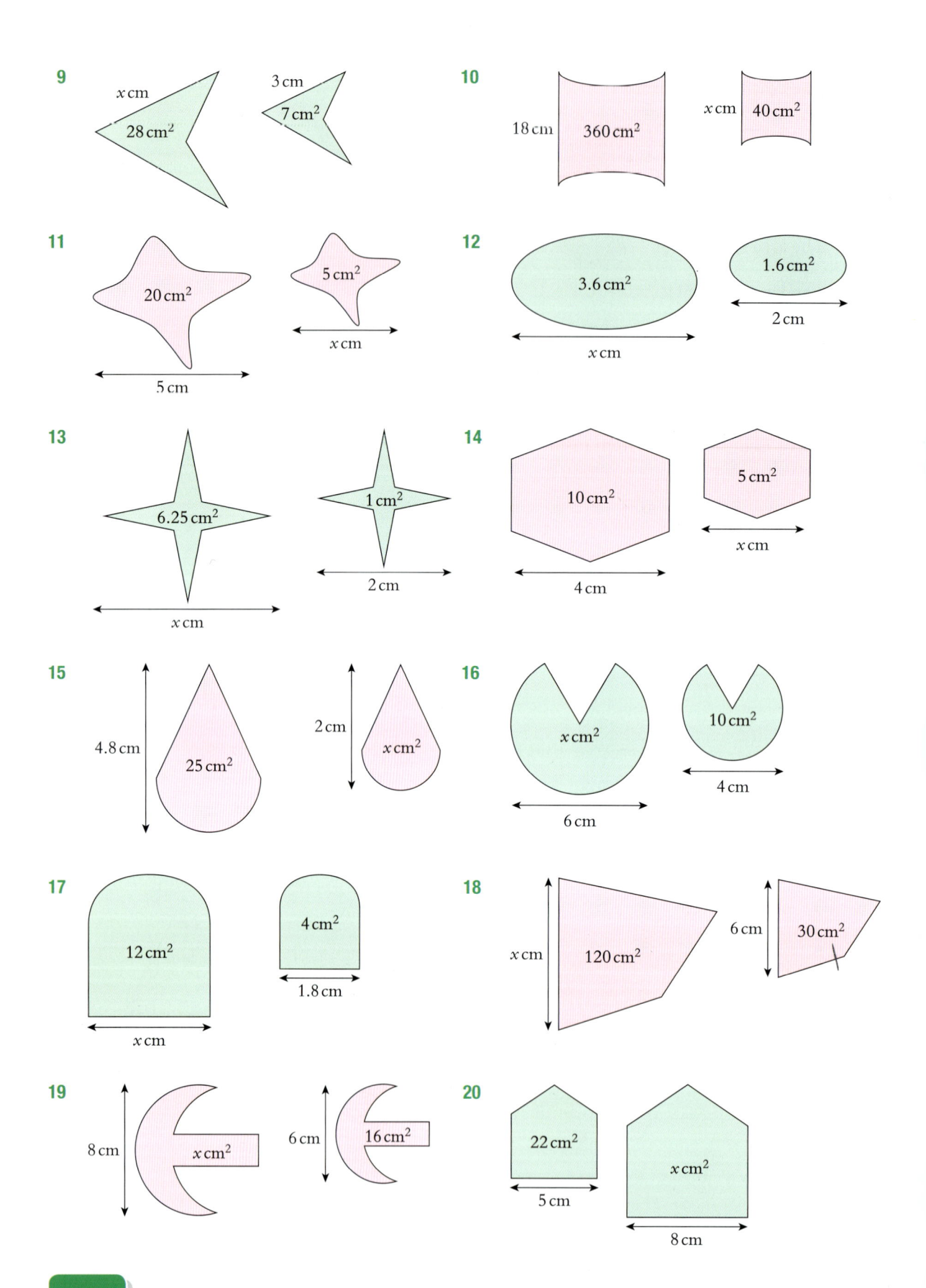

10 18 cm, 360 cm², x cm, 40 cm²

11 20 cm², 5 cm, 5 cm², x cm

12 3.6 cm², x cm, 1.6 cm², 2 cm

13 6.25 cm², x cm, 1 cm², 2 cm

14 10 cm², 4 cm, 5 cm², x cm

15 4.8 cm, 25 cm², 2 cm, x cm²

16 x cm², 6 cm, 10 cm², 4 cm

17 12 cm², x cm, 4 cm², 1.8 cm

18 x cm, 120 cm², 6 cm, 30 cm²

19 8 cm, x cm², 6 cm, 16 cm²

20 22 cm², 5 cm, x cm², 8 cm

EXAMPLE

Area of △CDE = 9 cm².
a Calculate the area of △ABC
b Calculate the area of ABDE

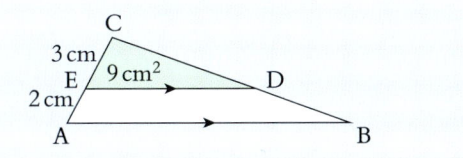

a △ABC and △CDE are similar.
The length scale factor $k = \dfrac{5}{3}$

The area scale factor $k^2 = \left(\dfrac{5}{3}\right)^2$

Area of △ABC = $\left(\dfrac{5}{3}\right)^2 \times 9 = 25$ cm²

b Area of ABDE = area of △ABC − area of △CDE = 25 − 9 = 16 cm²

EXERCISE 5.15

1 Calculate the area of △ABC and the area of ABDE for each of these diagrams.

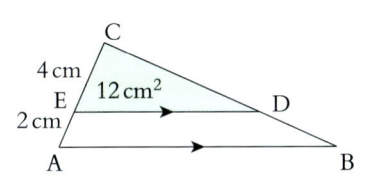

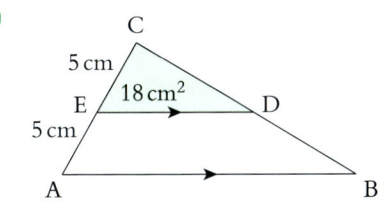

a (4 cm, 2 cm, 12 cm²)

b (5 cm, 5 cm, 18 cm²)

2 Find the value of x in each of these diagrams.

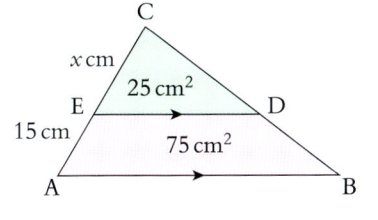

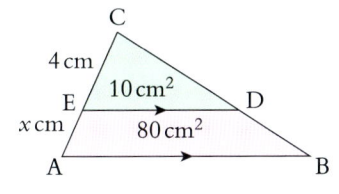

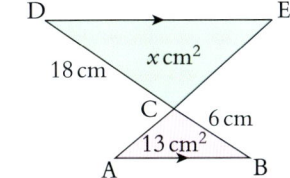

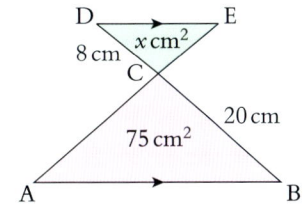

a (x cm, 15 cm, 25 cm², 75 cm²)

b (4 cm, x cm, 10 cm², 80 cm²)

c (18 cm, 6 cm, x cm², 13 cm²)

d (8 cm, 20 cm, x cm², 75 cm²)

KEY WORDS
length scale factor
area scale factor

Areas of similar shapes   249

Volumes of similar objects

THIS SECTION WILL SHOW YOU HOW TO:
- Find lengths and volumes of similar objects using scale factors

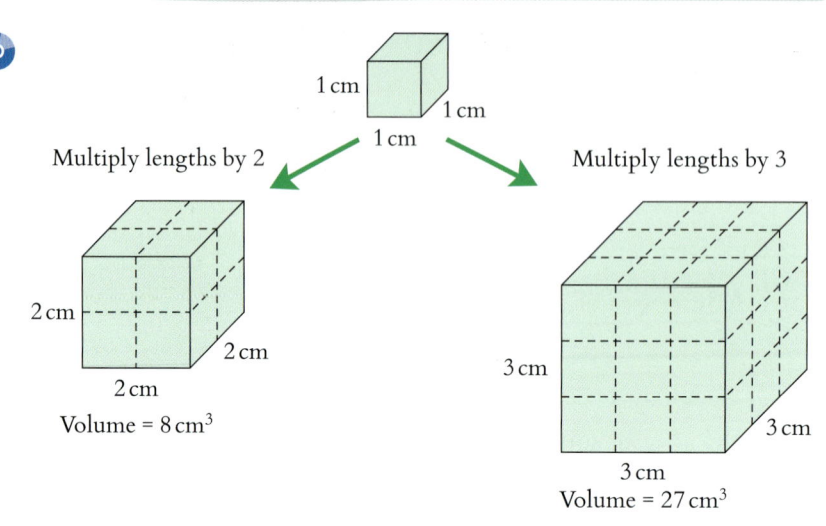

Multiply lengths by 2 ← 1 cm cube → Multiply lengths by 3

2 cm cube, Volume = 8 cm³

3 cm cube, Volume = 27 cm³

The diagram shows that:
- If the length scale factor is 2, then the **volume scale factor** is 8.
- If the length scale factor is 3, then the volume scale factor is 27.

General rule: If the length scale factor is k, then the volume scale factor is k^3.

EXAMPLE

The two solid hexagonal prisms are similar.
The volume of the smaller prism is 24 cm³
Calculate the volume of the larger prism.

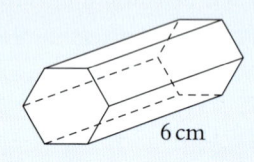

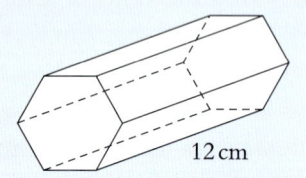

(smaller prism: 6 cm; larger prism: 12 cm)

The length scale factor $k = \dfrac{12}{6} = 2$
The volume scale factor $k^3 = 2^3$
So the volume of the larger prism = $2^3 \times 24 = 192$ cm³.

EXAMPLE

The two cones are similar.
The volume of the larger cone is 54 cm³.
The volume of the smaller cone is 16 cm³.
Find the height of the smaller cone.

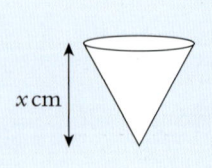

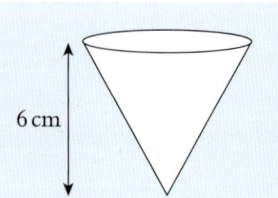

(smaller cone: x cm; larger cone: 6 cm)

The volume scale factor $k^3 = \dfrac{16}{54} = \dfrac{8}{27}$ and the length scale factor $k = \sqrt[3]{\dfrac{8}{27}} = \dfrac{2}{3}$

So the height of the smaller cone = $\dfrac{2}{3} \times 6 = 4$ cm.

250 UNIT 5

EXERCISE 5.16

In questions 1 to 10 each pair of objects are similar.
The volume of each object is written inside the object.
Find the value of x for each question.

1 Cuboid 9 cm³, 3 cm; similar cuboid x cm³, 6 cm.

2 Cylinder x cm³, 5 cm; similar cylinder 600 cm³, 10 cm.

3 Shape x cm³, 6 cm; similar shape 8 cm³, 2 cm.

4 Cone 40 cm³, 5 cm; similar cone x cm³, 7.5 cm.

5 Shape 189 cm³, 6 cm; similar shape x cm³, 4 cm.

6 Shape 100 cm³, 6 cm; similar shape x cm³, 10 cm.

7 Frustum 1200 cm³, 12 cm; similar frustum 150 cm³, x cm.

8 Pyramid 54 cm³, x cm; similar pyramid 16 cm³, 6 cm.

9 Bipyramid 810 cm³, 12 cm; similar bipyramid x cm³, 8 cm.

10 Triangular prism 162 cm³, 30 cm; similar prism 48 cm³, x cm.

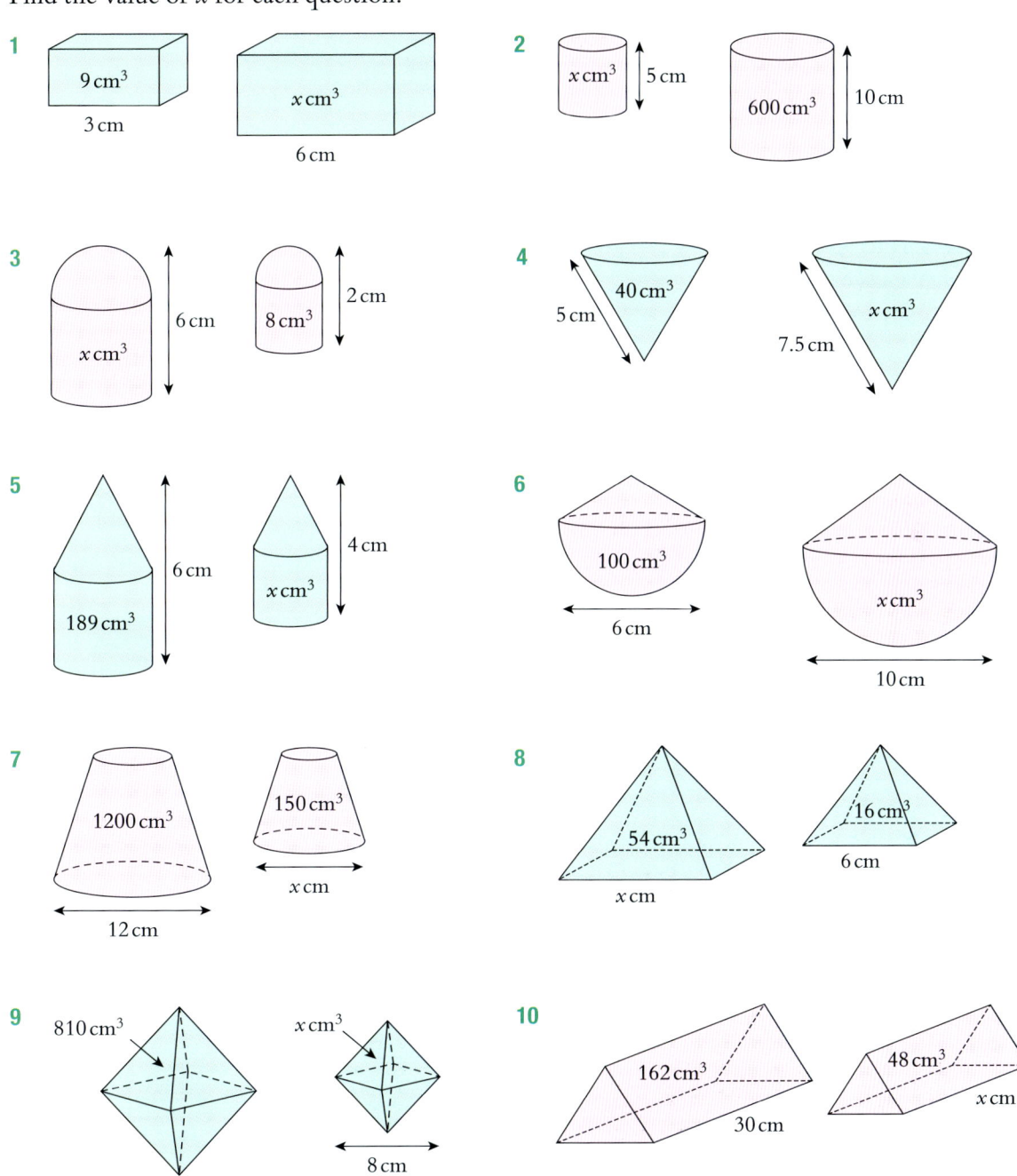

11 The two bottles are similar in shape.
The larger bottle holds 100 mℓ of perfume.
Calculate how many millilitres of perfume the smaller bottle holds.

12 The two bars of chocolate are similar in shape.
The smaller bar has a mass of 150 grams.
Calculate the mass of the larger chocolate bar.

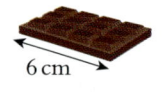

13 The two vases are similar in shape.
The largest vase holds 1.5 litres of water when full.
How much water does the smaller vase hold when full?

14 The two buckets are similar in shape.
The larger bucket holds 10 litres of water when full.
How much water does the smaller bucket hold when full?

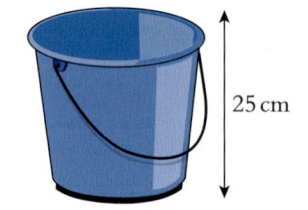

15 The two eggs are similar in shape.
Calculate the length of the smaller egg.

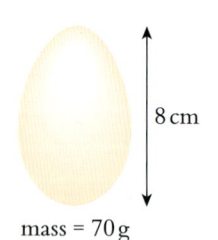

mass = 50 g

mass = 70 g

16 The two shampoo bottles are similar in shape.
Calculate the height of the larger bottle.

500 mℓ

300 mℓ

252 UNIT 5

17 A cylindrical cup has a height of 6 cm and a base radius of 4 cm.
    A large cylindrical jug is a similar shape to the cup.
    The jug is full of mango juice.
    The cup can be filled with mango juice from the jug exactly 125 times.
    Calculate the height and radius of the jug.

18 Two vases are similar in shape.
    The ratio of their heights is 3:2
    a The volume of the larger vase is 810 cm³.
       Calculate the volume of the smaller vase.
    b The surface area of the smaller vase is 240 cm².
       Calculate the surface area of the larger vase.

19 Two similar objects have a surface area of 4 cm² and 9 cm².
    The mass of the smaller object is 200 grams.
    Find the mass of the larger object.

20 Two similar cones have surface areas in the ratio 9:16
    a Find the ratio of their lengths.
    b Find the ratio of their volumes.

21 The two Russian dolls are similar.
    a The surface area of the larger doll is 150 cm².
       Calculate the surface area of the smaller doll.
    b The volume of the larger doll is 120 cm³.
       Calculate the volume of the smaller doll.

22 A model plane is made on a scale of 1:50
    Copy and complete the table.

	Real plane	Model plane
Wingspan	11.23 m	22.46 cm
Length of plane	9.12 m	……. cm
Wing area	22.48 m²	……. cm²

23 A model car is made on a scale of 1:40
    Copy and complete the table.

	Real car	Model car
Length of car	3.723 m	……. cm
Area of windscreen	0.4835 m²	……. cm²
Storage space in boot	160 litres	……. ml

KEY WORDS
volume scale factor

Volumes of similar objects   253

Grouped frequency tables

THIS SECTION WILL SHOW YOU HOW TO
- Find the modal group from a grouped frequency table
- Estimate the mean from a grouped frequency table

Large sets of data are often shown in a **grouped frequency table**.
The **modal class** is the group with the highest frequency.
In a grouped frequency table you do not know the individual data values so you can only estimate the value of the mean.
The **mid-value** of the groups is used to estimate the sum of the figures in each group.
For example, if you have 5 people whose heights, h cm, are in the group $164 < h \leq 168$ you can estimate the sum of their heights by calculating $5 \times 166 = 830$ cm.

EXAMPLE

The table shows the time, t minutes, taken by 100 people to complete a Sudoku puzzle.
a Write down the modal class.
b Calculate an estimate of the mean time taken.

Time (t minutes)	Frequency
$0 < t \leq 10$	20
$10 < t \leq 20$	32
$20 < t \leq 30$	41
$30 < t \leq 40$	5
$40 < t \leq 50$	2

a The highest frequency is 41, so the modal class is $20 < t \leq 30$.
b The mid-value of the group $0 < t \leq 10$ is 5, the mid-value of the group $10 < t \leq 20$ is 15 etc.
Two extra columns need to be added to the table for the calculations.

➡ **NOTE:** to find the mid-value of the group $10 < t \leq 20$ you calculate $\dfrac{10 + 20}{2} = \dfrac{30}{2} = 15$

Time (t minutes)	Frequency (f)	mid-value (x)	($f \times x$)
$0 < t \leq 10$	20	5	$20 \times 5 = 100$
$10 < t \leq 20$	32	15	$32 \times 15 = 480$
$20 < t \leq 30$	41	25	$41 \times 25 = 1025$
$30 < t \leq 40$	5	35	$5 \times 35 = 175$
$40 < t \leq 50$	2	45	$2 \times 45 = 90$
	Total = 100		Total = 1870

The estimated mean time = $\dfrac{\text{total time}}{\text{total number of people}}$

$= \dfrac{1870}{8.7} = 18.7$ minutes

EXAMPLE

The table shows the number of points scored by a rugby team in a series of 50 games.
Estimate the mean number of points scored per game.

Points	Frequency
5 – 15	6
16 – 20	13
21 – 25	26
26 – 36	5

The mid-value of the group 5 – 15 is $\frac{5+15}{2} = 10$

The mid-value of the group 16 – 20 is $\frac{16+20}{2} = 18$ etc.

Two extra columns need to be added to the table for the calculations.

Points	Frequency (f)	Mid-value(x)	(f × x)
5 – 15	6	10	6 × 10 = 60
16 – 20	13	18	13 × 18 = 234
21 – 25	26	23	26 × 23 = 598
26 – 36	5	31	5 × 31 = 155
	Total = 50		Total = 1047

The estimated mean number of points = $\frac{\text{total number of points}}{\text{total number of games}}$

$= \frac{1047}{50} = 20.94$ points.

EXERCISE 5.17

1 Write down the mid value for each of the following groups.

    a $40 < v \leq 50$ **b** $4 < t \leq 4.5$ **c** $9 < h \leq 15$ **d** $2.3 < x \leq 2.8$

    e 5 to 7   **f** 3 to 8   **g** 3.9 to 4.2   **h** 15 to 28

2 The table shows the speeds of 200 cars on a road.

Speed (v km/h)	$30 < v \leq 40$	$40 < v \leq 50$	$50 < v \leq 60$	$60 < v \leq 70$
Number of cars	37	85	76	2

    a Write down the modal class.

    b Copy and complete the table below and use it to calculate an estimate of the mean speed.

Speed (v km/h)	Number of cars (f)	Mid-value (x)	(f × x)
$30 < v \leq 40$	37	35	37 × 35 =
$40 < v \leq 50$	85		
$50 < v \leq 60$	76		
$60 < v \leq 70$	2		
	Total =		Total =

Grouped frequency tables

3 The table shows the heights of 150 students in a school.
   a Write down the modal class.
   b Calculate an estimate of the mean height.

Height (h cm)	Frequency
$120 < h \le 130$	3
$130 < h \le 140$	19
$140 < h \le 150$	36
$150 < h \le 160$	57
$160 < h \le 170$	22
$170 < h \le 180$	11
$180 < h \le 190$	2

4 The temperature of a fridge is checked regularly. The results are shown in the table.
   a Write down the modal class.
   b Calculate an estimate of the mean temperature.

Temperature (t°C)	Frequency
$2.5 < t \le 3$	4
$3 < t \le 3.5$	25
$3.5 < t \le 4$	13
$4 < t \le 4.5$	5
$4.5 < t \le 5$	3

5 The depth of water in a lake is recorded each day for 100 days.
   a Write down the modal class.
   b Calculate an estimate of the mean depth.

Depth (d metres)	Frequency
$10 < d \le 10.5$	8
$10.5 < d \le 11$	49
$11 < d \le 11.5$	38
$11.5 < d \le 12$	4
$12 < d \le 12.5$	1

6 Fifty new-born babies are weighed. The results are shown in the table.
   a Write down the modal class.
   b Calculate an estimate of the mean mass.
   c Calculate the percentage of new-born babies that had a mass greater than 4 kg.

Mass (m kg)	Frequency
$1 < m \le 2$	3
$2 < m \le 3$	18
$3 < m \le 4$	25
$4 < m \le 5$	3
$5 < m \le 6$	1

7 One hundred people are asked to record how many times they visit a doctor in one year.
   a Write down the modal class.
   b Calculate an estimate of the mean number of visits.

Number of visits	Frequency
0 to 2	58
3 to 5	24
6 to 8	15
9 to 14	3

8 Two hundred students record the number of hours of homework they do in a week.
   a Calculate an estimate of the mean number of hours of homework.
   b A student is chosen at random. Calculate the probability that they do more than 5 hours of homework in a week.

Hours (h hours)	Frequency
$0 < h \leq 2$	6
$2 < h \leq 5$	76
$5 < h \leq 9$	108
$9 < h \leq 15$	10

9 The table shows the lengths of 150 journeys made by a taxi driver in one week.
   a Calculate an estimate of the mean length of a journey.
   b Calculate the percentage of journeys that are over 4 miles in length.

Length (x miles)	Frequency
$0 < x \leq 1$	35
$1 < x \leq 2$	62
$2 < x \leq 3$	38
$3 < x \leq 4$	12
$4 < x \leq 5$	3

10 The table shows the times taken by 100 students to complete a test.
   a Write down the modal class.
   b Calculate an estimate of the mean time.
   c The data is regrouped to give the following table.

Time (t minutes)	Frequency
$0 < t \leq 60$	p
$60 < t \leq 120$	q
$120 < t \leq 150$	20

Write down the values of p and q.
   d Use the table in part c to calculate a new estimate of the mean time.

Time (t minutes)	Frequency
$0 < t \leq 30$	2
$30 < t \leq 60$	6
$60 < t \leq 90$	40
$90 < t \leq 120$	32
$120 < t \leq 150$	20

KEY WORDS
grouped frequency table
modal group
mid-value

Grouped frequency tables

Unit 5 Examination questions

1 A holiday in Europe was advertised at a cost of €245.
The exchange rate was $1 = €1.06.
Calculate the cost of the holiday in dollars, giving your
answer correct to the nearest cent. [2]

Cambridge IGCSE Mathematics 0580, Paper 21 Q5, June 2008

2 In January Sunanda changed £25 000 into dollars when the exchange rate was $1.96 = £1.
In June she changed the dollars back into pounds when the exchange rate was $1.75 = £1.
Calculate the profit she made, giving your answer in pounds (£). [3]

Cambridge IGCSE Mathematics 0580, Paper 21 Q11, June 2009

3 Michel changed $600 into pounds (£) when the exchange rate was £1 = $2.40.
He later changed all the pounds back into dollars when
the exchange rate was £1 = $2.60.

How many dollars did he receive? [2]

Cambridge IGCSE Mathematics 0580, Paper 21 Q2, June 2010

4 $f: x \mapsto 5 - 3x$

(a) Find $f(-1)$. [1]
(b) Find $f^{-1}(x)$. [2]
(c) Find $ff^{-1}(8)$. [1]

Cambridge IGCSE Mathematics 0580, Paper 2 Q15, November 2006

5 $$f(x) = \frac{x+3}{x}, \; x \neq 0.$$

(a) Calculate $f\left(\frac{1}{4}\right)$. [1]

(b) Solve $f(x) = \frac{1}{4}$. [2]

Cambridge IGCSE Mathematics 0580, Paper 2 Q11, June 2006

6 $f(x) = (x-1)^3 \quad g(x) = (x-1)^2 \quad h(x) = 3x + 1$

   (a) Work out $fg(-1)$. [2]

   (b) Find $gh(x)$ in its simplest form. [2]

   (c) Find $f^{-1}(x)$. [2]

Cambridge IGCSE Mathematics 0580, Paper 21 Q20, June 2010

7 $f(x) = x^3 - x^2 + 6x - 4$ and $g(x) = 2x - 1$.
Find

   (a) $f(-1)$, [1]

   (b) $gf(x)$, [2]

   (c) $g^{-1}(x)$. [2]

Cambridge IGCSE Mathematics 0580, Paper 21 Q18, June 2008

8 $\qquad f(x) = 4x + 1 \quad g(x) = x^3 + 1 \quad h(x) = \dfrac{2x+1}{3}$

   (a) Find the value of $gf(0)$. [2]

   (b) Find $fg(x)$. Simplify your answer. [2]

   (c) Find $h^{-1}(x)$. [2]

Cambridge IGCSE Mathematics 0580, Paper 21 Q22, November 2009

9 Factorise

   (a) $4x^2 - 9$, [1]

   (b) $4x^2 - 9x$, [1]

   (c) $4x^2 - 9x + 2$. [2]

Cambridge IGCSE Mathematics 0580, Paper 2 Q19, June 2006

10 A cylindrical glass has a radius of 3 centimetres and a height of 7 centimetres. A large cylindrical jar full of water is a similar shape to the glass. The glass can be filled with water from the jar exactly 216 times.
Work out the radius and height of the jar. [3]

Cambridge IGCSE Mathematics 0580, Paper 21 Q10, June 2008

11 A statue two metres high has a volume of five cubic metres.
A similar model of the statue has a height of four centimeters.

   (a) Calculate the volume of the model statue in cubic centimeters. [2]

   (b) Write your answer to **part (a)** in cubic metres. [1]

Cambridge IGCSE Mathematics 0580, Paper 2 Q13, November 2006

Examination questions

12 A company makes two models of television.
Model *A* has a rectangular screen that measures 44 cm by 32 cm.
Model *B* has a larger screen with these measurements increased in the ratio 5 : 4.

(a) Work out the measurements of the larger screen. [2]

(b) Find the **fraction** $\dfrac{\text{model } A \text{ screen area}}{\text{model } B \text{ screen area}}$ in its simplest form. [1]

Cambridge IGCSE Mathematics 0580, Paper 2 Q14, June 2006

13 Two similar vases have heights which are in the ratio 3 : 2.

(a) The volume of the larger vase is 1080 cm³.
Calculate the volume of the smaller vase. [2]

(b) The surface area of the smaller vase is 252 cm².
Calculate the surface area of the larger vase. [2]

Cambridge IGCSE Mathematics 0580, Paper 21 Q18, June 2009

14 (a)

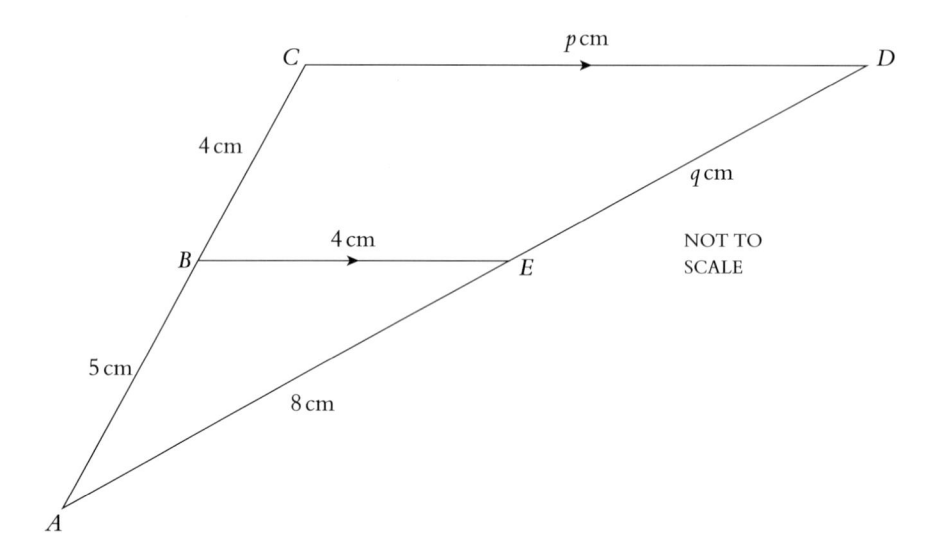

In the diagram triangles *ABE* and *ACD* are similar.
BE is parallel to *CD*.
$AB = 5$ cm, $BC = 4$ cm, $BE = 4$ cm, $AE = 8$ cm, $CD = p$ cm and $DE = q$ cm.
Work out the values of *p* and *q*. [4]

(b) A spherical balloon of radius 3 metres has a volume of 36π cubic metres.
It is further inflated until its radius is 12 m.
Calculate its new volume, leaving your answer in terms of π. [2]

Cambridge IGCSE Mathematics 0580, Paper 2 Q22, June 2006

15 [The surface area of a sphere of radius r is $4\pi r^2$ and the volume is $\frac{4}{3}\pi r^3$.]

(a) A solid metal sphere has a radius of 3.5 cm.
One cubic centimetre of the metal has a mass of 5.6 grams.
Calculate
- **(i)** the surface area of the shere, [2]
- **(ii)** the volume of the shere, [2]
- **(iii)** the mass of the sphere. [2]

(b)

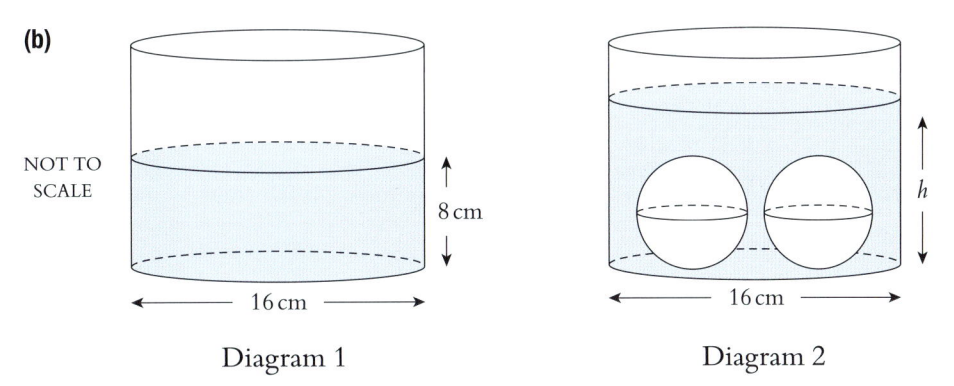

Diagram 1   Diagram 2

Diagram 1 shows a cylinder with a **diameter** of 16 cm.
It contains water to a depth of 8 cm.
Two spheres identical to the sphere in **part (a)** are placed in the water.
This is shown in Diagram 2.
Calculate h, the new depth of water in the cylinder. [4]

(c) A difference metal sphere has a mass of 1 kilogram.
One cubic centimeter of this metal has a mass of 4.8 grams.
Calculate the radius of this sphere. [3]

Cambridge IGCSE Mathematics 0580, Paper 4 Q4, November 2007

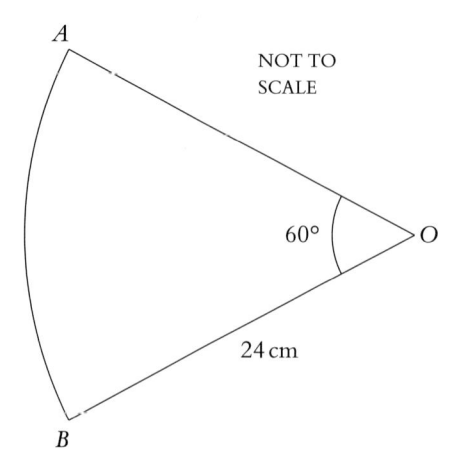

(a) The sector of a circle, centre O, radius 24 cm, has angle $AOB = 60°$.
   Calculate
   (i) the length of the arc AB, [2]
   (ii) the area of the sector OAB cm^2. [2]

(b) The points A and B of the sector are joined together to make a hollow cone as shown in diagram. The arc AB of the sector becomes the circumference of the base of the cone.

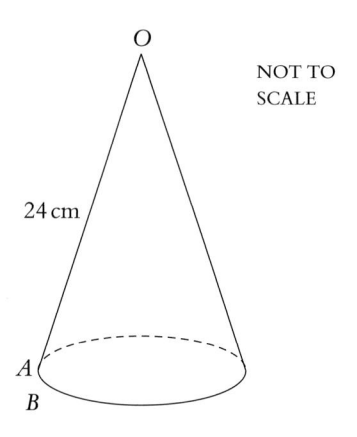

   Calculate
   (i) the radius of the base of the cone, [2]
   (ii) the height of the cone, [2]
   (iii) the volume of the cone,
        [The volume, V, of a cone of radius r and height h is $V = \frac{1}{3}\pi r^2 h$.] [2]

(c) A different cone, with radius x and height y, has a volume W.
   Find, **in terms of W**, the volume of
   (i) a similar cone, with both radius and height 3 times larger, [1]
   (ii) a cone of radius $2x$ and height y. [1]

Cambridge IGCSE Mathematics 0580, Paper 4 Q7, November 2009

17 Fifty students are timed when running one kilometre.

The results are shown in the table.

Time (t minutes)	$4.0 < t \leq 4.5$	$4.5 < t \leq 5.0$	$5.0 < t \leq 5.5$	$5.5 < t \leq 6.0$	$6.0 < t \leq 6.5$	$6.5 < t \leq 7.0$
Frequency	2	7	8	18	10	5

(a) Write down the modal time interval. [1]
(b) Calculate an estimate of the mean time. [4]

Cambridge IGCSE Mathematics 0580, Paper 4 Q8 a b, November 2009

18 Kristina asked 200 people how much they drink in one day.
The table shows her results.

Amount of water (x litres)	Number of people
$0 < x \leq 0.5$	8
$0.5 < x \leq 1$	27
$1 < x \leq 1.5$	45
$1.5 < x \leq 2$	50
$2 < x \leq 2.5$	39
$2.5 < x \leq 3$	21
$3 < x \leq 3.5$	7
$3.5 < x \leq 4$	3

(a) Write down the modal interval. [1]
(b) Calculate an estimate of the mean. [4]

Cambridge IGCSE Mathematics 0580, Paper 4 Q6 a b, June 2007

Percentages 3

> **THIS SECTION WILL SHOW YOU HOW TO**
> - Carry out calculations involving reverse percentages
> - find simple and compound interest

Reverse percentages

Sometimes you are told the final value after a percentage change and you have to calculate the original value.

This process is referred to as carrying out a **reverse percentage** calculation.

> **Reminder:**
> If a value is increased by 18% the multiplying factor is 1.18   (1 + 0.18)
> If a value is decreased by 13% the multiplying factor is 0.87   (1 − 0.13)

EXAMPLE

During 2010 the population of a small village increased by 4%.
At the end of the year the population was 468.
Calculate the population of the village at the beginning of the year.

original population × 1.04 = final population   *multiplying factor is 1.04   (= 1 + 0.04)*

$$x \times 1.04 = 468$$

$$x = \frac{468}{1.04}$$

$$x = 450$$

The population at the beginning of the year was 450.

➡ **CHECK:**
   450 × 1.04 = 468 ✓

EXAMPLE

Benito pays $159.29 for a surfboard in a sale.
The original price had been reduced by 15%.
Calculate the original price of the surfboard.

original price × 0.85 = final price   *multiplying factor is 0.85   (= 1 − 0.15)*

$$x \times 0.85 = 159.29$$

$$x = \frac{159.29}{0.85}$$

$$x = 187.40$$

The original price of the surfboard was $187.40

➡ **CHECK:**
   187.40 × 0.85 = 159.29 ✓

EXERCISE 6.1

1. In 2006 the population of Singapore was 4.5×10^6
   This was 50% greater than it was in 1990.
   Calculate the population of Singapore in 1990.

2. Carla makes a profit of 15% when she sells a mobile phone for $92
   Calculate how much Carla paid for the mobile phone.

3. In 2010 a train ticket cost $22
   This is 10% more than the cost of the ticket in 2005.
   Calculate the price of the ticket in 2005.

4. Sunita pays $240 for a bike in a sale.
   The original price had been reduced by 20%.
   Calculate the original price of the bike.

5. Gregor travels to work by train.
   One day, the train is slow and the journey takes 42 minutes.
   This is 10% more than the normal time taken.
   Calculate how long the journey normally takes.

6. The rent for a house is increased by 20% to $19 800
   Calculate the original rent.

7. The price of bananas is reduced by 25% to $1.74 per kilogram.
   Calculate the original price.

8. The cost of a holiday is reduced by 15% to $3009
   Calculate the original cost of the holiday.

9. The cost of a cinema ticket is increased by 12% to $9.52
   Calculate the original cost of the cinema ticket.

10. The cost of renting a car is reduced by 4% to $124.80
    Calculate the original cost of renting the car.

11. The price of gold increased by 2.5% to $45.92 per gram.
    Calculate the original price.

12. The temperature decreased by 25% to 16.2 °C
    Calculate the original temperature.

13. Zoe paid $5628 tax in 2010.
    This was 5% more than the amount of tax that she paid in 2009.
    Calculate the amount of tax that Zoe paid in 2009.

Simple interest

With **simple interest**, the interest earned is not reinvested.
This means that the amount of interest earned each year is unchanged.

> **EXAMPLE**
>
> Richard invests $400 at a rate of 5% per year simple interest.
> Calculate the amount Richard has after 3 years.
>
> Interest earned in first year = 5% of $400 = 0.05 × 400 = $20
> Interest earned in 3 years = 3 × 20 = $60
> Value of investment after 3 years = $400 + $60 = $460

Compound interest

With **compound interest**, the interest earned each year is reinvested.
This means that the amount of interest earned each year increases.
The following example shows three methods for finding the final value of an investment.

> **EXAMPLE**
>
> Richard invests $400 at a rate of 5% per year compound interest.
> Calculate the amount Richard has after 3 years.
>
> **Method 1**
>   Interest in first year = 0.05 × 400 = $20
>   Value of investment after 1 year = 400 + 20 = $420
>
>   Interest in second year = 0.05 × 420 = $21
>   Value of investment after 2 years = 420 + 21 = $441
>
>   Interest in third year = 0.05 × 441 = $22.05
>   Value of investment after 3 years = 441 + 22.05 = $463.05
>
> **Method 2** (Quicker method)
>   Value of investment after 1 year = 400 × 1.05 = $420   multiplying factor is 1.05 (= 1 + 0.05)
>   Value of investment after 2 years = 420 × 1.05 = $441
>   Value of investment after 3 years = 441 × 1.05 = $463.05
>
> **Method 3** (Quickest method!)
>   Value of investment after 3 years = 400×1.05^3 = $463.05

> **EXAMPLE**
>
> Simone invests $200 at a rate of 4% per year compound interest.
> Calculate the amount Simone has after 5 years.
>
> Value of investment after 5 years
> = 200×1.04^5
> = $243.33 (to the nearest cent)

EXERCISE 6.2

1. Amelia invests $80 at a rate of 15% per year, simple interest.
   Calculate the amount of interest Amelia earns in 5 years.

2. Karl invests $700 at a rate of 2% per year, simple interest.
   Calculate the total amount Karl has after 4 years.

3. Barbara invests $250 at a rate of 6% per year, simple interest.
   Calculate the total amount Barbara has after 3 years.

4. Sebastian invests $450 at a rate of 3.5% per year, simple interest.
   Calculate the total amount Sebastian has after 2 years.

5. Jack invests $500 at a rate of 3% per year, compound interest.
   Calculate the total amount Jack has after 2 years.

6. Omar invests $2000 at a rate of 2.5% per year, compound interest.
   Calculate the amount Omar has after 3 years.

7. Eduard invests $700 at a rate of 2% per year, compound interest.
   Calculate the amount Eduard has after 4 years.

8. Anelie invests $400 at a rate of 5% per year, compound interest.
   Calculate the amount of interest Anelie earns in 3 years.

9. Ferdinand invests $50 at a rate of r% per year, simple interest.
   After 4 years he has a total amount of $56.
   Calculate the value of r.

10. Hanna invests $400 at a rate of r% per year, simple interest.
    After 5 years she has a total amount of $700.
    Calculate the value of r.

11. Ferdinand invests $720 at a rate of 5% per year, compound interest.
    How many years will it be before his investment is worth more than $1000?

12. Robert bought his car 4 years ago.
    Each year the value of the car has depreciated by 15%.
    The car is now worth $6264.
    Calculate how much the car was worth when Robert bought it.

13. In 2009 the manatee population in Florida was estimated at 3800.
    The population is declining at a rate of approximately 1.1% per year.
    Estimate what the manatee population will be in the year in 2020.

> **KEY WORDS**
> reverse percentage
> simple interest
> compound interest

Speed, distance and time

THIS SECTION WILL SHOW YOU HOW TO
- Calculate speed, distance and time

Time

You need to be able to convert a time in minutes to a fraction (or decimal) of an hour.
To do this you must divide by 60.
You also need to be able to change a fraction (or decimal) of an hour into minutes.
To do this you must multiply by 60.

EXAMPLE

a Write 21 minutes as a fraction of an hour.
b Write 36 minutes as a decimal of an hour.

a 21 minutes = $\frac{21}{60}$ hour

   = $\frac{7}{20}$ hour

b 36 minutes = $\frac{36}{60}$ hour

   = 0.6 hours

EXAMPLE

a Change $\frac{5}{12}$ hour into minutes.
b Change 0.9 hours into minutes.

a $\frac{5}{12}$ hour = $\frac{5}{12}$ × 60 minutes

   = 25 minutes

b 0.9 hours = 0.9 × 60 minutes

   = 54 minutes

Speed

Speed is a measure of how fast an object is travelling.
When the speed is constant, the formula connecting speed, distance and time is

$$\text{speed} = \frac{\text{distance}}{\text{time}}$$

If the distance is measured in metres and the time is measured in seconds then the speed is measured in metres per second (m/s or ms^{-1}).
If the distance is measured in kilometres and the time is measured in hours then the speed is measured in kilometres per hour (km/h or kmh^{-1}).

EXAMPLE

A train takes 2 hours 55 minutes to travels 350 km at a constant speed.
Find the speed of the train in km/h.

speed = $\frac{\text{distance}}{\text{time}}$ = $\frac{350}{2\frac{11}{12}}$

= 120 km/h

➡ **NOTE:** you must change the 55 minutes into hours.

55 minutes = $\frac{55}{60}$ hour = $\frac{11}{12}$ hour

So, 2 hours 55 minutes = $2\frac{11}{12}$ hour

268 UNIT 6

EXAMPLE

Change 36 km/h into m/s.

$$36 \text{ kilometres/hour} = 36\,000 \text{ metres/hour}$$
$$= \frac{36\,000}{60} \text{ metres/minute}$$
$$= \frac{36\,000}{60 \times 60} \text{ metres/second}$$
$$= 10 \text{ m/s}$$

EXERCISE 6.3

1. Write these time intervals as fractions of an hour.
   - a 20 minutes
   - b 30 minutes
   - c 10 minutes
   - d 15 minutes
   - e 25 minutes
   - f 32 minutes
   - g 45 minutes
   - h 4 minutes
   - i 54 minutes
   - j 24 minutes
   - k 40 minutes
   - l 48 minutes

2. Change these time intervals into minutes.
   - a 0.5 hours
   - b 0.05 hours
   - c 0.25 hours
   - d 0.35 hours
   - e 0.8 hours
   - f 0.7 hours
   - g 0.1 hours
   - h 0.45 hours
   - i $\frac{2}{5}$ hour
   - j $\frac{3}{4}$ hour
   - k $\frac{5}{12}$ hour
   - l $\frac{3}{10}$ hour

3. Calculate the speed, in km/h, of an object that moves:
   - a 500 km in 8 hours
   - b 152 km in 4 hours
   - c 65 km in 5 hours
   - d 24 km in $\frac{1}{2}$ hours
   - e 6 km in 20 minutes
   - f 25 km in $2\frac{1}{2}$ hours
   - g 504 km in 10 h 30 min
   - h 76 km in 3 h 10 min
   - i 348 km in 2 h 54 min

4. Calculate the speed, in m/s, of an object that moves:
   - a 80 m in 4 s
   - b 125 m in 5 s
   - c 600 m in 12 s
   - d 72000 m in 1 h
   - e 5400 m in 3 min
   - f 49.68 km in 2 h 18 min

5.
   - a Change 72 km/h into m/s
   - b Change 12 km/h into m/s
   - c Change 360 km/h into m/s
   - d Change 20 m/s into km/h
   - e Change 35 m/s into km/h
   - f Change 41 m/s into km/h

6. A train of length 200 m passes through a tunnel of length 740 m.
   It takes 47 seconds to pass completely through the tunnel.
   Calculate the speed of the train in m/s.

7. Light travels a distance of 6×10^9 metres in 20 seconds.
   Calculate the speed of light in m/s.

8. Sound travels a distance of 2.06×10^4 metres in one minute
   Calculate the speed of sound in m/s.

The formula speed = $\frac{\text{distance}}{\text{time}}$ can be rearranged to give

distance = speed × time   or   time = $\frac{\text{distance}}{\text{speed}}$

This triangle is a useful memory aid for remembering the three formulae. To obtain the formula for finding S put your finger over the letter S.

This gives you S = $\frac{D}{T}$

Similarly, the formula for finding D is D = S × T and the formula for finding T is T = $\frac{D}{S}$.

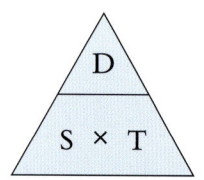

EXAMPLE

A car travels 200 km at a constant speed of 60 km/h. Calculate the time taken in hours and minutes.

time = $\frac{\text{distance}}{\text{speed}} = \frac{200}{60} = 3\frac{1}{3}$ hours

Time taken = 3 hours 20 minutes.

➡ **NOTE:** $\frac{1}{3}$ hour = $\frac{1}{3}$ × 60 minutes
= 20 minutes

Average speed

You can calculate the average speed for a journey using

average speed = $\frac{\text{total distance travelled}}{\text{total time taken}}$

EXAMPLE

A train travels 100 km at 80 km/h and then a further 150 km at 100 km/h.
a   Calculate the total time taken in hours and minutes.
b   Calculate the average speed for the whole journey.

a   time = $\frac{\text{distance}}{\text{speed}}$ time = $\frac{\text{distance}}{\text{speed}}$

   = $\frac{100}{80}$ = $\frac{150}{100}$

   = 1.25 hours   = 1.5 hours

Total time = 1.25 + 1.5 = 2.75 hours
= 2 hours 45 minutes

➡ **NOTE:** 0.75 hour = 0.75 × 60 minutes
= 45 minutes

b   Average speed = $\frac{\text{total distance travelled}}{\text{total time taken}}$

   = $\frac{100 + 150}{1.25 + 1.5}$

   = $\frac{250}{2.75}$

   = 90.9 km/h (to 3 s.f)

EXERCISE 6.4

1. Find the time taken, in hours and minutes, for the following journeys.
   - a 70 km at 35 km/h
   - b 150 km at 60 km/h
   - c 16 km at 32 km/h
   - d 18 km at 8 km/h
   - e 34 km at 12 km/h
   - f 53 km at 20 km/h

2. Find the distance travelled, in km, for the following journeys.
   - a 28 km/h for 2 hours
   - b 36 km/h for 6 hours
   - c 90 km/h for $3\frac{1}{2}$ hours
   - d 50 km/h for 30 minutes
   - e 60 km/h for 10 minutes
   - f 75 km/h for 4 minutes

3. A train travels 24 000 m in 40 minutes.
   Find the speed of the train in km/h.

4. A bird flies for 30 minutes at a speed of 7 m/s.
   Find the distance, in metres, travelled by the bird.

5. A car is travelling at a speed of 32 m/s.
   Calculate the time taken, in seconds, for the car to travel one kilometre.

6. A cyclist travels a distance of 91 km in 2 hours 12 minutes.
   Calculate the speed of the cyclist in km/h.

7. An aircraft travels at 1200 km/h for 2 hours and then 1500 km/h for 4 hours.
   - a Calculate the total distance travelled.
   - b Calculate the average speed for the whole journey.

8. A long distance runner runs at 5 m/s for 30 minutes and then at 4 m/s for one hour.
   - a Calculate the total distance travelled in km.
   - b Calculate the average speed of the runner in m/s.

9. A car travels at 80 km/h for 4 hours and then a speed of 100 km/h for a further 2 hours.
   - a Calculate the total distance travelled.
   - b Calculate the average speed for the whole journey.

10. A train of length 200 metres passes completely through a tunnel of length 250 metres in 15 seconds.
    Calculate the speed of the train.

11. Zavia drives at a speed of x km/h for one hour.
    She then drives at a speed of $(x + 20)$ km/h for the next two hours.
    The total distance travelled is 280 km.
    Calculate the value of x.

KEY WORDS
speed

Speed, distance and time   271

Sets and Venn diagrams

THIS SECTION WILL SHOW YOU HOW TO
- Use set language and notation
- Use Venn diagrams

The language of sets

A **set** is a collection of items.
Let A = The set of days in the week that begin with the letter T.
Then in set notation A = {Tuesday, Thursday}.

The items in the set are called **elements** or **members** of the set.
The list of elements is enclosed by { } and commas are used to separate the elements.

The **number of elements** in the set A is denoted by n(A).
In the above example: n(A) = 2.

The symbol $\in$ means 'is a member of'.
The symbol $\notin$ means 'is not a member of'.
In the above example Tuesday $\in$ A and Wednesday $\notin$ A.

The symbol $\emptyset$, or { }, is used to represent the **empty set**.
For example, if B = {days in the week that begin with the letter P} then B = $\emptyset$.

The **universal set** is denoted by the symbol $\mathscr{E}$.
The universal set contains all the elements being considered in a particular problem.
The sets
$\mathscr{E}$ = {days of the week} and
A = {Tuesday, Thursday} can be shown on a Venn diagram.

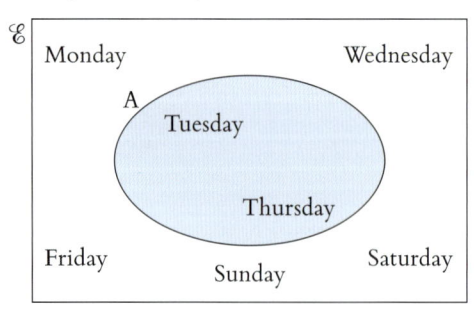

The set A has been shaded on the diagram.

EXERCISE 6.5

1. List these sets.
   a. {square numbers less than 50}
   b. {prime numbers less than 25}
   c. {months of the year}
   d. {factors of 12}
   e. {prime factors of 25}
   f. {even numbers less than 15}
   g. {odd numbers between 20 and 30}
   h. {prime numbers between 150 and 200}

2. List these sets.
   a. {factors of 49}
   b. {cube numbers between 100 and 200}
   c. {prime factors of 100}
   d. {factors of 20}
   e. {prime factors of 36}
   f. {months of the year beginning with J}

3. Describe these sets by a rule.
   a. {1, 4, 9, 16}
   b. {1, 3, 5, 7, …}
   c. {March, May}
   d. {4, 8, 12, 16, 20, 24, …}
   e. {1, 8, 27, 64, …}
   f. {red, blue, green, yellow, …}

4. Describe these sets by a rule.
   a. {2, 3, 5, 7, 11, …}
   b. {1, 2, 4, 5, 10, 20}
   c. {2, 4, 8, 16, 32}
   d. {(3, 4, 5), (6, 8, 10), (5, 12, 13), …}

5. Which of these statements are true?
   a. $y = 2x + 1 \notin$ {linear graphs}
   b. $y = x^3 + 1 \in$ {non-linear graphs}
   c. $1 \in$ {prime numbers}
   d. parallelogram $\in$ {triangles}
   e. kite $\notin$ {quadrilaterals}
   f. trapezium $\notin$ polygons

6. Which of these statements are true?
   a. $5 \in$ {factors of 56}
   b. $3 \in$ {factors of 4350}
   c. $2 \notin$ {factors of $2x^2 - 4x$}
   d. $x \in$ {factors $x^2 - 3x$}

7. Which of these statements are true?
   a. $x - 2 \in$ {factors of $x^2 - 2x$}
   b. $x + 2 \in$ {factors of $x^2 + 2x - 8$}
   c. $x + 5 \notin$ {factors of $x^2 - 25$}
   d. $3x + 2 \in$ {factors of $6x^2 + x - 2$}

8. Which of these statements are true?
   a. {odd numbers divisible by 2} = ∅
   b. {even numbers divisible by 3} = ∅
   c. {prime numbers between 48 and 52} = ∅
   d. {months of the year beginning with the letter B} = ∅
   e. {common factors of 23 and 29} = ∅
   f. {answers to the equation $x^2 = -4$} = ∅

Venn diagrams

The **complement** of set A is the set of all elements not in A
The complement of set A is denoted by A'.

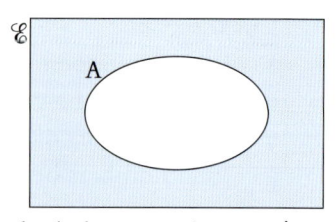

Shaded region shows A'

The **intersection** of A and B is the set of elements that are in both A and B
The intersection of A and B is denoted by $A \cap B$.

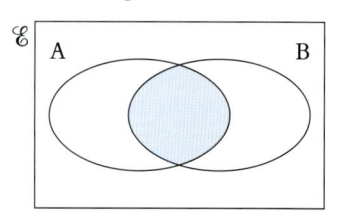

Shaded region shows $A \cap B$

The **union** of A and B is the set of elements that are in A or B or both.
The union of A and B is denoted by $A \cup B$.

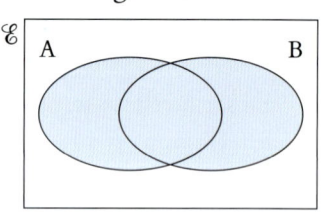

Shaded region shows $A \cup B$

Representing sets on a Venn diagram
When the expressions are more complicated you many need to use some diagrams for your working before deciding on your answer.

EXAMPLE

On a Venn diagram shade the regions **a** $A \cap B'$ **b** $A \cup B'$

Working

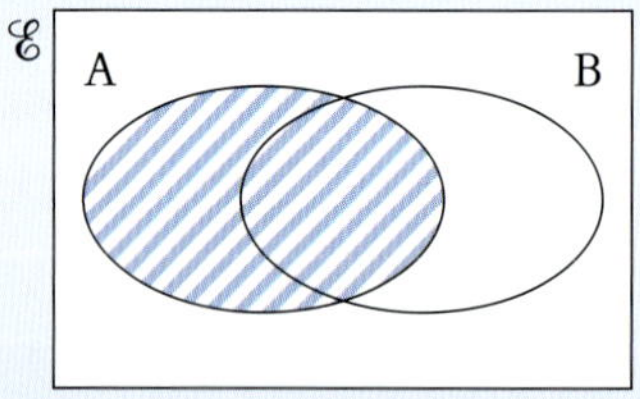

shade set A in one direction

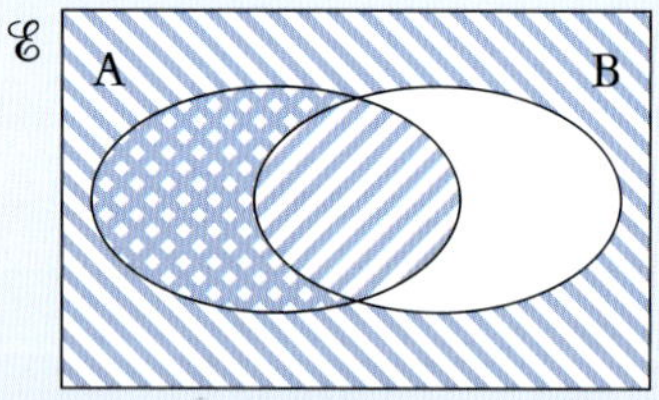

shade set B' in the opposite direction

a $A \cap B'$ is the region that is in both A and B' so you need the region with double shading.

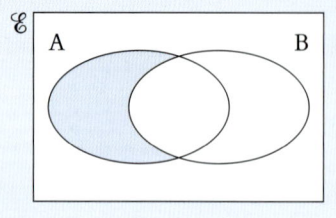

b $A \cup B'$ is the region that is in A or B' or both so you need all of the shaded regions.

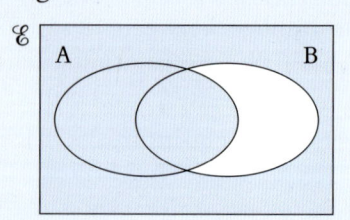

shaded region shows $A \cap B'$

shaded region shows $A \cup B'$

EXERCISE 6.6

1 On copies of this diagram shade the following regions.
- **a** B'
- **b** $A \cup B$
- **c** $A \cap B$
- **d** $A' \cap B$
- **e** $A' \cap B'$
- **f** $(A' \cup B)'$
- **g** $(A \cap B') \cup (A' \cap B)$
- **h** $(A \cap B) \cup (A \cup B)'$

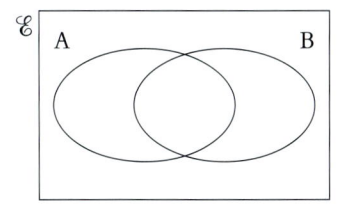

> If $\mathcal{E} = \{1, 2, 3, 4, 5\}$, $A = \{1, 3, 5\}$ and $B = \{3, 5\}$ then B is called a **proper subset** of A.
> This is written as $B \subset A$.
> On a Venn diagram the set B is drawn inside the set A.

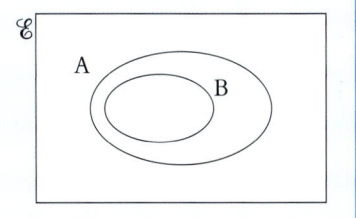

2 On copies of this diagram shade the following regions.
- **a** A'
- **b** $A \cup B$
- **c** $A \cap B$
- **d** $A' \cap B$
- **e** $A' \cap B'$
- **f** $(A \cup B)'$
- **g** $A \cup B'$
- **h** $(A' \cup B)'$
- **i** $(A \cap B) \cup (A \cup B)'$

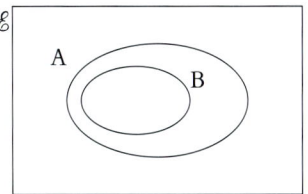

> If $\mathcal{E} = \{1, 2, 3, 4, 5\}$, $A = \{1, 2\}$ and $B = \{3, 5\}$ then A and B do not intersect. $A \cap B = \emptyset$.
> On a Venn diagram the sets A and B are drawn so that they do not overlap.

3 On copies of this diagram shade the following regions.
- **a** B'
- **b** $A \cup B$
- **c** $(A \cap B)'$
- **d** $A \cap B'$
- **e** $A' \cap B'$
- **f** $(A \cup B)'$

4 On copies of this diagram shade the following regions.
- **a** $A \cap B \cap C$
- **b** $(A \cap B \cap C)'$
- **c** $A \cup B \cup C$
- **d** $(A \cup B \cup C)'$
- **e** $A \cup (B \cap C)$
- **f** $A \cap (B \cup C)$
- **g** $A' \cap B' \cap C'$
- **h** $(A \cap B)' \cap C$
- **i** $(A \cup C)' \cap (A \cup B)'$

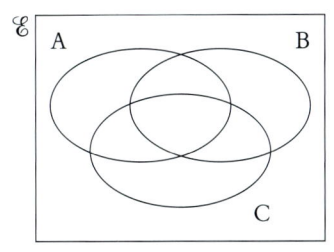

Sets and Venn diagrams

You might be given a Venn diagram and then be asked to use set notation to describe the shaded region.

$A \cap B'$

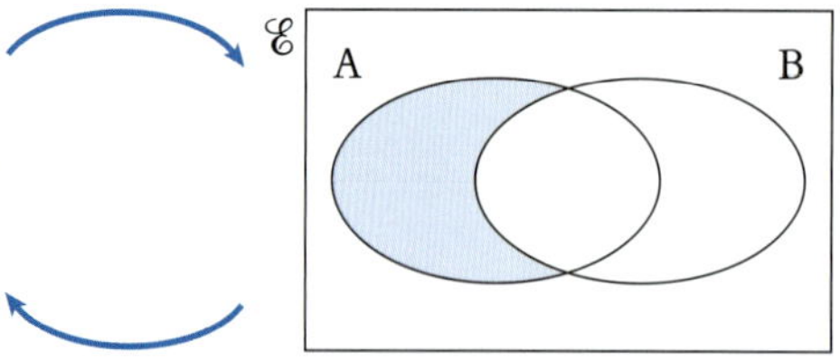

EXAMPLE

Describe the shaded regions using set notation.

a 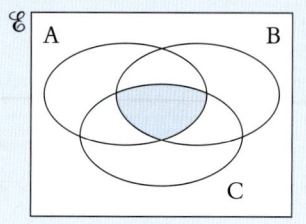
b 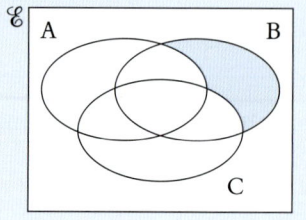
c

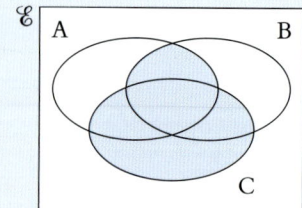

---

a
The shaded region is in A, B and C. 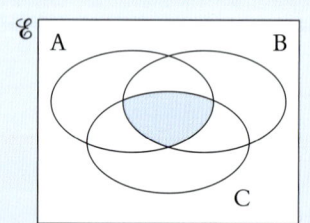
The region is $A \cap B \cap C$.

b 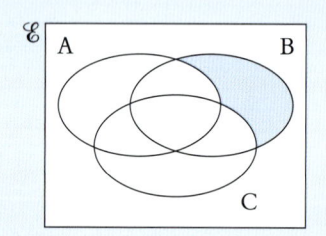
The shaded region is in B and outside $A \cup C$.
The region is $B \cap (A \cup C)'$.

c 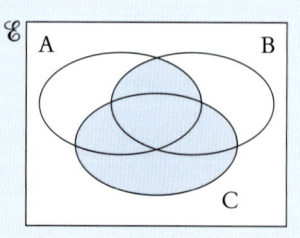
The shaded region contains C and $A \cap B$.
The region is $C \cup (A \cap B)$.

EXERCISE 6.7

1 Describe the shaded regions using set notation.

a, b, 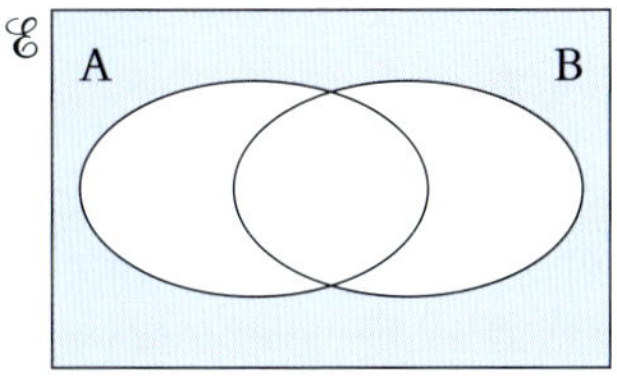 c, 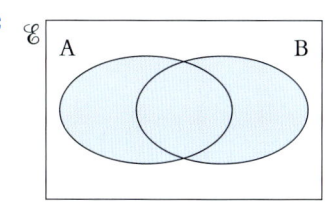 d, 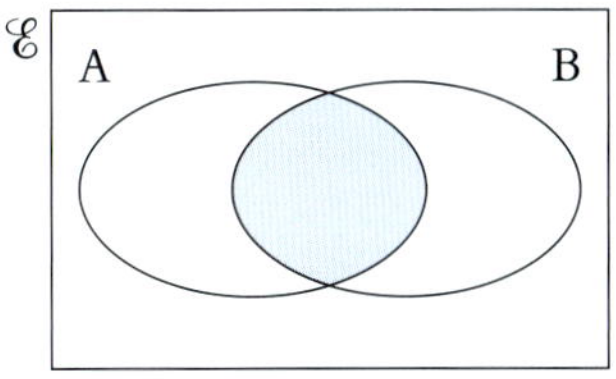 e, 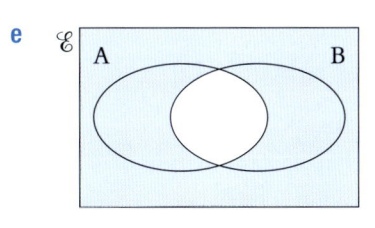 f

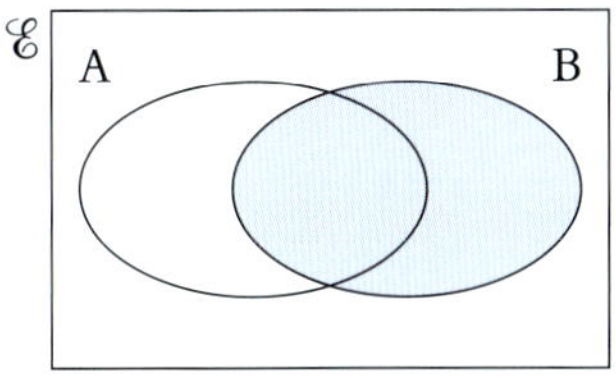

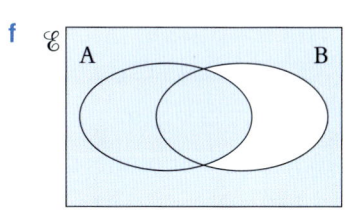

2 Describe the shaded regions using set notation.

a, b, 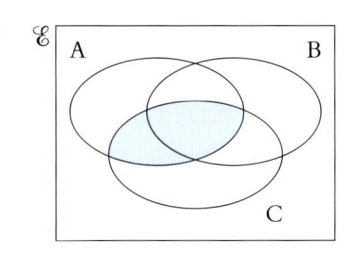 c, 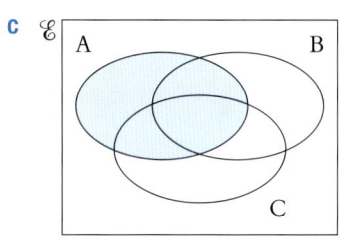 d, 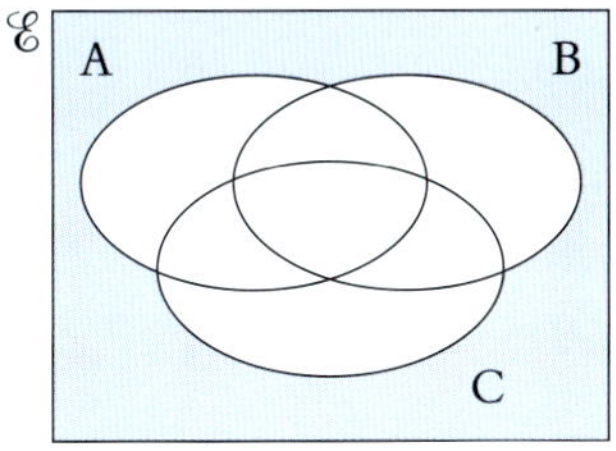 e, 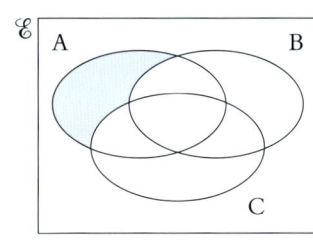 f, 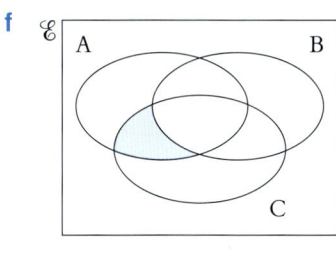 g, 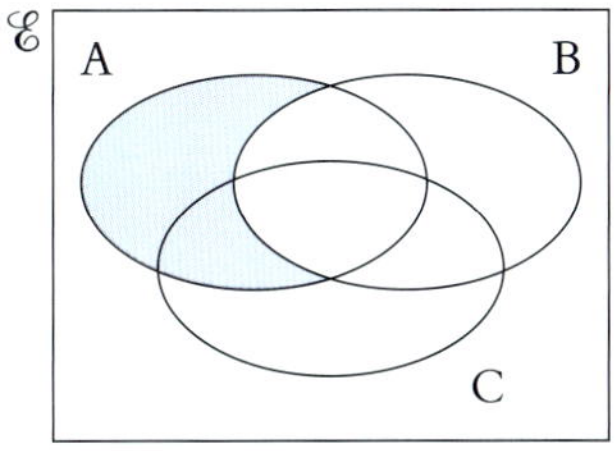 h, 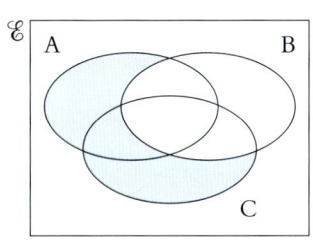 i

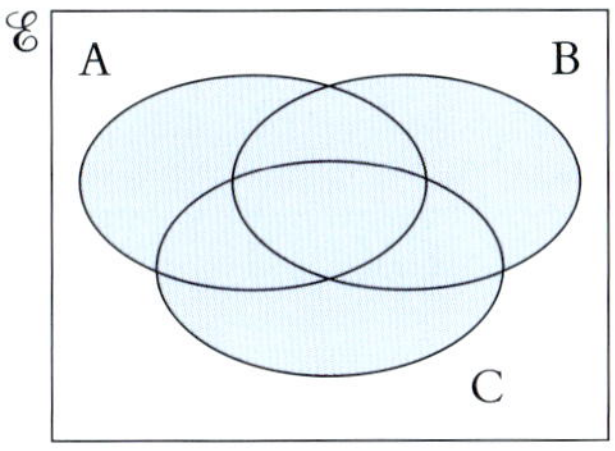

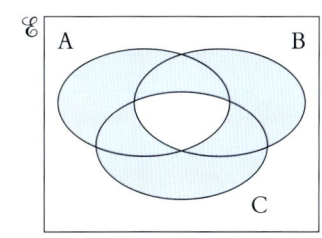

3 Describe the shaded regions using set notation.

a, b, 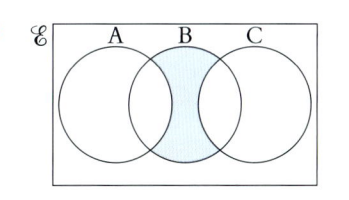 c

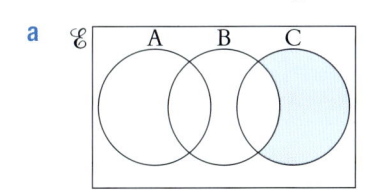

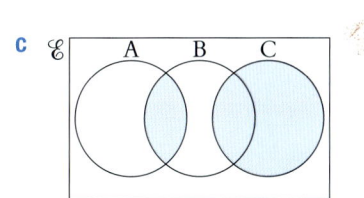

Further set language

If A = {1, 3, 5} and B = {1, 3, 5} then A is called a **subset** of B.
'A is a subset of B' means that all the elements of set A are contained in set B.
This is written as A ⊆ B
[C ⊄ D means C is **not a subset** of D]

If A = {3, 5} and B = {1, 3, 5} then A is called a **proper subset** of B.
'A is a proper subset of B' means that all the elements of set
A are contained in set B but there is at least one element of set B that is not in set A.
This is written as A ⊂ B
[C ⊄ D means C is **not a proper subset** of D]

A = {$x : x$ is a multiple of 4} means
'A is the set of numbers x such that x is a multiple of 4' so A = {4, 8, 12, 16, …}
B = {$(x, y): y = x + 1$} means 'B is the set of points (x, y) that lie on the line $y = x + 1$'.

Listing sets and illustrating on a Venn diagram.

EXAMPLE

ℰ = {1, 2, 3, 4, 5, 6, 7, 8, 9, 10}
A = {2, 4, 6, 8}   B = {1, 4, 9}   C = {1, 2, 5, 7}
a List A′   **b** List B ∩ C   **c** List A ∩ B ∩ C   **d** List A′ ∪ B
e Draw a Venn diagram to illustrate the sets ℰ, A, B and C.

a A′ means every element in ℰ that is not in A.
A′ = {1, 3, 5, 7, 9, 10}

b B ∩ C means the elements that are common to both B and C.
B ∩ C = {1}

c A ∩ B ∩ C means the elements that are common to all three sets.
A ∩ B ∩ C = ∅ or A ∩ B ∩ C = { }

d A′ ∪ B means the elements that are in A′ or B or both.
A′ = {1, 3, 5, 7, 9, 10} and B = {1, 4, 9}
A′ ∪ B = {1, 3, 4, 5, 7, 9, 10}

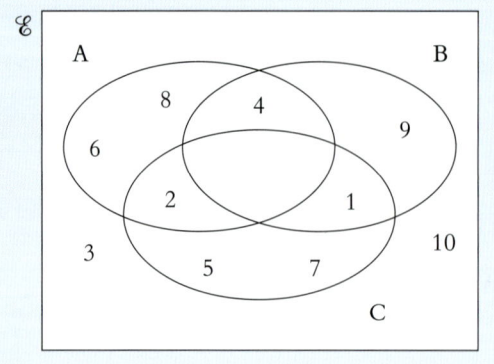

e

EXERCISE 6.8

1 List the following sets.
  - **a** A = {$x : x$ is a multiple of 3}
  - **b** B = {$x : x$ is a factor of 20}
  - **c** C = {$x : x$ is an odd number}
  - **d** D = {$x : x$ is a prime number less than 20}
  - **e** E = {$x : x$ is a prime factor of 36}
  - **f** F = {$x : 2x + 1 = 5$}
  - **g** G = {$x : x^2 = 25$}
  - **h** H = {$x : x^3 = 27$}
  - **i** I = {$x : 2x + 3 = 2x - 5$}

2 List the following sets. 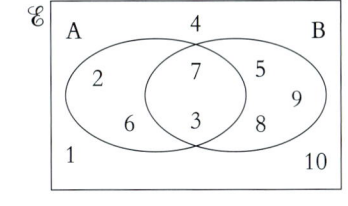
  - **a** A
  - **b** B
  - **c** A′
  - **d** B′
  - **e** A ∪ B
  - **f** A ∩ B
  - **g** (A ∪ B)′
  - **h** A ∩ B′

3 List the following sets. 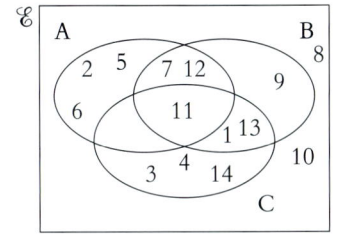
  - **a** A′
  - **b** B′
  - **c** A ∩ B ∩ C
  - **d** B ∪ C
  - **e** (A ∩ B ∩ C)′
  - **f** A′ ∩ B
  - **g** (A ∪ C)′
  - **h** B ∩ C′

4 ℰ = {$x : x$ is an integer and $1 \leq x \leq 10$}
A = {factors of 10}   B = {multiples of 4}   C = {prime numbers}
List the following sets.
  - **a** A
  - **b** B
  - **c** C
  - **d** A ∩ B ∩ C
  - **e** A ∩ B
  - **f** A ∩ C
  - **g** B ∩ C
  - **h** (A ∪ B ∪ C)′
  - **i** Draw a Venn diagram to illustrate the sets ℰ, A, B and C.

5 ℰ = {a, b, c, d, e, f, g}
A = {a, c, g}   B = {b, c, d}
List the following sets.
  - **a** A′
  - **b** B′
  - **c** A ∩ B
  - **d** A ∪ B
  - **e** A′ ∩ B
  - **f** A ∩ B′
  - **g** (A ∩ B)′
  - **h** A′ ∩ B′
  - **i** Draw a Venn diagram to illustrate the sets ℰ, A and B.

6 ℰ = {triangles}   E = {equilateral triangles}   I = {isosceles triangles}
Draw a Venn diagram to illustrate these sets.

7 ℰ = {quadrilaterals}   P = {parallelograms}   R = {rectangles}
Draw a Venn diagram to illustrate these sets.

Using sets to solve problems

You can also use Venn diagrams to show the number of elements in each set. This is particularly useful when solving problems.

EXAMPLE

$n(P) = 35$, $n(Q) = 31$, $n(R) = 32$,
$n(P \cap Q) = 13$, $n(P \cap R) = 16$, $n(Q \cap R) = 7$,
$n(P \cap Q \cap R) = 4$ and $n(P \cup Q \cup R)' = 6$.
Draw a Venn diagram to find $n(\mathcal{E})$.

The first pieces of information to go on the Venn diagram are
$n(P \cap Q \cap R) = 4$ and $n(P \cup Q \cup R)' = 6$.

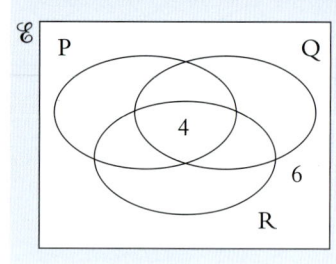

[Venn diagram with 4 in centre intersection and 6 outside the circles]

$n(P \cap Q) = 13$ $13 - 4 = 9$
$n(P \cap R) = 16$ $16 - 4 = 12$
$n(Q \cap R) = 7$ $7 - 4 = 3$

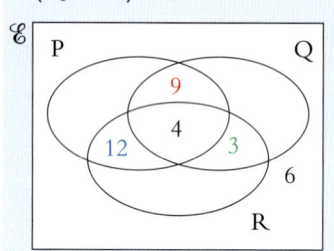

[Venn diagram with 9, 4, 12, 3, 6]

$n(P) = 35$ $35 - (9 + 12 + 4) = 10$
$n(Q) = 31$ $31 - (9 + 3 + 4) = 15$
$n(R) = 32$ $32 - (12 + 3 + 4) = 13$

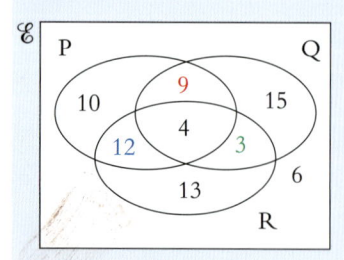

[Venn diagram with 10, 9, 15, 12, 4, 3, 13, 6]

Now add all the numbers on the Venn diagram together to find $n(\mathcal{E})$.
$n(\mathcal{E}) = 10 + 9 + 15 + 12 + 4 + 3 + 13 + 6 = 72$

EXERCISE 6.9

1 The Venn diagram shows information about the number of elements in the sets.

   For example n(A) = 5 + 8 = 13.
   Find
   a n(A′)   b n(B)   c n(B′)
   d n(A ∩ B)   e n(A ∪ B)   f n(A ∪ B)′
   g n(A′ ∩ B)   h n(A ∩ B′).

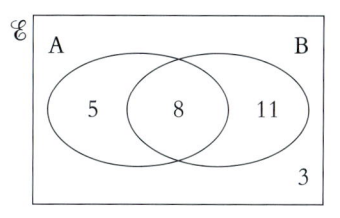

2 n(A) = 18, n(B) = 13 and n(A ∩ B) = 3. Draw a Venn diagram to find n(A ∪ B).

3 n(A) = 12, n(B) = 11 and n(A ∩ B) = 4. Draw a Venn diagram to find n(A ∪ B).

4 n(A) = 35, n(B) = 33 and n(A ∩ B) = 13. Draw a Venn diagram to find n(A ∪ B).

5 n(A) = 9, n(B) = 11 and n(A ∪ B) = 13. Draw a Venn diagram to find n(A ∩ B).

6 n(A) = 65, n(B) = 57 and n(A ∪ B) = 114. Draw a Venn diagram to find n(A ∩ B).

7 n(ℰ) = 27, n(A) = 12, n(B) = 9 and n(A ∩ B) = 4.
   Draw a Venn diagram to find   a n(A ∪ B)   b n(A ∪ B)′   c n(A ∩ B′).

8 n(ℰ) = 40, n(A) = 25, n(B) = 22 and n(A ∪ B)′ = 3.
   Draw a Venn diagram to find   a n(A ∩ B)   b n(A ∩ B′).

9 n(ℰ) = 50, n(A) = 26, n(B) = 22 and n(A ∩ B) = 12.
   Draw a Venn diagram to find   a n(A ∪ B)   b n(A ∪ B)′   c n(A′ ∩ B).

10 n(A) = 15, n(A ∩ B) = 10 and n(A′ ∩ B) = 7.
   Draw a Venn diagram to find   a n(B)   b n(A ∪ B).

11 n(A) = 11, n(B) = 15, n(C) = 10, n(A ∩ B) = 5, n(A ∩ C) = 4,
   n(B ∩ C) = 7 and n(A ∩ B ∩ C) = 3.
   Draw a Venn diagram to find n(A ∪ B ∪ C).

12 n(A) = 16, n(B) = 12, n(C) = 9, n(B ∩ C) = 3, n(A ∩ C) = 5,
   n(A ∪ B) = 22 and n(A ∩ B ∩ C) = 2.
   Draw a Venn diagram to find n(A ∪ B ∪ C).

13 n(A) = 17, n(B) = 13, n(C) = 16, n(A ∩ B) = 10, n(A ∩ C) = 6, n(B ∩ C) = 7,
   n(A ∩ B ∩ C) = 4 and n(A ∪ B ∪ C)′ = 8.
   Draw a Venn diagram to find n(ℰ).

EXAMPLE

There are 29 students in a class.
22 study biology (B), 19 study chemistry (C) and 3 study neither biology or chemistry.
How many students study both biology and chemistry?

Method 1
3 students study neither biology or chemistry so the number 3 goes outside the sets B and C.

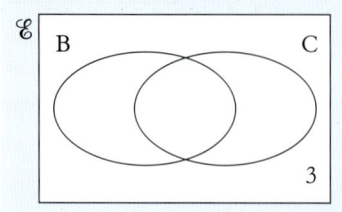

$29 - 3 = 26$ so there are now 26 students left to put on the Venn diagram.
$22 + 19 = 41$ and $41 - 26 = 15$
The number that goes in the intersection is 15 and the remaining numbers can now be filled in

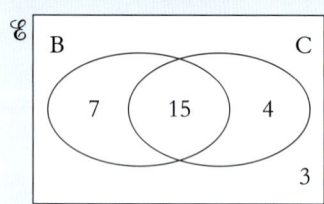

The number of students who study both biology and chemistry is 15.

Method 2 (Using algebra)
The number 3 goes outside the sets B and C on the Venn diagram.
Let x = the number of students that study both biology and chemistry.

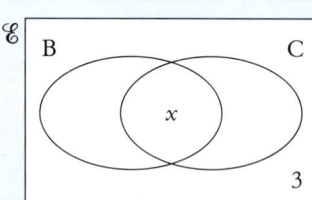

The number of students who study biology and not chemistry is $22 - x$.
The number of students who study chemistry and not biology is $19 - x$.
So the Venn diagram becomes

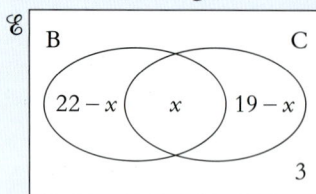

There are a total of 29 students.
So $(22 - x) + x + (19 - x) + 3 = 29$ $44 - x = 29$ $x = 15$
The number of students who study both biology and chemistry is 15.

282 UNIT 6

EXERCISE 6.10

1. There are 30 students in a class.
   17 study geography (G), 16 study history (H) and 6 study neither geography or history.
   Copy and complete the Venn diagram to show this information.
   How many students study both geography and history?

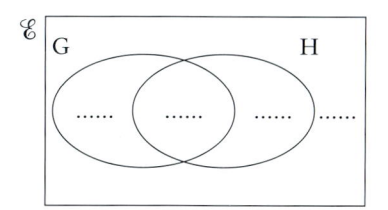

2. There are 34 students in a class.
   20 like apple juice, 28 like orange juice and 2 students don't like either drink.
   How many students like both apple and orange juice?

3. 50 people were asked if they had ever been to Mauritius or the Seychelles on holiday.
   11 had been to Mauritius, 7 had been to the Seychelles and 35 had been to neither.
   How many had been to both?

4. 26 students go on an activity holiday.
   15 have sailed before, 8 have rock climbed before and 7 have done neither.
   How many have done both before?

5. There are 27 students in a class.
   22 have scientific calculators, 7 have graphical calculators and 2 students have neither.
   How many students have both a scientific calculator and a graphical calculator?

6. 100 students do a mathematics test.
   89 had a pair of compasses, 87 had a protractor and 86 had a ruler.
   78 students had a ruler and protractor.
   81 had a pair of compasses and a protractor.
   79 had a pair of compasses and a ruler.
   74 students had all three pieces of equipment.
   How many students had
   a none of the three pieces of equipment,
   b only a protractor.

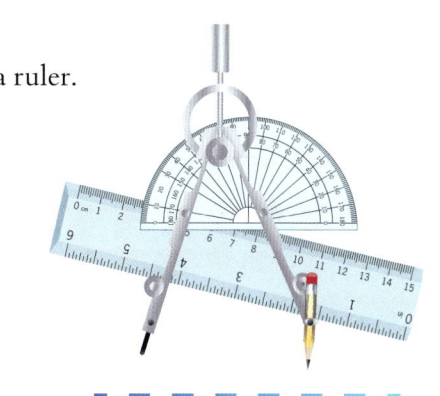

7. A group of students are given three problems to solve. (Problem A, B and C.)
   13 students solve problem A, 17 solve problem B and 7 solve problem C.
   9 solve problems A and B.
   6 solve problems B and C.
   5 solve problems A and C.
   4 solve all three problems.
   2 solve none of the three problems.
   How many students are there altogether.

KEY WORDS

set, element
member, empty set
universal set
Venn diagram
complement
intersection
union
subset

Indices 2

THIS SECTION WILL SHOW YOU HOW TO
- Use fraction indices

Reminder: $a^m \times a^n = a^{m+n}$ $a^m \div a^n = a^{m-n}$ $(a^m)^n = a^{mn}$

$a^0 = 1$ $a^{-n} = \dfrac{1}{a^n}$

Fractional indices

Consider $\sqrt{16} \times \sqrt{16}$ and $16^{\frac{1}{2}} \times 16^{\frac{1}{2}}$

$\qquad\qquad = 4 \times 4 \qquad\qquad\quad = 16^{\frac{1}{2}+\frac{1}{2}} = 16^1$

$\qquad\qquad = 16 \qquad\qquad\qquad\; = 16$

$\qquad\qquad\qquad\text{so}\;\; 16^{\frac{1}{2}} = \sqrt{16}$

The index $\dfrac{1}{2}$ means 'square root' $a^{\frac{1}{2}} = \sqrt{a}$

The index $\dfrac{1}{3}$ means 'cube root' $a^{\frac{1}{3}} = \sqrt[3]{a}$

This can be written more generally as:
The index $\dfrac{1}{n}$ means 'nth root' $a^{\frac{1}{n}} = \sqrt[n]{a}$

EXAMPLE

Find the value of   **a** $16^{\frac{1}{2}}$ **b** $27^{\frac{1}{3}}$ **c** $32^{-\frac{1}{5}}$

a $16^{\frac{1}{2}} = \sqrt{16} = 4$ **b** $27^{\frac{1}{3}} = \sqrt[3]{27} = 3$ **c** $32^{-\frac{1}{5}} = \dfrac{1}{32^{\frac{1}{5}}} = \dfrac{1}{\sqrt[5]{32}} = \dfrac{1}{2}$

$8^{\frac{2}{3}}$ can be written as $\left(8^{\frac{1}{3}}\right)^2 = \left(\sqrt[3]{8}\right)^2 = 2^2 = 4$ similarly $8^{-\frac{2}{3}} = \dfrac{1}{8^{\frac{2}{3}}} = \dfrac{1}{\left(\sqrt[3]{8}\right)^2} = \dfrac{1}{2^2} = \dfrac{1}{4}$

In general: $a^{\frac{m}{n}} = \left(\sqrt[n]{a}\right)^m$

EXAMPLE

Find the value of   **a** $125^{\frac{4}{3}}$ **b** $16^{-\frac{3}{4}}$

a $125^{\frac{4}{3}} = \left(125^{\frac{1}{3}}\right)^4 = \left(\sqrt[3]{125}\right)^4 = 5^4 = 625$ **b** $16^{-\frac{3}{4}} = \dfrac{1}{16^{\frac{3}{4}}} = \dfrac{1}{\left(\sqrt[4]{16}\right)^3} = \dfrac{1}{2^3} = \dfrac{1}{8}$

EXERCISE 6.11

1 Find the value of

a $9^{\frac{1}{2}}$ b $64^{\frac{1}{2}}$ c $27^{\frac{1}{3}}$ d $81^{\frac{1}{2}}$ e $64^{\frac{1}{3}}$ f $8^{\frac{1}{3}}$

g $100^{\frac{1}{2}}$ h $81^{\frac{1}{4}}$ i $144^{\frac{1}{2}}$ j $169^{\frac{1}{2}}$ k $400^{\frac{1}{2}}$ l $8000^{\frac{1}{3}}$

2 Find the value of

a $9^{-\frac{1}{2}}$ b $4^{-\frac{1}{2}}$ c $27^{-\frac{1}{3}}$ d $25^{-\frac{1}{2}}$ e $81^{-\frac{1}{4}}$ f $125^{-\frac{1}{3}}$

g $8^{-\frac{1}{3}}$ h $64^{-\frac{1}{3}}$ i $100^{-\frac{1}{2}}$ j $36^{-\frac{1}{2}}$ k $121^{-\frac{1}{2}}$ l $16^{-\frac{1}{4}}$

3 Find the value of

a $25^{\frac{3}{2}}$ b $8^{\frac{4}{3}}$ c $4^{\frac{3}{2}}$ d $9^{\frac{3}{2}}$ e $100^{\frac{5}{2}}$ f $1^{\frac{2}{3}}$

g $8^{\frac{2}{3}}$ h $9^{\frac{5}{2}}$ i $1000^{\frac{2}{3}}$ j $169^{\frac{3}{2}}$ k $81^{\frac{3}{4}}$ l $8^{\frac{4}{3}}$

4 Find the value of

a $25^{-\frac{3}{2}}$ b $8^{-\frac{4}{3}}$ c $16^{-\frac{3}{2}}$ d $4^{-\frac{3}{2}}$ e $100^{-\frac{3}{2}}$ f $8^{-\frac{2}{3}}$

g $9^{-\frac{5}{2}}$ h $64^{-\frac{5}{6}}$ i $1000^{-\frac{5}{3}}$ j $64^{-\frac{2}{3}}$ k $81^{-\frac{3}{4}}$ l $1^{-\frac{4}{3}}$

5 Simplify the following. Write your answers in the form x^n.

a $\left(x^3\right)^{\frac{2}{3}}$ b $x^{\frac{1}{2}} \times x^{1\frac{1}{2}}$ c $x^{\frac{1}{4}} \times x^{\frac{1}{4}}$ d $\left(x^{-4}\right)^{\frac{1}{2}}$

e $\left(x^{\frac{1}{3}}\right)^3$ f $x^{2\frac{1}{2}} \times x^{-\frac{1}{2}}$ g $x^{-\frac{1}{2}} \times x^{-\frac{1}{2}}$ h $x^{3\frac{1}{2}} \times x^{\frac{1}{2}}$

i $\left(x^{\frac{3}{4}}\right)^4$ j $x^{\frac{2}{3}} \div x^{-\frac{1}{3}}$ k $\left(x^4\right)^{-\frac{1}{2}}$ l $\left(x^{-\frac{1}{4}}\right)^{-8}$

m $x^{1\frac{2}{5}} \div x^{-\frac{3}{5}}$ n $x^{\frac{1}{4}} \times x^{-\frac{1}{4}}$ o $x^{\frac{1}{3}} \times x^{\frac{1}{2}}$ p $x^{\frac{1}{3}} \div x^{\frac{1}{2}}$

6 Simplify

a $3x^{0.5} \times 5x^{0.5}$ b $8x^{0.5} \times 2x^{-0.5}$ c $12x^{1.5} \div 2x^{-2.5}$ d $15x^{\frac{2}{3}} \div 5x^{\frac{1}{3}}$

e $\dfrac{8x^{0.5} \times 3x^{1.5}}{4x^{0.5}}$ f $\dfrac{5x^{0.25} \times 6x^{1.25}}{2x^{-0.5} \times 5x^3}$ g $\dfrac{8x^{-3}}{2x \times 2x^{0.5}}$ h $\dfrac{5x^{0.5} + 3x^{0.5}}{2x^{-5}}$

7 Simplify

a $\dfrac{15x^6 y^{0.5}}{3x^9 y^{2.5}}$ b $\left(2x^3 y^{\frac{3}{2}}\right)^6$ c $\sqrt{25x^6 y^{\frac{1}{2}}}$ d $\sqrt{4x^8 y^6} \times \sqrt[3]{8x^3 y^{-6}}$

Indices 2

EXAMPLE

Find the value of n in these equations. **a** $2^n = 64$ **b** $3^n = \dfrac{1}{81}$

a $2^n = 64$ *first write 64 as a power of 2* $64 = 2 \times 2 \times 2 \times 2 \times 2 \times 2 = 2^6$
$2^n = 2^6$
$n = 6$

b $3^n = \dfrac{1}{81}$ *first write $\dfrac{1}{81}$ as a power of 3* $\dfrac{1}{81} = \dfrac{1}{3 \times 3 \times 3 \times 3} = \dfrac{1}{3^4} = 3^{-4}$
$3^n = 3^{-4}$
$n = -4$

EXAMPLE

Find the value of n in these equations. **a** $5^{n+1} = 125$ **b** $2^{2n-1} = 32$.

a $5^{n+1} = 125$ *first write 125 as a power of 5* $125 = 5 \times 5 \times 5 = 5^3$
$5^{n+1} = 5^3$
$n + 1 = 3$
$n = 2$

b $2^{2n-1} = 32$ *first write 32 as a power of 2* $32 = 2 \times 2 \times 2 \times 2 \times 2 = 2^5$
$2^{2n-1} = 2^5$
$2n - 1 = 5$
$2n = 6$
$n = 3$

EXAMPLE

Find the value of n in these equations. **a** $9^{n+1} = 3^5$ **b** $8^{n-1} = 2^6$.

a $9^{n+1} = 3^5$ *the base numbers are different so replace 9 with 3^2*
$\left(3^2\right)^{n+1} = 3^5$ *use the rule $\left(a^m\right)^n = a^{mn}$*
$3^{2n+2} = 3^5$
$2n + 2 = 5$
$2n = 3$
$n = 1.5$

b $8^{n-1} = 2^6$ *the base numbers are different so replace 8 with 2^3*
$\left(2^3\right)^{n-1} = 2^6$ *use the rule $\left(a^m\right)^n = a^{mn}$*
$2^{3n-3} = 2^6$
$3n - 3 = 6$
$3n = 9$
$n = 3$

EXERCISE 6.12

1 Find the value of n in these equations.
- **a** $2^n = 32$
- **b** $3^n = 9$
- **c** $4^n = 4$
- **d** $2^n = 1$
- **e** $10^n = 100$
- **f** $5^n = 125$
- **g** $2^n = 256$
- **h** $3^n = 27$
- **i** $2^n = 2$
- **j** $4^n = 64$
- **k** $3^n = 81$
- **l** $2^n = 16$

2 Find the value of n in these equations.
- **a** $2^n = \dfrac{1}{16}$
- **b** $3^n = \dfrac{1}{9}$
- **c** $2^n = \dfrac{1}{128}$
- **d** $4^n = \dfrac{1}{16}$
- **e** $6^n = \dfrac{1}{36}$
- **f** $4^n = \dfrac{1}{64}$
- **g** $2^n = \dfrac{1}{8}$
- **h** $3^n = \dfrac{1}{81}$
- **i** $3^n = \dfrac{1}{3}$
- **j** $10^n = \dfrac{1}{1000}$
- **k** $5^n = \dfrac{1}{625}$
- **l** $2^n = \dfrac{1}{2}$

3 Find the value of n in these equations.
- **a** $2^{2n} = 64$
- **b** $2^{n+1} = 16$
- **c** $2^{n-1} = 8$
- **d** $3^{n+2} = 81$
- **e** $4^{2n} = 64$
- **f** $10^{3n} = 1\,000\,000$
- **g** $2^{n-2} = 128$
- **h** $2^{2n+1} = 8$
- **i** $2^{2n-1} = 32$
- **j** $3^{2n+1} = 27$
- **k** $5^{n-1} = 125$
- **l** $7^{2n+1} = 49$

4 Find the value of n in these equations.
- **a** $2^{n+1} = \dfrac{1}{2}$
- **b** $2^{n-1} = \dfrac{1}{4}$
- **c** $3^{2n} = \dfrac{1}{27}$
- **d** $4^{2n} = \dfrac{1}{16}$
- **e** $10^{n+1} = \dfrac{1}{1000}$
- **f** $5^{2n+1} = \dfrac{1}{625}$
- **g** $2^{3n+2} = \dfrac{1}{128}$
- **h** $10^{2n} = \dfrac{1}{1\,000\,000}$

5 Find the value of x in these equations.
- **a** $32^x = 2$
- **b** $8^x = 2$
- **c** $9^x = 3$
- **d** $16^x = 4$
- **e** $10000^x = 10$
- **f** $27^x = 3$
- **g** $125^x = 5$
- **h** $169^x = 13$
- **i** $\left(\dfrac{1}{2}\right)^x = 128$
- **j** $\left(\dfrac{1}{3}\right)^x = 27$
- **k** $\left(\dfrac{1}{5}\right)^{2x} = 25$
- **l** $\left(\dfrac{1}{2}\right)^{2x} = 2$

6 Find the value of n in these equations.
- **a** $8^{2n+1} = 2^4$
- **b** $9^{n+1} = 3^6$
- **c** $4^{2n+3} = 2^6$
- **d** $25^{2n-1} = 5^4$
- **e** $3^{n+1} = 9^{n+1}$
- **f** $8^{2n-3} = 2^{n+6}$
- **g** $25^{n+2} = 5^{n+1}$
- **h** $1000^{2n} = 10^{n+10}$

KEY WORDS
fractional indices

Solving quadratic equations by factorisation

THIS SECTION WILL SHOW YOU HOW TO
- Use factorisation to solve quadratic equations

A **quadratic equation** is an equation that can be written in the form $ax^2 + bx + c = 0$ where a, b and c are numbers and $a \neq 0$.

Examples of quadratic equations are: $\quad 3x^2 - 5x + 4 = 0 \quad 2x^2 - 6x = 3 \quad x^2 - 9 = 0$

A very important concept that is used when solving quadratic equations by factorisation is:

If $a \times b = 0$ then $a = 0$ or $b = 0$

EXAMPLE

Solve these equations **a** $x^2 - 3x = 0$ **b** $x^2 = 10x$

a $x^2 - 3x = 0$ factorise
$x(x - 3) = 0$ the product of x and $(x - 3)$ is 0 so one of them must be 0
$x = 0$ or $x - 3 = 0$
$x = 0$ or $x = 3$ ➡ CHECK: $0^2 - 3 \times 0 = 0$ ✓ $3^2 - 3 \times 3 = 0$ ✓

b $x^2 = 10x$ rearrange so that all the terms are on one side
$x^2 - 10x = 0$ factorise
$x(x - 10) = 0$
$x = 0$ or $x - 10 = 0$
$x = 0$ or $x = 10$ ➡ CHECK: $0^2 = 10 \times 0$ ✓ $10^2 = 10 \times 10$ ✓

EXAMPLE

Solve these equations **a** $x^2 = 25$ **b** $x^2 - 5x + 6 = 0$ **c** $2x^2 + x = 21$

a $x^2 = 25$ rearrange so that all the terms are on one side
$x^2 - 25 = 0$ factorise
$(x + 5)(x - 5) = 0$
$x + 5 = 0$ or $x - 5 = 0$
$x = -5$ or $x = 5$ ➡ CHECK: $(-5)^2 = 25$ ✓ $5^2 = 25$ ✓

➡ **NOTE:** part a could also be done by simply saying that if $x^2 = 25$ then $x = \pm 5$.

b $x^2 - 5x + 6 = 0$
$(x - 2)(x - 3) = 0$ factorise
$x - 2 = 0$ or $x - 3 = 0$
$x = 2$ or $x = 3$ ➡ CHECK: $2^2 - (5 \times 2) + 6 = 0$ ✓ $3^2 - (5 \times 3) + 6 = 0$ ✓

c $2x^2 + x = 21$ rearrange so that all the terms are on one side
$2x^2 + x - 21 = 0$ factorise
$(2x + 7)(x - 3) = 0$
$2x + 7 = 0$ or $x - 3 = 0$
$2x = -7$ or $x = 3$
$x = -3\frac{1}{2}$ or $x = 3$ ➡ CHECK: $2 \times \left(-3\frac{1}{2}\right)^2 + \left(-3\frac{1}{2}\right) = 21$ ✓
$\qquad\qquad\qquad\qquad\qquad\qquad\qquad 2 \times 3^2 + 3 = 21$ ✓

EXERCISE 6.13

1. Solve these equations.
   a. $x^2 = 16$
   b. $x^2 = 36$
   c. $x^2 = 100$
   d. $x^2 = \frac{1}{4}$
   e. $2x^2 = 50$
   f. $3x^2 = 27$
   g. $5x^2 = 180$
   h. $3x^2 = 363$
   i. $9x^2 = 4$
   j. $16x^2 = 9$
   k. $25x^2 = 16$
   l. $9x^2 = 1$
   m. $x^2 - 81 = 0$
   n. $y^2 - 1 = 0$
   o. $x^2 - 400 = 0$
   p. $4x^2 - 9 = 0$

2. Is it possible to solve the equation $x^2 + 16 = 0$? Explain your answer.

3. Solve these equations.
   a. $x^2 + 6x = 0$
   b. $x^2 - 8x = 0$
   c. $y^2 - 10y = 0$
   d. $a^2 + 2a = 0$
   e. $x^2 - 5x = 0$
   f. $y^2 - 12y = 0$
   g. $4x^2 + 9x = 0$
   h. $3y^2 - y = 0$
   i. $2x^2 + 13x = 0$
   j. $x^2 - 25x = 0$
   k. $x^2 + 40x = 0$
   l. $x^2 + 3x = 0$
   m. $x^2 = 4x$
   n. $y^2 = 15y$
   o. $a^2 = 11a$
   p. $b^2 = b$

4. Solve these equations.
   a. $x^2 + 5x + 6 = 0$
   b. $x^2 - 5x - 6 = 0$
   c. $x^2 + x - 6 = 0$
   d. $a^2 - 9a + 20 = 0$
   e. $x^2 - 5x - 14 = 0$
   f. $y^2 + 12y + 27 = 0$
   g. $x^2 - x - 20 = 0$
   h. $y^2 - 10y + 16 = 0$
   i. $x^2 + 7x - 30 = 0$
   j. $x^2 - 16x + 64 = 0$
   k. $x^2 - 10x + 25 = 0$
   l. $x^2 + 4x + 3 = 0$
   m. $x^2 + 4x - 21 = 0$
   n. $x^2 - 6x + 9 = 0$
   o. $x^2 - 9x + 8 = 0$
   p. $x^2 + 4x - 45 = 0$

5. Solve these equations.
   a. $2x^2 - 5x - 3 = 0$
   b. $3x^2 - 5x + 2 = 0$
   c. $5y^2 + 2y - 3 = 0$
   d. $2x^2 + 13x - 7 = 0$
   e. $4x^2 - 12x + 5 = 0$
   f. $6x^2 + 13x - 5 = 0$
   g. $10x^2 - 13x + 4 = 0$
   h. $6x^2 + 29x - 5 = 0$
   i. $6x^2 + 11x - 10 = 0$
   j. $4x^2 - 16x + 15 = 0$
   k. $7x^2 + 33x - 10 = 0$
   l. $12x^2 - 7x - 10 = 0$

6. Farah was asked to solve the equation $x^2 - 12x + 32 = 12$.

   $x^2 - 12x + 32 = 12$
   $(x - 8)(x - 4) = 12$
   $x - 8 = 12$ or $x - 4 = 12$
   $x = 20$ or $x = 16$

   She has got it wrong.
   Copy out the question and correct her working.

7. Solve these equations.
   a. $x^2 = 5x - 6$
   b. $x^2 = 8x - 12$
   c. $x^2 = 18 - 3x$
   d. $(x + 3)^2 - 1 = 0$
   e. $8x - x^2 = 16$
   f. $7 + 6x = x^2$
   g. $x^2 - 12x + 37 = 2$
   h. $2x = x^2 - 15$
   i. $x(x + 3) = 18$

   HINT
   rearrange so that the equations are in the form $ax^2 + bx + c = 0$

8. Solve these equations.
   a. $x = \dfrac{6}{x - 5}$
   b. $x = \dfrac{20}{x + 1}$
   c. $\dfrac{x + 2}{8} = \dfrac{10}{x + 4}$
   d. $x - 1 = \dfrac{12}{x}$

Solving quadratic equations by factorisation

Problems involving quadratic equations

> **EXAMPLE**
>
> The sum of the squares of three consecutive integers is 110. Find the three numbers.
>
> Let the numbers be x, $x + 1$ and $x + 2$
> $x^2 + (x + 1)^2 + (x + 2)^2 = 110$ expand: $(x + 1)^2 = x^2 + 2x + 1$ and
> $x^2 + x^2 + 2x + 1 + x^2 + 4x + 4 = 110$ $(x + 2)^2 = x^2 + 4x + 4$
> $3x^2 + 6x + 5 = 110$ rearrange to the form $ax^2 + bx + c = 0$
> $3x^2 + 6x - 105 = 0$ take out a common factor of 3
> $3(x^2 + 2x - 35) = 0$ factorise
> $3(x - 5)(x + 7) = 0$
> $x - 5 = 0$ or $x + 7 = 0$
> $x = 5$ or $x = -7$
> The three numbers are 5, 6 and 7 or −7, −6 and −5.

> **EXAMPLE**
>
> Find the points of intersection of the graphs of $y = x^2 + 4x - 3$ and $y = x + 1$.
>
> You need to find the points where the curve and the line meet.
> This will happen when the y coordinates are equal.
> $x^2 + 4x - 3 = x + 1$ rearrange to the form $ax^2 + bx + c = 0$
> $x^2 + 3x - 4 = 0$ factorise
> $(x + 4)(x - 1) = 0$
> $x + 4 = 0$ or $x - 1 = 0$
> $x = -4$ or $x = 1$ substitute to find y
> when $x = -4$, $y = -4 + 1 = -3$
> when $x = 1$, $y = 1 + 1 = 2$ write the two coordinates
> The points of intersection are (−4, −3) and (1, 2).

> **EXAMPLE**
>
> The sides of the right-angled triangle shown are in cm.
> Find   **a**   the value of x
>   **b**   the lengths of the sides of the triangle.
>
> **a**   Using Pythagoras' theorem:
> $(3x + 8)^2 + (2x + 1)^2 = (4x + 1)^2$ expand the brackets
> $9x^2 + 48x + 64 + 4x^2 + 4x + 1 = 16x^2 + 8x + 1$ collect like terms
> $13x^2 + 52x + 65 = 16x^2 + 8x + 1$ rearrange to the form $ax^2 + bx + c = 0$
> $3x^2 - 44x - 64 = 0$ factorise
> $(3x + 4)(x - 16) = 0$
> $3x + 4 = 0$ or $x - 16 = 0$
> $3x = -4$ or $x = 16$ x cannot be negative as this would give
> $x = -1\frac{1}{3}$ or $x = 16$ negative lengths
> So $x = 16$.
>
> **b**   The lengths of the sides of the triangle are 33 cm, 56 cm and 65 cm.

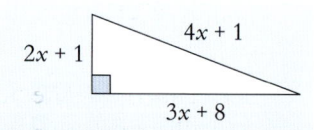

EXERCISE 6.14

1. **a** If x is subtracted from x^2 the answer is 42. Find the two possible values for x.
   b If x is added to x^2 the answer is 306. Find the two possible values for x.

2. **a** I think of a number. I then square the number and add twice the original number. My answer is 99. Find the two possible values for the original number.
   b I think of a number. I then square the number and subtract twice the original number. My answer is 120. Find the two possible values for the original number.

3. The product of two numbers is 270. The two numbers are x and $x + 3$. Find the two possible values for x.

4. For this right-angled triangle
   a Show that $2x^2 + x^2 - 210 = 0$
   b Hence find the sides of the triangle.

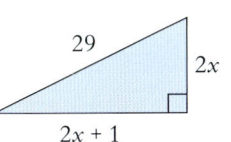

   (Right-angled triangle with hypotenuse 29, vertical side $2x$, horizontal side $2x + 1$.)

5. The two rectangles have the same area. Find the value of x.

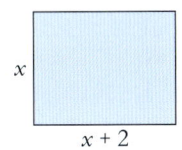

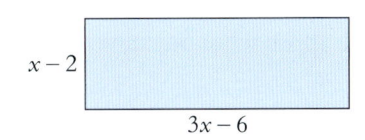

   (Rectangle 1: sides x and $x + 2$. Rectangle 2: sides $x - 2$ and $3x - 6$.)

6. Rectangle A is 12 cm² bigger than rectangle B. Find the value of x.

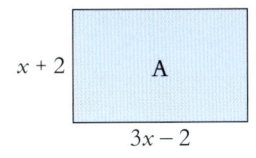

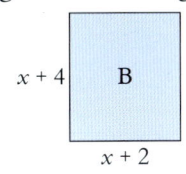

   (Rectangle A: sides $x + 2$ and $3x - 2$. Rectangle B: sides $x + 4$ and $x + 2$.)

7. The lengths of the sides of two squares differ by 3 cm and the sum of their areas is 317 cm². Find the length of the sides of the two squares.

8. The area of the trapezium is 60 cm². Find the value of x.

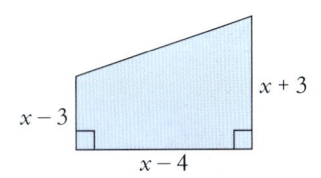

   (Trapezium with parallel sides $x - 3$ and $x + 3$, and perpendicular distance $x - 4$.)

9. A stone is catapulted upwards.
   The height, h metres, of the stone above the ground after t seconds is given by the formula $h = 10t - 2t^2$.
   Find the values of t when the stone is 12 meters above the ground.

10. Find the points of intersection of the graphs of $y = x^2 - x - 12$ and $y = x + 3$.

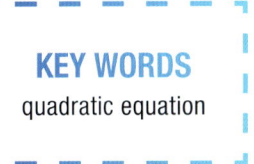

KEY WORDS
quadratic equation

Solving quadratic equations by factorisation

Reciprocal graphs

THIS SECTION WILL SHOW YOU HOW TO
- Draw reciprocal graphs

The **reciprocal** of x is $\frac{1}{x}$. The graph of $y = \frac{1}{x}$ is called a reciprocal graph.

Other examples of reciprocal graphs are: $y = \frac{2}{x} + 1$, $y = 2x - \frac{5}{x}$ and $y = \frac{6}{x+2}$

To draw the graph of $y = \frac{6}{x}$ for $-6 \leq x \leq 6$ you must first make a table of values.

x	-6	-5	-4	-3	-2	-1	0	1	2	3	4	5	6
y	-1	-1.2	-1.5	-2	-3	-6		6	3	2	1.5	1.2	1

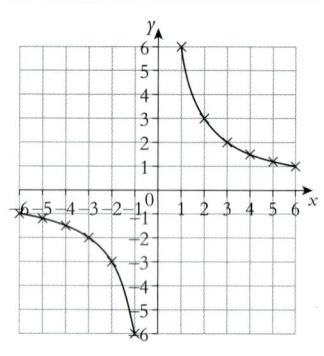

There is no value for y when $x = 0$ because $\frac{6}{0}$ is undefined.

The graph consists of two separate curves, you must not join them together.
The curve approaches but never touches the axes.
The lines that the curve approaches are called **asymptotes**.

➡ **NOTE:** you do not need to know the word asymptote in the examination.

EXAMPLE

Complete the table and draw the graph of $y = \frac{4}{x-2} + 1$

x	-6	-4	-2	0	1	1.5	2	2.5	3	4	6	8	10
y		0.33			-3			9				1.67	

Write down the equations of the two asymptotes.

x	-6	-4	-2	0	1	1.5	2	2.5	3	4	6	8	10
y	0.5	0.33	0	-1	-3	-7		9	5	3	2	1.67	1.5

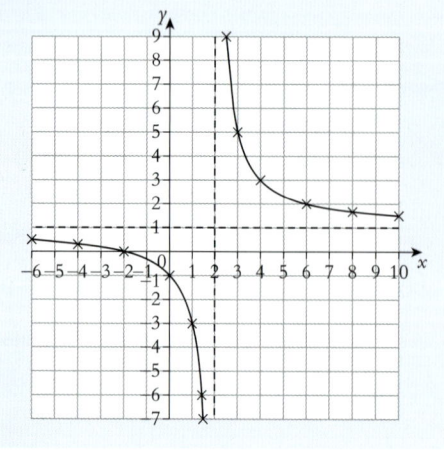

Note there is no y value when $x = 2$
The asymptotes are represented by dashed lines.
The equations of the asymptotes are $x = 2$ and $y = 1$.

EXERCISE 6.15

For each of these functions make a table of values and draw the graph of the function between the stated x values.

1. $y = \dfrac{5}{x}$ $-5 \leq x \leq 5$
2. $y = -\dfrac{5}{x}$ $-5 \leq x \leq 5$
3. $y = \dfrac{4}{x} + 2$ $-5 \leq x \leq 5$
4. $y = \dfrac{5}{x-3}$ $-2 \leq x \leq 8$
5. $y = \dfrac{6}{x} + x + 1$ $-6 \leq x \leq 6$
6. $y = x^2 + \dfrac{6}{x}$ $-4 \leq x \leq 4$
7. $y = \dfrac{12}{x^2}$ $-6 \leq x \leq 6$
8. $y = \dfrac{4}{x^2} + x - 2$ $-6 \leq x \leq 6$

9. Amir wants to make a rectangular enclosure next to a farm building.
   He wants the enclosure to have an area of 500 m².
   He will need fencing for three of the sides of the enclosure.

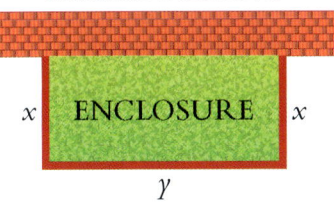

   a. By considering the area of the enclosure explain why $y = \dfrac{500}{x}$
   b. Write down an expression in terms of x and y for the total length, T, of fencing that is needed.
   c. Show that $T = 2x + \dfrac{500}{x}$
   d. Copy and complete this table of values for x and T.

x	5	10	15	20	30	40	50
T	110						110

   e. Draw the graph of $T = 2x + \dfrac{500}{x}$
   f. Use your graph to estimate the minimum length of fencing that is needed and the value of x that gives the minimum length of fencing.

KEY WORDS

reciprocal
asymptote

Circle theorems

THIS SECTION WILL SHOW YOU HOW TO
- Use the properties of tangents and chords to solve problems
- Use the circle theorems to solve problems

Tangents

A straight line can intersect a circle in three possible ways. It can be:

A DIAMETER   A CHORD   A TANGENT

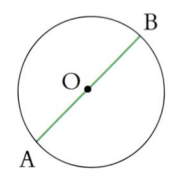

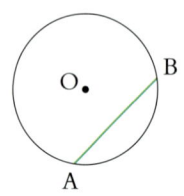

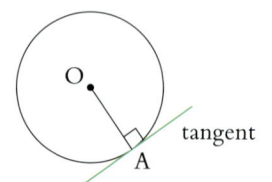

2 points of intersection   2 point of intersection   1 point of intersection

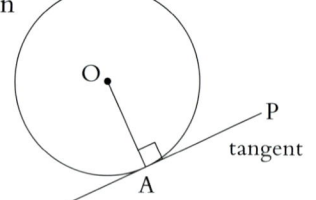

PROPERTY 1
A **tangent** to a circle 'touches' the circle at one point.
The angle between a tangent and a radius is a right angle.
$O\hat{A}P = 90°$.

PROPERTY 2
The two tangents drawn from a point P outside a circle are equal in length.
AP = BP.

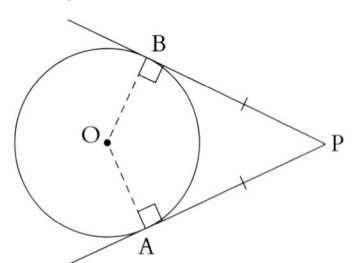

EXAMPLE

In the diagram, O is the centre of the circle radius 5 cm. AP is a tangent to the circle at A and AP = 15 cm.
a Calculate the length of OP and the size of angle AOP.
b Calculate the shaded area.

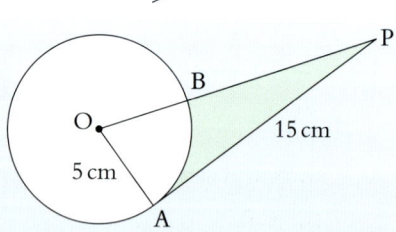

---

a $O\hat{A}P = 90°$ tangent and radius at right angles

$15^2 + 5^2 = OP^2$ $\tan A\hat{O}P = \dfrac{15}{5}$

$\quad OP^2 = 250$ $A\hat{O}P = \tan^{-1}\left(\dfrac{15}{5}\right)$

$\quad OP = 15.8$ cm $A\hat{O}P = 71.6°$

b Area of triangle AOP = $\dfrac{1}{2} \times$ base $\times$ height = $\dfrac{1}{2} \times 15 \times 5 = 37.5$ cm^2

Area of sector OAB = $\dfrac{71.6}{360} \times \pi \times 5^2 = 15.6$ cm^2

Shaded area = area of triangle AOP − area of sector OAB = 37.5 − 15.6 = 21.9 cm^2

EXERCISE 6.16

1. O is the centre of the circle and AP is a tangent to the circle.
   Calculate the value of x for each diagram.

   a) [circle with radius OA = 4 cm, OP = 10 cm, AP = x]

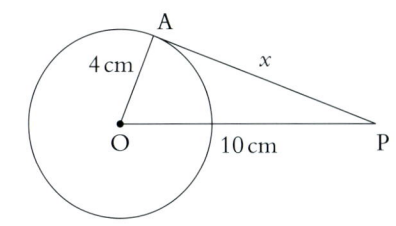

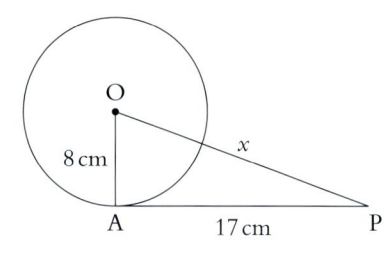

   b) [circle with OA = 8 cm, AP = 17 cm, OP = x]

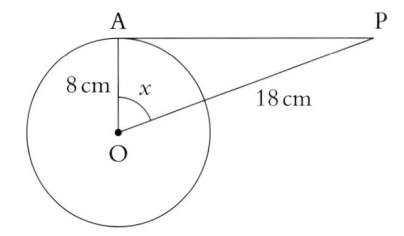

   c) [circle with OA = 8 cm, AP = 18 cm, angle AOP = x]

   d) [circle with OA = 3 cm, AP = 9 cm, angle AOP = x]

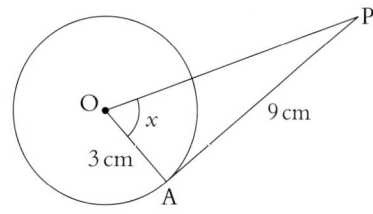

2. O is the centre of the circle.
   AP is a tangent to the circle at A.
   OP = 12 cm and AP = 10 cm.
   Calculate:
   a. the radius of the circle,
   b. angle AOP,
   c. the area of triangle OAP,
   d. the shaded area.

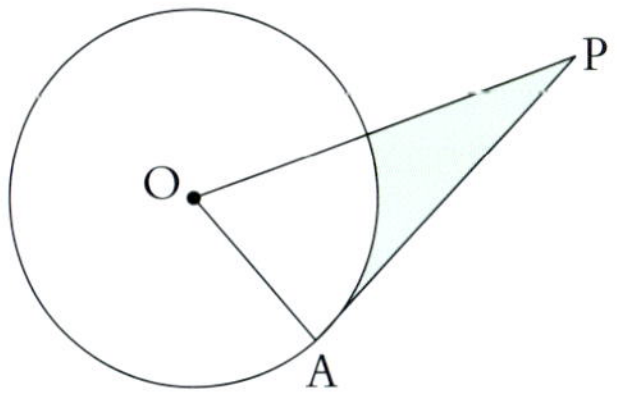

3. O is the centre of the circle, radius 6 cm.
   AP and BP are tangents to the circle.
   AP = 8 cm.
   Calculate:
   a. angle AOP,
   b. angle AOB,
   c. the area of quadrilateral OAPB,
   d. the area of sector OAB,
   e. the shaded area.

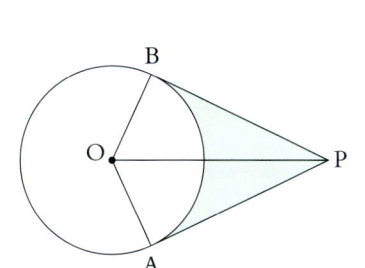

4. ABC is an equilateral triangle.
   AB, BC and AC are tangents to the circle at the points P, Q and R.
   O is the centre of the circle, radius 8 cm.
   Calculate:
   a. the length PO,
   b. the length PC,
   c. the length PA,
   d. the area of triangle ABC,
   e. the shaded area.

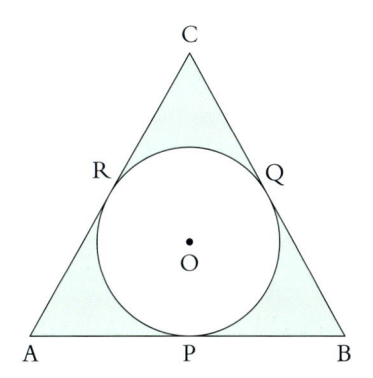

Circle theorems 295

Chords and segments

A straight line joining two points on the circumference of a circle is called a **chord**.

A chord divides a circle into two **segments**.

Symmetry properties of chords

PROPERTY 1

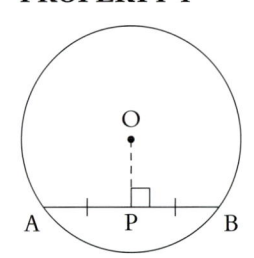

The perpendicular line from the centre of a circle to a **chord** bisects the chord.

AP = PB.

➡ **NOTE:** triangle OAB is isosceles.

PROPERTY 2

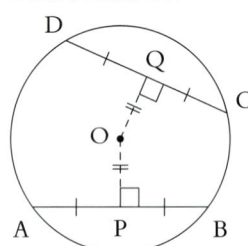

If two chords AB and CD are the same length then they will be the same perpendicular distance from the centre of the circle.

If AB = CD then OP = OQ.

EXAMPLE

A chord AB lies on a circle of radius 12 cm. The chord is 18 cm long. M is the midpoint of AB.
a   Calculate the length of OM and angle AOB.
b   Calculate the shaded area.

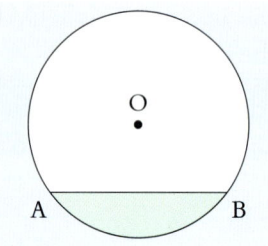

---

a

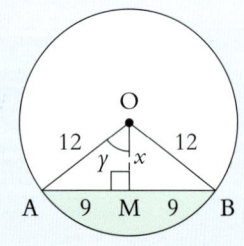

Using Pythagoras

$x^2 + 9^2 = 12^2$
$x^2 + 81 = 144$
$x^2 = 63$
$x = 7.937...$

OM is 7.94 cm (to 3 s.f.)

Let $A\hat{O}M = y$

$\sin y = \dfrac{9}{12}$

$y = \sin^{-1}\left(\dfrac{9}{12}\right)$

$y = 48.59...$

$A\hat{O}B = 2 \times 48.59...$

$A\hat{O}B = 97.2°$ (to 1 d.p.)

b   Area of triangle AOB = $\dfrac{1}{2}$ × base × height = $\dfrac{1}{2}$ × 18 × 7.937 = 71.44 cm²

Area of sector AOB = $\dfrac{97.2}{360}$ × π × 12² = 122.12 cm²

Area of shaded segment = area of sector − area of triangle
Area of shaded segment = 122.12 − 71.44 = 50.7 cm²

EXERCISE 6.17

1. A chord AB lies on a circle of radius 5 cm.
   The chord is 2 cm from the centre of the circle.
   Calculate the length of the chord.

2. A chord AB lies on a circle of radius 8 cm.
   The chord is 12 cm long.
   Calculate the distance of the chord from the centre of the circle.

3. A chord AB is 24 cm long.
   The distance of the chord from the centre of the circle is 5 cm.
   Calculate the radius of the circle.

4. O is the centre of the circle, radius 15 cm.
   The chord AB is 24 cm long.
   a Calculate the distance of the chord from O.
   b Calculate the area of triangle OAB.

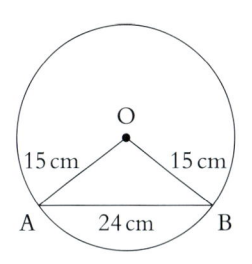

5. O is the centre of the circle, radius 5 cm.
   $A\hat{O}B$ = 140°. M is the midpoint of AB.
   Calculate:
   a the length of OM,
   b the length of AM,
   c the length of the chord AB,
   d the area of triangle OAB,
   e the area of sector OAB,
   f the area of the shaded segment.

6. O is the centre of the circle, radius 12 cm.
   The chords AB and CD are parallel.
   AB = 20 cm and CD = 15 cm.
   Calculate the distance between the two chords.

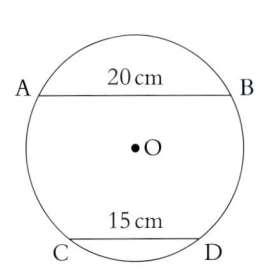

7. O is the centre of the circle, radius 10 cm.
   AB = 16 cm and AC = BC.
   M is the midpoint of AB.
   Calculate:
   a the length of OM,
   b the length of CM,
   c the area of triangle ABC.

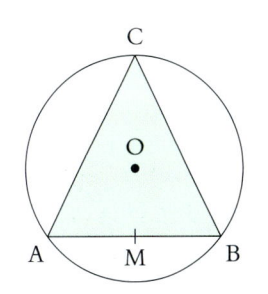

Circle theorems

EXAMPLE

Find the values of x and y.
Give reasons for your answers.

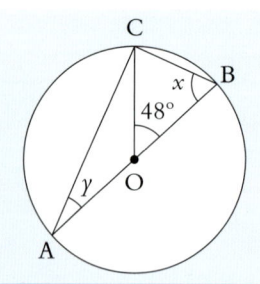

$x = \dfrac{180 - 48}{2} = 66°$ △BOC is isosceles

$A\hat{O}C = 180 - 48 = 132°$ angles on a straight line

$y = \dfrac{180 - 132}{2} = 24°$ △AOC is isosceles

EXERCISE 6.18

Find the size of each lettered angle.

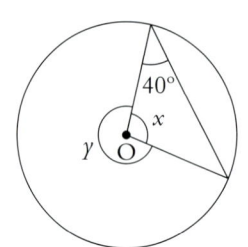

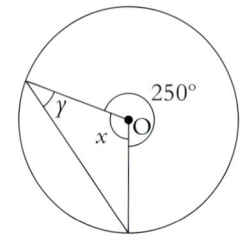

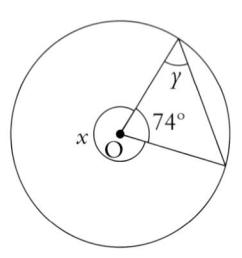

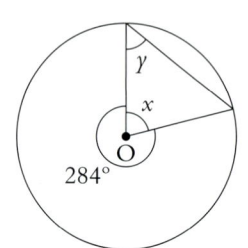

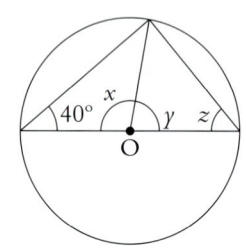

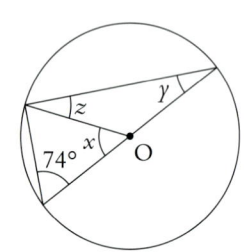

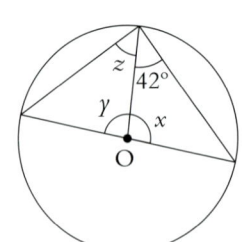

1. x, y, 120°

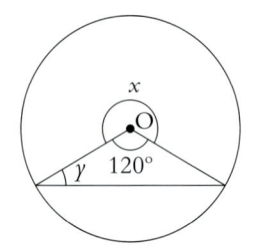

2. 40°, x, y

3. 250°, y, x

4. y, 74°, x

5. y, x, 284°

6. 40°, x, y, z

7. z, y, 74°, x

8. z, 42°, y, x

9. 47°, y, x

298   UNIT 6

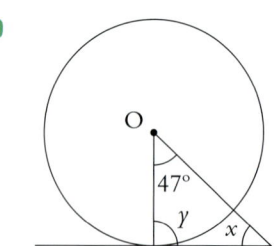

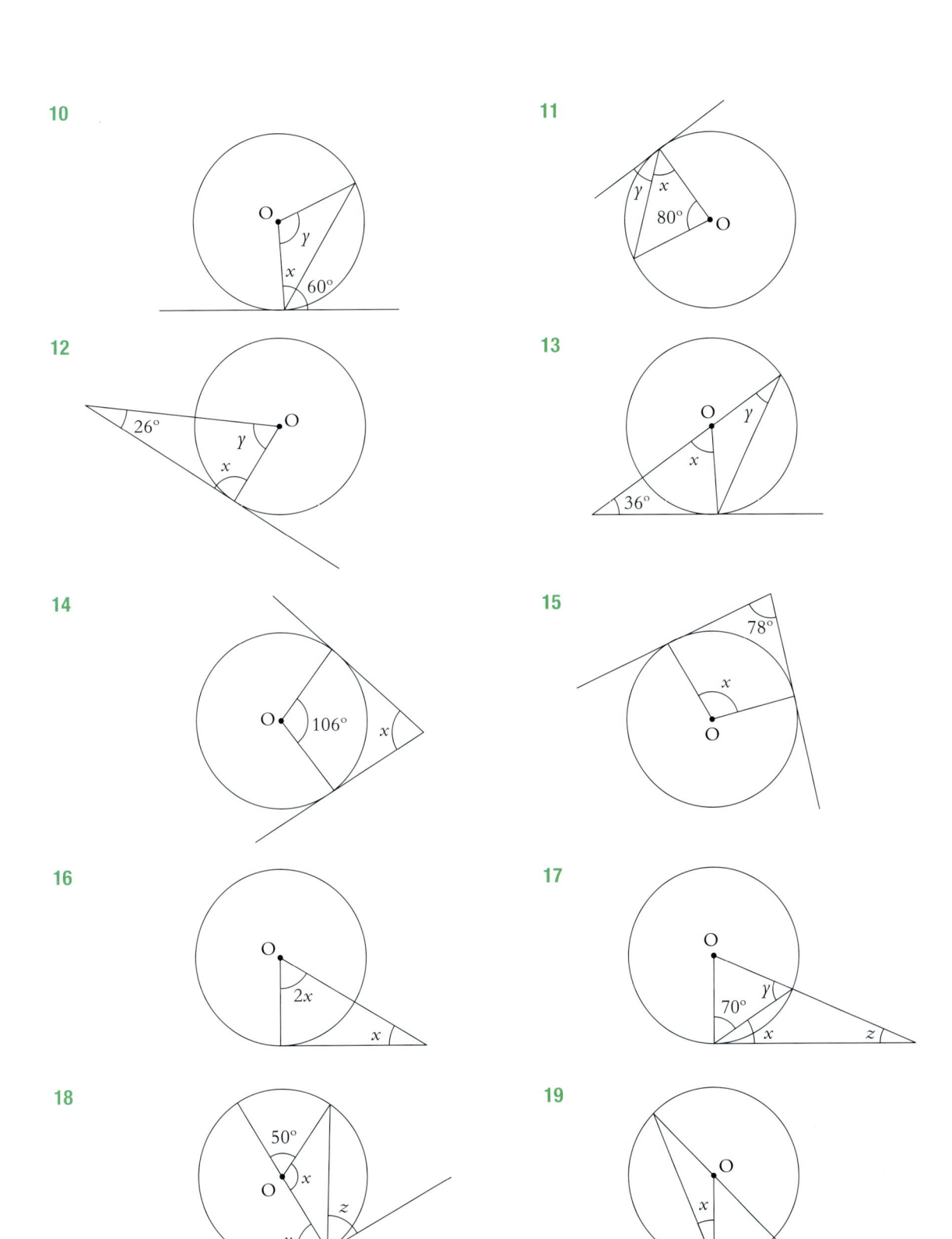

10

11

12

13

14

15

16

17

18

19

THEOREM 1

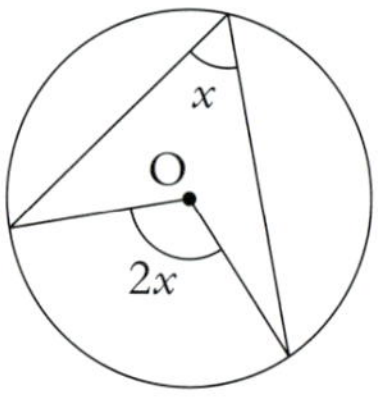

The angle at the centre is twice the angle at the circumference.

THEOREM 2

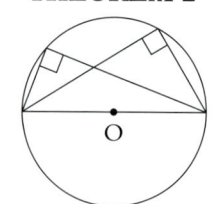

An angle in a semi-circle is always a right angle.

Proof of theorem 1

Let $O\hat{C}A = a$ and $O\hat{C}B = b$
$C\hat{A}O = a$ and $C\hat{B}O = b$ *isosceles triangles*
$A\hat{O}D = 2a$ and $B\hat{O}D = 2b$ *exterior angles of triangles*
$A\hat{C}B = a + b$ and $A\hat{O}B = 2a + 2b$
So $A\hat{O}B = 2 \times A\hat{C}B$

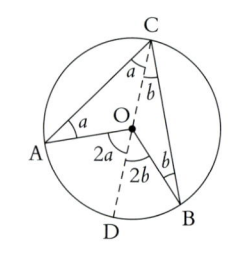

Proof of theorem 2

Angle at centre = 180°
Angle at centre = 2 × angle at circumference
$180° = 2 \times a$
$a = 90°$

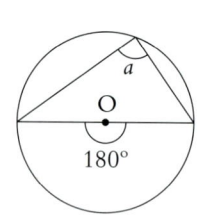

EXAMPLE

Find the value of x for each of these diagrams.

a   **b** 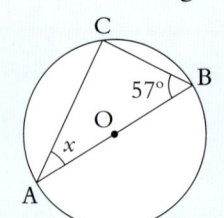   **c** 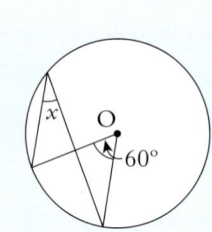   **d**

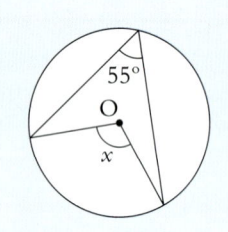

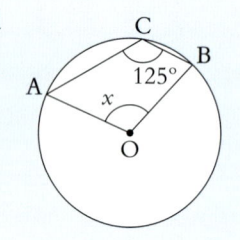

a $x = 2 \times 55°$
     $x = 110°$

b $x = 180° - (90° + 57°)$
     $x = 33°$

c $60° = 2 \times x$
     $x = 30°$

d   (reflex) $A\hat{O}B = 2 \times 125 = 250°$
     $x = 360 - 250 = 110°$

EXAMPLE

Find the values of x and y.
Give reasons for your answers.

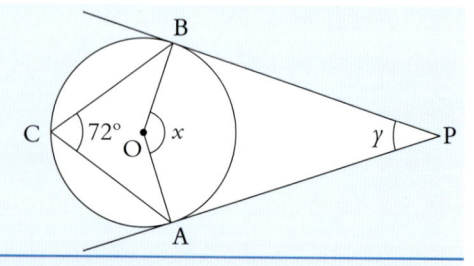

$x = 2 \times 72 = 144°$ angle at centre = 2 × angle at circumference
$O\hat{B}P = O\hat{A}P = 90°$ tangent and radius at right angles
$y = 360 - (90 + 90 + 144) = 36°$ angles in a quadrilateral add up to 360°

EXERCISE 6.19

Find the value of the letters in each of the following diagrams.

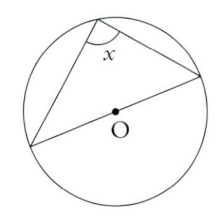

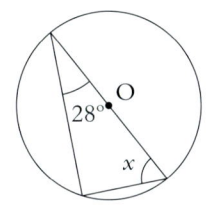

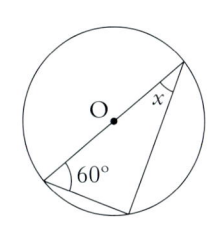

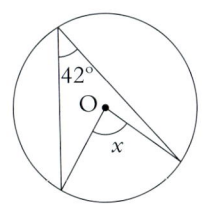

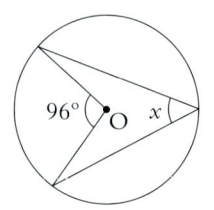

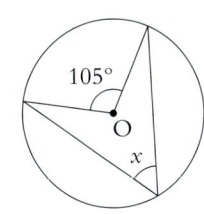

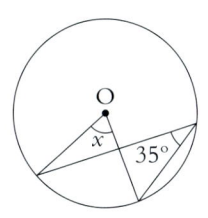

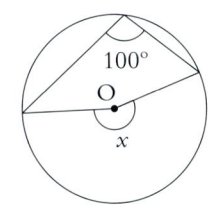

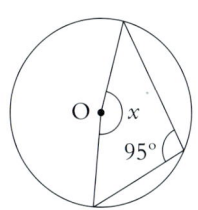

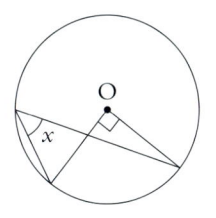

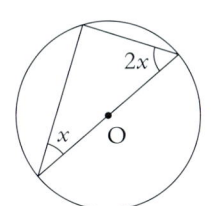

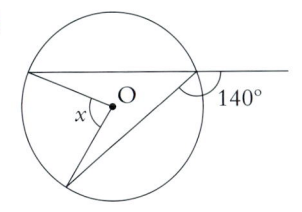

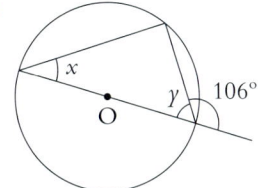

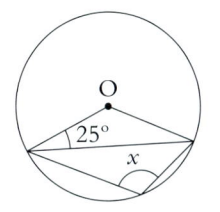

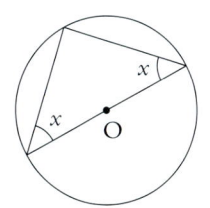

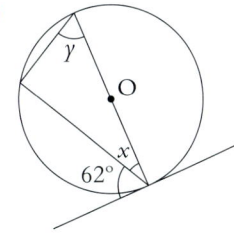

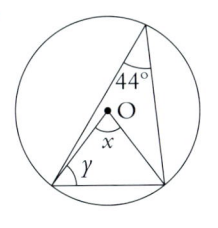

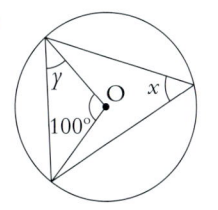

1.
2.
3.
4.
5.
6.
7.
8.
9.
10.
11.
12.
13.
14.
15.
16.
17.
18.

Circle theorems

THEOREM 3

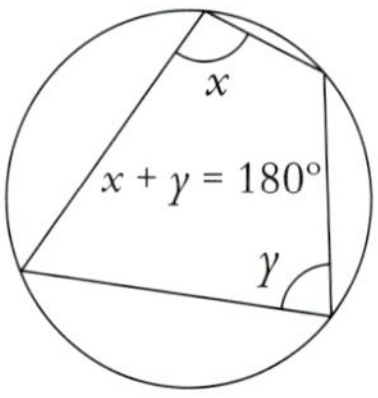

Opposite angles of a **cyclic quadrilateral** add up to 180°.

THEOREM 4

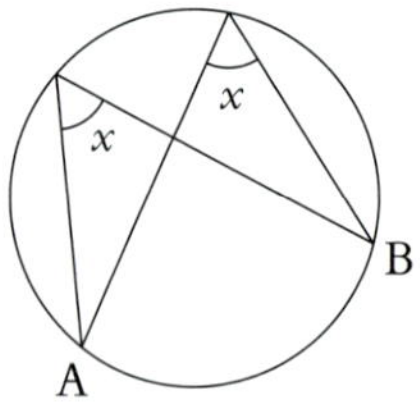

Angles from the same arc in the same segment are equal.

➡ **NOTE:** all four vertices of a cyclic quadrilateral lie on the circumference of a circle.

Proof of theorem 3

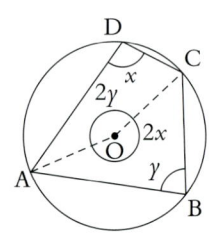

Let $A\hat{D}C = x$ and $A\hat{B}C = y$
(reflex) $A\hat{O}C = 2x$ angle at centre = 2 × angle at circumference
(obtuse) $A\hat{O}C = 2y$ angle at centre = 2 × angle at circumference
$2x + 2y = 360°$ angles at a point
So $x + y = 180°$

Proof of theorem 4

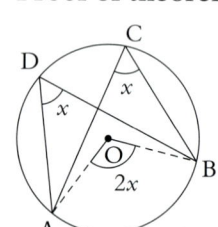

Let $A\hat{D}B = x$
Then $A\hat{O}B = 2x$ angle at centre = 2 × angle at circumference
So $A\hat{C}B = x$ angle at centre = 2 × angle at circumference

EXAMPLE

Find the value of x and y for each of these diagrams.

a   **b**

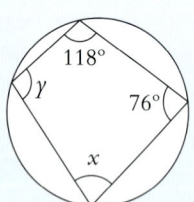

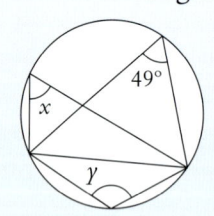

---

a $x + 118° = 180°$ and $y + 76° = 180°$ opposite angles in a cyclic quadrilateral
      $x = 62°$ $y = 104°$

b $x = 49°$ angles from the same arc in the same segment
     $y + 49° = 180°$ opposite angles in a cyclic quadrilateral
          $y = 131°$

EXERCISE 6.20

Find the value of the letters in each of the following diagrams.

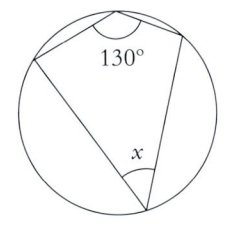

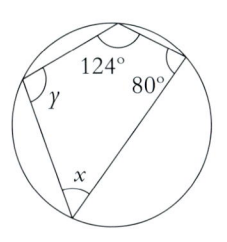

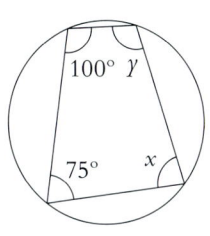

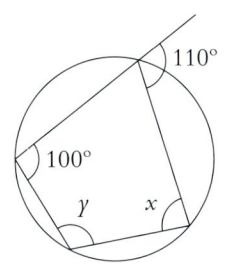

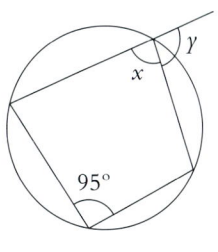

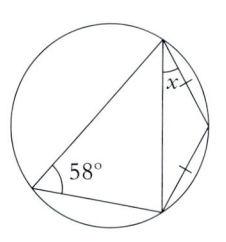

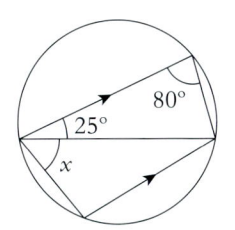

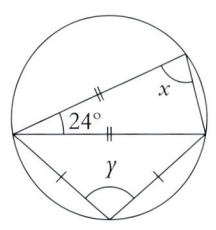

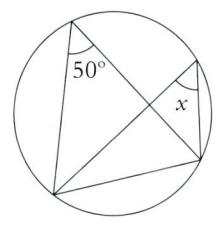

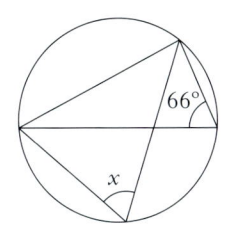

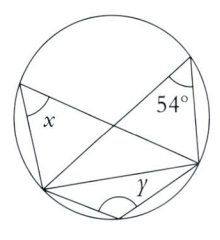

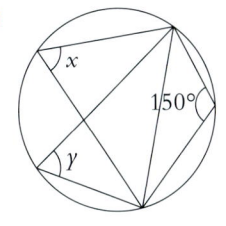

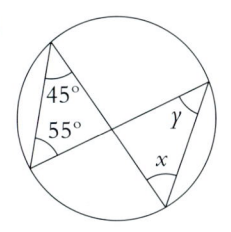

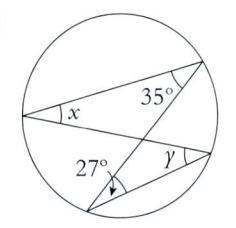

Summary of circle theorems:

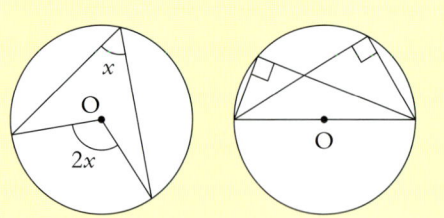

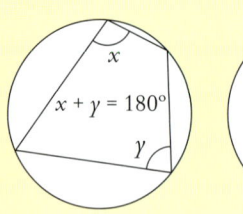

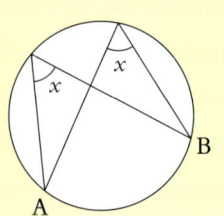

You need to learn all these rules together with the properties of tangents and chords.

EXAMPLE

Find the values of x and y.
Give reasons for your answers.

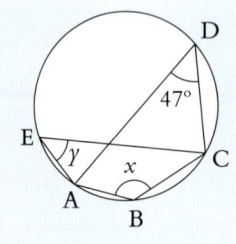

$x = 133°$ opposite angles of a cyclic quadrilateral
$y = 47°$ angles from the same arc in the same segment

EXAMPLE

Find the values of x, y and z.
Give reasons for your answers.

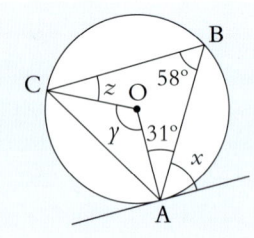

$x = 59°$ tangent and radius at right angles
$y = 116°$ angle at centre = 2 × angle at circumference

To find z you need to use quadrilateral OABC.
First find the angle at O.

reflex $\widehat{AOC} = 244°$ angles at a point add up to 360°
$z = 360 - (58 + 31 + 244)$ angles in a quadrilateral add up to 360°
$z = 27°$

304   UNIT 6

EXERCISE 6.21

Find the value of the letters in each of the following diagrams.

1

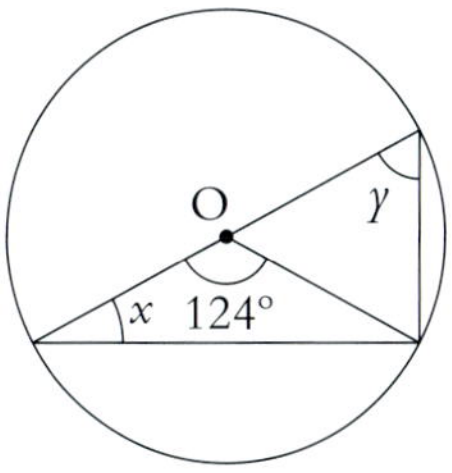

2

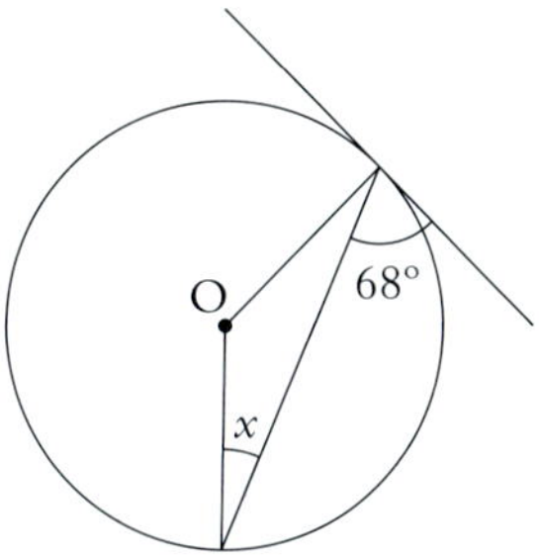

3

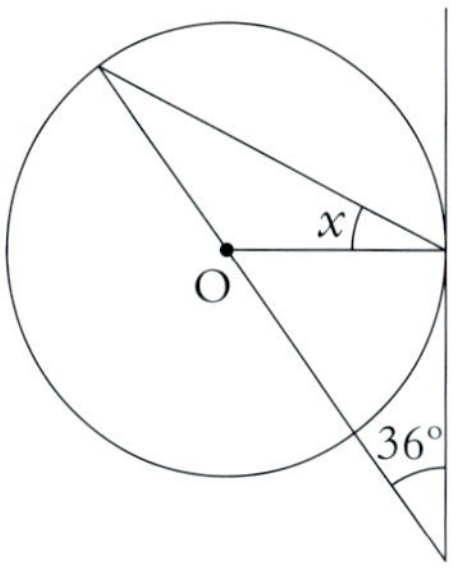

4

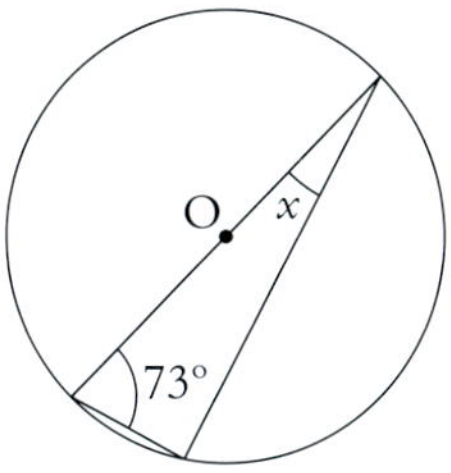

5

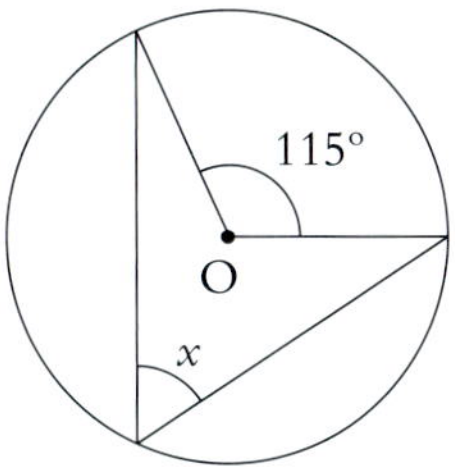

6

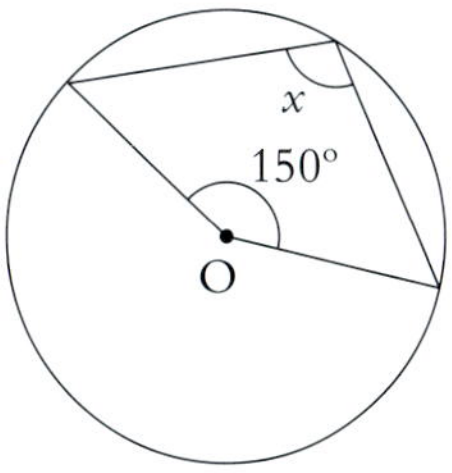

7

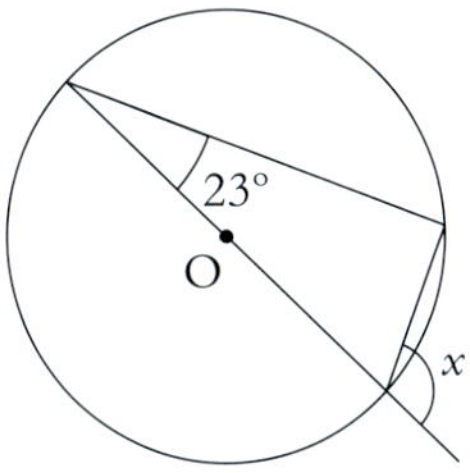

8

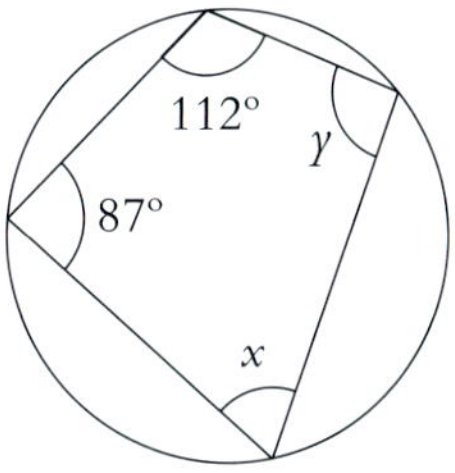

9

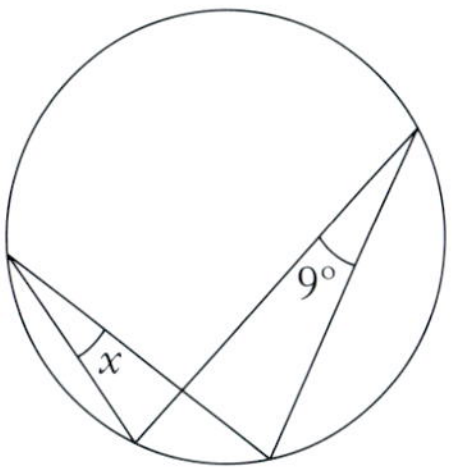

10

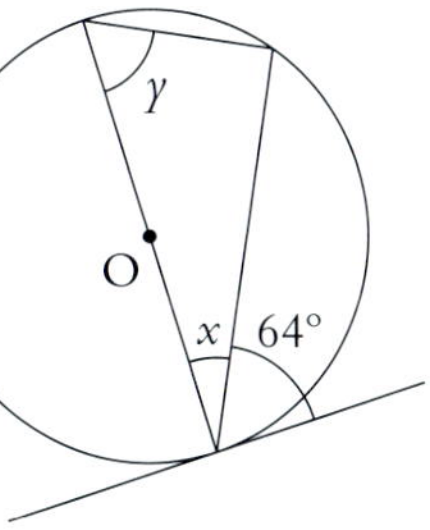

11

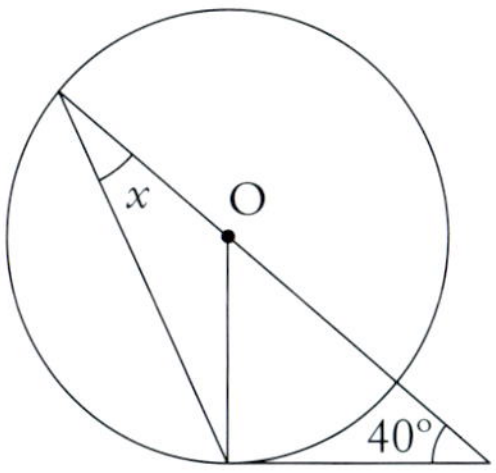

12

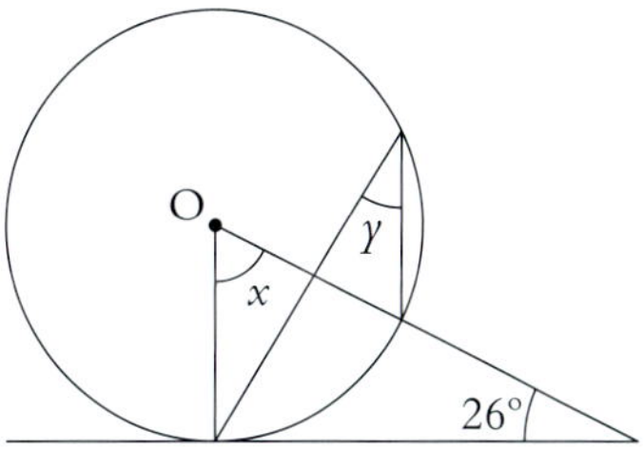

13

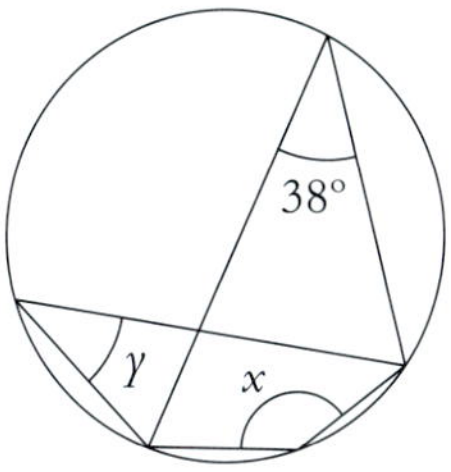

14

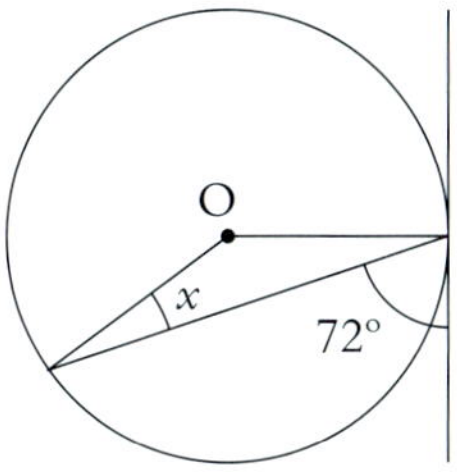

15

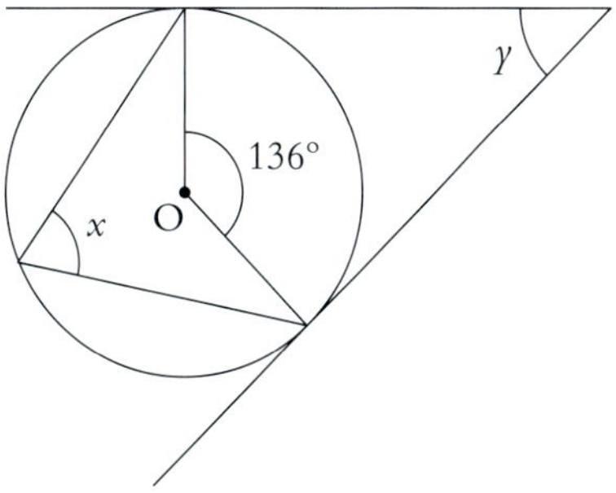

UNIT 6

16.
17.
18.
19.
20.
21.
22.
23.
24.
25.
26.
27.
28.
29.
30.

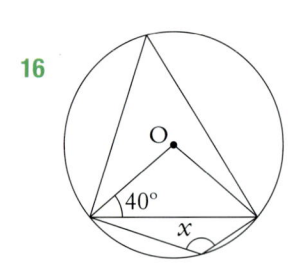

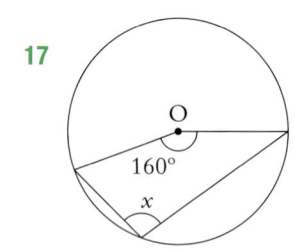

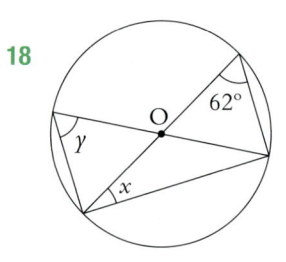

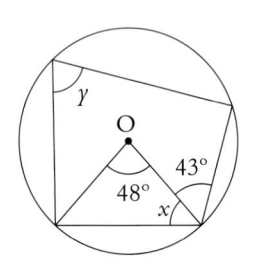

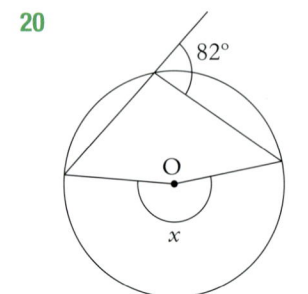

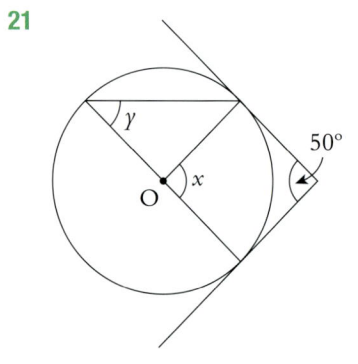

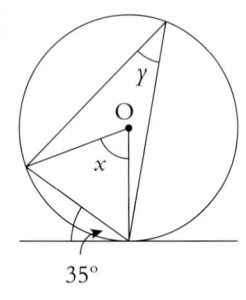

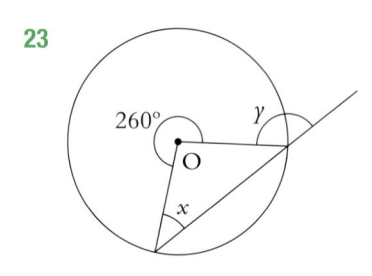

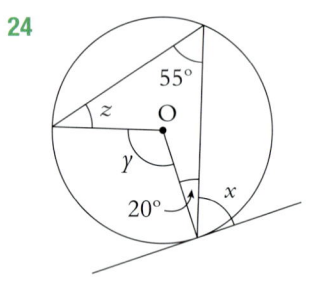

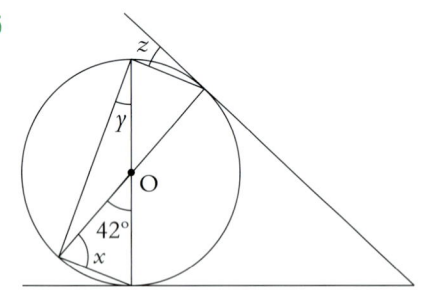

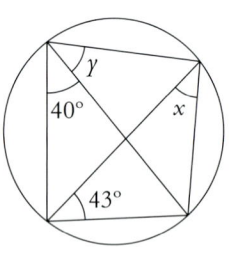

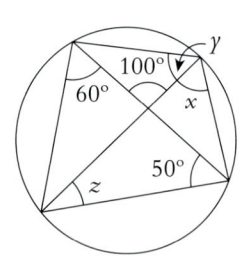

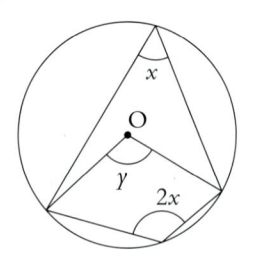

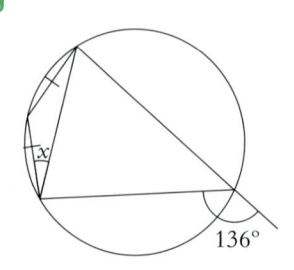

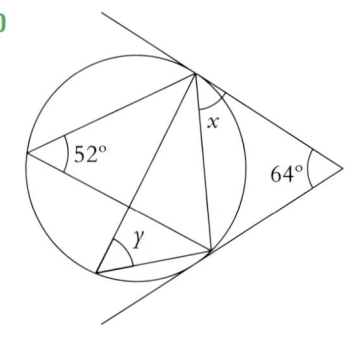

For questions **31–36** explain how you calculate each variable using the various circle theorems etc.

31

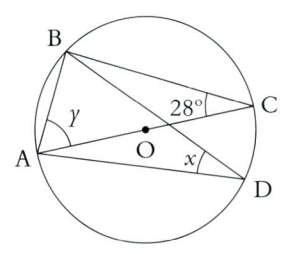

32

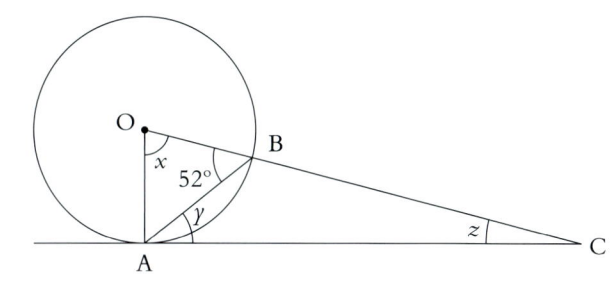

33

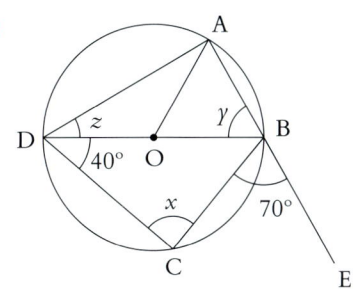

34

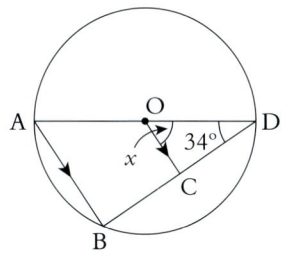

35

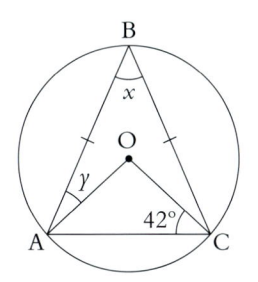

36

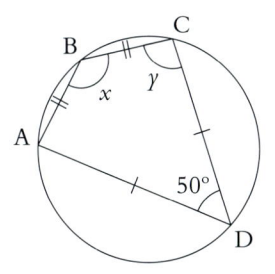

37 Use your answers to question **36** to make a general statement about cyclic kites.

CHALLENGE 1
Prove that $\widehat{BCD} = x$.

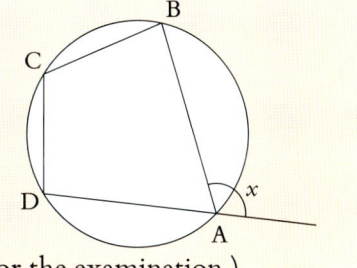

This means that the exterior angle of a cyclic quadrilateral is equal to the opposite interior angle.
(You do not have to learn this rule for the examination.)

CHALLENGE 2
Prove that $\widehat{ACB} = x$.

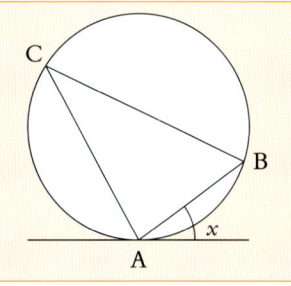

This rule is called The Alternate Segment Theorem.
(You do not have to learn this rule for the examination.)

KEY WORDS
tangent
chord
segment
cyclic quadrilateral

Circle theorems 307

Stretches and shears

> **THIS SECTION WILL SHOW YOU HOW TO**
> - Recognise and use stretches and shears.

Stretches

In a **stretch**, the points on an object move at right angles to a fixed line.

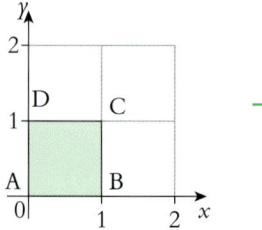

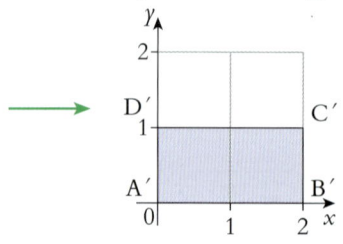

In this example ABCD has been stretched to give A′B′C′D′.
The points on the y-axis have not moved, so the y-axis (or $x = 0$) is called the **invariant line**.

The perpendicular distance of each point from the invariant line has doubled, so the **stretch factor** is 2.

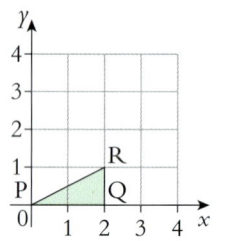

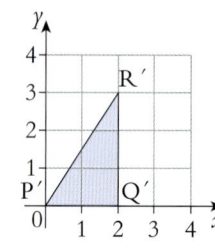

Here PQR has been transformed onto P′Q′R′ by:
- a stretch
- stretch factor 3
- with the x-axis as the invariant line

The following diagram shows a stretch where the invariant line is not the x- or y-axis.

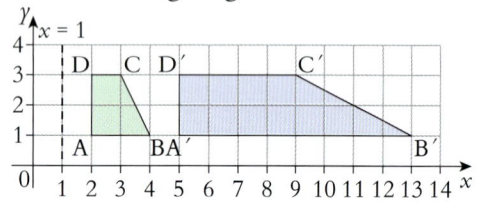

➡ **NOTE:** if the stretch factor is 4 then area of image = 4 × area of object

$A'B' = 4 \times AB$ and $D'C' = 4 \times DC$. So the stretch factor is 4.
The perpendicular distance of each point from the line $x = 1$ has quadrupled.
So the invariant line is $x = 1$.

EXAMPLE

A(1, 1), B(4, 1) and C(3, 3).
Draw the image of triangle ABC after a stretch, stretch factor −2 and with the y-axis invariant.

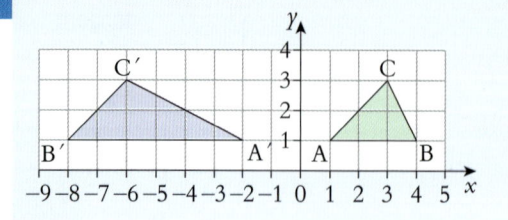

➡ **NOTE:** if the scale factor is negative the stretch is on the opposite side of the invariant line.

EXERCISE 6.22

1. Copy the diagram.
   Transform shape P by a stretch, stretch factor 3, with the y-axis invariant.

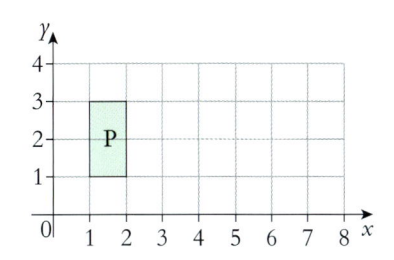

2. Copy the diagram.
   Transform shape P by a stretch, stretch factor 2, with the x-axis invariant.

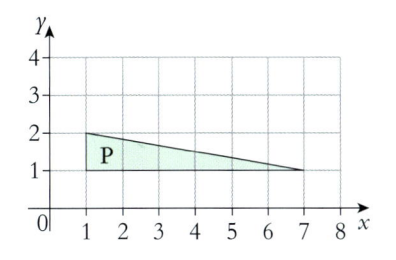

3. Describe fully the stretch that takes shape A onto shape B for each of these.

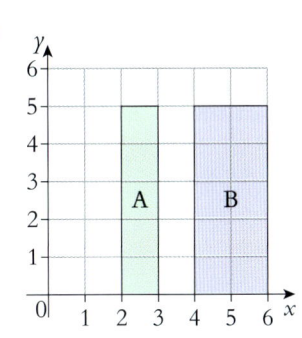

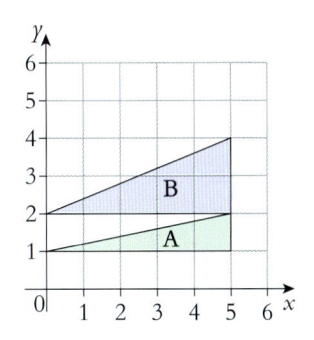

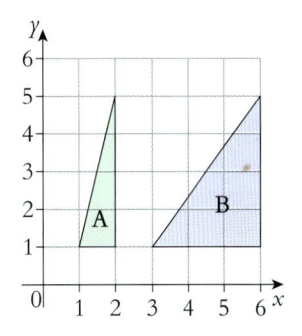

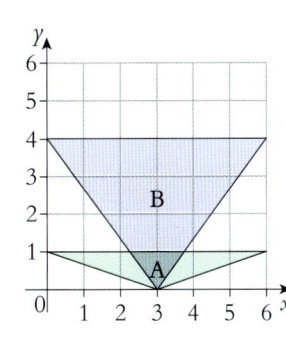

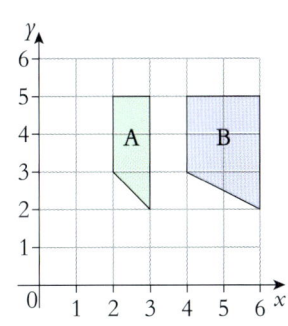

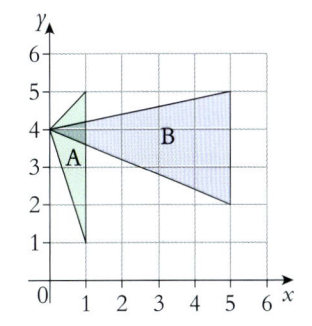

   a
   b
   c
   d
   e
   f

4. Copy the diagram.
   Transform shape P by a stretch, stretch factor $\frac{1}{2}$, with the x-axis invariant.

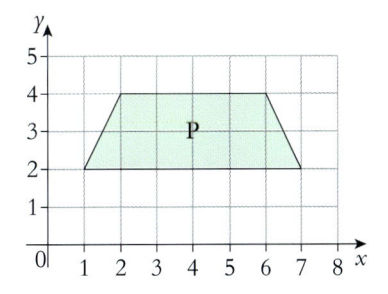

5 Describe fully the stretch that takes shape A onto shape B for each of these:

a   b 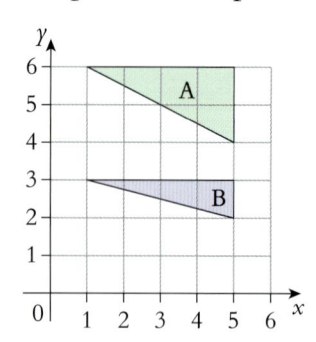   c

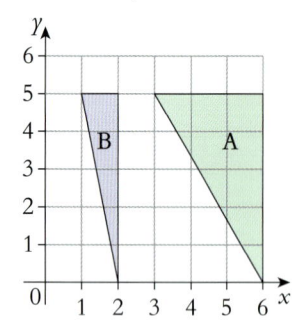

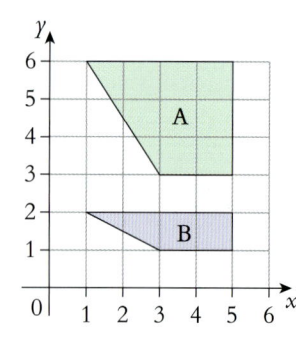

6 Copy the diagram.
Transform shape P by a stretch, stretch factor −3, with the y-axis invariant.

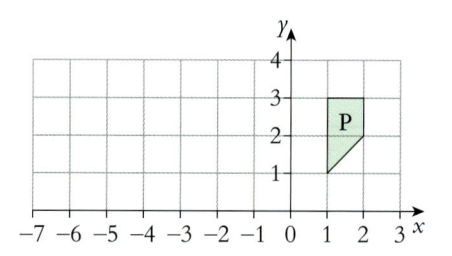

7 Describe fully the stretch that takes shape A onto shape B for each of these:

a   b 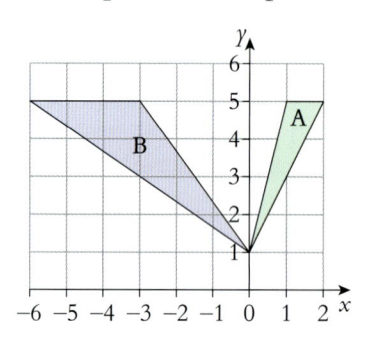   c

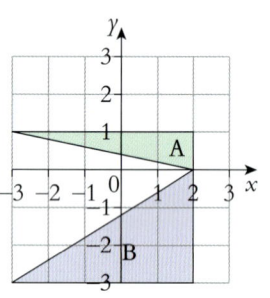

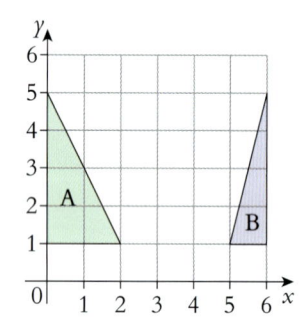

8 Copy the diagram.
Transform shape P by a stretch, stretch factor 2, with $x = 2$ invariant.

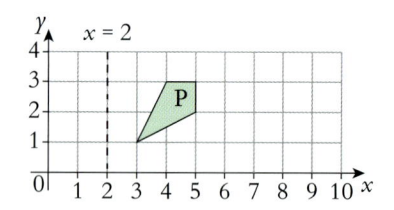

9 Describe fully the stretch that takes shape A onto shape B for each of these.

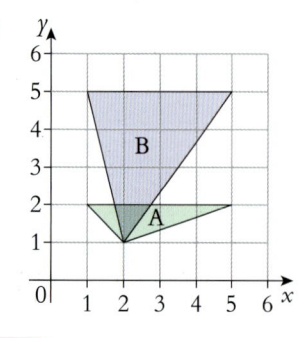

a   b 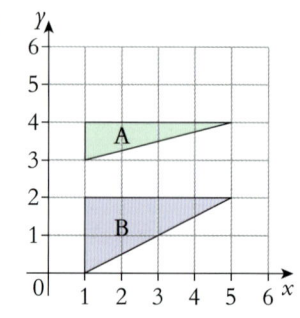   c

310   UNIT 6

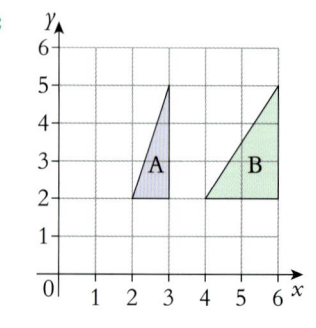

d

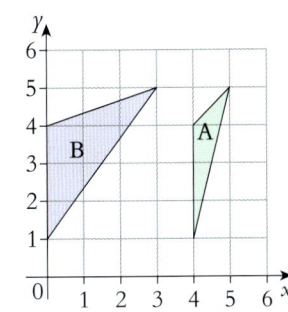

e

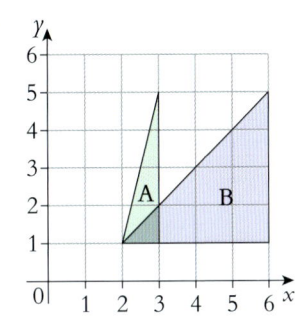

f

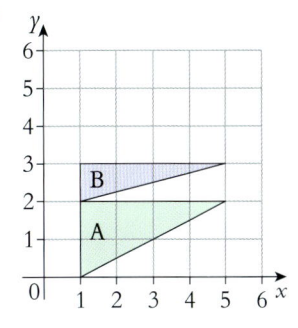

10 Copy the diagram.
  a Draw the image of shape P after a stretch, stretch factor 2 with the x-axis invariant. Label the image Q.
  b Draw the image of shape Q after a stretch, stretch factor 2 with the y-axis invariant. Label the image R.
  c Describe fully the single transformation that moves shape P onto shape R.

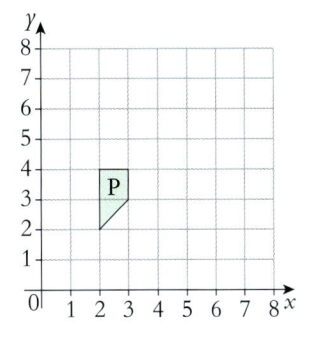

11 Describe fully the single transformation that moves
  a triangle A onto triangle B,
  b triangle B onto triangle C,
  c triangle A onto triangle C.

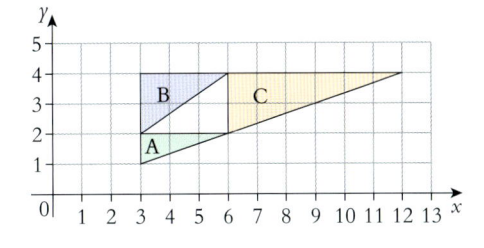

12 Describe fully the single transformation that moves
  a triangle A onto triangle B,
  b triangle B onto triangle C,
  c triangle A onto triangle C.

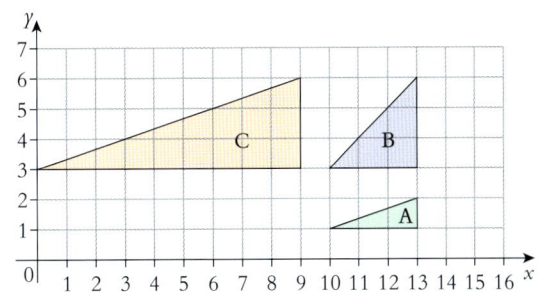

13 Copy the diagram.
Transform shape P by a stretch, stretch factor 2, with $y = x$ invariant.

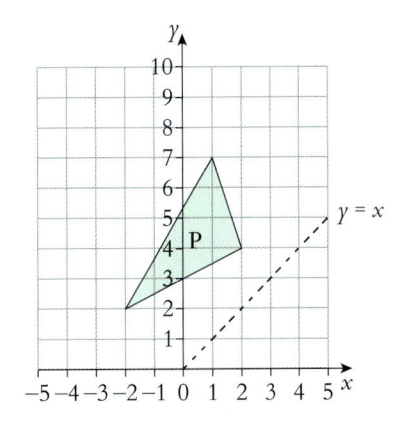

Stretches and shears 311

14 Describe fully the stretch that takes shape A onto shape B for each of these

a **b** 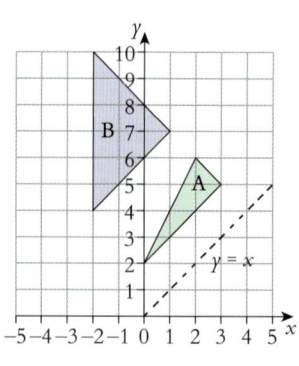 **c**

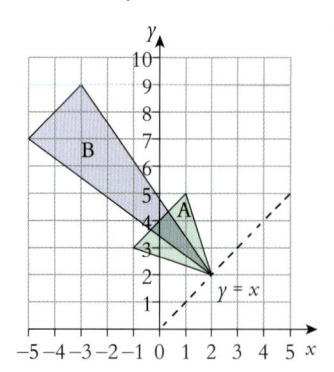

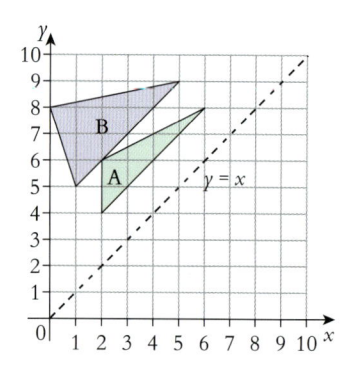

Shears

In a shear, all the points on an object move parallel to a fixed line (called the invariant line).
A shear does not change the area of a shape.
To calculate the shear factor use

$$\text{shear factor} = \frac{\text{distance moved by a point}}{\text{perpendicular distance of point from the invariant line}}$$

To calculate the distance moved by a point use

$$\text{distance moved by a point} = \text{shear factor} \times \text{perpendicular distance of point from the invariant line}$$

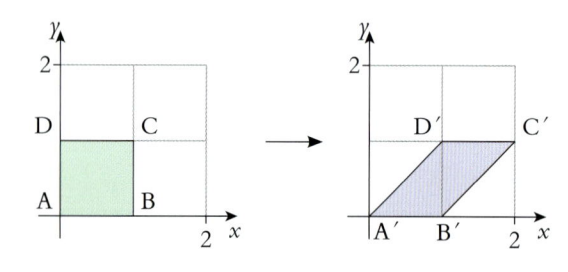

Here ABCD has been sheared to give A′B′C′D′.
The points on the x-axis have not moved,
so the x-axis (or $y = 0$) is called the invariant line.
DD′ = 1 and distance of D from invariant line = 1.
So, shear factor = $\frac{1}{1}$ = 1.

EXAMPLE

A(0, 0), B(2, 0) and C(1, 2).
Transform triangle ABC by a shear, shear factor 3 with the x axis invariant.

A and B are on the x-axis (the invariant line) so
they do not move.
CC′ = 3 × distance of C from the invariant line
CC′ = 3 × 2 = 6   so   C′ = (7, 2)

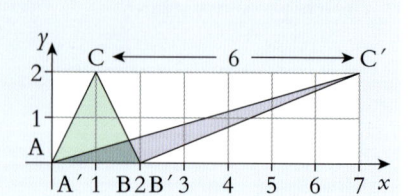

312   UNIT 6

EXAMPLE

Describe fully the single transformation that takes shape ABCD onto shape A'B'C'D'.

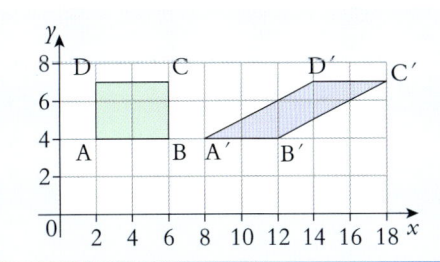

The transformation is a shear.
For the invariant line you find:
- where the lines AD and A'D' intersect
- where the lines BC and B'C' intersect

The lines intersect on the line $y = 1$.
So the invariant line is $y = 1$.
For the shear factor:
- DD' = 12
- distance of D from the invariant line = 6

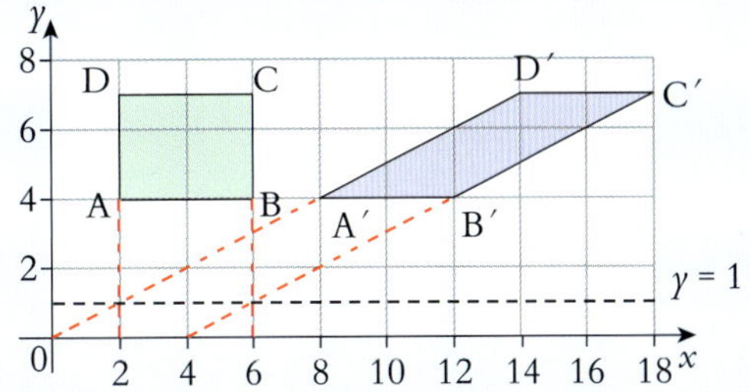

So the shear factor = $\frac{12}{6}$ = 2. The transformation is a shear, shear factor 2 with $y = 1$ invariant.

EXERCISE 6.23

1. Copy the diagram.
   Transform shape P by a shear, shear factor 3, with the x-axis invariant.

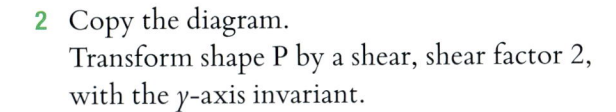

2. Copy the diagram.
   Transform shape P by a shear, shear factor 2, with the y-axis invariant.

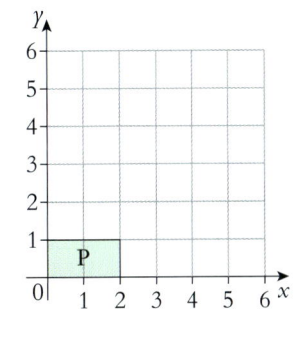

3. Describe fully the shear that takes shape A onto shape B for each of these.

   a   b 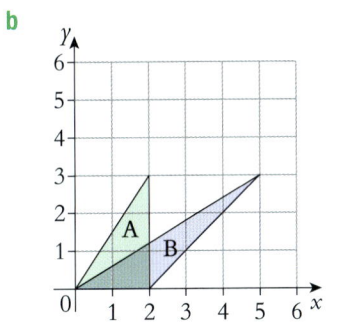   c

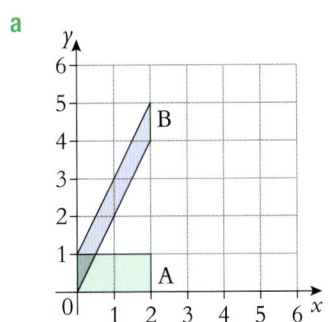

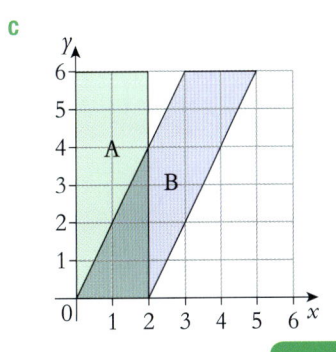

Stretches and shears

d, **e**, **f**, **g**, **h**, **i**

4 Copy the diagram.
Transform shape P by a shear, shear factor 2, with $y = 1$ invariant.

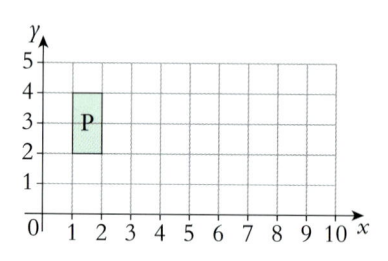

5 Describe fully the shear that takes shape A onto shape B for each of these.

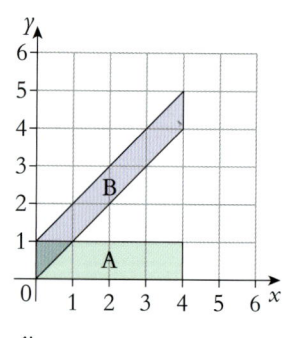

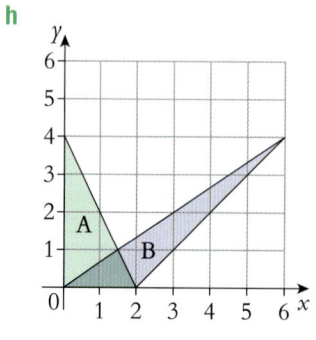

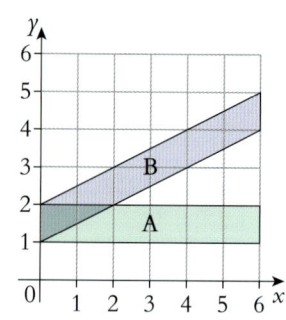

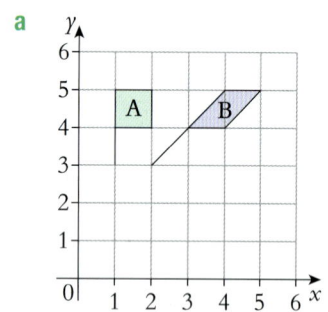

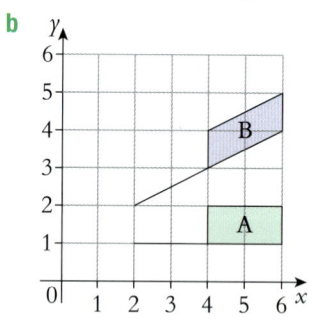

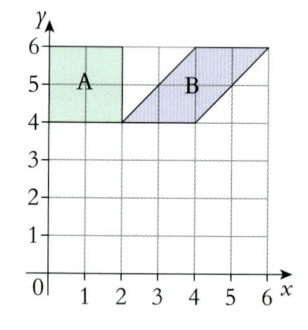

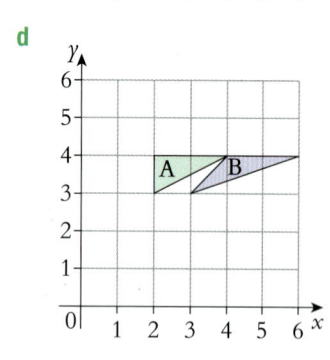

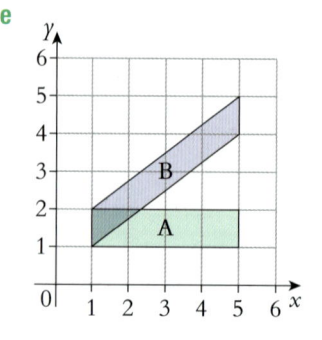

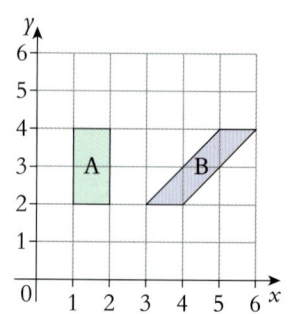

a, **b**, **c**, **d**, **e**, **f**

6 Copy the diagram.
   Transform shape P by a shear, shear factor 2, with $y = 2$ invariant.

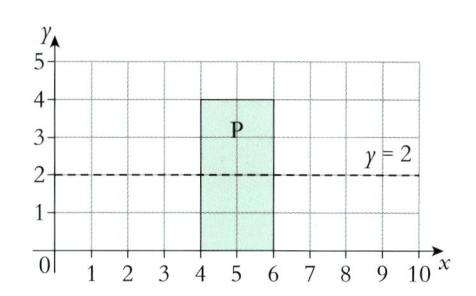

7 Describe fully the shear that takes shape A onto shape B for each of these.
   a 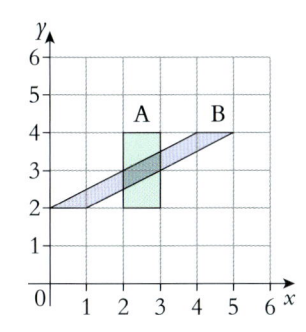
   b 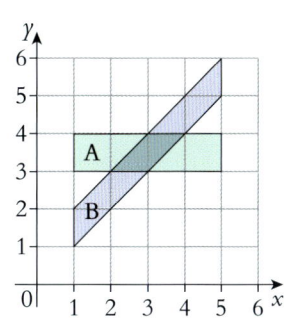
   c

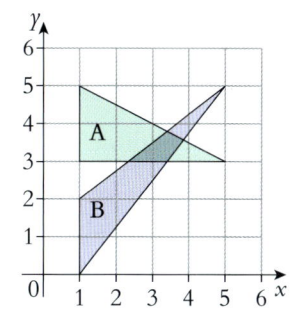

8 Copy the diagram.
   Transform shape P by a shear, shear factor −2, with the x-axis invariant.

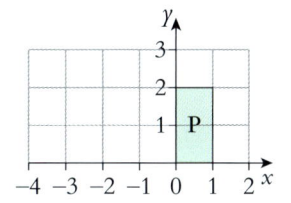

9 Describe fully the shear that takes shape A onto shape B for each of these
   a 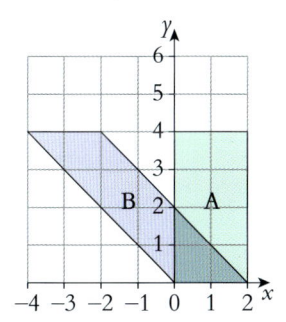
   b 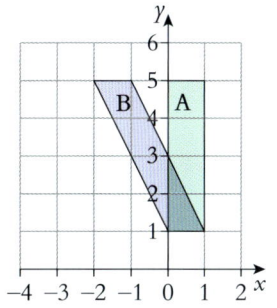
   c

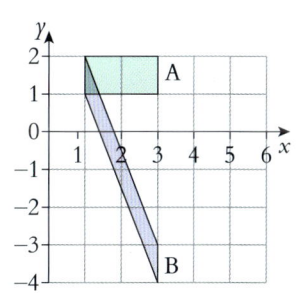

10 Describe fully the single transformation that moves
   a triangle A onto triangle B,
   b triangle B onto triangle C,
   c triangle A onto triangle C.

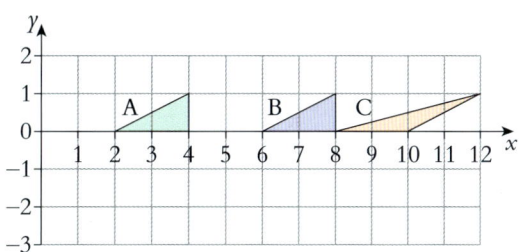

KEY WORDS
Stretch
stretch factor
shear
shear factor
invariant line

Stretches and shears

Probability 2

THIS SECTION WILL SHOW YOU HOW TO
- Use tree diagrams to represent outcomes of compound events
- Use tree diagrams to calculate probabilities of compound events

Combined independent events

Two events A and B are **independent** if one event happening has no effect on whether or not the other event happens. If A and B are independent events, then

P(A happening and B happening) = P(A happening) × P(B happening)

P(A and B) = P(A) × P(B)

EXAMPLE

A fair coin is tossed and a fair three-sided spinner is spun.

Find the probability of getting a head on the coin and a 2 on the spinner.

P(head and 2) = P(H and 2) = P(H) × P(2) = $\frac{1}{2} \times \frac{1}{3} = \frac{1}{6}$

Tree diagrams for independent events

A **tree diagram** is a clear way of showing all the possible outcomes of combined events. They can also be used to calculate probabilities of combined events.

You must draw your tree diagram clearly and write the probabilities on each branch.

EXAMPLE

A bag contains 3 red balls and 2 yellow balls.
A ball is chosen at random from the bag, its colour noted and it is then replaced in the bag.
A second ball is then chosen at random from the bag.
a Draw a tree diagram to show all the possible outcomes.
b Find the probability that **i** both balls are red   **ii** both balls are yellow
   iii different coloured balls are chosen.

a

First choice	Second choice	Outcome	Probability
R ($\frac{3}{5}$)	R ($\frac{3}{5}$)	RR	$\frac{3}{5} \times \frac{3}{5} = \frac{9}{25}$
	Y ($\frac{2}{5}$)	RY	$\frac{3}{5} \times \frac{2}{5} = \frac{6}{25}$
Y ($\frac{2}{5}$)	R ($\frac{3}{5}$)	YR	$\frac{2}{5} \times \frac{3}{5} = \frac{6}{25}$
	Y ($\frac{2}{5}$)	YY	$\frac{2}{5} \times \frac{2}{5} = \frac{4}{25}$

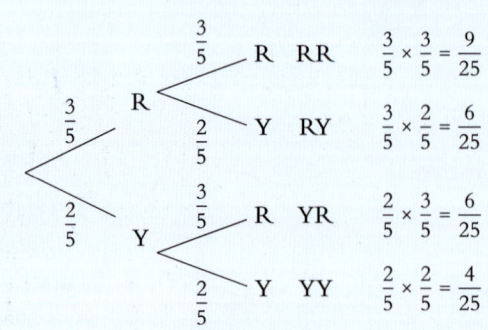

➡ **NOTE:** multiply along the branches.

➡ **CHECK:** the probabilities of all the routes through the diagram should add to 1.

➡ **NOTE:** if you need more than one of the outcomes add the probabilities together.

b
i P(RR) = $\frac{9}{25}$ **ii** P(YY) = $\frac{4}{25}$ **iii** P(different colours) = P(RY) + P(YR) = $\frac{6}{25} + \frac{6}{25} = \frac{12}{25}$

EXERCISE 6.24

1 The probability that an archer hits a target when he shoots an arrow is 0.7. The archer shoots two arrows at a target. 
   a Copy and complete the tree diagram.
   (H means he hits the target and M means he misses the target.)

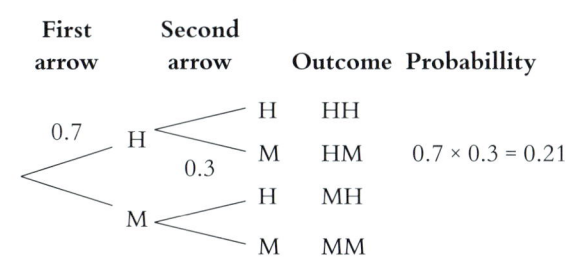

   |First arrow | Second arrow | Outcome | Probabillity |

   ```
                  H    HH
         0.7  H <
                  M    HM    0.7 × 0.3 = 0.21
              0.3
                  H    MH
              M <
                  M    MM
   ```

   b Find the probability that
   i both arrows hit the target,
   ii neither arrow hits the target,
   iii only one arrow hits the target,
   iv at least one arrow hits the target.

2 The circular board is divided into five coloured sections. When the spinner is spun it is equally likely to stop in any of the five sections. The spinner is spun twice.
Draw a tree diagram to show all the possible outcomes.
Find the probability that
   a both colours are blue,   **b** both colours are yellow,
   c the first colour is blue and the second is yellow.

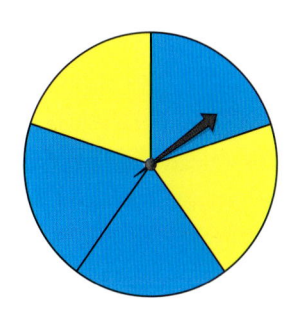

3 Bag A contains 2 green marbles and 1 yellow marble.
Bag B contains 3 green marbles and 2 yellow marbles.
One marble from each bag is selected at random.
Draw a tree diagram to show all the possible outcomes.
Find the probability of selecting
   a two greens,   **b** two yellows,   **c** one of each colour.

4 The probability that an egg on a production line is cracked is 0.01
Two eggs are selected at random from the production line.
Find the probability that,
   a both eggs are cracked,
   b neither egg is cracked,
   c exactly one of the eggs is cracked,
   d at least one egg is cracked.

Probability 2

5 An unbiased six-sided dice is rolled twice.

   a Copy and complete the tree diagram. ($\bar{6}$ means 'not getting a 6' when the dice is rolled.)

First roll	Second roll	Outcome	Probabillity

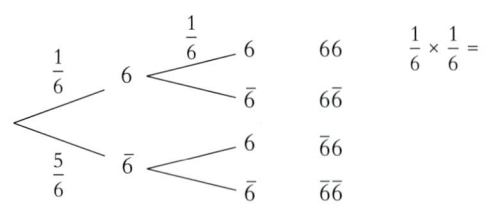

$\frac{1}{6}$ — 6 $\diagdown$ $\frac{1}{6}$ — 6   66 $\frac{1}{6} \times \frac{1}{6} =$

             $\bar{6}$ 6$\bar{6}$

$\frac{5}{6}$ — $\bar{6}$ $\diagdown$ 6 $\bar{6}$6

             $\bar{6}$ $\bar{6}\bar{6}$

   b Use the tree diagram to find the probability of obtaining
      i two sixes,   **ii** no sixes,   **iii** exactly one six.

6 A bag contains 5 green counters and 2 yellow counters.
Ali chooses a counter at random from the bag, he notes the colour and then he replaces it in the bag.
He then chooses a second counter at random.
Draw a tree diagram to show all the possible outcomes.
Find the probability that he chooses

   a two greens,   **b** two yellows,
   c two of the same colour,   **d** at least one green.

7 Three unbiased coins are thrown.

   a Copy and complete the tree diagram to show all the possible outcomes.
(H = head and T = tail)

First throw	Second throw	Third throw	Outcome	Probabillity

$\frac{1}{2}$ H — $\frac{1}{2}$ H — $\frac{1}{2}$ H   HHH $\frac{1}{2} \times \frac{1}{2} \times \frac{1}{2} = \frac{1}{8}$

                        T   HHT

            T   H
                   T

$\frac{1}{2}$ T — H — H
                 T
            T — H
                 T

   b Use the tree diagram to find the probability of getting
      i three tails,   **ii** no tails,   **iii** exactly one head,   **iv** at least two tails.

8 The probability that a biased coin lands on a head is $\frac{2}{3}$.
The coin is thrown three times.
Draw a tree diagram to show all the possible outcomes.
Use the tree diagram to find the probability of getting

   a three heads,   **b** three tails,   **c** exactly one tail,   **d** at least one head.

9 Three unbiased six-sided dice are rolled.

Find the probability of obtaining
- **a** three sixes,
- **b** no sixes,
- **c** exactly one six,
- **d** exactly two sixes.

10 Kurt posts three letters on the same day.
The probability that a letter sent by post will arrive the next day is 0.8
Find the probability that
- **a** all three letters arrive the next day,
- **b** no letters arrive the next day,
- **c** only one letter arrives the next day.

11 A bag contains 5 green counters, 3 red counters and 2 yellow counters.
Ali chooses a counter at random from the bag, he notes the colour and then he replaces it in the bag.
He then chooses a second counter at random.
- **a** Copy and complete the tree diagram to show all the possible outcomes.

First choice	Second choice	Outcome	Probability
$\frac{5}{10}$ G	$\frac{5}{10}$ G	GG	$\frac{5}{10} \times \frac{5}{10} = \frac{25}{100}$
	R	GR	
	Y		
$\frac{3}{10}$ R	G		
	R		
	Y		
$\frac{2}{10}$ Y	G		
	R		
	Y		

- **b** Find the probability that he chooses
  - **i** two greens,
  - **ii** two reds,
  - **iii** two yellows,
  - **iv** two of the same colour,
  - **v** two different colours.

12

Colour	Red	Blue	Yellow
Probability	$\frac{1}{4}$	$\frac{1}{3}$	$\frac{5}{12}$

The circular board is divided into three unequal sections.
The table shows the probabilities of the spinner landing on each of the three colours. The spinner is spun three times.
Draw a tree diagram to show all the possible outcomes.
Find the probability of getting
- **a** three reds,
- **b** three blues,
- **c** three yellows,
- **d** one of each colour.

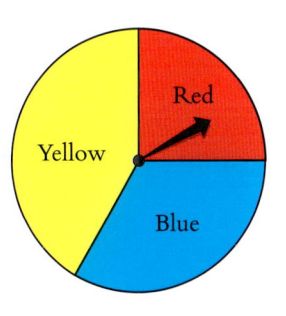

Tree diagrams for dependent events

Two events A and B are **dependent** if one event happening has an effect on whether or not the other event happens.

> **EXAMPLE**
>
> A bag contains 3 red balls and 2 yellow balls.
> A ball is chosen at random from the bag.
> The ball is **not** replaced in the bag.
> A second ball is then chosen at random from the bag. 
> **a** Draw a tree diagram to show all the possible outcomes.
> **b** Find the probability that
>   **i** both balls are red,
>   **ii** both balls are yellow,
>   **iii** different coloured balls are chosen.
>
> ---
>
> **a** If the first ball chosen is a red then the probabilities for the second ball are:
>
> $P(R) = \dfrac{2}{4}$ because there are now **two** red balls and only **four** balls to chose from
>
> $P(Y) = \dfrac{2}{4}$ because there are **two** yellow balls and only **four** balls to chose from  
>
> If the first ball chosen is a yellow then the probabilities for the second ball are:
>
> $P(R) = \dfrac{3}{4}$ because there are **three** red balls and only **four** balls to chose from
>
> $P(Y) = \dfrac{1}{4}$ because there is now **one** yellow ball and only **four** balls to chose from 
>
> You can now put the probabilities on a tree diagram.
>
> | First choice | Second choice | Outcome | Probabillity |
> |---|---|---|---|
> | $\tfrac{3}{5}$ R | $\tfrac{2}{4}$ R | RR | $\tfrac{3}{5} \times \tfrac{2}{4} = \tfrac{6}{20}$ |
> | | $\tfrac{2}{4}$ Y | RY | $\tfrac{3}{5} \times \tfrac{2}{4} = \tfrac{6}{20}$ |
> | $\tfrac{2}{5}$ Y | $\tfrac{3}{4}$ R | YR | $\tfrac{2}{5} \times \tfrac{3}{4} = \tfrac{6}{20}$ |
> | | $\tfrac{1}{4}$ Y | YY | $\tfrac{2}{5} \times \tfrac{1}{4} = \tfrac{2}{20}$ |
>
> **b i** $P(RR) = \dfrac{6}{20} = \dfrac{3}{10}$ **ii** $P(YY) = \dfrac{2}{20} = \dfrac{1}{10}$
>
> **iii** $P(\text{different colours}) = P(RY) + P(YR) = \dfrac{6}{20} + \dfrac{6}{20} = \dfrac{12}{20} = \dfrac{3}{5}$

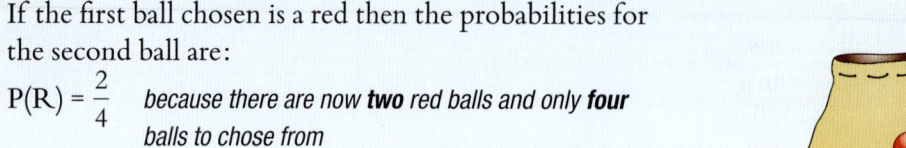

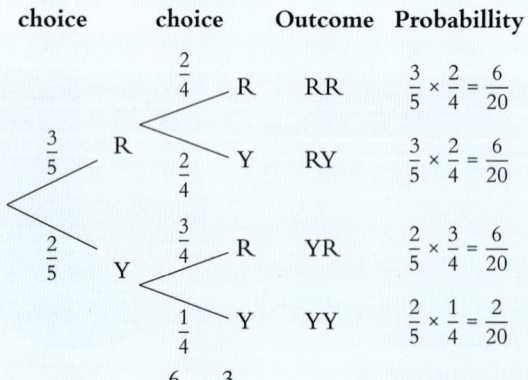

EXERCISE 6.25

1. A bag contains 5 orange balls and 2 blue balls.
   Raju chooses a ball at random from the bag.
   It is **not** replaced in the bag.
   He then chooses a second ball at random from the bag.
   Find the probability that he chooses
   - a two orange balls,
   - b two blue balls,
   - c two balls the same colour,
   - d two different coloured balls.

2. Shona has 3 one-dollar coins and 1 half-dollar coin in her purse.
   She chooses a coin at random from the purse.
   It is **not** replaced in the purse.
   She then chooses a second coin at random from the purse.
   Find the probability that she chooses
   - a 2 one-dollar coins,
   - b 2 half-dollar coins,
   - c at least 1 one-dollar coin.

3. A box contains 4 blue crayons and 2 yellow crayons.
   Kurt chooses a crayon at random from the box.
   It is **not** replaced in the box.
   He then chooses a second crayon at random from the box.
   Find the probability that he chooses
   - a two blue crayons,
   - b two yellow crayons,
   - c two different colour crayons,
   - d at least one yellow crayon.

4. A bag contains 6 orange sweets and 3 yellow sweets.
   Priya chooses a sweet at random and eats it.
   She then chooses a second sweet at random.
   Find the probability that she chooses
   - a two orange sweets,
   - b two yellow sweets,
   - c two different coloured sweets,
   - d at least one orange sweet.

5. A bag contains 4 blue marbles, 3 red marbles and 2 yellow marbles.
   Ali chooses a marble at random from the bag.
   It is **not** replaced in the bag.
   He then chooses a second marble at random from the bag.
   Find the probability that he chooses
   - a two blues,
   - b two reds,
   - c two yellows,
   - d two of the same colour,
   - e two different colours.

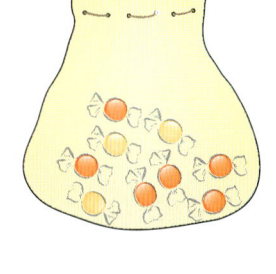

Probability 2

6 A ball is chosen at random from bag A, the colour is noted and then the ball is placed in bag B.
A ball is then chosen at random from bag B.
Find the probability of choosing
  a two blacks
  b two whites
  c two different colours
  e at least one black.

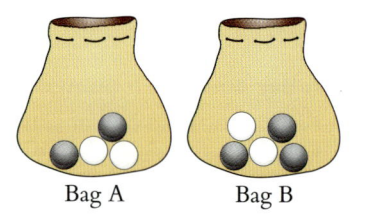

7 Omar either walks or cycles to school.
The probability that he cycles to school is 0.2.
If Omar cycles, the probability that he is late is 0.1.
If Omar walks, the probability he is late is 0.3.
  a Copy and complete the tree diagram.
  (C = cycle   W = walk   L = late   and $\bar{L}$ = not late)

  Outcome   Probabillity

  0.2 C 0.1 L   CL   0.2 × 0.1 = 0.02
           $\bar{L}$ C$\bar{L}$
       W 0.3 L   WL
           $\bar{L}$ W$\bar{L}$

  b Calculate the probability that Omar will be
    i late to school,   **ii** not late to school.

8 a Jacob is asked to select one shape at random from either bag A or bag B.
The probability that he chooses from bag A is $\frac{2}{5}$.
  i Copy and complete the tree diagram.

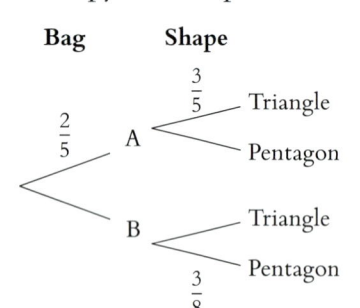

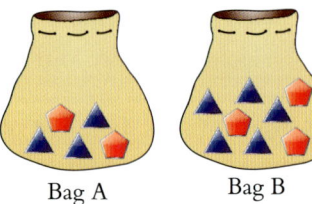

  Bag   Shape

  $\frac{2}{5}$ A $\frac{3}{5}$ Triangle
                  Pentagon
       B   Triangle
                  Pentagon
           $\frac{3}{8}$

  ii Calculate the probability that Jacob chooses a triangle.
  iii Calculate the probability that Jacob chooses a pentagon.
b Eduardo is asked to select 2 shapes at random from bag A or two shapes at random from bag B. (He does not replace the first shape before selecting the second shape.)
The probability that he chooses from bag A is $\frac{1}{3}$.
  i Calculate the probability that Eduardo chooses two triangles.
  ii Calculate the probability that Eduardo chooses two different shapes.

Further probability

EXAMPLE

The probability that Anna hits the bullseye on a dartboard is 0.1.
Find the probability that she hits the bullseye for the first time on her fourth attempt.

She must miss on her first three attempts and hit on her fourth attempt.
Let M = miss and H = hit.
P(M) = 0.9 and P(H) = 0.1.
P(M M M H) = P(M) × P(M) × P(M) × P(H)
           = 0.9 × 0.9 × 0.9 × 0.1 = 0.0729

EXERCISE 6.26

1 The circular board is divided into eight numbered sections.
   When the arrow is spun it is equally likely to stop in any of the eight sections.
   The spinner is spun several times until it lands on the number 2.
   Find the probability that this happens on the fourth spin.

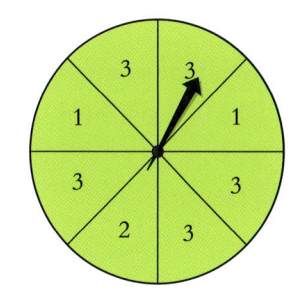

2 A fair coin is tossed repeatedly until it lands on a head.
   Find the probability that this happens on the sixth toss.

3 A six-sided dice is unbiased.
   It is numbered 1, 2, 3, 4, 5 and 6 on its faces.
   The dice is rolled several times until it lands on the number 5.
   Find the probability that this happens on the fifth roll.

4 The probability that Robert will pass his driving test on his first attempt is $\frac{2}{3}$.
   If he fails, the probability that he passes on any future attempt is $\frac{3}{4}$.
   Find the probability that he passes on his fifth attempt.

5 The bag contains 5 black marbles and 4 white marbles.
   Marbles are selected at random from the bag until a white marble is selected.
   (The marbles are **not** replaced.)
   Find the probability that this happens on the third selection.

KEY WORDS
tree diagram
independent events
dependent events

Probability 2

Unit 6 Examination questions

1 In 2004 Colin had a salary of $7200.

   (a) This was an increase of 20% on his salary in 2002.
Calculate his salary in 2002. [2]

   (b) In 2006 his salary increased to $8100.
Calculate the percentage increase from 2004 to 2006. [2]

Cambridge IGCSE Mathematics 0580, Paper 2 Q16, June 2006

2 Zainab borrows $198 from a bank to pay for a new bed.
The bank charges compound interest at 1.9% per month.
Calculate how much **interest** she owes at the end of 3 months.
Give your answer correct to 2 decimal places. [3]

Cambridge IGCSE Mathematics 0580, Paper 21 Q14, November 2009

3 Each year a school organizes a concert.

   (a) **(i)** In 2004 the cost of organising the concert was $385.
In 2005 the cost was 10% less than in 2004.
Calculate the cost in 2005. [2]

       (ii) The cost of $385 in 2004 was 10% more than the cost in 2003.
Calculate the cost in 2003. [2]

   (b) **(i)** In 2006 the number of tickets sold was 210.
The ratio
    Number of adult tickets : Number of student tickets was 23 : 19.
How many adult tickets were sold? [2]

       (ii) Adult tickets were $2.50 each and student tickets were $1.50 each.
Calculate the **total** amount **received** from selling the tickets. [2]

       (iii) In 2006 the cost of organising the concert was $410.
Calculate the percentage profit in 2006. [2]

   (c) In 2007, the number of tickets sold was again 210.
Adult tickets were $2.60 each and student tickets were $1.40 each.
The total amount received from selling the 210 tickets was $480.
How many student tickets were sold? [4]

Cambridge IGCSE Mathematics 0580, Paper 4 Q1, November 2007

4 Johan invested $600 for 3 years at 4% per year **compound** interest.
Calculate the final amount he had after three years. [3]

Cambridge IGCSE Mathematics 0580, Paper 21 Q9, November 2010

5 Alaric invests $100 at 4% per year **compound interest**.
  (a) How many dollars will Alaric have after 2 years? [2]
  (b) After x years, Alaric will have y dollars.
  He knows a formula to calculate y.
  The formula is $y = 100 \times 1.04^x$

x (Years)	0	10	20	30	40
y (Dollars)	100	p	219	q	480

  Use this formula to calculate the values of p and q in the table. [2]
  (c) Using a scale of 2 cm to represent 5 years on the x-axis and 2 cm
  to represent $50 on the y-axis, draw an x-axis for $0 \leq x \leq 40$ and
  y-axis for $0 \leq y \leq 500$.
  Plot the five points in the table and draw a smooth curve through them. [5]
  (d) Use your graph to estimate
   (i) how many dollars Alaric will have after 25 years, [1]
   (ii) how many years, to the nearest year, it takes for Alaric to have $200. [1]
  (e) Beatrice invests $100 at 7% per year **simple interest**.
   (i) Show that after 20 years Beatrice has $240. [2]
   (ii) How many dollars will Beatrice have after 40 years? [1]
   (iii) On the **same grid**, draw a graph to show how the $100
   which Beatrice invests will increase during the 40 years. [2]
  (f) Alaric first has more than Beatrice after n years.
  Use your graphs to find the value of n. [1]

Cambridge IGCSE Mathematics 0580, Paper 4 Q8, June 2008

6 A person in a car, travelling at 108 kilometres per hour, takes 1 second to go past
a building on the side of the road.
Calculate the length of the building in metres. [2]

Cambridge IGCSE Mathematics 0580, Paper 21 Q4, June 2010

7 The Canadian Maple Leaf train timetable from Toronto to Buffalo is show below.

Toronto	10 30
Oakville	10 52
Aldershot	11 07
Grimsby	11 41
St Catharines	11 59
Niagara Falls	12 24
Buffalo	13 25

  (a) How long does the journey take from Toronto to Buffalo? [1]
  (b) This journey is 154 kilometres. Calculate the average speed of the train. [2]

Cambridge IGCSE Mathematics 0580, Paper 2 Q8, November 2006

8 (a) Shade the region $A \cap B$ on a copy of the diagram

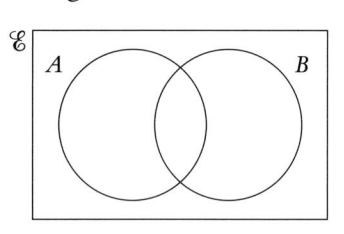

[1]

(b) Shade the region $(A \cup B)'$ on a copy of the diagram

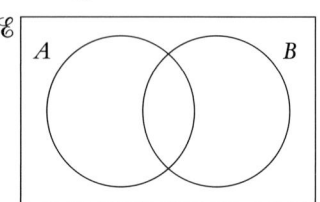

[1]

(c) Shade the complement of set B on a copy of the diagram

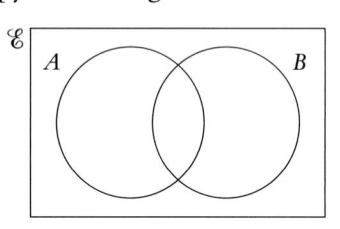

[1]

Cambridge IGCSE Mathematics 0580, Paper 2 Q11, November 2006

9 Shade the region required in a copy of each Venn Diagram.

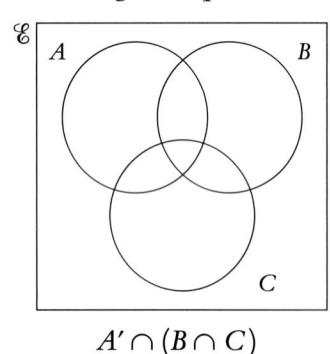

$A' \cap (B \cap C)$ 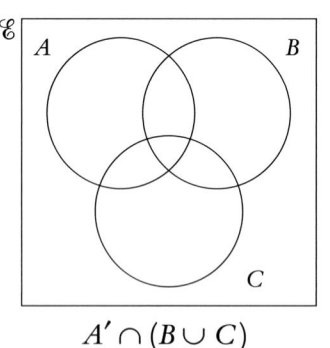 $A' \cap (B \cup C)$

[2]

Cambridge IGCSE Mathematics 0580, Paper 21 Q7, November 2009

10 $\mathscr{E} = \{1, 2, 3, 4, 5, 6, 7, 9, 11, 16\}$ $P = \{2, 3, 5, 7, 11\}$ $S = \{1, 4, 9, 16\}$ $M = \{3, 6, 9\}$

(a) Draw a Venn diagram to show this information. [2]
(b) Write down the value of $n(M' \cap P)$. [1]

Cambridge IGCSE Mathematics 0580, Paper 21 Q12, June 2008

11 In a survey of 60 cars, 25 use diesel, 20 use liquid hydrogen and 22 use electricity. No cars use all three fuels and 14 cars use both diesel and electricity. There are 8 cars which use diesel only, 15 cars which use liquid hydrogen only and 6 cars which use electricity only.
In the Venn diagram below

$\mathcal{E}$ = {cars in the survey},
D = {cars which use diesel},
L = {cars which use liquid hydrogen},
E = {cars which use electricity}.

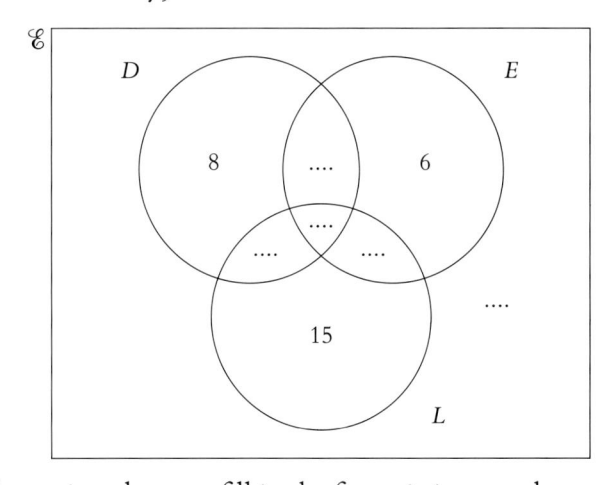

(a) Use the information above to fill in the five missing numbers in a copy of the Venn diagram. [4]
(b) Find the number of cars which use diesel but not electricity. [1]
(c) Find $n(D' \cap (E \cup L))$. [1]

Cambridge IGCSE Mathematics 0580, Paper 21 Q22, November 2010

12 Simplify

(a) $\left(\dfrac{p^4}{16}\right)^{0.75}$, [2]

(b) $3^2 q^{-3} \div 2^3 q^{-2}$. [2]

Cambridge IGCSE Mathematics 0580, Paper 21 Q16, June 2010

13 Simplify $\dfrac{5}{18}x^{\frac{3}{2}} \div \dfrac{1}{2}x^{-\frac{5}{2}}$. [2]

Cambridge IGCSE Mathematics 0580, Paper 21 Q4, November 2009

14 Simplify $(27x^3)^{\frac{2}{3}}$. [2]

Cambridge IGCSE Mathematics 0580, Paper 21 Q8, June 2008

15 Find the value of n in the following equations.

    (a) $2^n = 1024$ [1]

    (b) $4^{2n-3} = 16$ [2]

Cambridge IGCSE Mathematics 0580, Paper 21 Q14, November 2010

16 Find the value of n in each of the following statements.

    (a) $32^n = 1$ [1]

    (b) $32^n = 2$ [1]

    (c) $32^n = 8$ [1]

Cambridge IGCSE Mathematics 0580, Paper 2 Q7, November 2006

17 Solve the equations

    (a) $0.2x - 3 = 0.5x$, [2]

    (b) $2x^2 - 11x + 12 = 0$. [3]

Cambridge IGCSE Mathematics 0580, Paper 2 Q17, November 2006

18

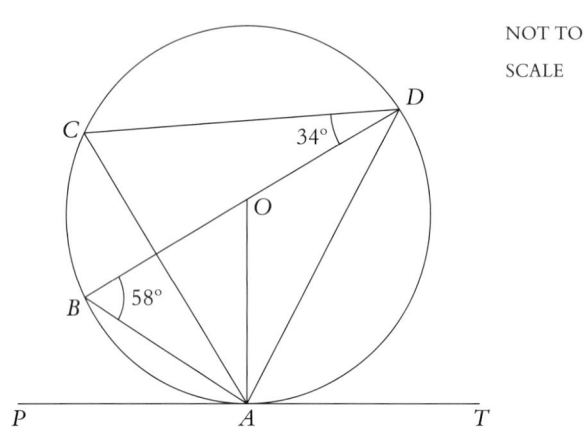

A, B, C and D lie on the circle, centre O.
BD is a diameter and PAT is the tangent at A.
Angle $ABD = 58°$ and angle $CDB = 34°$.

Find

    (a)   angle ACD, [1]

    (b)   angle ADB, [1]

    (c)   angle DAT, [1]

    (d)   angle CAO. [2]

Cambridge IGCSE Mathematics 0580, Paper 21 Q22, June 2009

19

Points A, B and C lie on a circle, centre O, with diameter AB.
BD, OCE and AF are parallel lines.
Angle $CBD = 68°$.

Calculate

(a) angle BOC, [2]
(b) angle ACE. [2]

Cambridge IGCSE Mathematics 0580, Paper 21 Q19, November 2009

20

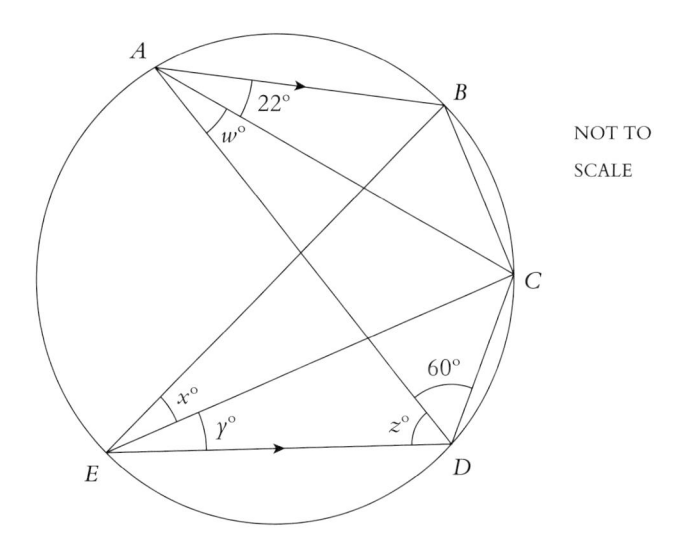

AD is a diameter of the circle $ABCDE$.
Angle $BAC = 22°$ and angle $ADC = 60°$.
AB and ED are parallel lines.
Find the values of w, x, y and z. [4]

Cambridge IGCSE Mathematics 0580, Paper 2 Q18, June 2006

21

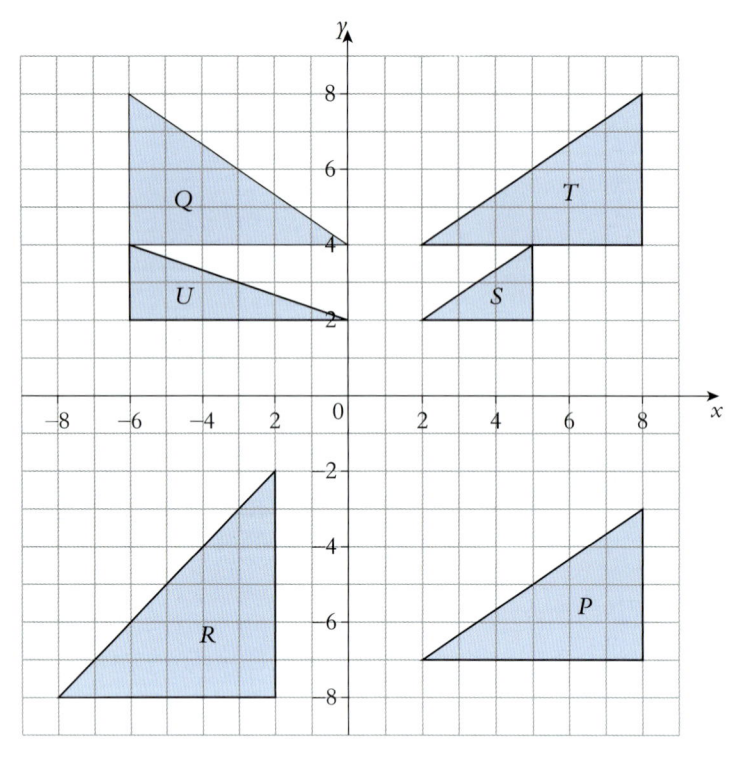

The diagram shows triangles P, Q, R, S, T and U.

(a) Describe fully the **single** transformation which maps triangle
- (i) T onto P, [2]
- (ii) Q onto T, [2]
- (iii) T onto R, [2]
- (iv) T onto S, [3]
- (v) U onto Q. [3]

Cambridge IGCSE Mathematics 0580, Paper 4 Q7 (a), November 2008

22

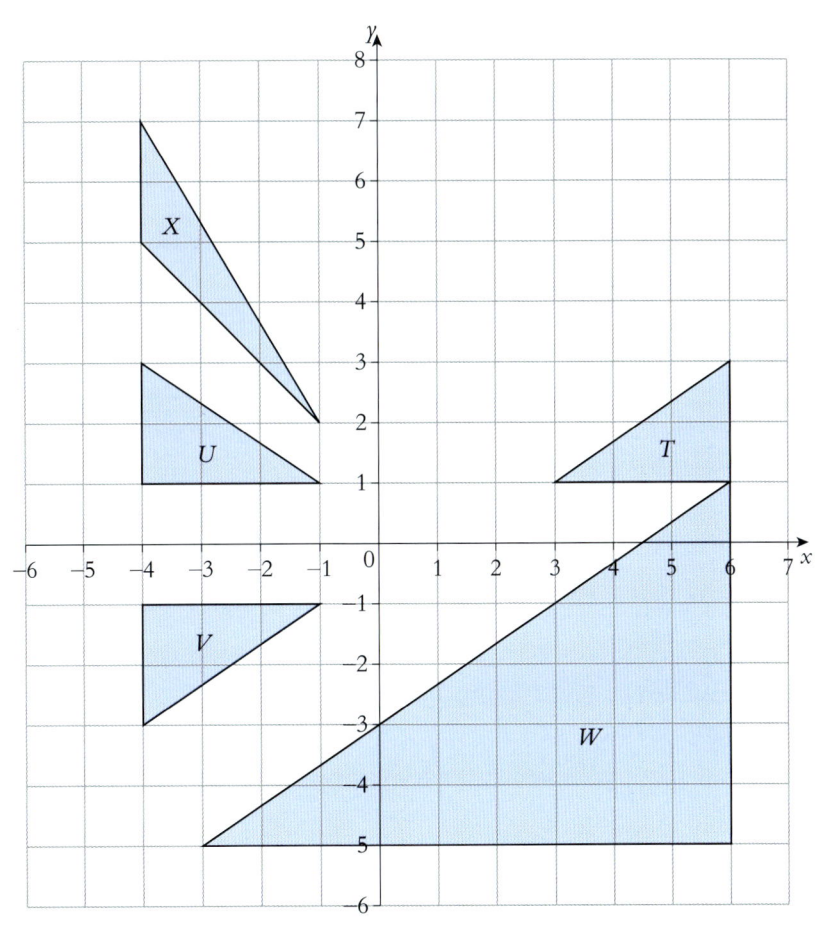

(a) Describe fully the **single** transformation which maps
- **(i)** triangle T onto triangle U, [2]
- **(ii)** triangle T onto triangle V, [3]
- **(iii)** triangle T onto triangle W, [3]
- **(iv)** triangle U onto triangle X. [3]

Cambridge IGCSE Mathematics 0580, Paper 4 Q2 (a), November 2009

23 Claude goes to school by bus.
The probability that the bus is late is 0.1.
If the bus is late, the probability that Claude is late to school is 0.8.
If the bus is not late, the probability that Claude is late to school is 0.05.
- **(i)** Calculate the probability that the bus is late and Claude is late to school. [1]
- **(ii)** Calculate the probability that Claude is late to school. [3]
- **(iii)** The school team lasts 56 days.
  How many days would Claude expect to be late? [1]

Cambridge IGCSE Mathematics 0580, Paper 4 Q2 (b), November 2007

24

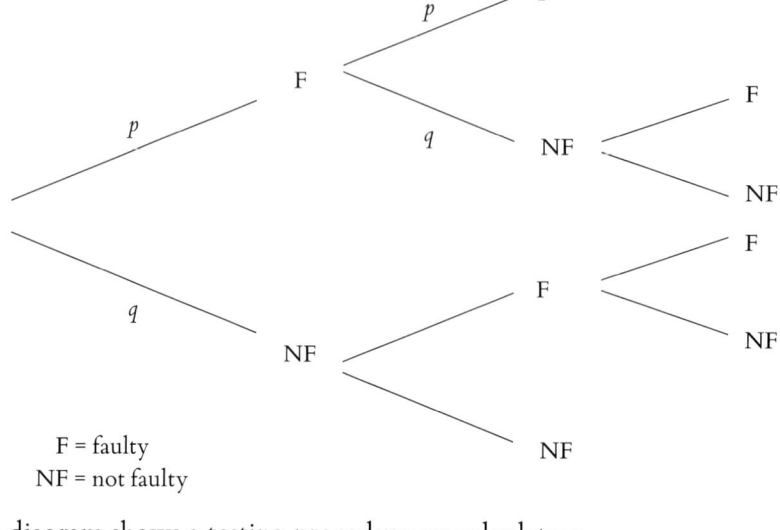

```
                    First              Second              Third
                    Calculator         Calculator          Calculator
```

```
                                              p     F
                                      F
                              p               q     NF ──── F
                                                        └── NF
                                                        
                                              F ──── F
                              q                        └── NF
                              NF
                                              NF
```

F = faulty
NF = not faulty

The tree diagram shows a testing procedure on calculators, taken from a large batch.

Each time a calculator is chosen at random, the probability that it is faulty (F) is $\frac{1}{20}$.

(a) Write down the values of p and q. [1]

(b) Two calculators are chosen at random.
Calculate the probability that
  (i) both are faulty, [2]
  (ii) exactly one is faulty. [2]

(c) If **exactly one** out of two calculators tested is faulty, then a third calculator is chosen at random.
Calculate the probability that exactly one of the first two calculators is faulty **and** the third one is faulty. [2]

(d) The whole batch of calculators is rejected
  either if the first two chosen are both faulty
  or if a third one needs to be chosen and it is faulty.
Calculate the probability that the whole batch is rejected. [2]

(e) In one month, 1000 batches of calculators are tested in this way.
How many batches are expected to be rejected? [1]

Cambridge IGCSE Mathematics 0580, Paper 4 Q8, June 2009

25

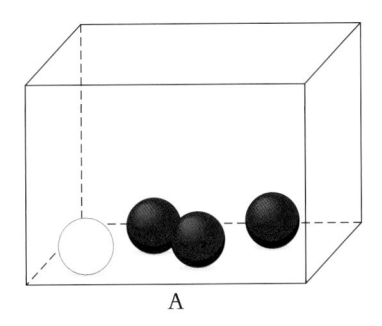

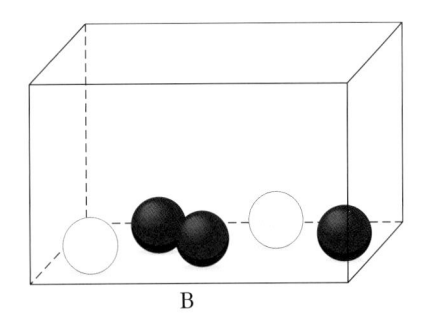

Box A contains 3 black balls and 1 white ball.
Box B contains 3 black balls and 2 white balls.

(a) A ball can be chosen at random from either box.
Copy and complete the following statement.
There is a greater probability of choosing a white ball from Box.....
Explain your answer. [1]

(b) Abdul chooses a box and then chooses a ball from this box at random.
The probability that he chooses box A is $\frac{2}{3}$.

(i) Complete a copy of the tree diagram by writing the four probabilities in the empty spaces.

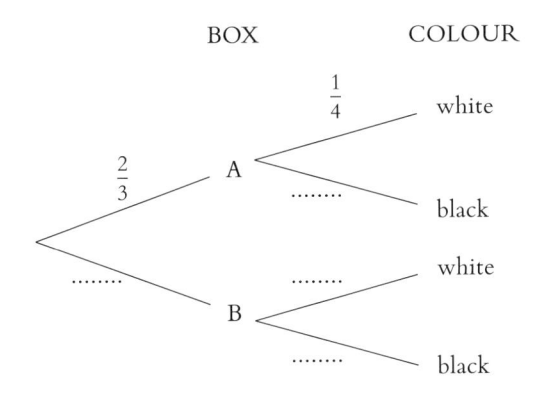

```
        BOX              COLOUR
                  1/4
                     ――― white
         2/3  A ＜
                  ......
                     ――― black
                  ......
                     ――― white
         ......  B ＜
                  ......
                     ――― black
```
[4]

(ii) Find the probability that Abdul chooses box A and a black ball. [2]
(iii) Find the probability that Abdul chooses a black ball. [2]

(c) Tatiana chooses a box and then chooses **two** balls from this box at random (without replacement).
The probability that she chooses box A is $\frac{2}{3}$.
Find the probability that Tatiana chooses two white balls. [2]

Cambridge IGCSE Mathematics 0580, Paper 41 Q4, June 2010

Distance-time graphs

THIS SECTION WILL SHOW YOU HOW TO
- Read information from a distance-time graph
- Find the speed from a distance-time graph

You can use travel graphs of distance against time to find out how the distance is changing. Speed is calculate using the formula:

$$\text{speed} = \frac{\text{distance travelled}}{\text{time taken}}$$

So the steepness of a **distance-time graph** represents the speed.

A straight line shows that the speed is constant (steady)

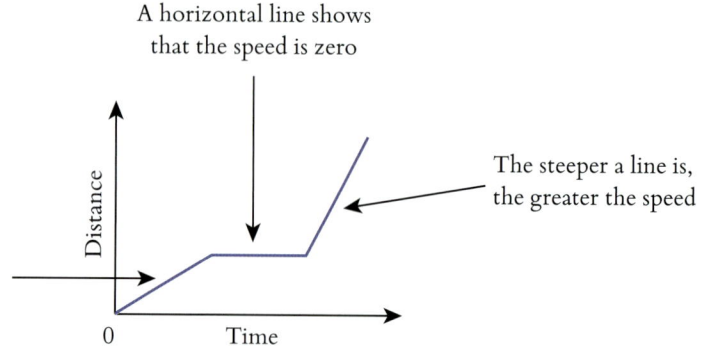

A horizontal line shows that the speed is zero

The steeper a line is, the greater the speed

EXAMPLE

Roxy leaves home at 8:00 am to go on a bike ride. She stops once for a rest. The travel graph shows information about her journey.

a How long does she stop for a rest?
b At what time does she arrive back home?
c How far does she travel in total?
d Calculate her average speed, in km/h, for the whole journey.
e Calculate her fastest speed, in km/h, for the whole journey.

a Four large squares on the horizontal axis represents one hour.
So, each large square represents 15 minutes. She stops for a rest for 15 minutes.
b She arrives back home at 10:30 am.
c Total distance travelled = 20 + 20 = 40 km.
d Average speed = $\dfrac{\text{total distance travelled}}{\text{total time taken}} = \dfrac{40}{2.5} = 16$ km/h
e Her fastest speed is when the graph is steepest. (This occurs between 8:00 and 8:30 am.)

Speed = $\dfrac{\text{distance travelled}}{\text{time taken}} = \dfrac{12}{0.5} = 24$ km/h

EXERCISE 7.1

1

The distance-time graph shows information about Paulo's car journey to work.
- **a** Calculate the speed of the car during the first stage of the journey.
  Give your answer in   **i** kilometres/minute,   **ii** kilometres/hour.
- **b** Calculate the speed of the car during the second stage of the journey.
  Give your answer in   **i** kilometres/minute,   **ii** kilometres/hour.
- **c** Calculate the speed of the car during the third stage of the journey.
  Give your answer in   **i** kilometres/minute,   **ii** kilometres/hour.
- **d** Calculate the average speed for the whole journey.
  Give your answer in kilometres/hour.

2

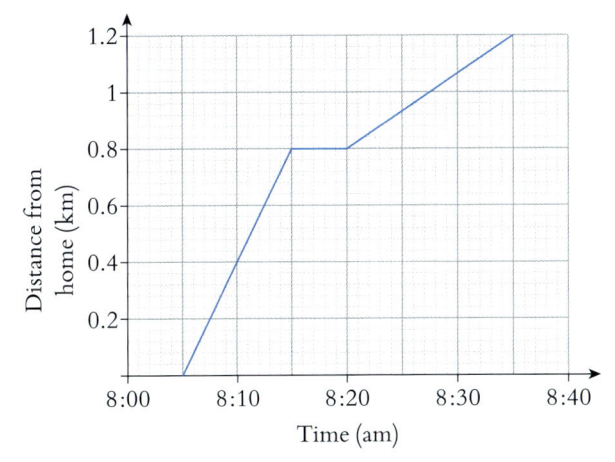

The travel graph shows information about Cara's walk to school each day.
She stops at a friend's house on the way.
- **a** How long does she stop at her friend's house?
- **b** Calculate her speed, in km/h, between 08:05 am and 08:15 am.
  Give your answer in   **i** kilometres/minute,   **ii** kilometres/hour.
- **c** Calculate her speed, in km/h, between 08:20 am and 08:35 am.
  Give your answer in   **i** kilometres/minute,   **ii** kilometres/hour.
- **d** Calculate her average speed, in km/h, for the whole journey.

Distance-time graphs

3

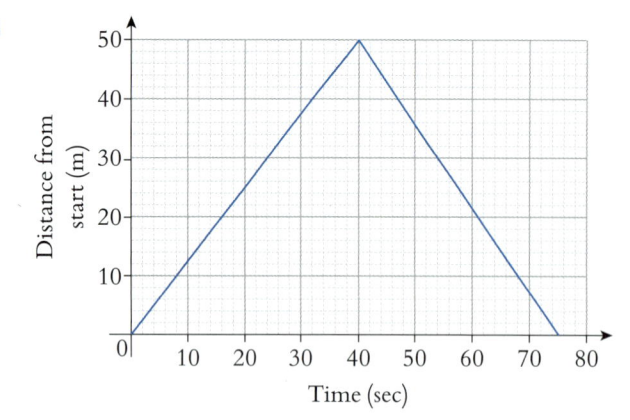

Henri swims in a 100-metre race. The swimming pool is 50 metres long.
He swims the first length in 40 seconds and the second length in 35 seconds.
  a Calculate his speed, in m/s, for the first length.
  b Calculate his speed, in m/s, for the second length.
  c Calculate his average speed, in m/s, for the total 100 metres.

4

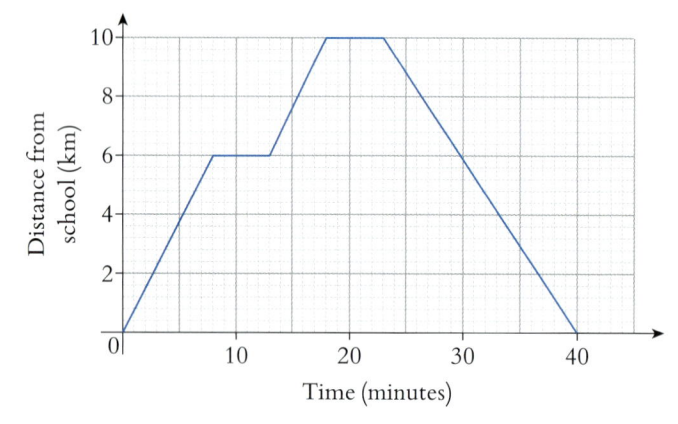

A school bus takes students home from school.
The bus stops at two villages. The bus then returns to school.
The distance-time graph shows information about the journey.
  a How far are the two villages from the school?
  b How long does the bus stop at each of the two villages?
  c Calculate the greatest speed of the bus.
     Give your answer in   i kilometres/minute,   ii kilometres/hour.
  d Calculate the average speed for the whole bus journey.
     Give your answer in   i kilometres/minute,   ii kilometres/hour.

5

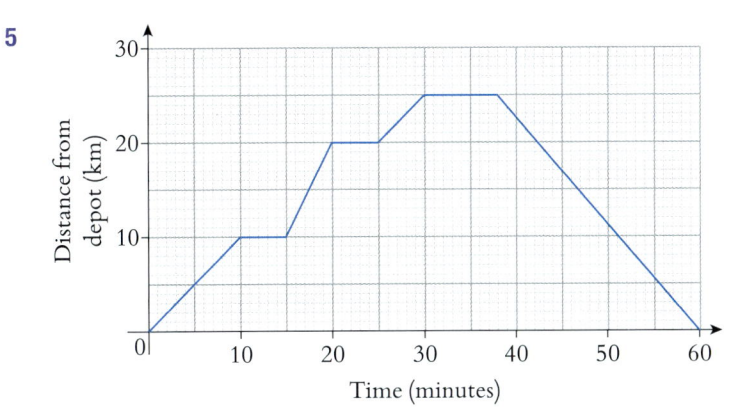

The distance-time graph shows information about the journey of a delivery van.
It starts at the depot and returns to the depot after 60 minutes.
- a How many times does the delivery van stop?
- b How many minutes did the delivery van stop altogether?
- c For which part of the journey did the delivery van travel fastest?
- d What is the total distance travelled by the delivery van?
- e Calculate the delivery van's average speed for the whole journey.

6 A train (P) leaves station A at 8 am and travels at a constant speed of 240 km/h towards station B.
The distance between the two stations is 180 km.
- a Show that the train arrives at station B at 8:45 am.

At 8:15 am a second train (Q) leaves station B and travels towards station A at a constant speed of 200 km/h.
- b Calculate the time that the train arrives at station A.
- c Copy and complete this travel graph to show the journey of the second train.

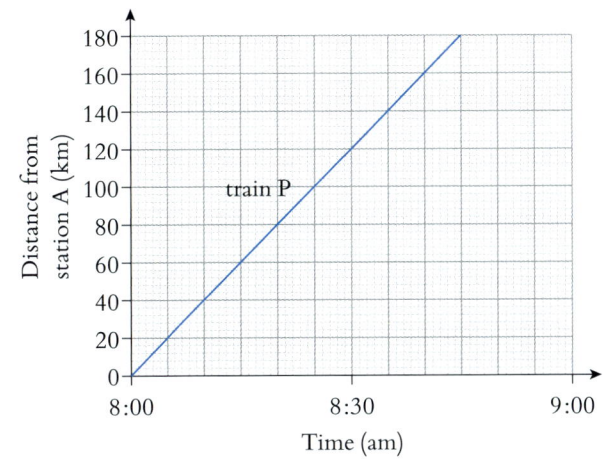

- d Use your graph to estimate the time at which the two trains pass each other.

> **KEY WORDS**
> distance-time graph

Distance-time graphs   337

Speed-time graphs

> **THIS SECTION WILL SHOW YOU HOW TO**
> - Find the acceleration and distance travelled from a speed-time graph

You can use travel graphs of speed against time to find out how the speed is changing. The rate of change of speed is called the **acceleration**.

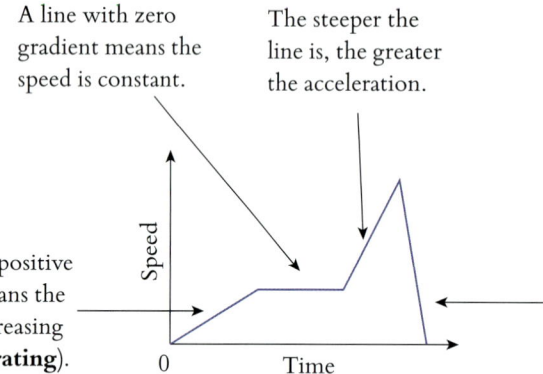

A line with zero gradient means the speed is constant.

The steeper the line is, the greater the acceleration.

A line with positive gradient means the speed in increasing (it is **accelerating**).

A line with negative gradient means the speed in decreasing (it is **decelerating**).

You must learn the following two rules:

> In a speed-time graph,
> acceleration = gradient of line = $\dfrac{\text{change in speed}}{\text{time}}$
>
> distance travelled = area under the graph

The units for acceleration are metres per second or m/s² (or ms⁻²).

EXAMPLE

The speed-time graph shows the speed of a car over 100 seconds.
a   Calculate the acceleration of the car over the first 20 seconds.
b   Calculate the acceleration of the car over the last 40 seconds.
c   Calculate the distance travelled by the car in the 100 seconds.

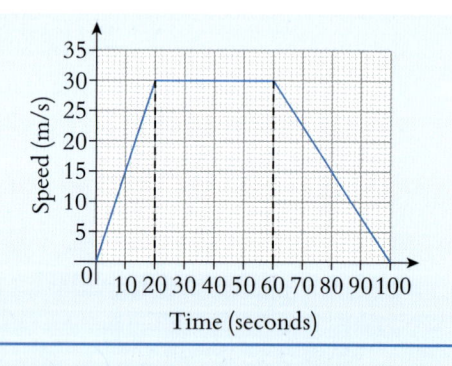

a   Acceleration = gradient of line = $\dfrac{30}{20}$ = 1.5 m/s²

b   Acceleration = gradient of line = $-\dfrac{30}{40}$ = −0.75 m/s²

➡ **NOTE:** this can also be called a deceleration of 0.75 m/s².

c   Distance travelled = area under the graph
    = $\left(\dfrac{1}{2} \times 20 \times 30\right) + (40 \times 30) + \left(\dfrac{1}{2} \times 40 \times 30\right)$
    = 2100 m

➡ **NOTE:** the quickest method for finding the area under this graph is to use
Area of trapezium = $\dfrac{1}{2}(a + b)h$
= $\dfrac{1}{2}(100 + 40) \times 30$ = 2100

338   UNIT 7

EXERCISE 7.2

1. The speed-time graph shows the speed of a car over the first 10 seconds of a journey. 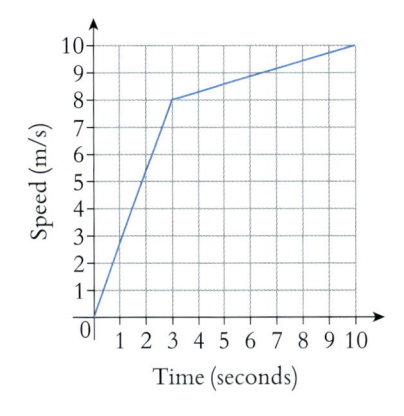
   a. Calculate the acceleration between 0 and 3 seconds.
   b. Calculate the acceleration between 3 and 10 seconds.
   c. Calculate the total distance travelled by the car in the first 10 seconds of its journey.

2. The speed-time graph shows the speed of a train over 60 seconds. 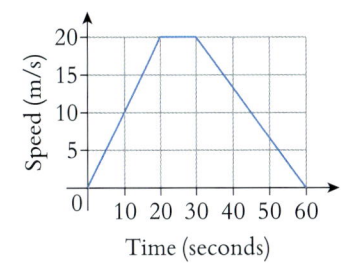
   a. Calculate the acceleration between 0 and 20 seconds.
   b. Calculate the deceleration between 30 and 60 seconds.
   c. Calculate the total distance travelled by the train.

3. The speed-time graph shows the speed of a cyclist over 60 seconds. 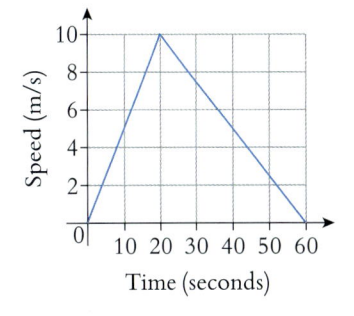
   a. Calculate the acceleration between 0 and 20 seconds.
   b. Calculate the deceleration between 20 and 60 seconds.
   c. Calculate the total distance travelled by the cyclist.
   d. Calculate the average speed of the cyclist over the 60 seconds.

4. The speed-time graph shows the speed of a car over the first 50 seconds of a journey. 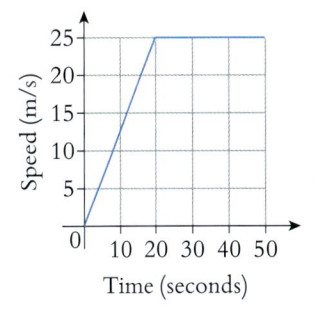
   a. Calculate the acceleration between 0 and 20 seconds.
   b. Calculate the distance travelled in the first 20 seconds.
   c. Calculate the total distance travelled by the car in the first 50 seconds of its journey.
   d. Calculate the average speed in the first 50 seconds of the journey.

Speed-time graphs

5 The speed-time graph shows the speed of a cyclist over the first 100 seconds of a journey. 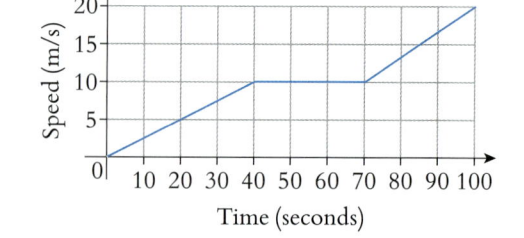
   a Calculate the acceleration between 0 and 40 seconds.
   b Calculate the acceleration between 70 and 100 seconds.
   c Calculate the distance travelled in the first 40 seconds.
   d Calculate the total distance travelled by the cyclist in the first 100 seconds of the journey.

6 The speed-time graph shows the first 60 seconds of a journey.

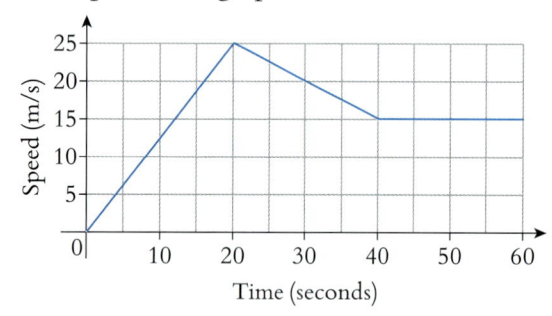

   a Calculate the acceleration between 0 and 20 seconds.
   b Calculate the acceleration between 20 and 40 seconds.
   c Calculate the distance travelled in the first 60 seconds of the journey.
   d Calculate the average speed in the first 60 seconds of the journey.

7 The speed-time graph shows the speed of a truck over a 3 minute journey. The truck travels a total distance of 2.5 km.

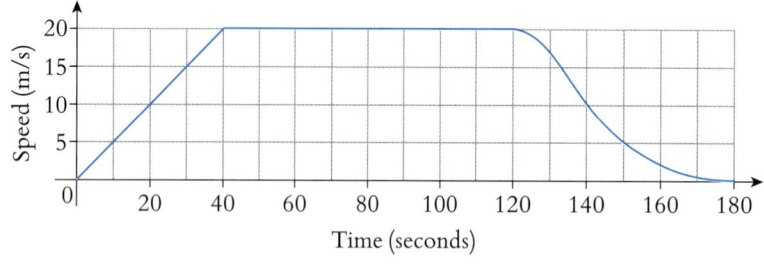

   a Calculate the acceleration between 0 and 40 seconds.
   b Calculate the distance travelled in the first 2 minutes of the journey.
   c Estimate the distance travelled by the truck while it was decelerating.

8 The speed-time graph shows the speed of a car and the speed of a cyclist over a period of 140 seconds.

   a Calculate the acceleration of the car during the first 40 seconds.
   b Calculate the deceleration of the car between 60 and 140 seconds.
   c Calculate the distance travelled by the cyclist in the 140 seconds.
   d Calculate the distance travelled by the car in the 140 seconds.
   e Calculate the distance travelled by the car while it was travelling faster than the cyclist.

9 The speed-time graph shows the speed for the first 7 minutes of a journey.

   Calculate the distance travelled, in kilometres, during the 7 minutes.

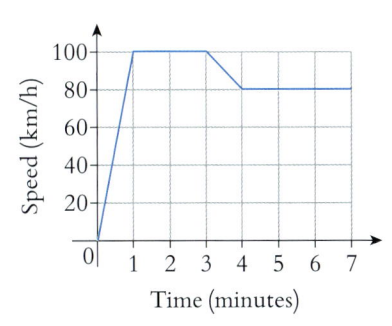

10 The speed-time graph shows the speed of a car over 80 seconds. The total distance travelled is 1.28 km.
   a Calculate the greatest speed of the car during the journey.
   b Calculate the average speed of the car for the journey.

11 The speed-time graph shows the speed of a motorbike for a journey lasting 35 seconds. The total distance travelled is 540 metres.
   a Calculate the greatest speed of the motorbike during the journey.
   b Calculate the acceleration between 0 and 10 seconds.

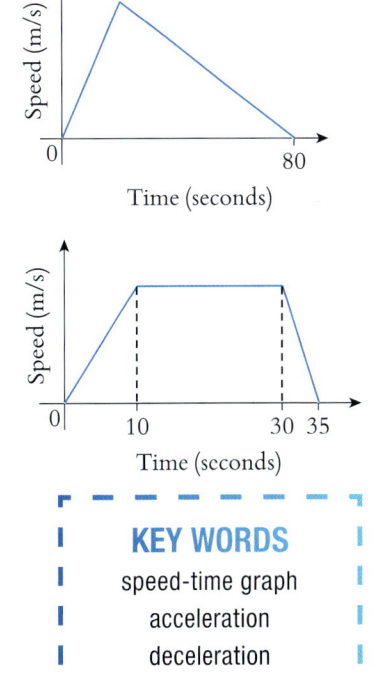

KEY WORDS
speed-time graph
acceleration
deceleration

Speed-time graphs 341

Rearranging formulae 2

THIS SECTION WILL SHOW YOU HOW TO
- Change the subject of a formula where the variable appears more than once

EXAMPLE

a Make x the subject of $ax + bx = c$

$ax + bx = c$ factorise
$x(a + b) = c$ divide both sides by $(a + b)$
$x = \dfrac{c}{a+b}$

b Make x the subject of $a(x - b) = c(x + d)$

$a(x - b) = c(x + d)$ expand the brackets
$ax - ab = cx + cd$ collect the x's on one side
$ax - cx = ab + cd$ factorise
$x(a - c) = ab + cd$ divide both sides by $(a - c)$
$x = \dfrac{ab+cd}{a-c}$

EXAMPLE

Make x the subject of $a = \dfrac{bx+c}{dx+e}$

$a = \dfrac{bx+c}{dx+e}$ multiply both sides by $(dx + e)$
$a(dx + e) = bx + c$ expand the brackets
$adx + ae = bx + c$ collect the x's on one side
$adx - bx = c - ae$ factorise
$x(ad - b) = c - ae$ divide both sides by $(ad - b)$
$x = \dfrac{c-ae}{ad-b}$

EXAMPLE

Make x the subject of $a = \sqrt{\dfrac{b-x}{x}}$

$a = \sqrt{\dfrac{b-x}{x}}$ square both sides
$a^2 = \dfrac{b-x}{x}$ multiply both sides by x
$a^2 x = b - x$ collect the x's on one side
$a^2 x + x = b$ factorise
$x(a^2 + 1) = b$ divide both sides by $(a^2 + 1)$
$x = \dfrac{b}{a^2+1}$

EXERCISE 7.3

Make x the subject of these formulae.

1. $px - qx = r$
2. $ax = b + cx$
3. $ax = bx + c$
4. $x = a + bx$
5. $x + a = bx + c$
6. $ax - b = cx + 5$
7. $3 - 2x = ax + b$
8. $a^2 + ax + bx = 0$
9. $a(x - 3) = b(x + 4)$
10. $7(ax + b) = 3(cx - d)$
11. $a = \dfrac{x-2}{x+3}$
12. $x = \dfrac{ax+b}{c}$
13. $e = \dfrac{dx+1}{2-3x}$
14. $\dfrac{ax+b}{2} = \dfrac{cx+d}{4}$
15. $\dfrac{x-4}{a} = \dfrac{x+3}{b}$
16. $a = \dfrac{\sqrt{x+bx}}{c}$
17. $a = \sqrt{\dfrac{2+x}{x}}$
18. $2\sqrt{\dfrac{x-p}{x}} = q$
19. $x^2 = a + bx^2$
20. $ax^2 + b = c - dx^2$
21. $f = \dfrac{xy}{x+y}$
22. $2a = \dfrac{a^2 x}{x+a}$

23. To make x the subject of the formula $a = \dfrac{b}{x} - c$ there are two possible methods:

Method 1

$a = \dfrac{b}{x} - c$ multiply by x

$ax = b - cx$ collect x's

$ax + cx = b$ factorise

$x(a + c) = b$ divide by $(a + c)$

$x = \dfrac{b}{a+c}$

Method 2

$a = \dfrac{b}{x} - c$ add c to both sides

$a + c = \dfrac{b}{x}$ multiply by x

$x(a + c) = b$ divide by $(a + c)$

$x = \dfrac{b}{a+c}$

a. $\dfrac{5}{x} + a = b$
b. $a - \dfrac{b}{x} = c$
c. $\dfrac{p}{x} + 5q = 2r$
d. $\dfrac{6f}{x} + g = 7h$
e. $\dfrac{a}{x} + 2 = \dfrac{b}{x} + 3c$
f. $\dfrac{a}{x} - b = c - \dfrac{d}{x}$

Rearranging formulae 2

Sequences

THIS SECTION WILL SHOW YOU HOW TO
- Write down and use the formula for the *n*th term of a sequence

Linear sequences
A **sequence** is a list of numbers or diagrams that are connected by a rule.

4, 7, 10, 13, … and … are sequences.
The numbers in a sequence are called the **terms** of the sequence.

The ***n*th term** is used to describe a general term in a sequence.

If the *n*th term = $5n + 3$
  1st term = $5 \times 1 + 6 = 11$
  2nd term = $5 \times 2 + 6 = 16$
  3rd term = $5 \times 3 + 6 = 21$
  4th term = $5 \times 4 + 6 = 26$
  The sequence is 11, 16, 21, 26, …

It is called a linear sequence because the differences between terms are all the same.
The rule to find the next term (the **term-to-term rule**) is +5 or add 5.

EXAMPLE

Find the *n*th term of these sequences
a 1, 5, 9, 13, … **b** 45, 38, 31, 24, …

a 1   5   9   13   The difference between the terms is 4.
   +4 +4 +4   The *n*th term formula will contain the expression $4n$.

Term position	1	2	3	4
Term	1	5	9	13

The rule is: multiply the term position by 4 and then subtract 3.

nth term = $4n - 3$

b 45   38   31   24   The difference between the terms is −7.
   −7 −7 −7   The *n*th term formula will contain the expression $-7n$

Term position	1	2	3	4
Term	45	38	31	24

The rule is: multiply the term position by −7 and then add 52.

nth term = $-7n + 52$ or nth term = $52 - 7n$

EXERCISE 7.4

1. Write down the next two terms for each of the following sequences.
   a 7, 11, 15, 19, …
   b 10, 15, 20, 25, …
   c 1, 2, 4, 8, 16, …
   d 8, 9.5, 11, 12.5, …
   e 20, 17, 14, 11, …
   f 2, 5, 12.5, 31.25, …
   g 108, 36, 12, 4, …
   h 100, 48, 22, 9, …
   i 1, 1, 2, 3, 5, 8, …
   j 1, 3, 7, 15, 31, …
   k −1, 4, 3, 7, 10, …
   l 0.01, 0.1, 1, 10, …

2. Write down the first 5 terms of the sequence whose nth term is
   a $n + 6$
   b $5n − 3$
   c $4n + 2$
   d $2n − 1$
   e $3n + 5$
   f $2n − 4$
   g $50 − 2n$
   h $16 − n$
   i $25 − 3n$
   j $0.5n + 2$
   k $30 − 0.5n$
   l $−4n + 7$

3.
   Diagram 1   Diagram 2   Diagram 3   Diagram 4

   Omar says the nth diagram has $n + 2$ squares.
   Is he correct? Explain your answer.

4.
   Diagram 1   Diagram 2   Diagram 3   Diagram 4
   1 cube   4 cubes   7 cubes   10 cubes
   How many cubes are in a diagram 5, b diagram 20, c diagram n?

5. Find the nth term for these linear sequences.
   a 7, 11, 15, 19, …
   b 0, 3, 6, 9, …
   c 8, 13, 18, 23, …
   d −4, −1, 2, 5, …
   e 2, 4.5, 7, 9.5, …
   f −8, −1, 6, 13, …
   g 1, 4, 7, 10, …
   h −1, 3, 7, 11, …
   i −2, 4, 10, 16, …
   j −5.5, −4, −2.5, −1, …
   k 8, 10.5, 13, 15.5, …
   l 6.02, 6.04, 6.06, 6.08, …

6. Find the nth term for these linear sequences.
   a 21, 19, 17, 15, …
   b 100, 96, 92, 88, …
   c 15, 8, 1, −6, …
   d 17, 14, 11, 8, …
   e 5, 3, 1, −1, …
   f 1.7, 1.3, 0.9, 0.5, …
   g −2, −5, −8, −11, …
   h 13, 11, 9, 7, …
   i 2, 1.5, 1, 0.5, …

7. nth term = $8n − 5$. Which term in the sequence has a value of 155?

8. nth term = $3n − 6$. Which term in the sequence has a value of 108?

9. nth term = $54 − 4n$. Which term in the sequence has a value of −46?

10. nth term = $\dfrac{7n − 3}{2}$. Which term in the sequence has a value of 219?

Non-linear sequences

The sequence 5, 7, 10, 14, … is called a non-linear sequence
+2 +3 +4

because the differences between the terms are not the same.
You should be able to recognise and use the following non-linear sequences:

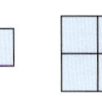

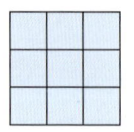

Square numbers   1, 4, 9, 16, … nth term = n^2

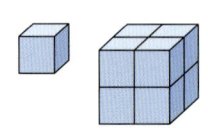

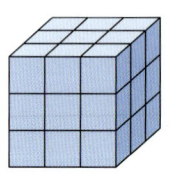

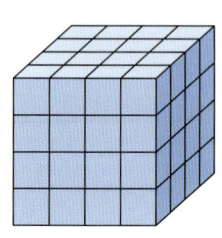

Cube numbers   1, 8, 27, 64, … nth term = n^3

Powers   For example 2, 4, 8, 16, … nth term = 2^n

EXAMPLE

Find the nth term of these sequences
a 1, 2, 4, 8, …   **b** 5, 8, 13, 20, …   **c** 2, 9, 28, 65, …

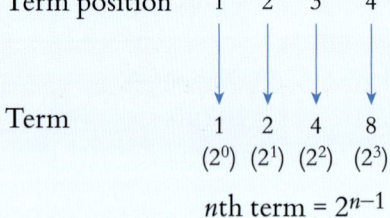

a Term position   1   2   3   4

Term   1   2   4   8
                 (2^0) (2^1) (2^2) (2^3)

The rule is: subtract 1 from the term position then do 2$^{(\text{term position} - 1)}$

nth term = 2^{n-1}

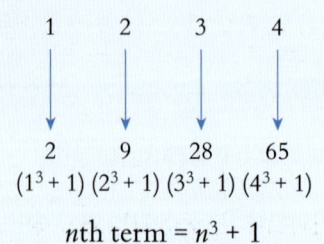

b Term position   1   2   3   4

Term   5   8   13 20
                 (1^2+4) (2^2+4) (3^2+4) (4^2+4)

The rule is: square the term position then add 4.

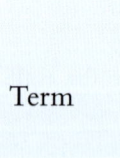

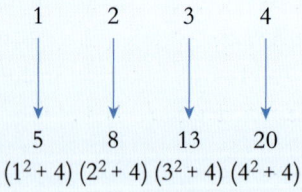

nth term = $n^2 + 4$

c Term position   1   2   3   4

Term   2   9   28 65
                 (1^3+1) (2^3+1) (3^3+1) (4^3+1)

The rule is: cube the term position and then add 1.

nth term = $n^3 + 1$

EXERCISE 7.5

1 Write down the first 5 terms of the non-linear sequence whose nth term is

   a $n^2 + 2$ **b** $n^2 - 3$ **c** $3n^2$ **d** $n(n-2)$

   e $n^3 + 1$ **f** $n^3 - 4$ **g** $\dfrac{n+2}{n}$ **h** $\dfrac{n(n+1)(n+2)}{6}$

2 Find the nth term for these non-linear sequences.

   a 1, 4, 9, 16, …   **b** 0, 3, 8, 15, …
   c 2, 8, 18, 32, …   **d** 3, 6, 11, 18, …
   e 11, 14, 19, 26, …   **f** 3, 12, 27, 48, …

HINT

these sequences are connected with square numbers

3 Find the nth term for these non-linear sequences.

   a 1, 8, 27, 64, 125, …   **b** 0, 7, 26, 63, 124, …
   c −1, 6, 25, 62, 123, …   **d** 2, 16, 54, 128, 250, …
   e 3, 10, 29, 66, 127, …

HINT

these sequences are connected with cube numbers

4 Find the nth term for these non linear sequences.

   a 10, 100, 1000, 10 000, …   **b** 2, 4, 8, 16, …
   c 3, 9, 27, 81, …   **d** 4, 8, 16, 32, …
   e 1, 3, 9, 27, …   **f** 0.1, 1, 10, 100, …

5 Find the nth term for the following sequences.

   a $1 \times 3, 2 \times 4, 3 \times 5, 4 \times 6, \ldots$ **b** $1 \times 2 \times 3, 2 \times 3 \times 4, 3 \times 4 \times 5, 4 \times 5 \times 6, \ldots$

   c $\dfrac{1}{6}, \dfrac{2}{7}, \dfrac{3}{8}, \dfrac{4}{9}, \ldots$ **d** $\dfrac{2}{5}, \dfrac{3}{10}, \dfrac{4}{17}, \dfrac{5}{26}, \ldots$

6 Write down the **next** term in each of the following sequences:

   a $128x^2, 64x^3, 32x^4, 16x^5, \ldots$ **b** $6x^3, 9x^5, 12x^7, 15x^9, \ldots$
   c $3x^2, 4x^3, 5x^4, 6x^5, \ldots$ **d** $x^6y^2, x^5y^3, x^4y^4, x^3y^5, \ldots$

7 The table shows the first four terms in three sequences A, B and C.

	Term 1	Term 2	Term 3	Term 4
Sequence A	1	8	27	64
Sequence B	4	7	10	13
Sequence C	−3	1	17	51

   a Find the nth term of sequence A.
   b Find the nth term of sequence B.
   c Use your answers to parts **a** and **b** to write down the nth term of sequence C.

Sequences

8 The diagram below shows how the fourth square pyramid number is made.

$1 + 4 + 9 + 16 = 30$

The fourth square pyramid number is 30.

Calculate the fifth and sixth square pyramid numbers.

9

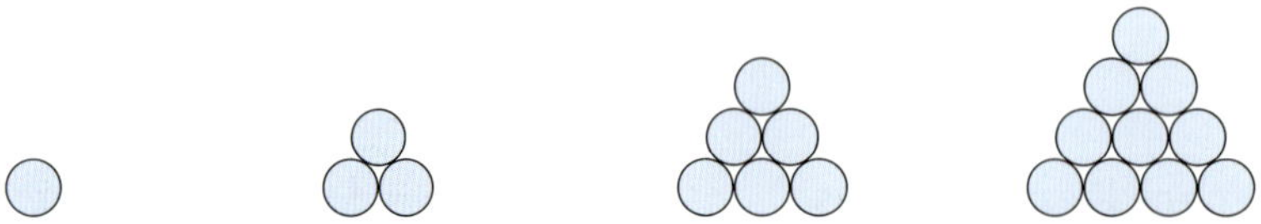

Diagram 1   Diagram 2   Diagram 3   Diagram 4

a How many circles are in diagram 4?
b Draw the next diagram in the sequence.
c Copy and complete the following table.

Diagram	1	2	3	4	5
Number of circles	$1 = \dfrac{1 \times 2}{2}$	$3 = \dfrac{2 \times 3}{2}$	$6 = \dfrac{3 \times 4}{2}$		

d How many circles are in diagram n?
e How many circles are in diagram 15?

10 The diagram shows four polygons. The diagonals for each polygon are shown.

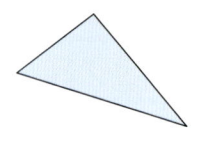

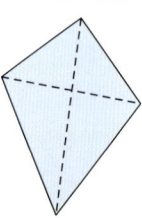

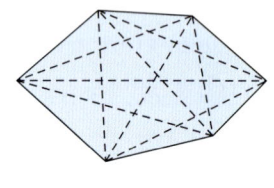

a Draw a polygon with seven sides. Show all the diagonals.
b Copy and complete the table.

Number of sides	3	4	5	6	7	8
Number of diagonals	0	2				

c How many diagonals are there in a 9-sided polygon?
d How many diagonals are there in a 20-sided polygon?
e Write down an expression for the number of diagonals in a polygon with n sides.

11 nth term $= 3n^2 + 7$. Which term in the sequence has a value of 3682?

12 nth term $= n^2 + 2n - 99$. Which term in the sequence has a value of 2501?

13 a Write down the next row for the number pattern.

b The sum of the numbers in row 6 is
$1 + 5 + 10 + 10 + 5 + 1 = 32 = 2^5$.
Copy and complete the table.

Row 1
Row 2
Row 3
Row 4
Row 5
Row 6

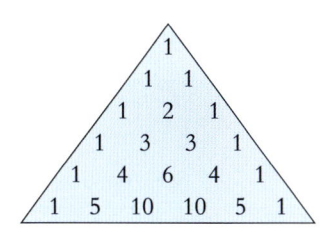

```
          1
         1 1
        1 2 1
       1 3 3 1
      1 4 6 4 1
     1 5 10 10 5 1
```

Row	1	2	3	4	5	6	7
Sum of numbers in row	$1 = 2^0$	$2 = 2^1$				$32 = 2^5$	

c Write down an expression for the sum of the numbers in row n.
d Which row has a sum of 32 768?

14

Year 1   Year 2   Year 3

The diagram shows the growth of a plant over three years.
Each year a flower is replaced by two stems and two flowers.

a Draw a diagram to show the plant in year 4.
b Copy and complete the table.

Year	1	2	3	4	5
Number of flowers	1	2	4		
Number of stems	1	3	7		

c How many flowers are there in year 6?
d How many flowers are there in year n?
e How many stems are there in year 6?
f How many stems are there in year n?

15 97, 75, 47, 40, …

The rule to find the next term in the sequence is: multiply together the digits of the last number and then add 12.
(To find the second term you work out $9 × 7 + 12 = 63 + 12 = 75$.)
What is the 100th number in the sequence?

Sequences

Further non-linear sequences

EXAMPLE

The nth term of the sequence 6, 11, 18, 27, ... is given by the formula
$$n\text{th term} = n^2 + bn + c$$
Find the values of b and c.

The 1st term is 6, so substitute $n = 1$ into the formula.
$$1^2 + b \times 1 + c = 6$$
$$b + c = 5$$
The 2nd term is 11, so substitute $n = 2$ into the formula.
$$2^2 + b \times 2 + c = 11$$
$$2b + c = 7$$
You now have a pair of simultaneous equations to solve.
$$2b + c = 7$$
$$b + c = 5$$
Subtracting gives $b = 2$ so $2 + c = 5$ so $c = 3$
$b = 2$ and $c = 3$

If you are asked to find the values of a, b and c when the nth term $= an^2 + bn + c$ for a given sequence then the problem is more complicated.
One method is to look at the differences in the terms.

For example   6   13   24   39   58

First difference   7   11   15   19

Second difference   4   4   4

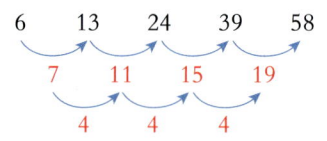

The second differences are all the same.
This means that it is a quadratic sequence.

A second difference of 2	⇒	the nth term will contain n^2
A second difference of 4	⇒	the nth term will contain $2n^2$
A second difference of 6	⇒	the nth term will contain $3n^2$

The sequence 6, 13, 24, 39, 58, ... has a second difference of 4 so the nth term will contain the term $2n^2$

So nth term $= 2n^2 + bn + c$
The 1st term is 6 $2 \times 1^2 + b \times 1 + c = 6$
                                  $b + c = 4$
The 2nd term is 13 $2 \times 2^2 + b \times 2 + c = 13$
                                  $2b + c = 5$
Simultaneous equations are $2b + c = 5$
                        and $b + c = 4$
Subtracting gives $b = 1$ so $1 + c = 4$ so $c = 3$
                                  $a = 2$, $b = 1$ and $c = 3$

EXERCISE 7.6

1 8, 14, 22, 32, ...
The nth term of the sequence is given by the formula nth term = $n^2 + bn + c$.
Find the values of b and c.

2 8, 16, 26, 38, 52, ...
The nth term of the sequence is given by the formula nth term = $n^2 + bn + c$.
Find the values of b and c.

3 3, 4, 7, 12, 19, ...
The nth term of the sequence is given by the formula nth term = $n^2 + bn + c$.
Find the values of b and c.

4 6, 15, 28, 45, 66, ...
The nth term of the sequence is given by the formula nth term = $2n^2 + bn + c$.
Find the values of b and c.

5 6, 16, 32, 54, 82, ...
The nth term of the sequence is given by the formula nth term = $3n^2 + bn + c$.
Find the values of b and c.

6 9, 20, 35, 54, 77, ...
The nth term of the sequence is given by the formula nth term = $an^2 + bn + c$.
Find the values of a, b and c.

7 4, 7, 14, 25, 40, ...
The nth term of the sequence is given by the formula nth term = $an^2 + bn + c$.
Find the values of a, b and c.

8 1, 11, 27, 49, 77, ...
The nth term of the sequence is given by the formula nth term = $an^2 + bn + c$.
Find the values of a, b and c.

9 4.5, 12, 20.5, 30, 40.5, ...
The nth term of the sequence is given by the formula nth term = $an^2 + bn + c$.
Find the values of a, b and c.

10 A sequence is made from sticks.

Diagram 1
4 sticks

Diagram 2
12 sticks

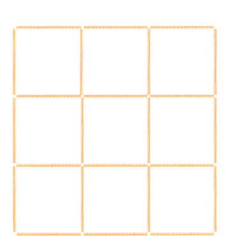

Diagram 3
24 sticks

How many sticks are needed to make:
  a diagram 4   b diagram 5   c diagram n?

11 A sequence is made from small squares.

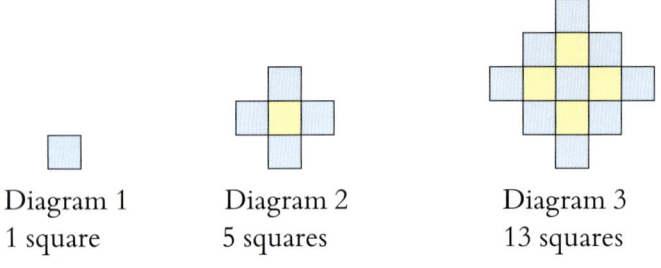

Diagram 1
1 square

Diagram 2
5 squares

Diagram 3
13 squares

How many small squares are needed to make:
  a diagram 4   b diagram 5   c diagram n?

12 The nth term of the sequence 2, 8, 20, 40, 70, …. is $\dfrac{n(n+1)(n+2)}{3}$.

Write down the nth term of the sequence 4, 16, 40, 80, 140, …

13 a Find the value of $1^2 + 2^2 + 3^2 + 4^2$.
   b The sum of the first n square numbers can be found using the formula
      $$1^2 + 2^2 + 3^2 + 4^2 + 5^2 + 6^2 + \ldots\ldots + n^2 = \dfrac{n(n+1)(2n+1)}{k}$$
      Use your answer to part **a** to find the value of k.

14 2, 21, 70, 161, 306, …

The nth term of the sequence is given by the formula nth term $= an^3 + 3n^2 + bn + 1$.
Find the values of a and b.

Further sequence notation

There are different ways of writing the terms of sequences, such as $u_1, u_2, u_3, u_4, \ldots$

EXAMPLE

A sequence is defined by $u_n = 3n + 5$.
Write down the first four terms of the sequence.

$u_1 = (3 \times 1) + 5 = 8$
$u_2 = (3 \times 2) + 5 = 11$
$u_3 = (3 \times 3) + 5 = 14$
$u_4 = (3 \times 4) + 5 = 17$

The first four terms of the sequence are 8, 11, 14 and 17.

EXAMPLE

$T_1 = 2 \qquad T_2 = 5 \qquad T_3 = 10 \qquad T_4 = 17$
Write down an expression for T_n

$T_1 = 1^2 + 1$
$T_2 = 2^2 + 1$
$T_3 = 3^2 + 1$
$T_4 = 4^2 + 1$
So $T_n = n^2 + 1$.

➡ **NOTE:** You should be able to recognise that the numbers 2, 5, 10 and 17 are connected to the square numbers 1, 4, 9 and 19

EXERCISE 7.7

1 Write down the first four terms of these sequences.
   a $u_n = 2n - 3$
   b $u_n = 5n + 1$
   c $u_n = 3n + 2$
   d $u_n = n^2$
   e $u_n = n^2 + 2$
   f $u_n = (n - 1)^2$
   g $u_n = n^3$
   h $u_n = n^3 - n^2$
   i $u_n = (n + 1)^3$
   j $u_n = 2^n$
   k $u_n = 3^{n-2}$
   l $u_n = 10^{n+3}$
   m $u_n = \dfrac{n}{n+2}$
   n $u_n = \dfrac{n+4}{n+5}$
   o $u_n = \dfrac{n^2}{2n+3}$

2 Find the formula for the nth term (u_n) for each of these sequences.
   a $u_1 = 4 \qquad u_2 = 9 \qquad u_3 = 14 \qquad u_4 = 19$
   b $u_1 = 8 \qquad u_2 = 11 \qquad u_3 = 14 \qquad u_4 = 17$
   c $u_1 = 20 \qquad u_2 = 17 \qquad u_3 = 14 \qquad u_4 = 11$
   d $u_1 = 0 \qquad u_2 = 3 \qquad u_3 = 8 \qquad u_4 = 15$
   e $u_1 = 1 \qquad u_2 = 3 \qquad u_3 = 9 \qquad u_4 = 27$
   f $u_1 = \dfrac{4}{7} \qquad u_2 = \dfrac{5}{9} \qquad u_3 = \dfrac{6}{11} \qquad u_4 = \dfrac{7}{13}$

KEY WORDS

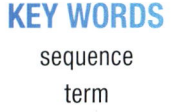

sequence
term
term-to-term rule
nth term

Exponential graphs

THIS SECTION WILL SHOW YOU HOW TO
- Recognise exponential graphs
- Draw exponential graphs
- Use exponential graphs in practical situations

Exponential functions are functions of the form $y = a^x$ where a is a positive constant. Examples of exponential functions include $y = 2^x$, $y = 3^x$, etc.
The graph of an exponential function has a characteristic shape.

EXAMPLE

a Draw the graph of $y = 2^x$ for $-3 \leq x \leq 3$.
b Use your graph to estimate the value of x when $2^x = 6$.
c By drawing a tangent to your graph estimate the gradient of the graph when $x = 1$.

a First make a table of values for x and y

when $x = -3$, $y = 2^{-3} = \dfrac{1}{2^3} = \dfrac{1}{8} = 0.125$ when $x = -2$, $y = 2^{-2} = \dfrac{1}{2^2} = \dfrac{1}{4} = 0.25$

when $x = -1$, $y = 2^{-1} = \dfrac{1}{2^1} = \dfrac{1}{2} = 0.5$ when $x = 0$, $y = 2^0 = 1$

when $x = 1$, $y = 2^1 = 2$ when $x = 2$, $y = 2^2 = 4$
when $x = 3$, $y = 2^3 = 8$

x	−3	−2	−1	0	1	2	3
y	0.125	0.25	0.5	1	2	4	8

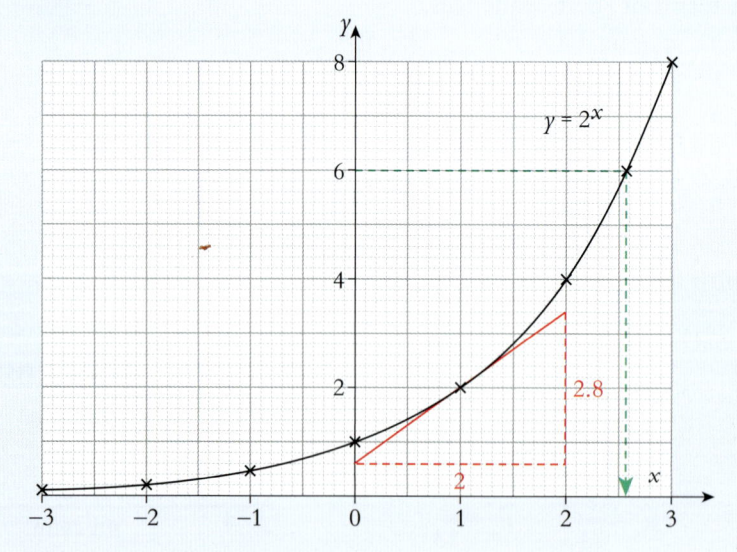

b When $2^x = 6$, $x \approx 2.6$

c When $x = 1$, gradient $\approx \dfrac{2.8}{2} = 1.4$

EXERCISE 7.8

For each of these functions make a table of values and draw the graph of the function between the stated x values.

1 $y = 3^x$ $-3 \leq x \leq 3$
2 $y = 3^{-x}$ $-3 \leq x \leq 3$

3 $y = 3^x + 1$ $-3 \leq x \leq 3$
4 $y = 3^x - 5$ $-3 \leq x \leq 3$

5 $y = 3^{x-2}$ $-1 \leq x \leq 5$
6 $y = 2^{-x}$ $-3 \leq x \leq 3$

7 $y = 2^x + 1$ $-3 \leq x \leq 3$
8 $y = 2^{x+1}$ $-4 \leq x \leq 2$

9 $y = 1 - 2^x$ $-3 \leq x \leq 3$
10 $y = \left(\dfrac{1}{2}\right)^x$ $-3 \leq x \leq 3$

11

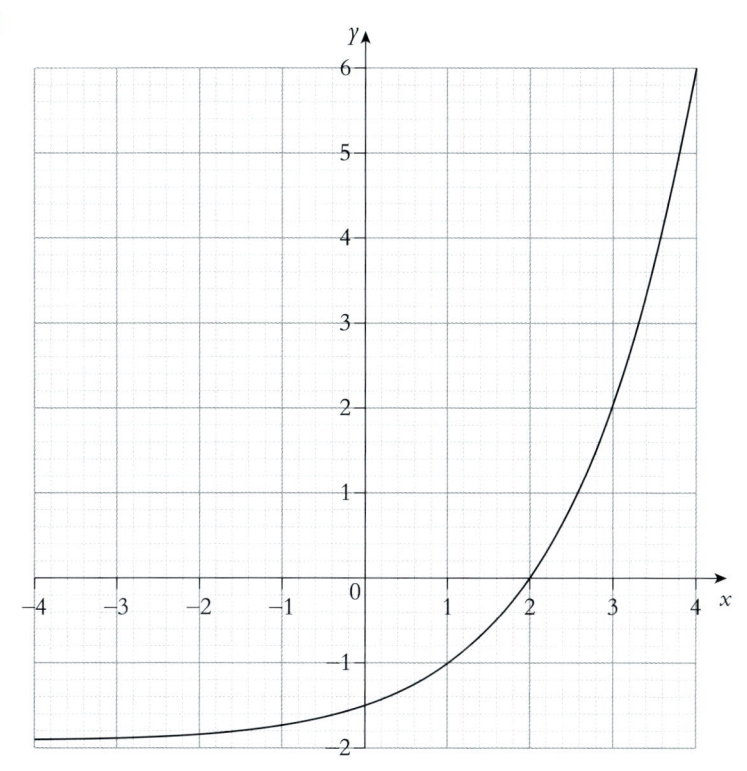

The diagram shows the accurate graph of $y = f(x)$
  a Use the graph to find
    i $f(3)$
    ii $f(0)$.
  b Use the graph to solve
    i $f(x) = 0$
    ii $f(x) = 4$.
  c Copy the graph and draw a tangent to your graph to estimate the gradient of the graph when $x = 3$

Practical applications of exponential graphs

Exponential curves are often used to model real-life situations.

EXAMPLE

A patch of mould is increasing in size.
The area (A cm^2) of the mould at time t days is given by the formula
$A = 0.9 \times (1.5)^t$.

a Use this formula to complete the table of values for t and A.

t	0	2	4	6	8	10
A	0.9					

b Draw the graph of $A = 0.9 \times (1.5)^t$
c Use your graph to estimate the number of days it will take for the area of the patch of mould to exceed 20 cm^2.

a when $t = 0$, $A = 0.9 \times 1.5^0 = 0.9$ when $t = 2$, $A = 0.9 \times 1.5^2 = 2.03$
   when $t = 4$, $A = 0.9 \times 1.5^4 = 4.56$ when $t = 6$, $A = 0.9 \times 1.5^6 = 10.3$
   when $t = 8$, $A = 0.9 \times 1.5^8 = 23.1$ when $t = 10$, $A = 0.9 \times 1.5^{10} = 51.9$

t	0	2	4	6	8	10
A	0.9	2.03	4.56	10.3	23.1	51.9

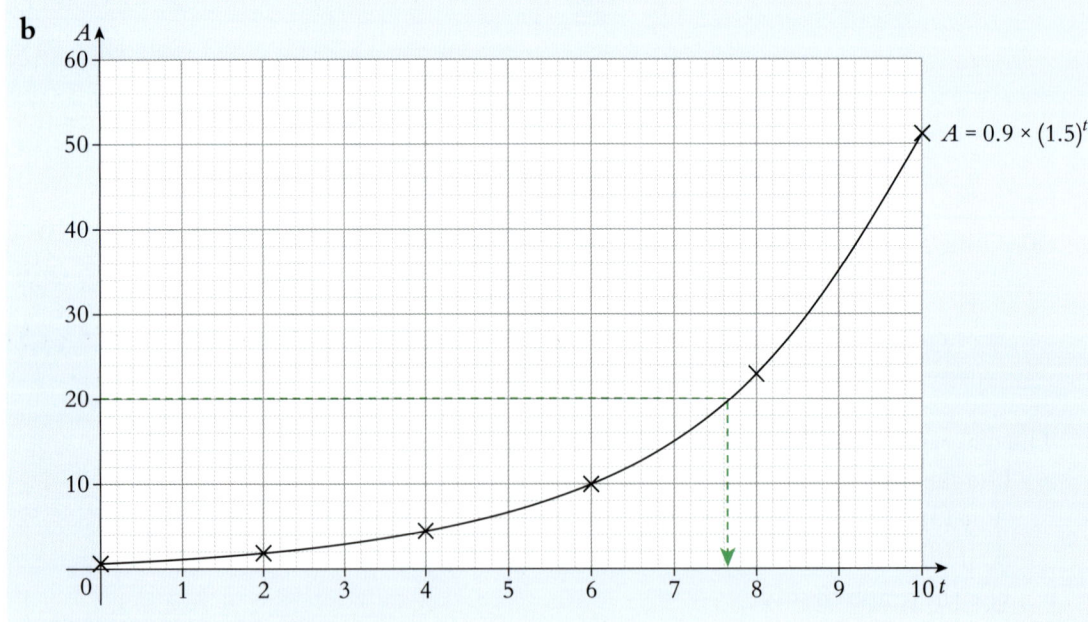

b

c When $A = 20$, $t \approx 7.65$ days.

EXERCISE 7.9

1 Kris invests $100 in a savings account.
   The number of dollars (y) in the savings account after x years is give by the formula
   $$y = 100 \times 1.1^x$$
   a Use this formula to copy and complete the table of values for x and y.

x	0	4	8	12	16	20
y	100	146				

   b Draw the graph of $y = 100 \times 1.1^x$
   c Use your graph to estimate how many years, to the nearest year, it takes for Kris to have $500.

2 A car is bought for $40 000. The value, $$y$, of the car after x years is given by the formula
   $$y = 40\,000 \times 0.9^x$$
   a Use this formula to copy and complete the table of values for x and y.

x	0	2	4	6	8	10
y	40 000	32 400				

   b Draw the graph of $y = 40\,000 \times 0.9^x$
   c Use your graph to estimate the number of years that it takes for the car to halve in value.

3 The population of lizards on an island is increasing.
   The population, P, over time t years, is given by the formula
   $$P = 250 \times (1.2)^t$$
   a Use this formula to copy and complete the table of values for t and P.

t	0	1	2	3	4	5	6
P	250						

   b Draw the graph of $P = 250 \times (1.2)^t$
   c Use your graph to estimate the number of years that it takes for the population to double.

4 The population of an endangered species of bird on an island is decreasing.
   The population, P, over time t years, is given by the formula
   $$P = 100 \times (0.75)^t$$
   a Use this formula to copy and complete the table of values for t and P.

t	0	1	2	3	4	5	6	7	8
P	100								

   a Draw the graph of $P = 100 \times (0.75)^t$
   b Use your graph to estimate the number of years that it takes for the population to fall below 20.

KEY WORDS
exponential

Matrices and transformations

THIS SECTION WILL SHOW YOU HOW TO
- Transform a shape using a matrix
- Find the matrix for a transformation

You can transform a shape using a matrix.

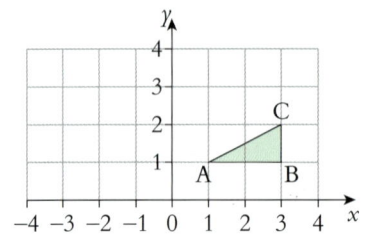

To transform △ABC using the **transformation matrix** $\begin{pmatrix} -1 & 0 \\ 0 & 1 \end{pmatrix}$

STEP 1

Write the coordinates of A, B and C as the column vectors $\begin{pmatrix} 1 \\ 1 \end{pmatrix}, \begin{pmatrix} 3 \\ 1 \end{pmatrix}$ and $\begin{pmatrix} 3 \\ 2 \end{pmatrix}$

and combine these column vectors to make the matrix $\begin{matrix} A & B & C \\ \begin{pmatrix} 1 & 3 & 3 \\ 1 & 1 & 2 \end{pmatrix} \end{matrix}$

STEP 2

Multiply the matrices $\begin{pmatrix} -1 & 0 \\ 0 & 1 \end{pmatrix} \begin{matrix} A & B & C \\ \begin{pmatrix} 1 & 3 & 3 \\ 1 & 1 & 2 \end{pmatrix} \end{matrix} = \begin{matrix} A' & B' & C' \\ \begin{pmatrix} -1 & -3 & -3 \\ 1 & 1 & 2 \end{pmatrix} \end{matrix}$

STEP 3

Plot the points A′(−1, 1), B′(−3, 1) and C′(−3, 2) on the graph.

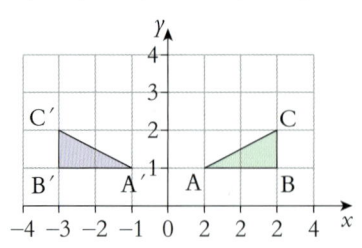

STEP 4

Describe the transformation.

The matrix $\begin{pmatrix} -1 & 0 \\ 0 & 1 \end{pmatrix}$ represents a reflection in the y-axis.

Reminder
- To describe a **reflection** you must state the mirror line.
- To describe a **rotation** you must state the direction of rotation, the angle of rotation and the centre of rotation.
- To describe an **enlargement** you must state the scale factor of the enlargement and the centre of enlargement.
- To describe a **stretch** you must state the stretch factor and the invariant line.
- To describe a **shear** you must state the shear factor and the invariant line.

EXAMPLE

A(1, 1), B(4, 1), C(3, 3) and D(1, 2).

Transform quadrilateral ABCD using the matrix $\begin{pmatrix} 0 & -1 \\ 1 & 0 \end{pmatrix}$.

Describe the transformation fully.

$$\begin{pmatrix} 0 & -1 \\ 1 & 0 \end{pmatrix} \begin{pmatrix} A & B & C & D \\ 1 & 4 & 3 & 1 \\ 1 & 1 & 3 & 2 \end{pmatrix} = \begin{pmatrix} A' & B' & C' & D' \\ -1 & -1 & -3 & -2 \\ 1 & 4 & 3 & 1 \end{pmatrix}$$

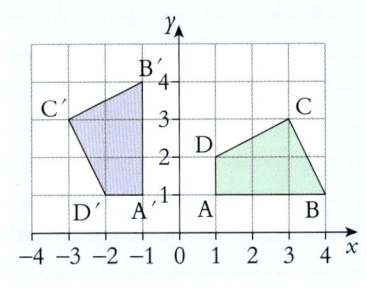

The transformation is a rotation, 90° anticlockwise about the point (0, 0).

EXERCISE 7.10

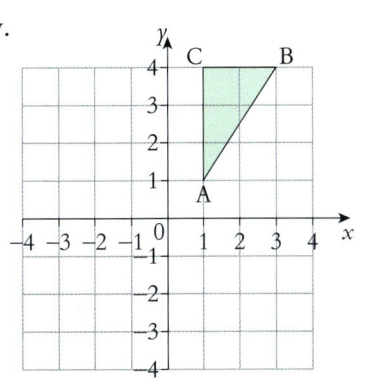

1 Transform triangle ABC using each of the matrices given below.
Label your image A'B'C'.
Describe fully the single transformation that moves ABC onto A'B'C'.
Draw a new diagram for each part of the question.

a $\begin{pmatrix} 0 & 1 \\ 1 & 0 \end{pmatrix}$ b $\begin{pmatrix} 1 & 0 \\ 0 & -1 \end{pmatrix}$ c $\begin{pmatrix} 0 & -1 \\ -1 & 0 \end{pmatrix}$

Matrices and transformations

2 Transform rectangle ABCD using each of the matrices given below.
Label your image A′B′C′D′.
Describe fully the single transformation that moves ABCD onto A′B′C′D′.
Draw a new diagram for each part of the question.

a $\begin{pmatrix} -1 & 0 \\ 0 & -1 \end{pmatrix}$ b $\begin{pmatrix} 0 & 1 \\ -1 & 0 \end{pmatrix}$ c $\begin{pmatrix} 0.8 & 0.6 \\ -0.6 & 0.8 \end{pmatrix}$

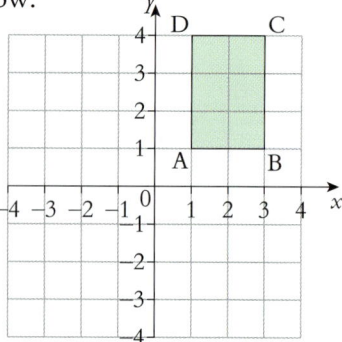

3 Plot the points A (2, 1), B(3, 4), C(1, 4) and D(2, 3) on a diagram.

Transform quadrilateral ABCD using the matrix $\begin{pmatrix} 2 & 0 \\ 0 & 2 \end{pmatrix}$
Label your image A′B′C′D′.
Describe fully the single transformation that moves ABCD onto A′B′C′D′.

4 Plot the points A(4, 1), B(3, 4), C(1, 4) and D(0, 2) on a diagram.

Transform quadrilateral ABCD using the matrix $\begin{pmatrix} 3 & 0 \\ 0 & 3 \end{pmatrix}$.
Label your image A′B′C′D′.
Describe fully the single transformation that moves ABCD onto A′B′C′D′.

5 Plot the points A(8, 2), B(4, 6) and C(2, 4) on a diagram.

Transform triangle ABC using the matrix $\begin{pmatrix} \frac{1}{2} & 0 \\ 0 & \frac{1}{2} \end{pmatrix}$.
Label your image A′B′C′.
Describe fully the single transformation that moves ABC onto A′B′C′.

6 Plot the points A(8, 2), B(4, 6), C(2, 6) and D(2, 4) on a diagram.

Transform quadrilateral ABCD using the matrix $\begin{pmatrix} -\frac{1}{2} & 0 \\ 0 & -\frac{1}{2} \end{pmatrix}$.
Label your image A′B′C′D′.
Describe fully the single transformation that moves ABCD onto A′B′C′D′.

7 Plot the points A(4, 0), B(3, 4), C(0, 2) and D(3, 2) on a diagram.

Transform quadrilateral ABCD using the matrix $\begin{pmatrix} -2 & 0 \\ 0 & -2 \end{pmatrix}$.
Label your image A′B′C′D′.
Describe fully the single transformation that moves ABCD onto A′B′C′D′.

8 Describe the transformation represented by the matrix $\begin{pmatrix} k & 0 \\ 0 & k \end{pmatrix}$.

9 Transform square ABCD using each of the matrices given below.
Label your image A′B′C′D′.
Describe fully the single transformation that moves ABCD onto A′B′C′D′.
Draw a new diagram for each part of the question.

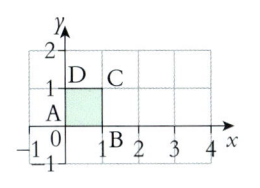

a $\begin{pmatrix} 1 & 1 \\ 0 & 1 \end{pmatrix}$ **b** $\begin{pmatrix} 1 & 2 \\ 0 & 1 \end{pmatrix}$ **c** $\begin{pmatrix} 1 & 3 \\ 0 & 1 \end{pmatrix}$

10 Describe the transformation represented by the matrix $\begin{pmatrix} 1 & k \\ 0 & 1 \end{pmatrix}$.

11 Transform square ABCD using each of the matrices given below.
Label your image A′B′C′D′.
Describe fully the single transformation that moves ABCD onto A′B′C′D′.
Draw a new diagram for each part of the question.

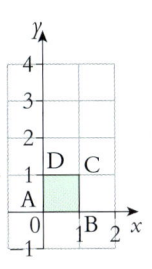

a $\begin{pmatrix} 1 & 0 \\ 1 & 1 \end{pmatrix}$ **b** $\begin{pmatrix} 1 & 0 \\ 2 & 1 \end{pmatrix}$ **c** $\begin{pmatrix} 1 & 0 \\ 3 & 1 \end{pmatrix}$

12 Describe the transformation represented by the matrix $\begin{pmatrix} 1 & 0 \\ k & 1 \end{pmatrix}$.

13 Transform square ABCD using each of the matrices given below.
Label your image A′B′C′D′.
Describe fully the single transformation that moves ABCD onto A′B′C′D′.
Draw a new diagram for each part of the question.

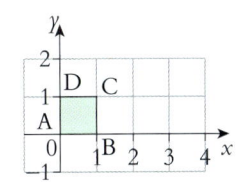

a $\begin{pmatrix} 2 & 0 \\ 0 & 1 \end{pmatrix}$ **b** $\begin{pmatrix} 3 & 0 \\ 0 & 1 \end{pmatrix}$ **c** $\begin{pmatrix} 4 & 0 \\ 0 & 1 \end{pmatrix}$

14 Describe the transformation represented by the matrix $\begin{pmatrix} k & 0 \\ 0 & 1 \end{pmatrix}$.

Matrices and transformations   361

15 Transform square ABCD using each of the matrices given below.
Label your image A′B′C′D′.
Describe fully the single transformation that moves ABCD onto A′B′C′D′.
Draw a new diagram for each part of the question.

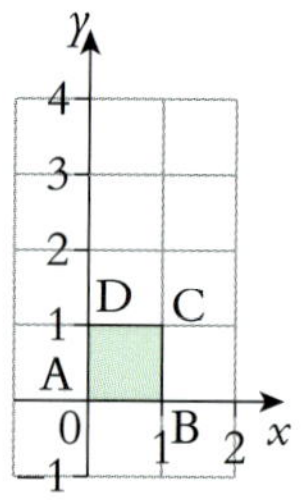

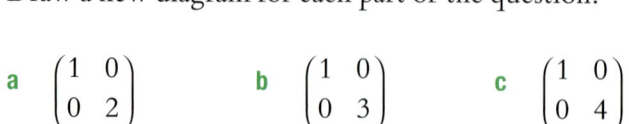

a $\begin{pmatrix} 1 & 0 \\ 0 & 2 \end{pmatrix}$ b $\begin{pmatrix} 1 & 0 \\ 0 & 3 \end{pmatrix}$ c $\begin{pmatrix} 1 & 0 \\ 0 & 4 \end{pmatrix}$

16 Describe the transformation represented by the matrix $\begin{pmatrix} 1 & 0 \\ 0 & k \end{pmatrix}$.

Finding the matrix for a transformation

$$\begin{pmatrix} a & b \\ c & d \end{pmatrix}\begin{pmatrix} 1 \\ 0 \end{pmatrix} = \begin{pmatrix} a \\ c \end{pmatrix} \text{ and } \begin{pmatrix} a & b \\ c & d \end{pmatrix}\begin{pmatrix} 0 \\ 1 \end{pmatrix} = \begin{pmatrix} b \\ d \end{pmatrix}$$

This means that the matrix $\begin{pmatrix} a & b \\ c & d \end{pmatrix}$ maps the vector $\begin{pmatrix} 1 \\ 0 \end{pmatrix} \rightarrow \begin{pmatrix} a \\ c \end{pmatrix}$

and the vector $\begin{pmatrix} 0 \\ 1 \end{pmatrix} \rightarrow \begin{pmatrix} b \\ d \end{pmatrix}$

So, if you know where the vectors $\begin{pmatrix} 1 \\ 0 \end{pmatrix}$ and $\begin{pmatrix} 0 \\ 1 \end{pmatrix}$ go you can write down the matrix for the transformation.

To do this you always transform the 'unit square'.

EXAMPLE

Find the transformation matrix for a reflection in the x-axis.

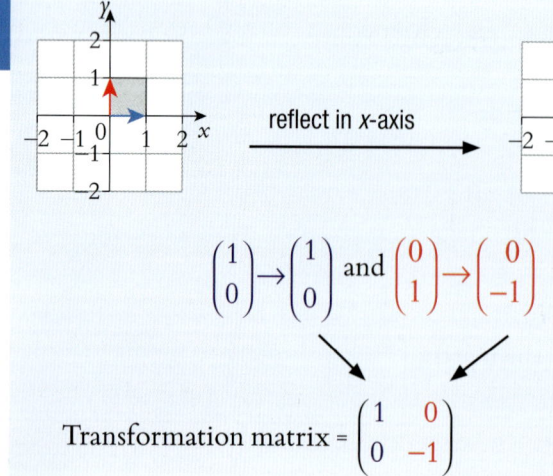

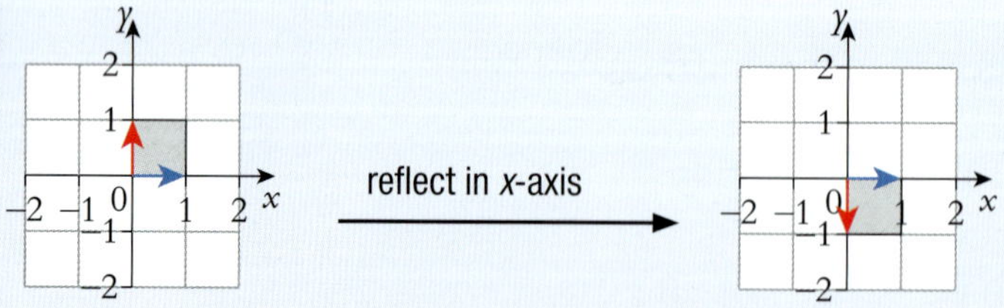

reflect in x-axis

$\begin{pmatrix} 1 \\ 0 \end{pmatrix} \rightarrow \begin{pmatrix} 1 \\ 0 \end{pmatrix}$ and $\begin{pmatrix} 0 \\ 1 \end{pmatrix} \rightarrow \begin{pmatrix} 0 \\ -1 \end{pmatrix}$

Transformation matrix = $\begin{pmatrix} 1 & 0 \\ 0 & -1 \end{pmatrix}$

EXAMPLE

Find the transformation matrix for a rotation of 180° about the point (0, 0).

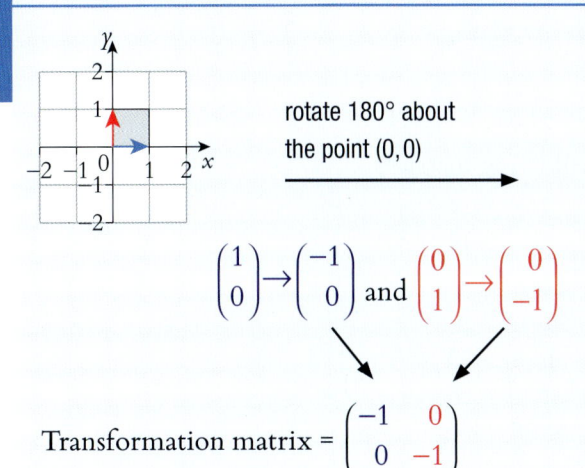

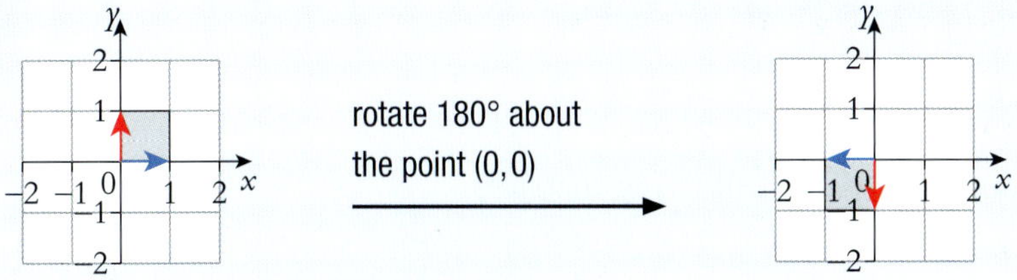

$$\begin{pmatrix}1\\0\end{pmatrix} \to \begin{pmatrix}-1\\0\end{pmatrix} \text{ and } \begin{pmatrix}0\\1\end{pmatrix} \to \begin{pmatrix}0\\-1\end{pmatrix}$$

Transformation matrix = $\begin{pmatrix}-1 & 0\\0 & -1\end{pmatrix}$

EXAMPLE

Find the transformation matrix for a shear, shear factor 2 with the x-axis invariant.

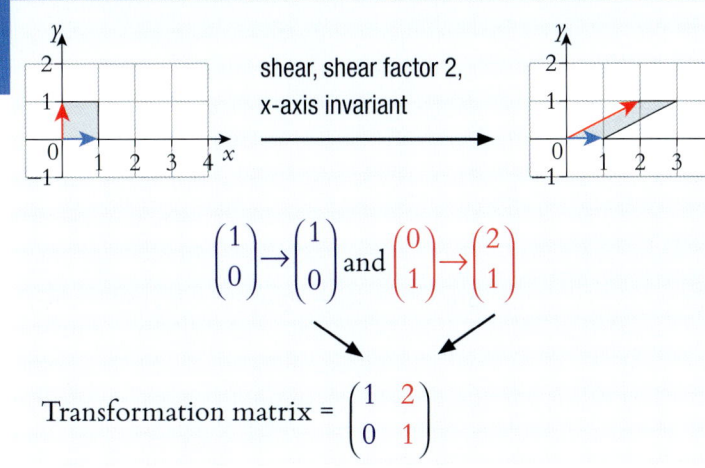

$$\begin{pmatrix}1\\0\end{pmatrix} \to \begin{pmatrix}1\\0\end{pmatrix} \text{ and } \begin{pmatrix}0\\1\end{pmatrix} \to \begin{pmatrix}2\\1\end{pmatrix}$$

Transformation matrix = $\begin{pmatrix}1 & 2\\0 & 1\end{pmatrix}$

EXERCISE 7.11

Find the matrix for each of the following transformations.

1. Rotation 90° clockwise about the point (0, 0).
2. Stretch with stretch factor 2 and the x-axis invariant.
3. Reflection in the x-axis.
4. Enlargement with scale factor 2, centre (0, 0).
5. Rotation 90° anticlockwise about the point (0, 0).
6. Enlargement with scale factor $\frac{1}{3}$, centre (0, 0).
7. Reflection in $y = x$.

Matrices and transformations

8 Shear with shear factor 4 and the *y*-axis invariant.

9 Stretch with stretch factor 3 and the *y*-axis invariant.

10 Reflection in *y* = −*x*.

Describe the transformation for each of the following matrices.

11 $\begin{pmatrix} 3 & 0 \\ 0 & 3 \end{pmatrix}$ 12 $\begin{pmatrix} 2 & 0 \\ 0 & 1 \end{pmatrix}$ 13 $\begin{pmatrix} 1 & 0 \\ 0 & -1 \end{pmatrix}$ 14 $\begin{pmatrix} \frac{1}{2} & 0 \\ 0 & \frac{1}{2} \end{pmatrix}$

15 $\begin{pmatrix} 0 & 1 \\ 1 & 0 \end{pmatrix}$ 16 $\begin{pmatrix} 1 & 5 \\ 0 & 1 \end{pmatrix}$ 17 $\begin{pmatrix} 4 & 0 \\ 0 & 4 \end{pmatrix}$ 18 $\begin{pmatrix} 0 & -1 \\ 1 & 0 \end{pmatrix}$

19 $\begin{pmatrix} 1 & 0 \\ 0 & 4 \end{pmatrix}$ 20 $\begin{pmatrix} -2 & 0 \\ 0 & -2 \end{pmatrix}$ 21 $\begin{pmatrix} 1 & 0 \\ 4 & 1 \end{pmatrix}$ 22 $\begin{pmatrix} 1 & 0 \\ 0 & -3 \end{pmatrix}$

23 Write down the matrix that maps
   a triangle A onto triangle B,
   b triangle B onto triangle C,
   c triangle A onto triangle C,
   d triangle C onto triangle A.

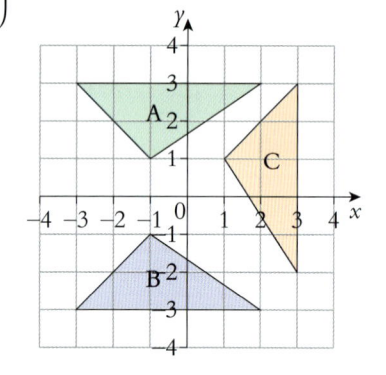

24 Write down the matrix that maps
   a triangle A onto triangle B,
   b triangle A onto triangle C,
   c triangle C onto triangle A,
   d triangle A onto triangle D.

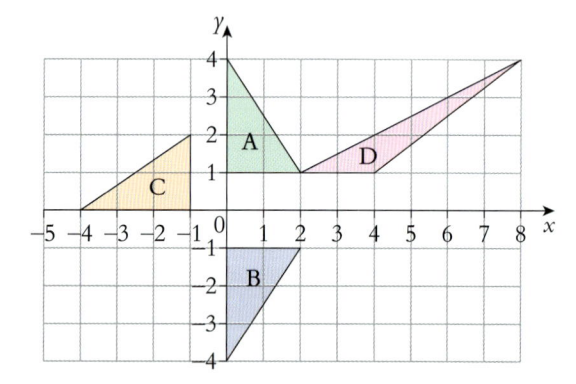

25 Write down the matrix that maps
   a triangle A onto triangle B,
   b triangle A onto triangle C,
   c triangle B onto triangle C,
   d triangle A onto triangle D,
   e triangle D onto triangle A.

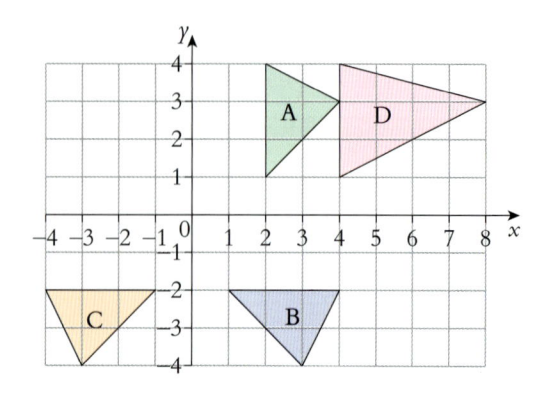

26 Write down the matrix that maps
  a triangle A onto triangle B,
  b triangle A onto triangle C,
  c triangle A onto triangle D,
  d triangle D onto triangle A,
  e triangle C onto triangle A.

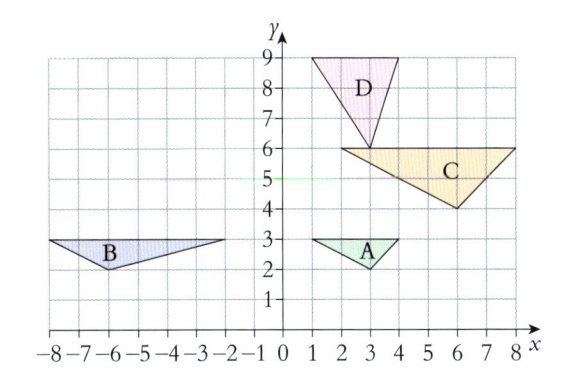

Combined transformations

EXAMPLE

A(2, 1), B(5, 1) and C(5, 3).

  a Transform $\triangle ABC$ using the matrix $\begin{pmatrix} 0 & 1 \\ 1 & 0 \end{pmatrix}$. Label this $A_1B_1C_1$.

  b Transform $\triangle A_1B_1C_1$ using the matrix $\begin{pmatrix} 0 & -1 \\ 1 & 0 \end{pmatrix}$. Label this $A_2B_2C_2$.

  c i Describe fully the single transformation which maps $\triangle ABC$ onto $\triangle A_2B_2C_2$.
     ii Find the 2 by 2 matrix which represents the transformation which maps $\triangle ABC$ onto $\triangle A_2B_2C_2$.

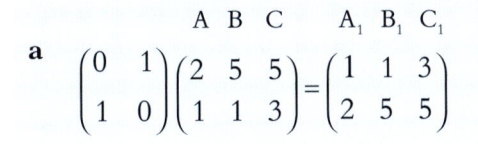

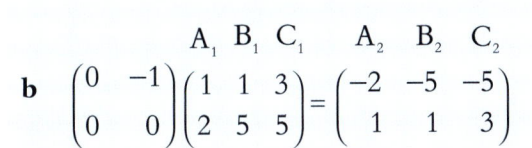

---

  a
$$\begin{matrix} & A & B & C \\ & & & \end{matrix}$$
$$\begin{pmatrix} 0 & 1 \\ 1 & 0 \end{pmatrix} \begin{pmatrix} 2 & 5 & 5 \\ 1 & 1 & 3 \end{pmatrix} = \begin{pmatrix} 1 & 1 & 3 \\ 2 & 5 & 5 \end{pmatrix} \begin{matrix} A_1 & B_1 & C_1 \end{matrix}$$

  b
$$\begin{pmatrix} 0 & -1 \\ 0 & 0 \end{pmatrix} \begin{pmatrix} 1 & 1 & 3 \\ 2 & 5 & 5 \end{pmatrix} = \begin{pmatrix} -2 & -5 & -5 \\ 1 & 1 & 3 \end{pmatrix}$$
with columns labelled $A_1\ B_1\ C_1$ and $A_2\ B_2\ C_2$.

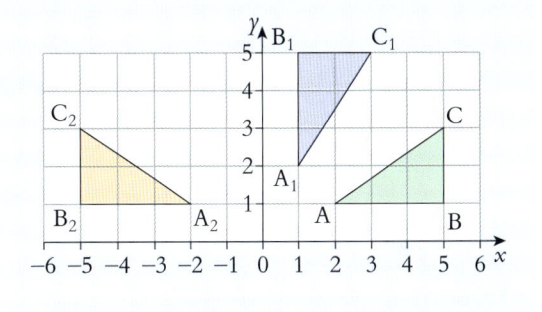

  c i The transformation which maps $\triangle ABC$ onto $\triangle A_2B_2C_2$ is a reflection in the y-axis.

     ii $\begin{pmatrix} 1 \\ 0 \end{pmatrix} \rightarrow \begin{pmatrix} -1 \\ 0 \end{pmatrix}$ and $\begin{pmatrix} 0 \\ 1 \end{pmatrix} \rightarrow \begin{pmatrix} 0 \\ 1 \end{pmatrix}$

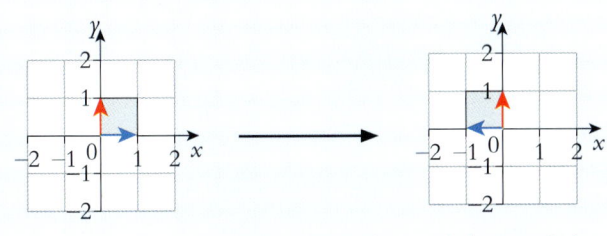

       Matrix = $\begin{pmatrix} -1 & 0 \\ 0 & 1 \end{pmatrix}$

Matrices and transformations

Notation

If you have a shape A and a transformation matrix R:
  R(A) means transform shape A using the matrix R.

If you have a shape A and two transformation matrices R and M:
  MR(A) means transform shape A using the matrix R
  then transform the image using the matrix M.

EXAMPLE

$R = \begin{pmatrix} -1 & 0 \\ 0 & -1 \end{pmatrix}$ and $E = \begin{pmatrix} -2 & 0 \\ 0 & -2 \end{pmatrix}$

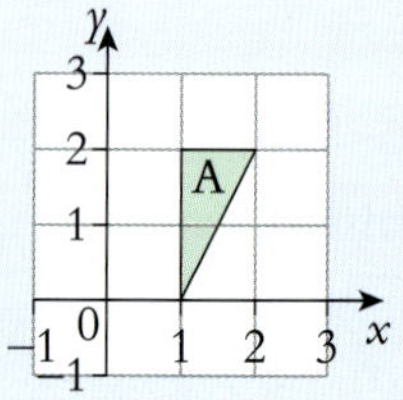

a Transform triangle A using the matrix R. Label this R(A). Describe the transformation represented by the matrix R.
b Transform triangle R(A) using the matrix E. Label this ER(A). Describe the transformation represented by the matrix E.
c **i** Describe fully the single transformation which maps triangle A onto ER(A).
  ii Find the 2 by 2 matrix which represents the transformation which maps triangle A onto ER(A).

---

a $\begin{pmatrix} -1 & 0 \\ 0 & -1 \end{pmatrix}\begin{pmatrix} 1 & 1 & 2 \\ 0 & 2 & 2 \end{pmatrix} = \begin{pmatrix} -1 & -1 & -2 \\ 0 & -2 & -2 \end{pmatrix}$

The matrix R represents a rotation of 180° about the point (0, 0).

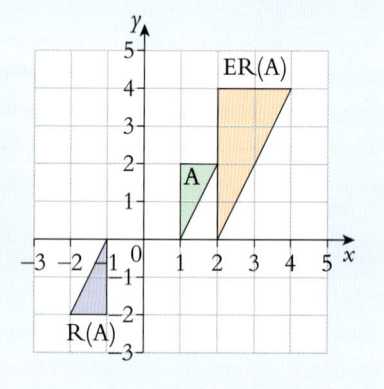

b $\begin{pmatrix} -2 & 0 \\ 0 & -2 \end{pmatrix}\begin{pmatrix} -1 & -1 & -2 \\ 0 & -2 & -2 \end{pmatrix} = \begin{pmatrix} 2 & 2 & 4 \\ 0 & 4 & 4 \end{pmatrix}$

The matrix E represents an enlargement, scale factor −2, centre (0, 0).

c **i** The transformation which maps triangle A onto ER(A) is an enlargement, scale factor 2, centre (0, 0).
  ii

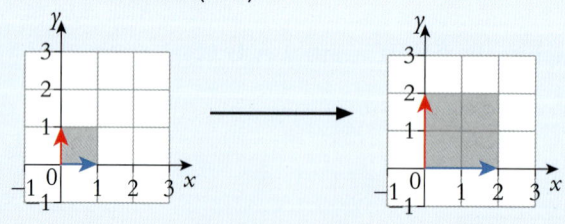

$\begin{pmatrix} 1 \\ 0 \end{pmatrix} \rightarrow \begin{pmatrix} 2 \\ 0 \end{pmatrix}$ and $\begin{pmatrix} 0 \\ 1 \end{pmatrix} \rightarrow \begin{pmatrix} 0 \\ 2 \end{pmatrix}$

Matrix $= \begin{pmatrix} 2 & 0 \\ 0 & 2 \end{pmatrix}$

EXERCISE 7.12

1. A(1, 1), B(4, 1) and C(2, 3).
   a. Transform △ABC using the matrix $\begin{pmatrix} 1 & 0 \\ 0 & -1 \end{pmatrix}$.
      Label this $A_1B_1C_1$.

   b. Transform $\triangle A_1B_1C_1$ using the matrix $\begin{pmatrix} -1 & 0 \\ 0 & 1 \end{pmatrix}$.
      Label this $A_2B_2C_2$.

   c. i. Describe fully the single transformation which maps △ABC onto $\triangle A_2B_2C_2$.
      ii. Find the 2 by 2 matrix which represents the transformation which maps △ABC onto $\triangle A_2B_2C_2$.

2. A(0, 2), B(1, 1) and C(3, 2).
   a. Transform △ABC using the matrix $\begin{pmatrix} 0 & 1 \\ -1 & 0 \end{pmatrix}$.
      Label this $A_1B_1C_1$.

   b. Transform $\triangle A_1B_1C_1$ using the matrix $\begin{pmatrix} -1 & 0 \\ 0 & -1 \end{pmatrix}$.
      Label this $A_2B_2C_2$.

   c. i. Describe fully the single transformation which maps △ABC onto $\triangle A_2B_2C_2$.
      ii. Find the 2 by 2 matrix which represents the transformation which maps △ABC onto $\triangle A_2B_2C_2$.

3. Rectangle X has vertices (1, 1), (2, 1), (2, 3) and (1, 3).

   a. Transform rectangle X using the matrix $R = \begin{pmatrix} 0 & -1 \\ 1 & 0 \end{pmatrix}$.
      Label this R(X).

   b. Describe fully the single transformation that maps
      i. X onto R(X)   ii. R(X) onto X.

   c. Find R^{-1}, the inverse of the matrix R.

   d. Transform R(X) by the matrix R^{-1}. Comment on your results.

4. Triangle A has vertices (−3, 1), (−9, 1) and (−7, 4).

   a. Transform triangle A using the matrix $R = \begin{pmatrix} 0 & 1 \\ -1 & 0 \end{pmatrix}$.
      Label this R(A).

   b. Transform triangle R(A) using the matrix $M = \begin{pmatrix} 0 & 1 \\ 1 & 0 \end{pmatrix}$.
      Label this MR(A).

   c. Describe fully the single transformation which maps triangle A onto MR(A).

> **KEY WORDS**
> transformation matrix

Cumulative frequency

THIS SECTION WILL SHOW YOU HOW TO
- Construct and use cumulative frequency diagrams
- Estimate and interpret the median, percentiles, quartiles and inter-quartile range

Cumulative frequency tables and diagrams

You can construct a **cumulative frequency** table from a grouped frequency table.

The grouped frequency table below shows information about the times taken, t seconds, for 25 students to solve a puzzle.

The cumulative frequency (cf) of the data is found by calculating the running total of the frequencies (f).

For example, to find the number of people taking twenty seconds or less ($t \leq 20$) you add 3 and 6. To find the number of people taking thirty seconds or less ($t \leq 30$) you add 3, 6 and 10.

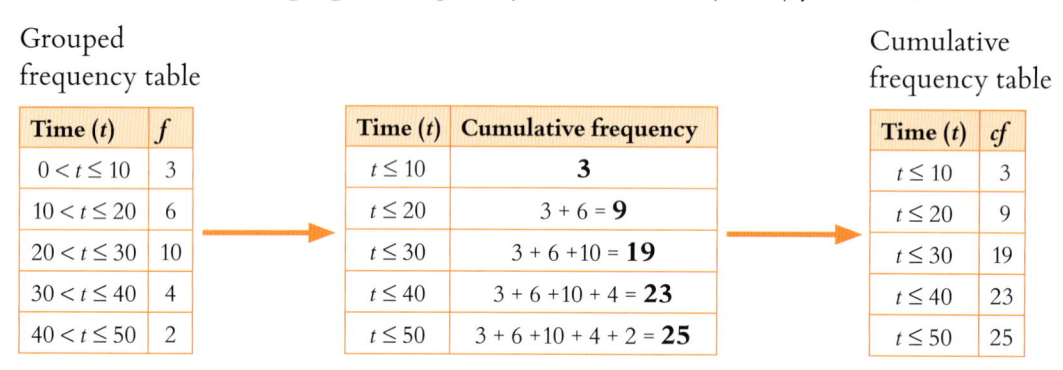

Grouped frequency table

Time (t)	f
$0 < t \leq 10$	3
$10 < t \leq 20$	6
$20 < t \leq 30$	10
$30 < t \leq 40$	4
$40 < t \leq 50$	2

Time (t)	Cumulative frequency
$t \leq 10$	3
$t \leq 20$	3 + 6 = **9**
$t \leq 30$	3 + 6 + 10 = **19**
$t \leq 40$	3 + 6 + 10 + 4 = **23**
$t \leq 50$	3 + 6 + 10 + 4 + 2 = **25**

Cumulative frequency table

Time (t)	cf
$t \leq 10$	3
$t \leq 20$	9
$t \leq 30$	19
$t \leq 40$	23
$t \leq 50$	25

To draw a cumulative frequency diagram (or c.f. graph), you must use the values from the cumulative frequency table.
So for the above example, you must plot (10, 3), (20, 9), (30, 19), (40, 23) and (50, 25).
It is important to note that the points are plotted at the upper end of the class.

Cumulative frequency diagram

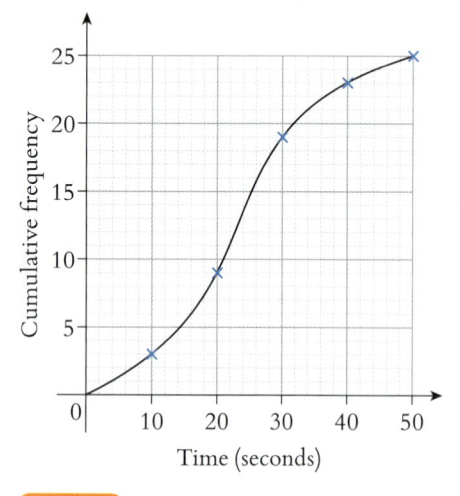

There are no students who took 0 seconds or less so the point (0, 0) can also be plotted. Draw a smooth curve through the points.

➡ **NOTE:** it is important to remember that the cumulative frequency diagram shows the number of students taking less than or equal to a particular time.

UNIT 7

A cumulative frequency diagram can be used to find estimates.
The green line shows how the curve can be used to estimate the number of students that took 26 seconds or less to solve the puzzle.

The number of students taking 26 seconds or less = 15

The red line shows how the curve can be used to estimate the number of students that took more than 35 seconds to solve the puzzle.

The number of students taking 35 seconds or less = 21
So, the number of people taking more than 35 seconds
= 25 − 21 = 4

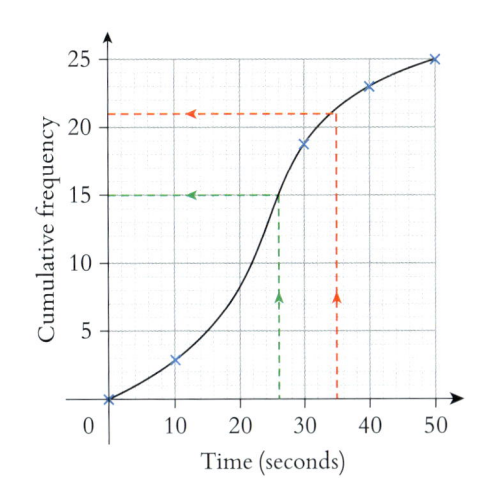

EXERCISE 7.13

1   The grouped frequency table shows information about the lengths of 50 telephone calls.

Time (t mins)	Frequency
$0 < t \leq 2$	15
$2 < t \leq 4$	12
$4 < t \leq 6$	10
$6 < t \leq 8$	8
$8 < t \leq 10$	5

Time (t mins)	Cumulative frequency
$t \leq 2$	15
$t \leq 4$	27
$t \leq 6$	
$t \leq 8$	
$t \leq 10$	

  a Copy and complete the cumulative frequency table.
  b Draw a cumulative frequency diagram to show the information.
  c Estimate the number of phone calls that took 7 minutes or less.
  d Estimate the number of phone calls that took more than 3 minutes.

2   The grouped frequency table shows information about the heights of 100 boys.

Height (h cm)	Frequency
$155 < h \leq 160$	7
$160 < h \leq 165$	24
$165 < h \leq 170$	35
$170 < h \leq 175$	22
$175 < h \leq 180$	12

Height (h cm)	Cumulative frequency
$h \leq 160$	7
$h \leq 165$	31
$h \leq 170$	
$h \leq 175$	
$h \leq 180$	

  a Copy and complete the cumulative frequency table.
  b Draw a cumulative frequency diagram to show the information.
  c Estimate the number of boys that are 162 cm or less in height.
  d Estimate the number of boys that are more than 173 cm in height.

Cumulative frequency   369

3 The table shows information about the marks, m %, achieved in an examination by a group of students.
  a Construct a cumulative frequency table for this data.
  b Draw a cumulative frequency diagram to show the data.
  c Estimate the number of students that scored 68% or less.
  d The pass mark was 48%.
     Estimate the number of students that passed the exam.

Mark (m %)	Frequency
$0 < m \leq 20$	5
$20 < m \leq 40$	10
$40 < m \leq 60$	14
$60 < m \leq 80$	17
$80 < m \leq 100$	4

4 60 newborn babies are weighed.
The table shows the results.
  a Construct a cumulative frequency table for this data.
  b Draw a cumulative frequency diagram to show the data.
  c Estimate the number of newborn babies with a mass of 3.2 kg or less.
  d Estimate the number of newborn babies with a mass greater than 3.7 kg.

Mass (m kg)	Frequency
$2 < m \leq 2.5$	6
$2.5 < m \leq 3$	10
$3 < m \leq 3.5$	17
$3.5 < m \leq 4$	13
$4 < m \leq 4.5$	8
$4.5 < m \leq 5$	6

5 The table shows information about the midday temperatures, t °C, recorded over fifty consecutive days.
  a Draw a cumulative frequency diagram to show the data.
  b Estimate the number of days with a temperature of 17°C or less.
  c Estimate the percentage of days when the temperature was over 21°C.

Temperature (t °C)	Frequency
$6 < t \leq 10$	2
$10 < t \leq 14$	4
$14 < t \leq 18$	20
$18 < t \leq 22$	15
$22 < t \leq 26$	9

6 The cumulative frequency table shows information about the speeds, v km/h, of sixty cars passing a checkpoint.

Speed (v km/h)	Cumulative frequency
$v \leq 30$	6
$v \leq 40$	24
$v \leq 50$	50
$v \leq 60$	60

Speed (v km/h)	Frequency
$0 < v \leq 30$	6
$30 < v \leq 40$	18
$40 < v \leq 50$	
$50 < v \leq 60$	

Copy and complete the grouped frequency table for this data.

7 The cumulative frequency table shows information about the marks, $m\%$, obtained in an examination by a group of eighty students.

Mark ($m\%$)	Cumulative frequency
$m \leq 20$	3
$m \leq 40$	21
$m \leq 60$	58
$m \leq 80$	79
$m \leq 100$	80

Mark ($m\%$)	Frequency
$0 < m \leq 20$	3
$20 < m \leq 40$	18
$40 < m \leq 60$	
$60 < m \leq 80$	
$80 < m \leq 100$	

Copy and complete the grouped frequency table for this data.

8 The cumulative frequency diagram shows information about the waiting times of forty patients at a doctor's surgery.
  a Estimate the number of patients that waited 24 minutes or less.
  b Estimate the number of patients that waited more than 36 minutes.
  c Use the information in the diagram to copy and complete the following two tables.

Time (t mins)	Cumulative frequency
$t \leq 10$	
$t \leq 20$	
$t \leq 30$	
$t \leq 40$	
$t \leq 50$	

Time (t mins)	Frequency
$0 < t \leq 10$	
$10 < t \leq 20$	
$20 < t \leq 30$	
$30 < t \leq 40$	
$40 < t \leq 50$	

  d Calculate an estimate of the mean waiting time.

9 Forty light bulbs were tested to see how long they would last. The cumulative frequency diagram shows information about how long they lasted.
  a Estimate the number of light bulbs that lasted 740 hours or less.
  b Estimate the percentage of light bulbs that lasted more than 960 hours.
  c Use the information in the diagram to complete the following two tables.

Time (t hours)	Cumulative frequency
$t \leq 700$	
$t \leq 800$	
$t \leq 900$	
$t \leq 1000$	
$t \leq 1100$	

Time (t hours)	Frequency
$600 < t \leq 700$	
$700 < t \leq 800$	
$800 < t \leq 900$	
$900 < t \leq 1000$	
$1000 < t \leq 1100$	

  d Calculate an estimate of the mean time.

Cumulative frequency

The median, quartiles and inter-quartile range

The median, **lower quartile (LQ)**, **upper quartile (UQ)** and **inter-quartile range (IQR)** for a set of data can be estimated using a cumulative frequency diagram.

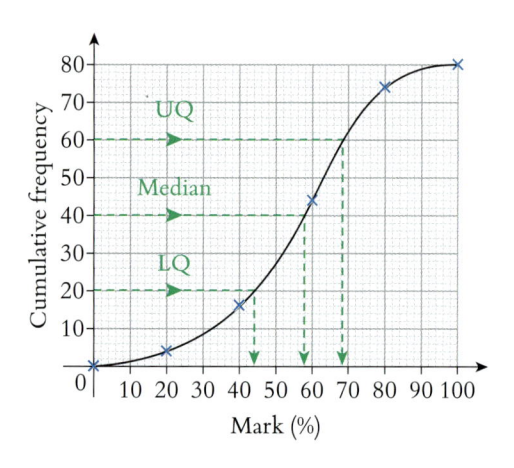

To find the median:
Draw a horizontal line from 40 $\left(\frac{1}{2} \text{ of } 80\right)$ on the cf axis to meet the curve.
The median value is where the vertical line from this point meets the horizontal axis.

To find the lower quartile:
Draw a horizontal line from 20 $\left(\frac{1}{4} \text{ of } 80\right)$ on the cf axis to meet the curve.
The LQ value is where the vertical line from this point meets the horizontal axis.

To find the upper quartile:
Draw a horizontal line from 60 $\left(\frac{3}{4} \text{ of } 80\right)$ on the cf axis to meet the curve.
The UQ value is where the vertical line from this point meets the horizontal axis.

The values are taken from the horizontal axis.

Median = 58%   Lower quartile = 44%   Upper quartile = 68%

The inter-quartile range is a measure of spread of the middle 50% of the data and is calculated using:

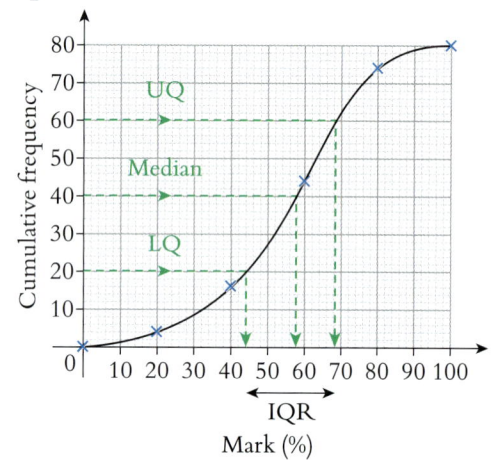

> Inter-quartile range = upper quartile − lower quartile

So for this cumulative frequency diagram:
Inter-quartile range = 68% − 44% = 24%

Percentiles
Percentiles divide the cumulative frequency into hundredths.
The lower quartile is the 25th percentile.
The median is the 50th percentile.
The upper quartile is the 75th percentile.

To find the 60th percentile for this diagram:
Draw a horizontal line from 48 (60% of 80) on the cumulative frequency axis to meet the curve.
The 60th percentile value is where the vertical line from this point meets the horizontal axis.

60th percentile = 62%

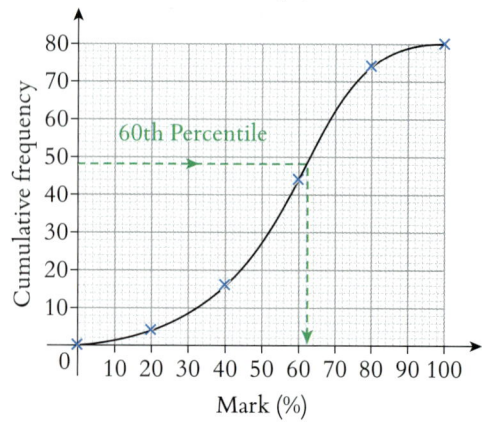

Comparing data sets

You can compare data sets using information obtained from a cumulative frequency diagram.
To do this you compare:
- The medians (this is a measure of average),
- The inter-quartile ranges (this is a measure of spread).

EXAMPLE

The table shows information about the examination results of a group of boys and girls.
Compare the boys' results with the girls' results.

	Median mark	IQR
Boys	56	24
Girls	52	35

The boys' median is greater than the girls' median, so on average the boys achieved higher marks.
The IQR for the girls is greater than the IQR for the boys, so the marks for the girls are more varied.

EXERCISE 7.14

1. For each of the cumulative frequency diagrams find estimates for
   - i the median,
   - ii the LQ,
   - iii the UQ,
   - iv the IQR,
   - v the 40th percentile.

a
[Cumulative frequency vs Height (cm)]

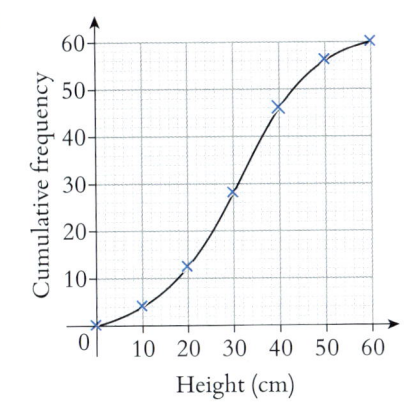

b
[Cumulative frequency vs Mark (%)]

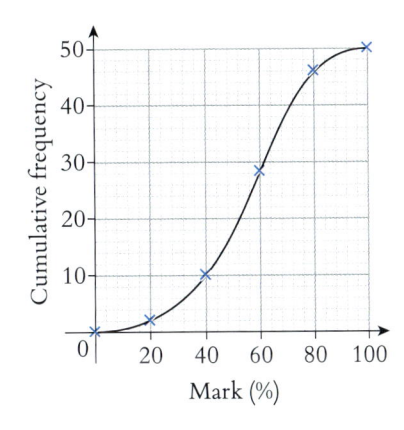

c
[Cumulative frequency vs Mass (kg)]

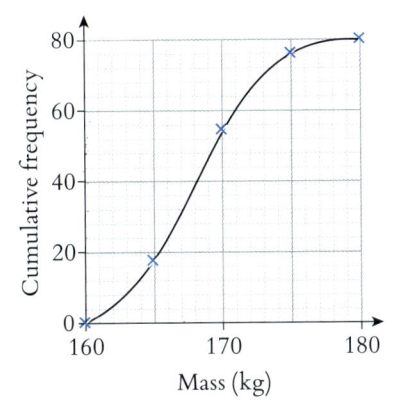

d
[Cumulative frequency vs Speed (km/h)]

Cumulative frequency 373

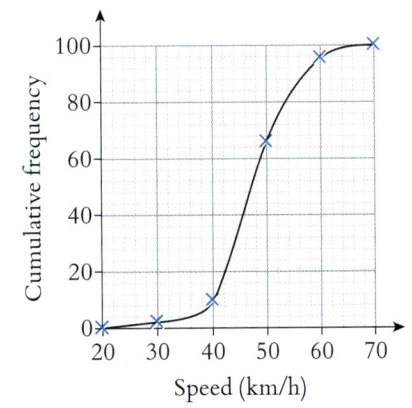

2 200 apples are weighed.
The table shows the results.
  a Draw a cumulative frequency diagram to show the data.
  b Use your cumulative frequency diagram to find
    i the median,
    ii the lower quartile,
    iii the upper quartile,
    iv the inter-quartile range,
    v the 68th percentile.

Mass (m kg)	Frequency
$100 < m \leq 110$	2
$110 < m \leq 120$	2
$120 < m \leq 130$	9
$130 < m \leq 140$	19
$140 < m \leq 150$	40
$150 < m \leq 160$	48
$160 < m \leq 170$	40
$170 < m \leq 180$	24
$180 < m \leq 190$	12
$190 < m \leq 200$	4

3 150 passengers on an aircraft had their baggage weighed.
The table shows the results.
  a Draw a cumulative frequency diagram to show the data.
  b Use your cumulative frequency diagram to find
    i the median,
    ii the lower quartile,
    iii the upper quartile,
    iv the inter-quartile range,
    v the 84th percentile.

Mass (m kg)	Frequency
$0 < m \leq 5$	10
$5 < m \leq 10$	20
$10 < m \leq 15$	38
$15 < m \leq 20$	35
$20 < m \leq 25$	25
$25 < m \leq 30$	12
$30 < m \leq 35$	10

4 60 potatoes are weighed.
The table shows the results.
  a Draw a cumulative frequency diagram to show the data.
  b Use your cumulative frequency diagram to find
    i the median,
    ii the lower quartile,
    iii the upper quartile,
    iv the inter-quartile range,
    v the 36th percentile.

Mass (m g)	Frequency
$200 < m \leq 250$	4
$250 < m \leq 300$	7
$300 < m \leq 350$	3
$350 < m \leq 400$	16
$400 < m \leq 450$	21
$450 < m \leq 500$	9

5

	Median time	IQR
10 to 20 year olds	150	20
60 to 70 year olds	300	28

The table shows information about the reaction times (milliseconds) of a group of 10 to 20 year olds and the reaction times of a group of 60 to 70 year olds.
Compare the two sets of data.

6

[Cumulative frequency diagram showing two curves labelled "paper 1" and "paper 2", with Mark (%) on the x-axis (0 to 100) and Cumulative frequency on the y-axis (0 to 80).]

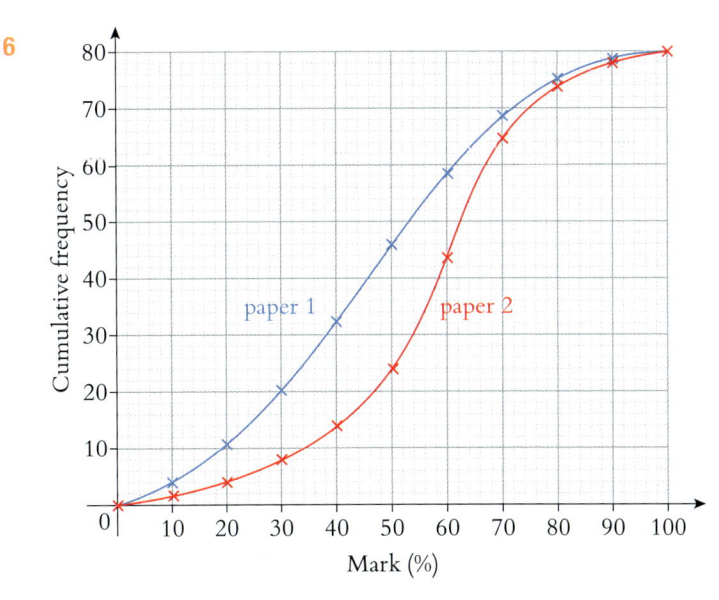

The cumulative frequency diagram shows information about the marks scored by a group of eighty students on Paper 1 and Paper 2 of a mathematics examination.

a   Use the cumulative frequency diagram to copy and complete the following table.

	Median	LQ	UQ	IQR
Paper 1				
Paper 2				

b   Compare the marks for Paper 1 and Paper 2.

7   In a survey, 100 shoppers leaving two different supermarkets were asked how much they had just spent. The results are shown in the two tables.

Supermarket A

Amount ($ x)	Frequency
$0 < x \leq 30$	2
$30 < x \leq 60$	25
$60 < x \leq 90$	47
$90 < x \leq 120$	16
$120 < x \leq 150$	10

Supermarket B

Amount ($ x)	Frequency
$0 < x \leq 30$	14
$30 < x \leq 60$	24
$60 < x \leq 90$	34
$90 < x \leq 120$	26
$120 < x \leq 150$	2

a   Draw a cumulative frequency diagram to show the information about the amounts spent in supermarket A and in supermarket B.

b   Use the cumulative frequency diagram to copy and complete the following table.

	Median	LQ	UQ	IQR
Supermarket A				
Supermarket B				

c   Compare the amounts of money spent in supermarkets A and B.

KEY WORDS
cumulative frequency
lower quartile
upper quartile
inter-quartile range
percentile

Cumulative frequency

Unit 7 Examination questions

1

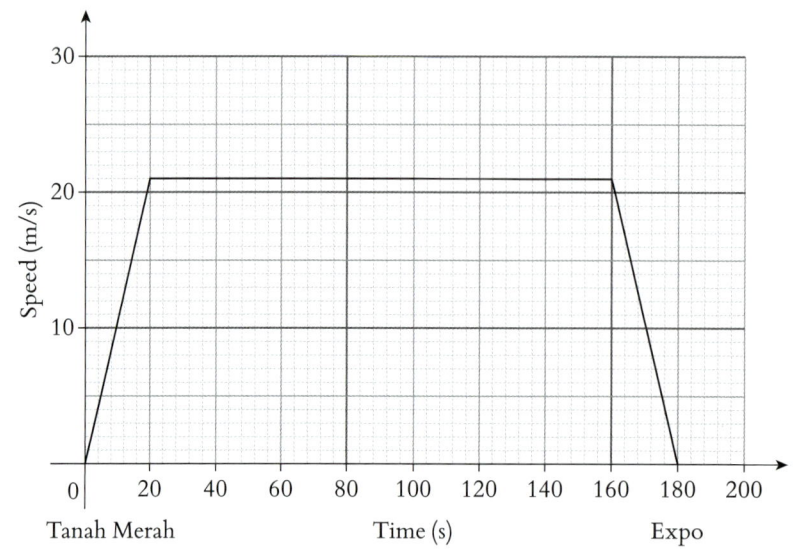

The graph shows the train journey between Tanah Merah and Expo in Singapore. Work out

(a) the acceleration of the train when it leaves Tanah Merah, [2]
(b) the distance between Tanah Merah and Expo, [3]
(c) the average speed of the train for the journey. [1]

Cambridge IGCSE Mathematics 0580, Paper 21 Q18, November 2008

2

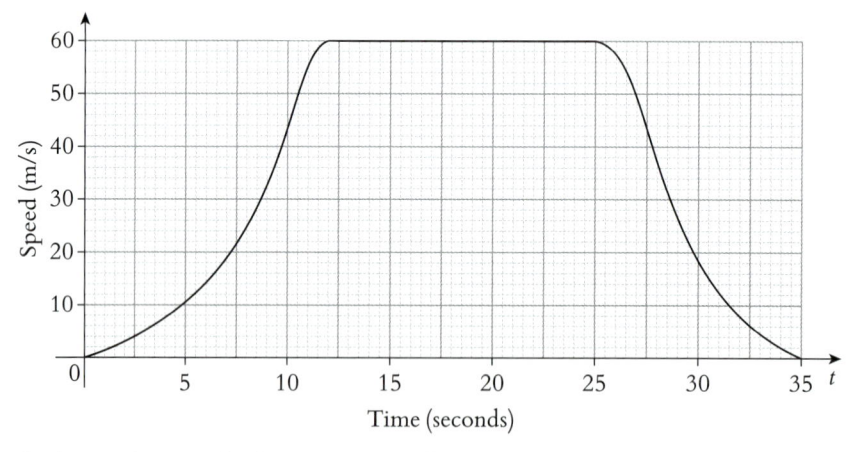

The graph shows the speed of a sports car after t seconds.
It starts from rest and accelerates to its maximum speed in 12 seconds.

(a) (i) Draw a tangent to a copy of the graph at $t = 7$. [1]
    (ii) Find the acceleration of the car at $t = 7$. [2]
(b) The car travels at its maximum speed for 13 seconds.
    Find the distance travelled by the car at its maximum speed. [2]

Cambridge IGCSE Mathematics 0580, Paper 21 Q18, November 2010

3 **(a)** Factorise $ax^2 + bx^2$. [1]
  (b) Make x the subject of the formula
   $ax^2 + bx^2 - d^2 = p^2$. [2]

Cambridge IGCSE Mathematics 0580, Paper 21 Q8, November 2008

4 For the sequence $5\frac{1}{2}$, 7, $8\frac{1}{2}$, 10, $11\frac{1}{2}$, ...

  (a) find an expression for the nth term, [2]
  (b) work out the 100th term. [1]

Cambridge IGCSE Mathematics 0580, Paper 2 Q10, June 2006

5 For each of the following sequences, write down the next term.

  (a) 2, 3, 5, 8, 13, ... [1]
  (b) x^6, $6x^5$, $30x^4$, $120x^3$, ... [1]
  (c) 2, 6, 18, 54, 162, ... [1]

Cambridge IGCSE Mathematics 0580, Paper 2 Q9, November 2006

6 **(a)** The formula for the nth term of the sequence
   1, 5, 14, 30, 55, 91, ... is $\frac{n(n+1)(2n+1)}{6}$.
   Find the 20th term. [1]
  (b) The nth term of the sequence 10, 17, 26, 37, 50, ... is $(n+2)^2 + 1$.
   Writer down the formula for the nth term of the sequence
   17, 26, 37, 50, 65, ... [1]

Cambridge IGCSE Mathematics 0580, Paper 21 Q4, June 2008

7 The table shows some terms of several sequences.

Team	1	2	3	4		8	
Sequence **P**	7	5	3	1		p	
Sequence **Q**	1	8	27	64		q	
Sequence **R**	$\dfrac{1}{2}$	$\dfrac{2}{3}$	$\dfrac{3}{4}$	$\dfrac{4}{5}$		r	
Sequence **S**	4	9	16	25		s	
Sequence **T**	1	3	9	27		t	
Sequence **U**	3	6	7	−2		u	

(a) Find the values of p, q, r, s, t and u. [6]
(b) Find the nth term of sequence
  (i) **P**, [1]
  (ii) **Q**, [1]
  (iii) **R**, [1]
  (iv) **S**, [1]
  (v) **T**, [1]
  (vi) **U**. [1]
(c) Which term in sequence **P** is equal to −777? [2]
(d) Which term in sequence **T** is equal to 177 147? [2]

Cambridge IGCSE Mathematics 0580, Paper 4 Q9, November 2007

8

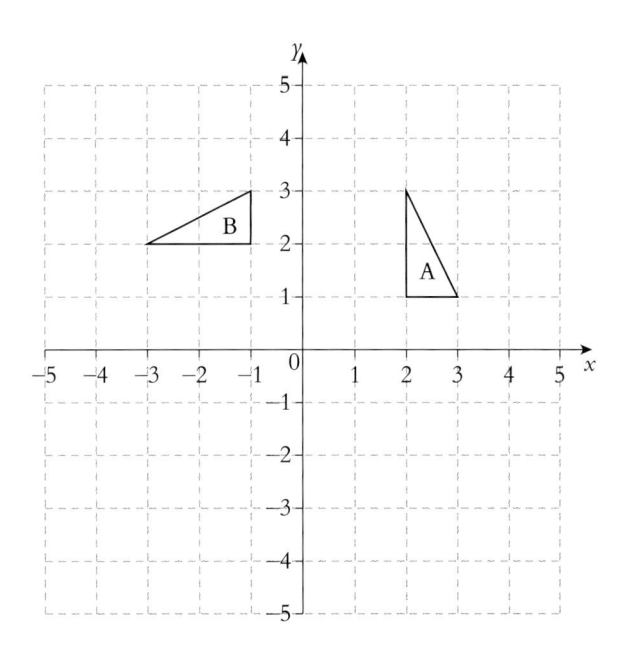

(a) A transformation is represented by the matrix $\begin{pmatrix} 0 & -1 \\ -1 & 0 \end{pmatrix}$.

    (i) On a copy of the grid above, draw the image of triangle A after this transformation. [2]

    (ii) Describe fully this transformation. [2]

(b) Find the 2 by 2 matrix representing the transformation which maps triangle A onto triangle B. [2]

Cambridge IGCSE Mathematics 0580, Paper 21 Q19, June 2008

9 (a) Draw x and y axes from 0 to 12 using a scale of 1 cm to 1 unit on each axis. [1]
    (b) Draw and label triangle T with vertices (8, 6), (6, 10), and (10, 12). [1]
    (c) Triangle T is reflected in the line $y = x$.
        (i) Draw the image of triangle T. Label this image P [2]
        (ii) Write down the matrix which represents this reflection. [2]

    (d) A transformation is represented by the matrix $\begin{pmatrix} \frac{1}{2} & 0 \\ 0 & \frac{1}{2} \end{pmatrix}$

        (i) Draw the image of triangle T under this transformation. Label this image Q. [2]
        (ii) Describe fully this single transformation. [3]
    (e) Triangle T is stretched with the y-axis invariant and a stretch factor of $\frac{1}{2}$. Draw the image of triangle T. Label this image R. [2]

Cambridge IGCSE Mathematics 0580, Paper 4 Q7, November 2007

10

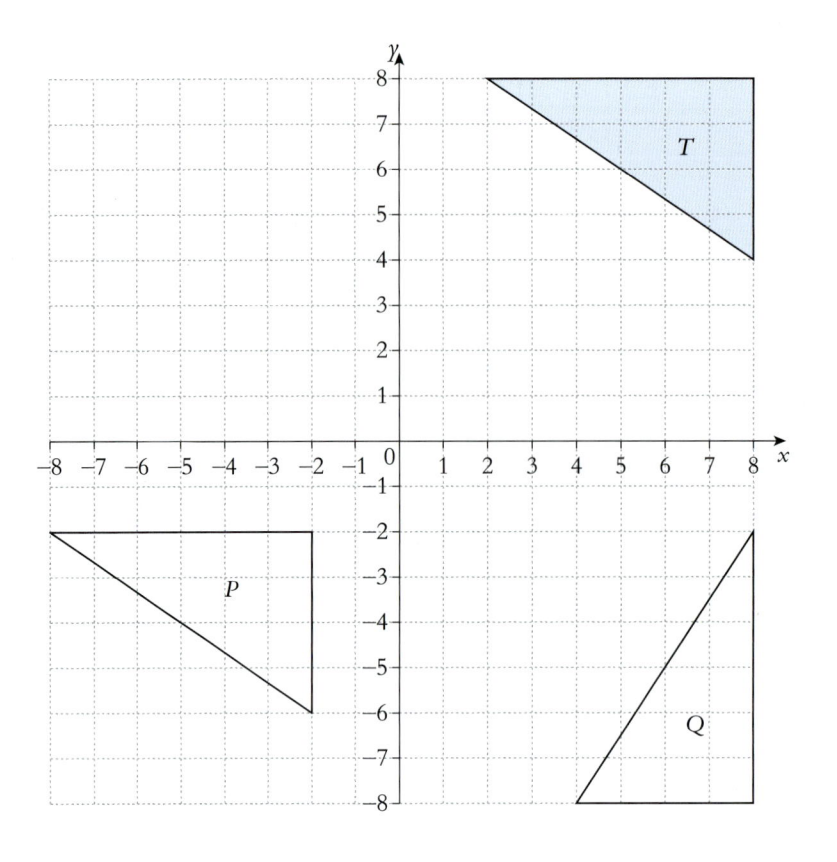

(a) On a copy of the grid, draw the enlargement of the triangle T, centre (0, 0), scale factor $\frac{1}{2}$. [2]

(b) The matrix $\begin{pmatrix} -1 & 0 \\ 0 & 1 \end{pmatrix}$ represents a transformation.

  (i) Calculate the matrix product $\begin{pmatrix} -1 & 0 \\ 0 & 1 \end{pmatrix}\begin{pmatrix} 8 & 8 & 2 \\ 4 & 8 & 8 \end{pmatrix}$. [2]

  (ii) On the grid, draw the image of the triangle T under this transformation. [2]
  (iii) Describe fully this **single** transformation. [2]

(c) Describe fully the **single** transformation which maps
  (i) triangle T onto triangle P, [2]
  (ii) triangle T onto triangle Q. [3]

(d) Find the 2 by 2 matrix which represents the transformation in **part (c) (ii)**. [2]

Cambridge IGCSE Mathematics 0580, Paper 41 Q3, June 2010

11 The number of hours that a group of 80 students spent using a computer in a week was recorded. The results are shown by the cumulative frequency curve.

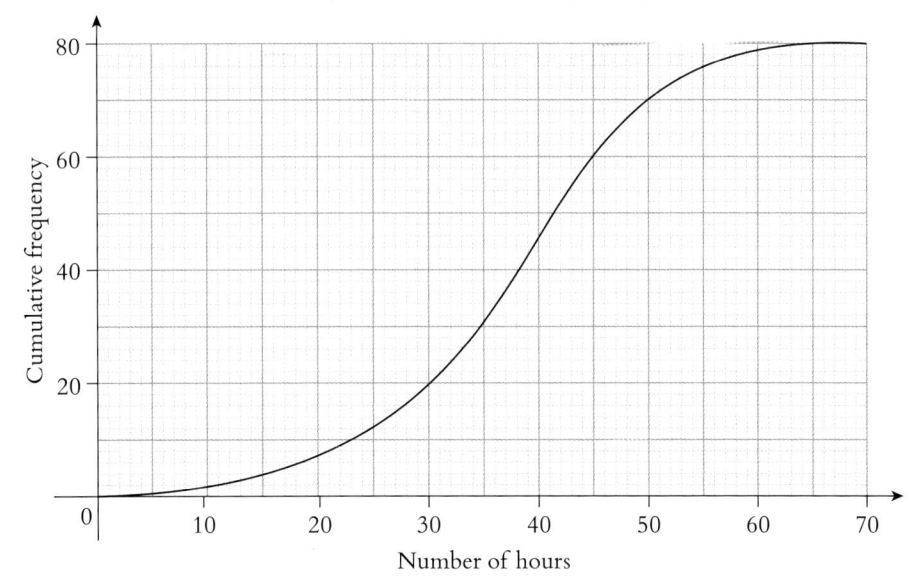

Use the cumulative frequency curve to find

(a) the median, [1]
(b) the upper quartile, [1]
(c) the interquartile range, [1]
(d) the number of students who spent more than 50 hours using a computer in a week. [2]

Cambridge IGCSE Mathematics 0580, Paper 21 Q20, November 2009

Rational and irrational numbers

> **THIS SECTION WILL SHOW YOU HOW TO**
> - Change a recurring decimal to a fraction
> - Identify rational and irrational numbers

Recurring decimals

Recurring decimals contain digits that are repeated over and over again.

Dots are used to show how the decimals recur.

$\frac{2}{3} = 0.666... = 0.\dot{6}$

$\frac{4}{11} = 0.363636... = 0.\dot{3}\dot{6}$

$\frac{5}{12} = 0.41666... = 0.41\dot{6}$

$\frac{2}{7} = 0.2857142857142... = 0.\dot{2}8571\dot{4}$

➡ **NOTE:** to change a fraction to a decimal divide the numerator by the denominator.
eg $\frac{2}{3} = 2 \div 3$

To change $0.\dot{2}\dot{3}$ to a fraction use the following method:

let $x = 0.232323...$ *multiply both sides by 100*
$100x = 23.232323...$ *write $x = 0.232323...$ underneath*
$x = 0.232323...$ *subtract the two equations*
$\overline{}$
$99x = 23$
$x = \frac{23}{99}$

➡ **NOTE:** if 1 digit recurs multiply by 10.
If 2 digits recur multiply by 100.
If 3 digits recur multiply by 1000.

➡ **CHECK:**
$23 \div 99 = 0.232323...$ ✓

Rational numbers

A **rational number** is a number that can be written in the form $\frac{a}{b}$ where a and b are integers.

Rational numbers include:

- All fractions, for example $\frac{4}{5}$
- All integers, for example $5 = \frac{5}{1}$
- All mixed numbers, for example $2\frac{1}{3} = \frac{7}{3}$
- All terminating decimals, for example $5.19 = \frac{519}{100}$
- All recurring decimals, for example $0.\dot{6} = \frac{2}{3}$
- All square roots that can be calculated exactly, for example $\sqrt{25} = 5 = \frac{5}{1}$

Irrational numbers

An **irrational number** is a number that *cannot* be written in the form $\frac{a}{b}$ where a and b are integers. Examples of irrational numbers are π, $\sqrt{7}$ and $\sqrt[3]{2}$.

EXERCISE 8.1

1 Write these fractions as recurring decimals.

    a $\dfrac{1}{3}$ **b** $\dfrac{5}{9}$ **c** $\dfrac{2}{11}$ **d** $\dfrac{7}{22}$ **e** $\dfrac{11}{18}$

    f $\dfrac{59}{90}$ **g** $\dfrac{5}{33}$ **h** $\dfrac{19}{111}$ **i** $\dfrac{2}{7}$ **j** $\dfrac{5}{13}$

2 Write these recurring decimals as fractions.

    a $0.\dot{3}$ **b** $0.\dot{7}$ **c** $0.\dot{3}\dot{6}$ **d** $0.\dot{6}\dot{8}$ **e** $0.\dot{1}\dot{4}$

    f $0.1\dot{4}\dot{5}$ **g** $0.2\dot{3}\dot{8}$ **h** $0.5\dot{2}$ **i** $0.6\dot{1}$ **j** $0.\dot{1}3\dot{6}$

3 $\sqrt{49}$ $\sqrt{36}$ $\sqrt{40}$ $\sqrt{\dfrac{4}{9}}$ $\sqrt{64}$

   From the list above, write down
    a a factor of 12   **b** an irrational number
    c a cube number   **d** a prime number
    e a number smaller than 1.

4 Which of these numbers are **not** rational numbers?

    $\dfrac{4}{7}$ π $7\dfrac{1}{11}$ $\sqrt{\dfrac{5}{8}}$ 0.003 $\sqrt{0.04}$ $\sqrt[3]{8}$

5 Which of these numbers are irrational?

    $\sqrt{1}$ $\sqrt{2}$ $\sqrt{3}$ $\sqrt{4}$ $\sqrt{5}$ $\sqrt{6}$

6 Is x rational or irrational for each of these right-angled triangles.

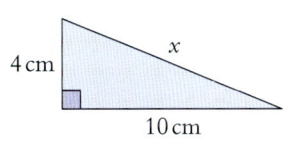

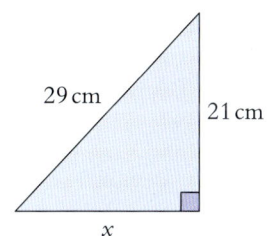

    a Right-angled triangle with legs 4 cm and 10 cm, hypotenuse x.

    b Right-angled triangle with hypotenuse 29 cm, one leg 21 cm, other leg x.

7 The circle has radius 5 cm.
   The circle touches the sides of the square.
   Calculate
    a the area of the circle,
    b the area of the square,
    c the shaded area.
   State whether each of your answers is rational or irrational.

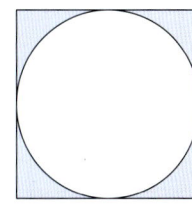

KEY WORDS
recurring decimal
rational
irrational

Rational and irrational numbers

Solving quadratic equations using the formula

THIS SECTION WILL SHOW YOU HOW TO
- Solve quadratic equations using the quadratic formula

Some quadratic equations do not factorise.
For example: $x^2 - 5x + 3 = 0$ does not factorise because you cannot find two numbers that multiply to give 3 and add to give -5.

> Quadratic equations of the form $ax^2 + bx + c = 0$ can be solved using the formula
> $$x = \frac{-b \pm \sqrt{b^2 - 4ac}}{2a}$$

In the formula:
- a is the coefficient of the x^2 term
- b is the coefficient of the x term
- c is the constant.

The $\pm$ symbol means 'plus or minus'.

➡ **NOTE:** the quadratic formula is not given to you in the examination. You must learn the formula carefully.

Factorising gives exact answers. If a question asks you to give your answer to 2 d.p., it is a hint that you need to use the formula.

EXAMPLE

Solve $2x^2 - 9x + 8 = 0$, giving your answers to two decimal places.

$2x^2 - 9x + 8 = 0$ first write down the values of a, b and c.

$a = 2 \quad b = -9 \quad c = 8$

$x = \dfrac{-b \pm \sqrt{b^2 - 4ac}}{2a}$ substitute the values into the formula

$x = \dfrac{-(-9) \pm \sqrt{(-9)^2 - 4 \times 2 \times 8}}{2 \times 2}$ $-(-9) = 9$ and $(-9)^2 = 81$

$x = \dfrac{9 \pm \sqrt{81 - 64}}{4}$

$x = \dfrac{9 \pm \sqrt{17}}{4}$

$x = \dfrac{9 + \sqrt{17}}{4}$ or $x = \dfrac{9 - \sqrt{17}}{4}$

$x = 3.28$ or $x = 1.22$ (2 d.p.)

UNIT 8

EXERCISE 8.2

1 Rafiu is asked to solve the equation $x^2 + 11x - 5 = 0$ correct to 3 s.f.
   He writes

   $$x = \frac{-11 \pm \sqrt{121 + 20}}{2}$$
   $$= \frac{-11 \pm 11.9}{2}$$
   $$= 0.450 \text{ or } -11.5$$

   The correct solution is $x = 0.437$ or $x = -11.4$
   Explain why his answers are not accurate.

2 Solve these equations using the quadratic formula.
   Where appropriate give your answers correct to 2 decimal places.
   a $x^2 + 2x - 24 = 0$ b $x^2 + 3x - 2 = 0$ c $x^2 - x - 7 = 0$
   d $a^2 - 2a - 2 = 0$ e $x^2 - 6x + 2 = 0$ f $y^2 + 3y - 14 = 0$
   g $x^2 + 4x - 7 = 0$ h $3y^2 - 8y + 2 = 0$ i $2x^2 - 5x - 6 = 0$
   j $3x^2 - 5x - 1 = 0$ k $x^2 + 10x - 400 = 0$ l $4x^2 + 7x - 5 = 0$
   m $2x^2 - 3x - 4 = 0$ n $x^2 + 2x - 5 = 0$ o $2x^2 + 6x - 5 = 0$
   p $x^2 + 10x + 11 = 0$

3 Use the quadratic formula to solve the equation $x^2 + 5x + 9 = 0$
   Explain what happens.

4 Use the quadratic formula to solve the equation $4x^2 - 20x + 25 = 0$
   Explain what happens.

5 Solve these equations.
   Give your answers correct to 2 decimal places.
   a $x(x + 2) = 12$ b $x(x + 2) = 9 - 2x$
   c $(x - 2)^2 = 6$ d $x(x + 8) = x - 3$
   e $2x^2 = x(x - 7) - 3$ f $(3x - 2)(x - 2) = 7$
   g $(3x)^2 = x + 4$ h $(x - 3)^2 + x = 28$

   HINT
   You need to expand the brackets and rearrange to the form $ax^2 + bx + c = 0$ before using the formula.

6 Solve these equations.
   Give your answers correct to 2 decimal places.
   a $x + 4 = \dfrac{9}{x}$
   b $x - \dfrac{11}{x} = 2$

   HINT
   You need to multiply both sides of the equation by x and then rearrange to the form $ax^2 + bx + c = 0$ before using the formula.

7 Solve these equations.
   Give your answers correct to 2 decimal places.
   a $1 + \dfrac{2}{x} = \dfrac{14}{x^2}$ b $x - 3 = \dfrac{5}{3x}$ c $x - 2 = \dfrac{1}{x + 5}$

Solving quadratic equations using the formula   385

The number of roots of a quadratic equation

The answers to an equation are called the **roots** of the equation. Consider solving the following three equations using the formula $x = \dfrac{-b \pm \sqrt{b^2 - 4ac}}{2a}$

$x^2 + 2x - 15 = 0$

$x = \dfrac{-2 \pm \sqrt{2^2 - 4 \times 1 \times -15}}{2 \times 1}$

$x = \dfrac{-2 \pm \sqrt{64}}{2}$

$x = 3$ or $x = -5$

two roots

$x^2 + 2x + 1 = 0$

$x = \dfrac{-2 \pm \sqrt{2^2 - 4 \times 1 \times 1}}{2 \times 1}$

$x = \dfrac{-2 \pm \sqrt{0}}{2}$

$x = -1$ (repeated)

one root

$x^2 + x + 9 = 0$

$x = \dfrac{-1 \pm \sqrt{1^2 - 4 \times 1 \times 9}}{2 \times 1}$

$x = \dfrac{-1 \pm \sqrt{-35}}{2}$

no roots

A quadratic equation can have two, one or no roots.

The part of the quadratic formula underneath the square root sign is called the **discriminant**.

➡ **NOTE:** you will not be tested on the use of a discriminant in the examination.

discriminant = $b^2 - 4ac$

The discriminant tells you how many roots there are for the equation.

$b^2 - 4ac$
- > 0 means there will be **two** roots
- $= 0$ means there will be **one** root
- < 0 means there will be **no** roots

EXAMPLE

Calculate the discriminant for the equation $x^2 - 3x + 4 = 0$. How many roots does the equation have?

$x^2 - 3x + 4 = 0$ $a = 1$ $b = -3$ $c = 4$

discriminant = $b^2 - 4ac = (-3)^2 - 4 \times 1 \times 4$

$= 9 - 16$

$= -7$

The discriminant is negative so there are no roots.

EXERCISE 8.3

1. State whether these equations will have two, one or no roots.
   a. $x^2 + 9x - 12 = 0$
   b. $x^2 - 6x + 9 = 0$
   c. $x^2 + 8x + 20 = 0$
   d. $x^2 - 3x - 11 = 0$
   e. $x^2 + 12x + 36 = 0$
   f. $x^2 + 2x - 17 = 0$
   g. $2x^2 + x + 2 = 0$
   h. $x^2 + 16 = 8x$
   i. $6x^2 - 19x + 15 = 0$
   j. $7x^2 = 6x + 7$

2. Find the values of k for which $x^2 + kx + 25 = 0$ has one root.

3. Find the values of k for which $kx^2 + 8x + 8 = 0$ has two roots.

EXERCISE 8.4

1. The area of the rectangle is 15 cm².
   Find the value of x.

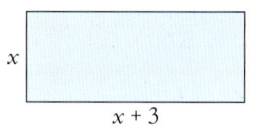

2. The area of the right-angled triangle is 6 cm².
   Find the value of x.

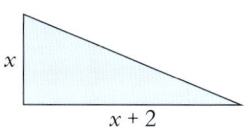

3. The area of the right-angled triangle is 38 cm².
   Find the value of x.

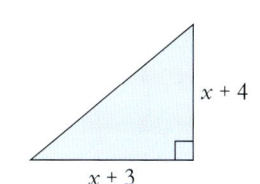

4. The area of the trapezium is 62 cm².
   Find the value of x.

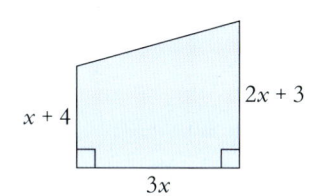

5. A ball is thrown upwards. The height, h metres, of the ball above the ground after t seconds is given by the formula $h = 2 + 15t - 5t^2$. Find the values of t when the ball is 4 metres above the ground.

6. The length of a rectangular field is 20 metres greater than its width, w metres.
   a. Write down an expression for the area, A m², of the field.
   b. The area of the field is 7000 m². Find the width and perimeter of the field.

7. The solid circular cylinder has a base radius of r cm and a height of 8 cm. The total surface area of the cylinder is 130π cm².
   a. Show that $r^2 + 8r - 65 = 0$.
   b. Solve the equation to find the value of r.

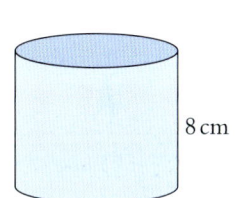

8. The two rectangles have the same area. Find the value of x.

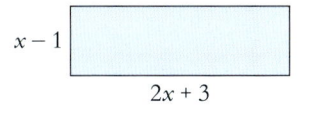

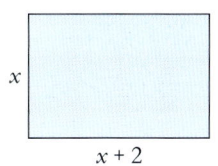

9. Find the points of intersection of the graphs of $y = x^2 + 3x - 2$ and $y = x - 1$.

10. Raju runs at a speed of $(x - 2)$ m/s for $(x + 2)$ seconds and the distance he travels is $(9x + 6)$ metres.
    a. Show that $x^2 - 9x - 10 = 0$.
    b. Solve the equation to find x.
    c. Write down the time taken and the distance travelled.

KEY WORDS
roots
discriminant

Solving quadratic equations using the formula

Further algebraic fractions

THIS SECTION WILL SHOW YOU HOW TO
- Add and subtract more complicated algebraic fractions
- Solve more complicated equations that involve algebraic fractions

EXAMPLE

Write $\dfrac{2}{x} + \dfrac{1}{x+2}$ as a single fraction.

$\dfrac{2}{x} + \dfrac{1}{x+2}$ 　　　　　the common denominator is $x(x+2)$

$= \dfrac{2(x+2)}{x(x+2)} + \dfrac{x}{x(x+2)}$ 　　　$\dfrac{2}{x} = \dfrac{2(x+2)}{x(x+2)}$ and $\dfrac{1}{x+2} = \dfrac{x}{x(x+2)}$

$= \dfrac{2(x+2) + x}{x(x+2)}$

$= \dfrac{2x+4+x}{x(x+2)} = \dfrac{3x+4}{x(x+2)}$

EXAMPLE

Write $\dfrac{3}{x-2} - \dfrac{4}{x+5}$ as a single fraction.

$\dfrac{3}{x-2} - \dfrac{4}{x+5}$ 　　　　the common denominator is $(x-2)(x+5)$

$= \dfrac{3(x+5)}{(x-2)(x+5)} - \dfrac{4(x-2)}{(x-2)(x+5)}$ 　　$\dfrac{3}{x-2} = \dfrac{3(x+5)}{(x-2)(x+5)}$ and $\dfrac{4}{x+5} = \dfrac{4(x-2)}{(x-2)(x+5)}$

$= \dfrac{3(x+5) - 4(x-2)}{(x-2)(x+5)}$

$= \dfrac{3x+15-4x+8}{(x-2)(x+5)} = \dfrac{23-x}{(x-2)(x+5)}$

EXAMPLE

Solve $\dfrac{2x+4}{x-1} = \dfrac{x+4}{x-2}$

$\dfrac{2x+4}{x-1} = \dfrac{x+4}{x-2}$ 　　　there is a single fraction on each side so you can cross multiply

$(2x+4)(x-2) = (x+4)(x-1)$ 　　expand the brackets
$2x^2 - 8 = x^2 + 3x - 4$ 　　　rearrange to the form $ax^2 + bx + c = 0$
$x^2 - 3x - 4 = 0$ 　　　　　　factorise
$(x-4)(x+1) = 0$
$x - 4 = 0$ 　or　 $x + 1 = 0$
　$x = 4$ 　or　 $x = -1$

EXAMPLE

Solve $\dfrac{x}{x+3} - \dfrac{2}{x-4} = 1$

Method 1

$\dfrac{x}{x+3} - \dfrac{2}{x-4} = 1$

$\dfrac{x(x-4)}{(x+3)(x-4)} - \dfrac{2(x+3)}{(x+3)(x-4)} = 1$

$\dfrac{x(x-4) - 2(x+3)}{(x+3)(x-4)} = 1$

$\dfrac{x^2 - 4x - 2x - 6}{(x+3)(x-4)} = 1$

$x^2 - 6x - 6 = (x+3)(x-4)$

$x^2 - 6x - 6 = x^2 - x - 12$

$6 = 5x$

$x = 1\dfrac{1}{5}$

Method 2

$\dfrac{x}{x+3} - \dfrac{2}{x-4} = 1$

multiply all 3 terms by $(x+3)(x-4)$

$(x+3)(x-4) \times \dfrac{x}{x+3} - (x+3)(x-4) \times \dfrac{2}{x-4}$

$= (x+3)(x-4) \times 1$

$x(x-4) - 2(x+3) = (x+3)(x-4)$

$x^2 - 4x - 2x - 6 = x^2 - x - 12$

$x^2 - 6x - 6 = x^2 - x - 12$

$6 = 5x$

$x = 1\dfrac{1}{5}$

EXERCISE 8.5

1 Write as a single fraction.

a $\dfrac{3}{x} + \dfrac{2}{x-5}$

b $\dfrac{6}{x+3} + \dfrac{5}{x+2}$

c $\dfrac{8}{x-2} - \dfrac{3}{x+7}$

d $\dfrac{8}{2x} - \dfrac{1}{3x+1}$

e $\dfrac{7}{2x+5} - \dfrac{2}{3x-2}$

f $\dfrac{6}{1-2x} + \dfrac{5}{x-4}$

g $\dfrac{x+3}{x+2} - \dfrac{x+2}{x+3}$

h $\dfrac{x+5}{x+1} - \dfrac{x-2}{x+3}$

i $\dfrac{2}{x(x+1)} + \dfrac{3}{x+1}$

j $\dfrac{1}{x^2+6x+8} + \dfrac{1}{x+4}$

k $\dfrac{2}{x^2-2x-3} - \dfrac{3}{x-3}$

l $\dfrac{x(x-y)}{x+y} + y$

2 Solve these equations.

a $\dfrac{x+3}{x-3} = \dfrac{x+2}{x+4}$

b $\dfrac{x-6}{x+7} = \dfrac{x+3}{x-2}$

c $\dfrac{x+2}{5x+2} = \dfrac{2x-1}{4x+1}$

3 Solve these equations.

a $\dfrac{6}{x-1} + \dfrac{5}{x+2} = 4$

b $\dfrac{7}{2x-3} - \dfrac{5}{2x+1} = 6$

c $\dfrac{8}{x+1} + \dfrac{20}{x+4} = 2$

d $\dfrac{3x}{x-2} - \dfrac{2}{x-3} = 4$

e $\dfrac{6}{x+3} - \dfrac{5}{2x+1} = 8$

f $\dfrac{4}{x-3} + \dfrac{9}{2x-5} = 7$

g $\dfrac{2x}{4-x} + \dfrac{8}{x} = 6$

h $\dfrac{2x}{6+2x} - \dfrac{8}{2x+1} = 3$

i $\dfrac{x}{x-4} - \dfrac{x}{x-3} = 1$

Further algebraic fractions

Variation

THIS SECTION WILL SHOW YOU HOW TO
- Use direct and inverse variation

Direct variation (linear relationships)

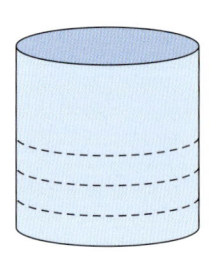

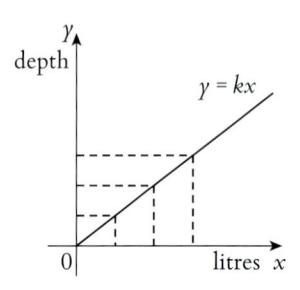

When a cylinder is filled with water, each litre of water increases the depth of water in the cylinder by the same amount.
The graph of the depth of the water against the number of litres in the cylinder is a straight line passing through the origin.
This type of relationship is known as **direct (linear) proportion** or **direct variation**.
(This means that if x is doubled then y also doubles.)

There are other ways of describing the relationship.
You need to be familiar with all of them.
- y is directly proportional to x
- y is proportional to x
- y varies directly as x
- y varies as x

The symbol for proportionality is $\propto$

> y is proportional to x
> $y \propto x$
> $y = kx$

k is called the **constant of proportionality**.
(It has the same value as the gradient of the graph.)

EXAMPLE

y is directly proportional to x and $y = 48$ when $x = 8$.
 a Find the formula for y in terms of x.
 b Find the value of y when $x = 2.5$
 c Find the value of x when $y = 120$

 a $y \propto x$ so $y = kx$
    substituting $y = 48$ and $x = 8$ into $y = kx$ gives $48 = k \times 8$ $k = 6$
    The formula is $y = 6x$
 b substituting $x = 2.5$ into $y = 6x$ gives $y = 6 \times 2.5$ $\Rightarrow$ $y = 15$
 c substituting $y = 120$ into $y = 6x$ gives $120 = 6 \times x$ $\Rightarrow$ $x = 20$

EXERCISE 8.6

1. y is proportional to x and $y = 30$ when $x = 6$
   a. Find the formula for y in terms of x.
   b. Find the value of y when $x = 8$
   c. Find the value of x when $y = 7$

2. y is proportional to x and $y = 56$ when $x = 7$
   a. Find the formula for y in terms of x.
   b. Find the value of y when $x = 3$
   c. Find the value of x when $y = 6$

3. y is proportional to x and $y = 4.5$ when $x = 9$
   a. Find the formula for y in terms of x.
   b. Find the value of y when $x = 2.4$
   c. Find the value of x when $y = 5$

4. y is proportional to x and $y = 6$ when $x = 18$
   a. Find the formula for y in terms of x.
   b. Find the value of y when $x = 4$
   c. Find the value of x when $y = 3.4$

5. y is proportional to x and $y = 0.6$ when $x = 6$
   a. Find the formula for y in terms of x.
   b. Find the value of y when $x = 8$
   c. Find the value of x when $y = 7$

6. If y is directly proportional to x, copy and complete this table.

x	2	5	
y		15	39

7. A stone is dropped from the top of a cliff. The speed of the stone, v m/s, varies directly with the time, t seconds, after it was dropped. After 2 seconds its speed is 19.6 m/s.
   a. Find the formula for v in terms of t.
   b. Find the speed when $t = 3$
   c. Find the time when $v = 24.5$

8. The extension of a spring, e cm, is proportional to the mass, m kg, that hangs from it.
   When a mass of 0.5 kg hangs from the spring its extension is 2.2 cm.
   a. Find the formula for e in terms of m.
   b. Find the extension when $m = 0.8$
   c. Find the mass when $e = 1.76$

9. The depth of oil, d m, in a tank varies directly with the volume, V m³, of oil that is in the tank. When $d = 0.9$, $V = 0.6$
   a. Find the formula for d in terms of V.
   b. Find the depth when $V = 1.2$
   c. Find the volume when $d = 0.57$

Variation

Direct variation (nonlinear relationships)

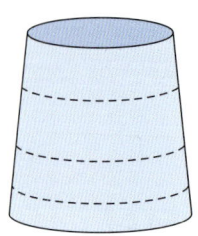

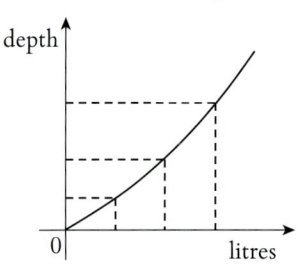

When the frustum of the cone is filled with water, each litre of water increases the depth of the water by different amounts.
A graph of the depth of the water against the number of litres in the frustum is a curve passing through the origin.

This relationship is known as a direct non-linear relationship.

> **EXAMPLE**
>
> y is proportional to the square of x and $y = 12$ when $x = 2$
> **a** Find the formula for y in terms of x.
> **b** Find the value of y when $x = 4$ **c** Find the value of x when $y = 3$

a $y \propto x^2$ so $y = kx^2$
substituting $y = 12$ and $x = 2$ into $y = kx^2$ gives $12 = k \times 2^2$
$12 = 4 \times k$
$k = 3$

The formula is $y = 3x^2$

b   substituting $x = 4$ into $y = 3x^2$ gives $y = 3 \times 4^2$
$y = 3 \times 16$ $y = 48$

c   substituting $y = 3$ into $y = 3x^2$ gives $3 = 3x^2$
$x^2 = 1$
$x = \pm 1$

➡ **NOTE:** there are two numbers that square to give the number 1.

> **EXAMPLE**
>
> The time for a pendulum to swing, t seconds, is proportional to the square root of the length, l m, of the pendulum.
> $l = 0.81$ when $t = 1.8$. Find the value of t when $l = 0.49$

You must first find the formula connecting t and l.
$t \propto \sqrt{l}$ so $t \propto k\sqrt{l}$
substituting $t = 1.8$ and $l = 0.81$ into $t = k\sqrt{l}$ gives $1.8 = k \times \sqrt{0.81}$
$1.8 = k \times 0.9$
$k = 2$

The formula is $t = 2\sqrt{l}$

substituting $l = 0.49$ into $t = 2\sqrt{l}$ gives $t = 2 \times \sqrt{0.49}$
$t = 2 \times 0.7$
$t = 1.4$

EXERCISE 8.7

1. y is proportional to the square of x and $y = 32$ when $x = 4$
   a Find the formula for y in terms of x. b Find the value of y when $x = 5$
   c Find the values of x when $y = 128$

2. y varies directly as the square of x and $y = 4.5$ when $x = 3$
   a Find the formula for y in terms of x. b Find the value of y when $x = 9$
   c Find the value of x when $y = 32$

3. y is proportional to the cube of x and $y = 108$ when $x = 3$
   a Find the formula for y in terms of x. b Find the value of y when $x = 2$
   c Find the value of x when $y = 256$

4. y is directly proportional to $(x + 3)^2$ and $y = 32$ when $x = 1$
   a Find the formula for y in terms of x. b Find the value of y when $x = 4$
   c Find the values of x when $y = 50$

5. y varies directly as $(x - 2)^2$ and $y = 20$ when $x = 4$
   a Find the formula for y in terms of x. b Find the value of y when $x = 8$

6. y is proportional to the square root of x and $y = 12$ when $x = 4$
   a Find the formula for y in terms of x. b Find the value of y when $x = 6.25$
   c Find the value of x when $y = 48$

7. If y varies directly as the cube of x, copy and complete this table.

x	1	2	
y		16	250

8. The breaking distance, d m, of a car is proportional to the square of the speed, v mph.
   $d = 6$ when $v = 20$
   a Find the formula for d in terms of v.
   b Find the value of d when $v = 40$

9. The energy (E) stored in a spring is proportional to the square of the extension (e).
   $E = 300$ when $e = 5$
   a Find the formula for E in terms of e.
   b Find the value of E when $e = 2.5$

10. The air resistance (R) to a car varies directly with the square of the speed (v).
    $v = 40$ when $R = 3200$
    a Find the formula for R in terms of v.
    b Find the value of R when $v = 35$

Inverse variation

When two variables are in **inverse proportion** (or **inverse variation**), one of the variables increases as the other decreases.

The number of tractors ploughing a field and the time taken are in inverse proportion.
If 1 tractor takes 6 hours then 2 tractors will take 3 hours and the following table can be obtained.

Number of tractors (x)	1	2	3	4	5	6
Time taken (y hours)	6	3	2	1.5	1.2	1

Notice that the product of corresponding x and y values in the table above always give the number 6. So $xy = 6$.
Dividing by x gives $y = \dfrac{6}{x}$.

> If two variables x and y are in inverse proportion, then the product of their values will stay the same.
> 'y is inversely proportional to x' can be written as $y \propto \dfrac{1}{x}$ or $y = \dfrac{k}{x}$ where k is the constant of proportionality.

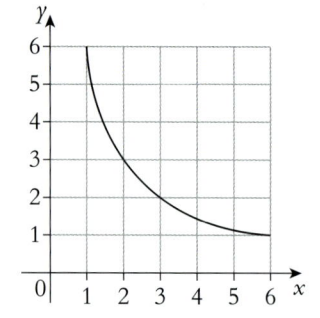

EXAMPLE

y is inversely proportional to x and $y = 8$ when $x = 2$
a Find the formula for y in terms of x
b Find the value of y when $x = 4$
c Find the value of x when $y = 0.5$

a $y \propto \dfrac{1}{x}$

$y = \dfrac{k}{x}$

$8 = \dfrac{k}{2}$

$k = 16$

The formula is $y = \dfrac{16}{x}$

b $y = \dfrac{16}{x}$

$y = \dfrac{16}{4}$

$y = 4$

c $y = \dfrac{16}{x}$

$0.5 = \dfrac{16}{x}$

$x = \dfrac{16}{0.5}$

$x = 32$

EXAMPLE

y is inversely proportional to x.
a Find the formula for y in terms of x.
b Find the values of a and b.

x	a	5	20
y	15	6	b

a $5 \times 6 = 30$ so $xy = 30$ so formula is $y = \dfrac{30}{x}$
b $a \times 15 = 30$ so $a = 2$
c $20 \times b = 30$ so $b = 1.5$

EXERCISE 8.8

1. y is inversely proportional to x and $y = 3$ when $x = 2$
   a Find the formula for y in terms of x.
   b Find the value of y when $x = 6$
   c Find the value of x when $y = 1.5$

2. y varies inversely as x and $y = 10$ when $x = 2.4$
   a Find the formula for y in terms of x.
   b Find the value of y when $x = 4$
   c Find the value of x when $y = 8$

3. y is inversely proportional to x and $y = 5$ when $x = 2$
   a Find the formula for y in terms of x.
   b Find the value of y when $x = 0.5$
   c Find the value of x when $y = 2.5$

4. y varies inversely as x and $y = 5$ when $x = 0.1$
   a Find the formula for y in terms of x.
   b Find the value of y when $x = 2$
   c Find the value of x when $y = 1$

5. If y varies inversely as x, copy and complete this table.

x	2	4	
y	18		6

6. If y is inversely proportional to x, copy and complete this table.

x	2.5	4	
y		25	20

7. The wave length, w, of a radio signal varies inversely as its frequency, f.
   $w = 1000$ when $f = 300$
   a Find the formula for w in terms of f.
   b Find the value of w when $f = 150$
   c Find the value of f when $w = 200$

8. The pressure, P, of a gas varies inversely as the volume, V, of the gas.
   $P = 1.5$ when $V = 200$
   a Find the formula for P in terms of V.
   b Find the value of P when $V = 150$
   c Find the value of V when $P = 2.5$

9. The time taken, t hours, to build a brick wall is inversely proportional to the number of men, n, that build the wall.
   $t = 12$ when $n = 2$
   a Find the formula for t in terms of n.
   b Find the value of t when $n = 3$
   c Find the value of n when $t = 4$

Further inverse variation

> **EXAMPLE**
>
> y varies inversely as $(x-2)^2$ and $y = 1$ when $x = 5$
> **a** Find the formula for y in terms of x
> **b** Find the value of y when $x = 4$
> **c** Find the values of x when $y = 9$
>
> **a** $y \propto \dfrac{1}{(x-2)^2}$ so $y = \dfrac{k}{(x-2)^2}$
>
> substituting $y = 1$ and $x = 5$ into $y = \dfrac{k}{(x-2)^2}$ gives $1 = \dfrac{k}{(5-2)^2}$
>
> $$1 = \dfrac{k}{9}$$
> $$k = 9$$
>
> The formula is $y = \dfrac{9}{(x-2)^2}$
>
> **b** substituting $x = 4$ into $y = \dfrac{9}{(x-2)^2}$ gives $y = \dfrac{9}{(4-2)^2}$
>
> $$y = \dfrac{9}{4}$$
> $$y = 2.25$$
>
> **c** substituting $y = 9$ into $y = \dfrac{9}{(x-2)^2}$ gives $9 = \dfrac{9}{(x-2)^2}$
>
> $9(x-2)^2 = 9$ multiply both sides by $(x-2)^2$
> $(x-2)^2 = 1$ divide both sides by 9
> $x^2 - 4x + 4 = 1$ expand brackets
> $x^2 - 4x + 3 = 0$ collect on one side
> $(x-3)(x-1) = 0$ factorise
> $x - 3 = 0$ or $x - 1 = 0$
> $x = 3$ or $x = 1$
>
> ➡ **CHECK:** $\dfrac{9}{(3-2)^2} = \dfrac{9}{1^2} = 9$ and $\dfrac{9}{(1-2)^2} = \dfrac{9}{(-1)^2} = \dfrac{9}{1} = 9$ ✓

EXERCISE 8.9

1 Match these cards.

a y varies inversely as the square of x	**A** $y \propto \dfrac{1}{\sqrt{x}}$
b y is proportional to the cube of x	**B** $y \propto \dfrac{1}{x^2}$
c y varies inversely as the cube of x	**C** $y \propto x^3$
d y varies inversely as the square root of x	**D** $y \propto x^2$
e y varies directly as the square of x	**E** $y \propto \dfrac{1}{x^3}$

2 y is inversely proportional to the square of x and $y = 4$ when $x = 3$
  a Find the formula for y in terms of x.
  b Find the value of y when $x = 2$
  c Find the value of x when $y = 1$

3 y varies inversely as the square root of x and $y = 10$ when $x = 4$
  a Find the formula for y in terms of x.
  b Find the value of y when $x = 16$
  c Find the value of x when $y = 4$

4 y is inversely proportional to $(x + 2)$ and $y = 3$ when $x = 3$
  a Find the formula for y in terms of x.
  b Find the value of y when $x = 8$
  c Find the value of x when $y = 3.75$

5 y varies inversely as $(x + 1)^2$ and $y = 4$ when $x = 1$
  a Find the formula for y in terms of x.
  b Find the value of y when $x = 3$
  c Find the values of x when $y = 16$

6 If y is inversely proportional to x^2, copy and complete this table.

x		2	5
y	50		2

7 The force of attraction, F newtons, between two magnets, is inversely proportional to the square of the distance, d cm, between them. $F = 5$ when $d = 2$
  a Find the formula for F in terms of d.
  b Find the value of F when $d = 0.5$
  c Find the value of d when $F = 0.2$

8 The intensity of sound (I decibels) varies inversely as the square of the distance (d m) from the source. For a jet engine the sound intensity is 140 decibels at a distance of 30 metres from the engine.
  a Find the formula for I in terms of d.
  b Find the value of I when $d = 100$
  c Find the value of d when $I = 2$

9 Match the four variation equations with the correct sketch graph.

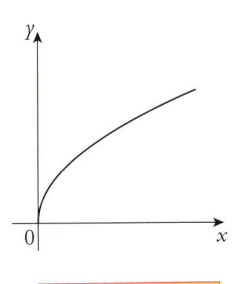

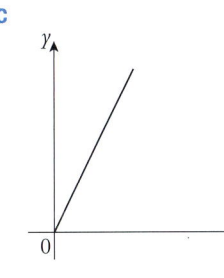

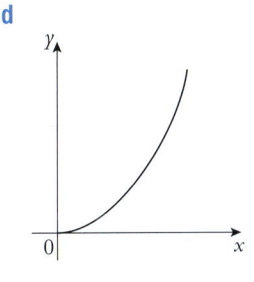

a **b** **c** **d**

A $y = \dfrac{k}{x}$ **B** $y = kx$ **C** $y = kx^2$ **D** $y = k\sqrt{x}$

> **KEY WORDS**
> direct proportion
> direct variation
> constant of proportionality
> inverse proportion
> inverse variation

Sine and cosine ratios up to 180°

THIS SECTION WILL SHOW YOU HOW TO
- Draw the sine and cosine graphs from 0° to 180°
- Solve trigonometric equations using the symmetry properties of the sine and cosine curves

The sine curve

You can use your calculator to find the y values, in the table below, for the graph of $y = \sin x$.

x	0	30	60	90	120	150	180
y	0	0.5	0.87	1	0.87	0.5	0

The graph of $y = \sin x$ can then be drawn using the values in the table.

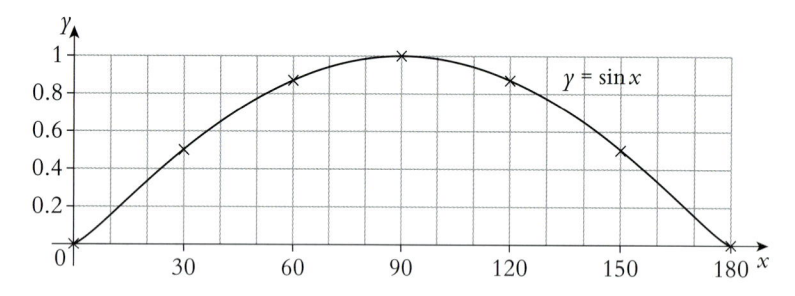

This graph starts at 0, increases to a maximum of 1 at 90° and then decreases to 0 at 180°.

The cosine curve

You can use your calculator to find the y values, in the table below, for the graph of $y = \cos x$.

x	0	30	60	90	120	150	180
y	1	0.87	0.5	0	−0.5	−0.87	−1

The graph of $y = \cos x$ can then be drawn using the values in the table.

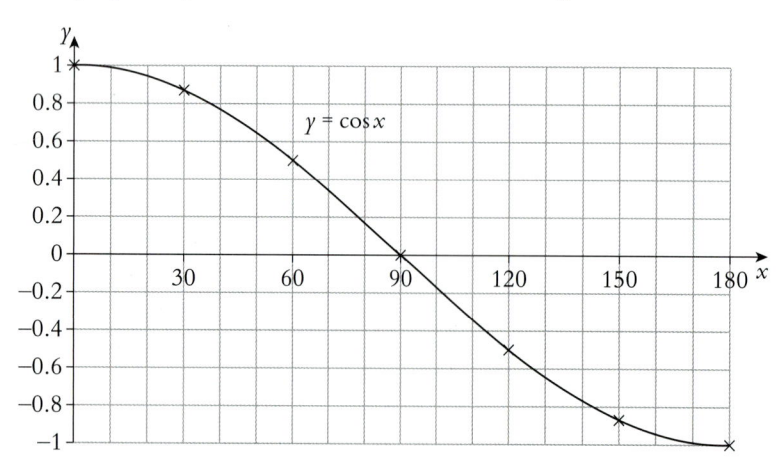

This graph starts at 1, decreases to 0 at 90° and then decreases further to −1 at 180°.
It is important to remember the shape of the sine and cosine curves.

Solving trigonometric equations

You are expected to be able to solve **trigonometric equations**, giving your answers in the range $0° \leq x \leq 180°$.

EXAMPLE

Solve the equation $\sin x = 0.4$ for $0° \leq x \leq 180°$.

You can find one answer to the equation $\sin x = 0.4$ using your calculator as follows
$x = \sin^{-1}(0.4)$
$x = 23.6°$
You must now check the sine curve
to see if there are any more answers.
The graph shows that there is another answer.
Using the symmetry of the curve,
the second answer is $x = 180° - 23.6° = 156.4°$
So the solution is $x = 23.6°$ or $x = 156.4°$

EXAMPLE

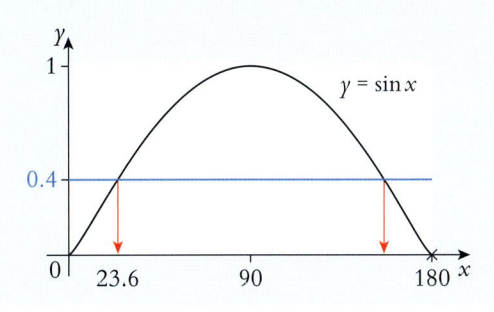

Solve the equation $\cos x = 0.624$ for $0° \leq x \leq 180°$.

Using a calculator
$x = \cos^{-1}(0.624)$
$x = 51.4°$
Now check the cosine curve
to see if there are any more answers.
There are no more answers.
So the only solution is $x = 51.4°$

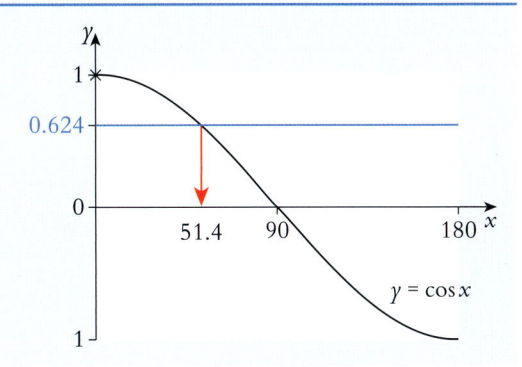

EXERCISE 8.10

Solve these equations for angles in the range $0° \leq x \leq 180°$

1. $\cos x = 0.2$
2. $\cos x = 0.7$
3. $\sin x = 0.88$
4. $\cos x = -0.35$
5. $\sin x = 0.2$
6. $\sin x = 1$
7. $\cos x = -1$
8. $\sin x = 0$
9. $\cos x = -2$
10. $\sin x = -0.75$
11. $\cos x = -0.76$
12. $\sin x = 0.81$
13. $\cos x = \frac{2}{7}$
14. $\sin x = \frac{3}{5}$
15. $\cos x = -\frac{3}{4}$
16. $5\sin x = 4$
17. $3\sin x = -1$
18. $8\cos x = -9$
19. $2 + \cos x = 2.5$
20. $1 - \sin x = 0.66$
21. $3 + 2\sin x = 4$
22. $(\cos x)^2 = \frac{1}{4}$
23. $(\sin x)^2 = \frac{4}{9}$
24. $2 + (\cos x)^2 = 2.9$

KEY WORDS
sine curve
cosine curve
trigonometric equation

Area of a triangle

> **THIS SECTION WILL SHOW YOU HOW TO**
> - Calculate the area of a triangle using $\frac{1}{2}ab\sin C$

You can use trigonometry to work out the area of a triangle.

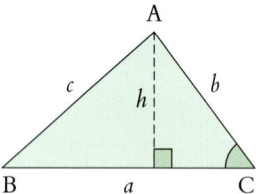

➡ **NOTE:**
angle A is opposite side a
angle B is opposite side b
angle C is opposite side c

Using the right-hand triangle $\sin C = \dfrac{h}{b}$ so $h = b\sin C$

Area of triangle = $\dfrac{1}{2}$ × base × perpendicular height = $\dfrac{1}{2} \times a \times h$

replacing h by $b\sin C$ gives area = $\dfrac{1}{2}ab\sin C$

> Area of triangle = $\dfrac{1}{2}ab\sin C$

You can use this formula when you know two sides and the angle between them.

EXAMPLE

Find the area of these triangles.

a) 5.6 cm, 35°, 8.7 cm

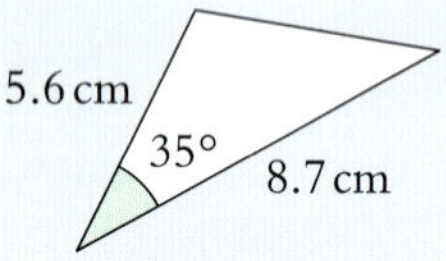

b) Equilateral triangle, 5 cm

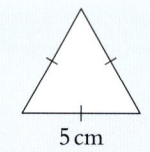

a Area = $\dfrac{1}{2} \times 8.7 \times 5.6 \times \sin 35°$
    = 14.0 cm² (to 3 s.f.)

b Equilateral triangle so each angle is 60°
    Area = $\dfrac{1}{2} \times 5 \times 5 \times \sin 60°$
    = 10.8 cm² (to 3 s.f.)

EXAMPLE

Find the area of the parallelogram.

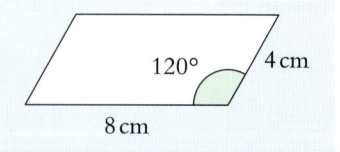

120°, 4 cm, 8 cm

Divide the parallelogram into two congruent triangles.

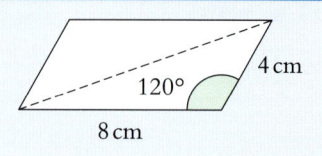

Area of triangle = $\dfrac{1}{2} \times 8 \times 4 \times \sin 120°$ = 13.856...

Area of parallelogram = 2 × 13.856 = 27.7 cm² (to 3 s.f.)

EXERCISE 8.11

1 Find the area of each triangle.

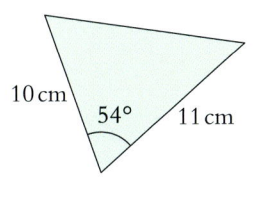

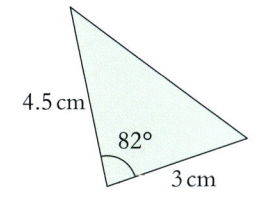

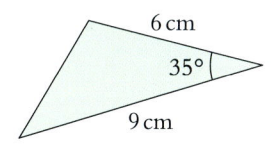

a) Triangle with sides 8 cm and 5 cm, included angle 130°.

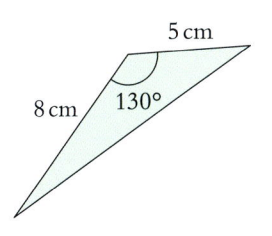

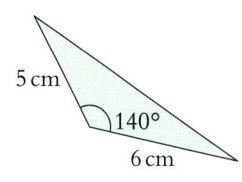

b) Triangle with sides 10 cm and 11 cm, included angle 54°.

c) Triangle with sides 5 cm and 6 cm, included angle 140°.

d) Triangle with sides 4.5 cm and 3 cm, included angle 82°.

e) Triangle with sides 6 cm and 9 cm, included angle 35°.

f) Triangle with sides 9.4 cm and 8.2 cm, included angle 58°.

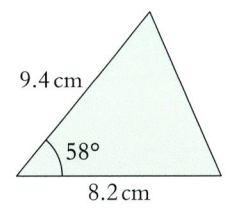

2 Find the area of each equilateral triangle.

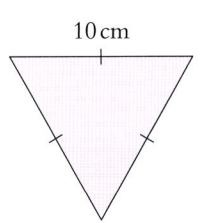

a) side 7 cm

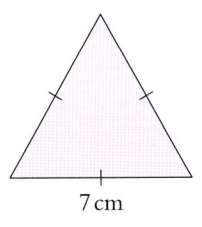

b) side 10 cm

c) side 8.2 cm

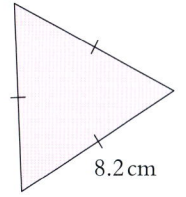

3 Find the area of each parallelogram.

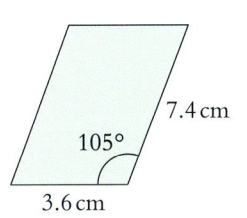

a) sides 15 cm and 12 cm, angle 72°

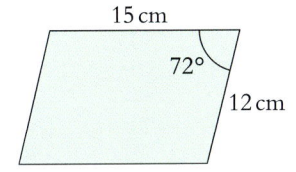

b) sides 7.4 cm and 3.6 cm, angle 105°

c) sides 10 cm and 7 cm, angle 58°

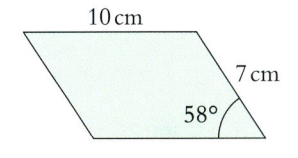

4 Find the area of each rhombus.

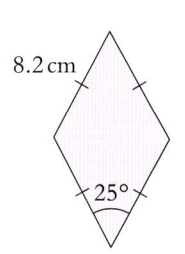

a) side 6 cm, angle 44°

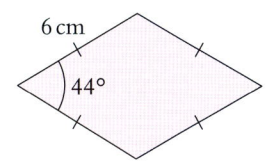

b) side 8.2 cm, angle 25°

c) side 4.6 cm, angle 100°

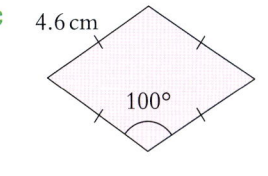

5 Find the area of a regular hexagon with sides of length 12 cm.

6   The area of the equilateral triangle is 40cm².
    Calculate the value of x.

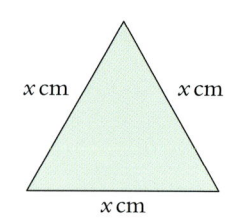

7   The area of the rhombus is 65 cm².
    Calculate the value of x.

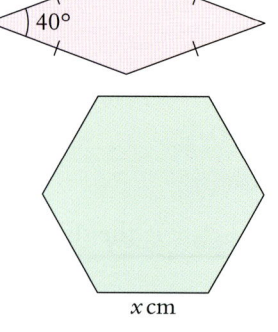

8   The area of the regular hexagon is 100 cm².
    Calculate the value of x.

9   The area of the triangle is 20 cm².
    Calculate the value of x.

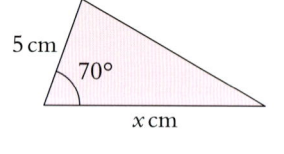

10 The area of the isosceles triangle is 82 cm².
    Calculate the value of x.

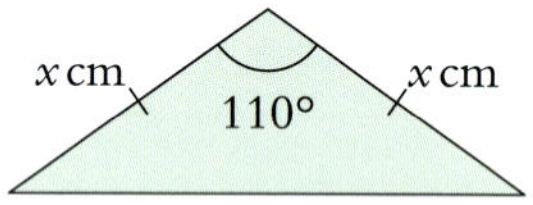

11 The regular tetrahedron has edges of length 10 cm.
    Calculate the total surface area of the tetrahedron.

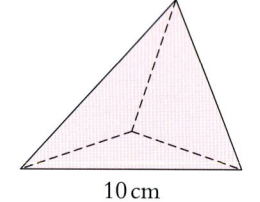

12 The regular octahedron has edges of length 6 cm.
    Calculate the total surface area of the octahedron.

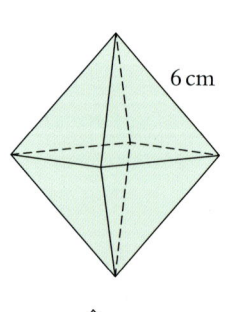

13 The triangle has an area of 18 cm².
    Calculate the two possible values of x.

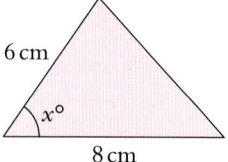

Area of a segment

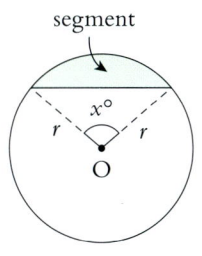

You can calculate the area of a segment of a circle using

Area of segment = area of sector − area of triangle

EXAMPLE

Find the area of the shaded segment.

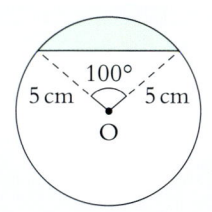

Area of segment = area of sector − area of triangle
$$= \frac{100}{360} \times \pi \times 5^2 - \frac{1}{2} \times 5 \times 5 \times \sin 100°$$
$$= 9.51 \text{ cm}^2 \text{ (to 3 s.f.)}$$

EXERCISE 8.12

1 Calculate the area of the shaded segment in each circle.

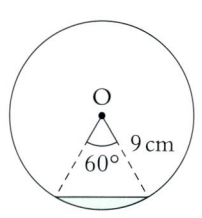

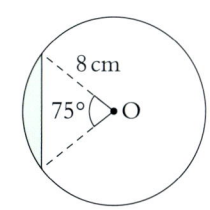

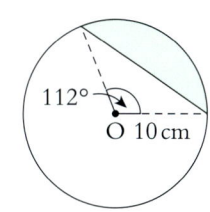

a 80°, 3 cm

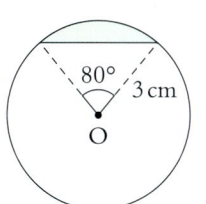

b 9 cm, 60°

c 120°, 4 cm

d 8 cm, 75°

e 112°, 10 cm

f 5 cm, 50°

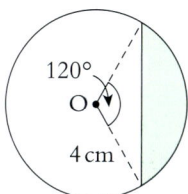

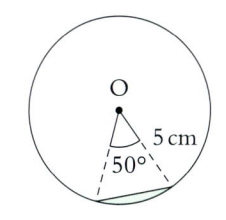

2 The area of the shaded segment is 100 cm². Calculate the value of r.

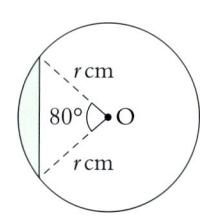

r cm, 80°, r cm

The sine and cosine rules

THIS SECTION WILL SHOW YOU HOW TO
- Find lengths and angles in triangles that are not right-angled

The sine rule

Using triangle AYC: Using triangle BYC:

$$\sin A = \frac{h}{b} \qquad \sin B = \frac{h}{a}$$

so $\quad h = b \sin A \qquad$ so $\quad h = a \sin B$

so $\quad b \sin A = a \sin B$

Rearranging gives $\dfrac{\sin A}{a} = \dfrac{\sin B}{b}$ or $\dfrac{a}{\sin A} = \dfrac{b}{\sin B}$

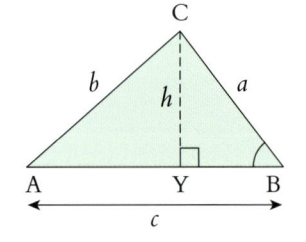

➡ **NOTE:**
angle A is opposite side a
angle B is opposite side b,
angle C is opposite side c

This rule is called the **sine rule** and it can be extended to include the third side and angle.

The sine rule: $\quad \dfrac{a}{\sin A} = \dfrac{b}{\sin B} = \dfrac{c}{\sin C}$

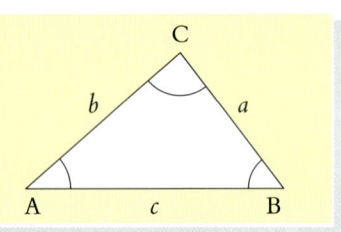

or $\quad \dfrac{\sin A}{a} = \dfrac{\sin B}{b} = \dfrac{\sin C}{c}$

The sine rule is used when you are working with two sides and two angles.

EXAMPLE

Find the value of x in each of these triangles.

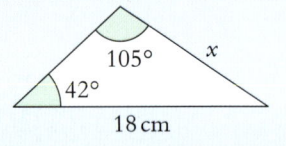

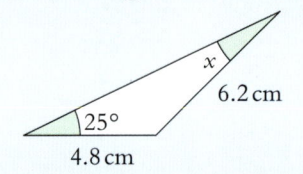

a) triangle with angles 42°, 105°, side 18 cm, side x

b) triangle with angle 25°, sides 4.8 cm, 6.2 cm, and x

➡ **NOTE:**
to find a side, use $\dfrac{a}{\sin A} = \dfrac{b}{\sin B}$

to find an angle, use $\dfrac{\sin A}{a} = \dfrac{\sin B}{b}$

a) $\dfrac{x}{\sin 42°} = \dfrac{18}{\sin 105°}$

$x = \sin 42° \times \dfrac{18}{\sin 105°}$

$x = 12.5$ cm (to 3 s.f.)

b) $\dfrac{\sin x}{4.8} = \dfrac{\sin 25°}{6.2}$

$\sin x = 4.8 \times \dfrac{\sin 25°}{6.2}$

$x = \sin^{-1}\left(4.8 \times \dfrac{\sin 25°}{6.2}\right)$

$x = 19.1°$ or $160.9°$ ⟵ This value is impossible because angles in a triangle add up to 180°
$x = 19.1°$ (to 1 d.p.) $(160.9° + 25° = 185.9°)$

404   UNIT 8

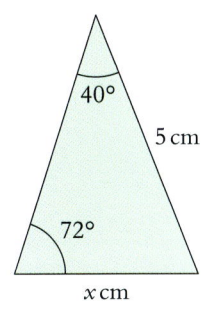

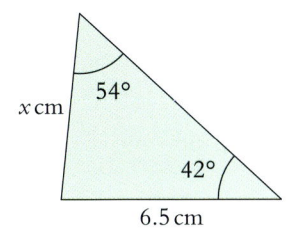

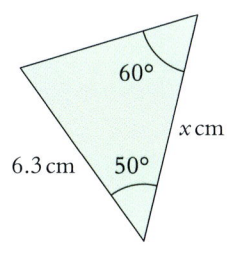

EXERCISE 8.13

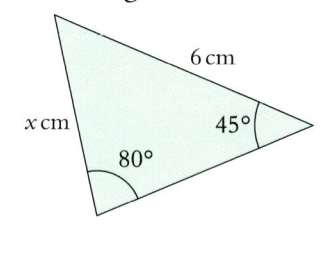

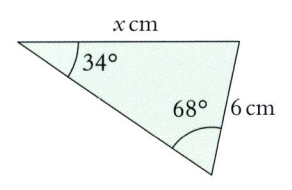

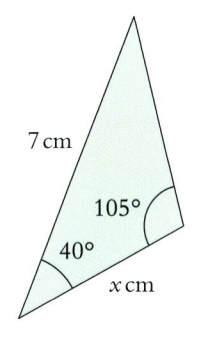

1 Find the value of x for each of these triangles.

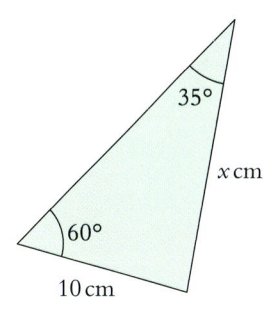

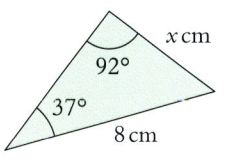

a) Triangle with angles 40° and 72°, side 5 cm opposite 72°, side x cm opposite 40°.

b) Triangle with angles 80° and 45°, side 6 cm opposite 80°, side x cm opposite 45°.

c) Triangle with angles 35° and 60°, side 10 cm opposite 35°, side x cm opposite 60°.

d) Triangle with angles 54° and 42°, side 6.5 cm opposite unknown angle, side x cm opposite 42°.

e) Triangle with angles 34° and 68°, side 6 cm opposite 34°, side x cm opposite the third angle.

f) Triangle with angles 92° and 37°, side 8 cm opposite 92°, side x cm opposite 37°.

g) Triangle with angles 60° and 50°, side 6.3 cm opposite 60°, side x cm opposite the third angle.

h) Triangle with angles 105° and 40°, side 7 cm opposite 40°, side x cm opposite 105°.

i) Triangle with angles 38° and 82°, side 12 cm opposite 82°, side x cm opposite 38°.

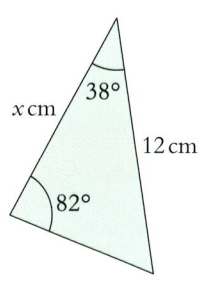

2 Find the value of x for each of these triangles.

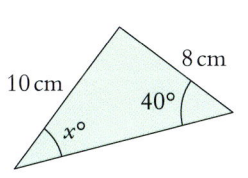

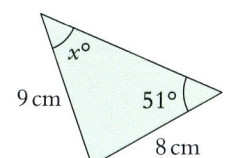

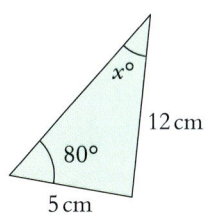

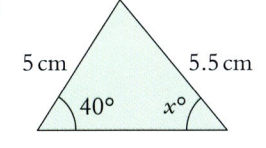

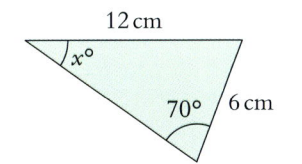

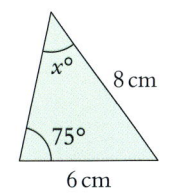

a) Triangle with side 10 cm, side 8 cm, included angle 40°, angle $x°$ opposite 8 cm.

b) Triangle with side 9 cm, side 8 cm, angle 51°, angle $x°$ opposite 8 cm.

c) Triangle with side 5 cm, side 12 cm, angle 80°, angle $x°$ opposite 5 cm.

d) Triangle with side 5 cm, side 5.5 cm, angle 40°, angle $x°$ opposite 5 cm.

e) Triangle with side 12 cm, side 6 cm, angle 70°, angle $x°$ opposite 6 cm.

f) Triangle with side 8 cm, side 6 cm, angle 75°, angle $x°$ opposite 6 cm.

The sine and cosine rules 405

g 13 cm, 15 cm, 82°, x°

h 20 cm, 15 cm, 100°, x°

i 8 cm, 65°, 11 cm, x°

3 Triangle ABC is isosceles.
Use the sine rule to calculate the value of x.

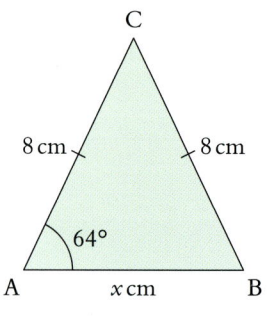

(Triangle with 8 cm, 8 cm, 64° at A, x cm at B)

4 ABCD is a quadrilateral.
  a Calculate the length of AC.
  b Calculate the size of angle CAD.

(Quadrilateral with D, 4.5 cm, C, 80°, 70°, 50°, A, 7 cm, B)

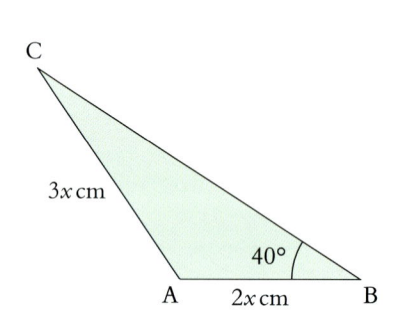

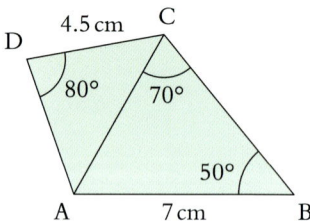

5 Calculate the size of angle ACB.

(Triangle with C, $3x$ cm, A, $2x$ cm, 40°, B)

6 The diagram shows three straight roads connecting points A, B and C.
The length of AB is 15 km.
  a Calculate the length of the road BC.
  b The bearing of A from B is 290°.
    i Find the bearing of B from A.
    ii Find the bearing of B from C.

(Diagram with A, 38°, C, 105°, 15 km, B)

7 The diagram shows three straight roads connecting points P, Q and R.
The length of PR is 8 km.
The bearing of R from P is 055°.
The bearing of Q from P is 080°.
The bearing of Q from R is 110°.
Calculate the length of PQ.

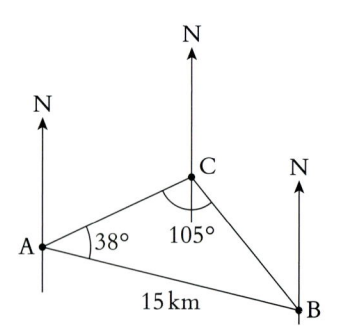

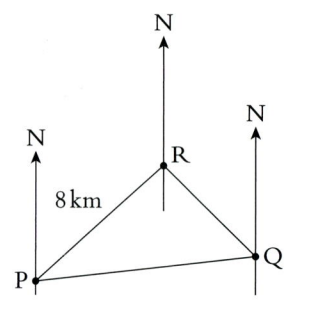

(Diagram with P, 8 km, R, Q)

Sine rule – the ambiguous case

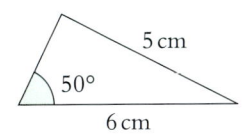

If you construct this triangle accurately, you will find there are two possible triangles that can be drawn.

ACCURATE CONSTRUCTIONS:

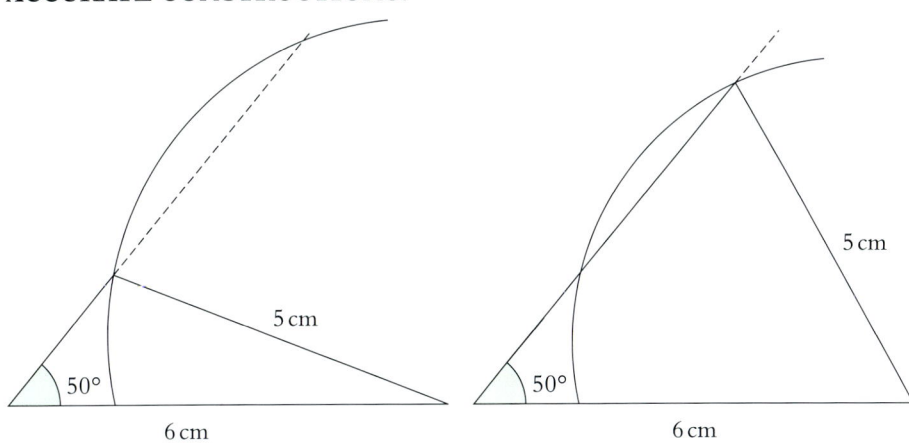

SOLVING USING SINE RULE

$$\frac{\sin x}{6} = \frac{\sin 50°}{5}$$

$$\sin x = 6 \times \frac{\sin 50°}{5}$$

$\sin x = 0.919...$

$x = 66.8°$ or $x = 180° - 66.8°$

$x = 66.8°$ or $x = 113.2°$

So the two possible triangles are

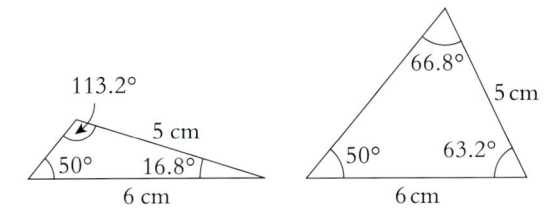

EXERCISE 8.14

Calculate the two possible values of x for each of these triangles. Check your calculations by constructing the triangles accurately.

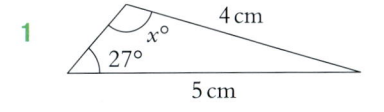

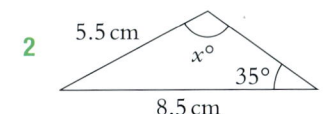

1.

2.

The cosine rule

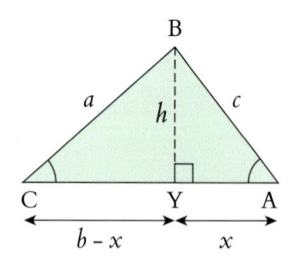

Using Pythagoras on triangle BYC:
$$h^2 = a^2 - (b-x)^2$$

Using Pythagoras on triangle AYB:
$$h^2 = c^2 - x^2$$

$a^2 - (b-x)^2 = c^2 - x^2$ *rearrange*
$a^2 = c^2 - x^2 + (b-x)^2$ *expand brackets*
$a^2 = c^2 - x^2 + b^2 - 2bx + x^2$ *collect like terms*
$a^2 = b^2 + c^2 - 2bx$

but, using △AYB $\cos A = \dfrac{x}{c}$ so $x = c \cos A$

so $a^2 = b^2 + c^2 - 2bc \cos A$

The cosine rule: $a^2 = b^2 + c^2 - 2bc \cos A$

or $\cos A = \dfrac{b^2 + c^2 - a^2}{2bc}$

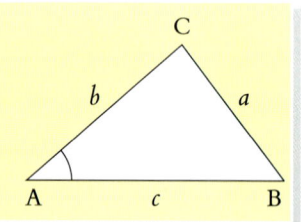

The cosine rule is used when you are working with three sides of a triangle and one angle.

Use $a^2 = b^2 + c^2 - 2bc \cos A$ to find the length of a side.

Use $\cos A = \dfrac{b^2 + c^2 - a^2}{2bc}$ to find the size of an angle.

EXAMPLE

Find the value of x in each of these triangles.

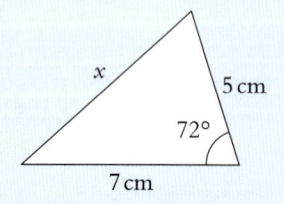

a

b

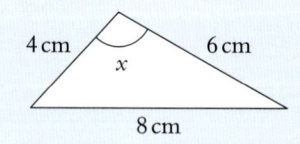

a Using the cosine rule:
$x^2 = 5^2 + 7^2 - 2 \times 5 \times 7 \times \cos 72°$
$x^2 = 52.368...$
$x = 7.24$ cm (to 3 s.f.)

b Using the cosine rule:
$\cos x = \left(\dfrac{4^2 + 6^2 - 8^2}{2 \times 4 \times 6} \right)$

$x = \cos^{-1} \left(\dfrac{4^2 + 6^2 - 8^2}{2 \times 4 \times 6} \right)$

$x = 104.5°$ (to 1 d.p.)

EXERCISE 8.15

1 Find the length of x for each of these triangles.

a 9 cm, 25°, 8 cm, x cm

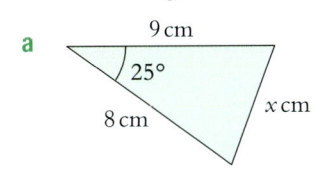

b 10 cm, x cm, 35°, 8 cm

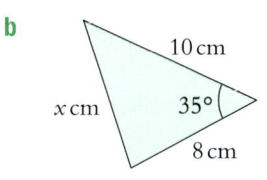

c 6 cm, 20°, 4 cm, x cm

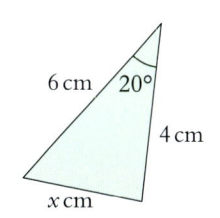

d x cm, 11 cm, 30°, 12 cm

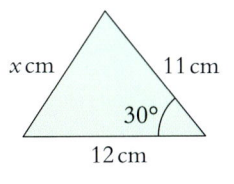

e 6 cm, x cm, 80°, 2 cm

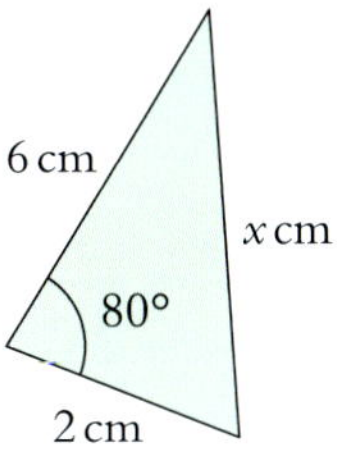

f x cm, 7.6 cm, 8 cm, 40°

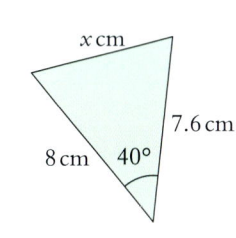

g 5 cm, x cm, 45°, 7.5 cm

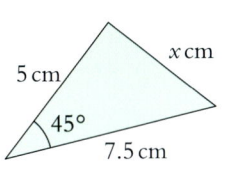

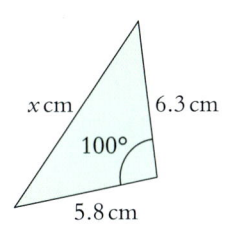

h x cm, 6.3 cm, 100°, 5.8 cm

i 7 cm, 30°, 10.2 cm, x cm

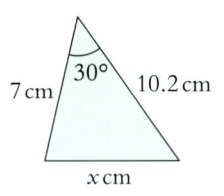

2 Find the value of x for each of these triangles.

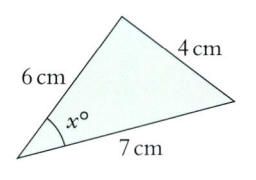

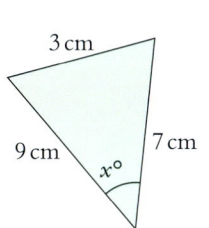

a 6 cm, 4 cm, $x°$, 7 cm

b 3 cm, 9 cm, 7 cm, $x°$

c 1.8 cm, 2 cm, $x°$, 2.5 cm

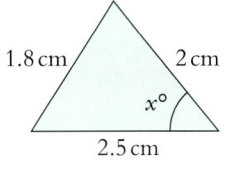

d 10 cm, $x°$, 8.3 cm, 5.6 cm

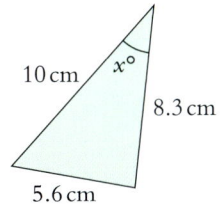

e 6 cm, 4 cm, $x°$, 4 cm

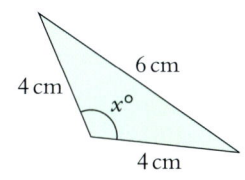

f 2 cm, $x°$, 3 cm, 1.8 cm

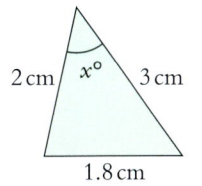

g 12 cm, 7.4 cm, $x°$, 7 cm

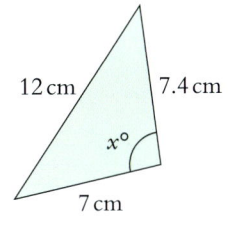

h 4 cm, $x°$, 5 cm, 2 cm

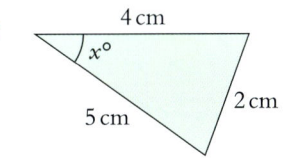

i $7y$, $3y$, $x°$, $5y$

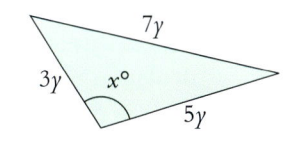

The sine and cosine rules   409

Solving problems using the sine and cosine rules

The sine and cosine rules are used to solve problems in triangles that are not right-angled.

Sometimes both of the rules have to be used.

Sometimes you have to choose carefully which rule is needed.

If no diagram is given, you must always start by sketching a clear diagram.

EXERCISE 8.16

1 Find the length of x for each of these triangles.

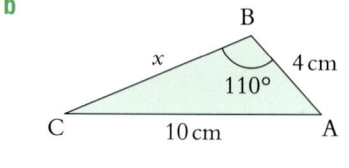

  a Triangle with C, B at top; CB = 15 cm, angle at B = 85°, side from B = 9 cm, x opposite at A.

  b Triangle with x from C to B, 4 cm from B to A, angle at A = 110°, CA = 10 cm.

  HINT
  you must first calculate angle ACB.

2 ABCD is a quadrilateral.
  a Find the length of BD.
  b Calculate the size of angle BCD.

  (DC = 2.4 cm, CB = 4.3 cm, DA = 3 cm, AB = 4 cm, right angle at A)

3 A, B and C are three points on horizontal ground.
  Point B is 5 km from A on a bearing of 120°, and C is 7 km from A on a bearing of 200°.
  a Find the size of angle BAC.
  b Calculate the length of BC.

4 In △ABC, AB = 6 cm, AC = 8 cm and BC = 4 cm.
  Calculate the size of the smallest angle.

  HINT
  the smallest angle is always opposite the shortest side.

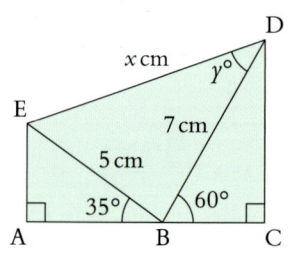

5 The lengths of the sides of a triangle are in the ratio 2 : 3 : 4.
  Calculate the size of the largest angle.

6 Calculate the value of x.

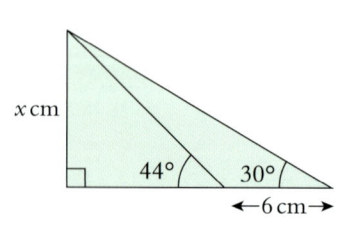

  (x cm side, angles 44° and 30°, base 6 cm)

7 a Find the size of angle DBE.
  b Calculate the area of triangle DBE
  c Calculate the value of x
  d Calculate the value of y
  e Calculate the angle of depression of E from D.

  (Figure: E to A with right angle at A, AB with 35° at B, BC with 60° at B, BD = 5 cm, DC = 7 cm, ED = x cm, angle at D = y°)

410   UNIT 8

8 Two sides of a triangle are 9.4 cm and 11.6 cm.
   The angle between the two sides is 65°.
   Calculate the perimeter of the triangle.

9 The parallelogram has sides of length 10 cm and 7 cm.
   The longest diagonal is 16 cm.
     a Calculate the length of the shortest diagonal.
     b Calculate the area of the parallelogram.

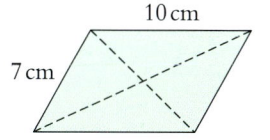

10 ABCD is a trapezium.
     a Calculate the length of BD.
     b Calculate the length of AD.
     c Calculate the area of the trapezium.

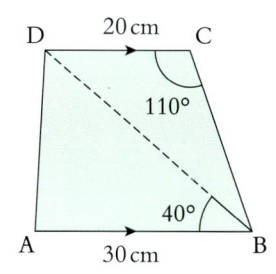

11 Calculate the area of the quadrilateral.

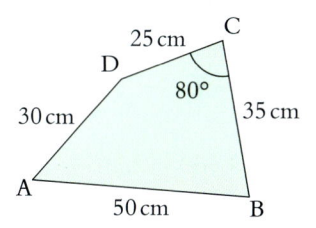

12 Triangle ABC has vertices A(1, 2), B(4, 6) and C(12, 0)
     a Calculate the length of
       i AB   **ii** BC   **iii** AC
     b Calculate the area of triangle ABC.

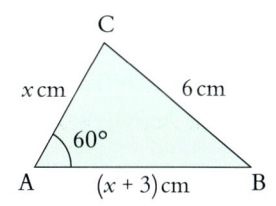

13 **a** Show that $x^2 + 3x - 27 = 0$.
     b Solve the equation to find the value of x.
     c Calculate the area of the triangle.

14 **a** Show that $3x^2 + 6x - 96 = 0$.
     b Solve the equation to find the value of x.
     c Calculate the area of the triangle.

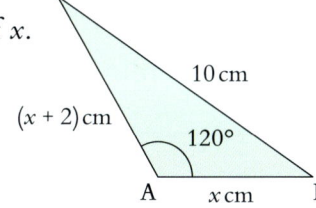

KEY WORDS
sine rule
cosine rule

The sine and cosine rules

Histograms

THIS SECTION WILL SHOW YOU HOW TO
• Draw and use a histogram

Grouped continuous data can be represented on a **histogram**.

In a histogram, the area of each bar is proportional to the frequency.

The bars can have different widths. (The width of a bar is called the **class width**.) The vertical axis represents the **frequency density**.

Frequency density = $\dfrac{\text{frequency}}{\text{class width}}$ or   frequency = frequency density × class width

EXAMPLE

The heights, h cm, of 200 plants are recorded. The results are shown in the frequency table. Draw a histogram to represent this information.

Height (h cm)	Frequency
$10 < h \leq 20$	20
$20 < h \leq 30$	40
$30 < h \leq 35$	40
$35 < h \leq 40$	45
$40 < h \leq 50$	25
$50 < h \leq 80$	30

Height (h cm)	Frequency	Class width	Frequency density
$10 < h \leq 20$	20	10	20 ÷ 10 = 2
$20 < h \leq 30$	40	10	40 ÷ 10 = 4
$30 < h \leq 35$	40	5	40 ÷ 5 = 8
$35 < h \leq 40$	45	5	45 ÷ 5 = 9
$40 < h \leq 50$	25	10	25 ÷ 10 = 2.5
$50 < h \leq 80$	30	30	30 ÷ 30 = 1

Add columns to the table to show the class widths and to calculate the frequency density.

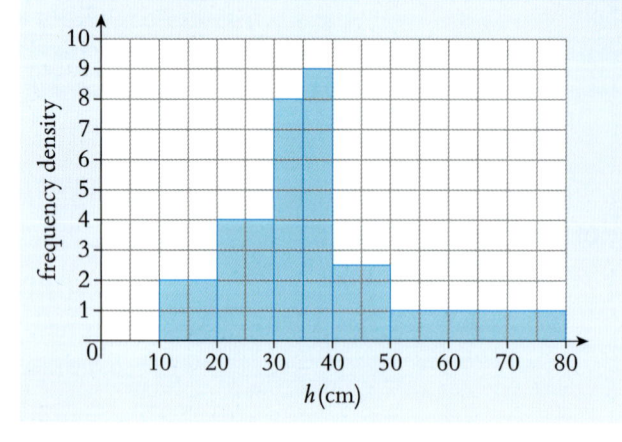

➡ **NOTE:** the vertical axis must be labelled frequency density.

412   UNIT 8

EXERCISE 8.17

1. The masses of 100 sweets are recorded.
   The results are shown in the frequency table.
   a Calculate the frequency density for each group.
   b Draw a histogram to represent this information.

Mass (m grams)	Frequency
$5 < m \leq 10$	5
$10 < m \leq 20$	35
$20 < m \leq 30$	45
$30 < m \leq 35$	15

2. The table shows the time, t hours, spent on homework each week by 100 students.
   a Calculate the frequency density for each group.
   b Draw a histogram to represent this information.

Time (t hours)	Frequency
$0 < t \leq 1$	7
$1 < t \leq 4$	39
$4 < t \leq 7$	42
$7 < t \leq 13$	12

3. The lengths of 85 phone calls are recorded.
   The results are shown in the table.
   a Calculate the frequency density for each group.
   b Draw a histogram to represent this information.

Time (t minutes)	Frequency
$0 < t \leq 1$	5
$1 < t \leq 3$	16
$3 < t \leq 7$	28
$7 < t \leq 9$	12
$9 < t \leq 15$	24

4. The heights of 200 students are measured.
   The results are shown in the table.
   a Calculate the frequency density for each group.
   b Draw a histogram to represent this information.

Height (h cm)	Frequency
$150 < h \leq 160$	24
$160 < h \leq 170$	58
$170 < h \leq 175$	63
$175 < h \leq 180$	45
$180 < h \leq 200$	10

5. The speeds of 130 cars passing a 70 km/h speed limit sign are recorded.
   The results are shown in the table.
   Use this information to draw a histogram.

Speed (v km/h)	Frequency
$50 < v \leq 60$	15
$60 < v \leq 65$	30
$65 < v \leq 70$	40
$70 < v \leq 75$	35
$75 < v \leq 90$	10

6. 200 students are given a puzzle to solve.
   The results are shown in the table.
   Use this information to draw a histogram.

Time (t minutes)	Frequency
$0 < t \leq 1$	4
$1 < t \leq 3$	45
$3 < t \leq 6$	90
$6 < t \leq 7$	26
$7 < t \leq 12$	35

Histograms

7 The table shows information about the time spent on the internet one evening by a group of 160 students. Draw a histogram to represent this information.

Time (t minutes)	Frequency
$0 < t \leq 40$	36
$40 < t \leq 70$	90
$70 < t \leq 150$	34

8 The histogram shows information about the lengths of 35 phone calls.

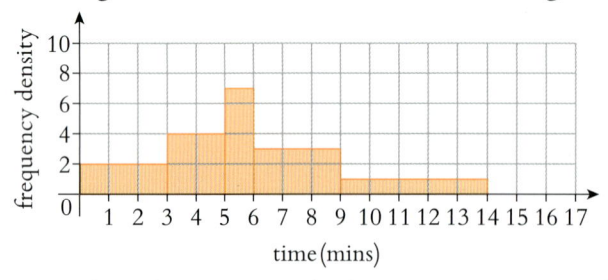

HINT

frequency = frequency density × class width

a Use the information in the histogram to copy and complete the following table.

Time (t mins)	$0 < t \leq 3$	$3 < t \leq 5$	$5 < t \leq 6$	$6 < t \leq 9$	$9 < t \leq 14$
Frequency					5

b Use the information in the table to calculate an estimate of the mean length of a phone call.

9 The histogram shows information about the masses of 25 newborn babies.

a Use the information in the histogram to copy and complete the following table.

Mass (m kg)	$1 < m \leq 2$	$2 < m \leq 3$	$3 < m \leq 3.5$	$3.5 < m \leq 4$	$4 < m \leq 5$
Frequency			6		

b Use the information in the table to calculate an estimate of the mean mass.

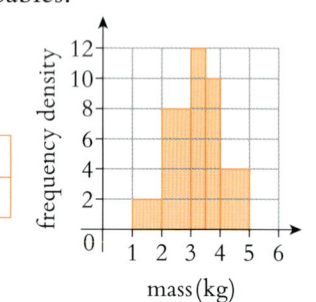

10 75 students do a sponsored swim to raise money for charity.
The table shows information about the amounts of money ($ x) raised.

Amount ($ x)	$0 < x \leq 10$	$10 < x \leq 20$	$20 < x \leq 35$	$35 < x \leq 50$
Frequency	15	40	15	5

When a histogram was drawn to show this information, the height of the column for the interval $20 < x \leq 35$ was 6 cm. Calculate the height of each of the other columns.

11 45 batteries are tested to see how long they last.
    The table shows the results

Number of hours (t)	$10 < t \leq 15$	$15 < t \leq 17$	$17 < t \leq 20$	$20 < t \leq 30$
Frequency	10	8	15	12

    When a histogram was drawn to show this information, the height of the column for the interval $10 < t \leq 15$ was 4 cm.
    Calculate the height of each of the other columns.

12 A group of students are asked how long they spent watching television one evening.
    The information is shown in the histogram.

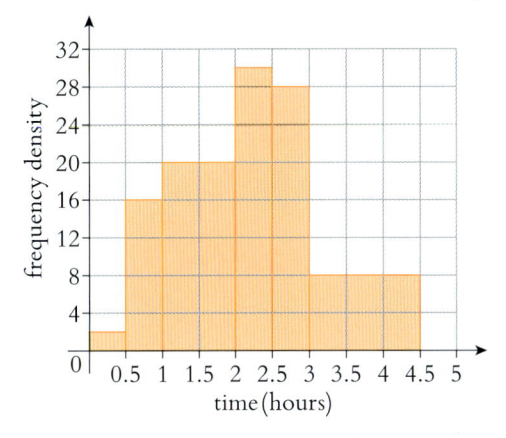

Time (t hours)	Frequency
$0 < t \leq 0.5$	1
$0.5 < t \leq 1$	
$1 < t \leq 2$	
$2 < t \leq 2.5$	
$2.5 < t \leq 3$	
$3 < t \leq 4.5$	

    a Use the information in the histogram to copy and complete the table.
    b Estimate how many students spent less than 1.5 hours watching television.
    c Use the information in the table to calculate an estimate of the mean number of hours.
    d A student is chosen at random.
       Find the probability that the student spent more than 2 hours watching television.

13 The frequency table and histogram are incomplete.
    They show information about the distances travelled to work by a group of workers.
    No worker travelled more than 28 km.

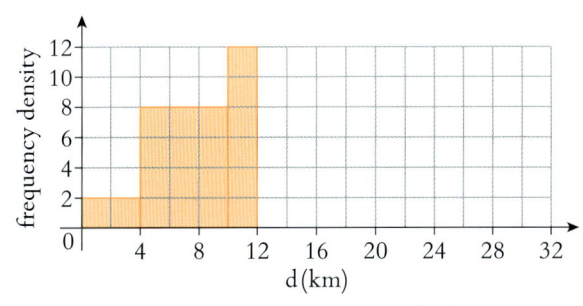

Distance (d km)	Frequency
$0 < d \leq 4$	8
$4 < d \leq 10$	
$10 < d \leq 12$	
$12 < d \leq 14$	20
$14 < d \leq 22$	40
$22 < d \leq 28$	18

    a Use the information given in the histogram to find the missing frequencies in the table.
    b Use the information given in the table to draw the missing bars on the histogram.
    c Calculate an estimate of the mean distance travelled.

    KEY WORDS
    histogram
    class width
    frequency density

Histograms   415

Unit 8 Examination questions

1 Solve the equation.
$$x^2 - 8x + 6 = 0$$
Show all your working and give your answers correct to 2 decimal places. [4]

Cambridge IGCSE Mathematics 0580, Paper 21 Q20, November 2010

2

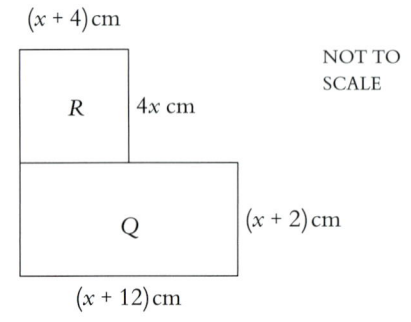

$(x + 4)$ cm

NOT TO SCALE

R $4x$ cm

Q $(x + 2)$ cm

$(x + 12)$ cm

 (a) (i) Write down an expression for the area of rectangle R. [1]
 (ii) Show that the total area of rectangles R and Q is
 $5x^2 + 30x + 24$ square centimetres. [1]
 (b) The total area of rectangles R and Q is 64 cm^2.
 Calculate the value of x correct to 1 decimal place. [4]

Cambridge IGCSE Mathematics 0580, Paper 2 Q20, November 2006

3

NOT TO SCALE

C

x cm

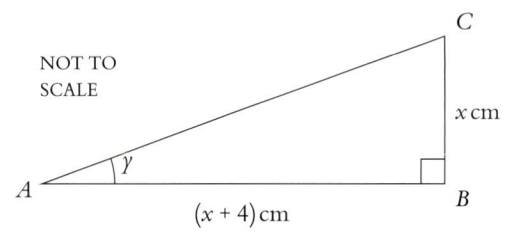

A y

$(x + 4)$ cm B

 (a) When the area of triangle ABC is 48 cm^2,
   (i) show that $x^2 + 4x - 96 = 0$, [2]
   (ii) solve the equation $x^2 + 4x - 96 = 0$. [2]
   (iii) write down the length of AB. [1]
 (b) When $\tan y = \dfrac{1}{6}$, find the value of x. [2]

 (c) When the length of AC is 9 cm,
   (i) show that $2x^2 + 8x - 65 = 0$, [2]
   (ii) solve the equation $2x^2 + 8x - 65 = 0$,
     (**Show your working** and give your answers correct to 2 decimal places.) [4]
   (iii) calculate the perimeter of triangle ABC. [1]

Cambridge IGCSE Mathematics 0580, Paper 4 Q2, November 2008

4 A packet of sweets contains chocolates and toffees.

  (a) There are x chocolates which have a total mass of 105 grams.
  Write down, in terms of x, the mean mass of a chocolate. [1]

  (b) There are $x + 4$ toffees which have a total mass of 105 grams.
  Write down, in terms of x, the mean mass of a toffee. [1]

  (c) The difference between the two mean masses in **parts a** and **b** is 0.8 grams.
  Write down an equation in x and show that it simplifies to $x^2 + 4x - 525 = 0$. [4]

  (d) (i) Factorise $x^2 + 4x - 525$. [2]
  (ii) Write down the solutions of $x^2 + 4x - 525 = 0$. [1]

  (e) Write down the total number of sweets in the packet. [1]

  (f) Find the mean mass of a sweet in the packet. [2]

Cambridge IGCSE Mathematics 0580, Paper 4 Q8, June 2007

5 (a) Solve the following equations.

  (i) $\dfrac{5}{w} = \dfrac{3}{w+1}$ [2]

  (ii) $(y + 1)^2 = 4$ [2]

  (iii) $\dfrac{x+1}{3} - \dfrac{x-2}{5} = 2$ [3]

  (b) (i) Factorise $u^2 - 9u - 10$. [2]
  (ii) Solve the equation $u^2 - 9u - 10 = 0$. [1]

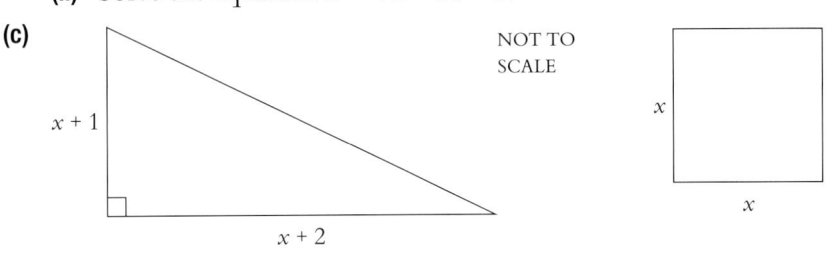

  (c)

  NOT TO SCALE

  Triangle with legs $x + 1$ and $x + 2$; square with side x.

  The area of the triangle is equal to the area of the square.
  All lengths are in centimetres.
  (i) Show that $x^2 - 3x - 2 = 0$. [3]
  (ii) Solve the equations $x^2 - 3x - 2 = 0$, giving your answers correct to 2 decimal places. Show all your working. [4]
  (iii) Calculate the area of one of the shapes. [1]

Cambridge IGCSE Mathematics 0580, Paper 41 Q9, June 2010

6 Write as a single fraction in its simplest form.

$$\frac{4}{2x+3} - \frac{2}{x-3}.$$ [3]

Cambridge IGCSE Mathematics 0580, Paper 21 Q11, November 2008

7 Simplify this fraction.

$$\frac{x^2 - 5x + 6}{x^2 - 4}$$ [4]

Cambridge IGCSE Mathematics 0580, Paper 21 Q16, November 2010

8 The braking distance, d, of a car is directly proportional to the square of its speed, v.
When $d = 5$, $v = 10$.
Find d when $v = 70$. [3]

Cambridge IGCSE Mathematics 0580, Paper 21 Q10, November 2009

9 A spray can is used to paint a wall.
The thickness of the paint on the wall is t. The distance of the spray can from the wall is d.
t is inversely proportional to the square of d.
$t = 0.2$ when $d = 8$.
Find t when $d = 10$. [3]

Cambridge IGCSE Mathematics 0580, Paper 21 Q13, June 2009

10 $\sin x° = 0.86603$ and $0 \leq x \leq 180$.
Find the two values of x. [2]

Cambridge IGCSE Mathematics 0580, Paper 21 Q6, November 2008

11

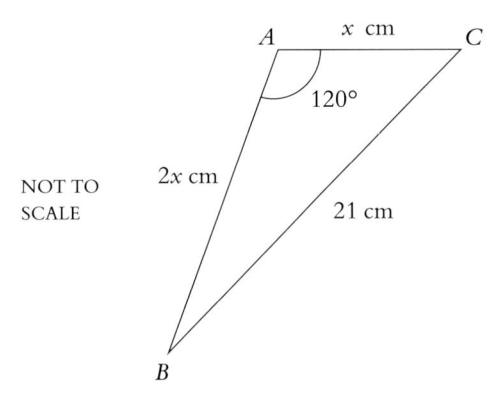

In triangle ABC, $AB = 2x$ cm, $AC = x$ cm, $BC = 21$ cm and angle $BAC = 120°$. Calculate the value of x. [3]

Cambridge IGCSE Mathematics 0580, Paper 21 Q11, June 2008

12 In triangle ABC, $AB = 6$ cm, $AC = 8$ cm and $BC = 21$ cm. Angle $ACB = 26.4°$. Calculate the area of the triangle ABC. [2]

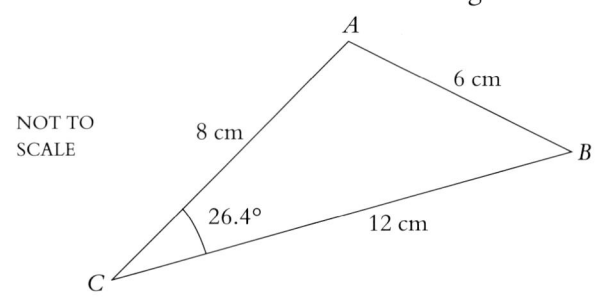

Cambridge IGCSE Mathematics 0580, Paper 2 Q5, June 2006

13

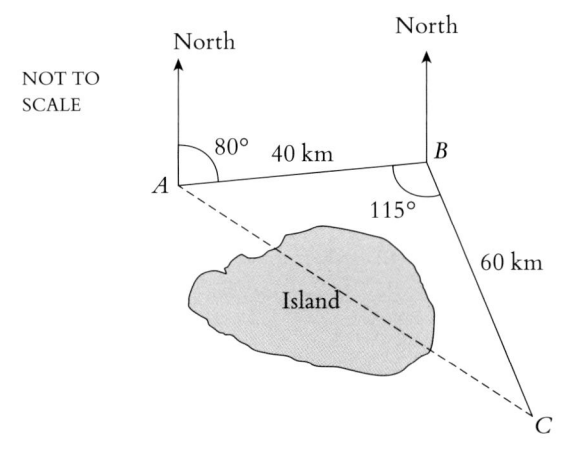

To avoid an island, a ship travels 40 kilometres from A to B and then 60 kilometres from B to C. The bearing of B from A is 080° and angle ABC is 115°.

(a) The ship leaves *A* at 11 55.
   It travels at an average speed of 35 km/h.
   Calculate, to the nearest minute, the time it arrives at *C*. [3]
(b) Find the bearing of
   (i) *A* from *B*, [1]
   (ii) *C* from *B*. [1]
(c) Calculate the straight line distance *AC*. [4]
(d) Calculate angle *BAC*. [3]
(e) Calculate how far *C* is **east** of *A*. [3]

Cambridge IGCSE Mathematics 0580, Paper 4 Q5, November 2008

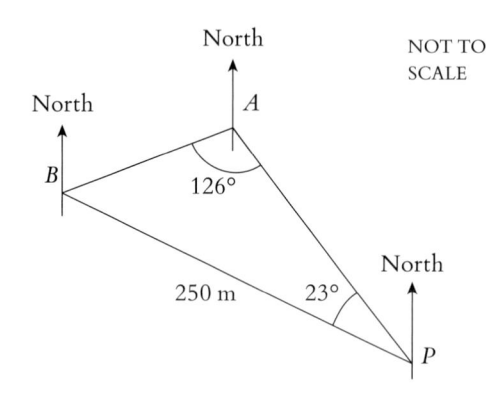

14

The diagram show three straight horizontal roads in a town, connecting points *P*, *A* and *B*.
PB = 250 m, angle *APB* = 23° and angle *BAP* = 126°.

(a) Calculate the length of the road AB. [3]
(b) The bearing of *A* from *P* is 303°.
   Find the bearing of
   (i) *B* from *P*, [1]
   (ii) *A* from *B*. [2]

Cambridge IGCSE Mathematics 0580, Paper 4 Q4, June 2009

15 The masses of 60 potatoes are measured.
The table shows the results.

Mass (m grams)	$10 < m \le 20$	$20 < m \le 40$	$40 < m \le 50$
Frequency	10	30	20

(a) Calculate an estimate of the mean. [4]

(b) On a copy of the grid, draw an accurate histogram to show the information in the table.

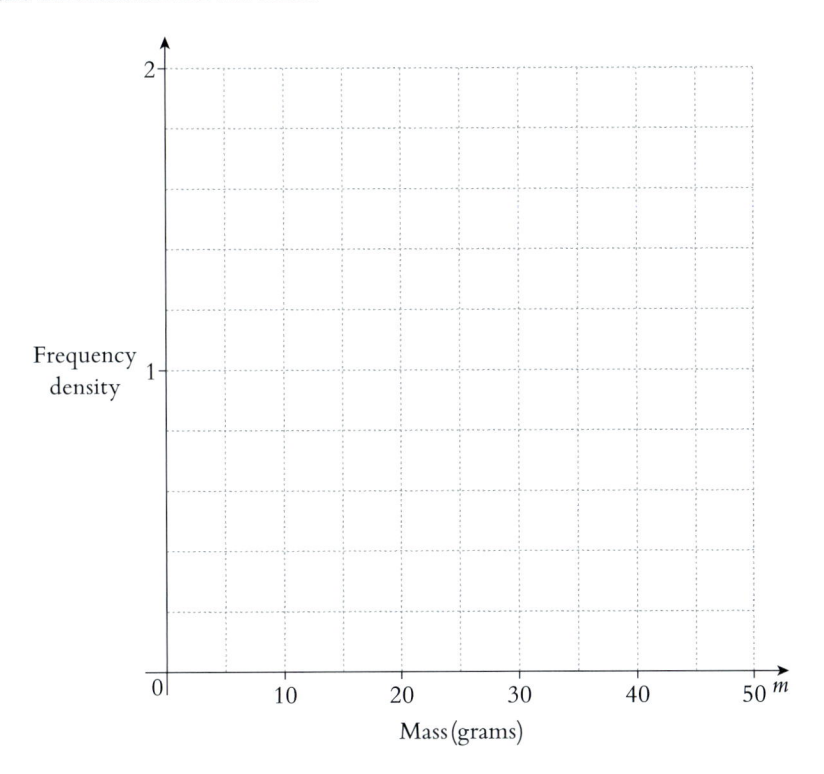

[3]

Cambridge IGCSE Mathematics 0580, Paper 41 Q6, June 2010

Upper and lower bounds

THIS SECTION WILL SHOW YOU HOW TO
- Find the upper and lower bounds of calculations

If the length of a nail is given as 6 cm correct to the nearest cm, you can find the **lower bound** and the **upper bound** for the length of the nail.

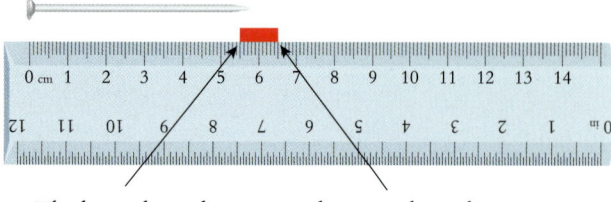

The lower bound is halfway between 5 cm and 6 cm. Lower bound = 5.5 cm.
The upper bound is halfway between 6 cm and 7 cm. Upper bound = 6.5 cm.
The range of possible values for the length of the nail can be written as:
$$5.5 \text{ cm} \leq \text{length} < 6.5 \text{ cm}$$

EXAMPLE

The mass of a parcel is 3.2 kg, correct to one decimal place.
Write down the upper and lower bounds for the mass of the parcel.

You need to think of the numbers (to 1 d.p.) directly below 3.2 and directly above 3.2.
These numbers are 3.1 and 3.3.
The lower bound is halfway between 3.1 and 3.2.
The upper bound is halfway between 3.2 and 3.3.
Lower bound = 3.15 cm   Upper bound = 3.25 cm.

Lower bound = 3.15 cm   Upper bound = 3.25 cm

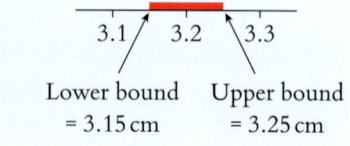

EXAMPLE

A rectangle has sides of length 5.7 cm and 3.4 cm, each correct to the nearest millimetre.
Calculate the upper and lower bounds for the area of the rectangle.

First write down the lower and upper bounds for the length and width of the rectangle.

Lower bound length = 5.65   Upper bound length = 5.75
Lower bound width = 3.35   Upper bound width = 3.45

So, the lower bound for the area = lower bound length × lower bound width
= 5.65 × 3.35
= 18.9275 cm²

and the upper bound for the area = upper bound length × upper bound width
= 5.75 × 3.45
= 19.8375 cm²

➡ **NOTE:** you must not round your answers when you are asked for a bound. You must write the full calculator display.

422   UNIT 9

EXAMPLE

$p = 8$ cm $\quad q = 2$ cm
The lengths of p and q have been given to the nearest centimetre.
Find the lower bound for each of the following calculations
  a $p + q$ b $p - q$ c $p \times q$ d $p \div q$

  a First write down the lower and upper bounds for the lengths of p and q.

   Lower bound of p = 7.5   Upper bound of p = 8.5
   Lower bound of q = 1.5   Upper bound of q = 2.5

   Lower bound of $(p + q)$ = lower bound of p + lower bound of q
   $\qquad\qquad\qquad\quad = 7.5 + 1.5$
   $\qquad\qquad\qquad\quad = 9$ cm

  b Lower bound of $(p - q)$ = lower bound of p − upper bound of q
   $\qquad\qquad\qquad\quad = 7.5 - 2.5$
   $\qquad\qquad\qquad\quad = 5$ cm

   ➡ **NOTE:** you must choose carefully which bounds to use in the calculation.

  c Lower bound of $(p \times q)$ = lower bound of p × lower bound of q
   $\qquad\qquad\qquad\quad = 7.5 \times 1.5$
   $\qquad\qquad\qquad\quad = 11.25$

  d Lower bound of $(p \div q)$ = lower bound of p ÷ upper bound of q
   $\qquad\qquad\qquad\quad = 7.5 \div 2.5$
   $\qquad\qquad\qquad\quad = 3$

EXERCISE 9.1

1 Write down the upper and lower bounds for each of these measurements.
  a 6 cm (to nearest cm)   b 32 min (to nearest minute)
  c 32 kg (to nearest kg)   d 9.2 cm (to nearest millimetre)
  e 7.7 cm (to 1 d.p.)   f 2.63 kg (to 3 s.f.)
  g 62.9 g (to 1 d.p.)   h 476 s (to nearest second)
  i 4.94 m (to 3 s.f.)   j 0.245 km (to nearest metre)

2 A bottle contains 250 ml of oil, correct to the nearest millilitre.
  Write down the range of possible values for the amount of oil,
  x millilitres, in the bottle.

3 A carton has a mass of 1.2 kg correct to 2 significant figures.
  Calculate the minimum possible total mass of 20 cartons.

4 The cost of making a necklace is $124 correct to the nearest dollar.
  Calculate the upper and lower bounds for the cost of making 150 necklaces.

Upper and lower bounds

5 When a bicycle wheel turns once, the bicycle travels 205 cm, correct to the nearest cm.
Calculate the upper and lower bounds for the distance travelled by the bicycle when the wheel has turned 150 times.

6 The length of each side of an equilateral triangle is 46 mm, correct to the nearest millimetre.
Calculate the upper and lower bounds for the perimeter of the triangle.

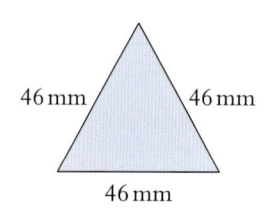

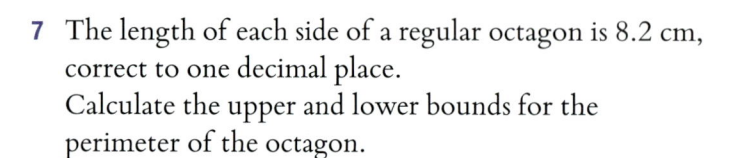

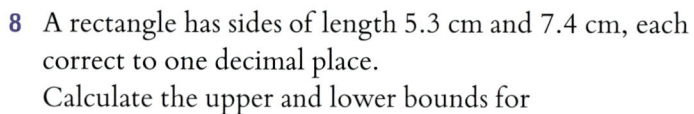

7 The length of each side of a regular octagon is 8.2 cm, correct to one decimal place.
Calculate the upper and lower bounds for the perimeter of the octagon.

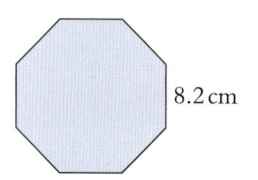

8 A rectangle has sides of length 5.3 cm and 7.4 cm, each correct to one decimal place.
Calculate the upper and lower bounds for
  a the perimeter of the rectangle,
  b the area of the rectangle.

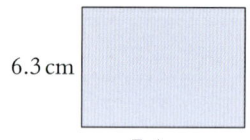

9 A rectangular photograph measures 22.4 cm by 18.6 cm, each correct to one decimal place.
Calculate the upper and lower bounds for
  a the perimeter of the photograph,
  b the area of the photograph.

10 The diagram shows a quadrilateral.
The lengths of the sides are given to the nearest cm.
Calculate the upper and lower bounds for the perimeter of the quadrilateral.

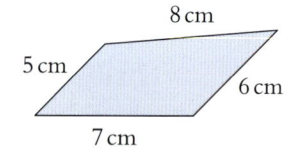

11 A cube has sides of length 4.7 cm correct to the nearest millimetre.
Calculate the upper and lower bounds for
  a the volume of the cube,
  b the surface area of the cube.

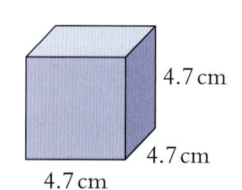

12 A cuboid has sides of length 8.1 cm, 5.2 cm and 3.5 cm correct to the nearest millimetre.
Calculate the upper and lower bounds for
  a the volume of the cuboid,
  b the surface area of the cuboid.

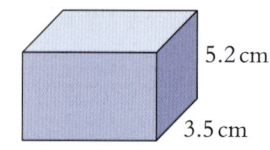

13 The lengths of the edges of the triangular prism are correct to the nearest centimetre.
    Calculate the upper and lower bounds for the volume of the prism.

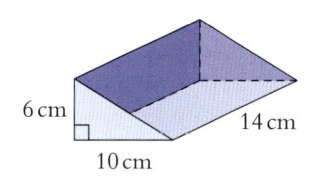

14 A truck can carry a maximum load of 8000 kg.
    The truck is loaded with packages each with a mass of 35 kg correct to two significant figures.
    Calculate the maximum number of packages that the truck can carry.

15 The area of a circle is 85 cm², correct to the nearest square centimetre.
    Calculate the lower bound for the radius, r, of the circle.
    (Write down all the numbers on your calculator display.)

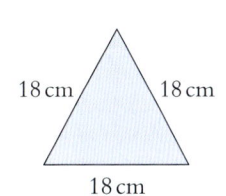

16 An equilateral triangle has sides of length 18 cm correct to the nearest centimetre.
    Calculate the lower bound for the area of the triangle.
    (Write down all the numbers on your calculator display.)

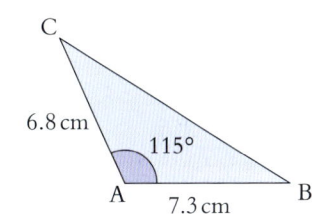

17 The lengths of AB and AC are correct to the nearest millimetre.
    Angle BAC is correct to the nearest degree.
    a   Calculate the lower bound for the length of BC.
    b   Calculate the upper bound for the area of the triangle.
    (In each case write down all the numbers on your calculator display.)

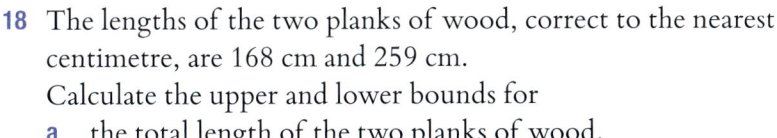

18 The lengths of the two planks of wood, correct to the nearest centimetre, are 168 cm and 259 cm.
    Calculate the upper and lower bounds for
    a   the total length of the two planks of wood.
    b   the difference in length of the two planks of wood.

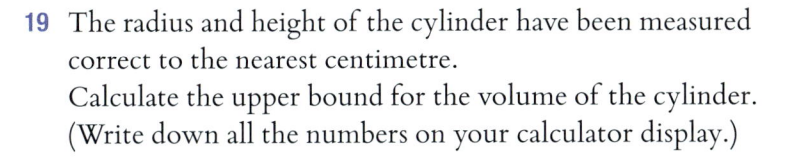

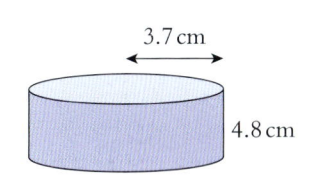

19 The radius and height of the cylinder have been measured correct to the nearest centimetre.
    Calculate the upper bound for the volume of the cylinder.
    (Write down all the numbers on your calculator display.)

20 The formula $a = \dfrac{v - u}{t}$ is used to find the acceleration for an object that increases its speed from u to v in a time t.
    If $u = 4.5$ m/s (to 1 d.p.), $v = 10.2$ m/s (to 1 d.p.) and $t = 3.6$ s (to 1 d.p.), calculate the upper bound for the acceleration.

KEY WORDS
upper bound
lower bound

Upper and lower bounds

Simultaneous equations 2

> **THIS SECTION WILL SHOW YOU HOW TO**
> - Solve simultaneous equations using the substitution method

Substitution method

In the substitution method you make x or y the subject of one of the equations and then substitute into the other equation.

EXAMPLE

Solve these simultaneous equations. $y = 3x - 10$
$x + y = -2$

$y = 3x - 10$ (1)
$x + y = -2$ (2)

Substitute for y from equation (1) into equation (2).

$x + (3x - 10) = -2$ *remove brackets and collect like terms*
$4x - 10 = -2$ *add 10 to both sides*
$4x = 8$ *divide both sides by 4*
$x = 2$

Substitute $x = 2$ in equation (1) and then solve to find y.

$y = 3 \times 2 - 10 = -4$

The solution is $x = 2$, $y = -4$

➡ **CHECK:** $-4 = 3 \times 2 - 10$ ✓
$2 + -4 = -2$ ✓

EXAMPLE

Solve these simultaneous equations. $2x + 3y = 0$
$x + 6y = 3$

$2x + 3y = 0$ (1)
$x + 6y = 3$ (2)

There is only one x in equation (2) so make x the subject of equation (2).

$x = 3 - 6y$

Substitute for x into equation (1).

$2(3 - 6y) + 3y = 0$ *remove brackets*
$6 - 12y + 3y = 0$ *collect like terms*
$6 - 9y = 0$ *add 9y to both sides*
$6 = 9y$ *divide both sides by 9*

$y = \dfrac{6}{9} = \dfrac{2}{3}$

Substitute $y = \dfrac{2}{3}$ in the equation $x = 3 - 6y$ to find x.

$x = 3 - \left(6 \times \dfrac{2}{3}\right) = 3 - 4 = -1$

The solution is $x = -1$, $y = \dfrac{2}{3}$

➡ **CHECK:** $(2 \times -1) + \left(3 \times \dfrac{2}{3}\right) = -2 + 2 = 0$ ✓

$-1 + \left(6 \times \dfrac{2}{3}\right) = -1 + 4 = 3$ ✓

EXERCISE 9.2

Solve these simultaneous equations

1.
   a) $x + 2y = 4$
   $x = 3y - 11$

   b) $x + 3y = 21$
   $y = 2x - 14$

   c) $5x + 2y = 1$
   $x = 3y + 7$

   d) $x + 2y = 11$
   $x = 2y + 5$

   e) $y = 6 + 2x$
   $4x + 2y = 16$

   f) $x = 3 - y$
   $2x - 3y = -19$

   g) $x = 2y + 10$
   $5x + 4y = 8$

   h) $y = 4x - 6$
   $2x - 3y = -7$

   i) $4x + 2y - 14 = 0$
   $x = y - 8$

   j) $5x - 3y = -17$
   $x = 3y - 25$

   k) $x = 18 + 7y$
   $x - 4y = 9$

   l) $y = 10x + 47$
   $2y - 5x = 34$

2.
   a) $y = 2x - 7$
   $y = x - 3$

   b) $x = 5y + 1$
   $x = 25 - 3y$

   c) $y = 3x - 9$
   $y = 4x + 14$

   d) $x = 3y + 5$
   $x = 2y - 4$

3.
   a) $2x + 3y = 0$
   $x - 2y = 14$

   b) $3x + y = 15$
   $5x - 3y = -3$

   c) $2x + y = -3$
   $3x - 2y = -8$

   d) $x - 5y = 18$
   $x + 3y = 2$

   e) $2x + 3y + 10 = 0$
   $x + 4y + 15 = 0$

   f) $x + 2y = 1$
   $3x - 2y = 19$

   g) $5x + 3y = -18$
   $2x + y = -7$

   h) $x + 2y = 11$
   $4y - x = -2$

   i) $3x + y = -1$
   $6x - 2y = -2$

   j) $x + 2y = 8$
   $3x + 4y - 6 = 0$

   k) $3x + 2y = 18$
   $x - 2y = -6$

   l) $3x + y = 0$
   $6x - 5y = 14$

4. Find the coordinates of the point where the lines $y = x + 2$ and $3x + 2y = 14$ intersect.

5. Find the coordinates of the point where the lines $y = 3x - 1$ and $2x - 5y = 18$ intersect.

6. A bag contains one-dollar coins and 50-cent coins. There are twice as many one-dollar coins as 50-cent coins and the total value of the coins is $42.50. How many one-dollar coins are there?

7. A magazine costs $2.50 more than a newspaper. The total cost of the magazine and the newspaper is $4.80. Find the cost of the magazine.

8. Adam is three times the age of Billy. The sum of their ages is 56 years. How old is Adam?

9. Sheena has some $10 and $5 notes. She has four times as many $10 notes as $5 notes and the total value of all the notes is $585. How many $5 notes does she have?

10. Theatre tickets cost either $30 or $42. One evening twice as many $30 tickets are sold as $42 tickets and the total money taken is $18 564. Find the total number of tickets that are sold.

KEY WORDS
substitution method

Linear programming

THIS SECTION WILL SHOW YOU HOW TO
- Solve problems using linear programming

EXAMPLE

A company plans to make two types of chair.
The machine time and craftsman's time needed for each type of chair are shown in the table.

	Chair A	Chair B
Machine time (hours)	2	3
Craftsman's time (hours)	4	2

The company has 30 hours of machine time and 32 hours of craftsman's time available.
The company wants to make at least 2 of chair A and at least 6 of chair B.
Let x be the number of chair A and y be the number of chair B.
Write down four inequalities in x and y.
Represent these four inequalities on a graph.
What is the largest number of type A chair that can be made?

Machine time gives $\quad 2x + 3y \leq 30$
Craftsman's time gives $\quad 4x + 2y \leq 32 \quad$ (this simplifies to $2x + y \leq 16$)
Number of chair A gives $\quad x \geq 2$
Number of chair B gives $\quad y \geq 6$
The boundary lines $2x + 3y = 30$, $2x + y = 16$, $x = 2$ and $y = 6$ need to be drawn.

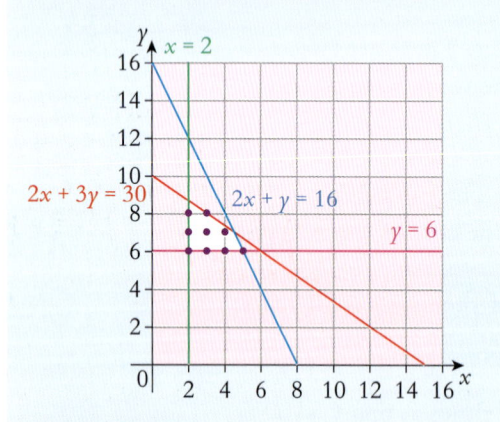

- The boundary lines are all **solid** lines in this example.
- Remember to shade the **unwanted** region.
- The unshaded region represents the required region.
- The red dots represent all the possible combinations of chairs in the required region (these are also shown in the table below).

Chair A (x)	2	2	2	3	3	3	4	4	5
Chair B (y)	6	7	8	6	7	8	6	7	6

The largest number of type A chair that can be made is 5.

➡ **NOTE:** the maximum or minimum numbers always occur at/near one of the corners of the required region.

EXERCISE 9.3

1. A car park has space for x cars and y motorcycles.
   The car park has an area of 900 m².
   Each car needs 15 m² of space and each motorcycle needs 3m² of space.
   There must be space for at least 40 cars.
   There must be space for at least 20 motorcycles.
   a Explain why $5x + y \leq 300$
   b Write down two more inequalities in x and y
   c Represent these three inequalities on a graph.
   d What is the largest possible number of cars in the car park?

2. Petra works in a hotel during the school holidays.
   In one week she spends x hours cleaning rooms and y hours waitressing.
   She spends no more than 15 hours working.
   She spends at least 6 hours waitressing.
   She spends at least twice as much time waitressing as she does cleaning.
   a Write down three inequalities in x and y.
   b Represent these three inequalities on a graph.
   c What is the largest possible number of hours she can spend cleaning rooms?

3. A builder plans to build some executive homes and some standard homes.
   He has 9000 m² of land available.
   He builds x executive homes and y standard homes.
   Each executive home requires 400 m² of land and each standard home requires 200 m² of land.
   The number of standard homes is to be less than the number of executive homes.
   There must be at least 10 standard homes.
   a Explain why $2x + y \leq 45$
   b Write down two more inequalities in x and y.
   c Represent these three inequalities on a graph.
   d What is the largest possible number of executive homes that can be built?

4. A school trip is planned for 100 students.
   The school uses x type A minibuses and y type B minibuses to transport the students.
   The type A minibus can carry 16 students and the type B minibus can carry 10 students.
   The school wishes to use no more than eight minibuses.
   The school wishes to use at least one type A minibus.
   The school wishes to use at least three type B minibuses.
   a Explain why $8x + 5y \geq 50$
   b Write down three more inequalities in x and y.
   c Represent these four inequalities on a graph.
   d What is the largest possible number of type A minibus that can be used?

Linear programming

EXAMPLE

An aircraft has 30 first class seats available and 220 economy class seats available.
A first class seat costs $240 and an economy class seat costs $120.
A flight will be cancelled if the money taken for the seats is less than $28 800.
Let x = the number of first class seats sold and y = the number of economy class seats sold.
a   Write down three inequalities in x and y.
b   Represent these three inequalities on a graph.
c   What is the minimum number of seats to be sold for the flight not to be cancelled?

a   First class seats $x \leq 30$
    Economy class seats $y \leq 220$
    Money taken $240x + 120y \geq 28800$ divide by 120 to simplify the inequality
                             $2x + y \geq 240$
    The three inequalities are $x \leq 30$, $y \leq 220$ and $2x + y \geq 240$.

b   The boundary lines $x = 30$, $y = 220$ and $2x + y = 240$ need to be drawn.
    To draw the line $2x + y = 240$ substitute $x = 0$ and $y = 0$ into the equation to find the axis crossing points.

x	0	120
y	240	0

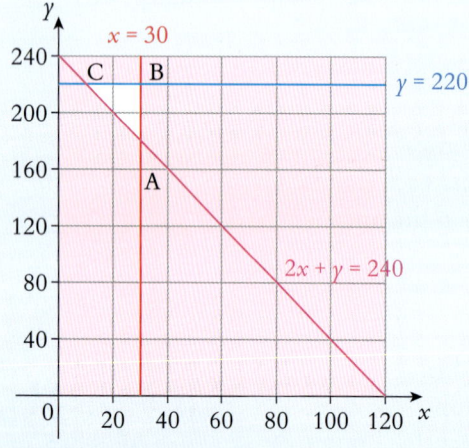

- The boundary lines are all **solid** lines in this example.
- Remember to shade the **unwanted** region.
- The unshaded region represents the required region.

c   The maximum or minimum occur at/near one of the corners of the required region.
    The three corners of the required region are A(30, 220), B(30, 180) and C(10, 220).

    At A number of seats = 30 + 220 = 250
    At B number of seats = 30 + 180 = 210
    At C number of seats = 10 + 220 = 230

    The minimum number of seats sold for the flight not to be cancelled is 210.

EXERCISE 9.4

1. A company makes model animals out of wood. They make giraffes and elephants.
   The carving time and sanding time needed for each model animal are shown below.

	Giraffe	Elephant
Carving time (hours)	4	3
Sanding time (hours)	1	1

   The company has 48 hours of carving time and 14 hours of sanding time available.
   The company wants to make at least 5 giraffes and at least 5 elephants.
   Let x be the number of giraffes and y be the number of elephants.
   a Write down four inequalities in x and y.
   b Represent these four inequalities on a graph.
   c The company makes a profit of $10 for each giraffe sold and a profit of $20 for each elephant sold.
      Calculate the greatest possible profit.

2. 4500 DVDs need to be packed into storage units.
   There are two sizes of storage units available.
   A large unit can hold 500 DVDs and a small unit can hold 300 DVDs.
   No more than 11 storage units are to be used.
   At least 2 of the storage units must be small.
   Let x be the number of large units used and y be the number of small units used.
   a Write down three inequalities in x and y.
   b Represent these three inequalities on a graph.
   c A large storage unit costs $10 and a small storage unit costs $8.
      Calculate the least possible cost of the storage units.

3. Lara is making some small cakes and biscuits to raise money for charity.
   She only has 600 g of flour and 580 g of sugar available.
   She has plenty of all the other ingredients.
   The amount of flour and sugar needed for each cake and biscuit are shown below.

	Small cake	Biscuit
Flour (g)	6	3
Sugar (g)	4	5

   Lara wants to make at least 50 small cakes and at least 40 biscuits.
   Let x be the number of small cakes and y be the number of biscuits made.
   a Write down four inequalities in x and y.
   b Represent these four inequalities on a graph.
   c Lara sells each cake for $1.50 and each biscuit for $1.
      Calculate the largest possible amount of money that Lara can make if she sells all her cakes and biscuits.

   KEY WORDS
   linear programming

Solving quadratic equations by completing the square

> **THIS SECTION WILL SHOW YOU HOW TO**
> - Complete the square
> - Solve quadratic equations by completing the square

Completing the square

If you expand the expressions $(x + b)^2$ and $(x - b)^2$ you will obtain the results:

$$(x + b)^2 = x^2 + 2bx + b^2 \quad \text{and} \quad (x - b)^2 = x^2 - 2bx + b^2$$

Rearranging these give you the following important results:

$$x^2 + 2bx = (x + b)^2 - b^2$$
$$x^2 - 2bx = (x - b)^2 - b^2$$

This is known as **completing the square**.

To complete the square for $x^2 + 6x$:

$$6 \div 2 = 3$$

$$x^2 + 6x = (x + 3)^2 - 3^2$$
$$= (x + 3)^2 - 9$$

➡ **CHECK:** $(x + 3)(x + 3) - 9 = x^2 + 3x + 3x + 9 - 9 = x^2 + 6x$ ✓

EXAMPLE

Complete the square for the expressions
a $x^2 - 12x$ b $x^2 - 7x$

a $x^2 - 12x$:

$$12 \div 2 = 6$$

$$x^2 - 12x = (x - 6)^2 - 6^2$$
$$= (x - 6)^2 - 36$$

➡ **CHECK:** $(x - 6)(x - 6) - 36$
$$= x^2 - 6x - 6x + 36 - 36$$
$$= x^2 - 12x \checkmark$$

b $x^2 - 7x$:

$$7 \div 2 = 3.5$$

$$x^2 - 7x = (x - 3.5)^2 - 3.5^2$$
$$= (x - 3.5)^2 - 12.25$$

➡ **CHECK:** $(x - 3.5)(x - 3.5) - 12.25$
$$= x^2 - 3.5x - 3.5x + 12.25 - 12.25$$
$$= x^2 - 7x \checkmark$$

EXERCISE 9.5

1 Complete the square for these expressions.
- **a** $x^2 - 2x$
- **b** $x^2 + 4x$
- **c** $x^2 - 10x$
- **d** $x^2 + 20x$
- **e** $x^2 - 16x$
- **f** $x^2 - 12x$
- **g** $x^2 + 18x$
- **h** $x^2 + 36x$
- **i** $x^2 + 22x$
- **j** $x^2 + 24x$
- **k** $x^2 - 3x$
- **l** $x^2 + 5x$
- **m** $x^2 + 7x$
- **n** $x^2 + 9x$
- **o** $x^2 - 20x$
- **p** $x^2 + x$

You may need to take out a numerical factor before completing the square.

$5x^2 - 20x \longrightarrow 5[x^2 - 4x] \longrightarrow 5[(x - 2)^2 - 4] \longrightarrow 5(x - 2)^2 - 20$

take 5 outside   complete the square   multiply both terms by 5

2 Complete the square for these expressions.
- **a** $5x^2 - 20x$
- **b** $3x^2 + 12x$
- **c** $4x^2 + 20x$
- **d** $3x^2 - 21x$
- **e** $2x^2 + 12x$
- **f** $2x^2 - 5x$
- **g** $3x^2 - 4x$
- **h** $2x^2 + 9x$

You may also have a constant term in your quadratic expression.

$x^2 + 12x - 3 \longrightarrow (x + 6)^2 - 6^2 - 3 \longrightarrow (x + 6)^2 - 39$

complete the square for $x^2 + 12x$
$x^2 + 12x = (x + 6)^2 - 6^2$

combine the numbers
$-6^2 - 3 = -39$

3 Write these expressions in the form $(x + p)^2 + q$.
- **a** $x^2 - 8x + 15$
- **b** $x^2 - 4x - 12$
- **c** $x^2 - 6x - 4$
- **d** $x^2 + 4x - 3$
- **e** $x^2 + 12x - 5$
- **f** $x^2 - 8x + 1$
- **g** $x^2 + 10x - 3$
- **h** $x^2 + 9x - 22$

4 $x^2 - 10x + 4 = (x - a)^2 + b$. Find the values of a and b.

5 $x^2 + 8x - 3 = (x + p)^2 + q$. Find the values of p and q.

6 $2x^2 - 8x + 3 = 2(x - p)^2 + q$. Find the values of p and q.

7

You can make x the subject of the equation $x^2 + 8x = a$ by completing the square.

$x^2 + 8x = a \longrightarrow (x + 4)^2 - 4^2 = a \longrightarrow (x + 4)^2 = a + 16 \longrightarrow x + 4 = \pm\sqrt{a + 16} \longrightarrow x = -4 \pm \sqrt{a + 16}$

Make x the subject of these equations. (You will need to complete the square first.)
- **a** $x^2 - 6x = b$
- **b** $x^2 + 10x = a + b$

You can solve quadratic equations by completing the square.

> **EXAMPLE**
>
> Solve the equation $x^2 + 4x - 3 = 0$, giving your answers to two decimal places.
>
> | | |
> |---|---|
> | $x^2 + 4x - 3 = 0$ | add 3 to both sides |
> | $x^2 + 4x = 3$ | complete the square for $x^2 + 4x$ |
> | $(x + 2)^2 - 2^2 = 3$ | |
> | $(x + 2)^2 - 4 = 3$ | add 4 to both sides |
> | $(x + 2)^2 = 7$ | square root both sides |
> | $x + 2 = \pm\sqrt{7}$ | subtract 2 from both sides |
> | $x = -2 \pm \sqrt{7}$ | |
> | $x = -2 + \sqrt{7}$ or $x = -2 - \sqrt{7}$ | |
> | $x = 0.65$ or $x = -4.65$ (2 d.p.) | |

The next example shows you what to do if the coefficient of x^2 is not 1.

> **EXAMPLE**
>
> Solve the equation $2x^2 - 8x + 5 = 0$, giving your answers to two decimal places.
>
> | | |
> |---|---|
> | $2x^2 - 8x + 5 = 0$ | divide both sides by 2 so that the coefficient of x^2 is 1 |
> | $x^2 - 4x + 2\frac{1}{2} = 0$ | subtract $2\frac{1}{2}$ from both sides |
> | $x^2 - 4x = -2\frac{1}{2}$ | complete the square for $x^2 - 4x$ |
> | $(x - 2)^2 - 2^2 = -2\frac{1}{2}$ | |
> | $(x - 2)^2 - 4 = -2\frac{1}{2}$ | add 4 to both sides |
> | $(x - 2)^2 = 1\frac{1}{2}$ | square root both sides |
> | $x - 2 = \pm\sqrt{1\frac{1}{2}}$ | add 2 to both sides |
> | $x = 2 \pm \sqrt{1\frac{1}{2}}$ | |
> | $x = 2 + \sqrt{1\frac{1}{2}}$ or $x = 2 - \sqrt{1\frac{1}{2}}$ | |
> | $x = 3.22$ or $x = 0.78$ (2 d.p.) | |

You can use completing the square to find the **minimum** value of a quadratic expression.

$$x^2 + 4x + 5 = \boxed{(x + 2)^2} + 1$$

➡ **NOTE:** this part of the expression is a square so it will always be ≥ 0 (positive). The smallest value it can be is 0.

The minimum value of $x^2 + 4x + 5$ is $0 + 1 = 1$.

EXERCISE 9.6

In this exercise give your answers where appropriate correct to 2 decimal places.

1. Use completing the square to solve these equations.
   a. $x^2 + 6x + 8 = 0$
   b. $x^2 + 4x + 4 = 0$
   c. $x^2 + 2x - 8 = 0$
   d. $x^2 - 12x + 34 = 0$
   e. $x^2 - 6x + 7 = 0$
   f. $x^2 + 14x - 3 = 0$
   g. $x^2 - 20x - 33 = 0$
   h. $x^2 - 10x + 15 = 0$
   i. $x^2 - 2x - 6 = 0$

2. Use completing the square to solve these equations.
   a. $x^2 + 3x + 2 = 0$
   b. $x^2 - 7x + 12 = 0$
   c. $x^2 + x - 6 = 0$
   d. $x^2 - 5x + 5 = 0$
   e. $x^2 + x - 10 = 0$
   f. $x^2 - 11x + 3 = 0$

3. Use completing the square to solve these equations.
   a. $3x^2 - 5x - 2 = 0$
   b. $2x^2 + 7x + 3 = 0$
   c. $2x^2 + 16x + 30 = 0$
   d. $2x^2 - 16x + 4 = 0$
   e. $3x^2 + 6x + 2 = 0$
   f. $2x^2 + 8x - 6 = 0$

4. Anna is trying to solve the equations $12 = \dfrac{3}{x} - x$ and $x(x - 2) = 12$.

   Copy the equations and correct her working.

   a.
   $12 = \dfrac{3}{x} - x$
   $12x = 3 - x$
   $13x = 3$
   $x = \dfrac{3}{13}$

   b.
   $x(x - 2) = 12$
   $x = 12$ or $x - 2 = 12$
   $x = 12$ or $x = 14$

5. Solve these equations.
   a. $x(x - 4) = 20$
   b. $x^2 - 7 = 2x$
   c. $10 + 3x = 2x^2$
   d. $x = 3 + \dfrac{1}{x}$
   e. $8 = \dfrac{2}{x} + 5x$
   f. $x = \dfrac{6}{x + 2}$

6. Use completing the square to find the minimum value of $x^2 - 4x + 11$.

7. Use completing the square to show that $x^2 + 6x + 14$ is never less than 5.

8. Use completing the square to find the minimum value of $2x^2 - 4x + 5$.

CHALLENGE:

Try solving the equation $ax^2 + bx + c = 0$ by completing the square.

If you do it correctly you should obtain the answer
$x = \dfrac{-b \pm \sqrt{b^2 - 4ac}}{2a}$.

KEY WORDS
completing the square
minimum

Solving quadratic equations by completing the square

Using graphs to solve equations

THIS SECTION WILL SHOW YOU HOW TO
- Use graphs to solve equations

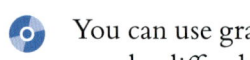

You can use graphs to find approximate solutions to equations that may be difficult to solve by other methods.

EXAMPLE

Draw the graph of $y = x^2 - 5x + 4$ for $0 \leq x \leq 5$.
Use your graph to solve the equations **a** $x^2 - 5x + 4 = 2$ **b** $x^2 - 6x + 6 = 0$

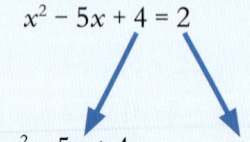

a $x^2 - 5x + 4 = 2$

$y = x^2 - 5x + 4$ $y = 2$

To solve the equation $x^2 - 5x + 4 = 2$ draw the curve $y = x^2 - 5x + 4$ and the line $y = 2$ and find the x coordinates of the points where $y = x^2 - 5x + 4$ and $y = 2$ intersect.
The answers to $x^2 - 5x + 4 = 2$ are
$x \approx 0.4$ and $x \approx 4.6$

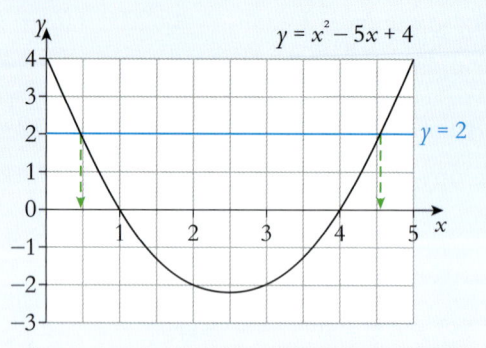

b To use the graph of $y = x^2 - 5x + 4$ to solve the equation $x^2 - 6x + 6 = 0$ you must first rearrange the equation into the form '$x^2 - 5x + 4 = ...$'

$x^2 - 6x + 6 = 0$ add x to both sides
$x^2 - 5x + 6 = x$ take 2 from both sides
$x^2 - 5x + 4 = x - 2$

$y = x^2 - 5x + 4$ $y = x - 2$

To solve the equation $x^2 - 6x + 6 = 0$ draw the curve $y = x^2 - 5x + 4$ and the line $y = x - 2$ and find the x coordinates of the points where $y = x^2 - 5x + 4$ and $y = x - 2$ intersect.
The answers to $x^2 - 6x + 6 = 0$ are
$x \approx 1.3$ and $x \approx 4.7$

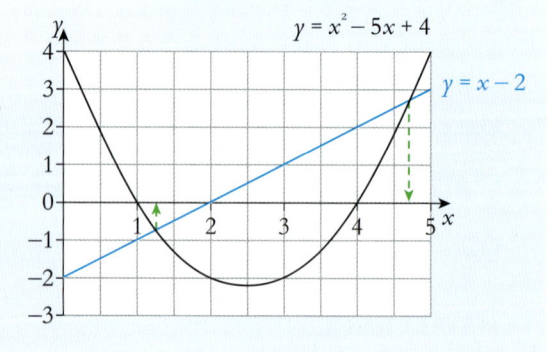

EXERCISE 9.7

1. $y = x^3 - 5x^2 + 10$ and $y = 2x$ are shown on the graph.
   Use the graph to solve the equation $x^3 - 5x^2 + 10x = 2x$

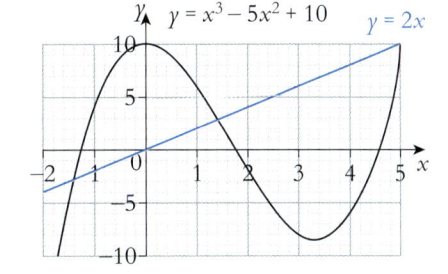

2. $y = \dfrac{4}{x}$ and $y = x - 2$ are shown on the graph.
   Use the graph to solve the equation $\dfrac{4}{x} = x - 2$

3. Draw the graph of $y = x^2$ for $-3 \leq x \leq 3$.
   Use your graph to solve these equations.
   a $x^2 = 6$ b $x^2 = x + 3$ c $x^2 - 2 = 2x$

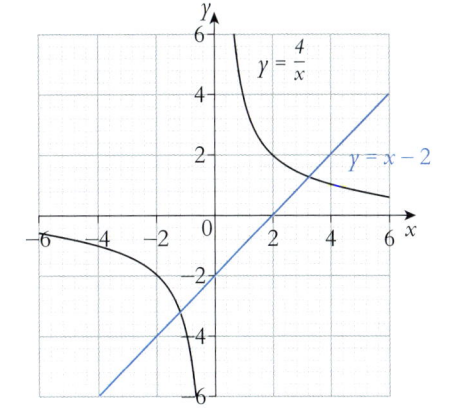

4. Draw the graph of $f(x) = x^2 - 3x$ for $-2 \leq x \leq 5$
   Use your graph to solve these equations.
   a $x^2 - 3x = 0$ b $x^2 - 3x = x + 2$
   c $x^2 - x - 4 = 0$

5. Draw the graph of $y = x^3 + 1$ for $-3 \leq x \leq 3$
   Use your graph to solve these equations.
   a $x^3 + 1 = 20$ b $x^3 + 1 = 5x$ c $x^3 + 10x - 9 = 0$

6. If the graph of $f(x) = 3^x$ has been drawn, what graph must also be drawn to solve these equations.
   a $3^x = 4$ b $3^x = 0.5$ c $3^x = 4x$ d $3^x - 2 = x$
   e $3^x - x^2 + 2 = 0$ f $x3^x = 1 + 3x$ g $1 - 3^x = 2x$ h $x^2 3^x = 1$

7. If the graph of $y = 3x^2 - \dfrac{1}{x}$ has been drawn, what graph must also be drawn to solve the equation $3x^3 - x^2 - 5x - 1 = 0$?

8. If the graph of $f(x) = \dfrac{2}{x^2}$ has been drawn, what graph must also be drawn to solve the equation $5x^3 - 5x^2 - 2 = 0$?

> **HINT**
>
> You must rearrange $3x^3 - x^2 - 5x - 1 = 0$ to the form '$3x^2 - \dfrac{1}{x} = ...$'

The roots of the equation f(x) = k

The graph of $f(x) = \dfrac{x^2}{2} - \dfrac{8}{x}$ is shown below.

The equation $\dfrac{x^2}{2} - \dfrac{8}{x} = k$ will have 3, 2 or 1 roots.

The number of roots depends on the value of k.

To solve the equation $\dfrac{x^2}{2} - \dfrac{8}{x} = 12$ the line $y = 12$ needs to be drawn.

To solve the equation $\dfrac{x^2}{2} - \dfrac{8}{x} = 6$ the line $y = 6$ needs to be drawn.

To solve the equation $\dfrac{x^2}{2} - \dfrac{8}{x} = -9$ the line $y = -9$ needs to be drawn.

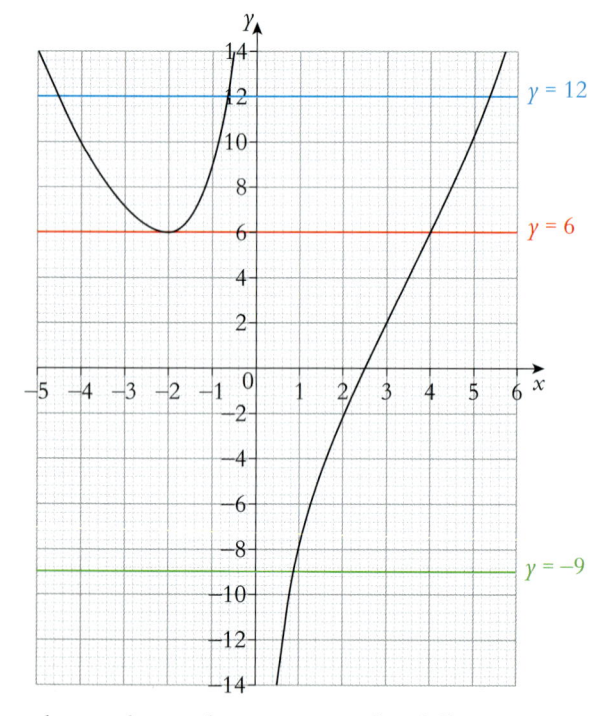

There are **three** roots to the equation $\dfrac{x^2}{2} - \dfrac{8}{x} = 12$
They are $x \approx -4.5$, $x \approx -0.7$ and $x \approx 5.2$

There are **two** roots to the equation $\dfrac{x^2}{2} - \dfrac{8}{x} = 6$
They are $x = -2$ and $x = 4$

There is **one** root to the equation $\dfrac{x^2}{2} - \dfrac{8}{x} = -9$
The root is $x \approx 0.9$

The results can be summarised as follows.

$\dfrac{x^2}{2} - \dfrac{8}{x} = k$

- $k > 6$ means there will be **three** roots
- $k = 6$ means there will be **two** roots
- $k < 6$ means there will be **one** root

EXERCISE 9.8

1. The graph of $f(x) = \dfrac{(x-2)^2}{2} + 1$ is shown.

   a. Use the graph to find the **two** roots of the equation $\dfrac{(x-2)^2}{2} + 1 = 5$

   b. For what value of k is there only one root of the equation $\dfrac{(x-2)^2}{2} + 1 = k$

   c. For what values of k are there no roots of the equation $\dfrac{(x-2)^2}{2} + 1 = k$

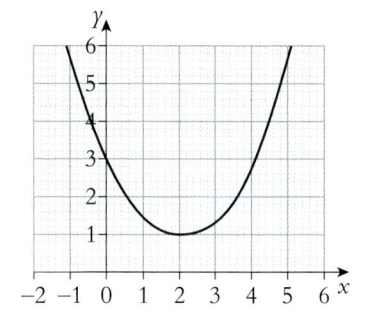

2. The graph of $f(x) = x + \dfrac{1}{2x} + 3$ is shown.

   a. Use the graph to find the **two** roots of the equation $x + \dfrac{1}{2x} + 3 = 6$

   b. There are three integer values of k for which the equation $x + \dfrac{1}{2x} + 3 = k$ has no roots.
   Write down these three values of k

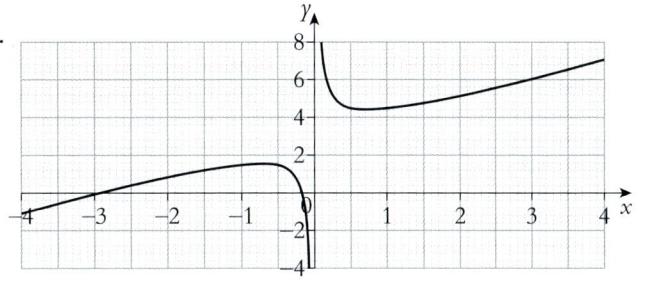

3. The graph of $f(x) = x^2 + \dfrac{2}{x}$ is shown.

   a. Use the graph to find the **three** roots of the equation $x^2 + \dfrac{2}{x} = 5$

   b. There is one value of k for which the equation $x^2 + \dfrac{2}{x} = k$ has two roots.
   Write down this value of k.

   c. For what values of k is there only one root of the equation $x^2 + \dfrac{2}{x} = k$

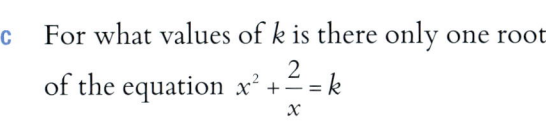

4. The graph of $y = x^3 - 3x + 1$ is shown.

   a. Use the graph to find the **three** roots of the equation $x^3 - 3x + 1 = 0$

   b. There are two values of k for which the equation $x^3 - 3x + 1 = k$ has two roots.
   Write down these two values of k

   c. For what values of k is there only one root of the equation $x^3 - 3x + 1 = k$.

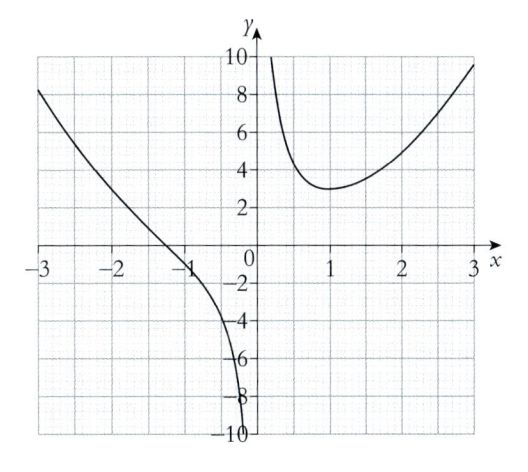

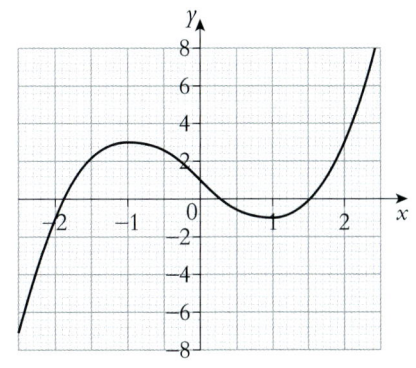

Using graphs to solve equations   439

Vectors and vector geometry

THIS SECTION WILL SHOW YOU HOW TO
- Use the correct notation for vectors
- Add and subtract vectors
- Solve problems in 2-D using vector geometry

A **vector** has both size and direction.
Examples of vectors are:
- displacement (5 km on a bearing of 300°)
- velocity (5 m/s due east)
- weight (20 N vertically downwards)

Vector notation

$\overrightarrow{AB}$ means the displacement from the point A to the point B.
A translation is described using a column vector.
A column vector describes the movement in both the x and y directions.

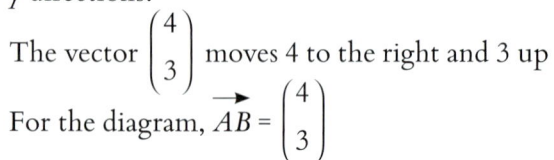

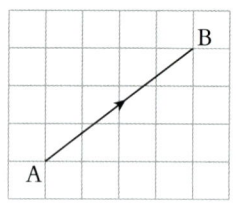

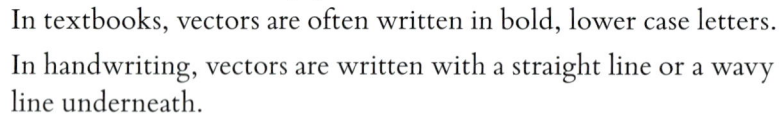

The vector $\begin{pmatrix} 4 \\ 3 \end{pmatrix}$ moves 4 to the right and 3 up

For the diagram, $\overrightarrow{AB} = \begin{pmatrix} 4 \\ 3 \end{pmatrix}$

In textbooks, vectors are often written in bold, lower case letters.

In handwriting, vectors are written with a straight line or a wavy line underneath.

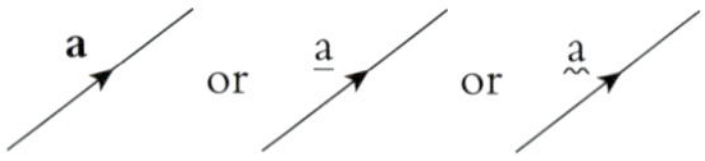

Magnitude of a vector

'The **magnitude** of the vector **a**' means 'the length of the vector **a**' and is denoted by |**a**|.
|**a**| is called the **modulus** of the vector **a**.

EXAMPLE

If $\mathbf{a} = \begin{pmatrix} -4 \\ 3 \end{pmatrix}$ find the magnitude of the vector **a**.

Using Pythagoras' theorem,
$|\mathbf{a}| = \sqrt{3^2 + (-4)^2}$
$= \sqrt{25}$
$= 5$

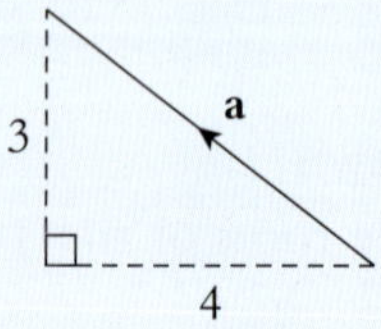

The magnitude of vector **a** is 5.

Two vectors are said to be equal if they are the same length and are in the same direction.

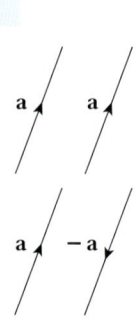

The vector −**a** is the same length as the vector **a** but is in the opposite direction.

EXERCISE 9.9

1 Draw these vectors on graph paper.

a $\begin{pmatrix} 5 \\ 1 \end{pmatrix}$ b $\begin{pmatrix} 4 \\ -2 \end{pmatrix}$ c $\begin{pmatrix} 0 \\ -4 \end{pmatrix}$ d $\begin{pmatrix} -5 \\ -3 \end{pmatrix}$ e $\begin{pmatrix} 6 \\ 0 \end{pmatrix}$

2 Write these vectors in column vector form.

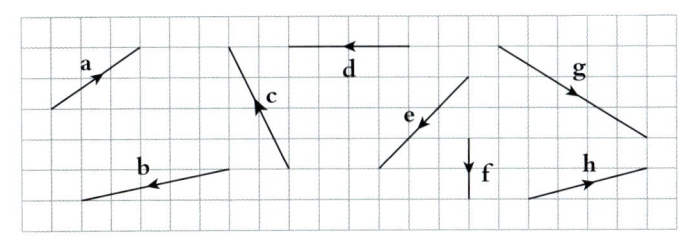

3 Draw the vectors $\begin{pmatrix} -4 \\ 2 \end{pmatrix}$ and $\begin{pmatrix} -8 \\ 4 \end{pmatrix}$.

What can you say about these two vectors?

4 Draw the vectors $\begin{pmatrix} 3 \\ 1 \end{pmatrix}$ and $\begin{pmatrix} -1 \\ 3 \end{pmatrix}$.

What can you say about these two vectors?

5 Find the magnitude of each of these vectors.

a $\begin{pmatrix} 6 \\ 8 \end{pmatrix}$ b $\begin{pmatrix} 5 \\ -12 \end{pmatrix}$ c $\begin{pmatrix} 0 \\ -4 \end{pmatrix}$ d $\begin{pmatrix} -15 \\ -8 \end{pmatrix}$ e $\begin{pmatrix} -20 \\ 21 \end{pmatrix}$

6 PQRSTU is a regular hexagon.
O is the centre of the hexagon.
$\overrightarrow{OP} = \mathbf{p}$ and $\overrightarrow{OQ} = \mathbf{q}$.
($\overrightarrow{TU} = \overrightarrow{OP} = \mathbf{p}$ because they are the same length and in the same direction.)

- **a** Write down all the vectors that are equal to **p**.
- **b** Write down all the vectors that are equal to **q**.
- **c** Write down all the vectors that are equal to −**p**.
- **d** Write down all the vectors that are equal to −**q**.

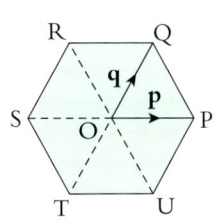

Adding and subtracting vectors

The vector **a** + **b** means the vector **a** followed by the vector **b**.
The vector **a** − **b** means the vector **a** followed by the vector −**b**.
The **resultant vector** is usually shown with a double arrow.
It is drawn from the starting point to the finishing point.

EXAMPLE

If $\mathbf{a} = \begin{pmatrix} 2 \\ 4 \end{pmatrix}$ and $\mathbf{b} = \begin{pmatrix} 5 \\ 1 \end{pmatrix}$. Draw a diagram to find the column vector **a** + **b**.

First draw the vector **a**.
Follow vector **a** by the vector **b**.
Counting the squares on the grid:

$\mathbf{a} + \mathbf{b} = \begin{pmatrix} 7 \\ 5 \end{pmatrix}$

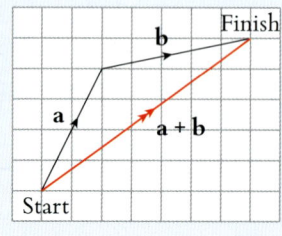

➡ NOTE: $\begin{pmatrix} 2 \\ 4 \end{pmatrix} + \begin{pmatrix} 5 \\ 1 \end{pmatrix} = \begin{pmatrix} 2+5 \\ 4+1 \end{pmatrix}$

$= \begin{pmatrix} 7 \\ 5 \end{pmatrix}$

EXAMPLE

If $\mathbf{a} = \begin{pmatrix} 2 \\ 4 \end{pmatrix}$ and $\mathbf{b} = \begin{pmatrix} 5 \\ 1 \end{pmatrix}$. Draw a diagram to find the column vector **a** − **b**.

First draw the vector **a**.
Follow vector **a** by the vector −**b**.
Counting the squares on the grid:

$\mathbf{a} - \mathbf{b} = \begin{pmatrix} -3 \\ 3 \end{pmatrix}$

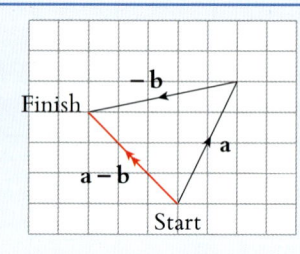

➡ NOTE: $\begin{pmatrix} 2 \\ 4 \end{pmatrix} - \begin{pmatrix} 5 \\ 1 \end{pmatrix} = \begin{pmatrix} 2-5 \\ 4-1 \end{pmatrix}$

$= \begin{pmatrix} -3 \\ 3 \end{pmatrix}$

Multiplication by a scalar

This vector can be written as 3**a**.

In the vector 3**a**, the number 3 is called a scalar.
(A scalar is a number.)

➡ NOTE: all these vectors are parallel.

442 UNIT 9

Parallel vectors

Two vectors are parallel if one vector can be written as a multiple of the other vector.
For example:

$\begin{pmatrix} 3 \\ -2 \end{pmatrix}$ and $\begin{pmatrix} 9 \\ -6 \end{pmatrix}$ are parallel and in the same direction because $\begin{pmatrix} 9 \\ -6 \end{pmatrix} = 3\begin{pmatrix} 3 \\ -2 \end{pmatrix}$

$\begin{pmatrix} 4 \\ -2 \end{pmatrix}$ and $\begin{pmatrix} -2 \\ 1 \end{pmatrix}$ are parallel and in opposite directions because $\begin{pmatrix} 4 \\ -2 \end{pmatrix} = -2\begin{pmatrix} -2 \\ 1 \end{pmatrix}$

Position vectors

The position vector of a point P means the displacement from the origin to the point P.
For this diagram, the position vector of A = $\overrightarrow{OA} = \begin{pmatrix} 2 \\ 3 \end{pmatrix}$

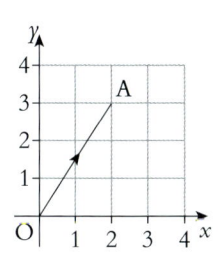

EXERCISE 9.10

1 $\mathbf{p} = \begin{pmatrix} 6 \\ 4 \end{pmatrix}$ Draw these vectors on squared paper.

    a $2\mathbf{p}$ b $\frac{1}{2}\mathbf{p}$ c $-\mathbf{p}$

    d $\frac{3}{2}\mathbf{p}$ e $-2\mathbf{p}$ f $3\mathbf{p}$

2 $\mathbf{p} = \begin{pmatrix} 6 \\ -3 \end{pmatrix}$ $\mathbf{q} = \begin{pmatrix} -2 \\ 4 \end{pmatrix}$ $\mathbf{r} = \begin{pmatrix} -1 \\ -2 \end{pmatrix}$

Write these vectors as column vectors.

    a $3\mathbf{q}$ b $5\mathbf{r}$ c $-4\mathbf{r}$

    d $-\frac{1}{2}\mathbf{q}$ e $\frac{1}{3}\mathbf{p}$ f $2\frac{1}{2}\mathbf{q}$

3 $\mathbf{p} = \begin{pmatrix} -2 \\ 1 \end{pmatrix}$ $\mathbf{q} = \begin{pmatrix} 6 \\ 2 \end{pmatrix}$ $\mathbf{r} = \begin{pmatrix} 0 \\ -4 \end{pmatrix}$

Write these as column vectors.

    a $\mathbf{p} + \mathbf{q}$ b $\mathbf{q} + \mathbf{r}$ c $\mathbf{r} - \mathbf{p}$ d $\mathbf{p} + \mathbf{q} - \mathbf{r}$

    e $2\mathbf{p} + 2\mathbf{q}$ f $7\mathbf{r} + \mathbf{p}$ g $\mathbf{q} - 3\mathbf{r}$ h $5\mathbf{p} + 4\mathbf{q} - 3\mathbf{r}$

    i $2\mathbf{p} + 3\mathbf{q}$ j $5\mathbf{q} - 2\mathbf{r}$ k $\mathbf{r} - 2\mathbf{p}$ l $3\mathbf{r} - \frac{1}{2}\mathbf{q}$

4 Which of these column vectors are parallel?

A $\begin{pmatrix} 5 \\ -10 \end{pmatrix}$ B $\begin{pmatrix} -4 \\ -7 \end{pmatrix}$ C $\begin{pmatrix} 5 \\ 1 \end{pmatrix}$ D $\begin{pmatrix} -1 \\ -3 \end{pmatrix}$

E $\begin{pmatrix} 20 \\ 4 \end{pmatrix}$ F $\begin{pmatrix} -10 \\ 20 \end{pmatrix}$ G $\begin{pmatrix} 3 \\ 9 \end{pmatrix}$ H $\begin{pmatrix} 12 \\ 21 \end{pmatrix}$

Vectors and vector geometry   443

5 Write down the position vectors of the points:
A, B, C, D, E, F, G and H.

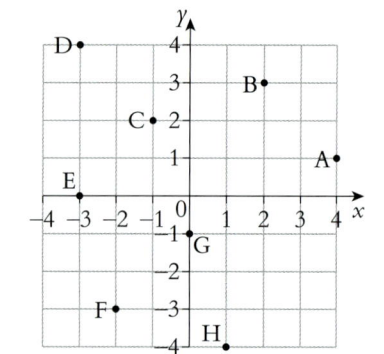

6 a Write as column vectors $\vec{AB}, \vec{BC}, \vec{CD}, \vec{DE}$ and $\vec{EA}$.
  b Write $\vec{AB} + \vec{BC} + \vec{CD} + \vec{DE} + \vec{EA}$ as a single column vector. Explain your answer.

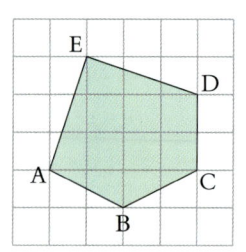

7 $\mathbf{p} = \begin{pmatrix} 5 \\ 6 \end{pmatrix}$ and $\mathbf{q} = \begin{pmatrix} 3 \\ 1 \end{pmatrix}$
  a Find as a single column vector, $3\mathbf{p} + 2\mathbf{q}$.
  b Calculate the value of $|3\mathbf{p} + 2\mathbf{q}|$.

8 $\mathbf{p} = \begin{pmatrix} 1 \\ 2 \end{pmatrix}$ and $\mathbf{q} = \begin{pmatrix} 5 \\ 21 \end{pmatrix}$
  a Find as a single column vector, $7\mathbf{p} + \mathbf{q}$.
  b Calculate the value of $|7\mathbf{p} + \mathbf{q}|$.

9 The diagram shows the position of two points A and B.
$\vec{AB} = \begin{pmatrix} p \\ q \end{pmatrix}$ where p and q are measured in kilometres.
  a Show that $p = 10.9$ correct to 3 s.f.
  b Calculate the value of q.

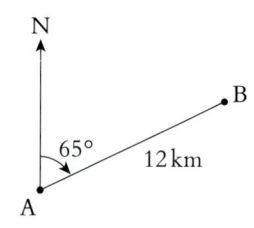

10 The diagram shows the position of two points A and B.
$\vec{AB} = \begin{pmatrix} p \\ q \end{pmatrix}$ where p and q are measured in kilometres.
Calculate the values of p and q.

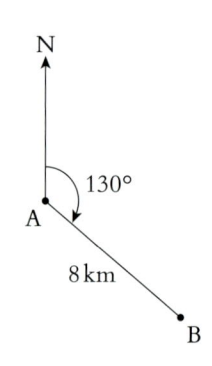

444 UNIT 9

11 The diagram shows the position of two points A and B.

   $\overrightarrow{AB} = \begin{pmatrix} p \\ q \end{pmatrix}$ where p and q are measured in kilometres.

   Calculate the values of p and q.

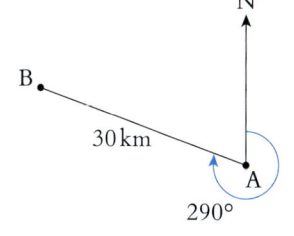

12 $\mathbf{p} = \begin{pmatrix} x \\ 4 \end{pmatrix}$ $\mathbf{q} = \begin{pmatrix} -2 \\ y \end{pmatrix}$ and $\mathbf{p} + \mathbf{q} = \begin{pmatrix} 5 \\ 3 \end{pmatrix}$

   Find the values of x and y.

13 $x \begin{pmatrix} 5 \\ 4 \end{pmatrix} + y \begin{pmatrix} 3 \\ -2 \end{pmatrix} = \begin{pmatrix} 3 \\ 20 \end{pmatrix}$

   Find the values of x and y.

HINT

Write down two simultaneous equations and solve.

14 ABCD is a parallelogram.
   A(1, 3), B(9, 4) and C(10, 7).
   a Write $\overrightarrow{AB}$ as a column vector.
   b Write $\overrightarrow{BC}$ as a column vector.
   c Write $\overrightarrow{AD}$ as a column vector.
   d Write down the position vector of the point D.

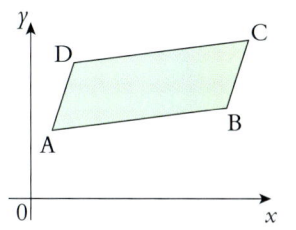

15 ABCD is a trapezium.
   A(4, 5), B(3, 2) and D(7, 7).
   $\overrightarrow{BC} = 2\overrightarrow{AD}$.
   a Write $\overrightarrow{AD}$ as a column vector.
   b Write $\overrightarrow{BC}$ as a column vector.
   c Find the coordinates of C.

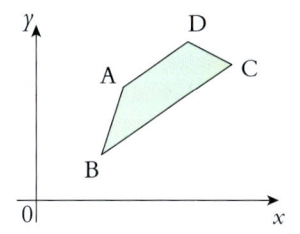

16 The position vector **r** is given by $\mathbf{r} = \begin{pmatrix} 3 \\ 5 \end{pmatrix} + t \begin{pmatrix} 2 \\ -1 \end{pmatrix}$.

   a Copy and complete the table below for the given values of t.

t	0	1	2	3	4
r				$\begin{pmatrix} 9 \\ 2 \end{pmatrix}$	

   b On a square grid, plot the five points given by the position vectors in the table.
   c What can you say about the five points?

Vectors and vector geometry

Vector geometry

EXAMPLE

OPQR is a parallelogram.
$\overrightarrow{OP} = \mathbf{p}$ and $\overrightarrow{OR} = \mathbf{r}$.
Find the following in terms of **p** and **r**

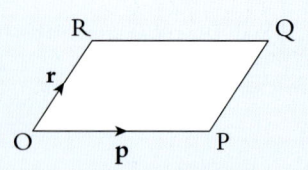

a $\overrightarrow{RQ}$,   b $\overrightarrow{QP}$,   c $\overrightarrow{OQ}$,   d $\overrightarrow{PR}$.

a $\overrightarrow{RQ} = \mathbf{p}$

b $\overrightarrow{QP} = -\mathbf{r}$

c $\overrightarrow{OQ} = \overrightarrow{OP} + \overrightarrow{PQ} = \mathbf{p} + \mathbf{r}$

➡ **NOTE:** the journey from O to Q is the 'same as' the journey from O to P, followed by the journey from P to Q.

d $\overrightarrow{PR} = \overrightarrow{PQ} + \overrightarrow{QR} = \mathbf{r} - \mathbf{p}$

Vectors are often used to prove geometric properties.
The next two examples illustrate this technique.

EXAMPLE

ABCD is a trapezium.
$\overrightarrow{DC} = 2\mathbf{p}$, $\overrightarrow{AB} = 6\mathbf{p}$ and $\overrightarrow{AD} = 2\mathbf{q}$.
E, F, G and H are the midpoints of the sides.
Find in terms of **p** and **q**

a $\overrightarrow{CB}$,   b $\overrightarrow{CF}$,   c $\overrightarrow{HE}$,
d $\overrightarrow{GF}$,   e $\overrightarrow{EF}$,   f $\overrightarrow{HG}$.

What can you say about quadrilateral EFGH?

a $\overrightarrow{CB} = \overrightarrow{CD} + \overrightarrow{DA} + \overrightarrow{AB}$
   $= -2\mathbf{p} - 2\mathbf{q} + 6\mathbf{p}$
   $= 4\mathbf{p} - 2\mathbf{q}$

b $\overrightarrow{CF} = \frac{1}{2}\overrightarrow{CB}$
   $= \frac{1}{2}(4\mathbf{p} - 2\mathbf{q})$
   $= 2\mathbf{p} - \mathbf{q}$

c $\overrightarrow{HE} = \overrightarrow{HA} + \overrightarrow{AE}$
   $= -\mathbf{q} + 3\mathbf{p}$
   $= 3\mathbf{p} - \mathbf{q}$

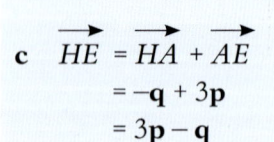

d $\overrightarrow{GF} = \overrightarrow{GC} + \overrightarrow{CF}$
   $= \mathbf{p} + (2\mathbf{p} - \mathbf{q})$
   $= 3\mathbf{p} - \mathbf{q}$

e $\overrightarrow{EF} = \overrightarrow{EB} + \overrightarrow{BF}$
   $= 3\mathbf{p} + (\mathbf{q} - 2\mathbf{p})$
   $= \mathbf{p} + \mathbf{q}$

f $\overrightarrow{HG} = \overrightarrow{HD} + \overrightarrow{DG}$
   $= \mathbf{q} + \mathbf{p}$
   $= \mathbf{p} + \mathbf{q}$

So $\overrightarrow{HE} = \overrightarrow{GF}$ and $\overrightarrow{EF} = \overrightarrow{HG}$.
This means that HE and GF are parallel and the same length.
Also, EF and HG are parallel and the same length.
So EFGH is a parallelogram.

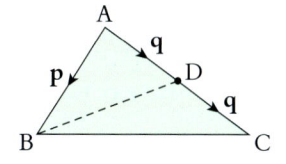

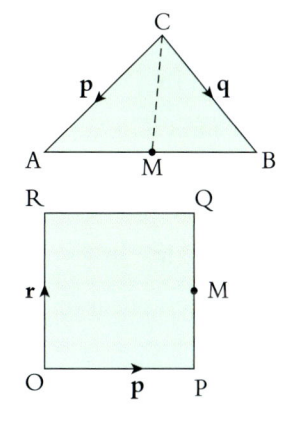

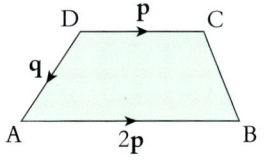

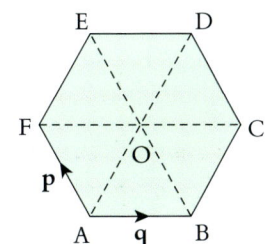

EXAMPLE

OPQ is a triangle.
X is the midpoint of OP.
Y is the midpoint of OQ.
$\overrightarrow{OP}$ = 2**p** and $\overrightarrow{OQ}$ = 2**q**.
Find in terms of **p** and **q**
  a $\overrightarrow{PQ}$,   **b** $\overrightarrow{XY}$.
What can you say about the lines PQ and XY?

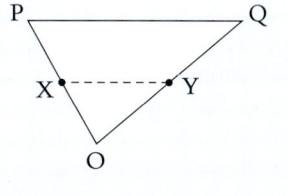

a $\overrightarrow{PQ} = \overrightarrow{PO} + \overrightarrow{OQ} = -2\mathbf{p} + 2\mathbf{q}$
b $\overrightarrow{XY} = \overrightarrow{XO} + \overrightarrow{OY} = -\mathbf{p} + \mathbf{q}$
So $\overrightarrow{PQ} = 2\overrightarrow{XY}$
This means that PQ and XY are parallel and that PQ is twice the length of XY.

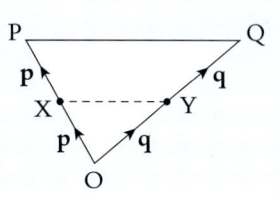

EXERCISE 9.11

1 ABC is a triangle. D is the midpoint of AC.
$\overrightarrow{AB}$ = **p** and $\overrightarrow{AD} = \overrightarrow{DC}$ = **q**.
Find in terms of **p** and **q**
  a $\overrightarrow{DA}$,   **b** $\overrightarrow{AC}$,   **c** $\overrightarrow{BD}$,   **d** $\overrightarrow{CB}$.

2 ABC is a triangle. M is the midpoint of AB.
$\overrightarrow{CA}$ = **p** and $\overrightarrow{CB}$ = **q**.
Find in terms of **p** and **q**
  a $\overrightarrow{AB}$,   **b** $\overrightarrow{AM}$,   **c** $\overrightarrow{BM}$,   **d** $\overrightarrow{CM}$.

3 OPQR is a square. M is the midpoint of PQ.
$\overrightarrow{OP}$ = **p** and $\overrightarrow{OR}$ = **r**.
Find in terms of **p** and **r**
  a $\overrightarrow{OQ}$,   **b** $\overrightarrow{MP}$,   **c** $\overrightarrow{OM}$,   **d** $\overrightarrow{MR}$.

4 ABCD is a trapezium.
$\overrightarrow{AB}$ = 2**p**, $\overrightarrow{DC}$ = **p** and $\overrightarrow{DA}$ = **q**.
Find in terms of **p** and **q**
  a $\overrightarrow{BA}$,   **b** $\overrightarrow{DB}$,   **c** $\overrightarrow{CA}$,   **d** $\overrightarrow{CB}$.

5 ABCDEF is a regular hexagon.
O is the centre of the hexagon.
$\overrightarrow{AF}$ = **p** and $\overrightarrow{AB}$ = **q**.
Find the following in terms of **p** and **q**.
  a $\overrightarrow{CD}$,   **b** $\overrightarrow{DE}$,   **c** $\overrightarrow{AO}$,
  d $\overrightarrow{CE}$,   **e** $\overrightarrow{FD}$,   **f** $\overrightarrow{DB}$.

Vectors and vector geometry

6 OABC is a parallelogram.

M is the midpoint of AB.
N is the midpoint of CB.
$\overrightarrow{OA}$ = **a** and $\overrightarrow{OC}$ = **c**.
Find in terms of **a** and **c**

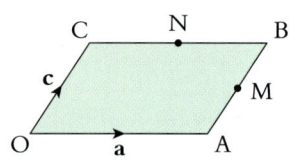

a $\overrightarrow{NB}$, b $\overrightarrow{BM}$, c $\overrightarrow{AC}$, d $\overrightarrow{MN}$.

What can you say about the lines AC and MN?

7 OABC is a parallelogram.
M is the point of intersection of the two diagonals.
$\overrightarrow{OA}$ = **a** and $\overrightarrow{OC}$ = **c**.
Find in terms of **a** and **c**

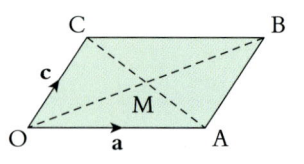

a $\overrightarrow{OB}$, b $\overrightarrow{CA}$, c $\overrightarrow{OM}$, d $\overrightarrow{CM}$.

8 ABCD is a parallelogram.
$\overrightarrow{AB}$ = **p** and $\overrightarrow{AD}$ = **q**.
E, F, G and H are the midpoints of the sides.
Find in terms of **p** and **q**

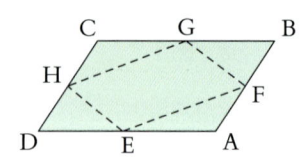

a $\overrightarrow{FG}$, b $\overrightarrow{EH}$, c $\overrightarrow{EF}$, d $\overrightarrow{HG}$.

What can you say about quadrilateral EFGH?

9 OABC and BCDE are two congruent parallelograms.
$\overrightarrow{OA}$ = **a** and $\overrightarrow{OC}$ = **c**.
M is the midpoint of BE and N is the midpoint of DE.
Find in terms of **a** and **c**

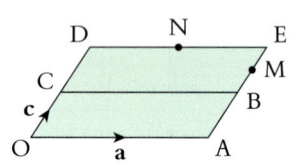

a $\overrightarrow{AM}$, b $\overrightarrow{AC}$, c $\overrightarrow{DM}$, d $\overrightarrow{MN}$.

What can you say about AC and MN?

10 ABCD is a trapezium.
BCDE is a parallelogram.
$\overrightarrow{EB}$ = **p**, $\overrightarrow{AE}$ = 3**p** and $\overrightarrow{ED}$ = **q**.
M is the midpoint of AD.
Find in terms of **p** and **q**

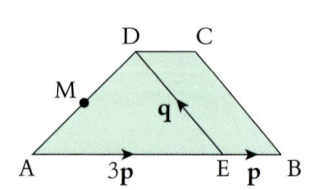

a $\overrightarrow{EC}$, b $\overrightarrow{EM}$, c $\overrightarrow{MC}$, d $\overrightarrow{AM}$.

Further vector geometry

Vector geometry can be used to prove that points are **collinear** (lie on a straight line).

If $\overrightarrow{AB} = k\overrightarrow{AC}$, then the points A, B and C are collinear.

(This is because the lines AB and AC must be parallel and the point A lies on both lines.)

Ratios can also be used in vector geometry.

If X is a point on the line AB such that $AX : XB = 2 : 3$

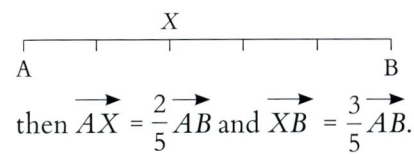

then $\overrightarrow{AX} = \frac{2}{5}\overrightarrow{AB}$ and $\overrightarrow{XB} = \frac{3}{5}\overrightarrow{AB}$.

The next example illustrates these techniques.

EXAMPLE

OPQR is a trapezium. 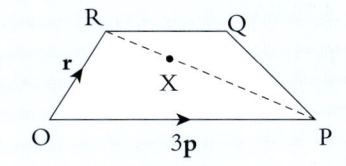
$\overrightarrow{OP} = 3\mathbf{p}$ and $\overrightarrow{OR} = \mathbf{r}$
$\overrightarrow{RQ} = \frac{1}{3}\overrightarrow{OP}$
X lies on RP such that $RX : XP = 1 : 3$

a Find in terms of **p** and **r**
   i $\overrightarrow{OQ}$,   **ii** $\overrightarrow{RP}$,   **iii** $\overrightarrow{RX}$,   **iv** $\overrightarrow{OX}$.

b What do your answers for $\overrightarrow{OQ}$ and $\overrightarrow{OX}$ tell you about the points O, X and Q?

---

a i $\overrightarrow{OQ} = \overrightarrow{OR} + \overrightarrow{RQ}$
$= \mathbf{r} + \mathbf{p}$

ii $\overrightarrow{RP} = \overrightarrow{RO} + \overrightarrow{OP}$
$= -\mathbf{r} + 3\mathbf{p}$

iii $\overrightarrow{RX} = \frac{1}{4}\overrightarrow{RP}$ ➡ **NOTE:** RX : XP = 1 : 3 means that $\overrightarrow{RX} = \frac{1}{4}\overrightarrow{RP}$.
$= \frac{1}{4}(-\mathbf{r} + 3\mathbf{p})$
$= \frac{3}{4}\mathbf{p} - \frac{1}{4}\mathbf{r}$

iv $\overrightarrow{OX} = \overrightarrow{OR} + \overrightarrow{RX} = \mathbf{r} + \frac{3}{4}\mathbf{p} - \frac{1}{4}\mathbf{r}$
$= \frac{3}{4}\mathbf{r} + \frac{3}{4}\mathbf{p}$
$= \frac{3}{4}(\mathbf{r} + \mathbf{p})$

b $\overrightarrow{OQ} = \mathbf{r} + \mathbf{p}$ and $\overrightarrow{OX} = \frac{3}{4}(\mathbf{r} + \mathbf{p})$
So $\overrightarrow{OX} = \frac{3}{4}\overrightarrow{OQ}$
This means that OX and OQ are parallel.
The point O lies on both of these line segments.
So the points O, X and Q are collinear.

EXERCISE 9.12

1 OPQ is a triangle and O is the origin.
$\overrightarrow{OP} = \mathbf{p}$ and $\overrightarrow{OQ} = \mathbf{q}$.
X is a point on PQ such that PX : XQ = 1 : 2
Find in terms of **p** and **q**
  a $\overrightarrow{PQ}$,   **b** $\overrightarrow{PX}$,   **c** the position vector of X.

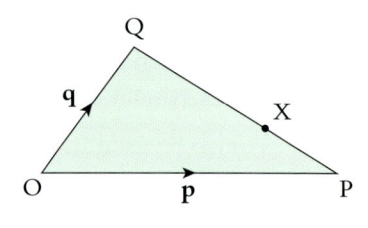

2 OPQ is a triangle and O is the origin.
$\overrightarrow{OP} = \mathbf{p}$ and $\overrightarrow{OQ} = \mathbf{q}$.
A is the midpoint on OQ.
B is a point on PQ such that PB : BQ = 3 : 1
Find in terms of **p** and **q**
  a $\overrightarrow{PQ}$,   **b** $\overrightarrow{PB}$,   **c** $\overrightarrow{AB}$,   **d** the position vector of B.

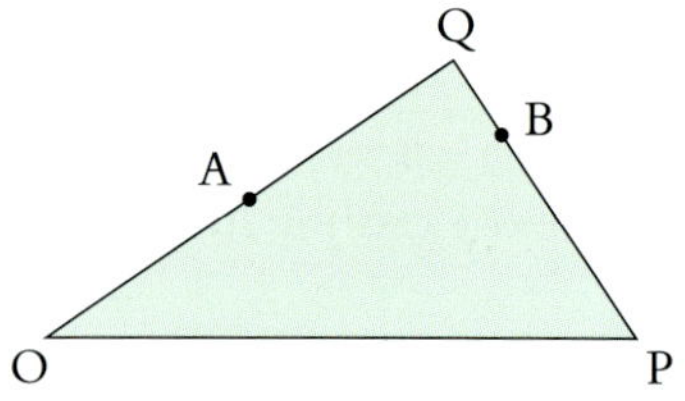

3 OPQ is a triangle and O is the origin.
$\overrightarrow{OP} = \mathbf{p}$ and $\overrightarrow{OQ} = \mathbf{q}$.
R is the midpoint on OQ.
S is a point on OP such that OS : SP = 1 : 2
T is a point on QP such that QT : TP = 3 : 2
Find in terms of **p** and **q**
  a $\overrightarrow{OR}$,   **b** $\overrightarrow{SR}$,   **c** $\overrightarrow{PQ}$,   **d** $\overrightarrow{PT}$.

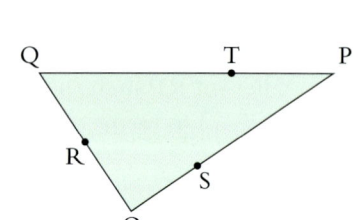

4 OPQR is a parallelogram and O is the origin.
$\overrightarrow{OP} = \mathbf{p}$ and $\overrightarrow{OR} = \mathbf{r}$.
X is a point on PQ such that PX : XQ = 1 : 2
Y is a point on RQ such that RY : YQ = 1 : 3
Find in terms of **p** and **r**
  a $\overrightarrow{PX}$,   **b** $\overrightarrow{RY}$,   **c** $\overrightarrow{XY}$.
  d the position vector of Y.

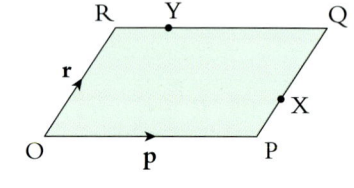

5 OABC is a trapezium.
$\overrightarrow{OA} = 3\mathbf{a}$, $\overrightarrow{CB} = 2\mathbf{a}$ and $\overrightarrow{OC} = \mathbf{c}$.
X is a point on AC such that AX : XC = 3 : 2
Find in terms of **a** and **c**
  a $\overrightarrow{AC}$,   **b** $\overrightarrow{AX}$,   **c** $\overrightarrow{OX}$,   **d** $\overrightarrow{OB}$.

What do your answers for $\overrightarrow{OX}$ and $\overrightarrow{OB}$ tell you about the points O, X and B?

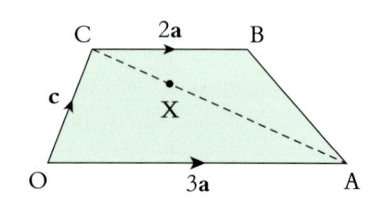

6 OPQR is a parallelogram and O is the origin.
$\overrightarrow{OP}$ = **p** and $\overrightarrow{OR}$ = **r**.
X is a point on PR such that PX : XR = 1 : 3
Find in terms of **p** and **r**
  a $\overrightarrow{PR}$, **b** $\overrightarrow{PX}$, **c** $\overrightarrow{XQ}$,
  d the position vector of X.

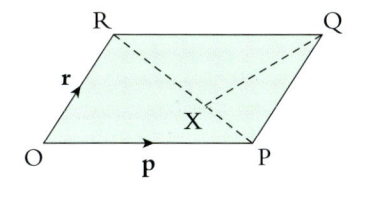

7 OPQR is a parallelogram and O is the origin.
$\overrightarrow{OP}$ = **p** and $\overrightarrow{OR}$ = **r**.
Y is a point on PQ such that PY : YQ = 2 : 3
The line RY is extended to the point X so that
$\overrightarrow{RX} = \frac{5}{3}\overrightarrow{RY}$.
Find in terms of **p** and **r**
  a $\overrightarrow{PY}$, **b** $\overrightarrow{RY}$, **c** $\overrightarrow{RX}$, **d** $\overrightarrow{OX}$.
Explain why O, P and X are collinear.

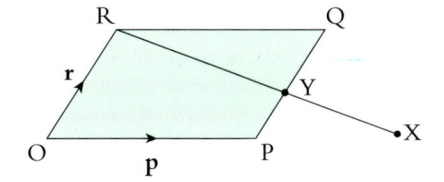

8 The position vector of A is $\begin{pmatrix} 1 \\ 5 \end{pmatrix}$. The position vector of B is $\begin{pmatrix} 4 \\ 3 \end{pmatrix}$

The position vector of C is $\begin{pmatrix} p \\ -3 \end{pmatrix}$. The position vector of D is $\begin{pmatrix} -7 \\ q \end{pmatrix}$

The points A, B, C and D are collinear. Find the values of p and q.

9 $\overrightarrow{OP}$ = **p** and $\overrightarrow{OQ}$ = **q**.
OA : AP = 1 : 2 and OB : BQ = 1 : 2
Prove that $\overrightarrow{PQ} = 3\overrightarrow{AB}$.

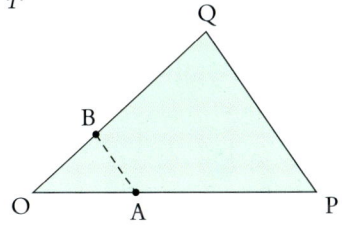

CHALLENGE:

OPQ is a triangle and O is the origin.
$\overrightarrow{OP}$ = **p** and $\overrightarrow{OQ}$ = **q**.
X is the midpoint of OP and Y is the midpoint of OQ.
V is a point on XQ such that
XV : VQ = 1 : 2.
  a Find the position vector of V.

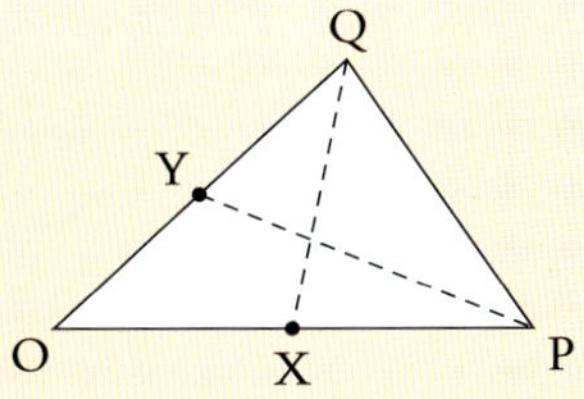

W is a point on YP such that YW : WP = 1 : 2.
  b Find the position vector of W.
What can you say about the points V and W?

KEY WORDS
vector
magnitude
modulus
resultant vector
scalar
position vector
collinear

Vectors and vector geometry   451

Probability 3

THIS SECTION WILL SHOW YOU HOW TO
- Use Venn diagrams to answer probability questions

Calculating probabilities from Venn diagrams

Probabilities can be calculated using information given on a Venn diagram.

EXAMPLE

In a survey, 100 adults are asked whether they like coffee (C) and whether they like tea (T).
The Venn diagram shows the results of the survey.

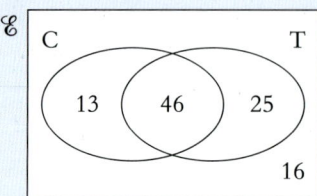

[Venn diagram: ℰ with two overlapping circles C and T. C-only: 13, intersection: 46, T-only: 25, outside: 16]

a A person is chosen at random.
Find the probability that the person does not like coffee.

b A person who likes tea is chosen at random.
Find the probability that this person also likes coffee.

c Two people are chosen at random.
Find the probability that they both like coffee.

---

a The number of people who do not like coffee = 25 + 16 = 41
You are choosing out of 100 people.

So, the probability that the person does not like coffee = $\dfrac{41}{100}$

b The number of people who like tea = 46 + 25 = 71
You are choosing out of 71 people and 46 of these also like coffee.

So, the probability that the person also likes coffee = $\dfrac{46}{71}$

c The probability the first person likes coffee = $\dfrac{13+46}{100} = \dfrac{59}{100}$

One person has been chosen so there are now only 99 to choose from.
Of these 99 people there are now only 58 that like coffee.

The probability the second person likes coffee = $\dfrac{58}{99}$

So, the probability that they both like coffee = $\dfrac{59}{100} \times \dfrac{58}{99} = \dfrac{1711}{4950} \approx 0.346$ (to 3 s.f.)

452 UNIT 9

EXERCISE 9.13

1. In a survey, 60 people are asked if they own a mobile phone (M) or a computer (C). The Venn diagram shows the results of the survey.

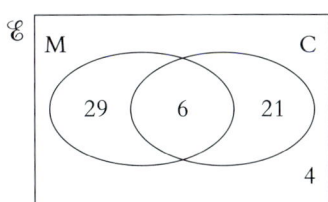

   [Venn diagram: M only = 29, M ∩ C = 6, C only = 21, outside = 4]

   a A person is chosen at random.
      i   Find the probability that the person owns a mobile phone.
      ii Find the probability that the person owns both a computer and a mobile phone.
      iii Find the probability that the person does not own a computer.
   b A person who owns a computer is chosen at random.
      Find the probability that this person does not own a mobile phone.
   c Two people are chosen at random. Find the probability that they both own a computer.

2. There are 100 students in a year group.
   50 study art (A), 29 study biology (B) and 13 study both subjects.
   a Show this information on a Venn diagram.
   b A student is chosen at random.
      i   Find the probability that the student studies biology but not art.
      ii Find the probability that the student studies neither biology nor art.
   c A student who studies art is chosen at random.
      Find the probability that this student does not study biology.
   d Two students are chosen at random. Find the probability that they both study art.

3. In a survey, 150 members of a youth club are asked whether they like singing (S) and whether they like dancing (D).
   The Venn diagram shows the results of the survey.

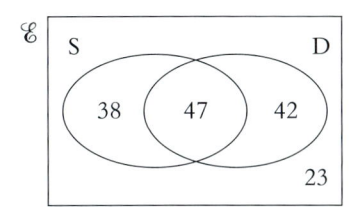

   [Venn diagram: S only = 38, S ∩ D = 47, D only = 42, outside = 23]

   a A member of the youth club is chosen at random.
      i   Find the probability that this person likes dancing.
      ii Find the probability that this person does not like singing.
      iii Find the probability that this person likes neither dancing nor singing.
   b A member of the youth club who likes dancing is chosen at random.
      Find the probability that this person likes singing.
   c Two members of the youth club are chosen at random.
      Find the probability that exactly one of the two members likes dancing.

Probability 3

4 In a survey, 100 students are asked if they like rock climbing (R), windsurfing (W) or swimming (S). The Venn diagram shows the results of the survey.

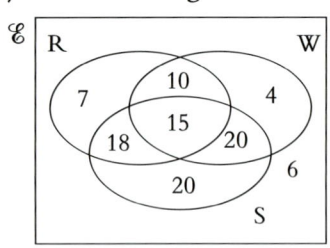

Venn diagram: R contains 7; R∩W = 10; W contains 4; R∩S = 18; R∩W∩S = 15; W∩S = 20; S contains 20; outside = 6.

  a One student is chosen at random.
   i Find the probability that the student likes swimming.
   ii Find the probability that the student likes rock climbing but not windsurfing.
   iii Find the probability that the student likes all three sports.
  b A student who likes rock climbing is chosen at random.
   Find the probability that this student does not like swimming.
  c Two students are chosen at random from those who like windsurfing.
   Find the probability that they both like exactly one other sport.

5 There are 35 students in a class.
  18 study geography (G), 22 study history and 5 study neither geography nor history.
  a Show this information on a Venn diagram.
  b A student is chosen at random.
   i Find the probability that the student studies both history and geography.
   ii Find the probability that the student studies history but not geography.
  c A student who studies history is chosen at random.
   Find the probability that this student studies geography.
  d Two students are chosen at random.
   Find the probability that only one of them studies history.

6 The brakes, steering and lights are tested on 200 cars.
  The Venn diagram shows the results of the survey.
  B = {cars with faulty brakes}   S = {cars with faulty steering}   L = {cars with faulty lights}

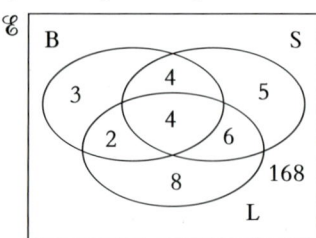

Venn diagram: B only = 3; B∩S = 4; S only = 5; B∩L = 2; B∩S∩L = 4; S∩L = 6; L only = 8; outside = 168.

  a One of the 200 cars is chosen at random.
   i Find the probability that the car has exactly one of the faults.
   ii Find the probability that the car has exactly two of the faults.
   iii Find the probability that the car has all three of the faults.
  b A car with faulty steering is chosen at random.
   Find the probability that this car also has faulty brakes.
  c Two cars are chosen at random. Find the probability that they both have faulty lights.

7 ℰ = {1, 2, 3, 4, 5, 6, 7, 8, 9, 10, 11, 12, 13, 14, 15, 16}
   A = {even numbers} and B = {multiples of 3}
   a Show these sets on a Venn diagram.
   b A number is chosen at random from the set ℰ.
      Find the probability that this number is in the set
      i   B   ii   A′   iii   A ∩ B   iv   A ∪ B   v   (A ∪ B)′

8 In a survey, 50 people are asked if they recycle glass (G), paper (P) or aluminium (A). The Venn diagram shows the results of the survey.

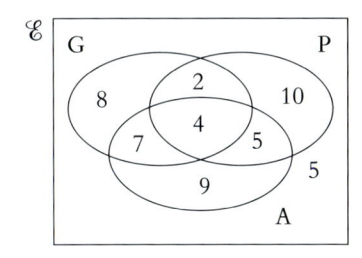

   a One of the 50 people is chosen at random.
      i   Find the probability that they recycle exactly one of the materials.
      ii Find the probability that they recycle exactly two of the materials.
      iii Find the probability that they recycle all three of the materials.
   b A person who recycles aluminium is chosen at random.
      Find the probability that this person also recycles paper.
   c Two people are chosen at random.
      Find the probability that exactly one of them recycles glass.

9 In a survey, a group of 50 students are asked whether they go to a gym (G) or whether they go swimming (S).
   16 students go to a gym.
   21 students go swimming.
   19 students neither go to a gym nor go swimming.
   a Show this information on a Venn diagram.
   b A student is chosen at random.
      Find the probability that the student goes to the gym and goes swimming.
   c A student who goes swimming is chosen at random.
      Find the probability that this student does not go to the gym.
   d Three students are chosen at random.
      Find the probability that all three students go to the gym.

Unit 9 Examination questions

1. A rectangle has sides of length 6.1 cm and 8.1 cm correct to 1 decimal place.
   Calculate the upper bound for the area of the rectangle as accurately as possible. [2]

   Cambridge IGCSE Mathematics 0580, Paper 21 Q7, November 2008

2. In 2005 there were 9 million bicycles in Beijing, correct to the nearest million.
   The average distance travelled by each bicycle in one day was
   6.5 km correct to one decimal place.
   Work out the upper bound for the **total** distance travelled by
   all the bicycles in one day. [2]

   Cambridge IGCSE Mathematics 0580, Paper 21 Q6, June 2009

3. Angharad sleeps for 8 hours each night, correct to the nearest 10 minutes.
   The total time she sleeps in the month of November (30 nights) is T hours.
   Between what limits does T lie? [2]

   Cambridge IGCSE Mathematics 0580, Paper 2 Q4, November 2006

4. The length of a side of a regular hexagon is 6.8 cm, correct
   to one decimal place.
   Find the smallest possible perimeter of the hexagon. [2]

   Cambridge IGCSE Mathematics 0580, Paper 21 Q8, November 2010

5. Solve the simultaneous equations
   $$2y + 3x = 6$$
   $$x = 4y + 16.$$ [3]

   Cambridge IGCSE Mathematics 0580, Paper 21 Q12, June 2009

6. A new school has x day students and y boarding students.
   The fees for a day student are $600 a term.
   The fees for a boarding student are $1200 a term.
   The school needs at least $720 000 a term.

   (a) Show that this information can be written as $x + 2y \geq 1200$. [1]
   (b) The school has a maximum of 900 students.
       Write down an inequality in x and y to show this information. [1]

(c) Draw two lines on a copy of the grid below and write the letter **R** in the region which represents these two inequalities.

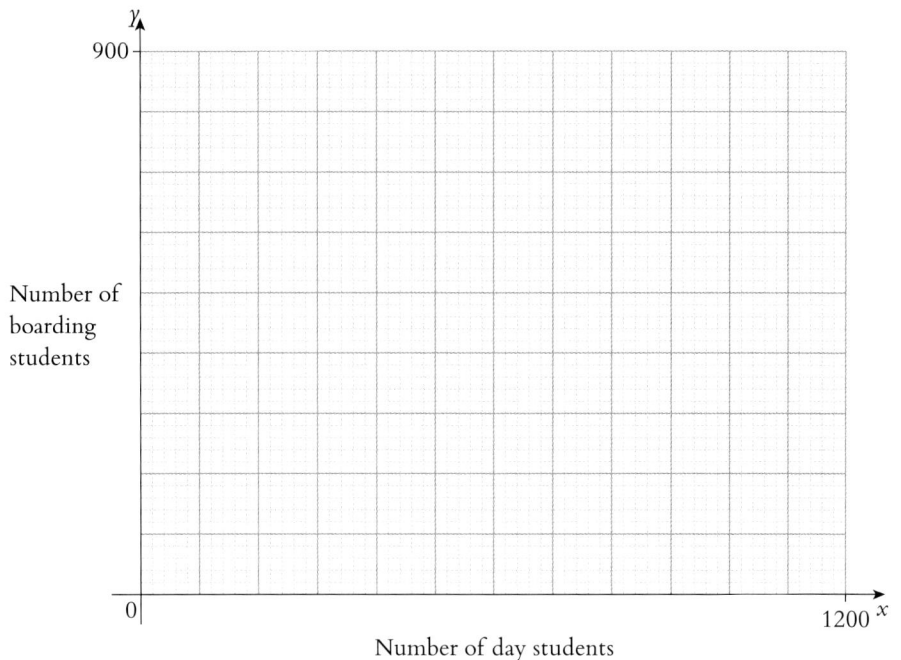

[Grid with y-axis labeled "Number of boarding students" going up to 900, and x-axis labeled "Number of day students" going up to 1200.]

[4]

(d) What is the least number of **boarding** students at the school? [1]

Cambridge IGCSE Mathematics 0580, Paper 21 Q20, November 2008

7 A company has a vehicle parking area of $1200 \, m^2$ with space for x cars and y trucks.
Each car requires $20 \, m^2$ of space and each truck requires $100 \, m^2$ of space.

(a) Show that $x + 5y \leq 60$. [1]
(b) There must also be space for
  (i) at least 40 vehicles, [1]
  (ii) at least 2 trucks. [1]
  Write down two more inequalities to show this information.

(c) One line has been drawn for you.
On a copy of the grid, show the three inequalities by drawing the other two lines and shading the **unwanted** regions.

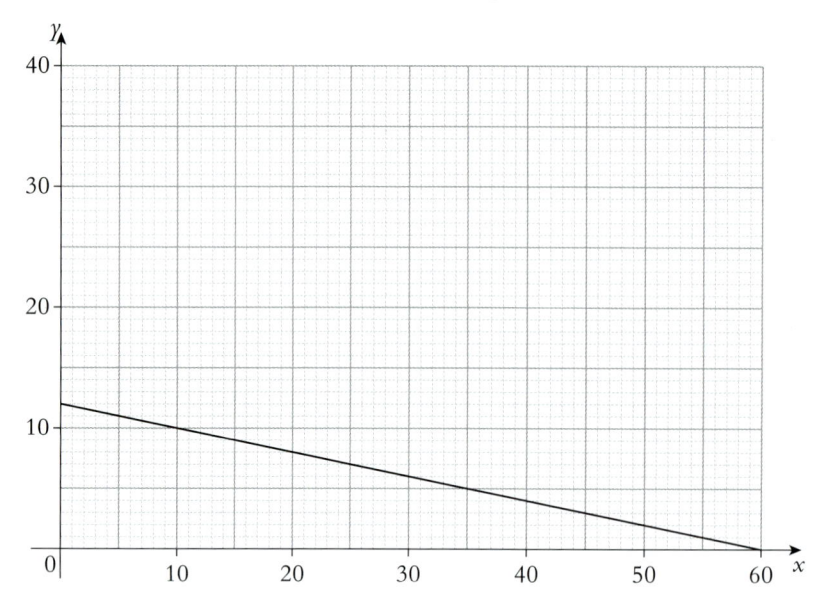

[4]

(d) Use your graph to find the largest possible number of trucks. [1]

(e) The company charges $5 for parking each car and $10 for parking each truck.
Find the number of cars and the number of trucks which give the company the greatest possible income.

Calculate this income. [3]

Cambridge IGCSE Mathematics 0580, Paper 41 Q10, June 2010

8 The table show some of the values of the function $f(x) = x^2 - \dfrac{1}{x}$, $x \neq 0$.

x	−3	−2	−1	−0.5	−0.2	0.2	0.5	1	2	3
y	9.3	4.5	2.0	2.3	p	−5.0	−1.8	q	3.5	r

(a) Find the values of p, q and r, correct to 1 decimal place. [3]
(b) Using a scale of 2 cm to represent 1 unit on the x-axis and 1 cm to represent 1 unit on the y-axis, draw an x-axis for $-3 \leq x \leq 3$ and $-6 \leq y \leq 10$.

Draw the graph of $y = f(x)$ for $-3 \leq x \leq -0.2$ and $0.2 \leq x \leq 3$. [6]

(c) (i) By drawing a suitable straight line, find the three values of x where $f(x) = -3x$. [3]

(ii) $x^2 - \dfrac{1}{x} = -3x$ can be written as $x^3 + ax^2 + b = 0$

Find the values of a and b. [2]

(d) Draw a tangent to the graph of $y = f(x)$ at the point where $x = -2$.

Use it to estimate the gradient of $y = f(x)$ when $x = -2$. [3]

Cambridge IGCSE Mathematics 0580, Paper 4 Q3, November 2008

9

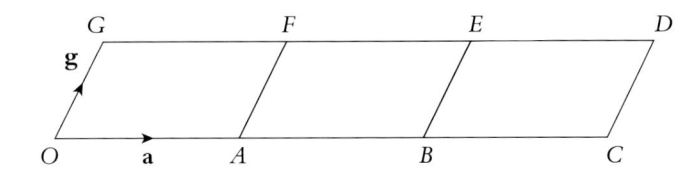

The diagram is made from three identical parallelograms.
O is the origin. $\overrightarrow{OA} = \mathbf{a}$ and $\overrightarrow{OG} = \mathbf{g}$.
Write down in terms of $\mathbf{a}$ and $\mathbf{g}$

(a) $\overrightarrow{GB}$, [1]
(b) the position vector of the centre of the parallelogram $BCDE$. [1]

Cambridge IGCSE Mathematics 0580, Paper 21 Q8, June 2009

10

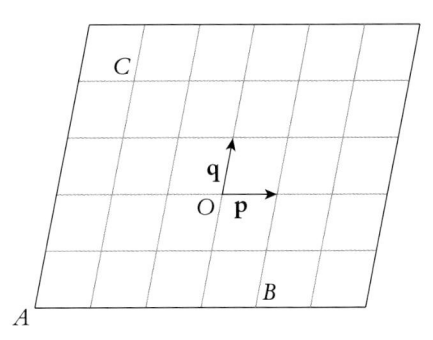

O is the origin. Vectors $\mathbf{p}$ and $\mathbf{q}$ are shown in the diagram.

(a) Write down in terms of $\mathbf{p}$ and $\mathbf{q}$, in their simplest form
  (i) the position vector of the point A, [1]
  (ii) $\overrightarrow{BC}$, [1]
  (iii) $\overrightarrow{BC} - \overrightarrow{AC}$. [2]
(b) If $|\mathbf{p}| = 2$, write down the value of $|\overrightarrow{AB}|$. [1]

Cambridge IGCSE Mathematics 0580, Paper 21 Q17, November 2008

11

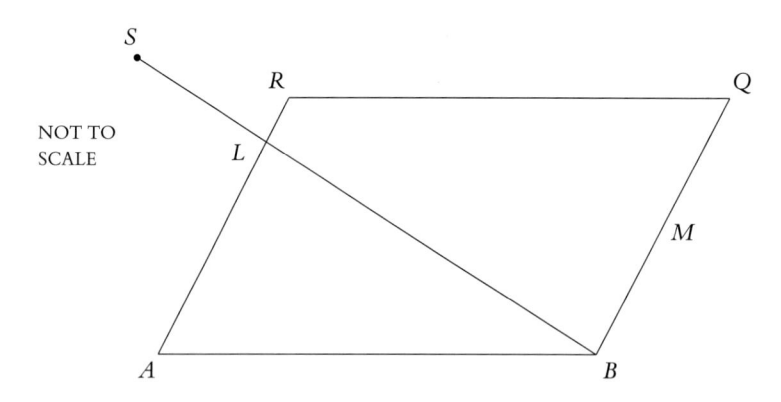

OPQR is a parallelogram.
O is the origin.
$\overrightarrow{OP}$ = **p** and $\overrightarrow{OR}$ = **r**.
M is the mid-point of PQ and L is on OR such that OL : LR = 2 : 1.
The line PL is extended to the point S.

(a) Find, in terms of **p** and **r**, in their simplest forms.
   (i) $\overrightarrow{OQ}$, [1]
   (ii) $\overrightarrow{PR}$, [1]
   (iii) $\overrightarrow{PL}$, [1]
   (iv) the position vector of M. [1]

(b) PLS is a straight line and PS = $\frac{3}{2}$PL.

   Find, in terms of **p** and / or **r**, in their simplest forms,
   (i) $\overrightarrow{PS}$, [1]
   (ii) $\overrightarrow{QS}$. [2]

(c) What can you say about the points Q, R and S? [1]

Cambridge IGCSE Mathematics 0580, Paper 4 Q9, June 2008

12 In a survey, 100 students are asked if they like basketball (*B*), football (*F*) and swimming (*S*). The Venn diagram shows the results.

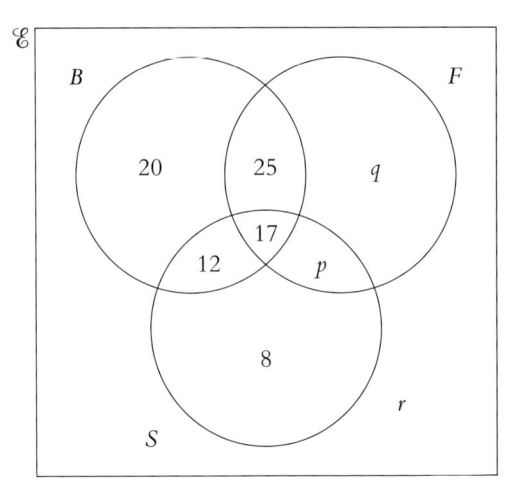

42 students like swimming.
40 students like exactly one sport.

(a) Find the values of *p*, *q* and *r*. [3]
(b) How many students like
  (i) all three sports, [1]
  (ii) basketball and swimming but not football [1]
(c) Find
  (i) n(B'), [1]
  (ii) n(($B \cup F) \cap S'$). [1]
(d) One student is chosen at random from the 100 students.
  Find the probability that the students.
  (i) only likes swimming, [1]
  (ii) likes basketball but not swimming. [1]
(e) Two students are chosen at random from those who like basketball.

  Find the probability that they each like exactly one other sport. [3]

Cambridge IGCSE Mathematics 0580, Paper 4 Q9, November 2008

Answers

Unit 1

Exercise 1.1
1. 17
2. 8
3. 2
4. 26
5. 6
6. 4
7. 19
8. 46
9. 3
10. 14
11. 48
12. 10
13. 50
14. 7
15. 44
16. 40
17. 196
18. 22
19. 34
20. 85
21. −1
22. 2
23. 2
24. −3
25. Answers as above.
26. ±4
27. −6
28. −36

Exercise 1.2
1. $(2 + 3) \times 4 + 5 = 25$
2. $2 \times 3 + 4 \times 5 = 26$
3. $2 + 3 \times (4 + 5) = 29$
4. $(2 + 3) \times 4^2 = 80$
5. $2 \times (3 + 4 \times 5) = 46$
6. $5 + 4 \times 3 − 2 = 15$
7. $(2 \times 3 + 4) \times 5 = 50$
8. $2 \times (3 + 4) \times 5 = 70$
9. $(2 + 3) \times (4 + 5) = 45$
10. $2 + 3 \times 4^2 = 50$

Exercise 1.3
1. −8
2. 10
3. −2
4. −3
5. −6
6. −18
7. 28
8. 28
9. −14
10. −57
11. −18
12. 0
13. −52
14. 25
15. 55
16. −15
17. 7
18. −47
19. −15
20. −5
21. 6
22. 9
23. −5
24. 53

Exercise 1.4
1. 60
2. −32
3. −32
4. 4
5. −11
6. 29
7. −10
8. 5
9. 121
10. −60
11. 64
12. 126
13. 60
14. 81
15. 225
16. −125
17. −216 000
18. 64
19. −3200
20. −1

Exercise 1.5
1.
   a. 1, 2, 5, 10
   b. 1, 3, 5, 15
   c. 1, 3, 9   d. 1, 17
   e. 1, 2, 3, 4, 5, 6, 10, 12, 15, 20, 30, 60
   f. 1, 2, 4, 5, 8, 10, 16, 20, 40, 80
   g. 1, 2, 4, 5, 10, 20, 25, 50, 100
   h. 1, 2, 4, 8, 16, 32, 64
   i. 1, 5, 25, 125
   j. 1, 2, 3, 5, 6, 9, 10, 15, 18, 30, 45, 90
2.
   a. 1, 2   b. 1, 5
   c. 1, 3, 9
   d. 1, 2, 4   e. 1, 5
   f. 1, 2, 3, 6
   g. 1, 2, 4, 5, 10, 20
   h. 1, 2, 3, 6   i. 1, 2, 3, 6
3.
   a. 2   b. 5   c. 18
   d. 9   e. 23
   f. 4   g. 15   h. 6   i. 4
4.
   a. 10, 20, 30, 40, 50, 60
   b. 6, 12, 18, 24, 30, 36
   c. 9, 18, 27, 36, 45, 54
   d. 18, 36, 54, 72, 90, 108
   e. 25, 50, 75, 100, 125, 150
   f. 40, 80, 120, 160, 200, 240
   g. 100, 200, 300, 400, 500, 600
   h. 12, 24, 36, 48, 60, 72
   i. 27, 54, 81, 108, 135, 162
   j. 121, 242, 363, 484, 605, 726
5.
   a. 24   b. 15   c. 18
   d. 56   e. 12   f. 24
   g. 42   h. 55   i. 120
6. 24 m

7. 3.30 am
8. 5 minutes

Exercise 1.6
1. 11, 17, 47
2. 3, 13, 23, 43, 53, 73
3.
   a. 50   b. 66   c. 189
   d. 70   e. 234   f. 132
   g. 4725   h. 15840
4.
   a. 2×5 b. $2 \times 3 \times 5^2$
   c. 3^4 d. $2^2 \times 3 \times 5$
   e. 2×37 f. $2^2 \times 5^2$
   g. 2×7^2 h. 2×5^3
   i. $2 \times 3 \times 5 \times 37$ j. $5^2 \times 11$
   k. $2^2 \times 3 \times 167$
   l. $2 \times 5 \times 13 \times 17$
5.
   a. 3, 7, 19
   b. 3, 15, 21
   c. 10, 15, 35
   d. 3, 10, 15
   e. 10, 19
6. 83, 89, 97
7. $713 = 23 \times 31$
8.
   a. 3 + 7   b. 3 + 11
   c. 2 + 23   d. 2 + 47
   e. 13 + 17   f. 3 + 17
   g. 7 + 31   h. 3 + 79
   i. 17 + 19   j. 7 + 41
9.
   a. 6, 168   b. 15, 315
   c. 14, 210   d. 5, 600
10.
   a. 6, 1260
   b. 7, 280
   c. 30, 1260

Exercise 1.7
1. 3^2
2.
   a. 1, 9, 49, 64, 289
   b. 1, 27, 64, 343, 512
3. 729

Exercise 1.8

1. $\frac{1}{2}$ 2. $\frac{5}{7}$ 3. $\frac{3}{4}$
4. $\frac{2}{5}$ 5. $\frac{5}{8}$ 6. $\frac{6}{7}$
7. $\frac{4}{7}$ 8. $\frac{4}{5}$ 9. $\frac{7}{8}$
10. $\frac{3}{11}$ 11. $\frac{17}{20}$ 12. $\frac{7}{12}$
13. $1\frac{3}{5}$ 14. $2\frac{1}{3}$ 15. $2\frac{1}{7}$
16. $1\frac{3}{17}$ 17. $6\frac{1}{2}$ 18. $4\frac{3}{4}$
19. $1\frac{5}{12}$ 20. $3\frac{2}{15}$ 21. $1\frac{7}{8}$
22. $1\frac{1}{21}$ 23. $3\frac{2}{5}$ 24. $3\frac{1}{3}$

Exercise 1.9

1. $3\frac{5}{6}$ 2. $5\frac{9}{20}$ 3. $7\frac{25}{42}$
4. $7\frac{7}{15}$ 5. $4\frac{7}{12}$ 6. $4\frac{7}{30}$
7. $4\frac{7}{40}$ 8. $15\frac{4}{9}$ 9. $2\frac{3}{10}$
10. $\frac{7}{24}$ 11. $3\frac{22}{63}$ 12. $2\frac{11}{40}$
13. $3\frac{11}{40}$ 14. $\frac{17}{30}$ 15. $3\frac{5}{12}$
16. $3\frac{29}{35}$ 17. $-2\frac{3}{10}$ 18. $6\frac{13}{15}$
19. $4\frac{1}{12}$ 20. -1 21. $-1\frac{4}{15}$
22. $1\frac{7}{15}$ 23. Answers as above.
24. $\frac{13}{35}$ 25. $\frac{13}{20}$ 26. $1\frac{9}{10}$
27. $6\frac{31}{35}$ 28. $-\frac{1}{12}$
29. a $\frac{13}{30}, \frac{3}{5}, \frac{2}{3}$ b $\frac{5}{9}, \frac{7}{12}, \frac{11}{36}$
   c $\frac{7}{10}, \frac{3}{4}, \frac{4}{5}$ d $\frac{5}{6}, \frac{8}{9}, \frac{11}{12}$

Exercise 1.10

1. $3\frac{2}{3}$ 2. $5\frac{1}{16}$ 3. $3\frac{29}{30}$
4. $19\frac{23}{28}$ 5. $11\frac{5}{18}$ 6. $20\frac{11}{25}$
7. $\frac{9}{14}$ 8. $2\frac{1}{2}$ 9. $1\frac{7}{10}$
10. $3\frac{29}{32}$ 11. $\frac{10}{27}$ 12. $1\frac{1}{20}$
13. $\frac{9}{25}$ 14. $6\frac{1}{4}$ 15. $\frac{27}{8000}$
16. $1\frac{7}{9}$ 17. $\frac{2}{5}$ 18. $1\frac{5}{8}$

19. Answers as above.
20. $1\frac{2}{3}$ 21. $1\frac{1}{2}$ 22. $\frac{15}{56}$
23. $2\frac{3}{11}$ 24. $3\frac{3}{4}$ litres

Exercise 1.11

1. a 5.3 b 8.8 c 6.0
   d 30.0 e 0.1 f 8.9
   g 4.2 h 16.3 i 644.0
   j 7.0
2. a 6.35 b 8.39 c 16.16
   d 4.21 e 5.70 f 0.04
   g 0.07 h 2.99 i 3.06
   j 7.10
3. a 7.222 b 6.162
   c 35.686 d 82.009
   e 24.789 f 3.000
   g 6.009 h 6.667
   i 102.103 j 5.010
4. a 3 b 8 c 10
   d 50 e 400 f 30 000
   g 50 000 h 0.002 i 0.09
   j 0.3
5. a 1.7 b 3.1 c 5700
   d 61 000 e 16 f 20
   g 100 h 0.0063
   i 0.058 j 0.00040
6. a 27.3 b 6.51 c 2590
   d 149 e 16.7 f 0.349
   g 0.0718 h 0.00808
   i 10.1 j 40 000
7. a 7 000 000 000
8. a 100 000 000
   b 110 000 000
   c 106 000 000
9. a 9000 b 8800 c 8850
10. a 0.06 b 0.064 c 0.0640
    d 0.1 e 0.06 f 0.064
11. a 40 000 b 40 000
    c 40 000 d 40 010
    e 40 008.6 f 40 008.63

Exercise 1.12

1. a 60 (62.72) b 16 (16.83)
   c 3000 (3555)
   d 2500 (2569)
   e 80 000 (81290)
   f 6 (5.695) g 5 (5.861)
   h 5 (4.404) i 16 (16.54)
2. a 10 (10.10) b 2 (2.081)
   c 3 (2.564)
3. Accurate answers given in brackets above.
4. $8 5. $9500

Exercise 1.13

1. $-x$ 2. $-3y$ 3. $6xy$
4. $-3xy$ 5. $7x - 9y$
6. $5p + 5q$ 7. $5xy - 3x$
8. $7x^2 - 15x$ 9. $7 + 4x$
10. $-7xy$ 11. $-3x^2 + 9$
12. $-6y^2 - 2y$ 13. $8xy - x$
14. $a - 4ab$ 15. $10ab - 6bc$
16. $x^4 + 5x^2 - 5$ 17. $5x^3 - x^2 + x$
18. $-2fg - 4gh$ 19. $4a^2b - 16ab^2$
20. $5cd^2 - 9c^2d$ 21. $6x + \frac{10}{x}$
22. $\frac{9}{x} - \frac{7}{y}$
23. $-2xy^2 + 12xy - 10x^2y + x^2 - y^2$
24. 3

Exercise 1.14

1. $7(2x + 3) = 14x + 21$
2. a $4x + 12$ b $6y + 12$
   c $4x + 20$ d $6 - 2a$
   e $7y - 28$ f $8x - 72$
   g $10x + 15$ h $18y + 24$
   i $35a + 42$ j $6x + 6y - 3$
   k $20a + 25b - 15$
   l $24p - 32q - 56r$ m $4x - 1$
   n $3x + 2$ o $5y - 2$
3. a $-3x - 12$ b $-2x + 12$
   c $-30 + 20x$ d $-16x - 40$
   e $-21x + 56$ f $-8x + 72$
   g $-5x - 4$ h $-2x + 7$
   i $-3x - 8$ j $-18x + 18y - 30$
   k $-12p - 16q + 20$
   l $-27x^2 - 9x + 6$
4. a $x^2 + 3x$ b $y^2 - 5y$
   c $7a - a^2$ d $10x - 6x^2$
   e $5y^2 + 40y$ f $6x^2 + 12xy$

5 **a** $5x + 30$ **b** $11x + 32$
   c $9y$ **d** $11x - 74$
   e $36x - 4y$ **f** $2x^2 - 11x$
   g $2x^2 + 6x$ **h** $2x^2 + 4x$
   i $38x - 9x^2$
   j $12xy - 8x - 15y$
   k $-4x^2 + 11xy + 21y^2$
6 $5(2x - 3) - 4(x - 6) =$
   $10x - 15 - 4x + 24 = 6x + 11$
7 **a** x **b** $2y - 11$
   c $23a - 10$ **d** $36x - 20$
   e $2x + 17$ **f** $11 - 8y$
   g $-h - 6g$ **h** $x^2 - 5x$
   i $11p^2 - 44p$
   j $63a + 6ab - 51a^2$
8 **a** $2x + 16$ **b** $27 - 3x$
   c $8 - 6x + 8y$ **d** $-1 - 5y$
   e $22 - 7y$ **f** $14x - 15$
   g $8y + x$ **h** $11x - 2x^2$

Exercise 1.15
1 **a** 6 **b** 9 **c** 7
   d $1\frac{1}{6}$ **e** $1\frac{1}{2}$ **f** $13\frac{1}{4}$
   g 3 **h** $-\frac{2}{3}$ **i** $\frac{3}{4}$
   j $1\frac{7}{8}$ **k** $-2\frac{1}{2}$ **l** $-2\frac{4}{5}$
   m $7\frac{2}{3}$ **n** $8\frac{1}{2}$ **o** $\frac{1}{2}$
   p $1\frac{5}{6}$ **q** 7 **r** $-2\frac{2}{5}$
   s 2 **t** $2\frac{4}{5}$
2 **a** 2 **b** 5 **c** 7
   d $3\frac{1}{2}$ **e** $\frac{1}{8}$ **f** $3\frac{4}{5}$
   g $\frac{1}{5}$ **h** $-6\frac{1}{2}$ **i** -5
   j 1 **k** $1\frac{1}{2}$ **l** 1
3 **a** 4 **b** 3 **c** 4
   d 5 **e** $6\frac{1}{2}$ **f** -4
   g $-\frac{1}{3}$ **h** $4\frac{1}{2}$ **i** $1\frac{2}{3}$
4 56, 57, 58
5 20

6 2
7 1.5 cm
8 5

Exercise 1.16
1 **a** 20 **b** 60 **c** 75
   d $6\frac{2}{3}$ **e** 45 **f** -28
   g 25 **h** $9\frac{1}{3}$ **i** -14
   j 22 **k** 24 **l** 20
2 **a** 7 **b** 11 **c** $3\frac{1}{4}$
   d $1\frac{1}{2}$ **e** $-2\frac{1}{2}$ **f** 5
   g $4\frac{1}{2}$ **h** $1\frac{7}{15}$
3 **a** 3 **b** $\frac{1}{2}$ **c** $1\frac{3}{4}$
   d $\frac{3}{10}$ **e** $\frac{1}{2}$ **f** 2
   g 5 **h** -2 **i** $1\frac{1}{2}$
   j -2 **k** $2\frac{1}{2}$ **l** -4
   m 3 **n** 3 **o** 1 **p** 0
4 **a** 0 **b** 3 **c** 4 **d** 1
   e $-\frac{8}{11}$ **f** -7 **g** 3
   h $-3\frac{7}{8}$
5 $12x + 4 = 12x - 5$ means that
   $4 = -5$ which is not possible
6 6

Exercise 1.17
1 $P = 1350 + 13n$
2 $Q = 100 - 9n$
3 $C = 4b + 7c$
4 $T = \frac{5r}{100} + \frac{6p}{100} + 4f$
5 $T = 30 + 4n$
6 $T = 45m + 20$

Exercise 1.18
1 -8 2 -23 3 -36
4 11 5 20 6 3
7 4 8 210 9 -2
10 16 11 -48 12 -480
13 -40 14 128 15 12

16 28 17 6 18 8
19 -1 20 -9 21 $-2\frac{1}{2}$
22 3 23 $3\frac{1}{2}$ 24 $-15\frac{2}{3}$
25 0 26 $-5\frac{4}{5}$ 27 -3
28 26

Exercise 1.19
1 **a** -10 **b** 5
   c -20 **d** 100
2 **a** 25 **b** 9 **c** 10
3 **a** 6 **b** 2.1 **c** 0.912
4 **a** 1 **b** 1.2 **c** 1.8
5 **a** 435 **b** 16
6 **a** 5 **b** 8
7 **a** 7 **b** 10
8 **a** 9 **b** 8

Exercise 1.20
1 **a** 1 **b** 4 **c** $-\frac{1}{3}$
   d $\frac{1}{4}$ **e** -2 **f** $\frac{2}{3}$
   g 0 **h** $-\frac{3}{2}$ **i** -1
   j 2
2 **a** 1 **b** $-\frac{9}{8}$ **c** $\frac{3}{2}$
   d $\frac{1}{3}$ **e** -4 **f** $\frac{9}{7}$
   g 0 **h** -11 **i** 4
   j -1 **k** $\frac{2}{5}$ **l** $\frac{1}{4}$
3 **a** 1 **b** 2 **c** $\frac{1}{2}$
   d 2 **e** $-\frac{1}{2}$ **f** infinite(∞)
   g 1 **h** 0
4 6
5 8
6 7
7 8
8 5
9 -3
10 2
11 11.25

Exercise 1.21
1 **a** $y = 2$ **b** $y = x$
   c $x = 4$ **d** $y = -x$

2 a

x	0	1	2
y	2	3	4

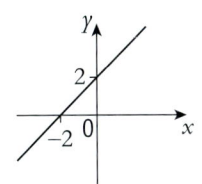

b

x	0	1	2
y	−5	−4	−3

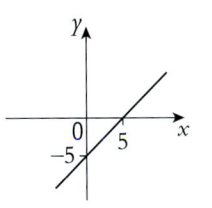

c

x	0	1	2
y	−2	0	2

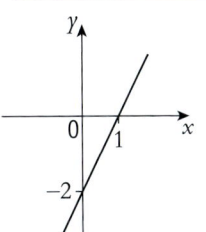

d

x	0	1	2
y	−4	−1	2

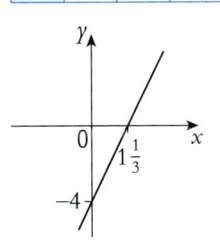

e

x	−2	0	2
y	−2	−1	0

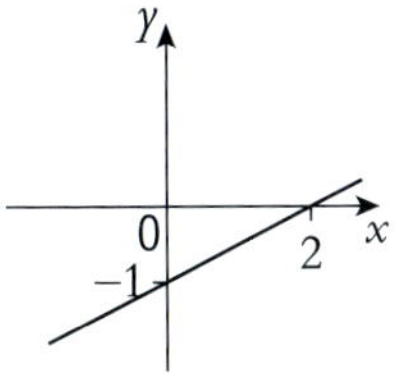

f

x	−3	0	3
y	−5	−4	−3

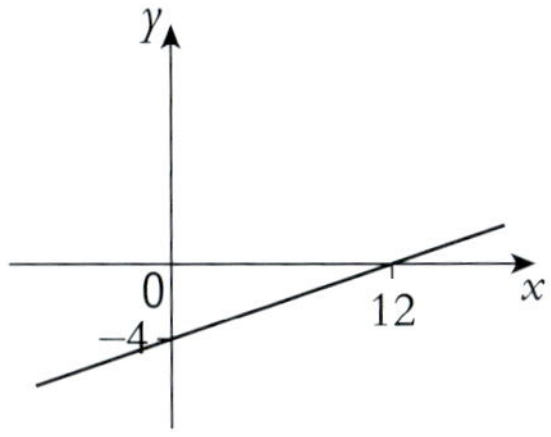

g

x	0	1	2
y	3	1	−1

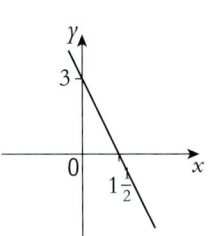

h

x	0	1	2
y	5	2	−1

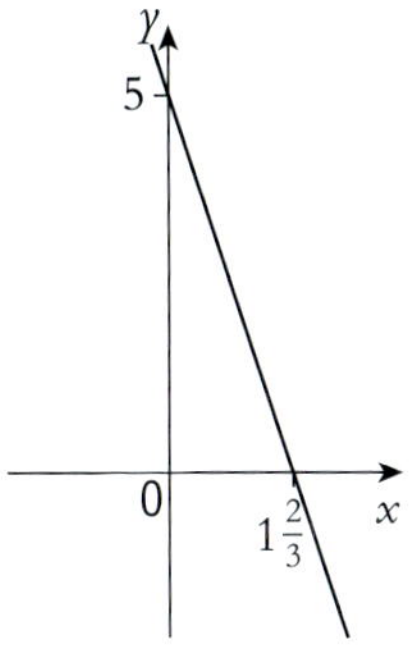

3 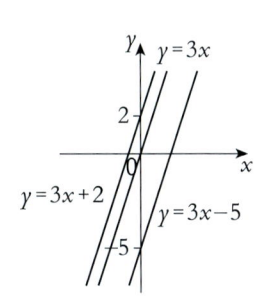 Parallel

4

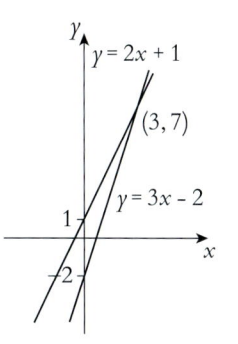

5

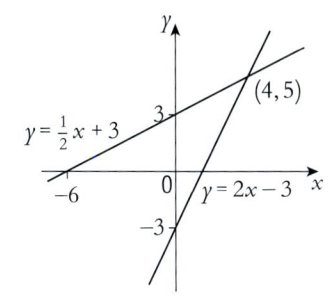

6 2

7 $-4\frac{1}{2}$

8 −14

Exercise 1.22

1

x	0	2
y	4	0

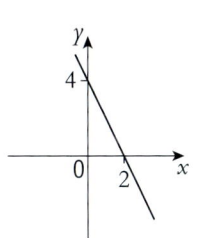

2

x	0	2
y	3	0

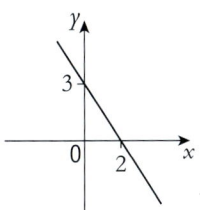

Answers 465

3

x	0	5
y	2	0

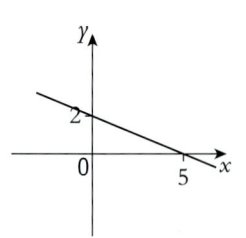

4

x	0	3
y	9	0

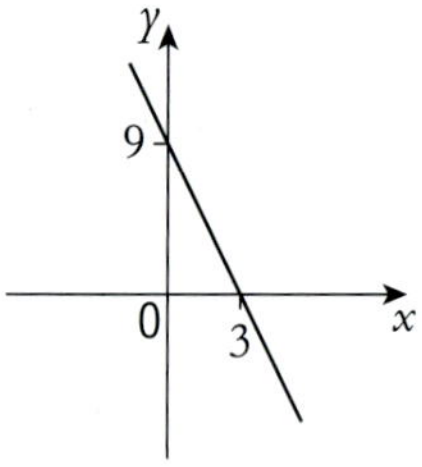

5

x	0	-6
y	-4	0

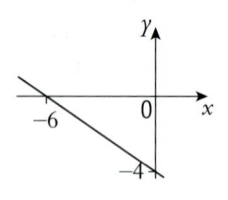

6

x	0	-5
y	-6	0

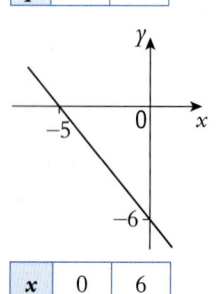

7

x	0	6
y	-3	0

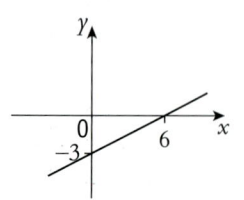

8

x	0	5
y	-4	0

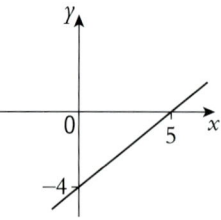

9

x	0	6
y	-4	0

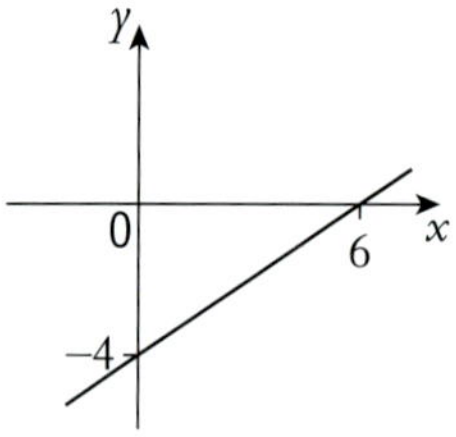

10

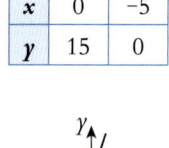

x	0	-5
y	15	0

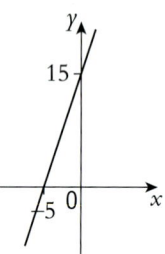

11

x	0	-6
y	10	0

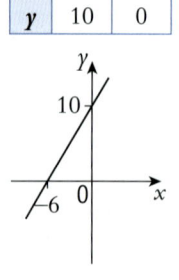

12

x	0	-2
y	12	0

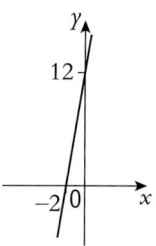

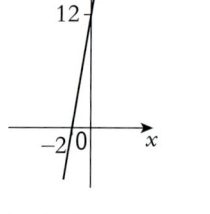

Exercise 1.23

1 $a = 138°$ $b = 42°$ $c = 138°$
2 $a = 150°$
3 $a = 55°$
4 $a = 36°$
5 $a = 49°$
6 $a = 25°$
7 $a = 68°$ $b = 44°$
8 $a = 18°$
9 $a = 50°$ $b = 65°$
10 $a = 40°$
11 $a = 40°$ $b = 35°$
12 $a = 30°$
13 $a = 40°$
14 $a = 120°$
15 $a = 55°$
16 $a = 50°$
17 $a = 40°$ $b = 40°$
18 $a = 45°$
19 $a = 100°$ $b = 130°$
20 $a = 65°$
21 $a = 60°$
22 $a = 60°$ $b = 70°$
23 $a = 73°$ $b = 45°$
24 $a = 109°$
25 33°, 66°, 99°, 162°
26 Several possible proofs, for example

$E\hat{F}B = F\hat{G}D$ (Corresponding angles)

$E\hat{F}B = 180° - B\hat{F}G$

$F\hat{G}D = 180° - F\hat{G}D$ (Angles on a straight line)

$B\hat{F}G + F\hat{G}D = 180°$

Exercise 1.24

1 a 4, 4 **b** 2, 2 **c** 0, 2

d 0, 1 **e** 2, 2 **f** 1, 1
2 a 1, 1 **b** ∞, ∞ **c** 0, 1
  d 2, 2 **e** 2, 2 **f** 8, 8
  g 0, 2 **h** 0, 4 **i** 0, 2
  j 6, 6 **k** 4, 4 **l** 0, 2
3 a 0, 2 **b** 2, 2 **c** 0, 4
4 a There are other possible answers.
  b There are other possible answers.
  c There are other possible answers.
5 a 1, 1 **b** 0, 2 **c** 2, 2
  d 0, 2
6 The four properties allow you to distinguish all quadrilaterals except a kite and an arrowhead.

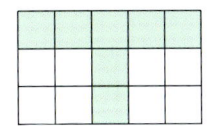

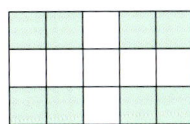

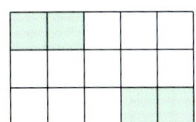

Exercise 1.25
1 (figures)
2 a 9 **b** ∞ **c** ∞
  d ∞ **e** 7

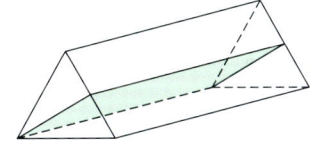

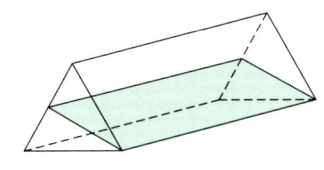

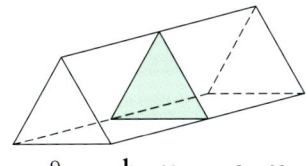

3 a 2 **b** 1
  c 0 **d** 5
4 (figure)

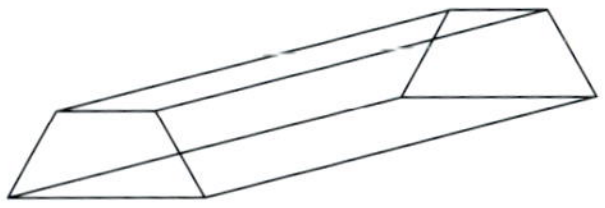

Exercise 1.26
1 No
  Sum of interior angles = 4 × 180° = 720°
2 124° **3** 140°
4 12 **5** 20°
6 1440° **7 a** 117°
  b 58° **c** 41°
  d 94° **e** 125° **f** 55°
8 a 108° **b** 135° **c** 67.5°
  d 22.5° **e** 36° **f** 81°
  g 117° **h** 36°
9 18

Exercise 1.27
1 He is incorrect Median = 5
2 a mean = 9 median = 10
    mode = 10 range = 5
  b mean = −2 median = −2
    no mode range = 15
  c mean = 58 median = 58.5
    no mode range = 15
  d mean = 3 median = 3
    mode = 3 range = 5
  e mean = $1\frac{1}{12}$ median = $1\frac{1}{12}$
    no mode range = $1\frac{1}{6}$
3 mean = 12 median = 13
  mode = 13 range = 9
4 mean = 10.336s
  median = 10.305s
  mode = 10.28s range = 0.6s
5 a Use median = 5 or mean = 4.5; mode = 0 or 8 which are extremal.
  b Use mode = 7 or median = 7; mean = 11.4 is sensitive to single large value.
  c Use median = 8 or mean = 8, though there are no data values near these averages. Mode not defined.
  d Use mean = 0 or median = 2 Mode = 3 is extremal.
6 a 659.2 kg **b** 7.14 m
7 1.75 m
8 a 69 **b** 7 **c** 13 **d** 56
9 68.2 **10** 61 **11** 37.1 cm

Exercise 1.28
1 a Incorrect Mode = 1
  b Incorrect Median = 1
2 a 1 **b** 1 **c** 1.6
3 mode = 1 median = 1
  mean = 1.2
4 mode = 1 median = 1
  mean = 1.43
5 mode = 7 median = 7
  mean = 7.025
6 a $x = 25, n = 41$ **b** 5

Examination Questions
1 a 1045.28 **b** 1000
2 a 590 **b** Neptune
3 53, 59
4 44
5 20
6 2
7 −170
8 a $500 + 170x$ **b** 11
9 a 6 **b** 0
10 a i 6 **ii** $4\frac{1}{2}$ **iii** 4.54
11 a i (figure) OR (figure)
  ii (figure)

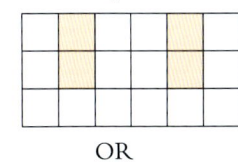

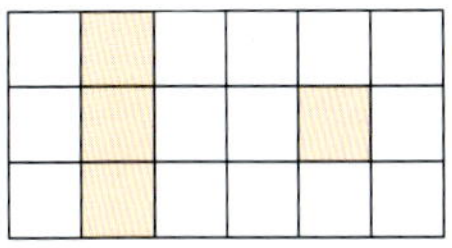

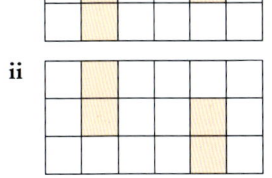

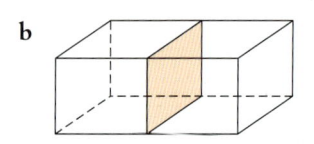

b Other answer possible

c 3

12 a 1   b 2.5   c 2.96
   d 2.9

13 a i 4   ii 5   iii 4.75
   b $\frac{190+3n}{40+n}$

Unit 2

Excercise 2.1

1 a $\frac{6}{25}$ b $\frac{4}{5}$ c $\frac{29}{50}$
  d $\frac{13}{20}$ e $\frac{3}{10}$

2 a 60%   b 35%   c 72%
  d 74%   e 87.5%

3 a 0.18   b 0.7   c 0.06
  d 0.47   e 0.025

4 a 28%   b 80%   c 8%
  d 120%   e 75.5%

5 a $12   b 312 km
  c 135 kg

6 30% of $60

7 a $64.96   b $89.28
  c $190.40   d $271.40
  d $451   f $198

8 a $54.40   b $272
  c $230   d $112
  e $482.50   f $359.28

9 $735

10 $883.50

11 $1277.50

12 $3510

13 $18

Excercise 2.2

1 a 2:3   b 3:4   c 1:5
  d 8:5   e 29:36   f 5:3
  g 8:7   h 7:11

2 a 1:5   b 1:20   c 1:4
  d 1:5 000 000   e 25:1

f 1:50   g 1:25   h 5:8
i 1:4

3 a 7:3   b 4:3   c 3:1
  d 1:5   e 2:5   f 26:11
  g 7:1   h 150:1

4 a 2:1   b 2:5   c 9:8
  d 3:5   e 1:4   f 9:10
  g 3:2   h 3:2

5 a 1:4:5   b 3:5:4   c 1:3:5
  d 8:3:7   e 21:45:77
  f 1:4:7   g 8:12:1
  h 10:15:4

6 a 1:4   b 1:$\frac{2}{3}$
  c 1:0.16   d 1:0.625
  e 1:500 000   f 1:400
  g 1:750 000   h 1:333 333$\frac{1}{3}$

7 3:1

8 6:15:11

9 3:7

10 1:1.6, 1:1.625, 1:1.615,
   1:1.619, 1:1.617, 1:1.618
   Tends to a constant
   ratio 1:1.61803399....

Excercise 2.3

1 a $56, $28   b 105 m, 175 m
  c $1200, $900
  d 187.2 km, 436.8 km
  e 112 kg, 70 kg
  c $23.20, $29

2 a $70, $140, $210
  b 225 m, 45 m, 90 m
  c $140, $60, $20
  d 900 km, 750 km, 1200 km
  e 8000 kg, 6000 kg, 10 000 kg
  f $269.60, $377.44, $431.36

3 $31.35

4 84°

5 36°

6 $\frac{4}{9}$

7 4:3

8 85

9 315

10 1161

11 115 cm

12 15

Excercise 2.4

1 a i 3.5 cm   ii 2.6 cm
     iii 1.6 cm
  b i 70 km   ii 52 km
     iii 32 km

2 40 km

3 a 15 km   b 32.5 km
  c 21 km   d 4.5 km

4 a 8 km   b 5 km
  c 7.6 km   d 1.4 km

5 a 2 cm   b 10 cm
  c 5 cm   d 8.6 cm

6 3.2 km   7 3 km

8 6 cm

9 27.2 km

10 4.8 cm

11 7.8 cm

Excercise 2.5

1 Incorrect $3^x \times 3^y = 3^{x+y}$

2 a x^9 b b^8 c y^{10}
  d c^{12} e a^{11} f p^8
  g y^9 h a^7

3 a $15x^6$ b $8b^4$ c $14y^{14}$
  d $25c^6$ e $6a^3$ f $70p^8$
  g $60y^6$ h $27a^6$

4 a $6x^6y^2$ b $30x^4y^3$
  c $10x^2y^3$ d $6x^4y^2$
  e $15x^3y^3$ f $35x^3y$
  g $4x^4y^2$ h $6a^8b^4$

5 a x^5 b b c a^3
  d x^5 e $6x^3$ f $2y^2$
  g $5b^7$ h $3c$

6 a $5y^2$ b $12ab$ c $\frac{x}{4}$
  d $\frac{y}{2}$ e $2a$ f $4abc$
  g a^2b^2c h $5x^2$ i x^5
  j y^7 k $6a^7$ l $3x^6$

7 a x^6 b a^{10} c b^{16}
  d y^{21} e $3a^6$ f $5x^{20}$
  g $8x^9$ h $81y^8$ i $8a^6$
  j $9y^8$ k $18x^4y^2$ l $40x^7$

8   a $27x^3y^6$ b $32a^{10}b^{15}$
    c $25x^8y^6$ d $1000x^{15}y^{21}$
9   a 4   b $13x^6$
    c $36x^3$ d $\dfrac{ab^3}{9}$
    e $4a^4bcd^2$ f $32x^{19}y^7$
    g $8x^9$ h $16x^4y^8$

Excercise 2.6

1 a $\dfrac{1}{3}$ b 1 c $\dfrac{1}{8}$ d $\dfrac{1}{27}$
   e 1   f $\dfrac{1}{1000}$ g $\dfrac{1}{16}$
   h $\dfrac{1}{7}$ i $\dfrac{1}{10000}$ j $\dfrac{1}{125}$
   k $\dfrac{1}{32}$ l $\dfrac{1}{81}$

2 a $\dfrac{1}{9}$ b $-\dfrac{1}{8}$ c $\dfrac{1}{36}$
   d $-\dfrac{1}{9}$ e $\dfrac{1}{16}$ f $-\dfrac{1}{1000}$

3 a $1\dfrac{1}{2}$ b $2\dfrac{1}{2}$ c $1\dfrac{1}{3}$
   d $1\dfrac{1}{5}$ e $3\dfrac{1}{2}$ f $1\dfrac{1}{4}$

4 a $2\dfrac{1}{4}$ b $6\dfrac{1}{4}$ c $1\dfrac{7}{9}$
   d $1\dfrac{11}{25}$ e $12\dfrac{1}{4}$ f $1\dfrac{9}{16}$
   g 8   h 1000

5 a a^4 b b^{-9} c c^{-2} d d^{-13}
   e d^3 f x^3 g y^4 h d^3
   i a^5 j b k 1 l d^{-4}

6 a x^{-2} b x^5 c $6x^9$ d x^{-5}

7 a $8x$ b $15x^{-3}$ c $6x^3$
   d $28x^{-5}$

8 a x^{-2} b y^{-6} c a^{-20} d b^{-8}
   e a^6 f y^2 g a^{10} h a

9 a $9a^{-2}$ b $125b^{-3}$ c $16c^{-8}$
   d $16d^{-4}$ e $\dfrac{1}{4}x^{-4}$ f $\dfrac{1}{27}y^{-6}$
   g $\dfrac{1}{25}x^{-6}$ h $\dfrac{1}{4}x^4$

10 a x b $\dfrac{1}{2}x^7$ c y^{-1}
   d $7x^{-2}$

11 $2a^2 + 3b^{-3}$

Excercise 2.7

1 $x \geq 1$ 2 $x < 2$ 3 $-1 \leq x \leq 2$
4 $-2 \leq x < 3$ 5 $x \leq 6$
6 $x \geq 4$ 7 $x < 7$ 8 $x > 2$
9 $x \geq 1\dfrac{1}{2}$ 10 $x < 6$ 11 $x \leq -1$

12 $x < 7$ 13 $x \geq 1$ 14 $x < 7\dfrac{1}{2}$
15 $x \leq -6$ 16 $x < 2$ 17 $x \geq 7$
18 $x < -\dfrac{1}{5}$ 19 $x > -2\dfrac{1}{2}$
20 $x \leq -4$ 21 $x \leq 1$ 22 $y > 8$
23 $x \geq 2$ 24 $x < 7$ 25 $x < \dfrac{1}{2}$
26 $x \geq -3$ 27 $x \leq 4$ 28 $x > 2$
29 $x \geq 3$ 30 $x < -7$ 31 $x \geq 9$
32 $x < 2$ 33 $x \geq 7$ 34 $x \leq 3$
35 $x \leq 2$ 36 $x < 1\dfrac{1}{2}$ 37 $x \leq 1$
38 $x \geq 18$ 39 $x > 4\dfrac{1}{2}$
40 $x \leq -3\dfrac{1}{5}$ 41 $x > 4\dfrac{2}{5}$
42 $x > 3$ 43 $x \leq -7$
44 $x > -5\dfrac{1}{2}$ 45 $x \leq -5\dfrac{1}{2}$
46 $-3 \leq x \leq 5$
    $-3, -2, -1, 0, 1, 2, 3, 4, 5$
47 $-3 \leq x < 4$ $-3, -2, -1, 0, 1, 2, 3$
48 $2 < x \leq 9$ $3, 4, 5, 6, 7, 8, 9$
49 $-2 < x \leq 4$ $-1, 0, 1, 2, 3, 4$
50 $2 < x < 4$ 3
51 $-2\dfrac{1}{2} \leq x < 5\dfrac{1}{2}$
    $-2, -1, 0, 1, 2, 3, 4, 5$
52 $0 < x \leq 4$ $1, 2, 3, 4$
53 $3 \leq x \leq 7$ $3, 4, 5, 6, 7$
54 $3 < x \leq 9$ $4, 5, 6, 7, 8, 9$
55 4   56 4   57 $x = 1, y = 1$

Excercise 2.8

1 a $\dfrac{4x}{5}$ b $\dfrac{5x}{7}$ c $\dfrac{x}{2}$
   d $\dfrac{x}{2}$ e $\dfrac{2x}{3}$ f $\dfrac{x}{2}$
   g $\dfrac{2x}{5}$ h $\dfrac{2x}{5}$ i $\dfrac{5x}{7}$

2 $\dfrac{y}{3} + \dfrac{2y}{5} = \dfrac{5y}{15} + \dfrac{6y}{15} = \dfrac{11y}{15}$

3 a $\dfrac{3x}{4}$ b $\dfrac{7x}{12}$ c $\dfrac{11x}{15}$
   d $\dfrac{17x}{20}$ e $\dfrac{x}{6}$ f $\dfrac{7x}{15}$
   g $\dfrac{x}{2}$ h $\dfrac{5x}{36}$ i $\dfrac{3x}{5}$

4 a $\dfrac{7x+6}{12}$ b $\dfrac{11x-13}{30}$
   c $\dfrac{17x+13}{35}$ d $\dfrac{37x+13}{35}$
   e $\dfrac{17x+33}{12}$ f $\dfrac{46x+64}{33}$

5 a $\dfrac{21x-2}{12}$ b $\dfrac{19x-1}{40}$

   c $\dfrac{9x-8}{30}$ d $\dfrac{21-37}{36}$
   e $\dfrac{12x+49}{15}$ f $\dfrac{19x-7.}{36}$

6 $\dfrac{34x+127}{60}$

7 $\dfrac{3x}{4} - \dfrac{2x}{3}$

Excercise 2.9

1 a 36 b 18 c 40
   d 12 e 40 f -36
   g 10 h 42

2 a 15 b 4 c 4 d 6
   e -4 f $\dfrac{2}{3}$ g 6 h -2

3 a 2 b 7 c 5 d 3
   e -3 f 6 g 4 h 8
   i -3 j -2 k $2\dfrac{1}{2}$
   l $\dfrac{1}{3}$

4 a 7 b 1 c 3 d 6
   e 4 f 1 g 5 h 7
   i -3 j -5 k $1\dfrac{1}{2}$
   l $\dfrac{1}{3}$

5 2

6 a $2 \times \left(\dfrac{x+2}{3} + \dfrac{4x+7}{5}\right) = 20$

   $5(x+2) + 3(4x+7) = 20 \times 3 \times 5 \div 2$
   $17x + 31 = 150$

   b $x = 7$; 3, 7

7 $5\dfrac{1}{2}$

Excercise 2.10

1 a 3, 2 b 2, -5 c $\dfrac{1}{2}$, 3
   d $\dfrac{2}{3}$, -1 e 2, 3 f -5, 4
   g $-\dfrac{1}{2}$, 7 h 3, $\dfrac{1}{2}$

2 a $1\dfrac{1}{2}$, 2 b 2, $-\dfrac{2}{3}$ c $\dfrac{2}{5}$, $-\dfrac{3}{5}$
   d $-\dfrac{2}{3}$, $1\dfrac{2}{3}$ e $-1\dfrac{1}{2}$, 3
   f $\dfrac{1}{3}$, -2 g 2, $-1\dfrac{1}{2}$
   h $-\dfrac{3}{7}$, $\dfrac{2}{7}$

3 a i $1\dfrac{1}{2}$ ii 3
      iii $y = 1\dfrac{1}{2}x + 3$
   b i $-\dfrac{1}{4}$ ii 4
      iii $y = -\dfrac{1}{4}x + 4$

Answers   469

c i $\frac{1}{2}$ ii -1
   iii $y = \frac{1}{2}x - 1$
d i $-\frac{5}{3}$ ii 5
   iii $y = -\frac{5}{3}x + 5$

4 a $y = -2$ b $y = 2x$
   c $y = 3x - 2$ d $y = \frac{1}{2}x + 6$
   e $3x + 4y = 26$ f $2y - x = 8$

5 a, b, 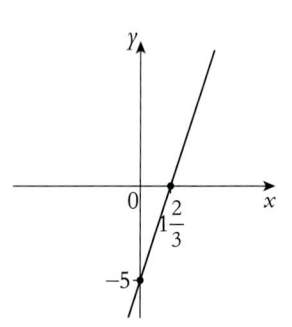 c, 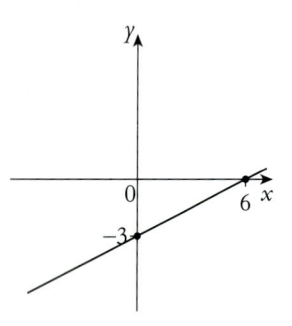 d, 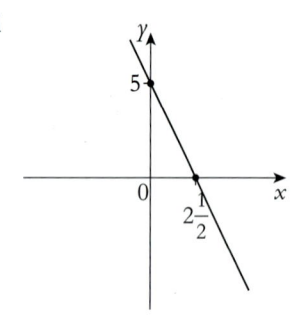 e, 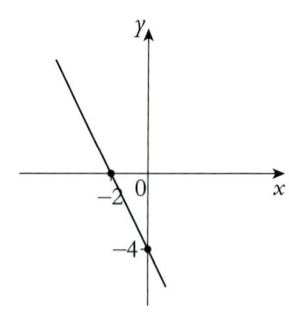 f, 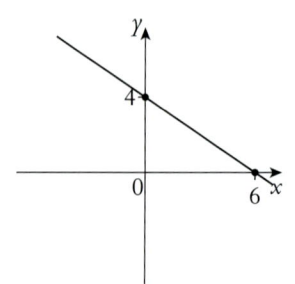 g, 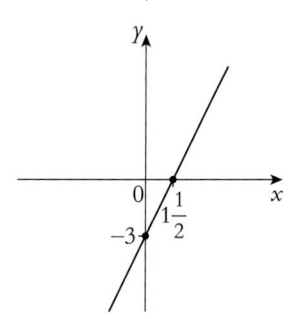 h (graphs)

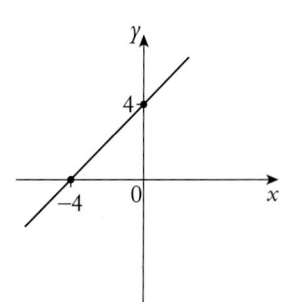

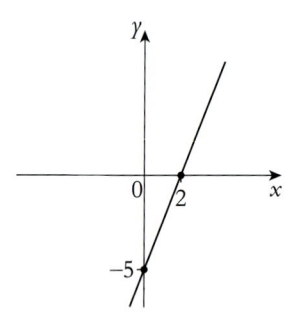

6 a E   b A, E   c B, C
   d D   e B, D

Excercise 2.11

1 a $y = \frac{1}{2}x + 6$ b $y = -2x - 3$
   c $y = x + 1$ d $y = 4x - 4$
   e $y = -2x - 2$ f $y = \frac{2}{3}x + 5$
   g $y = -\frac{3}{5}x - 10$ h $y = \frac{2}{5}x + 8$
   i $y = \frac{3}{4}x + 5$

2 a $y = \frac{1}{2}x + 2$ b $y = -3x - 4$
   c $y = x + 5$ d $y = -2x + 2$
   e $y = x - 5$ f $y = \frac{2}{3}x - \frac{1}{3}$
   g $y = -\frac{1}{5}x + 3\frac{1}{5}$ h $y = \frac{1}{4}x - 5$
   i $y = \frac{1}{2}x - 1\frac{1}{2}$

Excercise 2.12

1 a $(-3, 2)$ b $(4\frac{1}{2}, \frac{1}{2})$
   c $(2\frac{1}{2}, -1\frac{1}{2})$ d $(1\frac{1}{2}, 1\frac{1}{4})$
   e $(-2\frac{1}{4}, -1\frac{1}{4})$

2 a $(4, 3\frac{1}{2})$ b $(-2, 0)$
   c $(2\frac{1}{2}, -1)$ d $(-4, 5)$
   e $(-1\frac{1}{2}, 3\frac{1}{2})$ f $(4\frac{1}{2}, -4)$
   g $(-3\frac{1}{2}, -4\frac{1}{2})$
   h $(-3\frac{3}{4}, -2\frac{1}{4})$ i $(2\frac{1}{2}, \frac{3}{4})$

Excercise 2.13

1 a $y \leq 1$ b $x < -2$
   c $-1 \leq y < 3$ d $y \geq x$
   e $y > x - 2$ f $x + y \leq 2$
   g $y \geq -\frac{1}{2}x + 1$ h $y > 2x - 3$
   i $y < -3x + 3$

2 a, b (graphs)

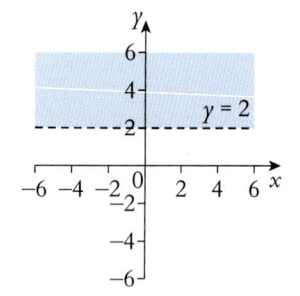

470 Answers

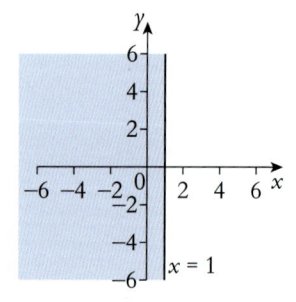

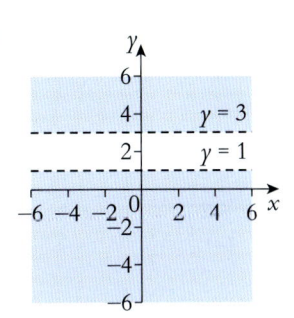

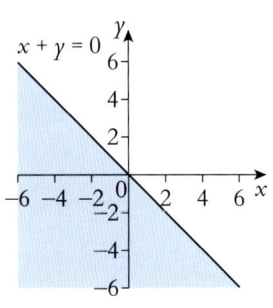

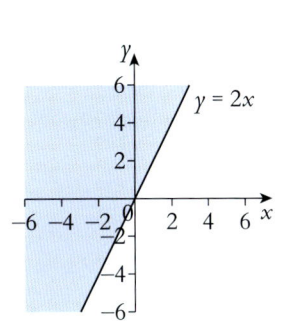

c

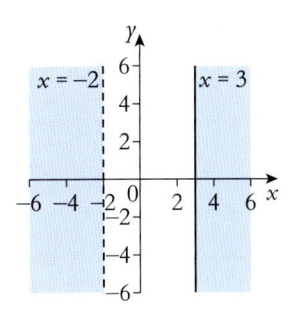

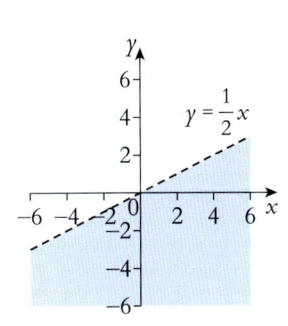

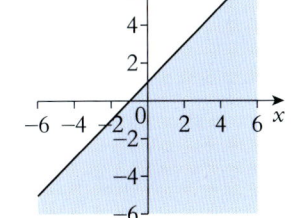

h

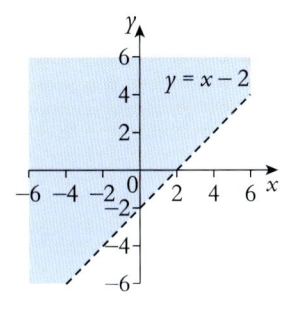

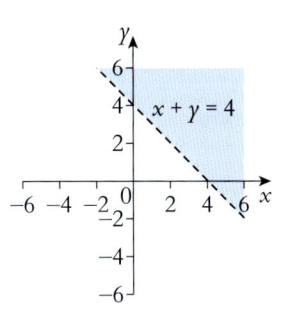

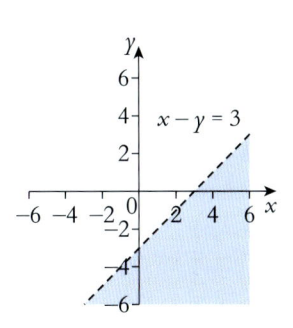

3 a

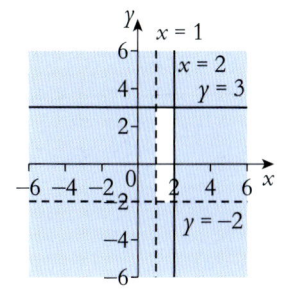

d

i

b

e

j

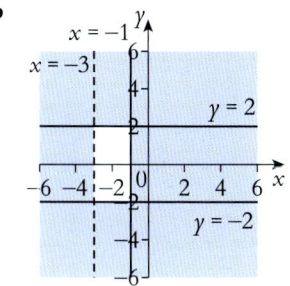

4 a $y \geq 2$, $y \leq 2x$
  b $y \leq 2$, $y \leq 2x$
  c $y \leq 2$, $y \geq 2x$

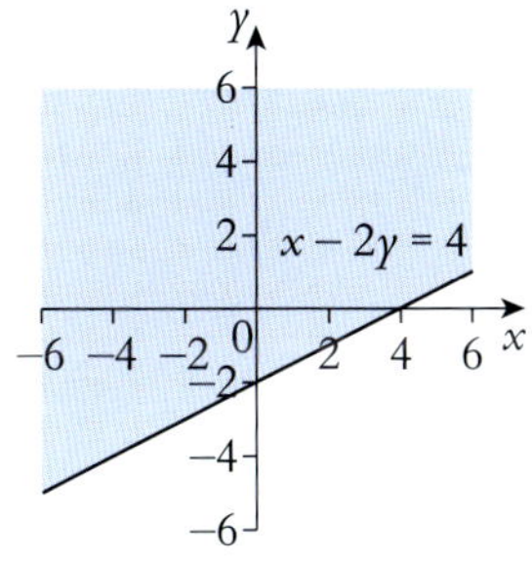

5 P and A, Q and C, R and B

6 a $x > -1$, $y > -1$, $x + y \leq 2$
  b $x > -1$, $y > 1$, $x + y \leq 3$
  c $x < -1$, $y > -1$, $y < x + 3$
  d $x > -1$, $x + y < 2$, $y \geq x - 2$
  e $x > -2$, $y \geq -3$, $y < \frac{1}{2}x + 1$
  f $y < 2$, $y > x$, $y < -2x - 2$
  g $x > -2$, $y \geq x$, $x + y < 3$
  h $y \leq 4$, $y > 2x$, $x + y > 0$
  i $x + 2y < 4$, $y \geq x - 2$, $y < 3x + 4$

f

k

g

l

7

Answers 471

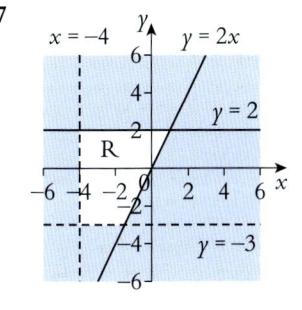

8

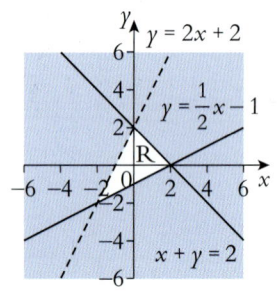

(graph showing $y = 2x + 2$, $y = \frac{1}{2}x - 1$, $x + y = 2$ with region R)

9

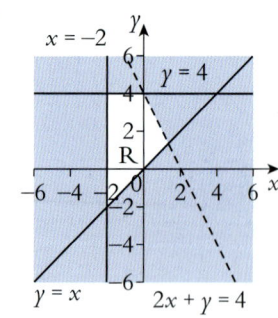

(graph showing $x = -2$, $y = 4$, $y = x$, $2x + y = 4$ with region R)

10 a

(graph with triangle region)

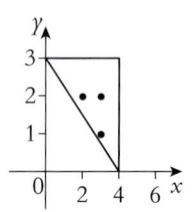

b $x < 4$, $y < 3$, $3x + 4y > 12$

c (2, 2), (3, 1), (3, 2)

Excercise 2.14

1 **a** 36 cm, 62 cm^2
   b 24 cm, 30 cm^2
   c 36 cm, 84 cm^2
   d 42 cm, 84 cm^2

2 **a** 45 cm^2 **b** 40 cm
   c 72 cm^2

3 **a** 18 cm^2 **b** 22.5 cm^2
   c 45 cm^2 **d** 66.5 cm^2

4 **a** 28 cm^2 **b** 30 cm^2
   c 160 cm^2 **d** 49.5 cm^2

5 8 cm

6 8.5

Excercise 2.15

1 **a** 8.06 cm   **b** 10.4 cm
   c 8.25 cm   **d** 6.62 m
   e 6.80 m   **f** 6.96 cm

2 $x^2 + 4^2 = 6^2$, $x^2 = 20$, $x = 4.47$ cm

3 **a** 6.93 cm   **b** 5.66 cm
   c 13.3 cm   **d** 8.08 cm
   e 19.0 cm   **f** 14.4 cm

4 38.1 km

5 4.77 m

6 **a** 3.54 cm   **b** 8.94 cm
   c 2.83 cm

7 7.81 cm

8 8.94 cm

9 10.6 cm

10 **a** 5 cm   **b** 22 cm
   c 26 cm^2 **c** 10 cm^2

11 **a** 4.33 cm   **b** 10.8 cm^2

12 6.63 cm

13 He is incorrect. $4^2 + 5^2 \neq 6^2$

14 Yes. $48^2 + 55^2 = 73^2$

15 5 miles

16 5.29 cm

17 17 cm

18 4.85 cm

19 0.828 cm

Excercise 2.16

1 **a** 5.39   **b** 3.61   **c** 5
   d 7.07

2 **a** 8.06   **b** 15.3   **c** 15.6
   d 8.54   **e** 11.7   **f** 13.9
   g 9.49   **h** 3.16   **i** 4.12

Excercise 2.17

1 **a** 5.39 m **b** 8.31 m
   c 12.3 m **d** 12.1 m

2 14.8 cm

3 11.1 cm

Excercise 2.18

Check students' drawings.

5 (5, 5.5)

7 15.2 cm

9 6.95 cm

10 (5, 6)

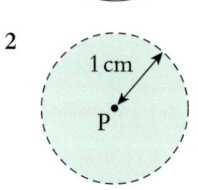

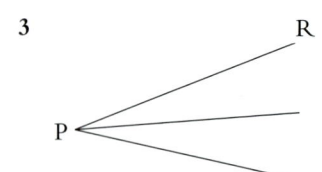

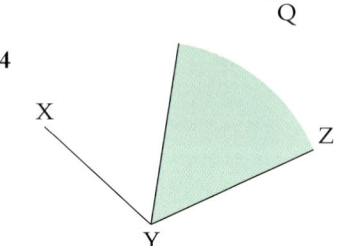

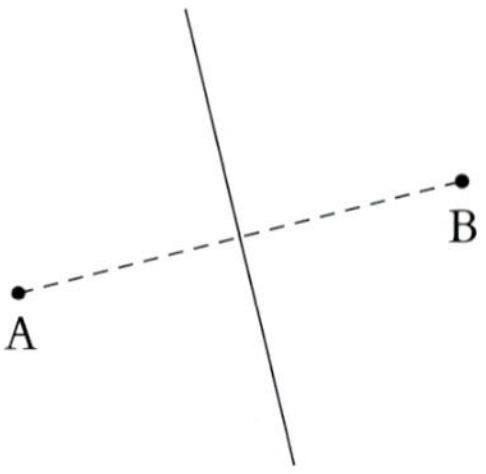

Excercise 2.19

1 *(circle with radius 2 cm, centre P)*

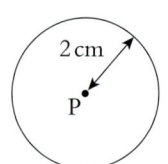

2 *(dashed circle with radius 1 cm, centre P)*

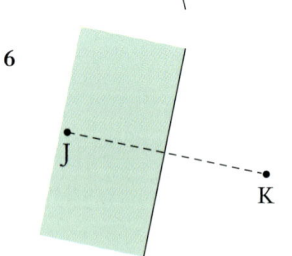

3 *(two rays from P to R and Q)*

4 *(sector with vertex Y, points X and Z)*

5 *(line segment with points A and B)*

6 *(rectangle with points J and K)*

7 *(semicircular region with points D and E)*

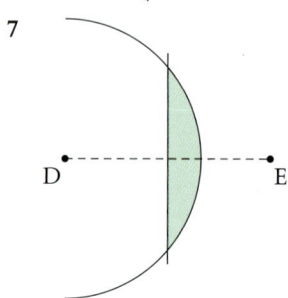

8
[figure: rounded square with inner square, gap 2]

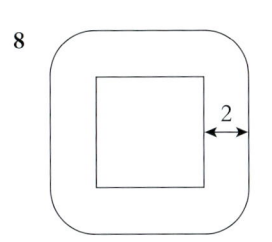

9
[figure: rounded triangle with inner triangle, gap 2]

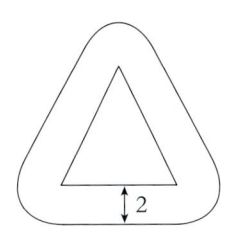

10
[figure: stadium shape with line of length 3 between two dots]

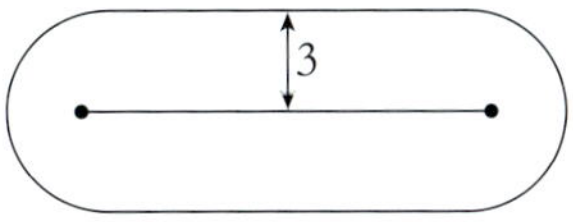

11
[figure: square ABCD with shaded triangle in upper region]

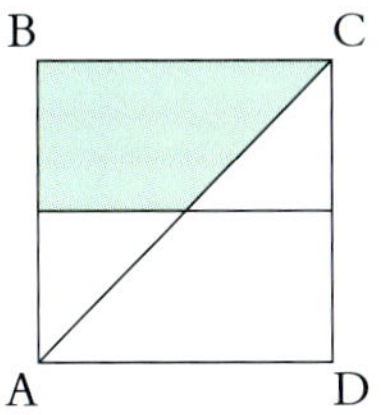

12
[figure: rectangle ABCD with point P and arc]

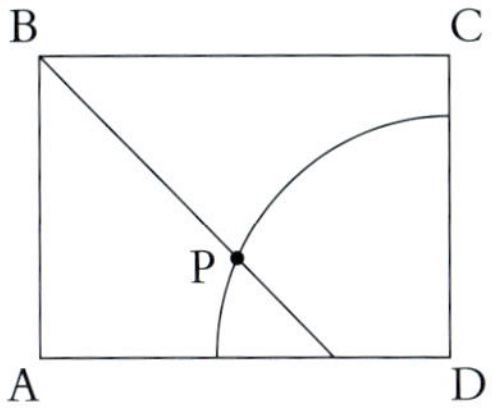

13
[figure: rectangle ABCD with shaded semicircle]

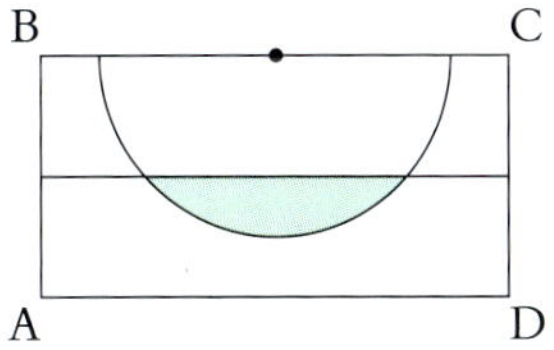

14
[figure: triangle with perpendiculars meeting at P]

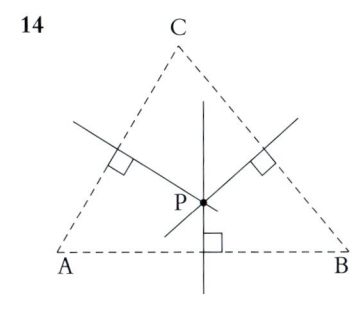

15
[figure: triangle QPR with arc construction]

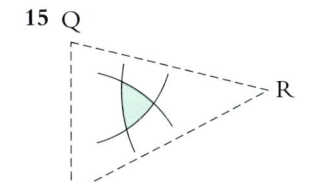

16
[figure: quadrilateral ABCD with shaded region]

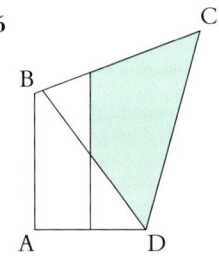

17
Lion
[figure: arc below point labelled Lion]

Electric fence
[figure: dashed line labelled Electric fence]

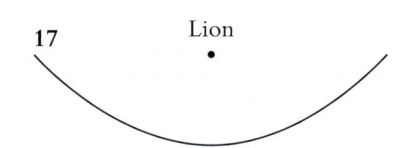

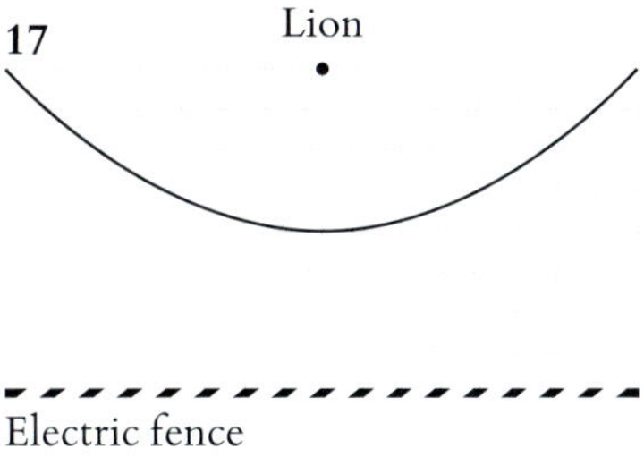

18
[figure: circle r = 6 with rectangle cut out, r = 2, side 4, 8]

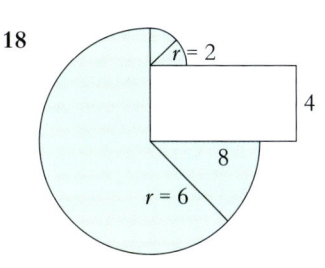

19 a
[figure: Locus of P, 36 cm above shaded strip]

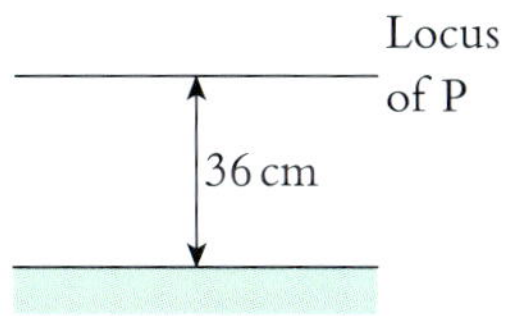

b
[figure: 61 cm, curve dipping to 11 cm above shaded strip]

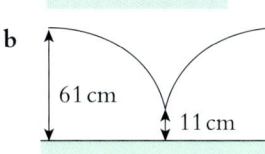

Excercise 2.20

1   a   18.8 cm, 28.3 cm²
    b   47.1 cm, 177 cm²
    c   15.1 cm, 18.1 cm²
    d   23.9 cm, 45.4 cm²
    e   126 cm, 1260 cm²
    f   29.5 cm, 69.4 cm²

2   a   163 cm   b   8.17 km
3   a   0.631 cm   b   3.96 m
4   a   30.9 cm²   b   30.9 cm²
    c   30.9 cm²   d   same
5   50.3 cm²
6   114 cm²

Excercise 2.21

1   15.4 cm, 14.1 cm²
2   25.0 cm, 38.5 cm²
3   13.4 cm, 9.42 cm²
4   22.9 cm, 34.8 cm²
5   23.4 cm, 28.1 cm²
6   30.3 cm, 44.6 cm²
7   24.6 cm, 11.4 cm²
8   40.6 cm, 51.8 cm²
9   31.4 cm, 21.5 cm²
10 25.1 cm, 25.1 cm²
11 14.9 cm, 15.8 cm²
12 28.6 cm
13 a   1.9544 cm, 10.04878 cm
    b   3.9088 cm, 20.09756 cm
    c   When the area quadruples, the lengths double.
14 a   2.523 cm, 9.010 cm
    b   7.569 cm, 27.03 cm
    c   When the area is multiplied by 9, the lengths treble.
15 35.5 cm²
16 7.78 cm
17 4 cm

Excercise 2.22

1   a   3.67 cm   b   19.5 cm
    c   6.14 cm   d   32.1 cm
2   a   6.68 cm²   b   22.3 cm²
    c   127 cm²   d   159 cm²
3   a   9.59 cm, 5.59 cm²
    b   65.1 cm, 61.1 cm²
    c   40.3 cm, 64.5 cm²
    d   39.2 cm, 49.7 cm²
4   76.4°
5   4.09 cm
6   a   40.9 cm   b   77.4 cm²
7   0.161 cm²

Answers   473

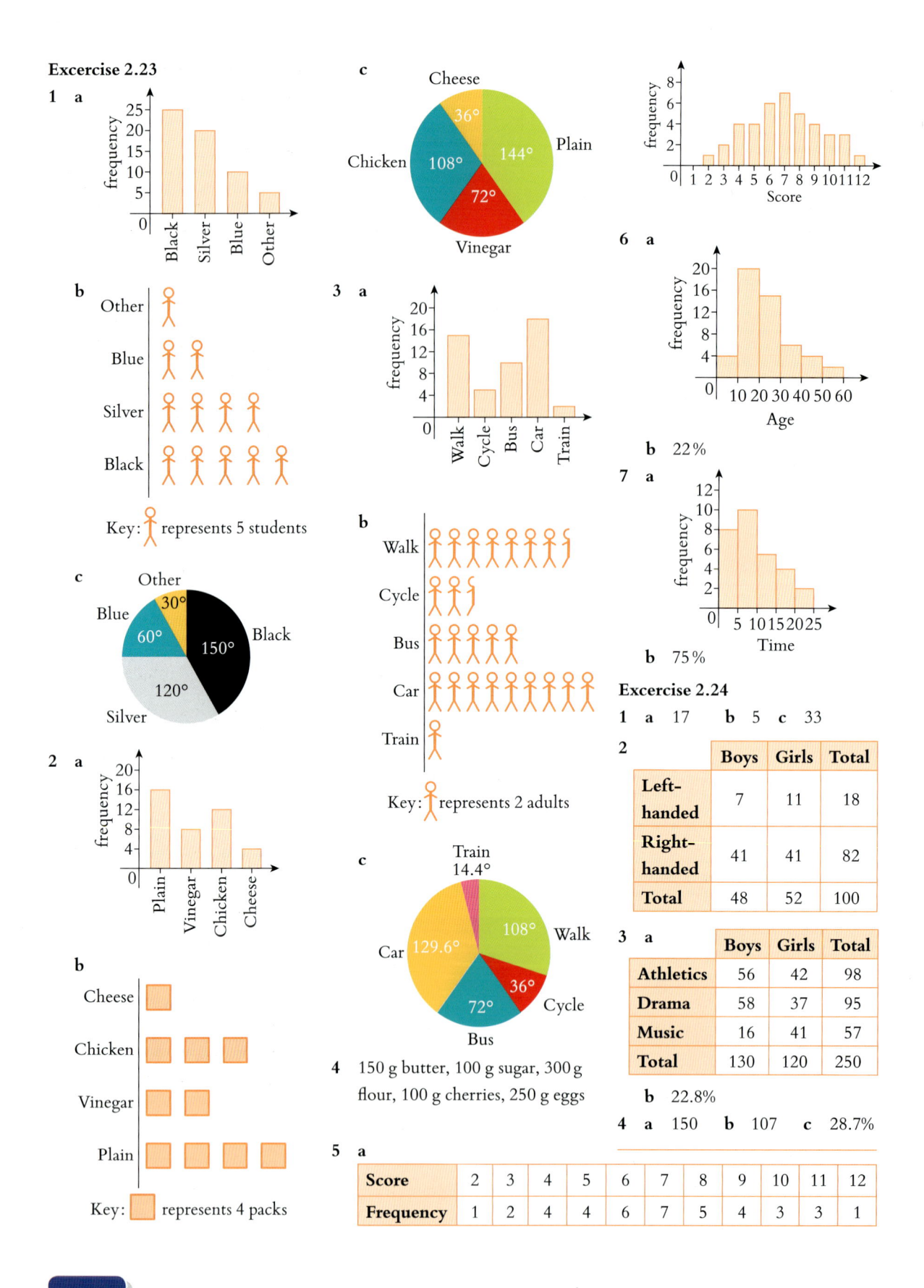

Excercise 2.23

1 a *(bar chart: Black 25, Silver 20, Blue 10, Other 5)*

b *(pictogram)*
Key: 👤 represents 5 students

c *(pie chart: Black 150°, Silver 120°, Blue 60°, Other 30°)*

2 a *(bar chart: Plain 16, Vinegar 8, Chicken 12, Cheese 4)*

b *(pictogram)*
Key: ▢ represents 4 packs

c *(pie chart: Cheese 36°, Plain 144°, Vinegar 72°, Chicken 108°)*

3 a *(bar chart: Walk 15, Cycle 5, Bus 10, Car 18, Train 2)*

b *(pictogram)*
Key: 👤 represents 2 adults

c *(pie chart: Walk 108°, Cycle 36°, Bus 72°, Car 129.6°, Train 14.4°)*

4 150 g butter, 100 g sugar, 300 g flour, 100 g cherries, 250 g eggs

5 a

Score	2	3	4	5	6	7	8	9	10	11	12
Frequency	1	2	4	4	6	7	5	4	3	3	1

(frequency diagram of scores)

6 a *(frequency diagram: Age 10–60)*

b 22%

7 a *(frequency diagram: Time 5–25)*

b 75%

Excercise 2.24

1 a 17   b 5   c 33

2

	Boys	Girls	Total
Left-handed	7	11	18
Right-handed	41	41	82
Total	48	52	100

3 a

	Boys	Girls	Total
Athletics	56	42	98
Drama	58	37	95
Music	16	41	57
Total	130	120	250

b 22.8%

4 a 150   b 107   c 28.7%

474 Answers

Examination Questions

1. $\dfrac{11x}{18}$
2. -5.2
3. **a** 3 **b** 8
4. $x \geq 0.8$
5. $x > -0.16$
6. $x < -23.5$
7. $x < -3$
8. 5
9. $y = \dfrac{1}{2}x + 5$
10. (4, 2)
11. **a** $y = 2x - 4$
    b (2, 0)
12. **a** Apply Pythagoras' theorem
    b 336 cm
13. 1.62 m^2
14. 24.3 m
15. 31.4 cm
16. **a** 53.4 cm^2 **b** 49.6 cm

Unit 3

Exercise 3.1

1. a, c, f, g
2. **a** 5×10^4 **b** 2×10^2
   c 7×10^5 **d** 5.33×10^3
   e 8.2×10^7 **f** 3.02×10^3
   g 6.66×10^5 **h** 3.636×10^1
3. **a** 480 000 **b** 370
   c 564 000 000 **d** 70 600 000
   e 20 000 **f** 1 100 000
   g 300.1 **h** 9 500 000 000
4. **a** 6×10^3 **b** 4.5×10^5
   c 3×10^4 **d** 4.1×10^6
   e 3.2×10^9 **f** 6.3×10^7
   g 4×10^3 **h** 2.5×10^7
   i 1.6×10^{13} **j** 5×10^4
   k 4.2×10^2 **l** 4.87×10^2
5. Answers as for question 4.
6. **a** 6.2×10^9 **b** 5.8×10^9
   c 1.1×10^{10} **d** 1.2×10^{18}
   e 3×10^1 **f** 3.6×10^{19}
   g 3.1×10^1 **h** 9×10^{10}
7. **a** 30 : 1 **b** 1 : 40 **c** 500 : 1
8. **a** 9.6×10^4 **b** 2.1×10^4
   c 1.46×10^6
9. **a** 2×10^4 **b** 3×10^2
   c 6×10^3 **d** 7×10^4
10. 2.13×10^4, 5.8×10^{13}, 2.04×10^{14}, 2.2×10^{14}
11. 1.49×10^8 12. 5 : 12
13. 1.2995×10^9

Exercise 3.2

1. **a** 1.5×10^{-2} **b** 6.7×10^{-4}
   c 9×10^{-4} **d** 6.06×10^{-4}
   e 3.2×10^6 **f** 5.2×10^{-4}
   g 4.4×10^{-1} **h** 5.308×10^7
   i 5.7×10^{-8} **j** 6.13×10^{-4}
   k 7.002×10^5 **l** 8.9×10^{-4}
2. **a** 0.000088 **b** 0.00000023
   c 0.006 **d** 0.0511
   e 99 000 **f** 0.00000104
   g 0.00068 **h** 600 000 000
   i 0.38 **j** 22
   k 408 000 **l** 0.0000095
3. **a** 6×10^{-2} **b** 7.2×10^{-5}
   c 5×10^{-8} **d** 6×10^{-4}
   e 3×10^{-9} **f** 6.4×10^{-7}
   g 3×10^{12} **h** 2.4×10^{-2}
   i 9×10^{-10} **j** 2×10^6
   k 2.7×10^3 **l** 2×10^{-8}
4. Answer as for question 3.
5. **a** 3×10^{-3} **b** 2×10^{-5}
   c 8×10^{-3} **d** 9×10^{-7}
6. **a** 6.3×10^{-8} **b** 5.7×10^{-8}
   c 1.86×10^{-7} **d** 1.8×10^{-16}
   e 2×10^1 **f** 3.6×10^{-15}
   g 9×10^{-18} **h** 1.5×10^{-6}
7. 8.35×10^{-22} grams
8. 1.87×10^{23}
9. 1.5×10^{-5} seconds
10. 2.625×10^{-5} kg

Exercise 3.3

1.

x	0	2	4
y	2	4	6

x	0	2	4
y	4	2	0

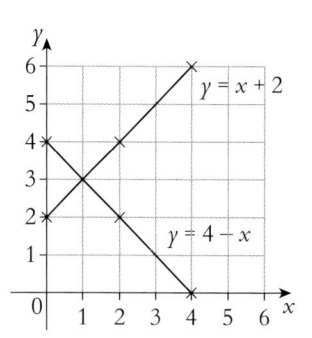

Graph showing $y = x + 2$ and $y = 4 - x$

$x = 1, y = 3$

2.

x	0	2	4
y	1	2	3

x	0	4
y	5	0

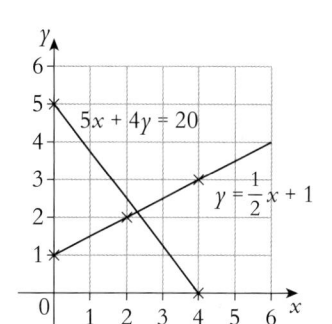

Graph showing $5x + 4y = 20$ and $y = \dfrac{1}{2}x + 1$

$x = 2\dfrac{2}{7}, y = 2\dfrac{1}{7}$

Exercise 3.4

1. **a** $x = 3, y = 4$
   b $x = 5, y = 2$
   c $x = 10.5, y = -2$
   d $x = 4, y = 4$
   e $x = 2, y = -1$
   f $x = -3, y = 4$
   g $x = -6, y = -2$
   h $x = 5, y = -2$

2 a $x = 2, y = 4$
    b $x = 9, y = 5$
    c $x = 2, y = 10$
    d $x = 7, y = 3$
    e $x = 5, y = -2$
    f $x = -3, y = 4$
    g $x = 7, y = -5$
    h $x = -2\frac{1}{2}, y = -6$

3 a $x = 6, y = 2$
    b $x = 3, y = 3$
    c $x = 5, y = -2$
    d $x = 1\frac{1}{2}, y = 3$
    e $x = -4, y = 2\frac{1}{2}$
    f $x = 2, y = 1\frac{1}{3}$
    g $x = -1, y = -2\frac{1}{4}$
    h $x = -4, y = 3\frac{1}{2}$

4 $x + y = 15, x - y = 4$;
    $x = 9\frac{1}{2}, y = 5\frac{1}{2}$

5 a 39, 15   b 15, −7

6 $x = 11, y = 2$

7 $x = 2.92, y = 1.8$

8 $0.85, $0.32

9 $1 = -2m + c, 4 = 4m + c$;
    $m = \frac{1}{2}, c = 2$

10 $39.68

Exercise 3.5

1 a $x = 3, y = 5$
    b $x = 4, y = 1$
    c $x = 6, y = 4$
    d $x = 3, y = 7$
    e $x = 5, y = -2$
    f $x = 10, y = -3$
    g $x = 1.625, y = 7.75$
    h $x = 3, y = -2.5$

2 a $x = 4, y = 7$
    b $x = 3, y = 2$
    c $x = 5, y = 3$
    d $x = 2, y = 2$
    e $x = 8, y = 6$
    f $x = 11, y = 4$
    g $x = 3, y = 9$
    h $x = 1, y = 7$

3 a $x = -\frac{1}{2}, y = 4$
    b $x = 3, y = -5$
    c $x = -2, y = 7$
    d $x = -3, y = -2$
    e $x = 2\frac{1}{2}, y = 3$
    f $x = 6, y = 1\frac{1}{2}$
    g $x = 3\frac{1}{2}, y = -5$
    h $x = -9, y = 4$

4 a $x = 4, y = 2$
    b $x = 2, y = 1$
    c $x = 4, y = 2$
    d $x = \frac{5}{7}, y = 4\frac{3}{7}$
    e $x = 2, y = -3$
    f $x = -1, y = 2$
    g $x = -2, y = 1$
    h $x = 1, y = 2$

5 a $x = 12, y = 15$
    b $x = 20, y = -16$
    c $x = -18, y = 10$
    d $x = 24, y = -6$

6 $x = 3, y = 7$

7 $x = 4, y = 2.5$

8 17

9 153

10 $1.85

11 $0.19

12 $a = 3, b = 5$

13 $a = 2, b = -3$

14 a $x = 3, y = -1$ b 28
    c 40

15 a $x = 1\frac{1}{2}, y = 4$ b 20
    c 25

16 a $x = 4, y = 3$ b 24
    c 27.7

Exercise 3.6

1 a $6(x + y)$ b $5xy(1 - 2y)$
    c $2(2x + 3y)$

2 a $4(x + 2)$ b $3(x - 2)$
    c $2(4x + 1)$ d $5(2x - 3)$
    e $7(y - 2)$ f $3(3y + 4)$
    g $4(5 - 2x)$ h $12(3 - y)$
    i $7(6 - x)$ j $2(9y - 2)$
    k $8(3x + 2)$ l $11(2y - 5)$

3 a $x(x + 5)$ b $x(x - 8)$
    c $y(y - 4)$ d $a(a + 9)$
    e $4x(x - 2)$ f $3y(y - 3)$
    g $2x(5x - 2)$ h $2y(4 - y)$
    i $5y(3y + 1)$ j $7x(2 - x)$
    k $2x(3 - 2x)$ l $3y(4y - 3)$

4 a $4(2\pi + 1)$ b $3(5 - 3\pi)$
    c $\pi r(r + l)$ d $2\pi r(r + h)$

5 a $3xy(2x - 1)$
    b $5x^4 y(2y - 1)$
    c $8xy(y^2 - 4x^2)$
    d $3x^2 y^2(5x - 4y^2)$
    e $3xy(4x - 5)$
    f $7pq(4q - 5p)$
    g $3ab(2a + 3)$
    h $7x^2 y(3x^3 - 4)$
    i $7ab^5(8a^3 b^2 - 5)$

6 a $3(a + b - c)$
    b $5(x + y + x^2)$
    c $a(2x - y - 3y^2)$
    d $x(x^2 + 4x - 2)$
    e $y^2(y^2 + 6y - 6)$
    f $pqr(pq - 1 + q)$
    g $4ab(5abc^2 - 2ab + 4c^2)$
    h $3x^2 y(5xy - 9 + 7y)$
    i $2p^3 q^3 r^2(4p^4 q^2 r + 3p - 2r)$

7 a $(x + 3)(y + 2)$
    b $(a + 2)(b - 4)$
    c $(p - 5)(q + 3)$
    d $(x - 3)(y - 5)$
    e $(x + y)^3$
    f $(a + b)^2 (a + b - 1)$
    g $(x - y)^2 (1 - x + y)$
    h $(x + y)(5 + x + y)$
    i $(x - y)^2 (x - y - 4)$

8 a $(y + 3)(x + 1)$
    b $(5 + x)(y - 2)$
    c $(4 + c)(a + b)$
    d $(p + 3)(q + 2r)$
    e $(3 - y)(x + 4)$
    f $(x - 4)(y - 3)$

Exercise 3.7

1. **a** $x - 2$ **b** $\dfrac{x+2}{4}$
   c $\dfrac{3x+4}{5}$

2. **a** $x + 2$ **b** $4y + 2$
   c $x + 2$ **d** $3y + 2$
   e $\dfrac{3x-6}{4}$ **f** $\dfrac{3x+2}{4}$
   g $\dfrac{3-4y}{2}$ **h** $\dfrac{6-4a}{5}$

3. **a** $x + 9$ **b** $y - 8$
   c $\dfrac{x-3}{x}$ **d** $1 - x^2$
   e $\dfrac{2x+y}{z}$ **f** $\dfrac{x+y}{z}$
   g $\dfrac{x+y}{y}$ **h** $\dfrac{x-2}{y}$

4. **a** 3 **b** 5 **c** $\dfrac{1}{6}$ **d** $\dfrac{1}{5}$
   e $\dfrac{2}{5}$ **f** $\dfrac{3x}{7}$ **g** $\dfrac{x}{z}$ **h** $\dfrac{2}{3}$

5. **a** x **b** $\dfrac{5}{x-5}$ **c** $\dfrac{x}{5}$
   d $\dfrac{2y}{5}$ **e** $\dfrac{x}{x-1}$ **f** $\dfrac{3}{(y-2)^2}$
   g x **h** $\dfrac{xy}{2}$

6. **a** $3\dfrac{1}{3}$ **b** $4\dfrac{2}{3}$
   c $\dfrac{x^2}{y^2}$ **d** $\dfrac{x+1}{y+1}$
   e $\dfrac{7(x+y)}{y}$ **f** $\dfrac{2f}{g}$

7. **a** $\dfrac{5xy}{12}$ **b** $\dfrac{9}{10}$

Exercise 3.8

1. **a** $x = \dfrac{p+q}{3}$ **b** $x = \dfrac{8b-a}{2}$
   c $x = (y-b)^2$

2. **a** $x = y - 3$ **b** $x = 5y + 6$
   c $x = 4 - y$ **d** $x = a + b$
   e $x = y + 8$ **f** $x = 5 - 2y$
   g $x = 6y - 8$ **h** $x = y - 10$

3. **a** $x = \dfrac{y}{6}$ **b** $x = \dfrac{y}{3}$
   c $x = \dfrac{y-4}{3}$ **d** $x = \dfrac{7-5y}{a}$
   e $x = 2p$ **f** $x = 15 + 5d$
   g $x = 2y - 16$
   h $x = 10a - 4ay$
   i $x = \dfrac{3y}{2}$ **j** $x = \dfrac{15y}{4}$
   k $x = \dfrac{7(a+b)}{3}$
   l $x = \dfrac{b(c+d)}{a}$

4. **a** $x = \dfrac{d-c}{2}$ **b** $x = \dfrac{y+2z}{5}$
   c $x = \dfrac{6-3y}{4}$ **d** $x = \dfrac{3b+c+d}{8}$
   e $x = \dfrac{d-b}{a}$
   f $x = \dfrac{c+cd}{b}$
   g $x = \dfrac{a+b+q}{p}$ **h** $x = \dfrac{y-8}{y}$

5. **a** $x = \dfrac{2}{y+z}$ **b** $x = \dfrac{p}{2q+3}$
   c $x = \dfrac{a-3y}{6}$ **d** $x = \dfrac{p+qr}{q}$

6. **a** $x = bc - a$ **b** $x = 3gh + 2f$
   c $x = \dfrac{cd - by}{a}$ **d** $x = \dfrac{p - mr}{q}$
   e $x = \dfrac{p}{q}$ **f** $x = \dfrac{g}{h}$
   g $x = \dfrac{p+q}{r}$ **h** $x = \dfrac{p-q}{5}$

7. **a** $x = \pm\sqrt{\dfrac{b}{a}}$ **b** $x = \pm\sqrt{\dfrac{q}{p}}$
   c $x = \pm\sqrt{\dfrac{c}{ab}}$ **d** $x = \pm\sqrt{\dfrac{bc}{a}}$
   e $x = \pm\sqrt{b-a}$
   f $x = \pm\sqrt{d+e}$
   g $x = \pm\sqrt{\dfrac{c-b}{a}}$
   h $x = \pm\sqrt{\dfrac{4r+5q}{p}}$
   i $x = \pm\sqrt{4(b+a)}$
   j $x = \pm\sqrt{a(d-bc)}$
   k $x = \pm\sqrt{g(f-3h)}$
   l $x = \pm\sqrt{2(3fg-g)}$

8. **a** $x = b^2 - a$ **b** $x = q^2r^2 + p$
   c $x = \dfrac{y+16}{2}$ **d** $x = \dfrac{z^2+4y}{4}$

9. They are both correct.

Exercise 3.9

1. **a** not congruent
   b congruent
   c congruent
   d not congruent
   e not congruent
   f congruent

2. **a** SAS **b** RHS **c** ASA
   d SAS **e** SSS **f** RHS

   For questions **3** to **5** other answers are possible.

3. SSS
5. SSS
6. RHS

Exercise 3.10

1. B and C
2. B and C
3. **a** $x = 12, y = 4$
   b $x = 3.6, y = 12.5$
   c $x = 15, y = 16.5$
   d $x = 4, y = 4$
   e $x = 8.75$
   f $x = 15, y = 17.5$

Exercise 3.11

1. **a** $x = 5, y = 12$
   b $x = 8.4, y = 2.3$
2. $w = 33\dfrac{1}{3}$
3. **a** 4 **b** 12 **c** 2 **d** $3\dfrac{1}{3}$
4. 54.5 m
5. **a** $x = 8, y = 20$
   b $x = 12, y = 5.5$
6. **a** $D = (5, 4), E = (14, 7)$
   b $(9\dfrac{1}{2}, 5\dfrac{1}{2})$
7. **a** $(-3, -1)$ **b** $(0, -3)$
   c $(-7\dfrac{1}{2}, 2)$

Answers 477

8 a $x = 4\frac{1}{2}, y = 7\frac{1}{2}$
   b $x = 30.45, y = 20.05$
9 a $x = 12, y = 10$
   b $x = 15, y = 7.5$

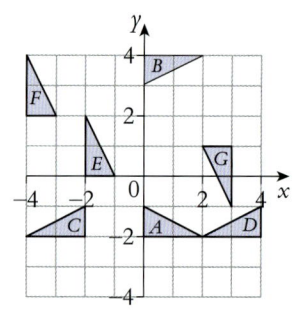

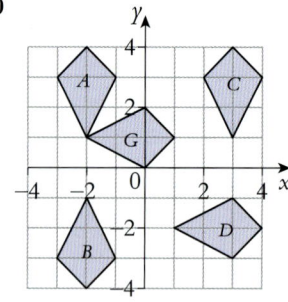

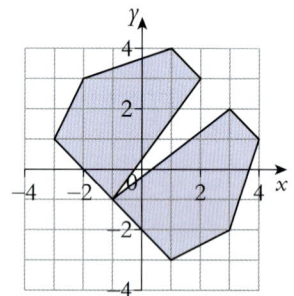

Exercise 3.12

1 a, b, 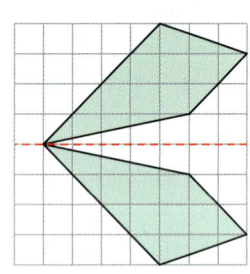 c, 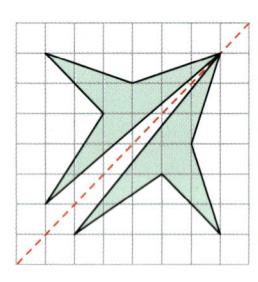 d, 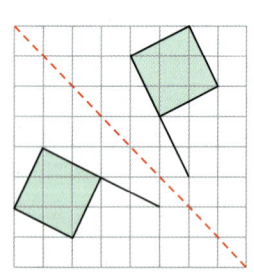 e, 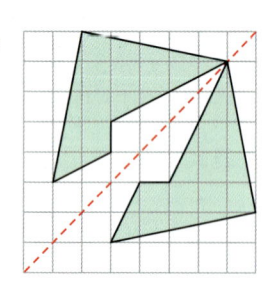 f (diagrams)

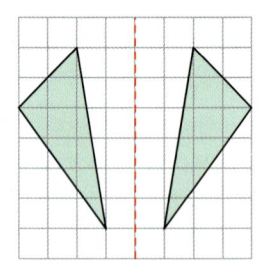

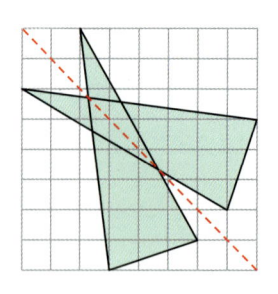

2 a $x = 6$ b $x = 10.5$
   c $x = 13$ d $x = 4$
   e $x = 6$ f $x = 11$
   g $x = 11$ h $x = 13$

3 a $y = 4.5$ b $y = 2$
   c $y = 1$ d $y = -1$
   e $y = 4$ f $y = -1$

4 a $x = 0$ b $y = 0$
   c $y = 1$

5 a $y = x$ b $x = 1$
   c $y = x$ d $y = 1$
   e $x = 0$

6 a $x = -1$ b $x = -1$
   c $y = 1$ d $y = 1$
   e $y = -x$ f $y = x$
   g $y = x$ h $y = -x$
   i $y = -x$ j $y = -x$

7 (diagram)

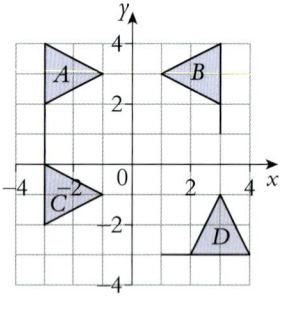

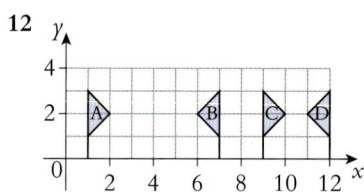

8 (diagram)

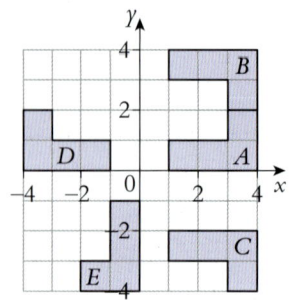

9 (diagram)

10 (diagram)

11 (diagram)

12 (diagram)
   d $x = 6\frac{1}{2}$

13 (diagram)
   d $y = 2$

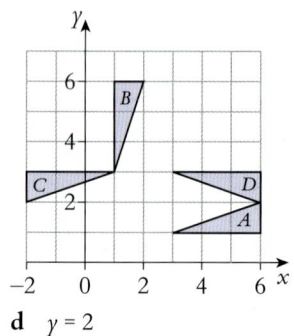

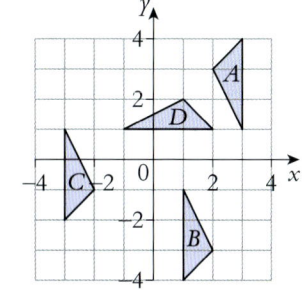

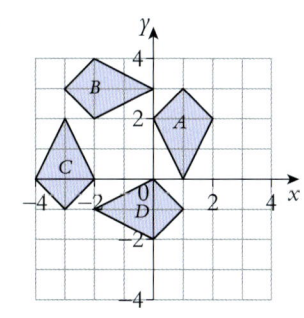

Exercise 3.13

1 a, b, 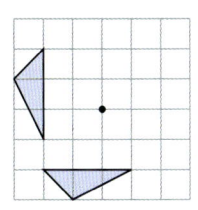 c, 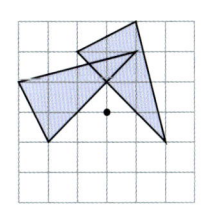 d, 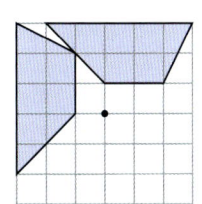 e, 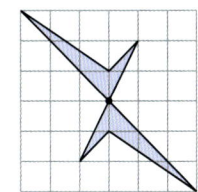 f (diagrams)

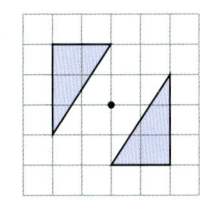

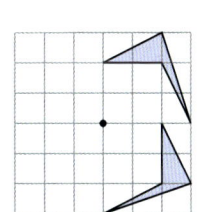

2 (diagram with triangles A, B, C, D on coordinate grid)

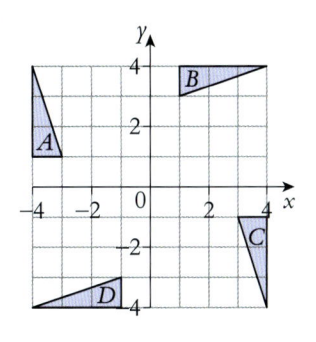

3 (diagram with shapes A, B, C, D on coordinate grid)

4 (diagram with shapes A, B, C, D on coordinate grid)

5 (diagram with shapes A, B, C, D on coordinate grid)

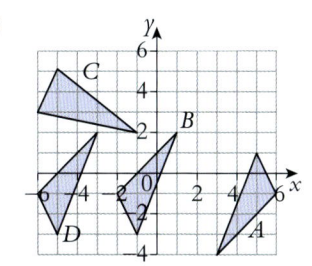

Exercise 3.14

1 a rotation, 90° clockwise about (0, 0)
   b rotation, 180° about (0, 0)
   c rotation, 90° anticlockwise about (0, 3)
   d rotation, 90° clockwise about (2, −1)
   e rotation, 90° anticlockwise about (2, 3)
   f rotation, 180° about $(0, -\tfrac{1}{2})$

2 a rotation, 180° about (3, 3)
   b rotation, 180° about (0, 0)
   c rotation, 90° anticlockwise about (−1, 2)
   d rotation, 90° clockwise about (−1, −1)
   e rotation, 90° anticlockwise about (1, −2)
   f rotation, 180° about (−1, 2)
   g rotation, 90° clockwise about (2, 1)
   h rotation, 90° anticlockwise about (1, −1)
   i rotation, 90° anticlockwise about (−2, −1)
   j rotation, 180° about (0.5, 0.5)
   k rotation, 90° anticlockwise about (−1, 2)
   l rotation, 90° clockwise about (2, 2)

Exercise 3.15

1 a $\begin{pmatrix} 2 \\ 4 \end{pmatrix}$ b $\begin{pmatrix} 3 \\ 3 \end{pmatrix}$ c $\begin{pmatrix} 5 \\ -1 \end{pmatrix}$

   d $\begin{pmatrix} 3 \\ 0 \end{pmatrix}$ e $\begin{pmatrix} 0 \\ -3 \end{pmatrix}$ f $\begin{pmatrix} -2 \\ 4 \end{pmatrix}$

2 a, b, 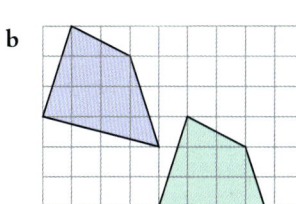 c (diagrams)

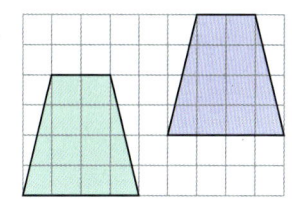

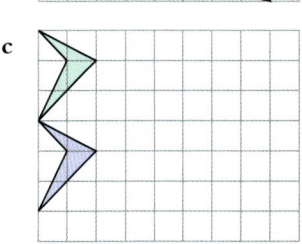

d [figure]

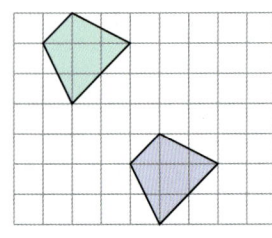

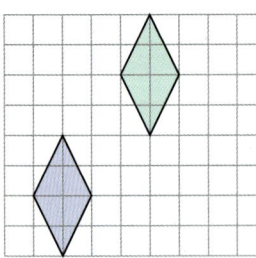

e [figure]

f [figure]

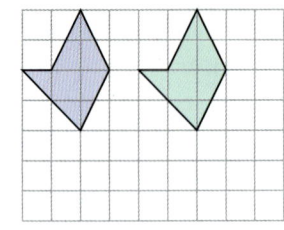

3 a (8, 8) **b** (5, −2)
  c (−4, −1) **d** (−9, 6)
  e (−4, −12) **f** (7, −8)

4 a translation $\begin{pmatrix} -4 \\ 1 \end{pmatrix}$

  b translation $\begin{pmatrix} -8 \\ -1 \end{pmatrix}$

  c translation $\begin{pmatrix} -9 \\ -4 \end{pmatrix}$

  d translation $\begin{pmatrix} -5 \\ -5 \end{pmatrix}$

  e translation $\begin{pmatrix} 2 \\ -5 \end{pmatrix}$

  f translation $\begin{pmatrix} 5 \\ -2 \end{pmatrix}$

  g translation $\begin{pmatrix} 5 \\ 1 \end{pmatrix}$

  h translation $\begin{pmatrix} 0 \\ -3 \end{pmatrix}$

  i translation $\begin{pmatrix} -7 \\ 0 \end{pmatrix}$

  j translation $\begin{pmatrix} 9 \\ 0 \end{pmatrix}$

  k translation $\begin{pmatrix} 10 \\ -4 \end{pmatrix}$

  l translation $\begin{pmatrix} 14 \\ -2 \end{pmatrix}$

5 a b [graph]

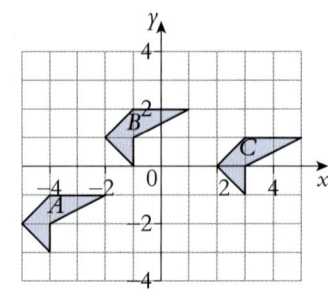

  c translation $\begin{pmatrix} 7 \\ 2 \end{pmatrix}$

6 a b [graph]

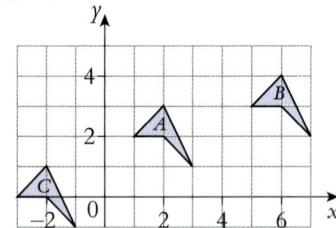

  c translation $\begin{pmatrix} 4 \\ -2 \end{pmatrix}$

7 $C = (1, 4)$ $B' = (6, −3)$
8 $A = (4, 1)$ $C = (−1, 3)$
  $D' = (−1, −2)$

Exercise 3.16

1 a reflection in $x = 3$
  b reflection in $y = x$
  c reflection in x-axis
  d rotation, 180° about (0, 0)
  e reflection in $y = −x$
  f translation $\begin{pmatrix} -6 \\ -1 \end{pmatrix}$
  g rotation, 90° anticlockwise about (1, 1)
  h reflection in $y = x$
  i rotation, 90° clockwise about (0, 0)
  j reflection in $y = −1$
  k reflection in y-axis
  l reflection in $x = 1$

2 a translation $\begin{pmatrix} 1 \\ 6 \end{pmatrix}$
  b rotation, 180° about (3.5, 2)
  c reflection in $x = 4$
  d rotation, 180° about (3, 7)
  e rotation, 90° clockwise about (5, 4)
  f translation $\begin{pmatrix} 5 \\ 3 \end{pmatrix}$
  g translation $\begin{pmatrix} 6 \\ -3 \end{pmatrix}$
  h reflection in $y = 5$
  i reflection in $x + y = 15$
  j translation $\begin{pmatrix} -3 \\ 4 \end{pmatrix}$
  k rotation, 90° clockwise about (8, 7)
  l rotation, 90° anticlockwise about (4, 1)

Exercise 3.17

1 a b c [graph]

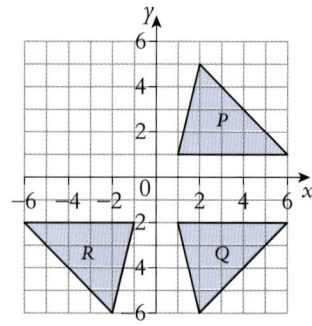

  d rotation, 180° about (0, 0)

2 a b c

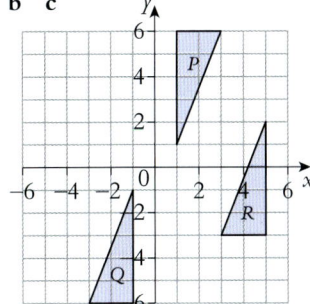

d rotation 180° about (3, 1.5)

3 a b c

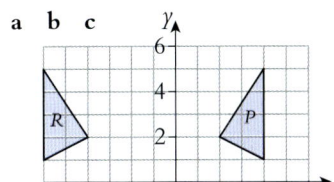

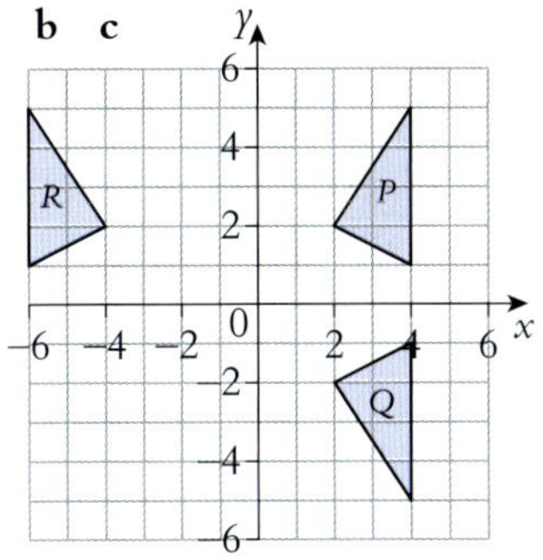

d reflection in $x = -1$

4 a b c

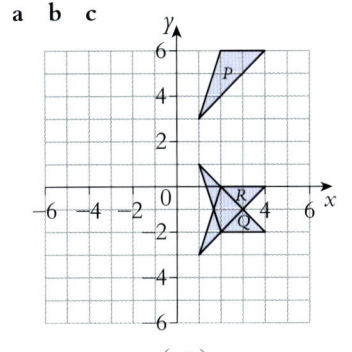

d translation $\begin{pmatrix} 0 \\ -6 \end{pmatrix}$

5 a b c

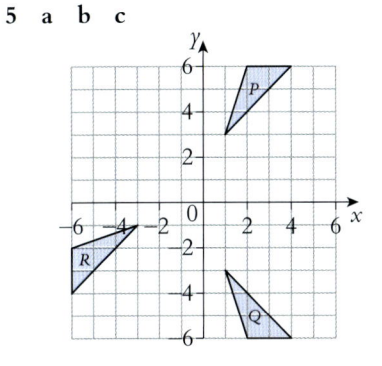

d reflection in $y = -x$

6 a b c

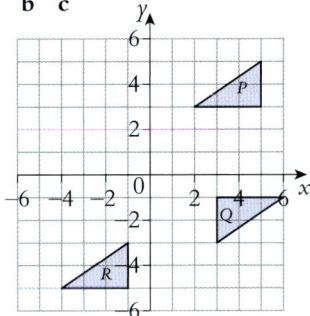

d translation $\begin{pmatrix} -6 \\ -8 \end{pmatrix}$

7 a translation $\begin{pmatrix} 6 \\ 10 \end{pmatrix}$

b translation $\begin{pmatrix} 3 \\ 10 \end{pmatrix}$

8 a rotation, 180° about (3, 2)

b translation $\begin{pmatrix} 0 \\ 6 \end{pmatrix}$

9 a translation $\begin{pmatrix} 4 \\ 2 \end{pmatrix}$

b translation $\begin{pmatrix} -6 \\ 10 \end{pmatrix}$

Exercise 3.18

1 translation $\begin{pmatrix} -2 \\ 4 \end{pmatrix}$

2 translation $\begin{pmatrix} 0 \\ -6 \end{pmatrix}$

3 translation $\begin{pmatrix} 5 \\ -7 \end{pmatrix}$

4 translation $\begin{pmatrix} 8 \\ 0 \end{pmatrix}$

5 reflection in x-axis
6 reflection in y-axis
7 reflection in $y = 3$
8 reflection in $x = -5$
9 reflection in $x + y = 4$

10 rotation, 90° anticlockwise about (0, 0)
11 rotation, 90° clockwise about (0, 0)
12 rotation, 180° about (0, 0)
13 rotation anticlockwise about (5, −2)
14 rotation 90° clockwise about (−4, 7)

Exercise 3.19

1 a SF 2, (6, 4)
  b SF 2, (2, 8)
  c SF 3, (2, 1)
  d SF 2, (0, 0)
  e SF 2, (6, 2)
  f SF 3, (−0.5, 7)

2 a enlargement, SF 2, centre (8, 3)
  b enlargement, SF 2, centre (12, 6)
  c enlargement, SF 2, centre (0, 7)
  d enlargement, SF 3, centre (13, 4)
  e enlargement, SF 2, centre (18, 3)
  f enlargement, SF 2, centre (22, 6)

Exercise 3.20

1 a

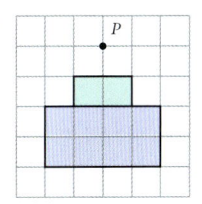

  b

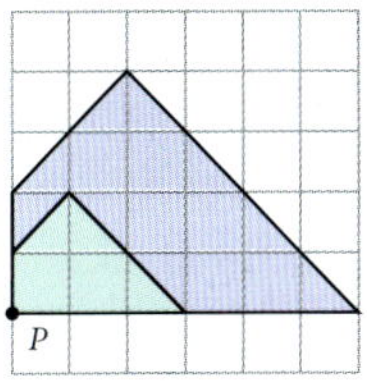

482 Answers

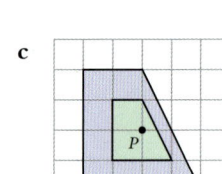

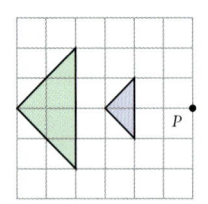

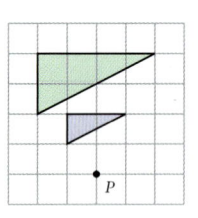

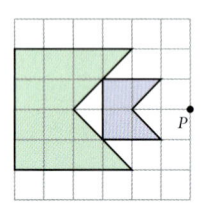

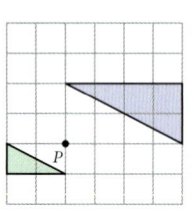

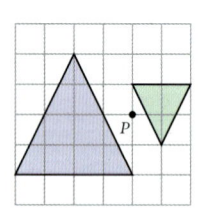

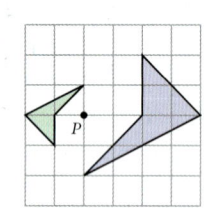

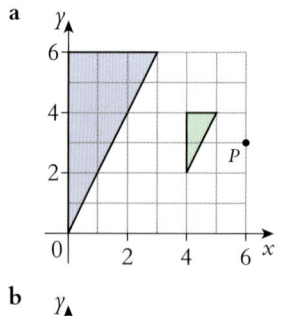

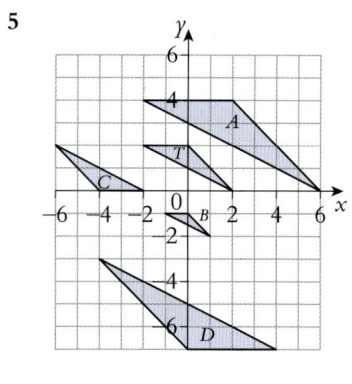

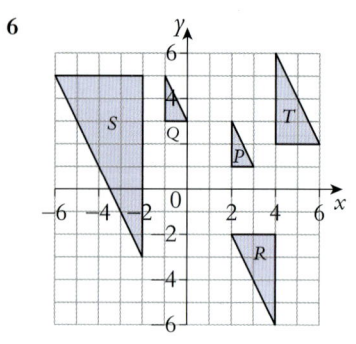

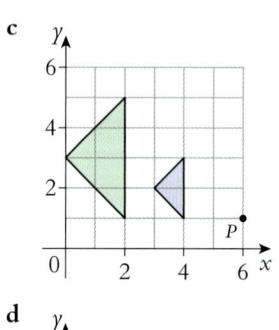

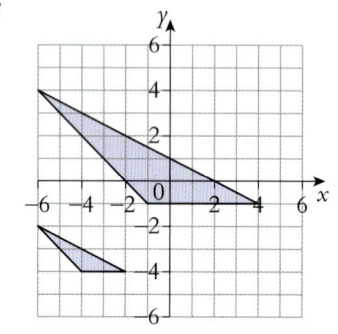

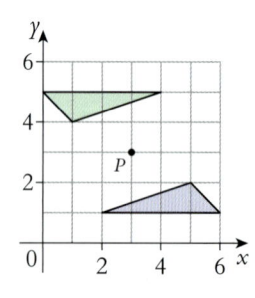

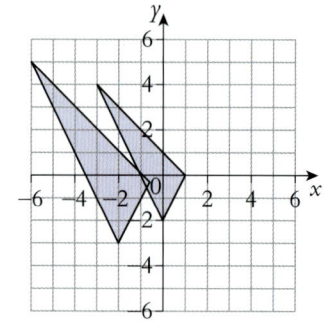

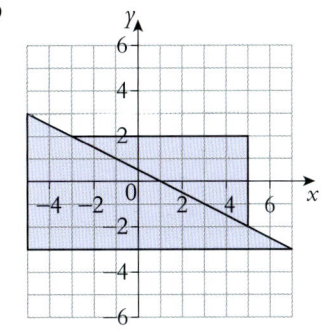

9

Exercise 3.21
1 a b c

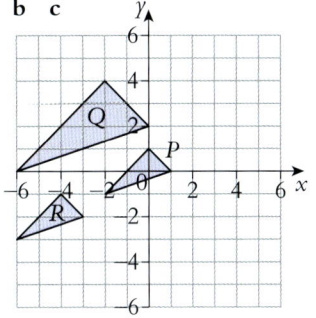

d translation $\begin{pmatrix} -4 \\ -2 \end{pmatrix}$

2 a b c

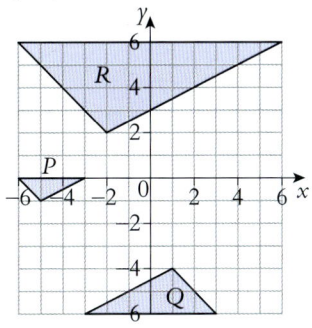

d enlargement, SF 4, centre $(-6, -2)$

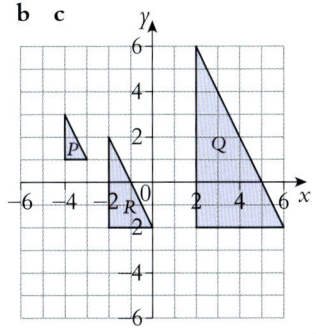

3 a b c

d enlargement, SF 2, centre $(-6, 4)$

Exercise 3.22
1 a 125 cm³   b 96 cm³
  c 120 000 cm³
2 a 180 cm³   b 192 cm³
  c 16 cm³
3 a 24 cm³   b 128 cm³
  c 2240 cm³ d 90 cm³
  e 450 cm³   f 540 cm³
4 a 3.14 m³   b 226 cm³
  c 236 cm³
5 a 151 cm³   b 490 cm³
  c 25 100 cm³
6 20 cm
7 a 7.875 m³
  b 7 875 000 cm³
  c 7875 litres
8 257 cm³
9 5.64 cm
10 306 cm
11 109.3 cm³
12 a 166 cm²   b 4157 cm³
   c 2.7 kg

Exercise 3.23
1 a 216 cm²   b 184 cm²
  c 928 cm²
2 a 240 cm²   b 88 cm²
  c 172 cm²
3 a 510 cm²   b 216 cm²
4 a 101 cm²   b 254 cm²
  c 176 cm²
5 a 75.4 cm² b 89.5 cm²
  c 50.3 cm²
6 a 234 cm²

Exercise 3.24
1 a $\frac{1}{8}$ b $\frac{1}{2}$ c $\frac{1}{4}$ d $\frac{1}{2}$
  e $\frac{1}{4}$ f 1
2 a $\frac{2}{5}$ b $\frac{2}{5}$ c $\frac{4}{5}$ d $\frac{1}{5}$
  e $\frac{4}{5}$ f 1

3 a $\frac{1}{10}$ b $\frac{1}{2}$ c $\frac{2}{5}$ d $\frac{9}{10}$
  e $\frac{3}{5}$ f $\frac{1}{2}$ g 0 h 1
4 a $\frac{1}{3}$ b $\frac{2}{3}$ c $\frac{4}{9}$ d $\frac{1}{9}$
5 a $\frac{17}{30}$ b $\frac{12}{25}$

Exercise 3.25
1 a

Number	1	2	3
Probability	$\frac{1}{4}$	$\frac{1}{8}$	$\frac{5}{8}$

  b i $\frac{3}{8}$ ii $\frac{7}{8}$ iii $\frac{3}{4}$ iv 0

2 a i $\frac{5}{8}$ ii $\frac{3}{8}$
  b i $\frac{4}{7}$ ii $\frac{3}{7}$
  c i $\frac{5}{7}$ ii $\frac{2}{7}$

3 a $\frac{1}{4}$ b green   c $\frac{5}{6}$

4 a 0.2   b 5   c 0.75

Exercise 3.26
1 a

	1	2	3	4	5
1	2	3	4	5	6
2	3	4	5	6	7
3	4	5	6	7	8
4	5	6	7	8	9
5	6	7	8	9	10

  b i $\frac{2}{25}$ ii $\frac{2}{25}$ iii 0
     iv $\frac{12}{25}$ v $\frac{1}{5}$ vi $\frac{11}{25}$

2 a

	1	2	3	4
1	2	3	4	5
2	3	4	5	6
3	4	5	6	7
4	5	6	7	8

  b i $\frac{1}{8}$ ii $\frac{1}{4}$ iii $\frac{1}{2}$ iv $\frac{3}{8}$

3 a

	2	4	6	8
1	2	4	6	8
2	4	8	12	16
3	6	12	18	24

b i $\frac{1}{6}$ **ii** 0

iii 1 **iv** $\frac{5}{12}$

Exercise 3.27

1 288
2 145
3 150
4 30
5 492
6 **a** 0.255, 0.42, 0.325
  b 234
7 **a** 0.18, 0.36, 0.27, 0.19
  b 162, 324, 243, 171

Examination Questions

1 4.496×10^9
2 1.152×10^{-2}
3 **a** 2.386×10^7
  b 2.34×10^{-2} (3 s.f.)
3 **a** 2.386×10^7
  b 2.34×10^{-2} (3 s.f.)
4 **a** 3×10^{11}
  b 5 000 000
6 **a** 2.67×10^{-2}
  b 0.0267
7 $x = 0.5, y = 3$
8 $x = 10, y = 3$
9 $x = 8, y = 5$
10 $x = 12, y = -10$
11 $2y(3x - z)$
12 $x = 2y - 14$
13 $d = \frac{2cw - 4w}{5}$
14 $y = (9 - 9x)^2$
15 **a** reflection in $y = x$

b

[graph showing triangles A, B, C on coordinate axes]

16 **a** G, E **b** A, B
17 **a b c d**

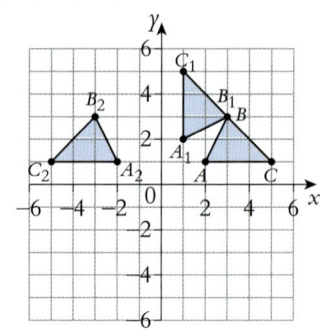

[graph showing triangles with vertices labeled A, B, B_1, B_2, C, C_1, C_2, A_1, A_2]

e reflection in the y-axis

18 **a i, ii b i**

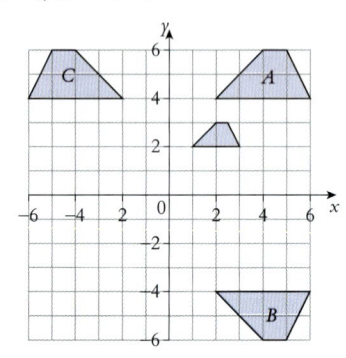

[graph showing trapezoids A, B, C on coordinate axes]

a iii Reflection in the y-axis

19 4.40 cm (3 s.f.)

Unit 4

Exercise 4.1

1	57.5 %	2	37 %	3	3.55 %
4	10 %	5	60 %	6	5 %
7	1.32 %	8	3.09 %	9	5 %
10	21.4 %	11	3.125 %		
12	7.42 %	13	16 %		

Exercise 4.2

1 **a** 1×4 **b** 1×1 **c** 1×3
  d 1×2 **e** 2×2 **f** 2×3
  g 2×1 **h** 2×4 **i** 3×1
  j 3×3 **k** 3×2 **l** 4×3

2 **a** $\begin{pmatrix} 7 & 4 \\ 4 & 13 \end{pmatrix}$ **b** $\begin{pmatrix} 3 & 4 \\ 2 & 3 \end{pmatrix}$

3 **a** $(8 \ 11 \ 15)$ **b** $(2 \ -5 \ 9)$

4 **a** $\begin{pmatrix} 12 & 4 & 6 \\ -2 & -2 & 11 \end{pmatrix}$

  b $\begin{pmatrix} -2 & -2 & 8 \\ -2 & 2 & 1 \end{pmatrix}$

5 **a** $\begin{pmatrix} 16 & 7 \\ -16 & 0 \\ 0 & 0 \\ 0 & -7 \end{pmatrix}$

  b $\begin{pmatrix} -4 & -3 \\ 0 & 0 \\ 4 & -10 \\ 14 & -1 \end{pmatrix}$

6 **a** $\begin{pmatrix} 7 & 10 \\ -10 & 2 \end{pmatrix}$

  b $\begin{pmatrix} 11 & -1 \\ -1 & -2 \end{pmatrix}$

  c $\begin{pmatrix} -1 & 3 \\ -5 & 6 \end{pmatrix}$

  d $\begin{pmatrix} -4 & 11 \\ -9 & 4 \end{pmatrix}$

  e $\begin{pmatrix} 13 & 8 \\ -8 & -2 \end{pmatrix}$

  f $\begin{pmatrix} 1 & 12 \\ -12 & 6 \end{pmatrix}$

  g $\begin{pmatrix} 9 & -10 \\ 6 & -2 \end{pmatrix}$

  h $\begin{pmatrix} -3 & -6 \\ 2 & 6 \end{pmatrix}$

484 Answers

7 a $\begin{pmatrix} 3 & -7 \\ 0 & 2 \\ 14 & 13 \end{pmatrix}$

b $\begin{pmatrix} 2 & 3 \\ 2 & 6 \\ 10 & -1 \end{pmatrix}$

c $\begin{pmatrix} -1 & -6 \\ 10 & 4 \\ 10 & -4 \end{pmatrix}$

d $\begin{pmatrix} 1 & -10 \\ -2 & -4 \\ 4 & 14 \end{pmatrix}$

e $\begin{pmatrix} 3 & 9 \\ -8 & 2 \\ 0 & 3 \end{pmatrix}$

f $\begin{pmatrix} 4 & -9 \\ -6 & -2 \\ 11 & 22 \end{pmatrix}$

g $\begin{pmatrix} 2 & 11 \\ -2 & 6 \\ 3 & -6 \end{pmatrix}$

h $\begin{pmatrix} -2 & -11 \\ 2 & -6 \\ -3 & 6 \end{pmatrix}$

8 $x = 7, y = -8$
9 $x = -7, y = -5$

Exercise 4.3

1 $(12\ -16)$

2 $\begin{pmatrix} 2 & -\frac{1}{2} \\ 0 & 1 \end{pmatrix}$

3 $\begin{pmatrix} 48 & -24 & -16 \\ 0 & 8 & -8 \end{pmatrix}$

4 $\begin{pmatrix} 6 & -3 & -18 & 15 \\ -3 & 0 & -6 & -3 \end{pmatrix}$

5 a $\begin{pmatrix} 4 & -6 \\ 2 & 2 \end{pmatrix}$ b $\begin{pmatrix} -20 & -30 \\ 0 & 10 \end{pmatrix}$

c $\begin{pmatrix} 15 & 0 \\ -6 & 3 \end{pmatrix}$ d $\begin{pmatrix} -2 & -3 \\ 0 & 1 \end{pmatrix}$

e $\begin{pmatrix} -6 & 9 \\ -3 & -3 \end{pmatrix}$ f $\begin{pmatrix} 19 & -6 \\ -4 & 5 \end{pmatrix}$

g $\begin{pmatrix} -35 & -30 \\ 6 & 7 \end{pmatrix}$ h $\begin{pmatrix} 2 & -9 \\ 2 & 3 \end{pmatrix}$

6 a (17) b (14) c (-72)
  d (15) e (8) f (6)
  g (44) h (-6) i (18)

7 $x = 9$
8 $x = 7$
9 $x = -2$

Exercise 4.4

1 a $\begin{pmatrix} 3 & 17 \\ 15 & 13 \end{pmatrix}$ b $\begin{pmatrix} 10 & 2 & 4 \\ 5 & 5 & 3 \end{pmatrix}$

c $\begin{pmatrix} 36 \\ 5 \end{pmatrix}$ d $\begin{pmatrix} 32 & 19 \\ 16 & 8 \end{pmatrix}$

e $\begin{pmatrix} 16 & 4 \\ 11 & 4 \end{pmatrix}$ f (19)

g $\begin{pmatrix} 36 & 0 \\ 10 & 6 \end{pmatrix}$

h $\begin{pmatrix} 19 & 3 & 7 & 11 \\ 11 & 6 & 5 & 4 \\ -9 & 0 & -3 & -6 \\ 34 & 12 & 14 & 16 \end{pmatrix}$

2 a $x = 3,\ y = 4$
  b $2 \times 4\ \ 3 \times 2$ is impossible because the numbers in the middle do not match

3 $\begin{pmatrix} 7 \\ 14 \\ 22 \end{pmatrix}$

Exercise 4.5

1 a 1 b 1 c 1 d -7
  e -1 f 2 g 4 h 0

2 $3\frac{1}{2}$
3 -1

4 a $\begin{pmatrix} -3 & 12 \\ -2 & 6 \end{pmatrix}$ b $\begin{pmatrix} 5 & -4 \\ 4 & -2 \end{pmatrix}$

5 a $\begin{pmatrix} -13 & 8 \\ -7 & -5 \end{pmatrix}$ b $\begin{pmatrix} -14 & -5 \\ 13 & -4 \end{pmatrix}$

c $\begin{pmatrix} 26 & 3 \\ 3 & 5 \end{pmatrix}$ d $\begin{pmatrix} 11 & 0 \\ 0 & 11 \end{pmatrix}$

6 a $\begin{pmatrix} 6 & -3 \\ -9 & 6 \end{pmatrix}$ b $\begin{pmatrix} 10 & 15 \\ 5 & 20 \end{pmatrix}$

c $\begin{pmatrix} 0 & -4 \\ -4 & -2 \end{pmatrix}$ d $\begin{pmatrix} 16 & 12 \\ -4 & 26 \end{pmatrix}$

e $\begin{pmatrix} 3 & 2 \\ -4 & -1 \end{pmatrix}$ f $\begin{pmatrix} -5 & 4 \\ -10 & 7 \end{pmatrix}$

g $\begin{pmatrix} 7 & -4 \\ -12 & 7 \end{pmatrix}$ h $\begin{pmatrix} 7 & 18 \\ 6 & 19 \end{pmatrix}$

i 1 j 5

7 a $\begin{pmatrix} 2 & -2 \\ 2 & 4 \end{pmatrix}$ b $\begin{pmatrix} 20 & -12 \\ 4 & -4 \end{pmatrix}$

c $\begin{pmatrix} -18 & 10 \\ -2 & 8 \end{pmatrix}$ d $\begin{pmatrix} 4 & -2 \\ 7 & -5 \end{pmatrix}$

e $\begin{pmatrix} 2 & -11 \\ 0 & -3 \end{pmatrix}$ f $\begin{pmatrix} 0 & -3 \\ 3 & 3 \end{pmatrix}$

g $\begin{pmatrix} 22 & -12 \\ 4 & -2 \end{pmatrix}$ h $\begin{pmatrix} 16 & -2 \\ 0 & 9 \end{pmatrix}$

i 3 j -2

8 $\begin{pmatrix} -5 & -3 \\ -2 & -1 \end{pmatrix}$

Exercise 4.6

1 a $\begin{pmatrix} 3 & -5 \\ -1 & 2 \end{pmatrix}$ b $\begin{pmatrix} 3 & -4 \\ -2 & 3 \end{pmatrix}$

c $\frac{1}{2}\begin{pmatrix} 2 & -2 \\ -4 & 5 \end{pmatrix}$ d $\begin{pmatrix} -2 & -1 \\ -1 & -1 \end{pmatrix}$

e $\begin{pmatrix} -1 & -2 \\ 2 & 3 \end{pmatrix}$ f $\frac{1}{2}\begin{pmatrix} 1 & 1 \\ 4 & 6 \end{pmatrix}$

g $-\frac{1}{2}\begin{pmatrix} -7 & -4 \\ 3 & 2 \end{pmatrix}$ h $\begin{pmatrix} 4 & 1 \\ -3 & -1 \end{pmatrix}$

Answers 485

i $\frac{1}{2}\begin{pmatrix} -1 & 1 \\ 7 & -9 \end{pmatrix}$ j $\frac{1}{2}\begin{pmatrix} 4 & 5 \\ 2 & 3 \end{pmatrix}$

k $\begin{pmatrix} 0.5 & 1 \\ 1 & 4 \end{pmatrix}$ l $\begin{pmatrix} -1 & 2 \\ -1 & 1 \end{pmatrix}$

2 Determinant = 0
3 $x = -5$
4 $x = 1\frac{1}{2}$
5 $Y = \begin{pmatrix} 1 & -3 \\ -2 & 7 \end{pmatrix}$
6 $X = -\frac{1}{14}\begin{pmatrix} 3 & -1 \\ -8 & -2 \end{pmatrix}$
7 $C = \begin{pmatrix} -1 & 3 \\ 2 & -5 \end{pmatrix}$

8 a $\begin{pmatrix} 18 & 24 \\ 4 & 4 \end{pmatrix}$ b $\begin{pmatrix} 13 & 16 \\ 4 & 5 \end{pmatrix}$

c $\begin{pmatrix} 19 & 24 \\ 6 & 7 \end{pmatrix}$ d $\begin{pmatrix} 117 & 144 \\ 36 & 45 \end{pmatrix}$

e $\begin{pmatrix} 55 & 68 \\ 17 & 21 \end{pmatrix}$ f -1

g $\begin{pmatrix} -1 & 4 \\ 1 & -3 \end{pmatrix}$ h $\begin{pmatrix} 1 & 0 \\ 0 & 1 \end{pmatrix}$

9 $Y = \begin{pmatrix} 2 & -2 \\ -3 & -3 \end{pmatrix}$

10 $X = \begin{pmatrix} 1 & 2 \\ -3 & 4 \end{pmatrix}$

Exercise 4.7

1 $x^2 + 13x + 30$ 2 $x^2 + 7x + 12$
3 $x^2 + 10x + 21$ 4 $x^2 + 6x + 8$
5 $x^2 + 6x + 5$ 6 $x^2 + 15x + 56$
7 $x^2 + 8x + 12$ 8 $x^2 + 13x + 42$
9 $x^2 + 8x + 16$ 10 $x^2 + 19x + 60$
11 $x^2 + 18x + 72$ 12 $2x^2 + 13x + 15$
13 $6x^2 + 19x + 10$

Exercise 4.8

1 a $x^2 + 4x + 3x + 12$
     $= x^2 + 7x + 12$
  b $x^2 - 6x + 4x - 24$
     $= x^2 - 2x - 24$
  c $x^2 - 4x - 5x + 20$
     $= x^2 - 9x + 20$
  d $x^2 + 4x + 4x + 16 = x^2 + 8x + 16$

2 a $x^2 + 3x + 2$
  b $x^2 + 12x + 35$
  c $x^2 + 11x + 24$ d $x^2 + 6x + 8$
  e $x^2 + 9x + 20$ f $x^2 + 4x + 3$
  g $x^2 + 12x + 27$
  h $x^2 + 11x + 30$
  i $x^2 + 7x + 12$ j $x^2 + 5x + 6$
  k $x^2 + 10x + 16$ l $x^2 + 5x + 4$

3 a $x^2 + x - 2$ b $x^2 + 4x - 12$
  c $x^2 + 3x - 18$ d $x^2 + 3x - 4$
  e $x^2 - 2x - 63$
  f $x^2 + 6x - 16$
  g $x^2 - 4x - 32$
  h $x^2 - 5x - 14$
  i $x^2 - 3x - 10$
  j $x^2 + 5x - 24$
  k $x^2 - 2x - 3$
  l $36 - 5x - x^2$

4 a $x^2 - 9$ b $x^2 - 64$
  c $x^2 - 36$ d $x^2 - 100$
  e $x^2 - 16$ f $x^2 - 81$
  g $x^2 - 1$ h $144 - x^2$

5 a $x^2 - 7x + 10$
  b $x^2 - 5x + 4$
  c $x^2 - 11x + 24$
  d $x^2 - 8x + 12$
  e $x^2 - 14x + 49$
  f $x^2 - 9x + 18$
  g $x^2 - 9x + 20$
  h $x^2 - 11x + 28$
  i $x^2 - 13x + 36$
  j $x^2 - 2x + 1$
  k $x^2 - 6x + 5$
  l $4 - 4x + x^2$

6 a $x^2 + 6x + 9$
  b $y^2 + 10y + 25$
  c $x^2 - 8x + 16$
  d $x^2 + 14x + 49$
  e $x^2 + 16y + 64$

  f $a^2 - 20a + 100$
  g $36 - 12x + x^2$
  h $4 + 4x + x^2$

7 a $2y^2 - 5y - 3$
  b $4x^2 + 5x - 6$
  c $5x^2 + 22x + 8$
  d $7y^2 + 33y - 10$
  e $8y^2 - 22y - 21$
  f $10x^2 - 49x - 33$
  g $15x^2 - 38x + 24$
  h $27y^2 - 96y + 20$
  i $24y^2 - 47y + 20$
  j $6a^2 - 5a - 21$
  k $3 - 13a + 12a^2$
  j $30x^2 - 3x - 9$

8 a $4x^2 - 1$ b $9x^2 - 4$
  c $25x^2 - 49$ d $25x^2 - 4$

9 a $4x^2 - 4x + 1$
  b $25x^2 + 20x + 4$
  c $9x^2 + 12xy + 4y^2$
  d $9y^2 - 48y + 64$

10 a $2x^2 + 3x + 19$
   b $8x^2 + 12x - 38$
   c $x^2 - 13x - 3$
   d $14x^2 - 12x + 3$
   e $13x^2 - 14x + 26$
   f $8x^2 - 8x + 10$
   g $32x^2 + 72x + 16$
   h $6x^2 - 48x - 73$

Exercise 4.9

1 a -5.5 b 2 c -2
2 a 35 b 23
3 a A and B b D and E
  c B and E d B and C
  e B and D f A and C
  g A and D h A and E
4 10
5 a $3x^2 + 9x - 30$
  b $2x^2 + 18x - 2$
6 answer given
7 a c^2 b $(a + b)^2$
  c $2ab$ d answer given

Exercise 4.10

1 a

x	-3	-2	-1	0	1	2	3
y	10	5	2	1	2	5	10

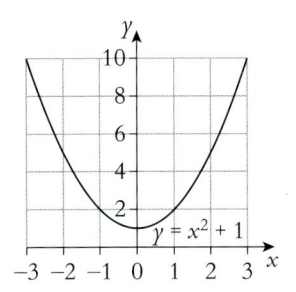

Graph of $y = x^2 + 1$

b

x	-3	-2	-1	0	1	2	3
y	7	2	-1	-2	-1	2	7

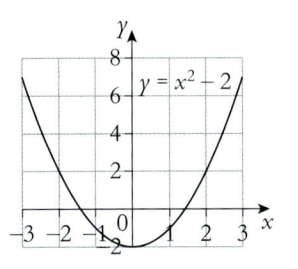

Graph of $y = x^2 - 2$

c

x	-3	-2	-1	0	1	2	3
y	-9	-4	-1	0	-1	-4	-9

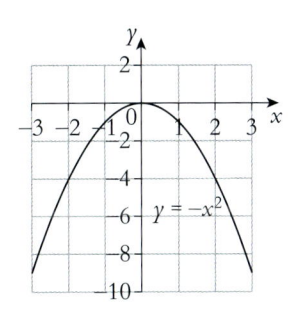

Graph of $y = -x^2$

d

x	-3	-2	-1	0	1	2	3
y	13	3	-3	-5	-3	3	13

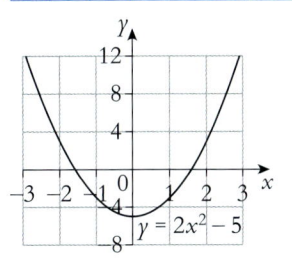

Graph of $y = 2x^2 - 5$

e

x	-3	-2	-1	0	1	2	3
y	2.5	0	-1.5	-2	-1.5	0	2.5

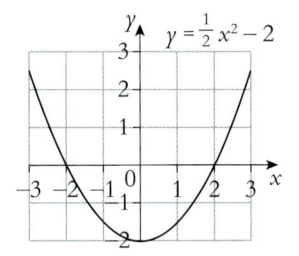

Graph of $y = \frac{1}{2}x^2 - 2$

f

x	-4	-3	-2	-1	0	1	2
y	8	3	0	-1	0	3	8

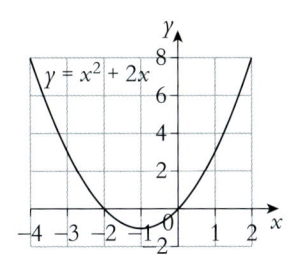

Graph of $y = x^2 + 2x$

g

x	-1	0	1	2	3	4
y	4	0	-2	-2	0	4

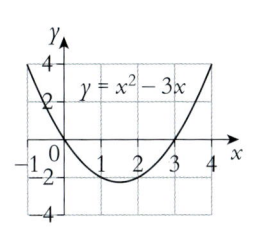

Graph of $y = x^2 - 3x$

h

x	-1	0	1	2	3	4
y	-4	0	2	2	0	-4

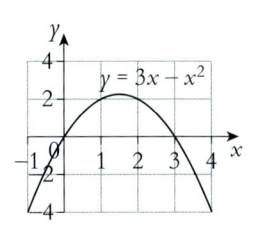

Graph of $y = 3x - x^2$

2 a

x	-1	0	1	2	3	4	5
y	8	3	0	-1	0	3	8

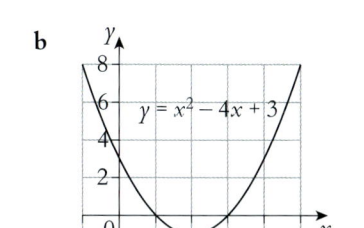

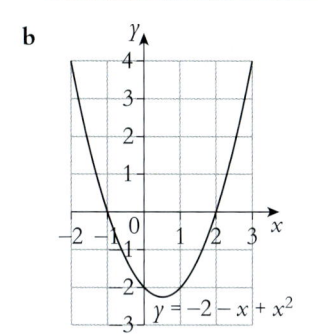

b Graph of $y = x^2 - 4x + 3$

c −0.8, 4.8

3 a

x	-2	-1	0	1	2	3
y	4	0	-2	-2	0	4

b Graph of $y = -2 - x + x^2$

c −1.6, 2.6

4 a

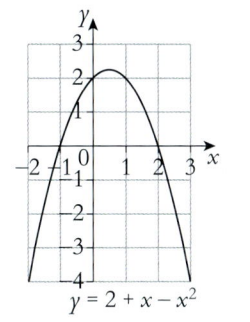

x	-2	-1	0	1	2	3
y	-4	0	2	2	0	-4

b Graph of $y = 2 + x - x^2$

c −0.6, 1.6

5 a

x	-1	0	1	2	3	4
y	8	1	-2	-1	4	13

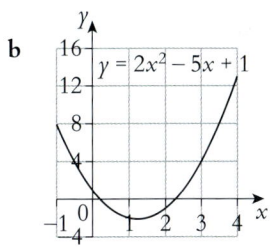

b Graph of $y = 2x^2 - 5x + 1$

c 0.2, 2.3

Exercise 4.11

1. **a** C **b** D **c** B **d** F
   e E **f** A
2. **a** 1.5 **b** 1 **c** 0.5 **d** 0
   e −0.5 **f** −1 **g** −1.5
3. **a** −4 **b** −2 **c** 0
   d 2 **e** 4

Exercise 4.12

1. **a** answer given

   b
x	0	0.2	0.4	0.6	0.8	1
V	0	0.24	0.56	0.96	1.44	2

   c graph of $V = x^2 + x$

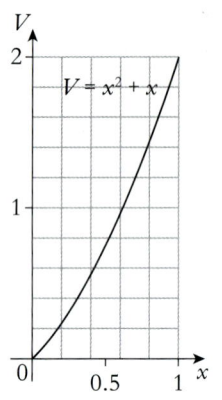

   d 0.75

2. **a**
t	0	1	2	3	4	5
h	0	20	30	30	20	0

   b graph of $h = 25t - 5t^2$

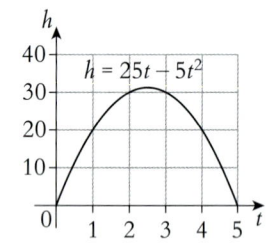

   c 31.25 m **d** 1.5, 3.5

3. **a**
x	0	10	20	30	40	50
y	0	8	12	12	8	0

   b graph of $y = x - 0.02x^2$

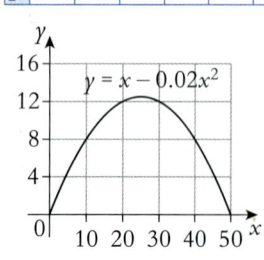

   c 12.5 m
   d 13.8 m $< x <$ 36.2 m

Exercise 4.13

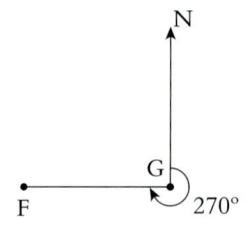

1. **a** diagram: bearing from A to B, 56°
   b diagram: bearing from F to G, 270°
   c diagram: bearing from M to N, 135°
   d diagram: bearing from P to Q, 340°

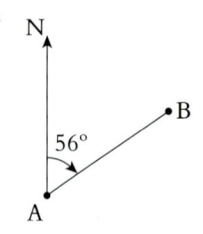

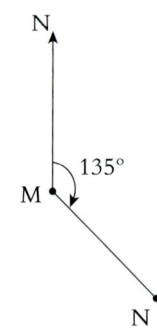

2. **a** 135° **b** 180°
   c 270° **d** 225°
   e 450° **f** 315°

3. **a** 085° **b** 135°
   c 010° **d** 300°

4. diagram with O, B, J; 77°, 220°, 1650 km, 1990 km

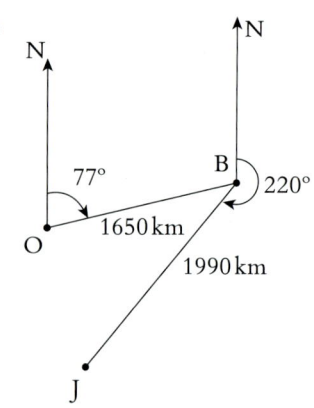

5. **a** diagram with A, B, C; 90°, 140°, 16 km, 12 km
   b 111°
   c 25.4 km

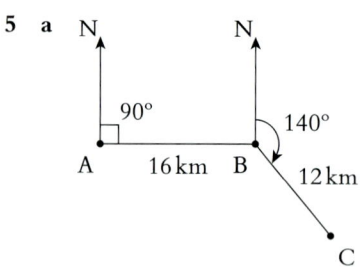

6. **a** 115° **b** 050° **c** 110°
   d 255° **e** 065° **f** 320°
   g 039° **h** 153°

7. 200°

8. **a** 232° **b** 302° **c** 067°
   d 115°

9. **a** 060° **b** 240° **c** 090°
   d 270° **e** 075° **f** 255°

10. **a** 300° **b** 305°

11. **a** 8 km **b** 035° **c** 215°

12. 13 km

13. diagram with P, L; 110°, 45°, Locus of points

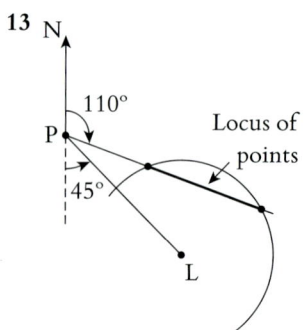

14. $x = 80, y = 60, z = 40$

15. $x = 83, y = 45, z = 52$

Exercise 4.14

All answers in cm to 3 s.f.

1. a 3.46   b 7.20   c 7.00
   d 9.63   e 13.7   f 7.85
   g 4.08   h 4.36   i 19.0
   j 14.3   k 12.6   l 8.31
   m 5.36   n 9.38   o 11.3
2. a $x = 8.66$, $y = 5.32$
   b $x = 10.9$, $y = 4.58$
   c $x = 4.69$, $y = 5.70$
   d $x = 6.71$, $y = 3.88$
   e $x = 6.40$, $y = 9.14$
   f $x = 13.2$, $y = 13.6$

Exercise 4.15

All answers to 3 s.f.

1. a 20.6°   b 38.7°   c 21.8°
   d 21.8°   e 18.4°   f 54.5°
   g 18.4°   h 16.7°   i 56.3°
2. a 54.5°   b 234°
3. $x = 18.4°$, $y = 36.9°$
4. a 65.2°   b 50.8°   c 43.8°
5. a $x = 36.9°$, $y = 29.2°$
   b $x = 38.7°$, $y = 19.3°$
   c $x = 23.6°$, $y = 10.9°$

Exercise 4.16

All answers in cm, unless stated otherwise, to 3 s.f.

1. a 6.51   b 3.35   c 5.73
   d 4   e 9.58   f 7.55
   g 14.5   h 8.67   i 1.75
   j 13.1   k 7.71   i 15.8
2. a 7.62   b 30.8   c 8.44
3. a 5.52, 12.9 cm^2
   b 9.38, 113 cm^2
   c 6.93, 27.7 cm^2
4. 3.47, 34.0 cm^2

Exercise 4.17

All answers to 3 s.f.

1. 44.4°   2 60°   3 44.4°
4. 23.6°   5 19.5°   6 48.2°
7. 65.4°   8 17.5°   9 52.0°
10. 25.4°   11 33.7°   12 50.6°
13. 67.1°   14 56.0°   15 51.8°

Exercise 4.18

All answers to 3 s.f.
All lengths in cm.

1. a 6.90   b 38.6°   c 31.1
   d 10.3   e 53.1   f 10.1
   g 5.20   h 10.5   i 60°
   j 11.8   k 14.8   l 9.97
   m 6.40   n 7.81   o 5.12
   p 37.9° q 66.4°   r 22.0°
2. a $x = 8.58$, $y = 30.1°$
   b $x = 4.52$, $y = 28.2°$
   c $x = 5.96$, $y = 13.3°$
3. a $x = 4.77$, $y = 57.8°$, $z = 5.63$
   b $x = 4.50$, $y = 5.41$, $z = 33.7°$
   c $x = 15.0$, $y = 26.6°$, $z = 8.94$
4. 60°
5. a $x = 4.31$, $y = 6.30$
   b $x = 11.0$, $y = 37.6°$
6. $x = 56.3°$, $y = 7.48$, $z = 15.5°$
7. a 41.6 cm^2 b 61.9 cm^2
   c 309 cm^2
8. a 1020 cm^2 b 2202 cm^3

Exercise 4.19

All answers to 3 s.f.

1. 114 m
2. a 2.41 m
   b 3.29 m
   c 2.13 m
3. a 56.4°   b 45.6°   c 41.1 m
4. 111 m
5. a 72.1 m   b 52.0 m
   c 102°   d 78°
   e 2450 m^2
6. a 11.8 cm   b 76.4 cm^2
7. a 56.8 m   b 0.123 m
8. a 8.68 km   b 49.2 km
9. a 1270 km   b 2720 km
10. a 826 km   b 929 km
    c 1240 km d 048°
    e 228°

Exercise 4.20

All answers to 3 s.f.

1. a 7.07, 8.66   b 35.3°
2. a 9.85, 11.0   b 26.9°
3. a 12.8   b 13.4   c 17.3°
4. a 5.66, 4.12, 4.58   b 55.6°
   c 51.8°
5. a 7.5, 9.60, 8.48
   b 38.7°
   c 90°
6. 15.9°

Exercise 4.21

All answers to 3 s.f.
All lengths in m.

1. 145
2. 9.99
3. 553
4. 828
5. 275
6. 3730
7. 45.3
8. 43.0
9. 4.06
10. a 7.03, 64.4
    b 9.96°
    c 6.26°

Exercise 4.22

1. a negative
   b zero
   c positive
   d zero
2. a

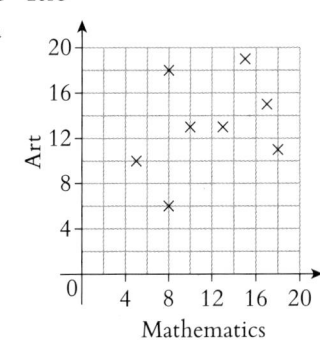

   b zero correlation

Answers   489

3   a

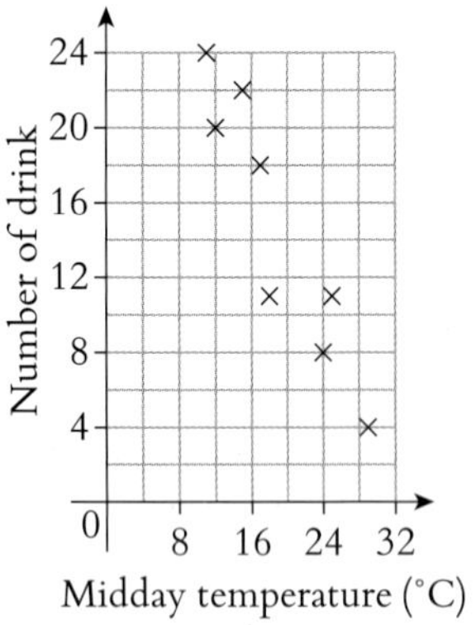

b   negative correlation

4   a

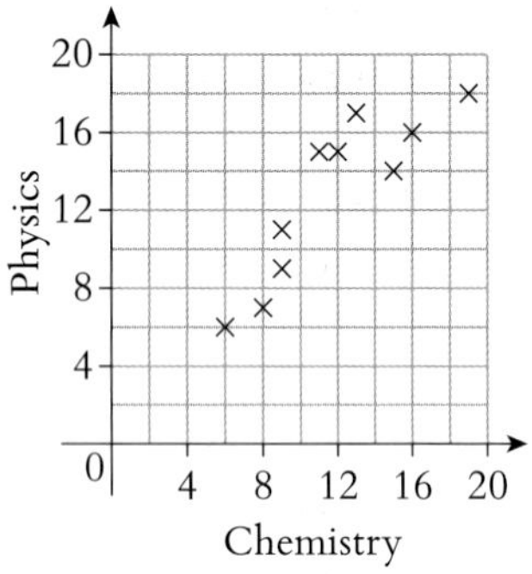

c   positive correlation

5   a

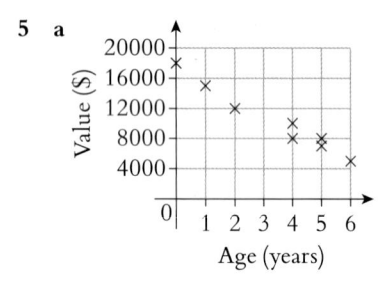

b   negative correlation

6   a   positive
    b   negative
    c   positive
    d   zero
    e   positive

Exercise 4.23

1   a   68 %
    b   25 %

2   a   8.3
    b   6.8

c

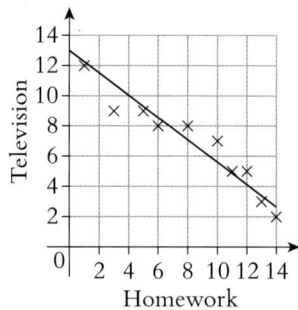

d   negative   e   9.5

3   a   21.05   b   176.2

c

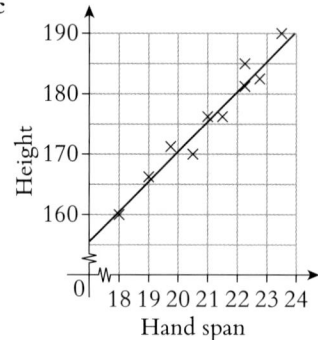

d   positive   e   174

Examination Questions

1   150%   2   161%

3   $\begin{pmatrix} 7 \\ 5 \\ 7 \end{pmatrix}$ 4 $a = 3, b = 4$

5   $\frac{1}{2}\begin{pmatrix} 5 & -3 \\ 4 & -2 \end{pmatrix}$ 6   5

7   a   (23)   b $x = 4, y = 6$
    c   determinant = 0

8   a   BA and CB   b $\begin{pmatrix} 8 & -24 \\ -4 & 16 \end{pmatrix}$
    c   determinant = 0

9   $48x - 36x^2$ or $12x(4 - 3x)$

10 a $105x^2$ b   22.6

11 a   24.7 m   b   11.5 m

12 a   i   5.74 cm   ii   6.32 cm
    b   132 cm²
    c   i   answer given   ii   50.8°
        iii   78.5°   iv   44.4°
        v   PHN or PMG

Unit 5

Exercise 5.1

1   $13.15
2   $30
3   $145
4   a   150 km   b   52.5 litres
5   1500 g flour, 3600 ml milk, 15 g salt, 15 eggs
6   16 seconds
7   2 hours 24 mins
8   7.5 days
9   6
10 a   18 hours   b   12
11 8
12 a   4 days   b   25

Exercise 5.2

1   a   i   320 Rf   ii   690 Rf
        iii   1040 Rf   iv   2560 Rf
    b   i $47   ii $86
        iii $19   iv $156

2   a   i   199 ARS   ii   1194 ARS
        iii   286.56 ARS
        iv   7960 ARS
    b   i $10.05   ii $20.60
        iii $125.63   iv $9.15

3   a   i   250.78 MYR
        ii   62.70 MYR
        iii   231.97 MYR
        iv   1567.40 MYR
    b   i $19.14   ii $5.74
        iii $95.70   iv $149.93

4   $75

Exercise 5.3

1   a   100   b   105
    c   65   d   82.5
2   a   20   b   33
    c   30   d   36
3   6250 kg
4   1 minute 57 seconds
5   560
6   105 kg
7   $1656

8 a 7.5 cm, 10 cm
  b i 75 cm² **ii** 48 cm²
  c 25 : 16
9 a 8 cm, 20 cm, 28 cm
  b 4480 cm³
  b 1890 cm³
  c 64 : 27

Exercise 5.4
1 a 11 **b** −9
  c 4 **d** 3
2 a −19 **b** 17
  c 3.5 **d** 6.5
3 a 10 **b** 0
  c 0 **d** 1.75
4 a 16 **b** −4
  c −4 **d** −5
5 a 66 **b** −6
  c 2.125 **d** 2
6 a 12.25 **b** 27.25
  c $\frac{7}{16}$ **d** 0.25
7 3
8 ±6
9 −1.75
10 −2.75
11 −3
12 0.5
13 ±3
14 −7

Exercise 5.5
1 a −6 **b** 2
2 a −29 **b** −37
3 a 1 **b** 25
4 a 3 **b** 7
5 a undefined **b** −0.5
6 a 9 **b** −6
  c 2.25 **d** 9 **e** 4
7 a −6 **b** 14
  c −7 **d** 14.5
  e 0.75
8 a D **b** A
  c B **d** C
9 a $x + 5$ **b** $x + 5$
  c $x - 6$

10 a $2x - 4$ **b** $2x - 9$
  c $4x + 3$
11 a $5x^2 + 15$ **b** $25x^2 + 3$
  c $25x$
12 a $\frac{3}{x-1}$
  b $\frac{3}{x-2} + 1$ or $\frac{x+1}{x-2}$
13 7
14 16.5
15 ±0.5

Exercise 5.6
1 a $x - 2$ **b** $x + 7$
  c $x - 22$ **d** $x + 0.5$
  e $\frac{x}{3}$ **f** $\frac{x}{8}$
  g $4x$ **h** $6x$
2 a $\frac{x+5}{2}$ **b** $\frac{x-8}{3}$
  c $\frac{x+2}{5}$ **d** $\frac{x-3}{4}$
  e $\frac{x-6}{3}$ **f** $\frac{x+28}{4}$
  g $\frac{x+20}{2}$ **h** $\frac{x-10}{2}$
  i $\frac{x-8}{10}$ **j** $\frac{15-x}{6}$
  k $\frac{28-x}{12}$ **l** $\frac{x+27}{6}$
  m $4x + 2$ **n** $5x - 6$
  o $\frac{4x+3}{2}$ **p** $\frac{5-6x}{3}$
3 a $\frac{5}{x+3}$ **b** $\frac{3}{x-2}$
  c $\frac{6}{x+7}$ **d** $\frac{5}{4-x}$
  e $\frac{2-3x}{x}$ **f** $\frac{4-x}{2x}$
  g $\frac{3+5x}{2x}$ **h** $\frac{3x-6}{2x}$
  i $\frac{1+3x}{x-2}$ **j** $\frac{5x+7}{1-2x}$
  k $\frac{8x+6}{x-1}$ **l** $\frac{3x-1}{4x-2}$
4 a $\sqrt[3]{x} - 3$ **b** $\sqrt[3]{x} + 5$
  c $\frac{\sqrt[3]{x}-7}{2}$ **d** $\frac{5-\sqrt[3]{x}}{3}$
5 a −1.5 **b** 2
  c −2.5
6 a 0.4 **b** 1.6
  c 0

7 1
8 −1
9 a 3 **b** 2.5
  c −29

Exercise 5.7
1 a $x(x + 5)$ **b** $x(x - 7)$
  c $3x(x - 5)$ **d** $6x(2 - 3x)$
2 a $(x + 3)(x + 1)$
  b $(x + 5)(x + 2)$
  c $(x + 1)(x + 1)$
  d $(x + 6)(x + 3)$
  e $(x + 4)(x + 2)$
  f $(x + 5)(x + 1)$
  g $(x + 2)(x + 3)$
  h $(x + 18)(x + 2)$
  i $(x + 5)(x + 5)$
  j $(x + 9)(x + 4)$
  k $(x + 4)(x + 5)$
  l $(x + 7)(x + 2)$
  m $(x + 7)(x + 3)$
  n $(x + 4)(x + 7)$
  o $(x + 21)(x + 1)$
  p $(x + 3)(x + 5)$
3 a $(x + 3)(x - 2)$
  b $(x - 6)(x + 1)$
  c $(x + 7)(x - 3)$
  d $(x - 8)(x + 2)$
  e $(x + 8)(x - 2)$
  f $(x - 16)(x + 1)$
  g $(x + 7)(x - 1)$
  h $(x - 8)(x + 5)$
  i $(x + 19)(x - 2)$
  j $(x - 11)(x + 3)$
  k $(x + 10)(x - 3)$
  l $(x - 5)(x + 3)$
  m $(x + 5)(x - 4)$
  n $(x - 20)(x + 1)$
  o $(x + 21)(x - 1)$
  p $(x - 7)(x + 1)$
4 a $(x - 5)(x - 2)$
  b $(x - 15)(x - 1)$
  c $(x - 3)(x - 7)$
  d $(x - 6)(x - 2)$

  e $(x - 10)(x - 1)$
  f $(x - 3)(x - 4)$
  g $(x - 7)(x - 1)$
  h $(x - 3)(x - 3)$
  i $(x - 5)(x - 6)$
  j $(x - 4)(x - 7)$
  k $(x - 21)(x - 1)$
  l $(x - 3)(x - 5)$
  m $(x - 7)(x - 2)$
  n $(x - 6)(x - 4)$
  o $(x - 13)(x - 1)$
  p $(x - 13)(x - 2)$
5 a $2(x + 4)(x + 2)$
  b $3(x - 4)(x + 3)$
  c $2(x - 5)(x + 1)$
  d $3(x - 11)(x - 1)$
  e $4(x + 3)(x + 1)$
  f $2(x + 9)(x + 1)$
  g $2(x - 3)(x - 3)$
  h $3(x - 9)(x + 4)$
  i $5(x - 5)(x + 1)$
  j $4(x + 3)(x + 5)$
  k $7(x + 2)(x + 1)$
  l $2(x + 7)(x - 5)$
6 a $x + 3$ b $x + 3$
  c $x + 9$ d $x + 11$
  e $x - 1$ f $x + 4$
  g $\dfrac{1}{x+24}$ h $\dfrac{x+3}{x-4}$
  i $\dfrac{x-8}{x-3}$ j $\dfrac{x+7}{x-5}$
  k $\dfrac{x}{x+6}$ l $\dfrac{x-1}{x}$

Exercise 5.8

1 a $(x - 9)(x + 9)$
  b $(y - 1)(y + 1)$
  c $(x - 8)(x + 8)$
  d $(a - 6)(a + 6)$
  e $(y - 10)(y + 10)$
  f $(4 - a)(4 + a)$
  g $(x - y)(x + y)$
  h $(12 - x)(12 + x)$
  i $(2x - 3)(2x + 3)$
  j $(4y - 1)(4y + 1)$
  k $(5x - 6)(5x + 6)$

  l $(3 - 4x)(3 + 4x)$
  m $(10x - 9y)(10x + 9y)$
  n $(5a - 7b)(5a + 7b)$
  o $(9x - 7y)(9x + 7y)$
  p $(11x - 10)(11x + 10)$
2 a $2(x + 5)(x - 5)$
  b $2(y + 2)(y - 2)$
  c $3(x + 1)(x - 1)$
  d $5(y + 4)(y - 4)$
  e $3(y + 8)(y - 8)$
  f $4(x + y)(x - y)$
  g $2(x + 11)(x - 11)$
  h $6(n + 10)(n - 10)$
  i $3(5y + 6)(5y - 6)$
3 $(x^2 + y^2)(x + y)(x - y)$
4 a $(2x + 1)(x + 5)$
  b $(2x + 3)(x + 1)$
  c $(2x + 1)(x + 7)$
  d $(2x + 11)(x + 1)$
  e $(3x + 2)(x + 1)$
  f $(3x + 1)(x + 5)$
  g $(3x + 1)(x + 3)$
  h $(3x + 7)(x + 1)$
  i $(2x - 1)(x + 3)$
  j $(3x + 5)(x - 1)$
  k $(3x - 1)(x - 7)$
  l $(5x + 3)(x - 1)$
  m $(7x - 2)(x - 1)$
  n $(5x - 2)(x + 1)$
  o $(11x + 1)(x - 2)$
  p $(2x - 13)(x - 1)$
5 a $(2x + 3)(x + 4)$
  b $(2x + 5)(x + 2)$
  c $(3x + 2)(x + 4)$
  d $(7x + 3)(x + 5)$
  e $(3x + 2)(2x + 1)$
  f $(6x + 5)(x + 1)$
  g $(4x + 1)(2x + 3)$
  h $(10x + 1)(x + 7)$
  i $(3x - 2)(x + 3)$
  j $(5x + 2)(x + 4)$
  k $(2x - 3)(x - 4)$
  l $(7x - 2)(x + 4)$
  m $(5x - 4)(x + 5)$

  n $(3x - 5)(x + 2)$
  o $(2x - 5)(x - 2)$
  p $(3x + 8)(x + 2)$
6 a $2(2x + 7)(x - 3)$
  b $3(2x + 1)(x + 3)$
  c $2(3x + 2)(x + 2)$
  d $4(3x + 1)(x + 2)$
  e $2y(4x + 1)(2x - 3)$
  f $2x(3x + 1)(x - 2)$
7 a $\dfrac{2x+1}{x}$
  b $\dfrac{3x-1}{x-2}$
  c $\dfrac{2x-5}{x+2}$ d $\dfrac{2x+3}{x+1}$
  e $\dfrac{x}{2x+3}$ f $\dfrac{2(x-5)}{3x-8}$

8 $3x(3x + 4)(3x + 2)^2$

Exercise 5.9

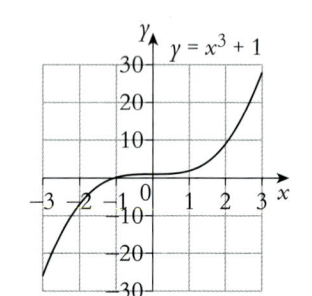

1 a

x	−3	−2	−1	0	1	2	3
y	−26	−7	0	1	2	9	28

[Graph of $y = x^3 + 1$]

 b

x	−3	−2	−1	0	1	2	3
y	29	10	3	2	1	−6	−25

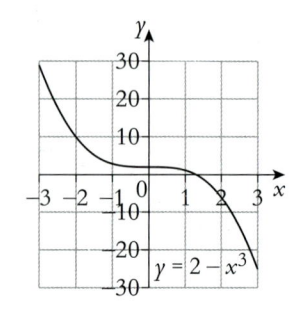

[Graph of $y = 2 - x^3$]

c

x	−3	−2	−1	0	1	2	3
y	27	8	1	0	−1	−8	−27

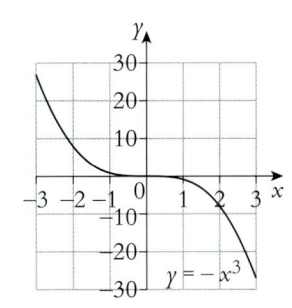

Graph of $y = -x^3$

d

x	−3	−2	−1	0	1	2	3
y	−33	−12	−3	0	3	12	33

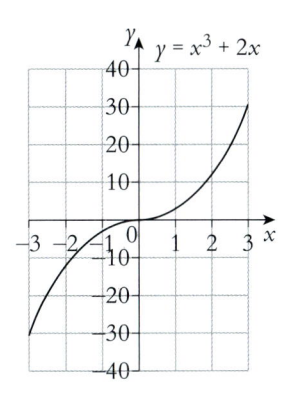

Graph of $y = x^3 + 2x$

e

x	−2	−1	0	1	2	3	4
y	20	4	0	2	4	0	−16

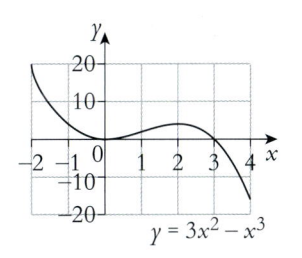

Graph of $y = 3x^2 - x^3$

f

x	−2	−1	0	1	2	3	4
y	−8	1	0	−5	−8	−3	16

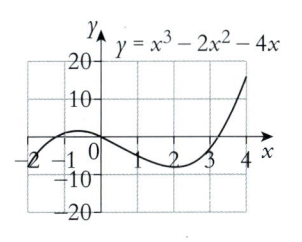

Graph of $y = x^3 - 2x^2 - 4x$

g

x	−1	0	1	2	3	4
y	−3	2	−1	−6	−7	2

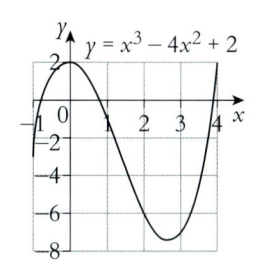

Graph of $y = x^3 - 4x^2 + 2$

h

x	−6	−5	−4	−3	−2	−1	0	1	2
y	2	14.5	18	15.5	10	4.5	2	5.5	18

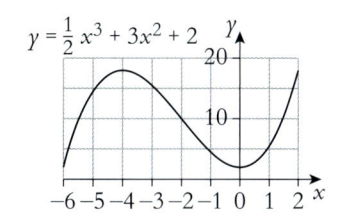

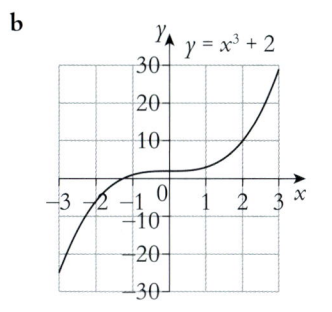

Graph of $y = \frac{1}{2}x^3 + 3x^2 + 2$

2 a

x	−3	−2	−1	0	1	2	3
y	−25	−6	1	2	3	10	29

b Graph of $y = x^3 + 2$

c 1.7

d 3

3 a

x	−2	−1	0	1	2	3	4
y	−6	3.5	6	4.5	2	1.5	6

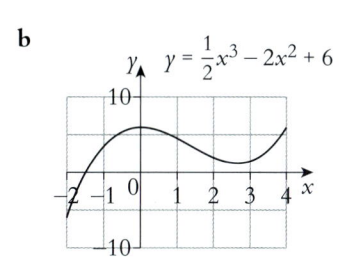

b Graph of $y = \frac{1}{2}x^3 - 2x^2 + 6$

c −0.9, 1.2, 3.7

d −2

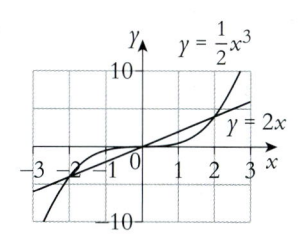

4 Graph of $y = \frac{1}{2}x^3$ and $y = 2x$

$(-2, -4)$ $(0, 0)$ $(2, 4)$

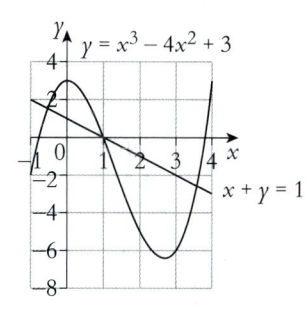

5 Graph of $y = x^3 - 4x^2 + 3$ and $x + y = 1$

$(-0.6, 1.6)$ $(1, 0)$ $(3.6, -2.6)$

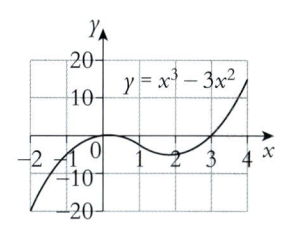

6 Graph of $y = x^3 - 3x^2$

$0 < x < 2$

Exercise 5.10

1 a answer given

b

x	0	0.2	0.4	0.6	0.8	1
v	0	0.096	0.448	1.152	2.304	4

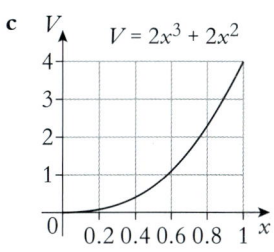

c Graph of $V = 2x^3 + 2x^2$

d 0.75

2 a answer given

b

x	0	1	2	3	4	5	6	7	8
V	0	252	384	420	384	300	192	84	0

Answers 493

c

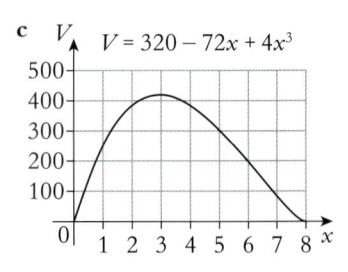

$V = 320 - 72x + 4x^3$

d 420 cm³, 2.9

Exercise 5.11

1. **a** 40 cm³ **b** 33 cm³ **c** 56 cm³ **d** 20 cm³
2. 128 cm³
3. 140 cm³
4. 14.4 cm³
5. 6 cm
6. 2.96×10^6 m³
7. 9.05×10^3 m³
8. 30.2 mm³
9. 224 cm³
10. 520 cm³
11. **a** 90 cm² **b** 246 cm² **c** 11.2 cm² **d** 154 cm² **e** 133 cm² **f** 412 cm²
12. **a** 173 cm² **b** 39.9 cm²
13. **a** 118 cm³ **b** 13.0 cm³

Exercise 5.12

1. **a** 117 cm³ **b** 56.5 cm³ **c** 236 cm³
2. 5.97 cm
3. 4.46 cm
4. 159 cm³
5. 687 cm³
6. **a** 660 cm³ **b** 775 cm³ **c** 1190 cm³
7. $\frac{7}{3} \pi r^2 h$
8. **a** 56.5 cm² **b** 126 cm² **c** 113 cm²
9. **a** 333 cm² **b** 163 cm² **c** 749 cm²
10. **a** 5.59 cm² **b** 54.5 cm² **c** 79.9 cm²
11. **a i** 43.4 cm² **ii** 2.3 cm **iii** 5.54 cm
    b i 282 cm² **ii** 6.42 cm **iii** 12.4 cm
    c i 1320 cm² **ii** 16.8 cm **iii** 18.5 cm
12. **a** 694 cm² **b** 1230 cm² **c** 860 cm²

Exercise 5.13

1. **a** 113 cm³ **b** 2140 cm³ **c** 776 cm³
2. **a** 113 cm² **b** 804 cm² **c** 408 cm²
3. **a** 134 cm³ **b** 8580 cm³ **c** 2090 cm³
4. **a** 151 cm² **b** 2410 cm² **c** 942 cm²
5. 4.69 cm
6. 6.31 cm
7. 221 cm³
8. 1.11 cm
9. 4.96 cm
10. **a** 1360 cm³ **b** 7800 cm³
11. **a** 641 cm² **b** 1990 cm²

Exercise 5.14

1. 20
2. 90
3. 27
4. 30
5. 11
6. 75
7. 27
8. 20
9. 6
10. 6
11. 2.5
12. 3
13. 5
14. 2.83
15. 4.34
16. 22.5
17. 3.12
18. 12
19. 28.4
20. 56.32

Exercise 5.15

1. **a** 27, 15 **b** 72, 54
2. **a** 15 **b** 8 **c** 117 **d** 12

Exercise 5.16

1. 72
2. 75
3. 216
4. 135
5. 56
6. 463
7. 6
8. 9
9. 240
10. 20
11. 12.5 m*l*
12. 4050 g
13. 0.633 *l*
14. 2.16 *l*
15. 7.15 cm
16. 23.7 cm
17. 30 cm, 20 cm
18. **a** 240 cm³ **b** 540 cm²
19. 675 g
20. **a** 3 : 4 **b** 27 : 64
21. **a** 96 cm² **b** 61.44 cm³
22. 18.24, 89.92
23. 9.3075, 3.02, 2.5

Exercise 5.17

1. **a** 45 **b** 4.25 **c** 12 **d** 2.55 **e** 6 **f** 5.5 **g** 4.05 **h** 21.5
2. **a** $40 < v \leq 50$

b

speed (v km/h)	f	x	$f \times x$
$30 < v \leq 40$	37	35	37 × 35 = 1295
$40 < v \leq 50$	85	45	85 × 45 = 3825
$50 < v \leq 60$	76	55	76 × 55 = 4180
$60 < v \leq 70$	2	65	2 × 65 = 130
	Total = 200		Total = 9430

mean = 47.15

3. **a** $150 < h \leq 160$ **b** 152.8 cm
4. **a** $3 < t \leq 3.5$ **b** 3.53 °C
5. **a** $10.5 < d \leq 11$ **b** 10.955 m
6. **a** $3 < m \leq 4$ **b** 3.12 kg **c** 8%
7. **a** 0 to 2 **b** 2.935
8. **a** 5.74 hrs **b** 0.59
9. **a** 1.74 miles **b** 2%
10. **a** $60 < t \leq 90$ **b** 93.6 mins **c** $p = 8, q = 72$ **d** 94.2 mins

Examination Questions

1. $231.13
2. £3000
3. $650
4. **a** 8 **b** $\frac{5-x}{3}$ **c** 8
5. **a** 13 **b** −4
6. **a** 27 **b** $9x^2$ **c** $\sqrt[3]{x} + 1$

494 Answers

7 a −14   b $2x^3 - 6x^2 + 12x - 9$
8 a $\dfrac{x+1}{2}$ b $4x^3 + 5$
   c $\dfrac{3x-1}{2}$
9 a $(2x + 3)(2x - 3)$
   b $x(4x - 9)$
   c $(4x - 1)(x - 2)$
10 18 cm, 42 cm
11 a 40 cm^3 b 0.00004
12 a 55 cm by 40 cm
    b $\dfrac{16}{25}$
13 a 320 cm^3 b 567 cm^2
14 a $p = 7.2$ $q = 6.4$ b 2304π
15 a i 154 cm^2 ii 180 cm^3
       iii 1006 g
    b 9.79   c 3.68 cm
16 a i 25.1   ii 302
    b i 4   ii 23.7
       iii 396
    c i 27 W   ii 4 W
17 a $5.5 < t \le 6.0$ b 5.67
18 a $1.5 < x \le 2$ b $1.7275 \, l$

Unit 6

Exercise 6.1
1 3×10^6 2 $80
3 $20   4 $300
5 38.2 minutes   6 $16 500
7 $2.32   8 $3540
9 $8.50   10 $130
11 $44.80   12 21.6°C
13 $5360

Exercise 6.2
1 $60   2 $756
3 $295   4 $481.50
5 $530.45   6 $2153.78
7 $757.70   8 $63.05
9 3   10 15
11 7 years   12 $12 000
13 3365

Exercise 6.3
1 a $\dfrac{1}{3}$ hr   b $\dfrac{1}{2}$ hr   c $\dfrac{1}{6}$ hr
   d $\dfrac{1}{4}$ hr   e $\dfrac{5}{12}$ hr   f $\dfrac{8}{15}$ hr
   g $\dfrac{3}{4}$ hr   h $\dfrac{1}{15}$ hr   i $\dfrac{9}{10}$ hr
   j $\dfrac{2}{5}$ hr   k $\dfrac{2}{3}$ hr   l $\dfrac{4}{5}$ hr
2 a 30 min   b 3 min   c 15 min
   d 21 min   e 48 min   f 42 min
   g 6 min   h 27 min   i 24 min
   j 45 min   k 25 min   l 18 min
3 a 62.5 km/hr   b 38 km/hr
   c 13 km/hr   d 48 km/hr
   e 18 km/hr   f 10 km/hr
   g 48 km/hr   h 24 km/hr
   i 120 km/hr
4 a 20 m/s   b 25 m/s
   c 50 m/s   d 20 m/s
   e 30 m/s   f 6 m/s
5 a 20 m/s   b $3\dfrac{1}{3}$ m/s
   c 100 m/s   d 72 km/h
   e 126 km/h
   f 147.6 km/h
6 20 m/s
7 3×10^8 m/s
8 343 m/s

Exercise 6.4
1 a 2 hours   b 2 hrs 30 mins
   c 30 mins
   d 2 hours 15 mins
   e 2 hours 50 mins
   f 2 hours 39 mins
2 a 56 km   b 216 km
   c 315 km   d 25 km
   e 10 km   f 5 km
3 36 km/h   4 12 600 m
5 31.25 s   6 41.4 km/h
7 a 8400 km   b 1400 km/h
8 a 23.4 km   b $4\dfrac{1}{3}$ m/s
9 a 520 km   b 86.7 km/h
10 30 m/s   11 80

Exercise 6.5
1 a {1, 4, 9, 16, 25, 36, 49}
   b {2, 3, 5, 7, 11, 13, 17, 19, 23}
   c {Jan, Feb, Mar, Apr, May, June, July, Aug, Sept, Oct, Nov, Dec}
   d {1, 2, 3, 4, 6, 12}
   e {5}
   f {2, 4, 6, 8, 12, 14}
   g {21, 23, 25, 27, 29}
   h {151, 157, 163, 167, 173, 179, 181, 191, 193, 197, 199}
2 a {1, 7, 49}   b {125}
   c {2, 5}
   d {1, 2, 4, 5, 10, 20}
   e {2, 3}
   f {Jan, June, July}
3 a {square numbers less than 17}
   b {odd numbers}
   c {months of the year beginning with M}
   d {multiples of 4}
   e {cube numbers}
   f {colours}
4 a {prime numbers}
   b {factors of 20}
   c {powers of 2 smaller than 33}
   d {Pythagorean triples}
5 a FALSE   b TRUE
   c FALSE   d FALSE
   e FALSE   f FALSE
6 a FALSE   b TRUE
   c FALSE   d TRUE
7 a TRUE   b FALSE
   c FALSE   d TRUE
8 a TRUE   b FALSE
   c TRUE   d TRUE
   e FALSE   f TRUE

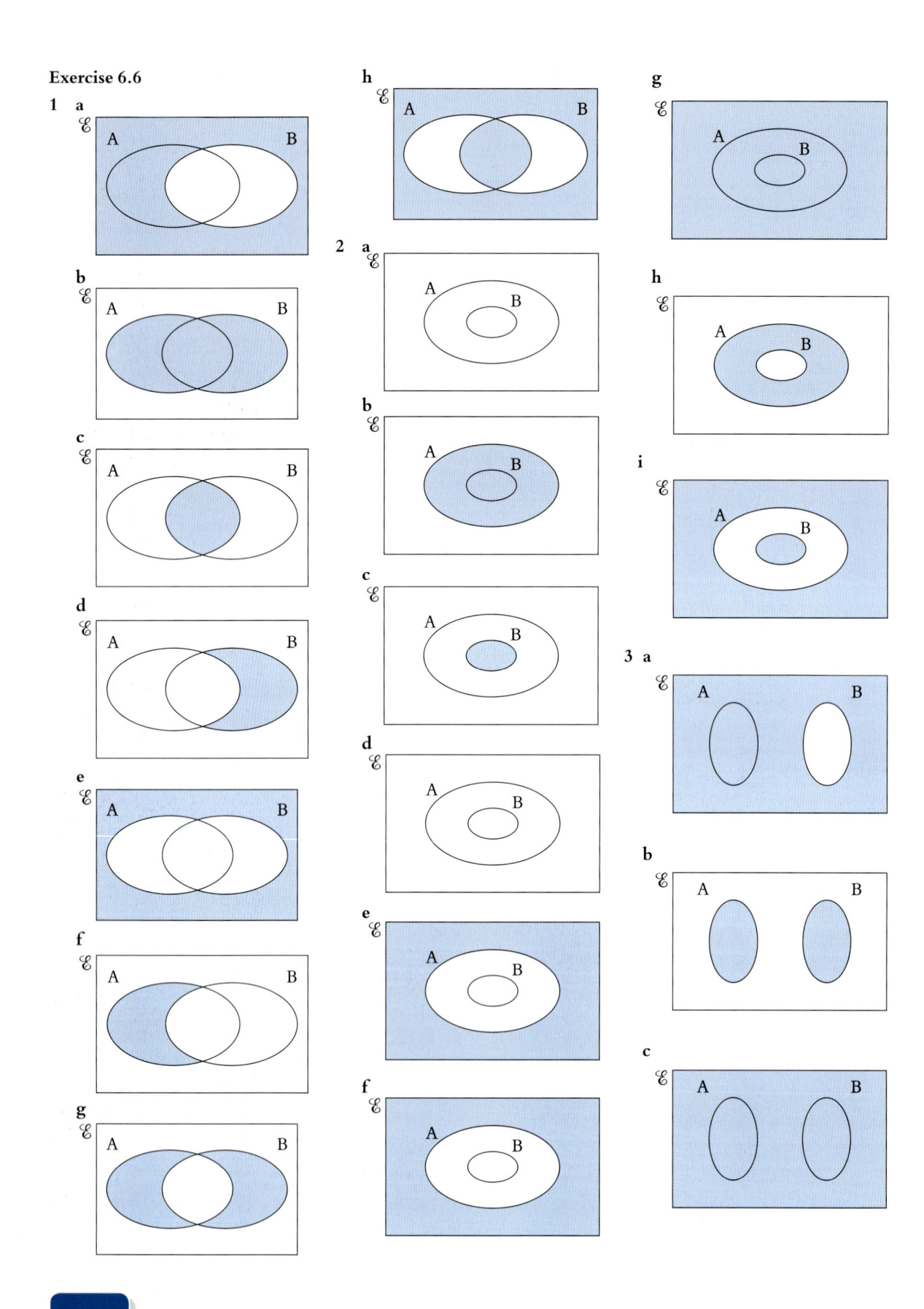

Exercise 6.6

1 a, b, c, d, e, f, g, h

2 a, b, c, d, e, f, g, h, i

3 a, b, c

d, **e**, 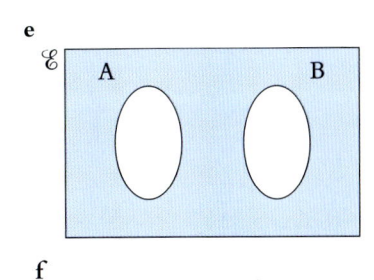 **f** (Venn diagrams)

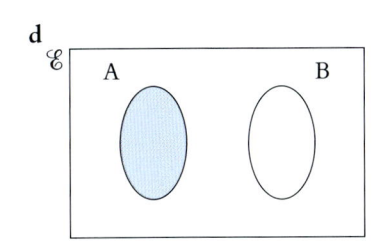

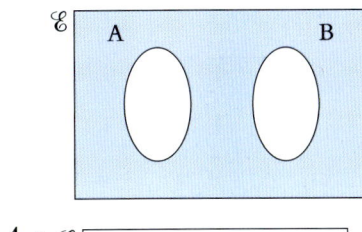

4 a, 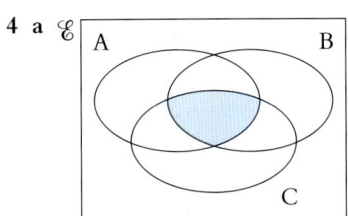 **b**, 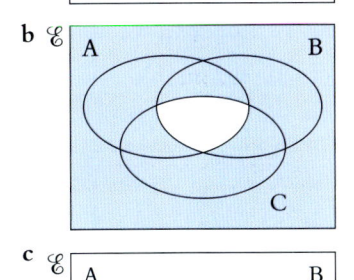 **c**, 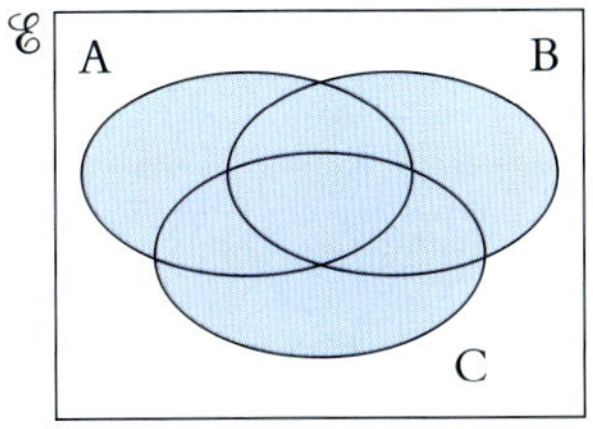 **d**, 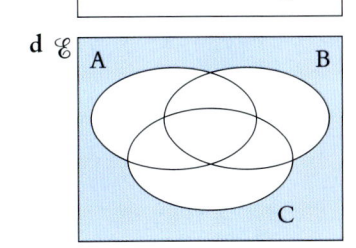 **e**, 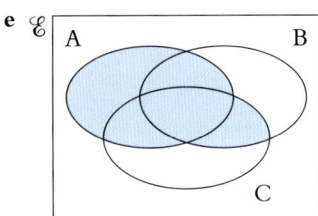 **f**, 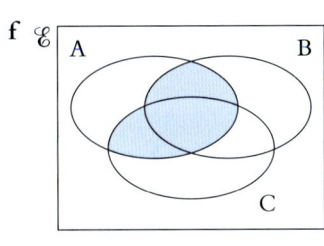 **g**, 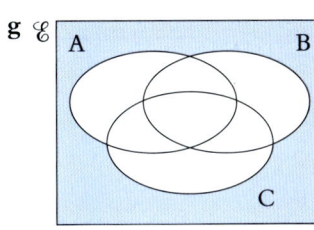 **h**, 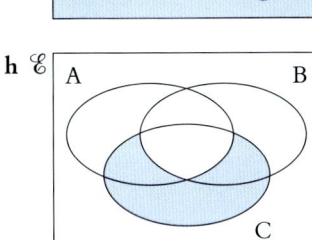 **i** (Venn diagrams)

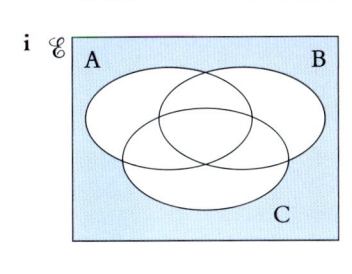

Exercise 6.7

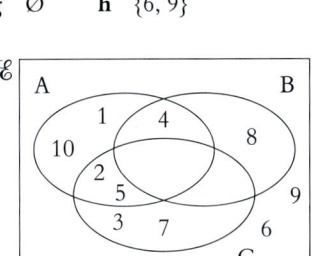

1. **a** B   **b** $(A \cup B)'$
   c $A \cup B$ **d** $A \cap B$
   e $(A \cap B)'$ **f** $A \cup (A \cup B)'$

2. **a** $A \cup B \cup C$ **b** $A \cap C$
   c A   **d** $(A \cup B \cup C)'$
   e $A \cap (B \cup C)'$ **f** $A \cap C \cap B'$
   g $A \cap B'$ **h** $(A \cup C) \cap B'$
   i $(A \cup B \cup C) \cap (A \cap B \cap C)'$

3. **a** $B' \cap C$ **b** $A' \cap B \cap C'$
   c $C \cup (A \cap B)$

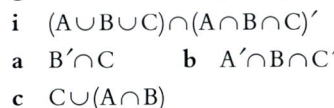

Exercise 6.8

1. **a** {3, 6, 9, 12,...}
   b {1, 2, 4, 5, 10, 20}
   c {1, 3, 5, 7,...}
   d {2, 3, 5, 7, 11, 13, 17, 19}
   e {2, 3}   **f** {2}
   g {−5, 5}   **h** {3}   **i** ∅

2. **a** {2, 3, 6, 7}   **b** {3, 5, 7, 8, 9}
   c {1, 4, 5, 8, 9, 10}
   d {1, 2, 4, 6, 10}
   e {2, 3, 5, 6, 7, 8, 9}
   f {3, 7}   **g** {1, 4, 10}
   h {2, 6}

3. **a** {1, 3, 4, 8, 9, 10, 13, 14}
   b {2, 3, 4, 5, 6, 8, 10, 14}
   c {11}
   d {1, 3, 4, 7, 9, 11, 12, 13, 14}
   e {1, 2, 3, 4, 5, 6, 7, 8, 9, 10, 12, 13, 14}
   f {1, 9, 13}   **g** {8, 9, 10}
   h {7, 9, 12}

4. **a** {1, 2, 4, 5, 10}
   b {4, 8}   **c** {2, 3, 5, 7}
   d ∅   **e** {4}   **f** {2, 5}
   g ∅   **h** {6, 9}
   i (Venn diagram with A, B, C containing: 1, 4, 10, 2, 5, 8, 3, 7, 6, 9)

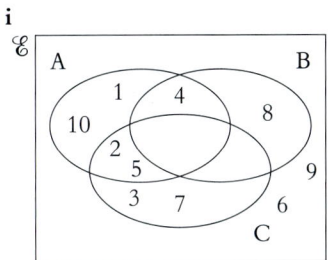

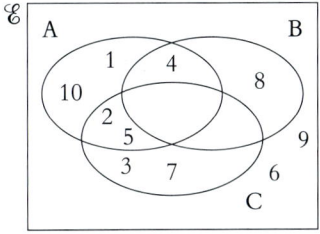

5. **a** {b, d, e, f}   **b** {a, e, f, g}
   c {c}   **d** {a, b, c, d, g}
   e {b, d}   **f** {a, g}
   g {a, b, d, e, f, g}
   h {e, f}
   i (Venn diagram with A, B containing: a, g, c, b, d, f, e)

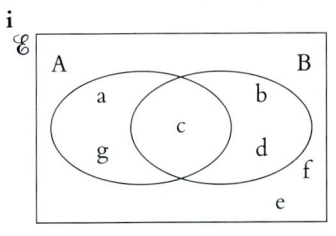

6 [Venn diagram: E inside I inside 𝓔]

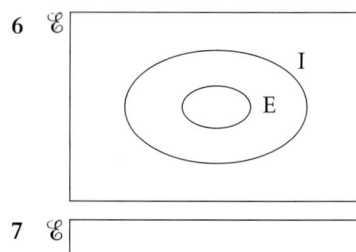

7 [Venn diagram: R inside P inside 𝓔]

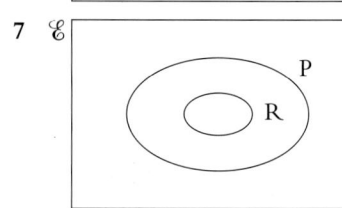

Exercise 6.9

1 a 14   b 19   c 8
   d 8   e 24   f 3
   g 11   h 5
2 28   3 19   4 55
5 7   6 8
7 a 17   b 10   c 8
8 a 10   b 15
9 a 36   b 14   c 10
10 a 17   b 22
11 23
12 25
13 35

Exercise 6.10

1 [Venn diagram with 𝓔, sets G and H overlapping; 8 in G only, 9 in intersection, 7 in H only, 6 outside, 9 outside]

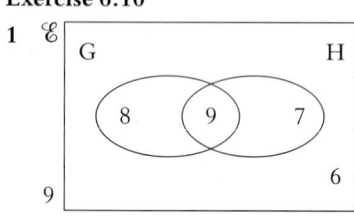

2 16
3 3
4 4
5 4
6 a 2   b 2
7 23

Exercise 6.11

1 a 3   b 8   c 3
   d 9   e 4   f 2
   g 10   h 3   i 12
   j 13   k 20   l 20
2 a $\frac{1}{3}$ b $\frac{1}{2}$ c $\frac{1}{3}$
   d $\frac{1}{5}$ e $\frac{1}{3}$ f $\frac{1}{5}$
   g $\frac{1}{2}$ h $\frac{1}{4}$ i $\frac{1}{10}$
   j $\frac{1}{6}$ k $\frac{1}{11}$ l $\frac{1}{2}$
3 a 125   b 16   c 8
   d 27   e 100 000 f 1
   g 4   h 243
   i 100   j 2197   k 27
   l 16
4 a $\frac{1}{125}$ b $\frac{1}{16}$ c $\frac{1}{64}$
   d $\frac{1}{8}$ e $\frac{1}{1000}$ f $\frac{1}{4}$
   g $\frac{1}{243}$ h $\frac{1}{32}$
   i $\frac{1}{100000}$ j $\frac{1}{16}$
   k $\frac{1}{27}$ l 1
5 a x^2 b x^2 c $x^{\frac{1}{2}}$
   d x^{-2} e x f x^2
   g x^{-1} h x^4 i x^3
   j x k x^{-2} l x^2
   m x^2 n 1   o $x^{\frac{5}{6}}$
   p $x^{-\frac{1}{6}}$
6 a $15x$ b 16   c $6x^4$
   d $3x^{\frac{1}{3}}$ e $6x^{1.5}$ f $3x^{-1}$
   g $2x^{-4.5}$ h $4x^{5.5}$
7 a $5x^{-3}y^{-2}$ b $64x^{18}y^9$
   c $5x^3y^{\frac{1}{4}}$ d $4x^5y$

Exercise 6.12

1 a 5   b 2   c 1
   d 0   e 2   f 3
   g 8   h 3   i 1
   j 3   k 4   l 4
2 a −4   b −2   c −7
   d −2   e −2   f −3
   g −3   h −4   i −1
   j −3   k −4   l −1
3 a 3   b 3   c 4
   d 2   e 1.5 f 2
   g 9   h 1   i 3
   j 1   k 4   l 0.5
4 a −2   d −1   c −1.5
   d −1   e −4   f −2.5
   g −3   h −3
5 a $\frac{1}{5}$ b $\frac{1}{3}$ c $\frac{1}{2}$
   d $\frac{1}{2}$ e $\frac{1}{4}$ f $\frac{1}{3}$
   g $\frac{1}{3}$ h $\frac{1}{2}$ i −7
   j −3   k −1   l −0.5
6 a $\frac{1}{6}$ b 2   c 0
   d 1.5   e −1   f 3
   g −3   h 2

Exercise 6.13

1 a ±4   b ±6   c ±10
   d $±\frac{1}{2}$ e ±5   f ±3
   g ±6   h ±11   i $±\frac{2}{3}$
   j $±\frac{3}{4}$ k $±\frac{4}{5}$ l $±\frac{1}{3}$
   m ±9   n ±1   o ±20
   p $±\frac{3}{2}$

2 No. You cannot find the square root of a negative number.

3 a 0, −6   b 0, 8   c 0, 10
   d 0, −2   e 0, 5   f 0, 12
   g 0, −2.25   h 0, $\frac{1}{3}$
   i 0, −6.5 j 0, 25   k 0, −40
   l 0, −3   m 0, 4   n 0, 15
   o 0, 11   p 0, 1
4 a −2, −3 b 6, −1   c 2, −3
   d 5, 4   e 7, −2   f −3, −9
   g 5, −4   h 8, 2   i 3, −10
   j 8   k 5   l −1, −3
   m 3, −7   n 3   o 8, 1
   p 5, −9
5 a $3, -\frac{1}{2}$ b $1, \frac{2}{3}$
   c 0.6, −1   d 0.5, −7
   e 2.5, 0.5 f $\frac{1}{3}, -2\frac{1}{2}$
   g 0.8, 0.5 h $\frac{1}{6}, -5$
   i $\frac{2}{3}, -2\frac{1}{2}$ j 2.5, 1.5
   k $\frac{2}{7}, -5$ l $1\frac{1}{4}, -\frac{2}{3}$

6 10, 2

7 **a** 3, 2 **b** 6, 2 **c** 3, −6
  d −2, −4 **e** 4 **f** 7, −1
  g 7, 5 **h** 5, −3 **i** 3, −6

8 **a** 6, −1 **b** 4, −5
  c 6, −12 **d** 4, −3

Exercise 6.14

1 **a** 7, −6 **b** 17, −18
2 **a** 9, −11 **b** 12, −10
3 15, −18
4 **a** $(2x+1)^2 + (2x)^2 = 29^2$
    $8x^2 + 4x + 1 = 841$
    $2x^2 + x − 210 = 0$
  b 20, 21, 29
5 6
6 4
7 11 cm, 14 cm
8 10
9 2 s, 3 s
10 (5, 8), (−3, 0)

Exercise 6.15

1

−5	−4	−3	−2	−1
−1	−1.25	−1.67	−2.5	−5

0	1	2	3	4	5
−	5	2.5	1.67	1.25	1

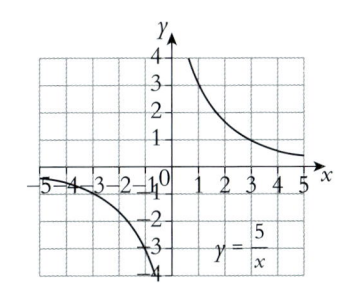

2

−5	−4	−3	−2	−1	0
1	1.25	1.67	2.5	5	−

1	2	3	4	5
−5	−2.5	−1.67	−1.25	−1

3

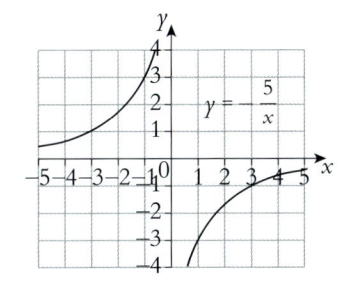

−5	−4	−3	−2	−1	0
1.2	1	0.67	0	−2	−

1	2	3	4	5
6	4	3.33	3	2.8

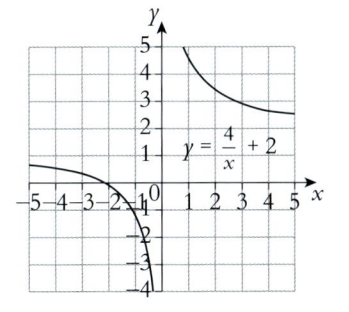

4

−2	−1	0	1	2
−1	−1.25	−1.67	−2.5	−5

3	4	5	6	7	8
−	5	2.5	1.67	1.25	1

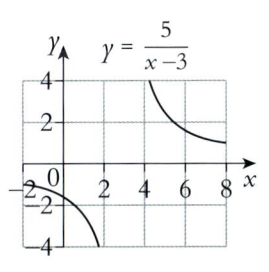

5

−6	−5	−4	−3	−2	−1
−6	−5.2	−4.5	−4	−4	−6

0	1	2	3	4	5	6
−	8	6	6	6.5	7.2	8

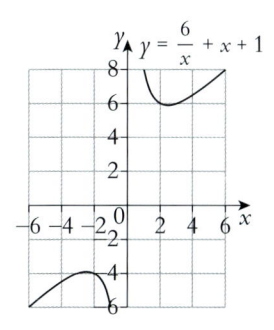

6

−4	−3	−2	−1	0
14.5	7	1	−5	−

1	2	3	4	−4
7	7	11	17.5	14.5

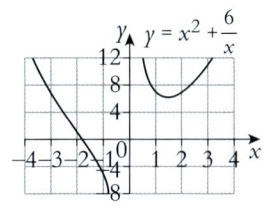

7

−6	−5	−4	−3	−2	−1
0.33	0.48	0.75	1.33	3	12

0	1	2	3	4	5	6
−	12	3	1.33	0.75	0.48	0.33

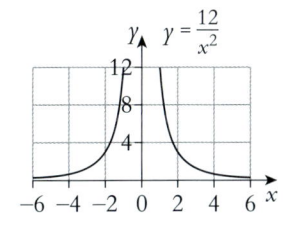

8

−6	−5	−4	−3	−2
−7.89	−6.84	−5.75	−4.56	−3

−1	0	1	2	3	4	5	6
1	−	3	1	1.44	2.25	3.16	4.11

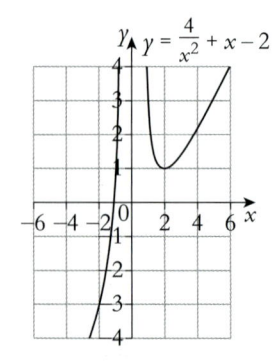

9 a answer given
   b $T = 2x + y$
   c answer given
   d

x	5	10	15	20	30	40	50
T	110	70	63.3	65	76.7	92.5	110

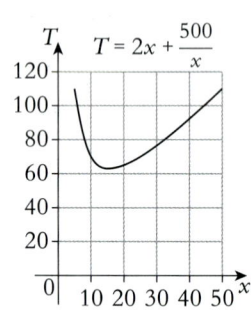

   e graph of $T = 2x + \dfrac{500}{x}$

   f minimum length = 63.2 m when $x = 15.8$

Exercise 6.16
1 **a** 9.17 cm   **b** 18.8 cm
   c 63.6°   **d** 71.6°
2 **a** 6.63 cm   **b** 56.4°
   c 33.2 cm²   **d** 11.5 cm²
3 **a** 53.1°   **b** 106°
   c 48 cm²   **d** 33.4 cm²
   e 14.6 cm²
4 **a** 8 cm   **b** 24 cm
   c 13.9 cm   **d** 333 cm²
   e 131 cm²

Exercise 6.17
1 9.17 cm
2 5.29 cm
3 13 cm
4 **a** 9 cm   **b** 108 cm²

5 **a** 1.71 cm   **b** 4.70 cm
   c 9.40 cm   **d** 8.03 cm²
   e 30.5 cm²   **f** 22.5 cm²
6 16.0 cm
7 **a** 6 cm   **b** 16 cm
   c 128 cm²

Exercise 6.18
1 $x = 240°$, $y = 30°$
2 $x = 100°$, $y = 260°$
3 $x = 110°$, $y = 35°$
4 $x = 286°$, $y = 53°$
5 $x = 76°$, $y = 52°$
6 $x = 100°$, $y = 80°$, $z = 50°$
7 $x = 32°$, $y = 16°$, $z = 16°$
8 $x = 96°$, $y = 84°$, $z = 48°$
9 $x = 43°$, $y = 90°$
10 $x = 30°$, $y = 120°$
11 $x = 50°$, $y = 40°$
12 $x = 90°$, $y = 64°$
13 $x = 54°$, $y = 27°$
14 $x = 74°$
15 $x = 102°$
16 $x = 30°$
17 $x = 20°$, $y = 70°$, $z = 50°$
18 $x = 130°$, $y = 90°$, $z = 65°$
19 $x = 22.5°$

Exercise 6.19
1 $x = 90°$
2 $x = 62°$
3 $x = 30°$
4 $x = 84°$
5 $x = 48°$
6 $x = 52.5°$
7 $x = 70°$
8 $x = 200°$
9 $x = 170°$
10 $x = 45°$
11 $x = 30°$
12 $x = 80°$
13 $x = 16°$, $y = 74°$
14 $x = 115°$
15 $x = 45°$
16 $x = 28°$, $y = 62°$
17 $x = 88°$, $y = 46°$

18 $x = 50°$, $y = 40°$

Exercise 6.20
1 $x = 50°$
2 $x = 56°$, $y = 100°$
3 $x = 80°$, $y = 105°$
4 $x = 123°$, $y = 56°$
5 $x = 80°$, $y = 110°$
6 $x = 85°$, $y = 95°$
7 $x = 29°$
8 $x = 100°$, $y = 55°$
9 $x = 78°$, $y = 102°$
10 $x = 50°$
11 $x = 66°$
12 $x = 54°$, $y = 126°$
13 $x = 30°$, $y = 30°$
14 $x = 55°$, $y = 45°$
15 $x = 27°$, $y = 35°$

Exercise 6.21
1 $x = 28°$, $y = 62°$
2 $x = 22°$
3 $x = 27°$
4 $x = 17°$
5 $x = 57.5°$
6 $x = 105°$
7 $x = 113°$
8 $x = 68°$, $y = 93°$
9 $x = 9°$
10 $x = 26°$, $y = 64°$
11 $x = 25°$
12 $x = 64°$, $y = 32°$
13 $x = 142°$, $y = 38°$
14 $x = 18°$
15 $x = 68°$, $y = 44°$
16 $x = 130°$
17 $x = 100°$
18 $x = 28°$, $y = 62°$
19 $x = 66°$, $y = 71°$
20 $x = 196°$
21 $x = 130°$, $y = 65°$
22 $x = 70°$, $y = 35°$
23 $x = 40°$, $y = 140°$
24 $x = 70°$, $y = 110°$, $z = 35°$
25 $x = 69°$, $y = 21°$, $z = 21°$
26 $x = 40°$, $y = 43°$

27 $x = 60°$, $y = 50°$, $z = 30°$
28 $x = 60°$, $y = 120°$
29 $x = 22°$
30 $x = 58°$, $y = 52°$

For questions **31–36** explanations are also required.

31 $x = 28°$, $y = 62°$
32 $x = 76°$, $y = 38°$, $z = 14°$
33 $x = 90°$, $y = 60°$, $z = 30°$
34 $x = 56°$
35 $x = 48°$, $y = 24°$
36 $x = 130°$, $y = 90°$

Exercise 6.22

1

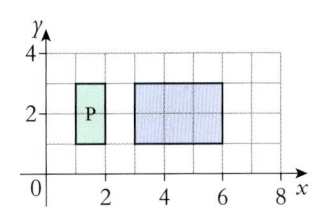

2

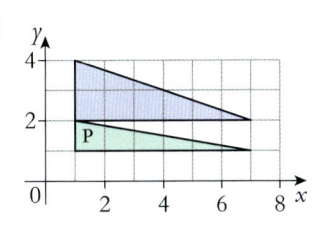

3 a stretch, SF 2, y-axis invariant
  b stretch, SF 2, x-axis invariant
  c stretch, SF 3, y-axis invariant
  d stretch, SF 4, x-axis invariant
  e stretch, SF 2, y-axis invariant
  f stretch, SF 5, y-axis invariant

4

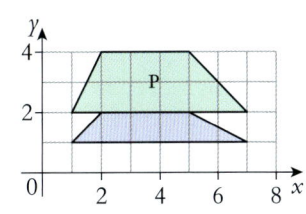

5 a stretch, SF $\frac{1}{3}$, y-axis invariant
  b stretch, SF $\frac{1}{2}$, x-axis invariant
  c stretch, SF $\frac{1}{3}$, x-axis invariant

6

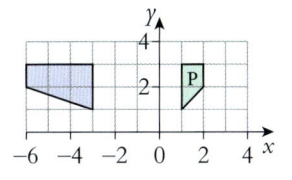

7 a stretch, SF –3, x-axis invariant
  b stretch, SF –3, y-axis invariant
  c stretch, SF –2, $x = 4$ invariant

8

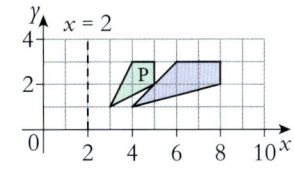

9 a stretch, SF 4, $y = 1$ invariant
  b stretch, SF 2, $y = 6$ invariant
  c stretch, SF 2, $x = 0$ invariant
  d stretch, SF 3, $x = 6$ invariant
  e stretch, SF 4, $x = 2$ invariant
  f stretch, SF $\frac{1}{2}$, $y = 4$ invariant

10 a b

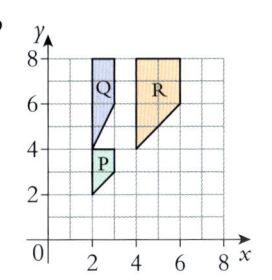

  c enlargement, SF 2, centre (0, 0)

11 a stretch, SF 2, x-axis invariant
  b stretch, SF 2, y-axis invariant
  c enlargement, SF 2, centre (0, 0)

12 a stretch, SF 3, x-axis invariant
  b stretch, SF 3, $x = 15$ invariant
  c enlargement, SF 3, centre (15, 0)

13

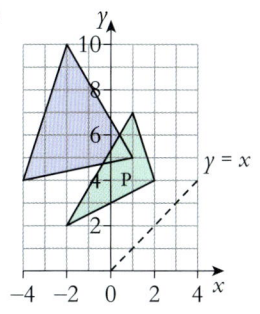

14 a stretch, SF 3, $y = x$ invariant
  b stretch, SF 3, $y = x$ invariant
  c stretch, SF 2, $y = x$ invariant

Exercise 6.23

1

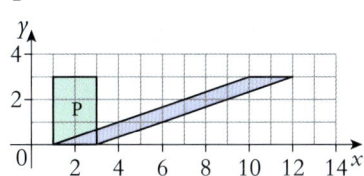

2

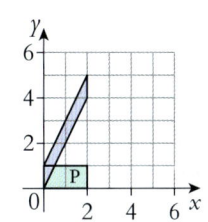

3 a shear, SF 2, y-axis invariant
  b shear, SF 1, x-axis invariant
  c shear, SF $\frac{1}{2}$, x-axis invariant
  d shear, SF $\frac{1}{2}$, x-axis invariant
  e shear, SF $\frac{1}{2}$, y-axis invariant
  f shear, SF 1, y-axis invariant
  g shear, SF 1, x-axis invariant
  h shear, SF $1\frac{1}{2}$, x-axis invariant
  i shear, SF $\frac{1}{2}$, y-axis invariant

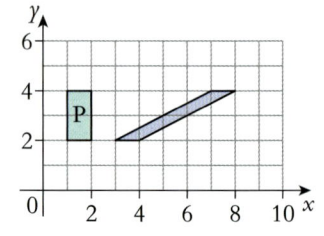

4

5
- **a** shear, SF 1, $y = 2$ invariant
- **b** shear, SF $\frac{1}{2}$, y-axis invariant
- **c** shear, SF 1, $y = 2$ invariant
- **d** shear, SF 1, $y = 2$ invariant
- **e** shear, SF $\frac{3}{4}$, $x = 1$ invariant
- **f** shear, SF 1, x-axis invariant

6

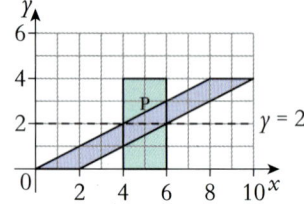

7
- **a** shear, SF 2, $y = 3$ invariant
- **b** shear, SF 1, $x = 3$ invariant
- **c** shear, SF $1\frac{1}{2}$, $x = 3$ invariant

8

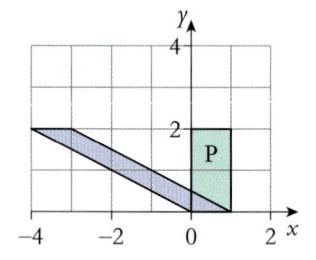

9
- **a** shear, SF -1, x-axis invariant
- **b** shear, SF $-\frac{1}{2}$, $y = 1$ invariant
- **c** shear, SF $-\frac{5}{2}$, $x = 1$ invariant

10 a translation $\begin{pmatrix} 4 \\ 0 \end{pmatrix}$
- **b** shear, SF 2, $y = -1$ invariant
- **c** shear, SF 2, $y = -3$ invariant

Exercise 6.24

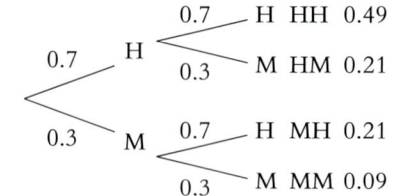

1 a (tree diagram: 0.7 H → 0.7 H HH 0.49; 0.3 M HM 0.21; 0.3 M → 0.7 H MH 0.21; 0.3 M MM 0.09)

b i 0.49 ii 0.09 iii 0.42 iv 0.91

2 a $\frac{9}{25}$ **b** $\frac{4}{25}$ **c** $\frac{6}{25}$

3 a $\frac{2}{5}$ **b** $\frac{2}{15}$ **c** $\frac{7}{15}$

4 a 0.0001 **b** 0.9801 **c** 0.0198 **d** 0.0199

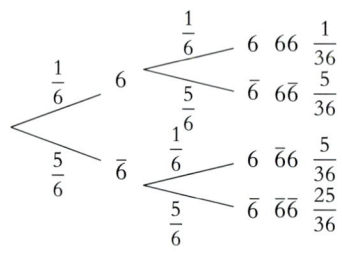

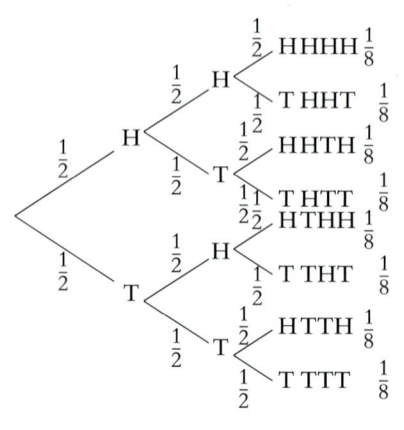

5 a (tree diagram with branches $\frac{1}{6}$, $\frac{5}{6}$ giving 6 66 $\frac{1}{36}$; 6 $\bar{6}$ $\frac{5}{36}$; $\bar{6}$ 6 $\frac{5}{36}$; $\bar{6}$ $\bar{6}$ $\frac{25}{36}$)

b i $\frac{1}{36}$ ii $\frac{25}{36}$ iii $\frac{5}{18}$

6 a $\frac{25}{49}$ **b** $\frac{4}{49}$ **c** $\frac{29}{49}$ **d** $\frac{45}{49}$

7 a (tree diagram HHHH $\frac{1}{8}$, THHT $\frac{1}{8}$, HHTH $\frac{1}{8}$, THTT $\frac{1}{8}$, HTHH $\frac{1}{8}$, THHT $\frac{1}{8}$, HTTH $\frac{1}{8}$, TTTT $\frac{1}{8}$)

b i $\frac{1}{8}$ ii $\frac{1}{8}$ iii $\frac{3}{8}$ iv $\frac{1}{2}$

8 a $\frac{8}{27}$ **b** $\frac{1}{27}$ **c** $\frac{4}{9}$ **d** $\frac{26}{27}$

9 a $\frac{1}{216}$ **b** $\frac{125}{216}$ **c** $\frac{25}{72}$ **d** $\frac{5}{72}$

10 a 0.512 **b** 0.008 **c** 0.096

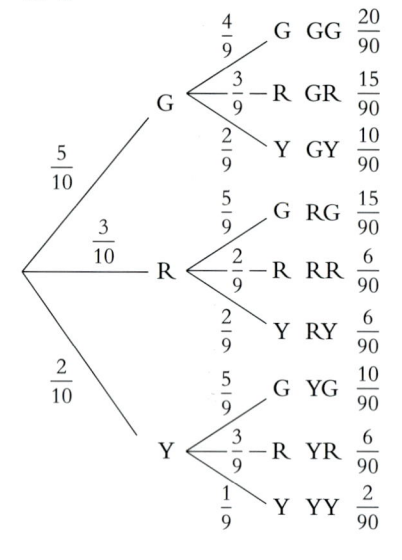

11 a (tree diagram: G $\frac{4}{9}$ GG $\frac{20}{90}$; $\frac{3}{9}$ R GR $\frac{15}{90}$; $\frac{2}{9}$ Y GY $\frac{10}{90}$; R $\frac{5}{9}$ G RG $\frac{15}{90}$; $\frac{2}{9}$ R RR $\frac{6}{90}$; $\frac{2}{9}$ Y RY $\frac{6}{90}$; Y $\frac{5}{9}$ G YG $\frac{10}{90}$; $\frac{3}{9}$ R YR $\frac{6}{90}$; $\frac{1}{9}$ Y YY $\frac{2}{90}$; with $\frac{5}{10}$ G, $\frac{3}{10}$ R, $\frac{2}{10}$ Y)

b i $\frac{2}{9}$ ii $\frac{1}{15}$ iii $\frac{1}{45}$ iv $\frac{14}{45}$ v $\frac{31}{45}$

12 a $\frac{1}{64}$ **b** $\frac{1}{27}$ **c** $\frac{125}{1728}$ **d** $\frac{5}{24}$

Exercise 6.25

1 a $\frac{10}{21}$ **b** $\frac{1}{21}$ **c** $\frac{11}{21}$ **d** $\frac{10}{21}$

2 a $\frac{1}{2}$ **b** 0 **c** 1

3 a $\frac{2}{5}$ **b** $\frac{1}{15}$ **c** $\frac{8}{15}$ **d** $\frac{3}{5}$

4 a $\frac{5}{12}$ b $\frac{1}{12}$ c $\frac{1}{2}$
   d $\frac{11}{12}$

5 a $\frac{1}{6}$ b $\frac{1}{12}$ c $\frac{1}{36}$
   d $\frac{5}{18}$ e $\frac{13}{18}$

6 a $\frac{1}{3}$ b $\frac{1}{4}$ c $\frac{5}{12}$
   d $\frac{3}{4}$

7 a

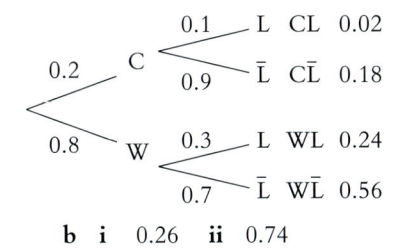

   Tree diagram: C (0.2) → L (0.1) CL 0.02, $\bar{L}$ (0.9) C$\bar{L}$ 0.18; W (0.8) → L (0.3) WL 0.24, $\bar{L}$ (0.7) W$\bar{L}$ 0.56

   b i 0.26   ii 0.74

8 a i

   Tree diagram: A ($\frac{2}{5}$) → triangle ($\frac{3}{5}$), pentagon ($\frac{2}{5}$); B ($\frac{3}{5}$) → triangle ($\frac{5}{8}$), pentagon ($\frac{3}{8}$)

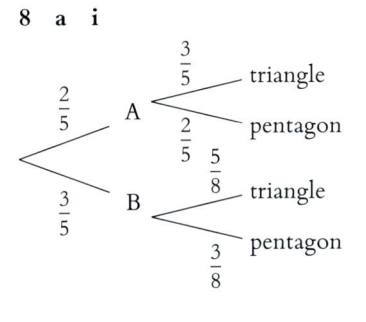

   ii $\frac{123}{200}$ iii $\frac{77}{200}$

   b i $\frac{71}{210}$ ii $\frac{39}{70}$

Exercise 6.26

1 $\frac{343}{4096}$

2 $\frac{1}{64}$

3 $\frac{625}{7776}$

4 $\frac{1}{256}$

5 $\frac{10}{63}$

Examination Questions

1 a 6000   b 12.5
2 11.50
3 a i $346.50
      ii $350
   b i 115
      ii $430
      iii 4.88%
   c 55
4 a 674.92
5 a $108.16
   b 148, 324
   c, e iii

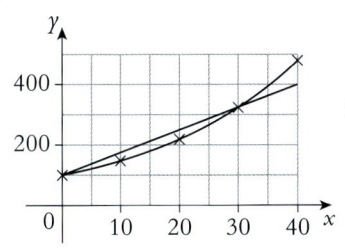

   [graph with y-axis up to 400, x-axis to 40, curve plotted through points]

   d i $267   ii 18
   e i answer given   ii $380
   f 28
6 30 m
7 a 2h 55 m   b 52.8 km/h
8 a [Venn diagram: A∩B shaded]
   b [Venn diagram: outside A∩B shaded]
   c [Venn diagram: A only and B only shaded, not intersection]

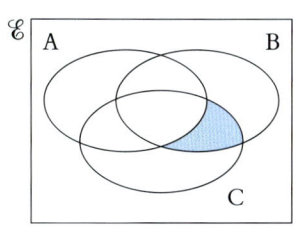

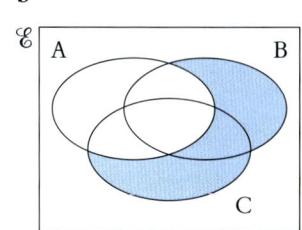

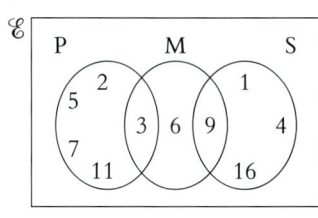

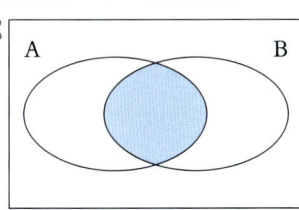

9 a [Venn diagram of A, B, C: region C∩B only, not A, shaded]
   b [Venn diagram of A, B, C: B only and C only regions shaded]

10 a [Venn diagram with sets P, M, S: P{5,2,7,11}, P∩M{3}, M{6}, M∩S{9}, S{1,4,16}]
   b 4

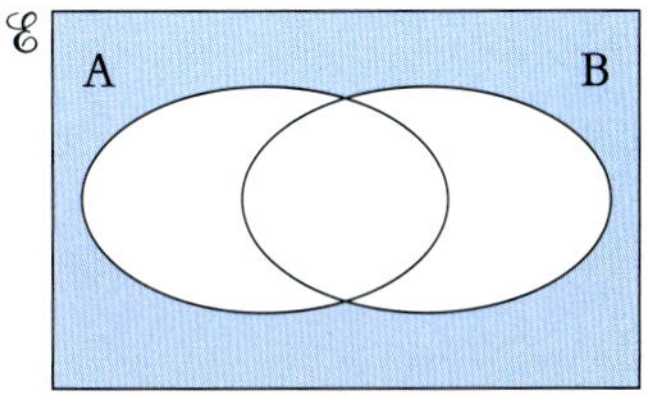

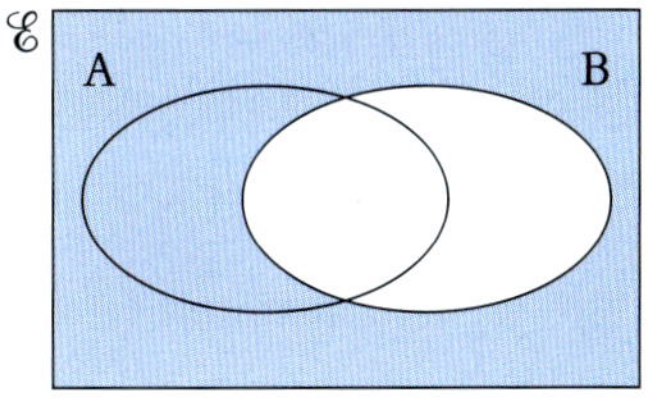

11 a [Venn diagram of D, E, L: 14 in D∩E, 0 in centre, 3 in D∩L, 2 in E∩L, 12 in L]
   b 11
   c 23

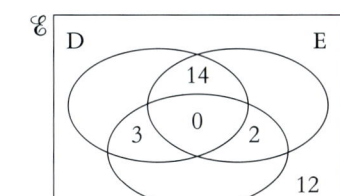

12 a $\frac{p^3}{8}$ b $\frac{9}{8}q^{-1}$

13 $1.25x^4$
14 $9x^2$
15 a 10   b 2.5
16 a 0   b 0.2
   c 0.6
17 a −10   b 4, 1.5

18 a 58   **b** 32   **c** 58
    d 24
19 a 44   **b** 158
20 30, 22, 30, 52
21 a i translation $\begin{pmatrix} 0 \\ -11 \end{pmatrix}$
    ii reflection in $x = 1$
    iii reflection in $y = -x$
    iv enlargement, SF 0.5, centre (2, 0)
    v stretch, SF 2, x-axis invariant
22 a i reflection in $x = 1$
    ii rotation, 180°, centre (1, 0)
    iii enlargement, SF 3, centre (6, 4)
    iv shear, SF -1, y-axis invariant
23 i 0.08   **ii** 0.125   **iii** 7
24 a $p = \frac{1}{20}$, $q = \frac{19}{20}$
    b i $\frac{1}{400}$ **ii** $\frac{38}{400}$
    c $\frac{38}{8000}$
    d $\frac{58}{8000}$
    e 7.25
25 a B   **b i** tree diagram: A ($\frac{2}{3}$) → white ($\frac{1}{4}$), black ($\frac{3}{4}$); B ($\frac{1}{3}$) → white ($\frac{2}{5}$), black ($\frac{3}{5}$)
    ii $\frac{1}{2}$ **iii** $\frac{7}{10}$
    c $\frac{1}{30}$

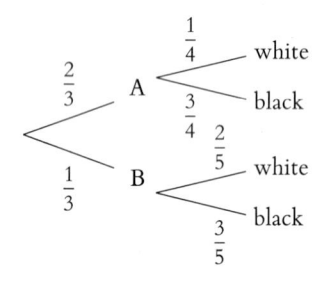

Unit 7

Exercise 7.1

1 a i 0.8 km/min
   ii 48 km/h
  b i 1.7 km/min
   ii 102 km/h
  c i 0.4 km/min
   ii 24 km/h
  d 64 km/h
2 a 5 mins
  b i 0.08 km/min
   ii 4.8 km/h
  c i 0.0267 km/min
   ii 1.6 km/h
  d 2.4 km/h
3 a 1.25 m/s   **b** 1.43 m/s
  c 1.33 m/s
4 a 6 km, 10 km
  b 5 min, 5 min
  c i 0.8 km/min
   ii 48 km/h
  d i 0.5 km/min
   ii 30 km/h
5 a 3   **b** 18 mins
  c between 15 and 20 mins
  d 50 km   **e** 50 km/h
6 a answer given   **b** 9:09 am
  c Distance-time graph showing train P and train Q between stations, crossing around 8:45
  d 8:32 am (approx)

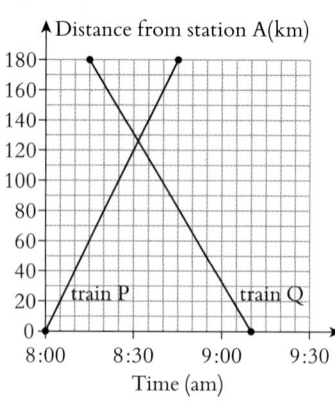

Exercise 7.2

1 a $2\frac{2}{3}$ m/s²   **b** $\frac{2}{7}$ m/s²
  c 75 m
2 a 1 m/s²   **b** $\frac{2}{3}$ m/s²
  c 700 m
3 a 0.5 m/s²   **b** 0.25 m/s²
  c 300 m   **d** 5 m/s
4 a 1.25 m/s²   **b** 250 m
  c 1000 m   **d** 20 m/s
5 a 0.25 m/s²   **b** $\frac{1}{3}$ m/s²
  c 200 m   **d** 950 m
6 a 1.25 m/s²   **b** -0.5 m/s²
  c 950 m   **d** 15.8 m/s
7 a 0.5 m/s²   **b** 2000 m
  c 225 m
8 a 0.8 m/s²   **b** 0.4 m/s²
  c 1120 m   **d** 2560 m
  e 2440 m
9 $9\frac{2}{3}$ km
10 a 32 m/s   **b** 16 m/s
11 a 19.6 m/s   **b** 1.96 m/s²

Exercise 7.3

1 $x = \dfrac{r}{p-q}$ **2** $x = \dfrac{b}{a-c}$
3 $x = \dfrac{c}{a-b}$ **4** $x = \dfrac{a}{1-b}$
5 $x = \dfrac{c-a}{1-b}$ **6** $x = \dfrac{5+b}{a-c}$
7 $x = \dfrac{3-b}{a+2}$ **8** $x = \dfrac{-a^2}{a+b}$
9 $x = \dfrac{3a+4b}{a-b}$ **10** $x = \dfrac{7b+3d}{3c-7a}$
11 $x = \dfrac{3a+2}{1-a}$ **12** $x = \dfrac{b}{c-a}$
13 $x = \dfrac{2e-1}{d+3e}$ **14** $x = \dfrac{d-2b}{2a-c}$
15 $x = \dfrac{3a+4b}{b-a}$ **16** $x = \dfrac{a^2c^2}{1+b}$
17 $x = \dfrac{2}{a^2-1}$ **18** $x = \dfrac{4p}{4-q^2}$
19 $x = \pm\sqrt{\dfrac{a}{1-b}}$ **20** $x = \sqrt{\dfrac{c-b}{a+d}}$

21 $x = \dfrac{fy}{y-f}$ **22** $x = \dfrac{2a}{a-2}$

23 a $x = \dfrac{5}{b-a}$ **b** $x = \dfrac{b}{a-c}$

  c $x = \dfrac{p}{2r-5q}$ **d** $x = \dfrac{6f}{7h-g}$

  e $x = \dfrac{a-b}{3c-2}$ **f** $x = \dfrac{a+d}{b+c}$

Exercise 7.4
1 a 23, 27 **b** 30, 35
  c 32, 64 **d** 14, 15.5
  e 8, 5
  f 78.125, 195.3125
  g $1\dfrac{1}{3}, \dfrac{4}{9}$ **h** 2.5, −0.75
  i 13, 21 **j** 63, 127
  k 17, 27 **l** 100, 1000

2 a 7, 8, 9, 10, 11
  b 2, 7, 12, 17, 22
  c 6, 10, 14, 18, 22
  d 1, 3, 5, 7, 9
  e 8, 11, 14, 17, 20
  f −2, 0, 2, 4, 6
  g 48, 46, 44, 42, 40
  h 15, 14, 13, 12, 11
  i 22, 19, 16, 13, 10
  j 2.5, 3, 3.5, 4, 4.5
  k 29.5, 29, 28.5, 28, 27.5
  l 3, −1, −5, −9, −13

3 nth term = $2n-1$. He is incorrect.

4 a 13 **b** 58 **c** $3n-2$

5 a $4n+3$ **b** $3n-3$
  c $5n+3$ **d** $3n-7$
  e $2.5n-0.5$ **f** $7n-15$
  g $3n-2$ **h** $4n-5$
  i $6n-8$ **j** $1.5n-7$
  k $2.5n+5.5$ **l** $0.02n+6$

6 a $23-2n$ **b** $104-4n$
  c $22-7n$ **d** $20-3n$
  e $7-2n$ **f** $2.1-0.4n$
  g $1-3n$ **h** $15-2n$
  i $2.5-0.5n$

7 20th

8 38th
9 25th
10 63rd

Exercise 7.5
1 a 3, 6, 11, 18, 27
  b −2, 1, 6, 13, 22
  c 3, 12, 27, 48, 75
  d −1, 0, 3, 8, 15
  e 2, 9, 28, 65, 126
  f −3, 4, 23, 60, 121
  g $3, 2, 1\dfrac{2}{3}, 1\dfrac{1}{2}, 1\dfrac{2}{5}$
  h 1, 4, 10, 20, 35

2 a n^2 **b** n^2-1 **c** $2n^2$
  d n^2+2 **e** n^2+10 **f** $3n^2$

3 a n^3 **b** n^3-1 **c** n^3-2
  d $2n^3$ **e** n^3+2

4 a 10^n **b** 2^n **c** 3^n
  d 2^{n+1} **e** 3^{n-1} **f** 10^{n-2}

5 a $n(n+2)$ **b** $n(n+1)(n+2)$
  c $\dfrac{n}{n+5}$ **d** $\dfrac{n+1}{(n+1)^2+1}$

6 a $8x^6$ **b** $18x^{11}$
  c $7x^6$ **d** x^2y^6

7 a n^3 **b** $3n+1$
  c n^3-3n-1

8 55, 91

9 a 10 **b** [triangle of circles image]
  c $10 = \dfrac{4 \times 5}{2}, 15 = \dfrac{5 \times 6}{2}$
  d $\dfrac{n(n+1)}{2}$ **e** 120

10 a [heptagon with diagonals image]
  b 5, 9, 14, 20
  c 27 **d** 170 **e** $\dfrac{n(n-3)}{2}$

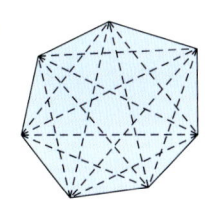

11 35th
12 50th
13 a 1, 6, 15, 20, 15, 6, 1

  b

Row	1	2	3
Sum	$1 = 2^0$	$2 = 2^1$	$4 = 2^2$

4	5	6	7
$8 = 2^3$	$16 = 2^4$	$32 = 2^5$	$64 = 2^6$

  c 2^{n-1}
  d 16

14 a [flower/tree diagram]

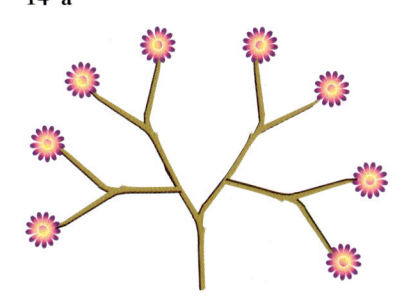

  b

Year	1	2	3	4	5
Flowers	1	2	4	8	16
Stems	1	3	7	15	31

  c 32 **d** 2^{n-1} **e** 63
  f $2^n - 1$

15 12

Exercise 7.6
1 $b=3, c=4$
2 $b=5, c=2$
3 $b=-2, c=4$
4 $b=3, c=1$
5 $b=1, c=2$
6 $a=2, b=5, c=2$
7 $a=2, b=-3, c=5$
8 $a=3, b=1, c=-3$
9 $a=0.5, b=6, c=-2$
10 a 40 **b** 60
  c $2n(n+1)$ or $2n^2+2n$

11 a 25   b 41
   c $(n-1)^2 + n^2$ or $2n^2 - 2n + 1$
12 $\dfrac{2n(n+1)(n+2)}{3}$
13 a 30   b $k = 6$
14 $a = 2, b = -4$

Exercise 7.7
1 a −1, 1, 3, 5
  b 6, 11, 16, 21
  c 5, 8, 11, 14
  d 1, 4, 9, 16
  e 3, 6, 11, 18
  f 0, 1, 4, 9
  g 1, 8, 27, 64
  h 0, 4, 18, 48
  i 8, 27, 64, 125
  j 2, 4, 8, 16
  k $\dfrac{1}{3}$, 0, 3, 9
  l 10 000, 100 000, 1 000 000, 10 000 000

2 a $U_n = 5n - 1$
  b $U_n = 3n + 5$
  c $U_n = 23 - 3n$
  d $U_n = n^2 - 1$
  e $U_n = 3^{n-1}$
  f $U_n = \dfrac{n+3}{2n+5}$

Exercise 7.8
1

x	−3	−2	−1	0	1	2	3
y	0.04	0.11	0.33	1	3	9	27

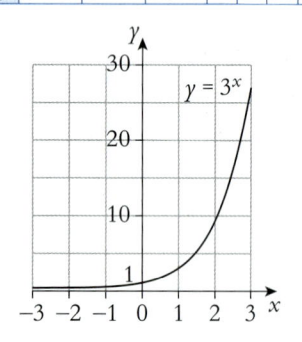

$y = 3^x$

2

x	−3	−2	−1	0	1	2	3
y	27	9	3	1	0.33	0.11	0.04

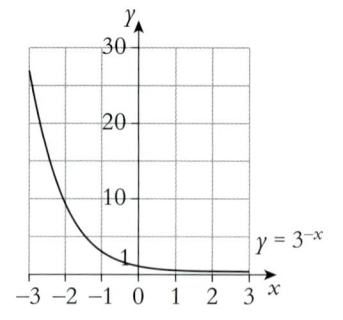

$y = 3^{-x}$

3

x	−3	−2	−1	0	1	2	3
y	1.04	1.11	1.33	2	4	10	28

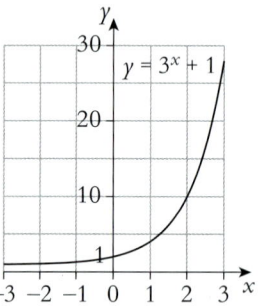

$y = 3^x + 1$

4

x	−3	−2	−1	0	1	2	3
y	−4.96	−4.89	−4.67	−4	−2	4	22

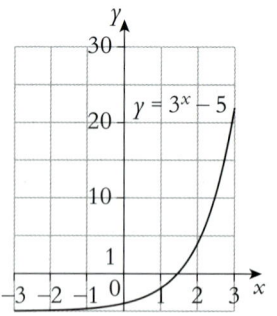

$y = 3^x - 5$

5

x	−1	0	1	2	3	4	5
y	0.04	0.11	0.33	1	3	9	27

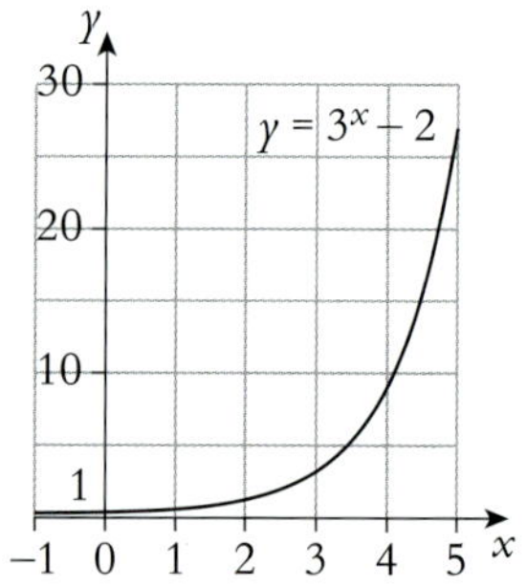

$y = 3^x - 2$

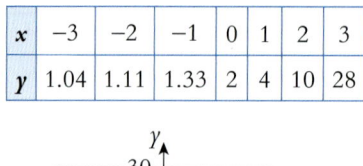

6

x	−3	−2	−1	0	1	2	3
y	8	4	2	1	0.5	0.25	0.13

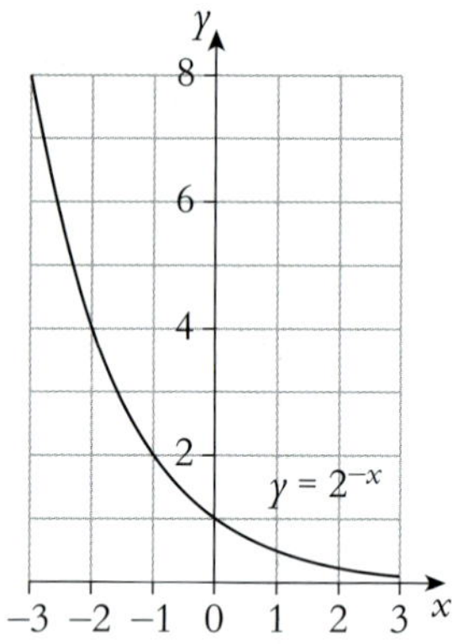

$y = 2^{-x}$

7

x	−3	−2	−1	0	1	2	3
y	1.13	1.25	1.5	2	3	5	9

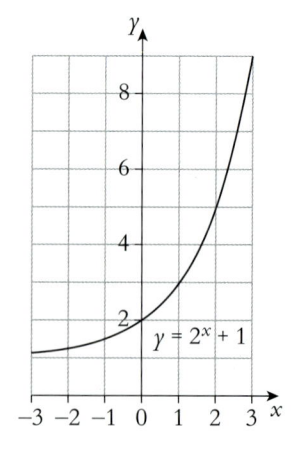

$y = 2^x + 1$

8

x	−4	−3	−2	−1	0	1	2
y	0.13	0.25	0.5	1	2	4	8

506

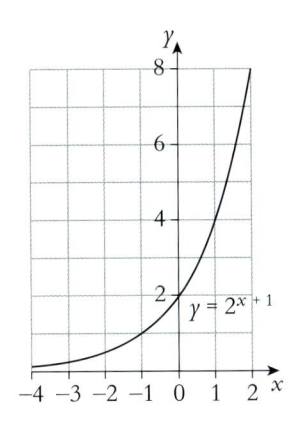

9

x	-3	-2	-1	0	1	2	3
y	0.87	0.75	0.5	0	-1	-3	-7

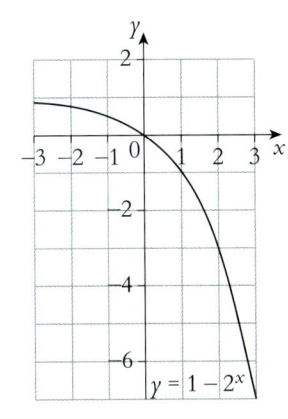

10

x	-3	-2	-1	0	1	2	3
y	8	4	2	1	0.5	0.25	0.13

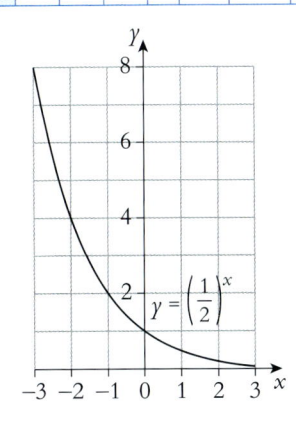

11 a i 2   ii -1.5
   b i 2   ii 3.6   c 2.8

Exercise 7.9

1 a 214, 314, 459, 673
   b (graph: $y = 100 \times 1.1^x$)
   c 17

2 a 26 200, 21 300, 17 200, 13 900
   b (graph: $y = 40\,000 \times 0.9^x$)
   c 6.6 years

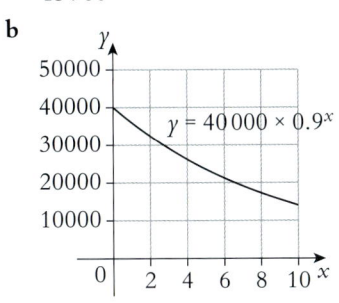

3 a 300, 360, 432, 518, 622, 746
   b (graph: $P = 250 \times (1.2)^t$)
   c 3.8 years

4 a 75, 56.3, 42.2, 31.6, 23.7, 17.8, 13.3, 10.0
   b (graph: $P = 100 \times (0.75)^t$)
   c 5.6 years

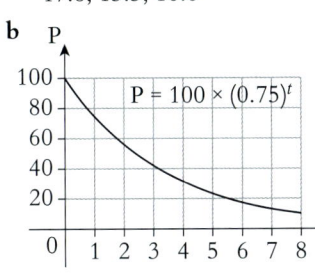

Exercise 7.10

1 a reflection in $y = x$
   b reflection in x-axis
   c reflection in $y = -x$
2 a rotation, 180° about (0, 0)
   b rotation, 90° clockwise about (0, 0)
   c rotation, 36.9° clockwise about (0, 0)
3 enlargement, SF 2, centre (0, 0)
4 enlargement, SF 3, centre (0, 0)
5 enlargement, SF $\frac{1}{2}$, centre (0, 0)
6 enlargement, SF $-\frac{1}{2}$, centre (0, 0)
7 enlargement, SF -2, centre (0, 0)
8 enlargement, SF k, centre (0, 0)
9 a shear, SF 1, x-axis invariant
   b shear, SF 2, x-axis invariant
   c shear, SF 3, x-axis invariant
10 shear, SF k, x-axis invariant
11 a shear, SF 1, y-axis invariant
    b shear, SF 2, y-axis invariant
    c shear, SF 3, y-axis invariant
12 shear, SF k, y-axis invariant
13 a stretch, SF 2, y-axis invariant
    b stretch, SF 3, y-axis invariant
    c stretch, SF 4, y-axis invariant
14 stretch, SF k, y-axis invariant
15 a stretch, SF 2, x-axis invariant
    b stretch, SF 3, x-axis invariant
    c stretch, SF 4, x-axis invariant
16 stretch, SF k, x-axis invariant.

Exercise 7.11

1 $\begin{pmatrix} 0 & -1 \\ -1 & 0 \end{pmatrix}$ 2 $\begin{pmatrix} 1 & 0 \\ 0 & 2 \end{pmatrix}$

3 $\begin{pmatrix} 1 & 0 \\ 0 & -1 \end{pmatrix}$ 4 $\begin{pmatrix} 2 & 0 \\ 0 & 2 \end{pmatrix}$

5 $\begin{pmatrix} 0 & -1 \\ -1 & 0 \end{pmatrix}$ 6 $\begin{pmatrix} \frac{1}{3} & 0 \\ 0 & \frac{1}{3} \end{pmatrix}$

7 $\begin{pmatrix} 0 & 1 \\ 1 & 0 \end{pmatrix}$ 8 $\begin{pmatrix} 1 & 0 \\ 4 & 1 \end{pmatrix}$

9 $\begin{pmatrix} 3 & 0 \\ 0 & 1 \end{pmatrix}$ 10 $\begin{pmatrix} 0 & -1 \\ -1 & 0 \end{pmatrix}$

11 enlargement, SF 3, centre (0, 0)
12 stretch, SF 2, y-axis invarient
13 reflection in the x-axis
14 enlargement, SF $\frac{1}{2}$, centre (0, 0)
15 reflection in $y = x$
16 shear, SF 5, x-axis invariant
17 enlargement, SF 4, centre (0, 0)
18 rotation, 90° anticlockwise about (0,0)
19 stretch, SF 4, x-axis invariant
20 enlargement, SF −2, centre (0,0)
21 shear, SF 4, y-axis invariant
22 stretch, SF −3, x-axis invariant
23 a $\begin{pmatrix} 1 & 0 \\ 0 & -1 \end{pmatrix}$ b $\begin{pmatrix} 0 & -1 \\ -1 & 0 \end{pmatrix}$

c $\begin{pmatrix} 0 & 1 \\ -1 & 0 \end{pmatrix}$ d $\begin{pmatrix} 0 & -1 \\ 1 & 0 \end{pmatrix}$

24 a $\begin{pmatrix} 1 & 0 \\ 0 & -1 \end{pmatrix}$ b $\begin{pmatrix} 0 & -1 \\ 1 & 0 \end{pmatrix}$

c $\begin{pmatrix} 0 & 1 \\ -1 & 0 \end{pmatrix}$ d $\begin{pmatrix} 1 & 2 \\ 0 & 1 \end{pmatrix}$

25 a $\begin{pmatrix} 0 & 1 \\ -1 & 0 \end{pmatrix}$ b $\begin{pmatrix} 0 & -1 \\ -1 & 0 \end{pmatrix}$

c $\begin{pmatrix} -1 & 0 \\ 0 & 1 \end{pmatrix}$ d $\begin{pmatrix} 2 & 0 \\ 0 & 1 \end{pmatrix}$

e $\begin{pmatrix} \frac{1}{2} & 0 \\ 0 & 1 \end{pmatrix}$

26 a $\begin{pmatrix} -2 & 0 \\ 0 & 1 \end{pmatrix}$ b $\begin{pmatrix} 2 & 0 \\ 0 & 2 \end{pmatrix}$

c $\begin{pmatrix} 1 & 0 \\ 0 & 3 \end{pmatrix}$ d $\begin{pmatrix} 1 & 0 \\ 0 & \frac{1}{3} \end{pmatrix}$

e $\begin{pmatrix} \frac{1}{2} & 0 \\ 0 & \frac{1}{2} \end{pmatrix}$

Exercise 7.12

1 a b

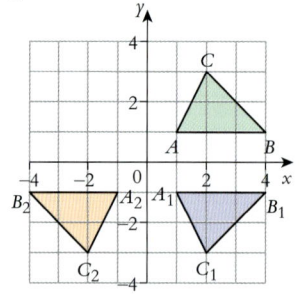

c i rotation, 180° about (0, 0)

ii $\begin{pmatrix} -1 & 0 \\ 0 & -1 \end{pmatrix}$

2 a b

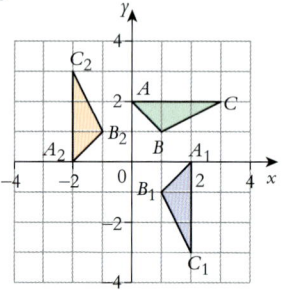

c i rotation, 90° anticlockwise about (0, 0)

ii $\begin{pmatrix} 0 & -1 \\ 1 & 0 \end{pmatrix}$

3 a

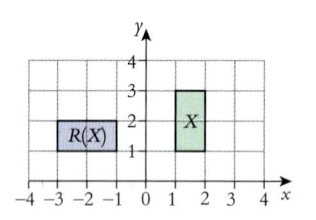

b i rotation, 90° anticlockwise about (0, 0)
  ii rotation, 90° clockwise about (0, 0)

c $\begin{pmatrix} 0 & 1 \\ -1 & 0 \end{pmatrix}$

d The inverse matrix R^{-1} transforms R(X) back to X.

4 a b

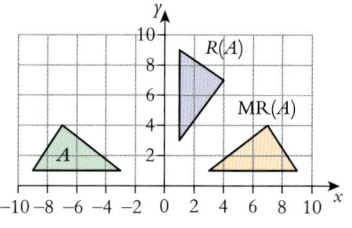

c reflection in the y-axis

Exercise 7.13

1 a

time	c.f.
$t \leq 2$	15
$t \leq 4$	27
$t \leq 6$	37
$t \leq 8$	45
$t \leq 10$	50

b

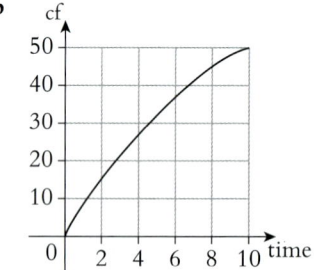

c 42   d 29

2 a

height	cf
$h \leq 160$	7
$h \leq 165$	31
$h \leq 170$	66
$h \leq 175$	88
$h \leq 180$	100

b

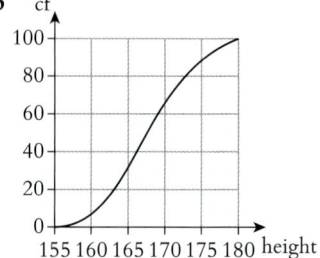

c 17   d 21

3 a

mark	cf
≤ 20	5
≤ 40	15
≤ 60	29
≤ 80	46
≤ 100	50

b cf graph vs mark

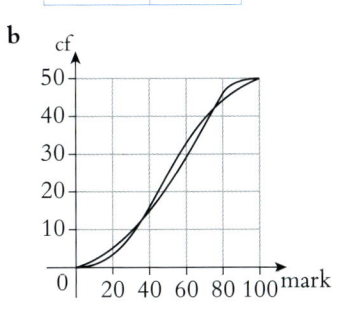

c 36 **d** 29

4 a

mass	c.f.
≤ 2.5	6
≤ 3	16
≤ 3.5	33
≤ 4	46
≤ 4.5	54
≤ 5	60

b cf graph vs mass

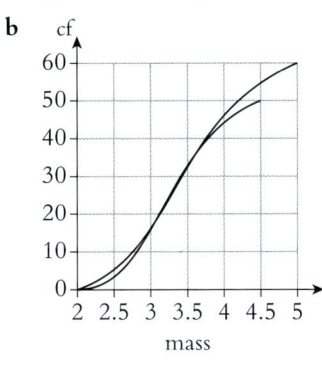

c 23 **d** 22

5 a cf graph vs temp

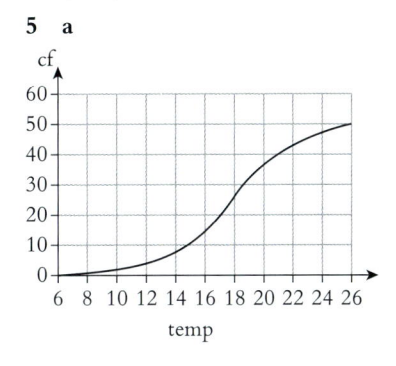

b 20 **c** 26%

6

speed	frequency
0 < v ≤ 30	6
30 < v ≤ 40	18
40 < v ≤ 50	26
50 < v ≤ 60	10

7

mark	frequency
0 < m ≤ 20	3
20 < m ≤ 40	18
40 < m ≤ 60	37
60 < m ≤ 80	21
80 < m ≤ 100	1

8 a 23 **b** 4

c

time	cf
t ≤ 10	4
t ≤ 20	16
t ≤ 30	32
t ≤ 40	38
t ≤ 50	40

time	frequency
0 < t ≤ 10	4
10 < t ≤ 20	12
20 < t ≤ 30	16
30 < t ≤ 40	6
40 < t ≤ 50	2

d 22.5 minutes

9 a 5 **b** 13%

c

time	cf
≤ 700	2
≤ 800	12
≤ 900	30
≤ 1000	38
≤ 1100	40

time	freq
600 < t ≤ 700	2
700 < t ≤ 800	10
800 < t ≤ 900	18
900 < t ≤ 1000	8
1000 < t ≤ 1100	2

d 845 hours

Exercise 7.14

1 a i 31 **ii** 22 **iii** 40
iv 18 **v** 28
b i 57 **ii** 44 **iii** 68
iv 24 **v** 52
c i 168 **ii** 165.5
iii 171 **iv** 5.5
v 167
d i 47 **ii** 44 **iii** 51
iv 7 **v** 45

2 a cf graph vs mass

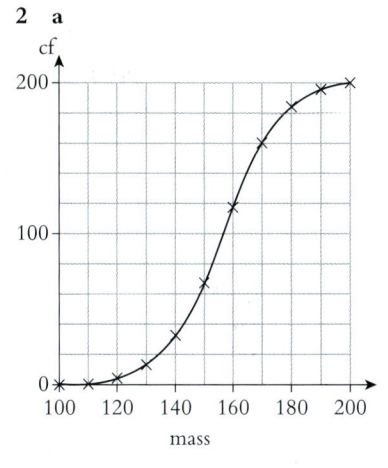

b i 156 **ii** 145
iii 168 **iv** 23
v 164

3 a cf graph vs mass

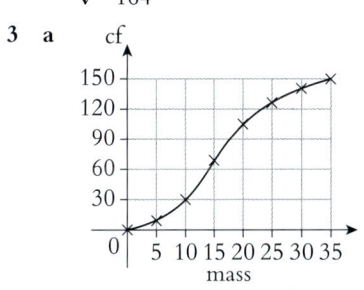

b i 16 **ii** 11 **iii** 22
iv 11 **v** 25

4 a cf graph vs mass

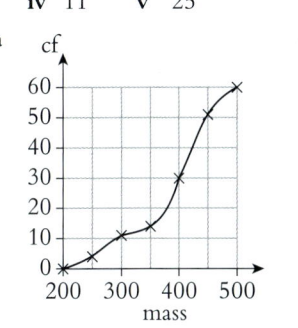

b i 400 **ii** 353
  iii 436 **iv** 83
  v 377

5 In general, the 10 to 20 year olds react quicker.
  The 10 to 20 year olds are less varied.

6 a

	Median	LQ	UQ	IQR
Paper 1	46	30	62	32
Paper 2	58	47	67	20

  b In general, the marks on paper 2 are higher.
  The marks on paper 2 are less varied.

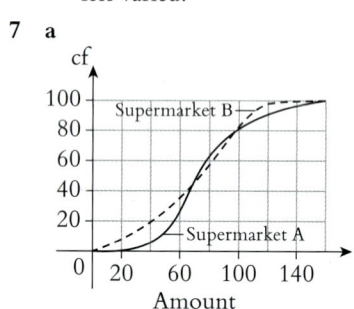

7 a

(cumulative frequency graph showing Supermarket A and Supermarket B against Amount)

  b

	Median	LQ	UQ	IQR
Super-maket A	75	58	92	34
Super-maket B	71	44	93	49

  c In general, shoppers spent more money in Supermarket A.
  The amounts spent in Supermarket A are less varied.

Examination Questions

1 a 1.05 **b** 3360 **c** 18.7
2 a ii 5 m/s² **b** 780 m
3 a $x^2(a+b)$ **b** $\pm\sqrt{\dfrac{p^2+d^2}{a+b}}$
4 a $1.5n+4$ **b** 154
5 a 21 **b** $360x^2$
  c 486

6 a 2870 **b** $(n+3)^2+1$
7 a $-7, 512, \dfrac{8}{9}, 81, 2187, -2106$
  b i $9-2n$ **ii** n^3
   iii $\dfrac{n}{n+1}$ **iv** $(n+1)^2$
   v 3^{n-1} **vi** $(n+1)^2-3^{n-1}$
  c 393 **d** 12

8 a i

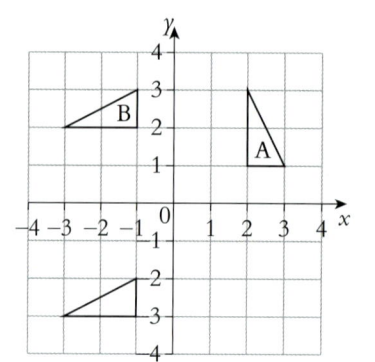

(diagram showing triangles A and B with reflections)

  ii reflection in $y=-x$
  b $\begin{pmatrix} 0 & -1 \\ 1 & 0 \end{pmatrix}$

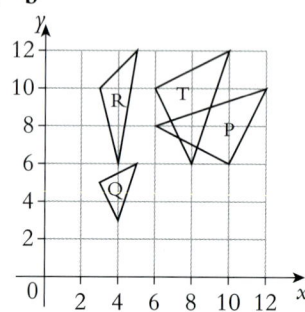

9 a b

(diagram showing triangles P, Q, R, T)

  c ii $\begin{pmatrix} 0 & 1 \\ 1 & 0 \end{pmatrix}$
  d ii enlargement, SF $\dfrac{1}{2}$, centre (0, 0)

10 a

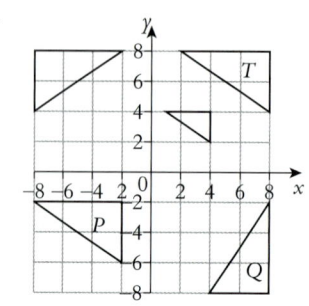

(diagram showing triangles P, Q, T)

  b i $\begin{pmatrix} -8 & -8 & -2 \\ 4 & 8 & 8 \end{pmatrix}$
   ii reflection in y-axis
  c i translation $\begin{pmatrix} -10 \\ -10 \end{pmatrix}$
   ii rotation, 90° clockwise about (0, 0)
  d $\begin{pmatrix} 0 & 1 \\ -1 & 0 \end{pmatrix}$

11 a 38 **b** 45 **c** 15
  d 10

Unit 8

Exercise 8.1

1 a $0.\dot{3}$ **b** $0.\dot{5}$ **c** $0.\dot{1}\dot{8}$
  d $0.3\dot{1}\dot{8}$ **e** $0.6\dot{1}$ **f** $0.6\dot{5}$
  g $0.\dot{1}\dot{5}$ **h** $0.\dot{1}7\dot{1}$
  i $0.\dot{2}8571\dot{4}$ **j** $0.\dot{3}8461\dot{5}$

2 a $\dfrac{1}{3}$ **b** $\dfrac{7}{9}$ **c** $\dfrac{4}{11}$
  d $\dfrac{68}{99}$ **e** $\dfrac{14}{99}$ **f** $\dfrac{145}{999}$
  g $\dfrac{238}{999}$ **h** $\dfrac{47}{90}$ **i** $\dfrac{11}{18}$
  j $\dfrac{3}{22}$

3 a $\sqrt{36}$ **b** $\sqrt{40}$ **c** $\sqrt{64}$
  d $\sqrt{49}$ **e** $\sqrt{\dfrac{4}{9}}$

4 $\pi, \sqrt{\dfrac{5}{8}}$

5 $\sqrt{2}, \sqrt{3}, \sqrt{5}, \sqrt{6}$

6 a irrational **b** rational
7 a 25π cm², irrational
  b 100 cm², rational
  c $100-25\pi$ cm², irrational

Exercise 8.2

1 The value of the square root was rounded too soon. Using $\sqrt{141}=11.8743...$ gives the correct answers.
2 a 4, −6 **b** 0.56, −3.56
  c 3.19, −2.19

d 2.73, −0.73
e 5.65, 0.35
f 2.53, −5.53
g 1.32, −5.32   h 2.39, 0.28
i 3.39, −0.89
j 1.85, −0.18
k 15.62, −25.62
l 0.54, −2.29
m 2.35, −0.85
n 1.45, −3.45
o 0.68, −3.68
p −1.26, −8.74

3 Impossible to solve because of the square root of a negative number

4 $x = 2.5$ There is only one (repeated) root of the equation.

5 a 2.61, −4.61
   b 1.61, −5.61
   c 4.45, −0.45
   d −0.46, −6.54
   e −0.46, −6.54
   f $3, -\frac{1}{3}$
   g 0.72, −0.61
   h 7.52, −2.52

6 a 1.61, −5.61
   b 4.46, −2.46

7 a 2.87, −4.87
   b 3.48, −0.48
   c 2.14, −5.14

Exercise 8.3
1 a 2   b 1   c 0   d 2
   e 1   f 2   g 0   h 1
   i 2   j 2
2 ±10
3 $k < 2$

Exercise 8.4
1 2.65   2 2.61   3 5.23
4 2.72   5 2.86, 0.14
6 a $A = w(w + 20)$
   b 74.26 m, 337 m

7 a answer given
   b 5
8 2.30
9 (0.414, −0.586), (−2.414, −3.414)
10 a answer given
   b 10   c 12, 96

Exercise 8.5
2 a $\frac{5(x-3)}{x(x-5)}$ b $\frac{11x+27}{(x+3)(x+2)}$
   c $\frac{5x+62}{(x-2)(x+7)}$ d $\frac{11x+4}{x(3x+1)}$
   e $\frac{17x-24}{(2x+5)(3x-2)}$
   f $\frac{-4x-19}{(1-2x)(x-4)}$
   g $\frac{2x+5}{(x+2)(x+3)}$
   h $\frac{9x+17}{(x+1)(x+3)}$
   i $\frac{2+3x}{x(x+1)}$ j $\frac{x+3}{(x+2)(x+4)}$
   k $\frac{-3x-1}{(x-3)(x+1)}$ l $\frac{x^2+y^2}{x+y}$

2 a −2.25   b −0.5   c $2, -\frac{1}{3}$

3 a $3, -1\frac{1}{4}$ b $2, -\frac{5}{6}$
   c 11, −2   d 5, 4
   e $-1, -2\frac{1}{16}$ f $4, 2\frac{5}{7}$
   g 2   h $-1\frac{1}{2}, -5\frac{1}{2}$
   i 6, 2

Exercise 8.6
1 a $y = 5x$ b 40   c 1.4
2 a $y = 8x$ b 24   c 0.75
3 a $y = 0.5x$ b 1.2   c 10
4 a $y = \frac{1}{3}x$ b $1\frac{1}{3}$ c 10.2
5 a $y = \frac{1}{10}x$ b 0.8   c 70
6
x	2	5	13
y	6	15	39

7 a $v = 9.8t$ b 29.4 m/s
   c 2.5 s
8 a $e = 4.4m$ b 3.52 cm
   c 0.4 kg
9 a $d = 1.5V$ b 1.8 m
   c 0.38 m³

Exercise 8.7
1 a $y = 2x^2$ b 50   c ±8
2 a $y = 0.5x^2$ b 40.5   c ±8
3 a $y = 4x^3$ b 32   c 4
4 a $y = 2(x+3)^2$ b 98
   c 2, −8
5 a $y = 5(x-2)^2$ b 180
6 a $y = 6\sqrt{x}$ b 15   c 64
7
x	1	2	5
y	2	16	250

8 a $d = 0.015v^2$ b 24
9 a $E = 12e^2$ b 75
10 a $R = 2v^2$ b 2450

Exercise 8.8
1 a $y = \frac{6}{x}$ b 1   c 4
2 a $y = \frac{24}{x}$ b 6   c 3
3 a $y = \frac{10}{x}$ b 20   c 4
4 a $y = \frac{1}{2x}$ b 0.25   c 0.5
5
x	2	4	6
y	18	9	6

6
x	2.5	4	5
y	40	25	20

7 a $w = \frac{300000}{f}$ b 2000
   c 1500
8 a $P = \frac{300}{V}$ b 2   c 120
9 a $t = \frac{24}{n}$ b 8   c 6

Exercise 8.9
1 a B   b C   c E
   d A   e D
2 a $y = \frac{36}{x^2}$ b 9   c ±6
3 a $y = \frac{20}{\sqrt{x}}$ b 1.25   c 25
4 a $y = \frac{15}{x+2}$ b 1.5   c 2
5 a $y = \frac{16}{(x+1)^2}$ b 1   c 1, −2

6

x	±1	2	5
y	50	12.5	2

7 a $F = \dfrac{20}{d^2}$ **b** 80 **c** 10

8 a $I = \dfrac{126000}{d^2}$ **b** 12.6
   c 251

9 a D **b** A **c** B **d** C

Exercise 8.10
1 78.5°   **2** 45.6°
3 61.6°, 118.4°   **4** 110.5°
5 11.5°, 168.5°   **6** 90°
7 180°   **8** 0°, 180°
9 no solution
10 no solution in range   **11** 139.5°
12 54.1°, 125.9°   **13** 73.4°
14 36.9°, 143.1°   **15** 138.6°
16 53.1°, 126.9°
17 no solution in range
18 no solution   **19** 60°
20 19.9°, 160.1°   **21** 30°, 150°
22 60°, 120°   **23** 41.8°, 138.2°
24 18.4°, 161.6°

Exercise 8.11
1 a 15.3 cm² **b** 44.5 cm²
  c 9.64 cm² **d** 6.68 cm²
  e 15.5 cm² **f** 32.7 cm²
2 a 21.2 cm² **b** 43.3 cm²
  c 29.1 cm²
3 a 171 cm² **b** 25.7 cm²
  c 59.4 cm²
4 a 25.0 cm² **b** 28.4 cm²
  c 20.8 cm²
5 374 cm²
6 9.61
7 10.1
8 6.20
9 8.51
10 13.2
11 173 cm²
12 125 cm²
13 48.6, 131.4

Exercise 8.12
1 a 1.85 cm² **b** 7.34 cm²
  c 9.83 cm² **d** 11.0 cm²
  e 51.4 cm² **f** 1.33 cm²
2 22.0

Exercise 8.13
1 a 3.38 **b** 4.31 **c** 15.1
  d 5.38 **e** 9.95 **f** 4.82
  g 6.84 **h** 4.16 **i** 10.5
2 a 30.9 **b** 43.7 **c** 24.2
  d 35.8 **e** 28.0 **f** 46.4
  g 59.1 **h** 47.6 **i** 41.2
3 7.01
4 a 5.71 cm **b** 51.0°
5 25.4°
6 a 9.56 km
  b i 110° **ii** 147°
7 13.1 km

Exercise 8.14
1 34.6, 145.4   **2** 62.4, 117.6

Exercise 8.15
1 a 3.81 cm **b** 5.74 cm
  c 2.63 cm **d** 6.03 cm
  e 5.99 cm **f** 5.35 cm
  g 5.31 cm **h** 9.27 cm
  i 5.42 cm
2 a 34.8 **b** 16.1 **c** 45.5
  d 34.1 **e** 97.2 **f** 35.6
  g 112.9 **h** 22.3 **i** 120

Exercise 8.16
1 a 12.8 cm **b** 7.90 cm
2 a 5 cm **b** 92.1°
3 a 80° **b** 7.86 km
4 29.0°
5 104.5°
6 8.61
7 a 85° **b** 17.4 cm²
  c 8.24 **d** 37.2 **e** 22.8°
8 32.4 cm
9 a 6.48 cm **b** 45.2 cm²
10 a 37.6 cm **b** 24.2 cm
  c 604 cm²

11 1020 cm²
12 a i 5 **ii** 10 **iii** 11.2
  b 25
13 a answer given **b** 3.91
  c 11.7 cm²
14 a answer given **b** 4.74
  c 13.9 cm²

Exercise 8.17
1 a 1, 3.5, 4.5, 3
  b

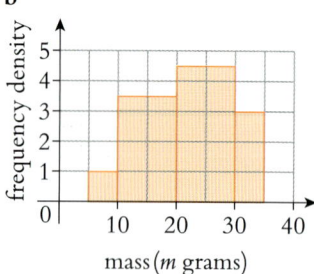

2 a 7, 13, 14, 2
  b

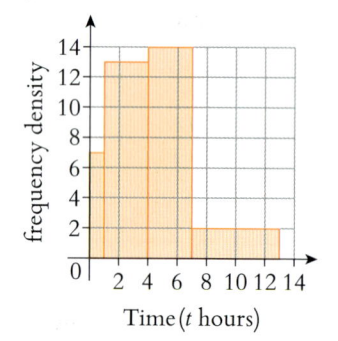

3 a 5, 8, 7, 6, 4
  b

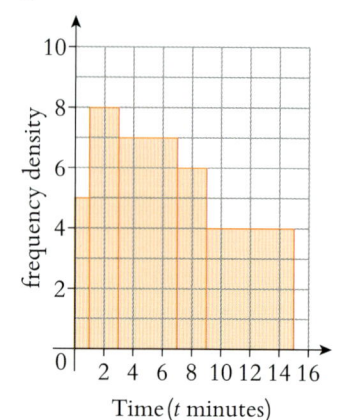

4 a 2.4, 5.8, 12.6, 9, 0.5

b *(histogram: Height (h cm), frequency density, bars around 150–200)*

5 *(histogram: Speed (v km/h), frequency density)*

6 *(histogram: Time (t minutes), frequency density, 0–12)*

7 *(histogram: Time (t minutes), frequency density, 0–140)*

8 a 6, 8, 7, 9, 5

b 5.84 min

9 a 2, 8, 6, 5, 4   **b** 3.17 kg

10 9 cm, 24 cm, 2 cm

11 8 cm, 10 cm, 2.4 cm

12 a 8, 20, 15, 14, 12

b 19

c 2.19 hours

d $\dfrac{41}{70}$

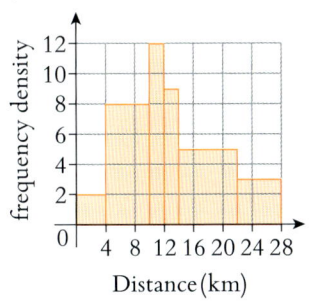

13 a 48, 24

b *(histogram: Distance (km), frequency density)*

c 12.9 km

Examination Questions

1 0.84, 7.16

2 a i $4x(x+4)$

ii answer given

b 1.1

3 a i answer given

ii −12, 8   **iii** 12 cm

b 0.8

c i answer given

ii −8.04, 4.04   **iii** 21.1 cm

4 a $\dfrac{105}{x}$ **b** $\dfrac{105}{x+4}$

c answer given

d i $(x+25)(x-21)$

ii −25, 21

e 46   **f** 4.57 g

5 a i −2.5   **ii** −3, 1

iii 9.5

b i $(u-10)(u+1)$

ii −1, 10

c i answer given

ii −0.56, 3.56

iii 12.7 cm^2

6 $\dfrac{-18}{(2x+3)(x-3)}$

7 $\dfrac{x-3}{x+2}$

8 245

9 0.128

10 60, 120

11 7.94

12 21.3

13 a 14:46

b i 260°   **iii** 145°

c 85.0 km   **d** 39.8°

e 73.8 km

14 a 121 km

b i 280°   **iii** 069°

15 a 32.5 g

b *(histogram: m, frequency density, 10–50)*

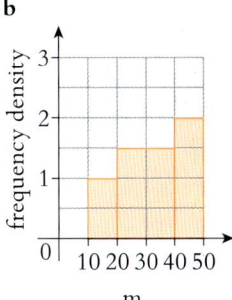

Unit 9

Exercise 9.1

1 a 5.5 cm, 6.5 cm

b 31.5 min, 32.5 min

c 31.5 kg, 32.5 kg

d 91.5 mm, 92.5 mm

e 7.65 cm, 7.75 cm

f 2.625 kg, 2.635 kg

g 62.85 g, 62.95 g

h 475.5 s, 476.5 s

i 4.935 m, 4.945 m

j 244.5 m, 245.5 m

2 $249.5 \leq x < 250.5$

3 23 kg

4 $18 675, $18 525

Answers 513

5 30 825 cm, 30 675 cm
6 139.5 mm, 136.5 mm
7 66 cm, 65.2 cm
8 a 25.6 cm, 25.2 cm
   b 39.8575 cm², 38.5875 cm²
9 a 82.2 cm, 81.8 cm
   b 418.6925 cm², 414.5925 cm²
10 28 cm, 24 cm
11 a 107.171875 cm³, 100.544625 cm³
    b 135.375 cm², 129.735 cm²
12 a 151.895625 cm³, 143.028375 cm³
    b 180.715 cm², 173.995 cm²
13 494.8125 cm³, 352.6875 cm³
14 231
15 5.186249645 cm
16 132.61014 cm²
17 a 11.77765262 cm
    b 22.90713754 cm²
18 a 428 cm, 426 cm
    b 92 cm, 90 cm
19 214.2664365 cm³
20 1.633802817 m/s²

Exercise 9.2

1 a $x = -2, y = 3$
   b $x = 9, y = 4$
   c $x = 1, y = -2$
   d $x = 8, y = 1\frac{1}{2}$
   e $x = \frac{1}{2}, y = 7$
   f $x = -2, y = 5$
   g $x = 4, y = -3$
   h $x = 2\frac{1}{2}, y = 4$
   i $x = -\frac{1}{3}, y = 7\frac{2}{3}$
   j $x = 2, y = 9$
   k $x = -3, y = -3$
   l $x = -4, y = 7$

2 a $x = 4, y = 1$
   b $x = 16, y = 3$
   c $x = -23, y = -78$
   d $x = -22, y = -9$
3 a $x = 6, y = -4$
   b $x = 3, y = 6$
   c $x = -2, y = 1$
   d $x = 8, y = -2$
   e $x = 1, y = -4$
   f $x = 5, y = -2$
   g $x = -3, y = -1$
   h $x = 8, y = 1\frac{1}{2}$
   i $x = -\frac{1}{3}, y = 0$
   j $x = -10, y = 9$
   k $x = 3, y = 4\frac{1}{2}$
   l $x = \frac{2}{3}, y = -2$
4 (2, 4)
5 (−1, −4)
6 34
7 $3.65
8 42
9 13
10 546

Exercise 9.3

1 a $15x + 3y \leq 900$
   b $x \geq 40, y \geq 20$
   c 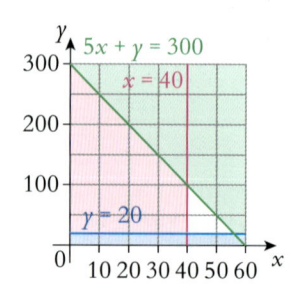

   d 56
2 a $x + y \leq 15, y \geq 6, y \geq 2x$
   b 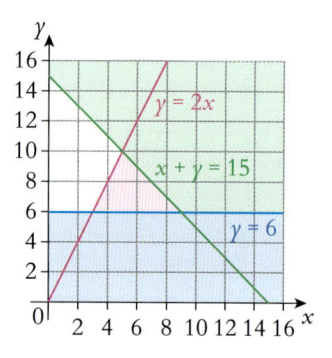

   c 5
3 a $400x + 200y \leq 9000$
   b $y < x, y \geq 10$
   c 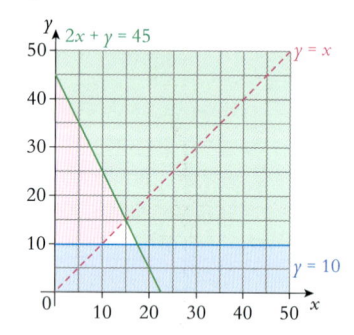

   d 17
4 a $16x + 10y \geq 100$
   b $x + y \leq 8 \quad x \geq 1 \quad y \geq 3$
   c 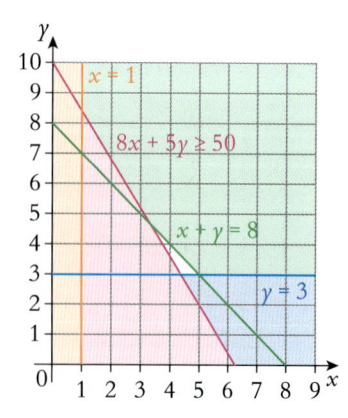

   d 5

Exercise 9.4

1. **a** $4x + 3y \leq 48$, $x + y \leq 14$, $x \geq 5$, $y \geq 5$
   b [graph showing lines $4x + 3y = 48$, $x = 5$, $x + y = 14$, $y = 5$]
   c $230

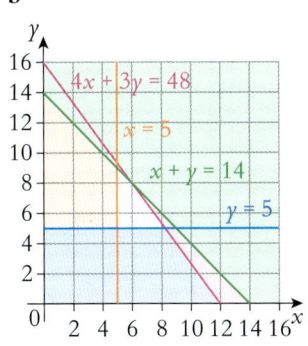

2. **a** $5x + 3y \geq 45$, $x + y \leq 11$, $y \geq 2$
   b [graph showing lines $5x + 3y = 45$, $x + y = 11$, $y = 2$]
   c $96

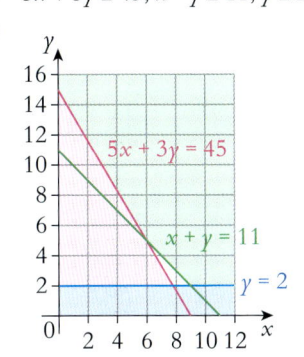

3. **a** $2x + y \leq 200$, $4x + 5y \leq 580$, $x \geq 50$, $y \geq 40$
   b [graph showing lines $x = 50$, $2x + y = 200$, $4x + 5y = 580$, $y = 40$]
   c $165

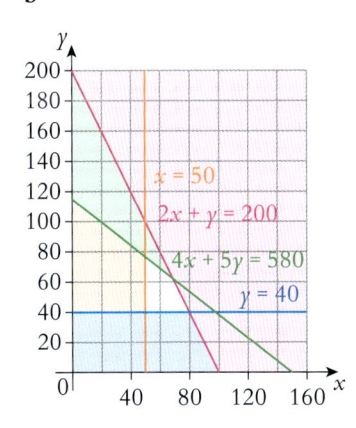

Exercise 9.5

1. **a** $(x - 1)^2 - 1$
   b $(x + 2)^2 - 4$
   c $(x - 5)^2 - 25$
   d $(x + 10)^2 - 100$
   e $(x - 8)^2 - 64$
   f $(x - 6)^2 - 36$
   g $(x + 9)^2 - 81$
   h $(x + 18)^2 - 324$
   i $(x + 11)^2 - 121$
   j $(x + 12)^2 - 144$
   k $(x - 1.5)^2 - 2.25$
   l $(x + 2.5)^2 - 6.25$
   m $(x + 3.5)^2 - 12.25$
   n $(x + 4.5)^2 - 20.25$
   o $(x - 10)^2 - 100$
   p $(x + 0.5)^2 - 0.25$

2. **a** $5(x - 2)^2 - 20$
   b $3(x + 2)^2 - 12$
   c $4(x + 2.5)^2 - 25$
   d $3(x - 3.5)^2 - 36.75$
   e $2(x + 3)^2 - 18$
   f $2(x - 1.25)^2 - 3.125$
   g $3(x - \frac{2}{3})^2 - 1\frac{1}{3}$
   h $2(x + 2.25)^2 - 10.125$

3. **a** $(x - 4)^2 - 1$
   b $(x - 2)^2 - 16$
   c $(x - 3)^2 - 13$
   d $(x + 2)^2 - 7$
   e $(x + 6)^2 - 41$
   f $(x - 4)^2 - 15$
   g $(x + 5)^2 - 28$
   h $(x + 4.5)^2 - 22.25$

4. $a = 5$, $b = -21$
5. $p = 4$, $q = -19$
6. $p = 2$, $q = -5$
7. **a** $x = 3 \pm \sqrt{b - 9}$
   b $x = -5 \pm \sqrt{a + b + 25}$

Exercise 9.6

1. **a** $-2, -4$ **b** -2
   c $2, -4$ **d** $7.41, 4.59$
   e $4.41, 1.59$
   f $0.21, -14.21$
   g $21.5, -1.53$
   h $8.16, 1.84$
   i $3.65, -1.65$

2. **a** $-1, -2$ **b** $4, 3$
   c $2, -3$ **d** $3.62, 1.38$
   e $2.70, -3.70$
   f $10.72, 0.28$

3. **a** $2, -\frac{1}{3}$ **b** $-\frac{1}{2}, -3$
   c $-3, -5$ **d** $7.74, 0.26$
   e $-0.42, -1.58$
   f $0.65, -4.65$

4. **a** $0.24, -12.24$
   b $4.61, -2.61$

5. **a** $6.90, -2.90$
   b $3.83, -1.83$
   c $3.11, -1.61$
   d $3.30, -0.30$
   e $1.29, 0.31$
   f $1.65, -3.65$

6. 7
7. answer given
8. 3

Exercise 9.7

1. $-1.41, 1.41, 5$
2. $-1.2, 3.2$
3. **a** $-2.45, 2.45$ **b** $-1.3, 2.3$
   c $-0.7, 2.7$
4. **a** $0, 3$ **b** $-0.45, 4.45$
   c $-1.6, 2.6$
5. **a** 2.7 **b** $2.1, 0.2, -2.3$
   c 0.84
6. **a** $y = 4$ **b** $y = 0.5$
   c $y = 4x$ **d** $y = x + 2$
   e $y = x^2 - 2$
   f $y = \frac{1}{x} + 3$
   g $y = 1 - 2x$
   h $y = \frac{1}{x^2}$
7. $y = x + 5$
8. $y = 5x - 5$

Exercise 9.8

1. **a** $-0.8, 4.8$ **b** $k = 1$
   c $k < 1$

2 a 0.2, 2.8
   b $k = 2$, $k = 3$, $k = 4$
3 a −2.4, 0.4, 2   b $k = 3$
   c $k < 3$
4 a −1.9, 0.3, 1.5
   b $k = -1$ and $k = 3$
   c $k > 3$ and $k < -1$

Exercise 9.9

1 a, b, 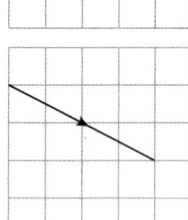 c, 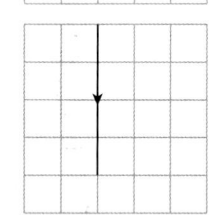 d, 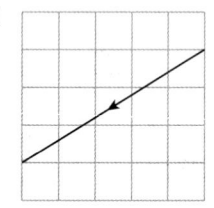 e (diagrams)

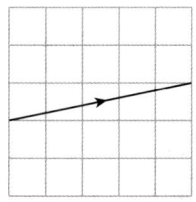

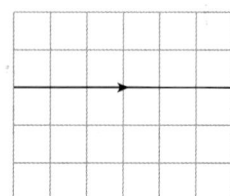

3 Parallel

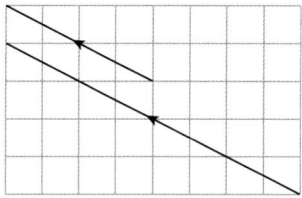

4 Perpendicular

5 a 10   b 13   c 4
   d 17   e 29

6 a $\vec{TU}, \vec{SO}, \vec{RQ}, \vec{OP}$
   b $\vec{UP}, \vec{TO}, \vec{SR}, \vec{OQ}$
   c $\vec{OS}, \vec{UT}, \vec{QR}, \vec{PO}$
   d $\vec{QO}, \vec{OT}, \vec{PU}, \vec{RS}$

Exercise 9.10

1 a, b, 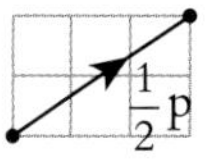 c, 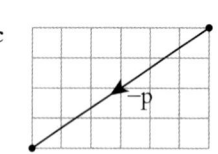 d, 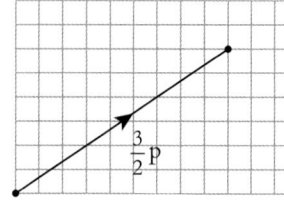 e, 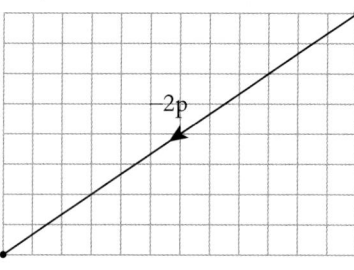 f (diagrams)

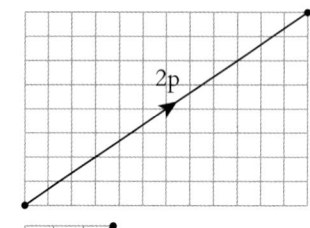

2 a $\begin{pmatrix} 3 \\ 2 \end{pmatrix}$ b $\begin{pmatrix} -5 \\ -1 \end{pmatrix}$ c $\begin{pmatrix} -2 \\ 4 \end{pmatrix}$
   d $\begin{pmatrix} -4 \\ 0 \end{pmatrix}$ e $\begin{pmatrix} -3 \\ -3 \end{pmatrix}$ f $\begin{pmatrix} 0 \\ -2 \end{pmatrix}$
   g $\begin{pmatrix} 5 \\ -3 \end{pmatrix}$ h $\begin{pmatrix} 4 \\ 1 \end{pmatrix}$

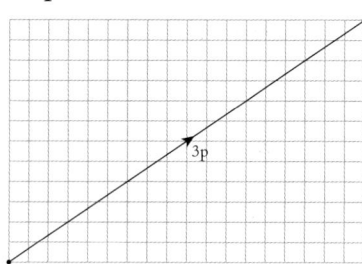

2 a $\begin{pmatrix} -6 \\ -10 \end{pmatrix}$ b $\begin{pmatrix} -5 \\ -10 \end{pmatrix}$ c $\begin{pmatrix} 4 \\ 8 \end{pmatrix}$
   d $\begin{pmatrix} 1 \\ -2 \end{pmatrix}$ e $\begin{pmatrix} 2 \\ -1 \end{pmatrix}$ f $\begin{pmatrix} -5 \\ 10 \end{pmatrix}$

3 a $\begin{pmatrix} 4 \\ 3 \end{pmatrix}$ b $\begin{pmatrix} 6 \\ -2 \end{pmatrix}$ c $\begin{pmatrix} 2 \\ -5 \end{pmatrix}$
   d $\begin{pmatrix} 4 \\ 7 \end{pmatrix}$ e $\begin{pmatrix} 8 \\ 6 \end{pmatrix}$
   f $\begin{pmatrix} -2 \\ -27 \end{pmatrix}$ g $\begin{pmatrix} 6 \\ 14 \end{pmatrix}$ h $\begin{pmatrix} 14 \\ 25 \end{pmatrix}$
   i $\begin{pmatrix} 14 \\ 8 \end{pmatrix}$ j $\begin{pmatrix} 30 \\ 18 \end{pmatrix}$ k $\begin{pmatrix} 4 \\ -6 \end{pmatrix}$
   l $\begin{pmatrix} -3 \\ -13 \end{pmatrix}$

4 C and E, B and H, G and D, A and F

5 a $\begin{pmatrix} 4 \\ 1 \end{pmatrix}$ b $\begin{pmatrix} 2 \\ 3 \end{pmatrix}$ c $\begin{pmatrix} -1 \\ 2 \end{pmatrix}$
   d $\begin{pmatrix} -3 \\ 4 \end{pmatrix}$ e $\begin{pmatrix} -3 \\ 0 \end{pmatrix}$ f $\begin{pmatrix} -2 \\ -3 \end{pmatrix}$
   g $\begin{pmatrix} 0 \\ -1 \end{pmatrix}$ h $\begin{pmatrix} 1 \\ -4 \end{pmatrix}$

6 a $\begin{pmatrix}2\\-1\end{pmatrix}, \begin{pmatrix}2\\1\end{pmatrix}, \begin{pmatrix}0\\2\end{pmatrix}, \begin{pmatrix}-3\\1\end{pmatrix}, \begin{pmatrix}-1\\-3\end{pmatrix}$

b $\begin{pmatrix}0\\0\end{pmatrix}$ Start and finish at the same place

7 a $\begin{pmatrix}21\\20\end{pmatrix}$ b 29

8 a $\begin{pmatrix}12\\35\end{pmatrix}$ b 37

9 a answer given
b 5.07
10 $p = 6.12, q = -5.14$
11 $p = -28.2, q = 10.3$
12 $x = 7, y = -1$
13 $x = 3, y = -4$

14 a $\begin{pmatrix}8\\1\end{pmatrix}$ b $\begin{pmatrix}1\\3\end{pmatrix}$ c $\begin{pmatrix}1\\3\end{pmatrix}$

d $\begin{pmatrix}2\\6\end{pmatrix}$

15 a $\begin{pmatrix}3\\2\end{pmatrix}$ b $\begin{pmatrix}6\\4\end{pmatrix}$ c (9, 6)

16 a

t	0	1	2	3	4
r	$\begin{pmatrix}3\\5\end{pmatrix}$	$\begin{pmatrix}5\\4\end{pmatrix}$	$\begin{pmatrix}7\\3\end{pmatrix}$	$\begin{pmatrix}9\\2\end{pmatrix}$	$\begin{pmatrix}11\\1\end{pmatrix}$

b

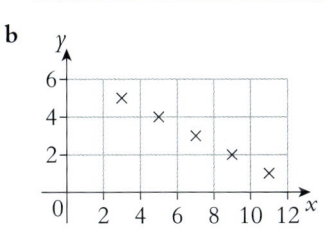

c The points lie in a straight line.

Exercise 9.11

1 a $-\mathbf{q}$ b $2\mathbf{q}$ c $-\mathbf{p}+\mathbf{q}$
  d $-2\mathbf{q}+\mathbf{p}$

2 a $-\mathbf{p}+\mathbf{q}$ b $-\frac{1}{2}\mathbf{p}+\frac{1}{2}\mathbf{q}$
  c $\frac{1}{2}\mathbf{p}-\frac{1}{2}\mathbf{q}$ d $\frac{1}{2}\mathbf{p}+\frac{1}{2}\mathbf{q}$

3 a $\mathbf{p}+\mathbf{r}$ b $\frac{1}{2}\mathbf{r}$ c $\mathbf{p}+\frac{1}{2}\mathbf{r}$
  d $\frac{1}{2}\mathbf{r}-\mathbf{p}$

4 a $\overrightarrow{BA} = -2\mathbf{p}$ b $\mathbf{q}+2\mathbf{p}$
  c $-\mathbf{p}+\mathbf{q}$ d $\mathbf{p}+\mathbf{q}$

5 a $\mathbf{p}$ b $-\mathbf{q}$ c $\mathbf{p}+\mathbf{q}$
  d $\mathbf{p}-\mathbf{q}$ e $2\mathbf{q}+\mathbf{p}$
  f $-\mathbf{q}-2\mathbf{p}$

6 a $\frac{1}{2}\mathbf{a}$ b $-\frac{1}{2}\mathbf{c}$ c $-\mathbf{a}+\mathbf{c}$
  d $\frac{1}{2}\mathbf{c}-\frac{1}{2}\mathbf{a}$
  parallel and AC = 2 MN

7 a $\mathbf{a}+\mathbf{c}$ b $-\mathbf{c}+\mathbf{a}$
  c $\frac{1}{2}\mathbf{a}+\frac{1}{2}\mathbf{c}$ d $\frac{1}{2}\mathbf{a}-\frac{1}{2}\mathbf{c}$

8 a $\frac{1}{2}\mathbf{p}+\frac{1}{2}\mathbf{q}$ b $\frac{1}{2}\mathbf{p}+\frac{1}{2}\mathbf{q}$
  c $-\frac{1}{2}\mathbf{q}+\frac{1}{2}\mathbf{p}$ d $\frac{1}{2}\mathbf{p}-\frac{1}{2}\mathbf{q}$
  parallelogram

9 a $\frac{3}{2}\mathbf{c}$ b $-\mathbf{a}+\mathbf{c}$
  c $\mathbf{a}-\frac{1}{2}\mathbf{c}$ d $\frac{1}{2}\mathbf{c}-\frac{1}{2}\mathbf{a}$
  parallel and AC = 2 MN

10 a $\mathbf{p}+\mathbf{q}$ b $-\frac{3}{2}\mathbf{p}+\frac{1}{2}\mathbf{q}$
   c $\frac{5}{2}\mathbf{p}+\frac{1}{2}\mathbf{q}$ d $\frac{1}{2}(3\mathbf{p}+\mathbf{q})$

Exercise 9.12

1 a $-\mathbf{p}+\mathbf{q}$ b $-\frac{1}{3}\mathbf{p}+\frac{1}{3}\mathbf{q}$
  c $\frac{2}{3}\mathbf{p}+\frac{1}{3}\mathbf{q}$

2 a $-\mathbf{p}+\mathbf{q}$ b $-\frac{3}{4}\mathbf{p}+\frac{3}{4}\mathbf{q}$
  c $\frac{1}{4}\mathbf{p}+\frac{1}{4}\mathbf{q}$ d $\frac{1}{4}\mathbf{p}+\frac{3}{4}\mathbf{q}$

3 a $\frac{1}{2}\mathbf{q}$ b $-\frac{1}{3}\mathbf{p}+\frac{1}{2}\mathbf{q}$
  c $-\mathbf{p}+\mathbf{q}$ d $-\frac{2}{5}\mathbf{p}+\frac{2}{5}\mathbf{q}$

4 a $\frac{1}{3}\mathbf{r}$ b $\frac{1}{4}\mathbf{p}$
  c $\frac{2}{3}\mathbf{r}-\frac{3}{4}\mathbf{p}$ d $\mathbf{r}+\frac{1}{4}\mathbf{p}$

5 a $-3\mathbf{a}+\mathbf{c}$ b $-\frac{9}{5}\mathbf{a}+\frac{3}{5}\mathbf{c}$
  c $\frac{6}{5}\mathbf{a}+\frac{3}{5}\mathbf{c}$ d $2\mathbf{a}+\mathbf{c}$
  collinear.

6 a $\mathbf{r}-\mathbf{p}$ b $\frac{1}{4}\mathbf{r}-\frac{1}{4}\mathbf{p}$
  c $\frac{1}{4}\mathbf{p}+\frac{3}{4}\mathbf{r}$ d $\frac{3}{4}\mathbf{p}+\frac{1}{4}\mathbf{r}$

7 a $\frac{2}{5}\mathbf{r}$ b $\mathbf{p}-\frac{3}{5}\mathbf{r}$
  c $\frac{5}{3}\mathbf{p}-\mathbf{r}$ d $\frac{5}{3}\mathbf{p}$
  $\overrightarrow{OX} = \frac{5}{3}\overrightarrow{OP}$

8 $p = 13, q = 10\frac{1}{3}$

9 answer given

Challenge

a $\frac{1}{3}p + \frac{1}{3}q$

b $\frac{1}{3}p + \frac{1}{3}q$

The points coincide.

Exercise 9.13

1 a i $\frac{7}{12}$ ii $\frac{1}{10}$ iii $\frac{11}{20}$

  b $\frac{7}{9}$ b $\frac{117}{590}$

2 a

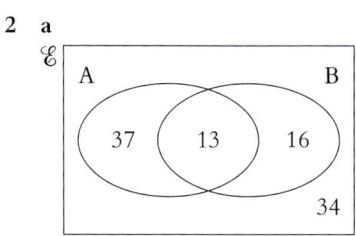

  b i $\frac{4}{25}$ ii $\frac{17}{50}$

  c $\frac{37}{50}$ d $\frac{49}{198}$

3 a i $\frac{89}{150}$ ii $\frac{13}{30}$ iii $\frac{23}{150}$

  b $\frac{47}{89}$ c 0.486

4 a i $\frac{73}{100}$ ii $\frac{1}{4}$ iii $\frac{3}{20}$

  b $\frac{17}{50}$ c $\frac{145}{392}$

5 a

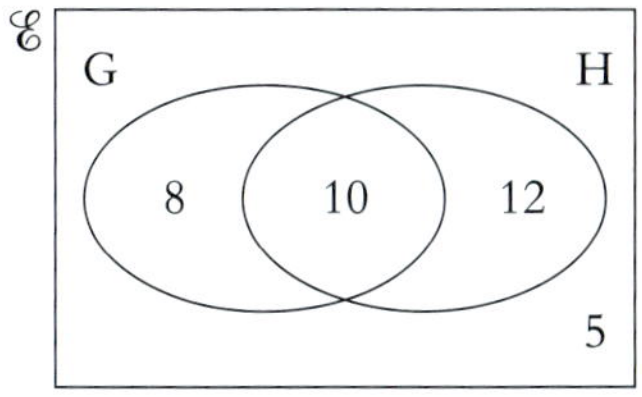

UNIT 9

Answers 517

b i $\frac{2}{7}$ ii $\frac{12}{35}$

  c $\frac{5}{11}$ **d** $\frac{286}{595}$

6 a i $\frac{2}{25}$ ii $\frac{3}{50}$ iii $\frac{1}{50}$

  b $\frac{8}{19}$ **c** $\frac{19}{1990}$

7 a

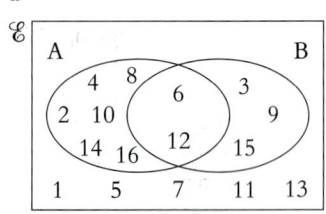

(Venn diagram: $\mathcal{E}$, sets A and B. A contains 4, 8, 2, 10, 14, 16; intersection 6, 12; B contains 3, 9, 15; outside 1, 5, 7, 11, 13)

  b i $\frac{5}{16}$ ii $\frac{1}{2}$ iii $\frac{1}{8}$

  iv $\frac{11}{16}$ v $\frac{5}{16}$

8 a i $\frac{27}{50}$ ii $\frac{7}{25}$ iii $\frac{2}{25}$

  b $\frac{9}{25}$ **c** $\frac{87}{175}$

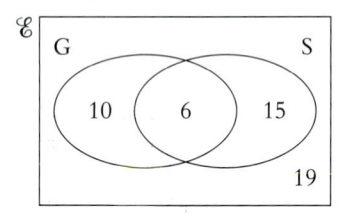

9 a

(Venn diagram: $\mathcal{E}$, sets G and S. G: 10; intersection 6; S: 15; outside 19)

  b $\frac{3}{25}$ **c** $\frac{5}{7}$ **d** $\frac{1}{35}$

Examination Questions

1 50.1225
2 62 225 000
3 $237.5 \leq T < 242.5$
4 40.5 cm
5 $x = 4$, $y = -3$
6 **a** $600x + 1200y \geq 720\,000$
   b $x + y \leq 900$
   c

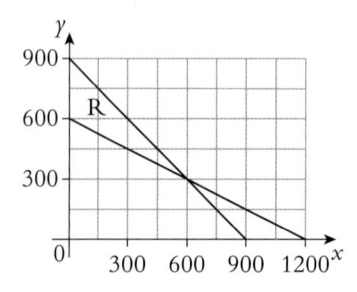

(graph with line R, axes y up to 900, x up to 1200)

   d 300
7 **a** $20x + 100y \leq 1200$
   b i $x + y \geq 40$
       ii $y \geq 2$
   c

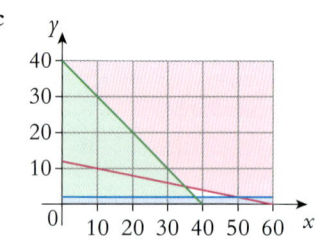

(graph, x to 60, y to 40, shaded regions)

   d 5
   e 50, 2, $270
8 **a** $p = 5.0$ $q = 0$ $r = 8.7$

b

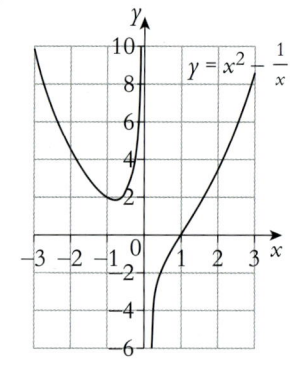

(graph of $y = x^2 - \frac{1}{x}$)

  c i $-2.88, -0.65, 0.53$
   ii $a = 3$, $b = -1$
  d -3.75

9 a $2\mathbf{a} - \mathbf{g}$ **b** $\frac{5}{2}\mathbf{a} + \frac{1}{2}\mathbf{g}$

10 a i $-3\mathbf{p} - 2\mathbf{q}$ ii $-3\mathbf{p} + 4\mathbf{q}$
    iii $-4\mathbf{p}$ **b** 8

11 a i $\mathbf{p} + \mathbf{r}$ ii $\mathbf{r} - \mathbf{p}$
    iii $-\mathbf{p} + \frac{2}{3}\mathbf{r}$ iv $\mathbf{p} + \frac{1}{2}\mathbf{r}$

  b i $\mathbf{r} - \frac{3}{2}\mathbf{p}$ ii $-\frac{3}{2}\mathbf{p}$

  c collinear

12 a $p = 5$, $q = 12$, $r = 1$
   b i 17 ii 12
   c i 26 ii 57
   d i $\frac{2}{25}$ ii $\frac{9}{20}$
   e $\frac{18}{73}$

Index

acceleration, 338
acute angles, 33
addition
    algebraic fractions, 64
    directed numbers, 3
    fractions, 11
    vectors, 442
adjacent sides, 194
algebra, 17–32, 58–75, 108–19, 170–89, 224–37, 272–93, 342–57, 383–97, 426–39
    see also matrix algebra
algebraic expressions, simplifying, 17–19
algebraic fractions, 388–9
    addition, 64
    manipulating, 64–7
    simplifying, 116
    subtraction, 64
alternate angles, 33
angles
    acute, 33
    alternate, 33
    at points, 33
    construction, 84
    corresponding, 33
    of depression, 208–9
    of elevation, 208–9
    interior, 39
    obtuse, 33
    properties, 33–5
    in quadrilaterals, 33
    reflex, 33
    right, 33
    on straight lines, 33
    in triangles, 33
    vertically opposite, 33
    see also exterior angles
arc length, 93–4
arcs, 93–4
area, 76–7
    circles, 90–5
    kites, 76
    parallelograms, 76
    rectangles, 76
    rhombuses, 76
    segments, 403
    similar shapes, 246–9
    trapeziums, 76
    triangles, 76, 400–3
    see also surface area
area scale factors, 246
asymptotes, 292
average speed, 270
averages, 41–3

bar charts, 96
bases, 58
bearings, 190–3
best fit, lines of, 212
biased dice, 156
BIDMAS rule, 2
bisectors
    construction, 84
    perpendicular, 84
boundary lines, 72–3
bounds
    lower, 422–5
    upper, 422–5
brackets
    double, expanding, 180–3
    expressions involving, 18

capacity, 150
centres
    of enlargements, 144
    of rotation, 132
charts
    bar, 96
    flow, 228
    pie, 96
    see also diagrams; graphs
chords, 294, 296
circle theorems, 294–307
circles, area, 90–5
circumference, 90–5
class width, 412
coefficients, 230
collinear points, 449
column matrices, 170
    multiplication, 172
column vectors, 136, 440
columns, 170
combined independent events, 316
combined transformations, 365

common denominators, 64
complements, 274
completing the square, quadratic equation solving, 432–5
composite functions, 226
compound interest, 266
cones
    surface area, 241
    volume, 241
congruency, 120
constant of proportionality, 390
constants, 230
constructions, geometrical, 84–5
continuous data, 96
    displaying, 97
    grouped, 412
conversions, currency, 220
correlations, and scatter diagrams, 210
corresponding angles, 33
corresponding sides, 123
cosine curves, 398
cosine ratio, 198, 200, 201, 398–9
cosine rule, 404–11
cross-multiplying, 22
cross-sections, 150
cube numbers, 9, 346
cubic graphs, 234–7
    applications, 236
cuboids, volume, 150
cumulative frequency, 368–75
cumulative frequency tables, 368
currency conversions, 220
curved surfaces, 154
curves
    cosine, 398
    gradients of, 186
    sine, 398
    tangents to, 186
cyclic quadrilaterals, 302
cylinders
    surface area, 154
    volume, 150, 151

data
    discrete, 96
    displaying, 96–9
    paired, 210
    types of, 96
    see also continuous data
data sets, comparing, 373
deceleration, 338
decimal places, 14–16
decimals, recurring, 382
denominators, 10
    common, 64
density, frequency, 412
dependent events, tree diagrams, 320
depression, angles of, 208–9
determinants, 176
diagrams
    possibility, 160
    see also charts; frequency diagrams; graphs; scatter diagrams; tree diagrams; Venn diagrams
diameters, 90, 294
dice
    biased, 156
    fair, 156
    unbiased, 156
difference of two squares, 232
direct proportion, 218–21, 390
direct variation, 390
directed numbers, 3–4
    addition, 3
    division, 4
    multiplication, 4
    subtraction, 3
discrete data, 96
discriminants, 386
distance, 268–71
    between points, 82
distance travelled, 338
distance–time graphs, 334–7
division
    directed numbers, 4
    fractions, 12
    indices, 58

double brackets, expanding, 180–3

elements, number of, 272
elevation, angles of, 208–9
elimination method, 109–11
empty sets, 272
enlargements, 144–9
   centres of, 144
equations
   roots of, 438
   solving, 66–7
      with graphs, 436–9
      of straight lines, 68–71
         finding, 70
   trigonometric, 399
   see also linear equations; quadratic equations; simultaneous equations
equilateral triangles, 33
estimation, 16
events, 156
   dependent, 320
   see also independent events
expanding, 114
   double brackets, 180–3
expected numbers, 162
experimental probability, 162
exponential graphs, 354–7
   applications, 356
expressions
   algebraic, 17–19
   factorising, 232
   involving brackets, 18
   simplifying, 17
   substitution into, 25
   see also quadratic expressions
exterior angles
   polygons, 39
   triangles, 33

factorising, 114–17, 230–3
   quadratic equation solving, 288–91
   quadratic expressions, 230, 288
factors, 5
   prime, 7
   shear, 312
   stretch, 308
   see also scale factors
fair dice, 156
fairness, 156

first, outside, inside, last (FOIL), 181
flow charts, 228
FOIL (first, outside, inside, last), 181
formulae
   constructing, 24
   quadratic equation solving, 384–7
   rearranging, 118–19, 342–3
   subjects of, 118
   substitution into, 25–7
   and variables, 24
fractional indices, 284
fractions
   addition, 11
   division, 12
   improper, 10
   language of, 10
   linear equations involving, solving, 22
   multiplication, 12
   rules for, 10–13
   subtraction, 11
   see also algebraic fractions
frequency
   cumulative, 368–75
   relative, 162
frequency density, 412
frequency diagrams, 97
   cumulative, 368–9
frequency tables, 44–5
   cumulative, 368
   grouped, 254–7, 368
frustums, 241
function notation, 224
functions, 224–9
   composite, 226
   inverse, 228
   quadratic, 184

geometrical constructions, 84–5
geometry, vector, 440–51
gradients, 68
   of curves, 186
   straight-line graphs, 28–32
graphical solutions, simultaneous equations, 108–9
graphs
   distance–time, 334–7
   equation solving, 436–9
   linear inequalities on, 72–5
   reciprocal, 292–3

regions, 72–3
speed–time, 338–41
see also charts; cubic graphs; diagrams; exponential graphs; quadratic graphs; straight-line graphs
greater than, 62
grouped continuous data, 412
grouped frequency tables, 254–7, 368

HCF (highest common factor), 5, 7
hemispheres
   surface area, 244
   volume, 244
hexagons, 153
highest common factor (HCF), 5, 7
histograms, 412–15
hypotenuse, 78, 194

identity matrices, 176
images, 128, 144
improper fractions, 10
independent events
   combined, 316
   tree diagrams, 316
indices, 58–61, 284–7, 346
   division, 58
   fractional, 284
   multiplication, 58
   negative, 60
   zero, 60
inequalities, linear, 62–3, 72–5
integers, 3
inter-quartile range (IQR), 372
intercepts, 68, 70
interest
   compound, 266
   simple, 266
interior angles, polygons, 39
intersections, 274
   points of, 294
invariant lines, 308
invariant points, 128
inverse functions, 228
inverse matrices, 178
inverse proportion, 218–21, 394
inverse transformations, 143
inverse variation, 394, 396

IQR (inter-quartile range), 372
IQs (upper quartiles), 372
irrational numbers, 382–3
isosceles triangles, 33

kites, area, 76

large numbers, 104
LCM (lowest common multiple), 5, 7
length scale factors, 246
less than, 62
like terms, 17
linear equations
   involving fractions, solving, 22
   solving, 20–3
linear inequalities
   on graphs, 72–5
   solving, 62–3
linear programming, 428–31
linear relationships, 390
linear sequences, 344
lines
   of best fit, 212
   boundary, 72–3
   invariant, 308
   midpoints of, 71
   mirror, 128
   parallel, 33, 68
   perpendicular, 33
   of symmetry, 36
   see also straight lines
loci, 86–9
losses, percentage, 168
lower bounds, 422–5
lower quartiles (LQs), 372
lowest common multiple (LCM), 5, 7
LQs (lower quartiles), 372

magnitude, vectors, 440
map scales, 52, 56–7
mapping, 224
matrices, 170–9
   identity, 176
   inverse, 178
   multiplication, 172, 174–5
   non-singular, 178
   order of, 170
   singular, 178
   and transformations, 358–67
   two by two, 176

zero, 176
see also column matrices;
row matrices;
transformation
matrices
matrix algebra, 170–9
mean, 41, 44, 254
median, 41, 44, 372
members, of sets, 272
mid-values, 254
midpoints, of lines, 71
minimum values, 434
mirror lines, 128
mixed numbers, 10
modal class, 254
mode, 41, 44
modulus, of vectors, 440
multiples, 5
multiplication
   column matrices, 172
   directed numbers, 4
   fractions, 12
   indices, 58
   matrices, 172, 174–5
   row matrices, 172

negative indices, 60
negative numbers, 3
non-linear relationships, 392
non-linear sequences, 346, 350
non-singular matrices, 178
notation
   function, 224
   sequence, 353
   transformation matrices, 366
   vectors, 440
nth terms, 344
numbers, 2–16, 50–7, 104–7, 168–9, 218–23, 264–71, 334–41, 382–3, 422–5
   cube, 9, 346
   expected, 162
   irrational, 382–3
   large, 104
   mixed, 10
   negative, 3
   positive, 3
   prime, 7
   rational, 382–3
   small, 106
   square, 9, 346
   see also directed numbers

numerators, 10, 64

objects, 128, 144
obtuse angles, 33
octahedrons, regular, 239
operations, order of, 2
opposite sides, 194
order
   of matrices, 170
   of operations, 2
outcomes, 156

paired data, 210
parallel lines, 33, 68
parallel vectors, 443
parallelograms, area, 76
percentage decrease, 168
percentage increase, 168
percentage losses, 168
percentage profits, 168
percentages, 50–1, 168–9, 264–7
   reverse, 264
percentiles, 372
perimeters, 76–7
perpendicular bisectors, 84
perpendicular lines, 33
pi (p), 90
pictograms, 96
pie charts, 96
planes, of symmetry, 38
points
   angles at, 33
   collinear, 449
   distance between, 82
   of intersection, 294
   invariant, 128
polygons, 39–40
   exterior angles, 39
   interior angles, 39
   regular, 39
position vectors, 443
positive numbers, 3
possibility diagrams, 160
powers *see* indices
prime factors, 7
   product of, 7
prime numbers, 7
primes, 7
prisms
   surface area, 154
   volume, 150, 151
probability, 156–63, 316–23, 452–6

calculating, from Venn diagrams, 452
   experimental, 162
problem-solving, using sets, 280
profits, percentage, 168
programming, linear, 428–31
proper subsets, 278
proportion
   direct, 218–21, 390
   inverse, 218–21, 394
proportionality, constant of, 390
pyramids
   surface area, 238
   truncated, 240
   volume, 238
Pythagoras' theorem, 78–83, 201
   applications, 204

quadrants, 92
quadratic equations
   roots, 386
   solving
      by completing the square, 432–5
      by factorisation, 288–91
      using formula, 384–7
quadratic expressions
   factorising, 230, 288
   minimum values, 434
quadratic functions, 184
quadratic graphs, 184–9
   applications, 188
quadrilaterals
   angles in, 33
   cyclic, 302
quartiles, 372

radii, 90
range, 41–3, 44
rational numbers, 382–3
ratios, 52–7
   decrease, 222–3
   dividing quantities into, 54–5
   increase, 222–3
   simplifying, 52
   tangent, 194
   trigonometric, 194–201, 398–9
   in vector geometry, 449
reciprocal graphs, 292–3
rectangles, area, 76

recurring decimals, 382
reflections, 128–31
reflex angles, 33
regions, graphs, 72–3
regular octahedrons, 239
regular polygons, 39
regular tetrahedrons, 240
relationships
   linear, 390
   non-linear, 392
relative frequency, 162
resultant vectors, 442
reverse percentages, 264
rhombuses, area, 76
right angles, 33
right-angled triangles, rules for, 201
roots
   of equations, 438
   quadratic equations, 386
rotational symmetry, 36
rotations, 128, 132–5
   centres of, 132
   describing, 134
rounding, 14
row matrices, 170
   multiplication, 172
rows, 170

scalars, multiplication by, 442
scale factors, 124, 149
   area, 246
   length, 246
   volume, 250
scales, map, 52, 56–7
scatter diagrams, 210–13
   and correlations, 210
sector area, 93–4
sectors, 93–4
segments, 296
   area, 403
semicircles, 92
sequence notation, 353
sequences, 344–53
   linear, 344
   non-linear, 346, 350
sets, 272–83
   empty, 272
   language of, 272, 278
   members of, 272
   problem-solving with, 280
   universal, 272
   on Venn diagrams, 274, 278
   see also subsets

shapes, 33–40, 76–95, 120–55, 190–209, 238–53, 294–315, 358–67, 398–411, 440–52
   see also similar shapes
shear factors, 312
shears, 308–15
sides
   adjacent, 194
   calculating, 194
   corresponding, 123
   opposite, 194
significant figures, 14–16
similar objects, volume, 250–3
similar shapes, 122
   area, 246–9
similar triangles, 120–7
simple interest, 266
simultaneous equations, 108–13, 426–7
   elimination method, 109–11
   graphical solution, 108–9
   substitution method, 426
sine curves, 398
sine ratio, 198, 200, 201, 398–9
sine rule, 404–11
singular matrices, 178
small numbers, 106
speed, 268–71, 334
   average, 270
speed–time graphs, 338–41
spheres
   surface area, 244
   volume, 244
square numbers, 9, 346
squares
   completing, 432–5
   difference of two, 232
standard form, 104–7
statistics, 41–5, 96–9, 156–63, 210–13, 254–7, 316–23, 368–75, 412–15, 452–5

straight lines
   angles on, 33
   equations of, 68–71
   finding, 70
straight-line graphs, 68–71
   gradients, 28–32
stretch factors, 308
stretches, 308–15
subjects, of formulae, 118
subsets, 278
   proper, 278
substitution
   into expressions, 25
   into formulae, 25–7
substitution method, 426
subtraction
   algebraic fractions, 64
   directed numbers, 3
   fractions, 11
   vectors, 442
surface area, 150–5, 238–45
   cones, 241
   cylinders, 154
   hemispheres, 244
   prisms, 154
   pyramids, 238
   spheres, 244
surfaces, curved, 154
symmetry, 36–8
   lines of, 36
   planes of, 38
   rotational, 36

tables
   two-way, 99
   see also frequency tables
tangent ratio, 194, 201
tangents, 294
   to curves, 186
term-to-term rule, 344
terms, 344
   like, 17
   nth, 344
tetrahedrons, regular, 240

theorems
   circle, 294–307
   see also Pythagoras' theorem
three-dimensional trigonometry, 206
time, 268–71
transformation matrices, 358, 362–3
   notation, 366
transformations, 128–43
   combined, 140–1, 365
   inverse, 143
   and matrices, 358–67
   translations, 128, 136–9
trapeziums, area, 76
tree diagrams
   dependent events, 320
   independent events, 316
triangles
   angles in, 33
   area, 76, 400–3
   construction, 84
   equilateral, 33
   exterior angles, 33
   isosceles, 33
   right-angled, 201
   similar, 120–7
trigonometric equations, solving, 399
trigonometric ratios, 194–201, 398–9
trigonometry, 194–207
   applications, 204
   three-dimensional, 206
truncated pyramids, 240
two-way tables, 99

unbiased dice, 156
unions, 274
units, of volume, converting, 150
universal sets, 272
upper bounds, 422–5
upper quartiles (IQs), 372

values
   mid-values, 254
   minimum, 434
variables, and formulae, 24
variation, 390–7
   direct, 390
   inverse, 394, 396
vector geometry, 440–51
   ratios in, 449
vectors, 440–51
   addition, 442
   column, 136, 440
   magnitude, 440
   modulus of, 440
   multiplication by scalars, 442
   notation, 440
   parallel, 443
   position, 443
   resultant, 442
   subtraction, 442
Venn diagrams, 272–83
   calculating probabilities from, 452
   representing sets on, 274, 278
vertically opposite angles, 33
volume, 150–5, 238–45
   cones, 241
   converting units of, 150
   cuboids, 150
   cylinders, 150, 151
   hemispheres, 244
   prisms, 150, 151
   pyramids, 238
   similar objects, 250–3
   spheres, 244
volume scale factors, 250

whole numbers, 3

y-intercepts, 68, 70

zero indices, 60
zero matrix, 176